Dust Explosions
in the Process Industries

With deep gratitude for their love and support, I dedicate this book to my wife Astrid and our children Kristian, Ragnar, Solveig and Jorunn, and to my mother and the memory of my father. The words in Isaiah 42.16 also gave me hope and courage.

#23584253

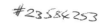

Dust Explosions in the Process Industries

Rolf K. Eckhoff

BUTTERWORTH
HEINEMANN

Butterworth–Heinemann Ltd
Linacre House, Jordan Hill, Oxford OX2 8DP

PART OF REED INTERNATIONAL BOOKS

OXFORD LONDON BOSTON
MUNICH NEW DELHI SINGAPORE SYDNEY
TOKYO TORONTO WELLINGTON

First published 1991

British Library Cataloguing in Publication Data
Eckhoff, Rolf K.
Dust explosions in the process industries.
I. Title
620.1

ISBN 0 7506 1109 X

Library of Congress Cataloguing in Publication Data
Eckhoff, Rolf K.
Dust explosions in the process industries / Rolf K. Eckhoff.
p. cm.
Includes index.
ISBN 0 7506 1109 X
1. Dust explosions. 2. Fire prevention. 3. Industrial accidents.
I. Title.
TH9446.D86E34 1991
604.7—dc20 91-597
CIP

Typeset by TecSet Ltd., Wallingston, Surrey.

Printed and bound in Great Britain by
Butler & Tanner Ltd, Frome and London

Contents

Foreword

Experience has shown all too clearly that ignition and explosion can occur wherever combustible dusts are handled or permitted to accumulate as a by-product of related activities. Despite reasonable precautions, accidents can and do happen; recognition of this universal hazard and the potential means for its control is widespread, as evidenced by the many individuals and groups worldwide performing research and developing codes and regulations.

The primary means of controlling and minimizing this recognized hazard are study, regulation and education; to accomplish this, specific knowledge must be generated and disseminated for the benefit of all interested people. Rolf Eckhoff has, in my estimation, prepared an outstanding book. It presents a detailed and comprehensive critique of all the significant phases relating to the hazard and control of a dust explosion, and offers an up to date evaluation of prevalent activities, testing methods, design measures and safe operating techniques.

The author is in an outstanding position to write this text, having spent a lifetime in research on dust and gas explosions. He assimilates information from worldwide contacts whilst retaining his independence of thought and the ability to see clearly through problems. His clear and concise language and thorough approach will benefit his fellow workers and all who read his book. His presentation of the mathematics, tables and figures is clear and striking. The inclusion of a comprehensive bibliography indicates not only his own thoroughness, but also the widespread nature of research into dust explosions throughout the world.

To my knowledge this book is the most complete compilation to date of the state of the art on industrial dust explosions.

John Nagy, Finleyville, PA, USA
(Formerly of the US Bureau of Mines)

Preface

The ambitious objective of this book is to provide an overview of the present state of the art. However, the amount of published information on dust explosions worldwide is vast, at the same time as much additional work was never printed in retrievable literature. Whilst I feel that I may have been able to cover some of the English/American and German open literature fairly well, most of the valuable research published in other languages has had to be left out simply because of the language barrier. Future attempts at summarizing some of this work in English should indeed be encouraged.

Although I have tried to give a reasonably balanced account, the book also reflects my personal research background. For example, other authors might perhaps not have included a separate chapter on the mechanics of dust deposits and dust particles. However, to me the important role of powder mechanics in dust explosions is evident. The book perhaps also reflects that most of my dust explosion research has been related to ignition, venting and testing.

The confrontation with the early research carried out by the pioneers in the UK, Germany, USA and other countries creates deep humility and admiration for the outstanding work performed by these people. Lack of sophisticated diagnostics did not prevent them from penetrating the logical structure of the problem and to draw long-lasting conclusions from their observations. It is a pity that much of this work seems to be neglected in more recent research. Too often mankind reinvents the wheel – this also applies to dust explosion research.

I would like to use this opportunity to thank Professor Emeritus H E Rose, Dr Sc, for bringing the existence of the phenomenon of dust explosions to my attention for the first time, and for giving me the opportunity to become acquainted with the subject, during my two years of study at King's College, London, 1966–68.

Many thanks also to Alv Astad and Helge Aas for their encouragement and active participation when dust explosion research, sponsored by Norwegian industry, was initiated at Chr. Michelsen Institute, Bergen, Norway, about 1970.

The Royal Norwegian Council for Scientific and Industrial Research (NTNF) has given valuable financial support to CMI's dust (and gas) explosion research from 1972 until today, and also allocated a generous special grant for the writing of this book. An additional valuable grant for the work with the book was given by the Swedish Fire Research Board (Brandforsk).

I am also deeply grateful to all the industrial companies, research institutions and colleagues in many countries, who made available to me and allowed me to make use of their photographs and other illustrations. A special thanks to Berufsgenossenschaftliches Institut für Arbeitssicherheit (BIA), in Germany for permission to translate and publish the tables in the Appendix.

I also wish to express my gratitude to those who have kindly read through sections of the draft manuscript and/or given constructive criticism and advice. John Nagy, Derek Bradley, Geoffrey Lunn, Bjørn Hjertager, Gisle Enstad, Dag Bjerketvedt, Ivar Ø. Sand

and Claus Donat should be mentioned specifically. Also my indebtedness goes to Chr. Michelsen Institute, which in its spirit of intellectual freedom coupled to responsibility, gave me the opportunity to establish dust and gas explosion research as an explicit activity of the institute. The institute also gave high priority to and allocated resources for the writing of this book, for which I am also most grateful.

This short preface does not allow me to mention the names of all the good people with whom I have had the privilege to work during my 20 years of dust and gas explosion research at CMI, and who deserve my sincere thanks. However, there is one exception – Kjell Fuhre, who worked with me from 1970 to 1988. I wish to thank him for having devoted his exceptional engineering talent to our experimental dust and gas explosion research, in laboratory scale as well as in full-size industrial equipment.

I wish to express a special thanks to Aaslaug Mikalsen, who, aided by more than 20 years of experience in interpreting my handwriting, was able to transform the untidy handwritten manuscript to a most presentable format on CMI's word processing system. Many thanks also to Per-Gunnar Lunde for having traced the majority of the drawings in the book.

Rolf K. Eckhoff

Chapter 1

Dust explosions – origin, propagation, prevention and mitigation: an overview

1.1
THE NATURE OF DUST EXPLOSIONS

1.1.1
THE PHENOMENON

1.1.1.1
What is an explosion?

The concept of *explosion* is not unambiguous. Various encyclopediae give varying definitions that mainly fall in two categories. The first focuses on the noise or 'bang' due to the sudden release of a strong pressure wave, or blast wave. The origin of this pressure wave, whether a chemical or mechanical energy release, is of secondary concern. This definition of an explosion is in accordance with the basic meaning of the word ('sudden outburst').

The second category of definitions is confined to explosions caused by sudden releases of chemical energy. This includes explosions of gases and dusts and solid explosives. The emphasis is then often put on the chemical energy release itself, and explosion is defined accordingly. One possible definition could then be 'An explosion is an exothermal chemical process that, when occurring at constant volume, gives rise to a sudden and significant pressure rise'.

In the present text the definition of an explosion will shift pragmatically between the two alternatives, focusing at either cause or effect, depending on the context.

1.1.1.2
What is a dust explosion?

The phenomenon named *dust explosions* is in fact quite simple and easy to envisage in terms of daily life experience. Any solid material that can burn in air will do so with a violence and speed that increases with increasing degree of sub-division of the material. Figure 1.1(a) illustrates how a piece of wood, once ignited, burns slowly, releasing its heat over a long period of time. When cut in small pieces, as illustrated in Figure 1.1(b), the combustion rate increases because the total contact surface area between wood and air has been increased. Also, the ignition of the wood has become easier. If the sub-division is continued right down to the level of small particles of sizes of the order of 0.1 mm or less, and the particles are suspended in a sufficiently large volume of air to give each particle enough space for its unrestricted burning, the combustion rate will be very fast, and the energy required for ignition very small. Such a burning dust cloud is a dust explosion. In

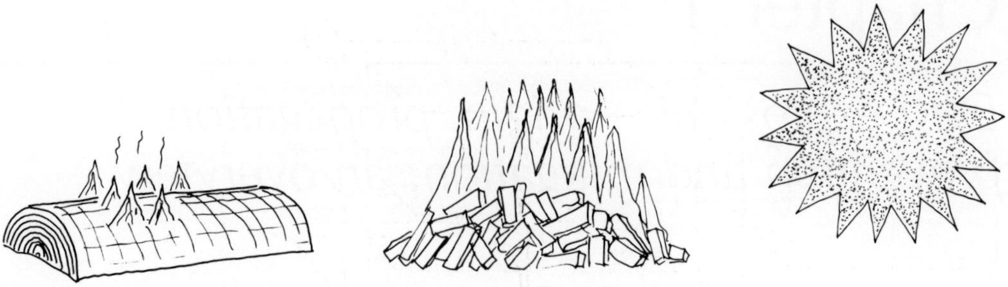

Figure 1.1 *Illustration of how the combustion rate of a given mass of combustible solid increases with increasing sub-division*

general, the dust cloud will be easier to ignite and burn more violently the smaller the dust particles are, down to some limiting particle size that depends on the type of dust material. If such an explosive combustion of a dust cloud takes place inside process equipment or work rooms, the pressure in the fully or partly enclosed explosion space may rise rapidly and the process equipment or building may burst, and life, limb and property can be lost.

1.1.1.3
Specific surface area – a convenient measure of dust fineness

The degree of subdivision of the solid can be expressed either in terms of a characteristic particle size, or as the total surface area per unit volume or unit mass of the solid. The latter characteristic is called the specific surface area of the subdivided solid.

Figure 1.2 illustrates the relationship between the particle size and the specific surface area. After sub-division of the original cube to the left into eight cubes of half the linear dimension of the original cube, the total surface area has increased by a factor of two, which indicates that the specific surface area is simply proportional to the reciprocal of the linear dimension of the cube. This can be confirmed by simply expressing the specific surface area S as the ratio between surface area and volume of one single cube of edge length x. One then finds

$$S = \frac{6x^2}{x^3} = \frac{6}{x} \tag{1.1}$$

This is also the specific surface area of a powder or dust consisting of monosized cubes of edge length x.

The same result applies to spheres of diameter x, because

$$S = \frac{\pi x^2}{(\pi/6) \times x^3} = \frac{6}{x} \tag{1.2}$$

For flake-shaped particles for which the thickness x is much smaller than the characteristic flake diameter, one has

$$S = \frac{2}{x} \tag{1.3}$$

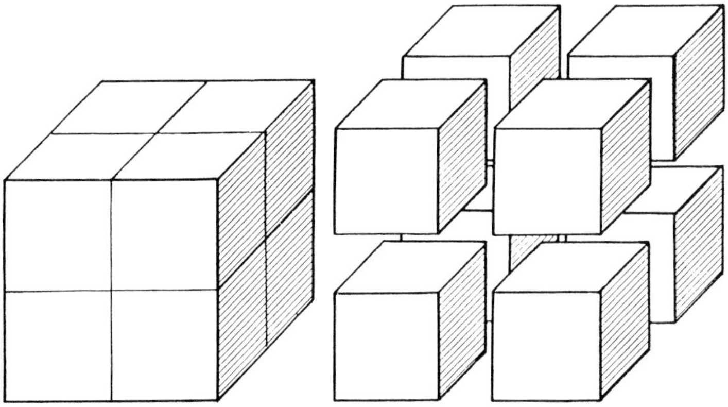

Figure 1.2 *Illustration of the increase of the specific surface area of a solid with increasing sub-division (From Hammond and Kaye, 1963)*

If a spherical particle of diameter, for example, 5 μm is compressed and deformed plastically to a thin flake of thickness, for example, 0.2 μm (flake diameter about 20 μm), Equations (1.2) and (1.3) show that the specific surface area increases by a factor of 8.3. This effect is utilized when producing highly reactive aluminium flakes from atomized (spherical) aluminium particles (see Section 1.3.2).

If the 'particles' are fibres of large length-to-diameter ratio, and the diameter is x, one gets

$$S = \frac{4}{x} \tag{1.4}$$

Fibrous dusts may either be natural, for example cellulose, or synthetic, such as flock materials. Ignitability and explosibility of synthetic flock materials were discussed by Schenk (1984).

In the case of polysized cubes or spheres, the specific surface area equals

$$S = 6\Sigma x_i^2 n_i / \Sigma x_i^3 n_i \tag{1.5}$$

where n_i is the number of particles in the size interval i in the sample considered.

As the particles get smaller, the inter-particle forces play an increasingly important role compared with gravity forces and in a given practical situation the dust in a dust cloud may not necessarily be dispersed into the small individual primary particles, but rather into larger agglomerates, or lumps. The effective particle size will therefore be larger and the effective specific surface area smaller than if the primary particles had been completely dispersed. This important aspect is discussed in Section 1.3.3 and in depth in Chapter 3. See also Section 7.4.2.

1.1.1.4
Factors influencing ignition sensitivity and explosion violence of dust clouds

Particle size/specific surface area of the dust is a central factor. However, there are other important factors too, and the comprehensive list may look as follows:

1. Chemical composition of the dust, including its moisture content.
2. Chemical composition, and initial pressure and temperature of the gas phase.

3. Distributions of particle sizes and shapes in the dust, determining the specific surface area of the dust in the fully dispersed state.
4. Degree of dispersion, or agglomeration, of dust particles, determining the effective specific surface area available to the combustion process in the dust cloud in the actual industrial situation.
5. Distribution of dust concentration in the actual cloud.
6. Distribution of initial turbulence in the actual cloud.
7. Possibility of generation of explosion-induced turbulence in the still unburnt part of the cloud. (Location of ignition source important parameter.)
8. Possibility of flame front distortion by other mechanisms than turbulence.
9. Possibility of significant radiative heat transfer (highly dependent on flame temperature, which in turn depends on particle chemistry).

Factors three and four have already been mentioned. These and other factors will be discussed in more detail in the subsequent sections. Factors one, two, three and nine can be regarded as basic parameters of the explosible dust cloud. Factors four to eight are, however, influenced by the actual industrial dust cloud generation process and explosion development. These in turn depend on the nature of the industrial process (flow rates, etc.) and the geometry of the system in which the dust cloud burns. The location of the ignition point is another parameter that can play an important role in deciding the course of the explosion.

In view of the wide spectrum of dust cloud concentrations, degrees of dust dispersion and turbulence, and of locations of potential ignition sources in industry, a correspondingly wide spectrum of possible dust cloud ignition sensitivities and combustion rates must be expected.

This complex reality of the process industry is also shared by laboratory experimentation and represents a constant challenge in the design of adequate experiments and interpretation of experimental results.

1.1.1.5 Previous books on the dust explosion problem

During the years several textbooks on the dust explosion hazard have been produced. One of the first ones was published in Germany by Beyersdorfer (1925), and he mentioned that his motivation for writing the book arose from three questions. The first, asked by most people, was: 'Are dust explosions really existing?' The second question, asked by the plant engineer, was: 'Why are we having so many dust explosions?' The final question was asked by the researcher: 'Why are we not having many more dust explosions?' Although out of date on some points, Beyersdorfer's pioneering book is still fascinating reading.

Almost half a century elapsed from the publication of Beyersdorfer's text till the next comprehensive book on dust explosions appeared. It should be mentioned though, that in the meantime some valuable summaries were published as parts of other books, or as reports. Examples are the reports by Verein deutscher Ingenieure (1957), and Brown and James (1962), and the sections on dust explosions in the handbook on room-explosions in general, edited by Freytag (1965). In his book on hazards due to static electricity, Haase (1972) paid attention also to the dust explosion problem. However, it was Palmer (1973) who produced the long desired updated comprehensive account of work in the Western world up to about 1970. In Eastern Europe, a book on the prevention of accidental dust explosions, edited by Nedin (1971), was issued in the USSR two years earlier. Cybulski's comprehensive account on coal dust explosions appeared in Polish in 1973, i.e. simultaneously with the publication of Palmer's book, and the English translation came two years later (Cybulski (1975)). In F. R. Germany, Bartknecht had conducted extensive research and testing related to dust explosions in coal mines as well as in the chemical process industries. This work was summarized in a book by Bartknecht (1978), which was subsequently translated to English. The book by Bodurtha (1980) on industrial explosion prevention and protection also contains a chapter on dust explosions.

Two years after, two further books were published, one by Field (1982) and one by Cross and Farrer (1982), each being quite comprehensive, but emphasizing different aspects of the dust explosion problem. In the subsequent year Nagy and Verakis (1983) published a book in which they summarized and analysed some of the extensive experimental and theoretical work conducted by US Bureau of Mines up to 1980 on initiation, propagation and venting of dust explosions. Three years later a book by Korolcenko (1986) was issued in the USSR, reviewing work on dust explosions published both in the West and in Eastern Europe. The year after, Bartknecht (1987) issued his second book, describing his extensive, more recent research and testing at Ciba-Geigy AG, related to dust explosion problems. The Institution of Chemical Engineers in the UK published a useful series of booklets reviewing the status of various aspects of the dust explosion problem, by Lunn (1984), Schofield (1984), Schofield and Abbott (1988), and Lunn (1988). The comprehensive book by Glor (1988) on electrostatic hazards in powder handling should also be specifically mentioned at this point. Valuable information on the same subject is also included in the book by Lüttgens and Glor (1989).

The proceedings of the international symposium on dust explosions, in Shenyang, P. R. China, published by North East University of Technology (1987), contains survey papers and special contributions from researchers from both Asia, America and Europe. EuropEx (1990) produced a collection of references to publications related to accidental explosions in general, including dust explosions. The collection is updated at intervals and contains references to standards, guidelines and directives, as well as to books and papers. Finally, attention is drawn to the proceedings of three conferences on dust explosions, in Nürnberg, published by the Verein deutscher Ingenieure (VDI) in 1978, 1984 and 1989 (see Verein deutscher Ingenieure in the reference list).

1.1.2
MATERIALS THAT CAN GIVE DUST EXPLOSIONS

Dust explosions generally arise from rapid release of heat due to the chemical reaction

$$\textit{fuel} + \textit{oxygen} \rightarrow \textit{oxides} + \textit{heat} \quad (1.6)$$

In some special cases metal dusts can also react exothermally with nitrogen or carbon dioxide, but most often oxidation by oxygen is the heat-generating process in a dust explosion. This means that only materials that are not already stable oxides can give rise to dust explosions. This excludes substances such as silicates, sulphates, nitrates, carbonates and phosphates and therefore dust clouds of Portland cement, sand, limestone, etc. cannot give dust explosions.

The materials that can give dust explosions include:

- Natural organic materials (grain, linen, sugar, etc.).
- Synthetic organic materials (plastics, organic pigments, pesticides, etc.).
- Coal and peat.
- Metals (aluminium, magnesium, zinc, iron, etc.).

The heat of combustion of the material is an important parameter because it determines the amount of heat that can be liberated in the explosion. However, when comparing the various materials in terms of their potential hazard, it is useful to relate the heat of combustion to the amount of oxygen consumed. This is because the gas in a given volume of dust cloud contains a limited amount of oxygen, which determines how much heat can be released in an explosion per unit volume of dust cloud. In Table 1.1 the heats of combustion of various substances, per mole of oxygen consumed, are given. Ca and Mg top the list, with Al closely behind. Si is also fairly high up on the list, with a heat of combustion per mole of oxygen about twice the value of typical natural and synthetic organic substances and coals.

Table 1.1 Heats of combustion (oxidation) of various substances per mole O_2 consumed

Substance	Oxidation product(s)	KJ/mole O_2
Ca	CaO	1270
Mg	MgO	1240
Al	Al_2O_3	1100
Si	SiO_2	830
Cr	Cr_2O_3	750
Zn	ZnO	700
Fe	Fe_2O_3	530
Cu	CuO	300
Sucrose	CO_2 and H_2O	470
Starch	CO_2 and H_2O	470
Polyethylene	CO_2 and H_2O	390
Carbon	CO_2	400
Coal	CO_2 and H_2O	400
Sulphur	SO_2	300

Table 1.1 is in accordance with the experience that the temperatures of flames of dusts of metals like Al and Si are very high compared with those of flames of organic dusts and coals.

The equations of state for ideal gases describes the mutual interdependence of the various parameters influencing the explosion pressure:

$$P = \frac{TnR}{V} \tag{1.7}$$

Here P is the pressure of the gas, T its temperature, V the volume in question, n the number of gas molecules in this volume, and R the universal gas constant. For constant volume, P is proportional to T and n. Normally the increase of T due to the heat developed in the burning dust cloud has the deciding influence on P , whereas the change of n only plays a minor role.

Combustion of metal dusts can cause the maximum possible relative reduction of n, by consuming all the oxygen in the formation of condensed metal oxides. If the gas is air and all the oxygen is consumed and all the nitrogen is left, n is reduced by about 20%.

In the case of organic dusts and coals, assuming that CO_2 (gas) and H_2O (gas) are the reaction products, the number of gas molecules per unit mass of dust cloud increases somewhat during combustion. This is because two H_2O molecules are generated per O_2 molecule consumed. Furthermore, in the case of organic substances containing oxygen, some H_2O and CO_2 is generated by decomposition of the solid material itself, without supply of oxygen from the air.

Consider as an example starch of composition $(C_6H_{10}O_5)_x$ suspended in air at the dust concentration that just consumes all the oxygen in the air to be completely transformed to CO_2 and H_2O (= stoichiometric concentration) 1 m^3 of air at normal ambient conditions contains about 8.7 moles of O_2 and 32.9 moles of N_2. When the starch is oxidized, all the O_2 is spent on transforming the carbon to CO_2, whereas the hydrogen and the oxygen in the starch are in just the right proportions to form H_2O by themselves. The 8.7 moles of O_2 is then capable of oxidizing 8.7/6 = 1.45 moles of $(C_6H_{10}C_5)$, i.e. about 235 g, which is the stoichiometric dust concentration in air at normal conditions. The reaction products will then be 8. 7 moles of CO_2 and 7.3 moles of H_2O. The total number of 41.6 moles of gas in the original 1 m^3 of dust cloud has therefore been transformed to 48.9 moles, i.e. an increase by 17.5%. In an explosion this will contribute to increasing the adiabatic constant-volume explosion pressure correspondingly.

It must be emphasized, however, that this formal consideration is not fully valid if the combustion of the organic particles also results in the formation of CO and char particles. This is discussed in greater detail in Chapter 4.

1.1.3 EXPLOSIBLE RANGE OF DUST CONCENTRATIONS – PRIMARY AND SECONDARY EXPLOSIONS

The explosive combustion of dust clouds, as illustrated in Figure 1.1(c) cannot take place unless the dust concentration, i.e. the mass of dust per unit volume of dust clouds, is within certain limits. This is analogous with combustion of homogeneous mixtures of gaseous fuels and air, for which the upper and lower flammability limits are well established. Figure 1.3 shows the explosible range for a typical natural organic material, such as maize starch, in air at normal temperature and atmospheric pressure.

The explosible range is quite narrow, extending over less than two orders of magnitude, from 50–100 g/m^3 on the lean side to 2–3 kg/m^3 on the rich one. As discussed in greater detail in Chapter 4, the explosibility limits differ somewhat for the various dust materials. For example, zinc powder has a minimum explosible concentration in air of about 500 g/m^3. Explosible dust clouds have a high optical density, even at the lower explosible limit. This is illustrated by the fact that the range of maximum permissible dust concentrations that are specified in the context of industrial hygiene in working atmospheres, are three to four orders of magnitude lower than minimum explosible dust concentrations. This means that the unpleasant dust concentration levels that can sometimes occur in the general working atmosphere of a factory, and which calls upon the attention of industrial hygiene authorities, are far below the concentration levels that can propagate dust flames.

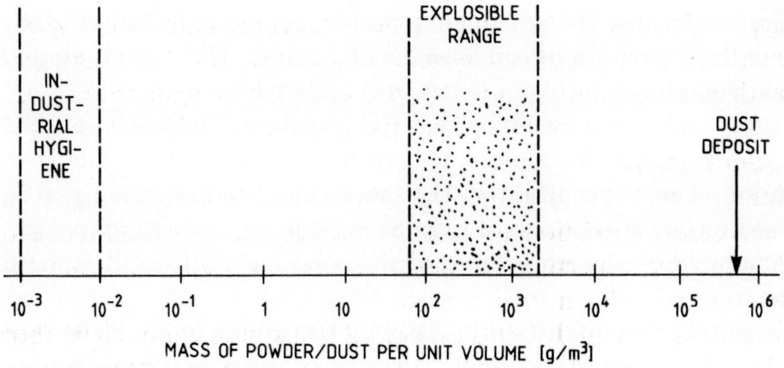

Figure 1.3 *Range of explosible dust concentrations in air at normal temperature and atmospheric pressure for a typical natural organic dust (maize starch), compared with typical range of maximum permissible dust concentrations in the context of industrial hygiene, and a typical density of deposits of natural organic dusts*

Therefore, the minimum explosible concentration corresponds to dust clouds of high optical densities, which are unlikely to occur regularly in work rooms of factories.

A visual impression of the density of explosible dust clouds is provided in Figure 1.4, which illustrates a cubical arrangement of cubical particles.

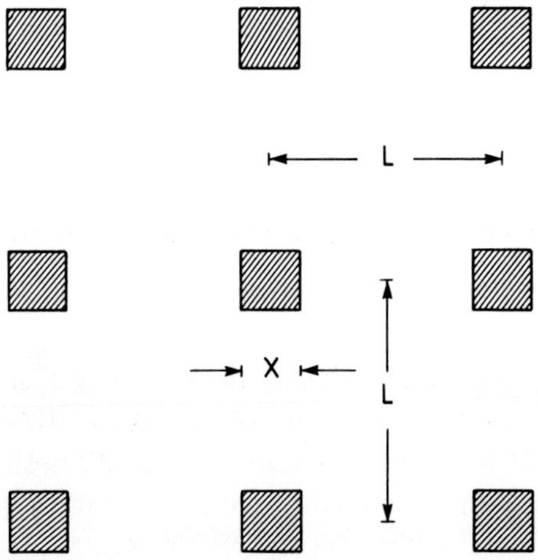

Figure 1.4 *Cubical dust particles of edge x arranged in space in a cubical pattern, with interparticle centre-to-centre distances of L.*

On average, there will be one cubical particle of volume x^3 per cube of dust cloud of volume L^3. If the particle density is ρ_p, the dust concentration equals

$$c = \rho_p (x/L)^3 \tag{1.8}$$

or, in a re-arranged form

$$L/x = (\rho_p/c)^{1/3} \tag{1.9}$$

For particles of density 1 g/cm^3, i.e. 10^6 g/m^3, a dust concentration of 50 g/m^3, corresponds to L/x = 27. For 500 g/m^3, which is a typical worst-case explosible concentration, L/x = 13. The actual density shown in Figure 1.4, of L/x = 4, corresponds to a very high dust concentration, 16 kg/m^3, which is well above the maximum explosible concentration for organic dusts (2–3 kg/m^3).

It is important to notice that the absolute inter-particle distance corresponding to a given dust concentration decreases proportionally with the particle size. For example, at a dust concentration of 500 g/m^3 and a particle density of 1 g/cm^3, L equals 1.3 mm for 100 μm particles, whereas it is only 13 μm for 1 μm particles.

Zehr (1965) quoted a rule of thumb by Intelmann, saying that if a glowing 25 W light bulb is observed through 2 m of dust cloud, the bulb cannot be seen at dust concentrations exceeding 40 g/m^3. This is illustrated in Figure 1.5. It follows from this that the dust clouds in which dust explosions are primarily initiated are normally found inside process

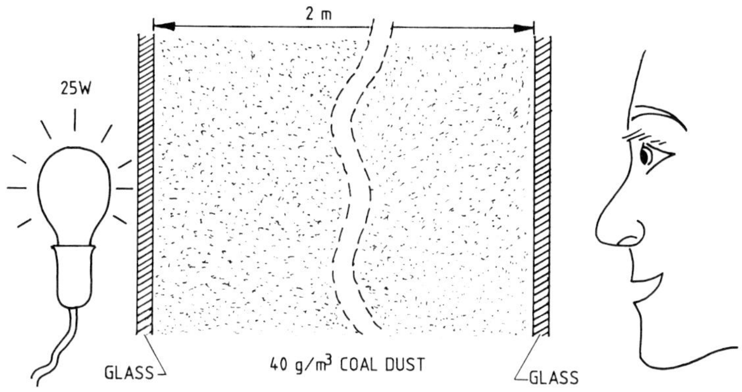

Figure 1.5 *A cloud of 40 g/m^3 of coal dust in air is so dense that a glowing 25 W light bulb can hardly be seen through a dust cloud of 2 m thickness*

equipment, such as mills, mixers, screens, dryers, cyclones, filters, bucket elevators, hoppers, silos, aspiration ducts and pipes for pneumatic transport of powders. Such explosions, initiated by some ignition source (see Section 1.1.4) are called *primary explosions*.

This reveals an important difference between primary dust and gas explosions. In the case of gases, the process equipment normally contains fuel only, with no air, and under such circumstances gas explosions inside process equipment are impossible. Therefore most primary gas explosions occur outside process equipment where gas from accidental leaks is mixed with air and explosible atmospheres generated.

One important objective of dust explosion control (see Section 1.4) is to limit primary explosions in process equipment to the process units in which they were initiated. A central concern is then to avoid *secondary explosions* due to entrainment of dust layers by the blast wave from the primary explosion. Figure 1.3 shows that there is a gap of two orders of magnitude between the maximum explosible dust concentration and the bulk density of dust layers and heaps. The consequence of this is illustrated in Figure 1.6. With reference to this figure it is seen that the simple relationship between the bulk density of the dust layer ρ_{bulk}, the layer thickness, h, the height, H, of the dust cloud produced from the layer, and the dust concentration, c, is:

$$c = \rho_{bulk} \frac{h}{H} \tag{1.10}$$

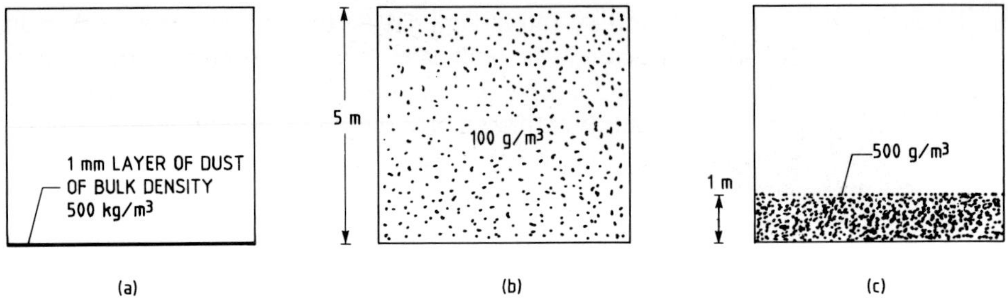

Figure 1.6 *Illustration of the potential hazard of even thin dust layers. A 1 mm layer of a dust of bulk density 500 kg/m³ will generate a cloud of average concentration 100 g/m³ if dispersed in a room of 5 m height. Partial dispersion up to only 1 m gives 500 g/m³*

If a dust layer of thickness h on the internal wall of a cylindrical duct of diameter D, is dispersed homogeneously over the whole tube cross-section, one has

$$c = \rho_{bulk} \frac{4h}{D} \tag{1.11}$$

In the case of a tube diameter of 0.2 m, typical of many dust extraction ducts in industry, a layer thickness of only 0.1 mm is sufficient for generating a dust concentration of 1000 g/m^3 with a dust of bulk density 500 kg/m^3.

In general, dispersable dust layers in process plants represent a potential hazard of extensive secondary dust explosions, which must be reduced to the extent possible. Figure 1.7 illustrates how secondary explosions in work rooms can be generated if preventive precautions are inadequate.

1.1.4
IGNITION SOURCES

1.1.4.1
Background

A combustible dust cloud will not start to burn unless it becomes ignited by a source of heat of sufficient strength. The most common ignition sources are

- Smouldering or burning dust.
- Open flames (welding, cutting, matches, etc.).

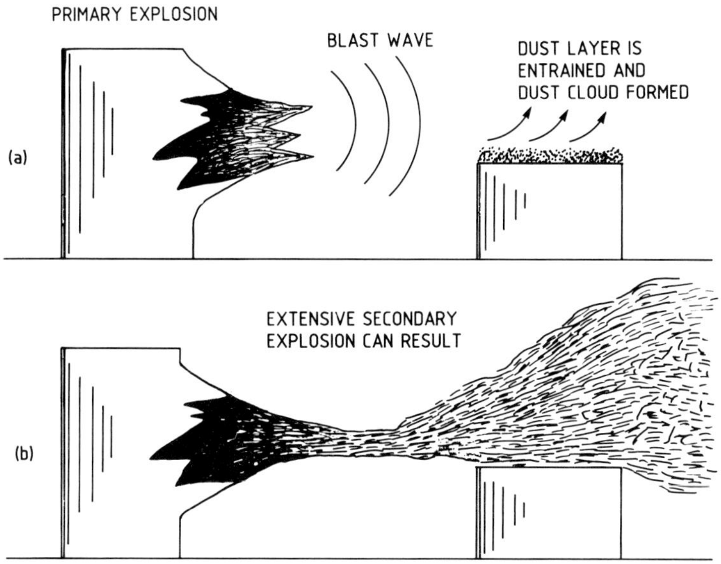

Figure 1.7 *Illustration of how the blast wave from a primary explosion entrains and disperses a dust layer, which is subsequently ignited by the primary dust flame*

- Hot surfaces (hot bearings, dryers, heaters, etc.).
- Heat from mechanical impacts.
- Electrical discharges and arcs.

Some of these sources will be discussed more extensively in Chapter 5, and only a brief outline will be given here.

There is considerable variation in the ignition sensitivity of various types of dusts. This will be discussed in Section 1.3. In order to quantify the ignition sensitivity of dust clouds and dust deposits, when exposed to various kinds of ignition sources, a range of laboratory-scale test methods have been developed, which will be described in Chapter 7.

1.1.4.2 Smouldering or burning dust

Experience has shown that combustible dusts, when deposited in heaps or layers, may under certain circumstances develop internal combustion and high temperatures. This is due to the porous structure of dust deposits, which gives oxygen access to the particle surface throughout the deposit, and also makes the heat conductivity of the deposit low. Consequently, heat developed due to comparatively slow initial oxidation at moderate temperatures inside the dust deposit, may not be conducted into the surroundings sufficiently fast to prevent the temperature in the reaction zone from rising. As long as oxygen is still available, the increased temperature increases the rate of oxidation, and the temperature inside the dust deposit increases even further. Depending on the permeability of the dust deposit and geometrical boundary conditions, the density difference between the hot combustion gases and the air of ambient temperature may create a draught that supplies fresh oxygen to the reaction zone and enhances the combustion process.

If a dust deposit containing such a hot reaction zone, often called a 'smouldering nest', is disturbed and dispersed by an air blast or a mechanical action, the burning dust can easily initiate a dust explosion if brought in contact with a combustible dust cloud. Sometimes the dust in the deposit that has not yet burnt, forms the explosible dust cloud.

The initial oxidation inside the deposit may sometimes be due to the dust or powder being deposited having a higher temperature than planned. However, some natural vegetable materials may develop initial spontaneous combustion even at normal ambient temperatures due to biochemical activity, if the content of fat and/or moisture is high.

In other cases the dust deposit or layer rests on a heated surface, which supplies the heat needed to trigger self-ignition in the dust. Such surfaces can be overheated bearings, heaters in workrooms, light bulbs, walls in dryers, etc. If the surface is not intended to be covered with dust, the dust deposit may prevent normal cooling by forming an insulating layer. This may give rise to an undesirable temperature rise in the surface, which will increase the probability of ignition of the dust further. In general the minimum temperature of the hot surface for the dust layer to self-ignite decreases with increasing thickness of the dust layer.

Figures 1.8, 1.9 and 1.10 illustrate various ways in which smouldering combustion in dust deposits can initiate dust explosions. The critical conditions for generation of

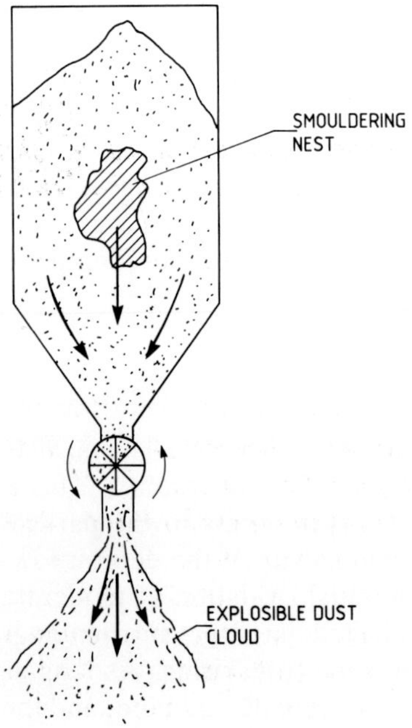

Figure 1.8 *A smouldering nest in a dust/powder deposit in a silo can initiate a dust explosion if the nest is discharged into an explosible dust cloud*

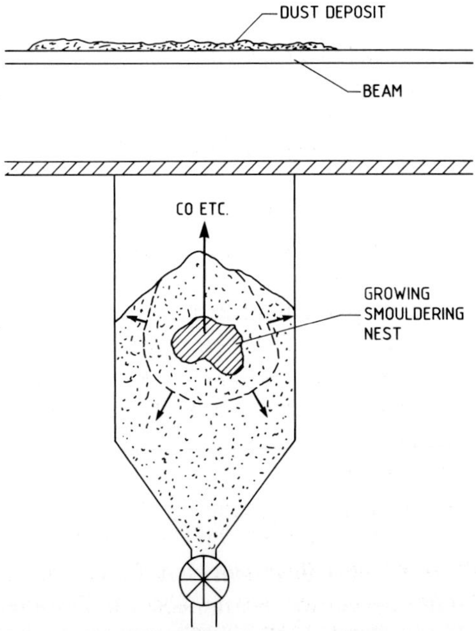

Figure 1.9 *Complex ignition sequence via gas explosion; due to limited supply of oxygen the smouldering nest develops CO and other combustible gases and creates an explosible mixture above the dust deposit. When the edge of the smouldering nest penetrates the top surface of the dust deposit, the gas ignites, and the gas explosion blows up the silo roof. Dust deposits in the room above the silo are dispersed and a major secondary dust explosion results*

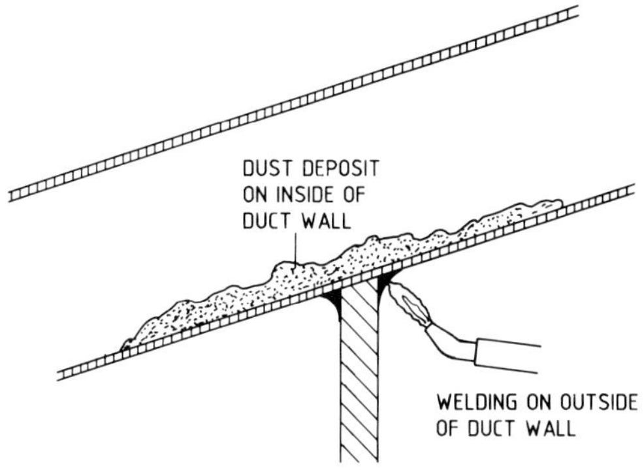

Figure 1.10 *A hidden dust deposit inside a duct can be brought to ignition by heat supplied to the duct wall from the outside*

smouldering nests are discussed in Chapter 5, whereas test methods for assessing the self-heating tendency of various dusts are described in Chapter 7.

It should be mentioned that van Laar (1981) found that burning cigarettes and cigars may give rise to smouldering fires in tapioca and soybean meal.

1.1.4.3
Open flames

The flames of welding and cutting burners are sufficiently powerful to initiate explosions in any dust cloud that is at all able to propagate a self-sustained flame. The cutting burner flame is particularly hazardous because it supplies excess oxygen to the working zone. If combustible dusts are dispersed in atmospheres containing more oxygen than in air, both ignition sensitivity and explosion violence increases compared with clouds in air (see Section 1.3.6). All codes and regulations for preventing dust explosions contain strict requirements to the safety precautions that have to be taken when performing hot work in areas containing dust.

Smoking should be prohibited in areas where combustible dusts exist. A burning wooden match develops about 100 J of thermal energy per second. This is more than sufficient for initiating explosions in most combustible dust clouds.

1.1.4.4
Hot surfaces

Besides igniting dust layers, hot surfaces can initiate dust explosions by direct contact between the dust cloud and the hot surface. However, the minimum hot surface temperatures needed for this are generally considerably higher (typically 400–500°C for organic dusts) than for ignition of dust layers. Further details are given in Chapters 5 and 7.

1.1.4.5
Heat from mechanical impacts

The literature on dust explosions is sometimes confusing when discussing ignition of dust clouds by heat from mechanical impacts. This is reflected in the use of terms such as 'friction' or 'friction sparks' when categorizing ignition sources. In order to clarify the situation, it seems useful to distinguish between friction and impact.

Friction is a process of fairly long duration whereby objects are rubbed against each other and heat is gradually accumulated. This produces hot surfaces, and in some cases inflammation, for example when an elevator or conveyor belt is slipping.

Impact is a short-duration interaction between two solid bodies under conditions of large transient mechanical forces. Small fragments of solid material may be torn off, and if made of metal, they may start burning in air due to the initial heat absorbed in the impact process. In addition, local 'hot spots' may be generated at the points of impact. In some cases the impact may occur repeatedly at one specific point, for example when a fixed object inside a bucket elevator is repeatedly hit by the buckets. This may gradually generate a hot spot of sufficient size and temperature to ignite the dust cloud directly.

A practical mechanical impact situation is illustrated in Figure 1.11. A steel bolt is accidentally entering the top of a large concrete silo during filling of the silo with maize starch. The bolt falls down into the nearly empty silo and hits the concrete wall near the silo bottom at a velocity of 25–30 m/s. Visible sparks are generated. A dense, explosible cloud of maize starch occupies the region where the impact occurs. Is ignition of the cloud probable? This problem will be discussed in detail in Chapter 5, but it should be indicated at this point that ignition by simple impacts where steel is the metal component, seem less likely than believed by many in the past. However, if the metal had been titanium or zirconium, ignition could have occurred in the situation illustrated in Figure 1.11.

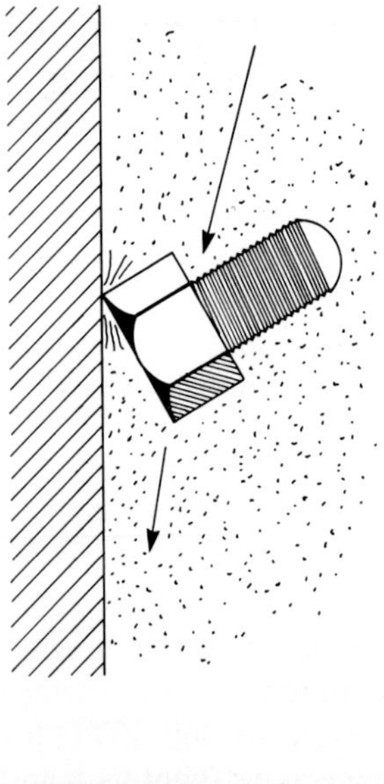

Figure 1.11 *A steel bolt is falling into a tall silo for maize starch and collides with the concrete silo wall at high velocity. Can the steel sparks generated initiate an explosion in the maize starch cloud in the silo?*

The thermite reaction ($2Al + Fe_2O_3 \rightarrow Al_2O_3 + 2Fe + heat$) is often mentioned as a potential ignition source from impacts involving aluminium and rust. However, if a lump of normal soft aluminium collides with a rusty steel surface, a thermite reaction will not necessarily take place. In fact, due to the softness of the aluminium, the result is often just a thin smear of aluminium on top of the rust . However, if this sandwich of aluminium and rust is given a hard blow by a third object, a thermite flash capable of igniting dust clouds can easily be produced. The same applies to a rusty surface that has been painted with aluminium paint, if the pigment content of the paint is comparatively high. (See also Chapters 5 and 7.)

1.1.4.6
Electric sparks and arcs; electrostatic discharges

It has been known since the beginning of this century that electric sparks and arcs can initiate dust explosions. The minimum spark energy required for ignition varies with the type of dust, the effective particle size distribution in the dust cloud, the dust concentration and turbulence, and the spatial and temporal distribution of the energy in the electric discharge or arc.

It was long thought that the electric spark energies needed for igniting dust clouds in air were generally much higher, by one or two orders of magnitude, than the minimum ignition energies for gases and vapours in air. However, it has now become generally accepted that many dusts can be ignited by spark energies in the range 1–10 mJ, i.e. very close to the typical range of gases and vapours.

It may be useful to distinguish between discharges caused by release of accumulated electrostatic charge, and sparks or arcs generated when live electric circuits are broken, either accidentally or intentionally (switches). In the latter case, if the points of rupture are separated at high speed, transient inductive sparks are formed across the gap as illustrated in Figure 1.12. If the current in the circuit prior to rupture is i and the circuit inductance L, the theoretical spark energy, neglecting external circuit losses, will be $1/2\ Li^2$. As an example, a current of 10 A and L equal to $10^{-5}\ H$ gives a theoretical spark energy of 0.5 mJ. This is too low for igniting most dust clouds in air. However, larger currents and/or inductances can easily give incendiary sparks. Sometimes rupture only results in a small gap of permanent distance. This may result in a hazardous stationary arc if the circuit is still live.

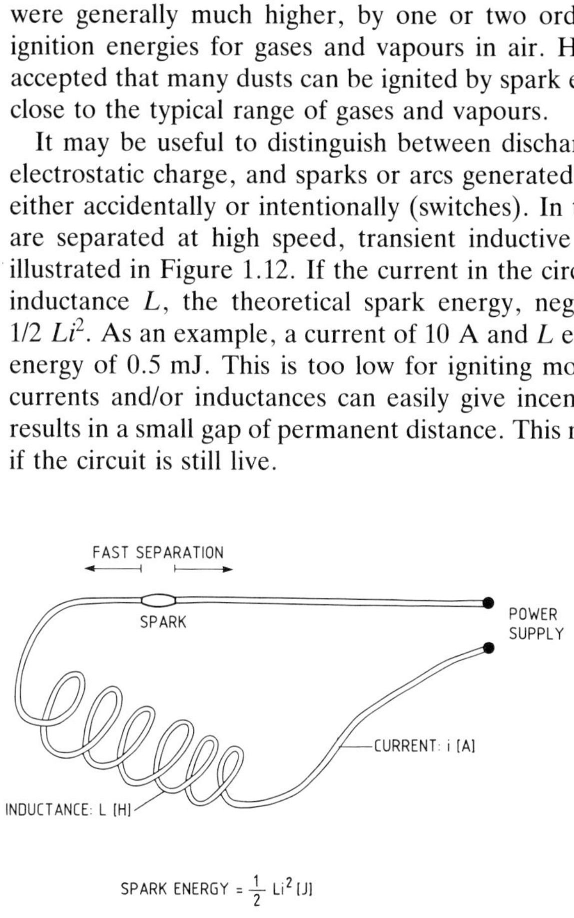

Figure 1.12 *Inductive spark or 'break flash' generated when a live electric circuit is suddenly broken and the points of rupture are separated at high speed*

Over the years the question of whether electrostatic discharges are capable of initiating dust explosions has been discussed repeatedly. The basic mechanism causing accumulation of electrostatic charges in industrial plants is transfer of charge between objects during rubbing. This occurs easily during handling and transport of powders and dusts where charge is exchanged between the powder/dust and the process equipment. The charge accumulated on process equipment or bulk powder can be released in various ways, depending on the circumstances. Glor (1988) and Lüttgens and Glor (1989) distinguished between six different types of electrostatic discharges, namely:

- Spark discharge.
- Brush discharge.
- Corona discharge.
- Propagating brush discharge.
- Discharge along the surface of powder/dust in bulk.
- Lightning-like discharge.

The differentiation between the various discharge types is not always straight-forward, but Glor's classification has turned out to be very useful when evaluating electrostatic hazards in practice in industry.

Spark discharges are by far the most hazardous type of the six with regard to initiating dust explosions in industry. Such discharges occur when the charge is accumulated on an electrically conducting, unearthed, object and discharged to earth across a small air gap. The gap distance must be sufficiently short to allow breakdown and spark channel formation at the actual voltage difference between the charged object and earth. On the other hand, in order for the spark to become incendiary, the gap distance must be sufficiently long to permit the required voltage difference to build up before break-down of the gap occurs. The theoretical spark energy, neglecting external circuit losses, equals $1/2\ CV^2$ where C is the capacitance of the un-earthed, charged process item with respect to earth, and V is the voltage difference. Figure 1.13 illustrates a practical situation that could lead to a dust explosion initiated by an electrostatic spark discharge.

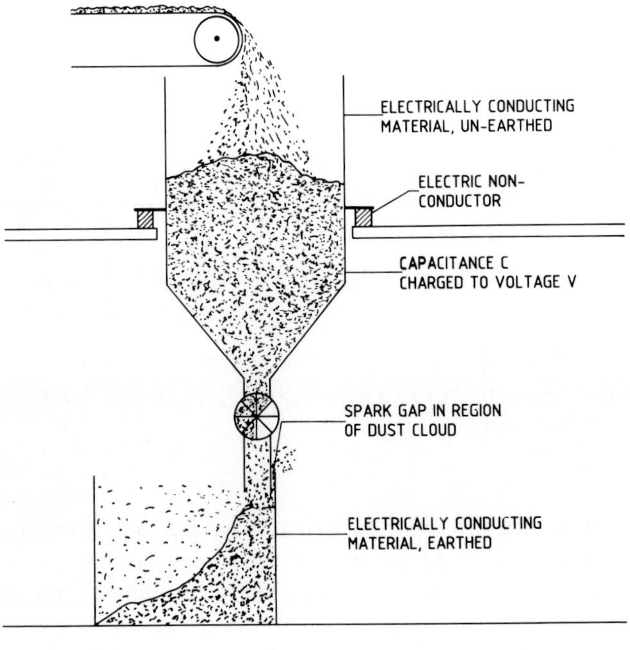

Figure 1.13 *Illustration of a practical situation that could lead to a dust explosion initiated by an electrostatic spark discharge*

Glor (1988) has given some typical approximate capacitance-to-earth values for objects encountered in the process industry. These have been incorporated in Table 1. 2 and used for estimating the maximum theoretical spark energy 1/2 CV^2 when discharging an object of capacitance C at a voltage V to earth.

Minimum electric spark energies (MIE) for ignition of dust clouds vary, as already mentioned, with dust type, particle size, etc. , but many dusts have MIE values well below the higher of the 1/2 CV^2 values in Table 1.2. However, it may not be appropriate to apply MIE values from standard tests directly to the electrostatic spark problem (see Chapter 5).

Turbulence in the dust cloud raises the effective MIE and therefore provides a safety factor. For example Yong Fan Yu (1985) was unable to ignite turbulent clouds of wheat grain dust in a container at the exit of a pneumatic transport pipe, even with soft electric sparks of energies of the order of 1 J.

Table 1.2 Maximum theoretical spark energies 1/2 CV^2 from discharge of various types of electrically conducting objects. Typical approximate capacitance values (From Glor, 1988)

Object	Capacitance [pF]	1/2 CV^2 (mJ) at various voltages 10 kV	20 kV	30 kV
Single screw	1	0.05	0.2	0.45
Flange (100 mm nominal size)	10	0.5	2	4.5
Shovel	20	1	4	9
Small container (bucket, 50 litres drum)	10-100	0.5-5	2-20	4.5-45
Funnel	10-100	0.5-5	2-20	4.5-45
Drum (~200 litres)	100-300	5-15	20-60	45-135
Person	100-300	5-15	20-60	45-135
Major plant items (large containers, reaction vessels)	100-1000	5-50	20-200	45-450
Road tanker	1000	50	200	450

Glor (1988) emphasized that, due to increasing use of non-conducting construction parts in modern industrial plants, the chance of overlooking un-earthed conducting items is high. Therefore the effort to ensure proper earthing of all conducting parts must be maintained, in particular in plants handling dusts of low MIE. According to Glor (1988) adequate earthing is maintained as long as the leak resistance to earth does not exceed 10^6 ohms for process equipment and 10^8 ohms for personnel. However, in practice, one aims for considerably lower resistances to earth.

Brush discharges occur between a single curved, earthed metal electrode (radius of curvature 5–50 mm) and a charged non-conducting surface (plastic, rubber, dust). Brush discharges can ignite explosible gas mixtures. However, according to Glor (1988), no ignition of a dust cloud by a brush discharge has yet been demonstrated, not even in sophisticated laboratory tests using very ignition sensitive dusts. It must be emphasized, however, that this does not apply if the powder/dust contains significant quantities of combustible solvents (see Section 1.3.9).

Corona discharges occur under the same conditions as brush discharges, but are associated with earthed electrodes of much smaller radii of curvature, such as sharp edges and needle tips. For this reason such discharges will occur at much lower field strengths than the brush discharges, and the discharge energies will therefore also be much lower. Consequently, the possibility of igniting dust clouds by corona discharges can be ruled out.

Propagating brush discharges can, however, initiate dust explosions. Such discharges, which will normally have much higher energies than ordinary brush discharges, occur if a double layer of charges of opposite polarity is generated across a thin sheet (< 8 mm thickness) of a non-conducting material (Glor (1988)). The reason for the high discharge energy is that the opposite charges allow the non-conductor surfaces to accumulate much higher charge densities than if the sheet had been charged on only one of the faces. Glor pointed out that in principle close contact of one of the faces of the sheet with earth is not necessary for obtaining a charged double layer. However, in practice earth on one side is the most common configuration. An example is illustrated in Figure 1.14. Powder is transported pneumatically in a steel pipe with an internal electrically insulating plastic coating. Due to the rubbing of the powder against the plastic, charge is accumulated on the internal face of the plastic coating. The high mobility of the electrons in the steel causes build-up of a corresponding charge of opposite polarity on the outer face of coating in contact with the steel. If a short passage between the two oppositely charged faces of the plastic coating is provided, either via a perforation of the coating or at the pipe exit, a propagating brush discharge can result.

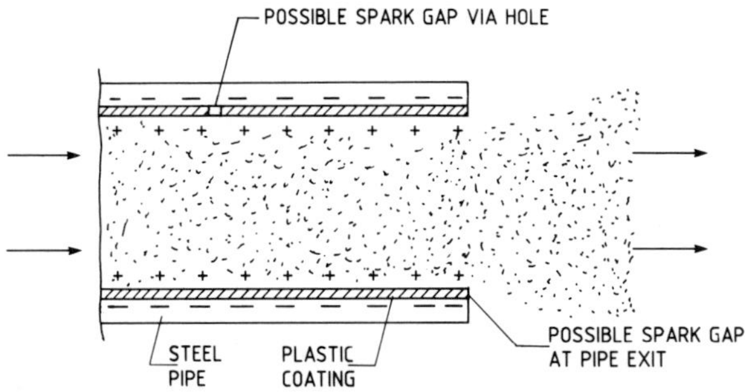

Figure 1.14 *Illustration of practical configuration of pneumatic powder transport that can lead to dust explosions initiated by propagating brush discharges*

Lüttgens (1985) and Lüttgens and Glor (1989) discussed a dust explosion in F. R. Germany that was initiated by a propagating brush discharge. Acrylic powder was transported pneumatically in a 50 mm diameter plastic pipe outdoors, and the earthed electrically conducting shield on the outer surface of the pipe was provided by rainwater and snow.

Glor (1988) identified five typical situations which may lead to propagating brush discharges during transport and handling of powders:

- High-velocity pneumatic transport of powder through an electrically insulating pipe, or a conductive pipe with an insulating internal coating.
- Use of inspection windows of glass or Plexiglass in pneumatic transport pipes.

- Continuous impact of powder particles onto an insulating surface (e.g. a coated dust deflector plate in the cyclone of a dust separator).
- Fast movement of conveyor or transmission belts made of an insulating material, or of a conductive material coated with an insulating layer of high dielectric strength.
- Filling of large containers or silos made of insulating material (e.g. flexible intermediate bulk containers) or of metallic containers or silos coated internally with an insulating layer of high dielectric strength.

Discharge along the surface of powder/dust in bulk may occur if non-conducting powders are blown or poured into a large container or silo. This is a fifth type of electrostatic discharge. When the charged particles settle in a heap in the container, very high space charge densities may be generated and luminous discharges may propagate along the surface of the powder heap, from its base to its top. However, theoretical calculations by Glor (1985) revealed that under realistic industrial conditions only very large particles, of 1–10 mm diameter, are likely to generate spark discharges due to this process. It further seems that very high specific electrical resistivity of the powder is also a requirement ($>10^{10}$ ohm•m) which probably limits this type of discharge to coarse plastic powders and granulates. Because of this large size, the particles generating the discharge are unlikely to give dust explosions, and therefore a possible explosion hazard must be associated with the simultaneous presence of an explosible cloud of an additional, fine dust fraction. The maximum equivalent spark energy for this type of discharge has been estimated to the order of 10 mJ, but still little is known about the exact nature and incendivity of these discharges. Glor (1988) pointed out that the probability of occurrence of such discharges increases with increasing charge-to-mass ratio in the powder, and increasing mass filling rate.

Lightning type discharge, which may in principle occur within an electrically insulating container with no conductive connection from the interior to the earth was the last type of discharge mentioned by Glor (1988) and Lüttgens and Glor (1989). However, as Glor stated, there is no evidence that lightning discharges have occurred in dust clouds generated in industrial operations. Thorpe *et al.* (1985) investigated the hazard of electrostatic ignition of dust clouds inside storage silos in a full-scale pneumatic conveying and storing facility. Sugar was used as test dust. They were able to draw some spark discharges from the charged dust cloud, but these were of low energy, and incapable of causing ignition. In fact, these spark discharges were not even able to ignite a propane/air mixture of minimum ignition energy less than 1 mJ.

Figure 1.15 gives an overall comparison of the equivalent energy ranges of the various electrostatic discharges discussed above, and typical MIE ranges for gases/vapours and dusts in air. The concept of 'equivalent energy', introduced by Gibson and Lloyd (1965), is defined as the energy of a spark discharge that has the same igniting power as the actual electrostatic discharge.

Further details of the generation and nature of the various types of electrostatic discharges are given by Glor (1988) and Lüttgens and Glor (1989). Some further details concerning electric sparks, and their ability to ignite dust clouds, are given in Chapter 5.

The Appendix gives some MIE values, determined by a standardized method, for various dusts.

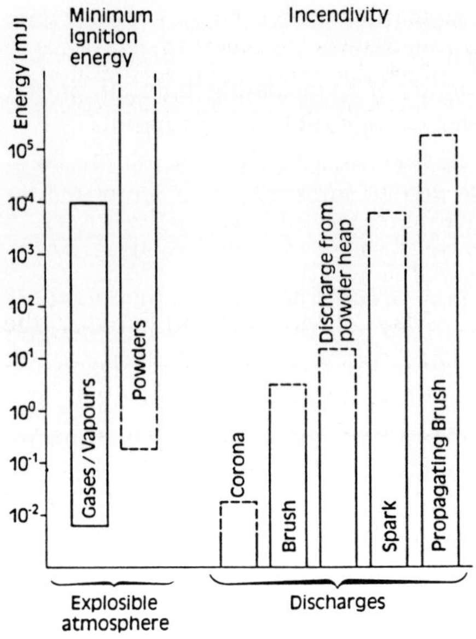

Figure 1.15 *Comparison of ranges of minimum ignition energies of dusts and gases in air, with the equivalent energies of various types of electrostatic discharges. The dotted parts of the columns represent approximate maximum and minimum limit ranges (From Glor, 1988)*

1.2 SIGNIFICANCE OF THE DUST EXPLOSION HAZARD: STATISTICAL RECORDS

1.2.1 RECORDING DUST EXPLOSIONS, AN ACTIVITY OF LONG TRADITIONS

Dust explosions have been a recognized threat to humans and property for a long time. One of the earliest comprehensive reports known is Count Morozzo's (1795) detailed analysis of an explosion in the flour warehouse of Mr. Giacomelli in Turin in 1785 (see full report in Chapter 2). It is interesting to observe that Morozzo also mentions even earlier incidents of violent combustion of clouds of flour in air.

However, at the time of Morozzo the coal mining industry was not fully aware of the important part played by coal dust in the serious coal mine explosions which had become quite common. Faraday and Lyell (1845) were probably some of the first scientists to realize the central role of coal dust in these explosions. In their report to Sir James Graham they discussed the fatal explosion in the Haswell coal mine near Durham, UK, on the 28th September 1844. It was concluded that the primary event was a methane/air ('firedamp') explosion initiated by a defective Davy lamp. However, the central role of the coal dust in developing the devastating main explosion was emphasized, based on a systematic analysis that is exemplary even today. Thus, in their report Faraday and Lyell stated:

> In considering the extent of the fire for the moment of explosion, it is not to be supposed that fire-damp is its only fuel; the coal dust swept by the rush of wind and flame from the floor, roof, and walls of the works would instantly take fire and burn, if there were oxygen enough in

> the air present to support its combustion; and we found the dust adhering to the face of the pillars, props, and walls in the direction of, and on the side towards the explosion, increasing gradually to a certain distance, as we neared the place of ignition. This deposit was in some parts half an inch, and in others almost an inch thick; it adhered together in a friable coked state; when examined with the glass it presented the fused round form of burnt coal dust, and when examined chemically, and compared with the coal itself reduced to powder, was found deprived of the greater portion of the bitumen, and in some instances entirely destitute of it. There is every reason to believe that much coal-gas was made from this dust in the very air itself of the mine by the flame of the fire-damp, which raised and swept it along; and much of the carbon of this dust remained unburnt only for want of air.

During the 150–200 years that have passed since the days of Morozzo and Faraday, the phenomenon of dust explosions has become fully accepted as a serious industrial hazard. Furthermore, since that time the expanding chemical and metallurgical industries have given birth to a steadily increasing number of new, finely divided combustible solid materials that have caused dust explosions to remain a significant hazard in many industries. As an important element in the constant efforts to fight the dust explosion hazard, actual accidents are carefully investigated. In some countries valuable statistical records are available, some of which will be discussed in the following.

1.2.2 DUST EXPLOSIONS IN USA 1900–1956

National Fire Protection Association published a report of important dust explosions in USA from 1900 to 1956 (NFPA 1957). The report gives informative details of a selection of 75 of the most serious and recent of the 1123 explosions recorded. The selection covers a wide range of dusts from all the categories wood, food and feed, metals, plastics, coal, paper and chemicals. In addition, each of the 1123 explosions is mentioned briefly individually by specifying the date, location, dust involved, probable ignition source, number of fatalities and injuries, and material losses.

Table 1.3 gives an overall summary of the consequences of explosions involving various dust categories. The table illustrates some interesting differences. For example, the metal dust explosions, representing 7.1% of the total number of explosions, were responsible for

Table 1.3 Dust explosions in USA 1900–1956: fatalities, injuries and material losses in a sample of 1123 accidental explosions

TYPE OF DUST	EXPLOSIONS		FATALITIES			INJURIES			MATERIAL LOSSES	
	Number	[%]	Number	[%]	per explosion	Number	[%]	per explosion	Million US$ *	per explosion
Wood and bark	162	14.5	38	5.6	0.23	160	9.0	0.99	11.4	0.070
Food and feed	577	51.4	409	60.5	0.71	1061	60.0	1.84	75.8	0.131
Metals	80	7.1	108	16.0	1.35	198	11.2	2.48	3.2	0.040
Plastics	61	5.4	44	6.5	0.72	121	6.8	1.98	3.7	0.061
Coal (not mines)	63	5.6	30	4.4	0.48	37	2.1	0.59	1.6	0.025
Paper	9	0.8	0	0.0	0.0	0	0.0	0.0	0.5	0.056
Others	171	15.2	47	7.0	0.27	193	10.9	1.13	4.3	0.025
All	1123	100.0	676	100.0		1770	100.0		100.5	

*Numerical value at year of explosion. Not inflated.
(Data from NFPA, 1957)

16% of all the fatalities and 11.2% of all the injuries, but only 3.2% of the material losses. The food and feed dust explosions also were responsible for higher percentages of fatalities and injuries than the 51.4% share of the number of explosions. Furthermore, food and feed gave by far the highest material loss per explosion. The pulverized coal dust explosions (not mining), on the contrary, gave lower percentages of both fatalities, injuries and material losses than their share of the total number of explosions.

1.2.3 DUST EXPLOSIONS IN F. R. GERMANY 1965–1985

Berufsgenossenschaftliches Institut für Arbeitssicherheit (Institute of Safety at Work of the Trade Unions) in F. R. Germany have conducted a programme of recording dust explosion accidents in F. R. Germany since the beginning of the 1960s. The first comprehensive report, covering 1965–1980 was published by Beck and Jeske (1982). A condensed version of the findings was given by Beck (1982). The comprehensive report contains a brief description of each explosion accident specifying the type of plant, the specific plant item, the type of dust, the likely ignition source, numbers of fatalities and injuries, and material losses. A further comprehensive report covering the explosions recorded from 1981 to 1985 was published by Jeske and Beck (1987), and the corresponding short version by Beck and Jeske (1988). Finally Jeske and Beck (1989) published an informative overview covering the whole span 1965–1985.

The total numbers of explosions recorded were 357 for 1965–1980 and 69 for 1981–1985. Beck and Jeske (1982) estimated the recorded explosions from 1965 to 1980 to about 15% of the total number of explosions that had actually occurred. The estimated number of actual dust explosions in F. R. Germany from 1965 to 1980 was therefore about 2400, i.e. about 160 per year. The number of explosions recorded per year for 1981–1985 was somewhat lower than for 1965–1980. However, because of the low percentage of recorded explosions it may not be justified to conclude that the annual number of accidental explosions dropped significantly after 1980.

Table 1.4 gives some data from F. R. Germany that can be compared directly with the older data from USA in Table 1.3. There are interesting differences in the distribution of the number of explosion accidents on the various dust categories. This may reflect both a change with time, from the first to the second part of this century, and differences between the structure of the industry in USA and in the F. R. Germany. One example is food and feed, which only represented 25% of all the explosions in F. R. Germany, whereas in USA the percentage was more than 50. However, the percentages of both fatalities and injuries for this dust group both in F. R. Germany and USA was higher than the percentage of explosions. On the other hand, the percentage of the explosions involving metal dusts was about twice as high in F. R. Germany as in USA. The higher percentage of both fatalities and injuries for metal dust explosions than the percentage of the number of explosions is, however, in agreement with the older data from USA. This probably reflects the extreme violence and temperatures of flames of metals like magnesium, aluminium and silicon.

Table 1.5 shows how the involvement of various categories of plant items in the explosions varies with dust type. This reflects differences between typical processes for producing, storing and handling the various categories of powders and dusts.

Table 1. 6 shows the frequencies of the various ignition sources initiating explosions in the same dust categories as used in Table 1.5. The category mechanical sparks may not be entirely unambiguous, and causes some problems with interpreting the data.

Table 1.4 Dust explosions in F. R. Germany 1965–1980: fatalities and injuries in a sample of 357 explosions

TYPE OF DUST	EXPLOSIONS		FATALITIES			INJURIES		
	Number	[%]	Number	[%]	per explosion	Number	[%]	per explosion
Wood	113	31.6	12	11.7	0.11	124	25	1.10
Food and feed	88	24.7	38	36.8	0.43	127	26	1.44
Metals	47	13.2	18	17.5	0.38	91	18.5	1.94
Plastics	46	12.9	18	17.5	0.39	98	20	2.13
Coal /peat	33	9.2	7	6.8	0.21	39	8	1.18
Paper	7	2.0	0	0.0	0.0	0	0	0.0
Others	23	6.4	10	9.7	0.43	13	2.5	0.56
All	357	100.0	103	100.0		492	100.0	

(From Beck, 1982)

Table 1.5 Dust explosions in F. R. Germany 1965–1985: frequencies in % of primary involvement of various plant items in a total of 426 dust explosions, and in the explosions of various categories of dusts

TYPE OF PLANT ITEM	TOTAL OF 426 EXPLOSIONS			Wood/ wood products	Coal/ peat	Food and feed	Plastics	Metals
	Number	% of total	% change 80/85					
Silos/bunkers	86	20.2	0	35.9	23.1	22.9	2	2
Dust collecting systems	73	17.2	+ 2.9	18.0	5.1	9.5	13.5	45.6
Milling and crushing plants	56	13.0	- 0.7	7.0	12.8	18.1	15.4	5.3
Conveying systems	43	10.1	0	4.7	5.1	26.7	17.3	2.0
Dryers	34	8.0	+ 0.4	10.2	2.0	7.6	9.6	2.0
Furnaces	23	5.4	+ 0.1	10.9	18.0	2.0	0	0
Mixing plants	20	4.7	+ 0.2	0	5.1	2.0	17.3	3.5
Grinding and polishing plants	19	4.5	0	3.9	0	0	2	22.8
Sieves/classifiers	12	2.8	- 0.3	4.7	0	2.8	0	3.5
Unknown/others	60	14.1	- 2.6	4.7	28.8	8.4	22.9	13.3
All	426	100.0	0	100.0	100.0	100.0	100.0	100.0

(From Jeske and Beck, 1989)

Table 1.7 gives an interesting correlation between the various plant items involved in the explosions, and the probable ignition sources. 'Mechanical sparks' are frequent ignition sources in dust collectors, mills and grinding plants, whereas smouldering nests are typical when the explosion is initiated in silos, bunkers and dryers. Apart from in dryers, spontaneous ignition was not very frequent. The distinction between smouldering nests and spontaneous heating may not always be obvious.

Electrostatic discharges were the dominating ignition source in mixing plants, but as Table 1.6 shows, electrostatic discharge ignition occurred almost solely with plastic dusts. Presumably mixers are quite frequent in plants producing and handling plastic dusts, and the combination of mixers and plastic dusts is favourable for generating electrostatic discharges.

Proust and Pineau (1989) showed that there is reasonably good agreement between the findings of Beck and Jeske for F. R. Germany and statistics of industrial dust explosions in UK from 1979–1984, as reported by Abbot (1988).

Table 1.6 Dust explosions in F. R. Germany: frequencies in % of initiation by various types of ignition sources iln a total of 426 explosions, and in the explosions of various categories of dusts

TYPE OF IGNITION SOURCE	TOTAL OF 426 EXPLOSIONS			Wood/ wood products	Coal/ peat	Food and feed	Plastics	Metals
	Number	% of total	% change 80/85					
Mechanical sparks	112	26.2	- 2.8	26.6	5.1	22.8	21.2	56.1
Smouldering nests	48	11.3	+ 1.5	19.5	20.5	5.7	9.6	0
Mechanical heating, Friction	38	9.0	0	9.4	5.1	12.4	9.6	3.5
Electrostatic discharges	37	8.7	0	2.3	0	6.7	34.6	5.3
Fire	33	7.8	- 0.6	14.8	12.8	4.8	2	2
Spontaneous ignition (self-ignition)	21	4.9	+ 0.4	3.1	15.4	6.7	2	3.5
Hot surfaces	21	4.9	- 0.4	5.5	10.3	2.8	3.9	3.5
Welding/cutting	21	4.9	+ 0.4	2.3	2.6	12.4	2	2
Electrical machinery	12	2.8	- 0.3	0	2.6	5.7	2	0
Unknown/not reported	68	16.0	+ 1.7	16.5*	25.6*	20.0*	13.1*	24.1*
Others	15	3.5	+ 0.1					
All	426	100.0	0	100.0	100.0	100.0	100.0	100.0

* This figure also includes 'Others'.

Table 1.7 Dust explosions in F. R. Germany 1965–1985: frequencies in % of various types of ignition sources of explosions initiated in various plant items

TYPE OF IGNITION SOURCE	All 426 explosions	Silos/ bunkers	Dust collectors/ separators	Mills and crushing plants	Conveying systems	Dryers	Mixing plants	Grinding plants	Sieves/ classifiers
Mechanical sparks	26.2	16.3	41.1	60.0	25.6	0	15.0	89.5	16.7
Smouldering nests	11.3	27.9	11.0	0	2.3	29.4	0	0	8.3
Mechanical heating Friction.	9.0	3.5	6.8	12.7	25.6	2.9	25.0	5.3	0
Electrostatic discharges	8.7	2.3	9.6	5.5	18.6	5.9	45.0	0	16.7
Fire	7.8	4.7	4.1	2	0	0	5.0	0	16.7
Spontaneous ignition (self-ignition)	4.9	2.3	2.7	0	4.7	14.7	0	0	8.3
Hot surfaces	4.9	11.6	0	3.6	2.3	23.5	0	0	0
Welding/cutting	4.9	5.8	2	0	4.7	2.9	5.9	0	0
Electrical machinery	2.8	2.3	2	0	0	0	0	0	0
Unknown/Others	19.5	23.3	20.7	16.2	16.2	20.7	4.1	5.2	33.3
All	100.0	100.0	100.0	100.0	100.0	100.0	100.0	100.0	100.0

(From Jeske and Beck, 1989)

1.2.4
RECENT STATISTICS OF GRAIN DUST EXPLOSIONS IN USA

Schoeff (1989) presented some statistical data that are shown in a slightly re-arranged form in Table 1.8. The data for 1900–1956 are from the same source as the data in Table 1.3. The alarming trend is that the annual number of explosions seems to increase rather than decrease. The annual number of fatalities is also higher for the last period 1979–1988 than for the previous one 1957–1975. The annual number of injuries for the last period is higher than for both previous periods. From 1957–1975 to 1979–1988 the annual estimated damage to facilities seems to have increased considerably more than what can be accounted for by inflation.

Table 1.8 Grain dust explosions in USA: Recent development

LOSS CATEGORY	1900-1956		1957-1975		1979-1988	
	Total	Per year	Total	Per year	Total	Per year
Number of explosions	490	8.6	192	10.1	202	20.2
Fatalities	381	6.8	68	3.6	54	5.4
Injuries	991	17.4	346	18.2	267	26.7
Estimated damage to facility, Mill. US$, not inflated	70	1.3	55	2.9	169	16.9

(Data from Schoeff, 1989).

It can be misleading to take the figures in Table 1.8 too far. However, the data do indicate that dust explosions remain a persistent threat to human life and limb, and to property. Therefore the efforts to fight the dust explosion hazard have to be continued.

1.3 DUST AND DUST CLOUD PROPERTIES THAT INFLUENCE IGNITABILITY AND EXPLOSION VIOLENCE

1.3.1 DUST CHEMISTRY INCLUDING MOISTURE

There are two aspects to consider, namely the thermodynamics of the explosion, and the kinetics. The thermodynamics is concerned with the amount of heat that is liberated during combustion, the kinetics with the rate at which the heat is liberated.

Dust chemistry influences both thermodynamics and kinetics, which are also to some extent coupled. Table 1.1 shows that there is a considerable difference between the amounts of heat developed per mole of oxygen consumed, for various groups of materials. Calcium, magnesium and aluminium top the list with 1100–1300 kJ/mole O_2. The lowest value is 300 kJ/mole O_2 for copper and sulphur. It would be expected that this difference is to some extent reflected in the maximum pressure of explosions, when performed adiabatically at constant volume. Zehr (1957) made some calculations of the maximum pressures to be expected under such conditions. In Figure 1.16 his results have been plotted against data from experiments in either 1 m^3 or 20 litre closed bombs, taken from Table A1 in the Appendix. For aluminium and magnesium Zehr only indicated that the theoretical values would be larger than 10 and 13.5 bar (g) respectively. Figure 1.16 suggests a fair correlation between theoretical and experimental data, with the theoretical results being somewhat higher than the experimental ones. This would be expected because of the idealized assumptions of stoichiometry and complete oxidation of all fuel, on which the calculations were based.

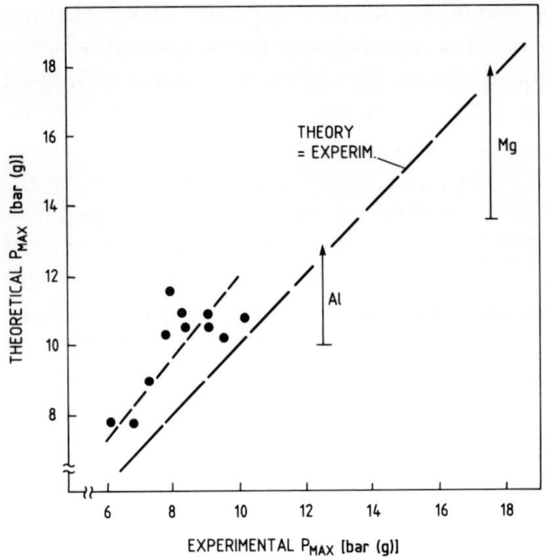

Figure 1.16 *Correlation between experimental P_{max} at constant volume from experiments in 1 m^3 or 20 litre closed vessels (Table A1, Appendix A), and theoretical P_{max} calculated by Zehr (1957).*

As discussed in detail in Chapter 7, the maximum rate with which the explosion pressure rises in closed-bomb experiments is a frequently used relative measure of the violence to be expected from explosions of a given dust.

Figure 1.17 shows how the maximum rates of pressure rise of starch (potato and maize starch) are systematically higher than for protein (two fish powders with fat removed) for the same specific surface area. The nitrogen compounds in the protein are probably in some way slowing down the combustion process.

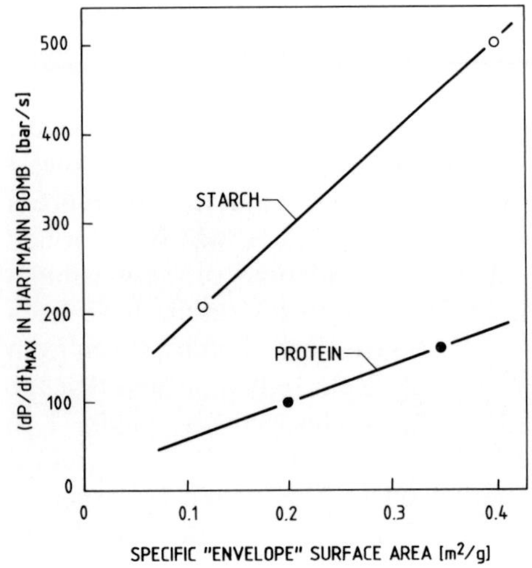

Figure 1.17 *Influence of chemistry (starch or protein) and specific surface area of natural organic materials on maximum rate of pressure rise in closed 1.2 litre Hartmann bomb (From Eckhoff, 1977/1978)*

Eckhoff (1977/1978) used the data in Figure 1.17 for producing an empirical equation, based on simple linear interpolation, for predicting maximum rates of pressure rise for natural organic dusts. Reasonable agreement with experiments was found for a range of food and feedstuffs dust, fish meals and cellulose.

Another example of the influence of dust chemistry on the explosion kinetics is shown in Figure 1.18. The heats of combustion of PVC and polyethylene are not very different. Closed-bomb experiments also give about the same maximum pressure for very small particle sizes. However, the chlorine in the PVC causes a quite dramatic drop in the rate of heat release as the median particle size increases beyond about 20 μm. Due to the very slow combustion, P_{max} for PVC also drops much faster as the particle size increases than for polyethylene. The retarding influence of chlorine on the combustion process most probably is of the same nature as that of the halogens in the halons, which were extensively used for explosion and fire suppression before the negative influence of such materials on the global environment was fully realized.

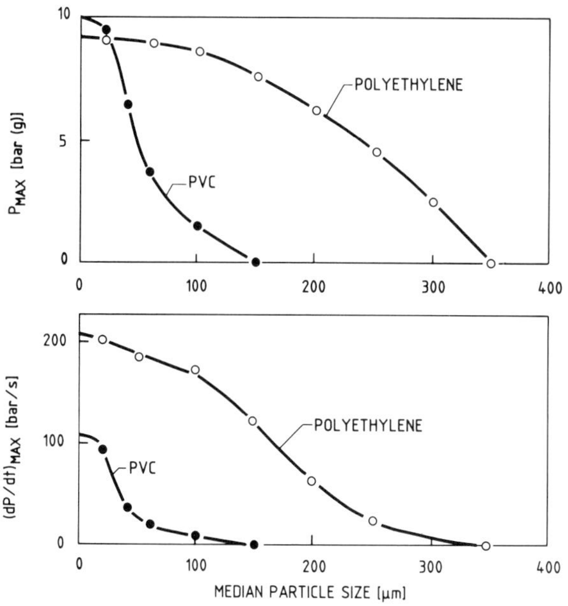

Figure 1.18 *Influence of chlorine in molecule of dust material on maximum explosion pressure and maximum rate of pressure rise in 1 m^3 standard ISO vessel, for various particle sizes (From Bartknecht, 1978)*

Moisture in the dust reduces both ignition sensitivity and explosion violence of dust clouds. Figure 1.19 illustrates the influence of dust moisture on the minimum electric spark ignition energy (MIE). The vertical axis is logarithmic, and it is seen that the effect is quite significant. If safety measures against electric spark ignition are based on MIE data for a finite dust moisture content, it is essential that this moisture content is not subsided in practice. The influence of dust moisture on the minimum ignition temperature of dust clouds is less marked. For example van Laar and Zeeuwen (1985) reported that flour of 14% moisture had a minimum ignition temperature of 470°C, whereas dry flour had 440°C. For starch the values were 400°C for the dry powder and 460°C with 13% moisture.

Figure 1.20 illustrates how the explosion violence is systematically reduced with increasing dust moisture content. The ignition delay characterizes the state of turbulence of the dust cloud at the moment of ignition in the sense that the turbulence intensity decreases as the ignition delay increases (see Chapter 4).

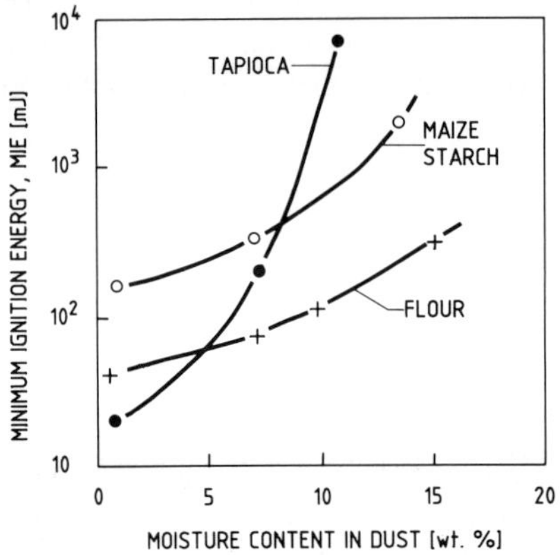

Figure 1.19 *Influence of dust moisture content on minimum electric spark ignition energy (MIE) for three dusts (From van Laar and Zeeuwen, 1985)*

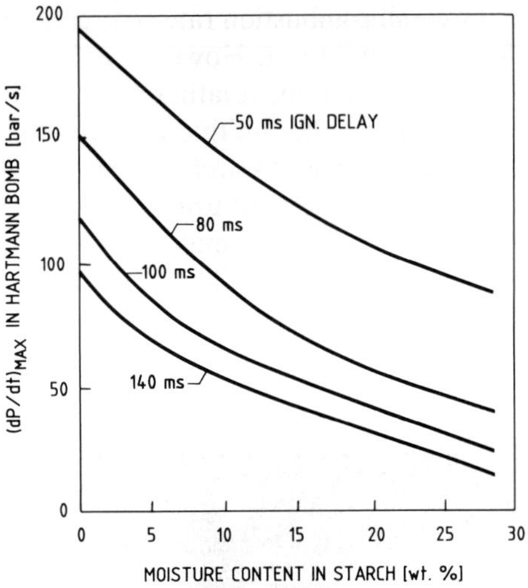

Figure 1.20 *Influence of moisture content in maize starch on maximum rate of pressure rise in Hartmann bomb for various ignition delays (time from dust dispersion to ignition) (From Eckhoff and Mathisen, 1977/1978)*

The specific role of moisture in reducing both ignition sensitivity and explosion violence of clouds of organic dusts is complex. First, evaporation and heating of water represents an inert heat sink. Secondly, the water vapour mixes with the pyrolysis gases in the preheating zone of the combustion wave, and makes the gas mixture less reactive. A third factor is that moisture increases the inter-particle cohesion of the dust and prevents dispersion into primary particles (see Chapter 3).

More detailed analyses of flame propagation in dust clouds of various materials are given in Chapter 4.

1.3.2 PARTICLE SIZE/SPECIFIC SURFACE AREA

Figure 1.17, in addition to illustrating the influence of dust chemistry on the dust cloud combustion rate, also shows a clear dependence on particle size/specific surface area for both materials. This is a general trend for most dusts. However, as discussed in detail in Chapter 4 for coal, this trend does not continue indefinitely as the particles get smaller. In the case of coal and organic materials pyrolysis/devolatilization always precedes combustion, which primarily occurs in the homogeneous gas phase. The limiting particle size, below which the combustion rate of the dust cloud does not increase any more, depends on the ratios between the time constants of the three consecutive processes: devolatilization, gas phase mixing and gas phase combustion. Particle size primarily influences the devolatilization rate. Therefore, if the gas phase combustion is the slowest of the three steps, increasing the devolatilization rate by decreasing the particle size does not increase the overall combustion rate. For coals it was found that the limiting particle diameter is of the order of 50 μm. However, for materials yielding gaseous pyrolysis products that are more reactive than volatiles from coal, e.g. due to unsaturated gaseous compounds, one would expect the limiting particle size to be smaller than for coal. For natural organic compounds such as starch and protein, the limiting particle diameter is probably not much smaller than about 10 μm, whereas for reactive dusts such as some organic dyes, it may well be considerably smaller.

Figures 1.21 and 1.22 show scanning electron microscope pictures of two typical natural organic dusts, namely a wood dust containing very irregular particle shapes, and maize starch having well-defined, nearly monosized and spherical particles.

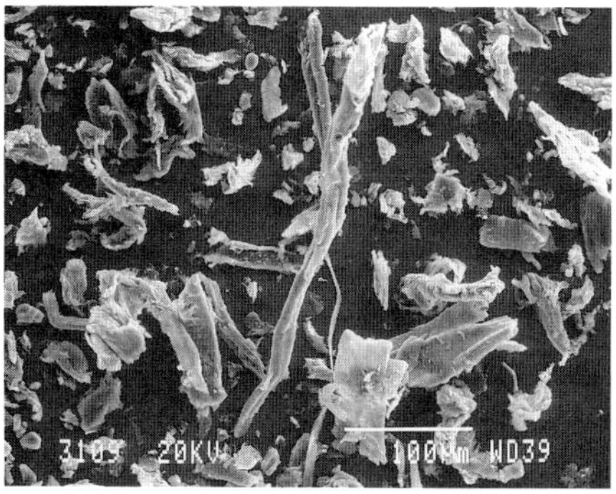

Figure 1.21 *Scanning electron microscope picture of wood dust (Courtesy of W. C. Wedberg)*

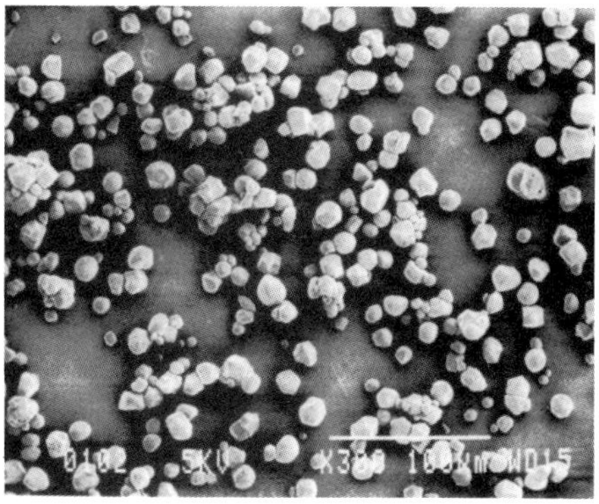

Figure 1.22 *Scanning electron microscope picture of native maize starch: typical particle size 10–15 μm (Courtesy of W. C. Wedberg)*

For metals, in particular those at the top of Table 1.1, the limiting particle size, below which there is no longer any increase in ignition sensitivity and explosion violence, is considerably smaller than for most organic materials. This is because these metals do not devolatilize or pyrolyze, but melt, evaporate and burn as discrete entities (see Chapter 4). Figure 1.23 shows how the combustion rate of clouds of aluminium dust in air increases systematically with the specific surface area of the dust, in agreement with the trend in Figure 1.17. However, the range of specific surface areas in Figure 1.23 is more than ten times that of Figure 1.17. For aluminium a specific surface area of 6.5 m^2/g corresponds to monosized spheres of diameter 0.34 μm, or flakes of thickness 0.11 μm, which is a more likely particle shape for the most violently exploding powders in Figure 1.23.

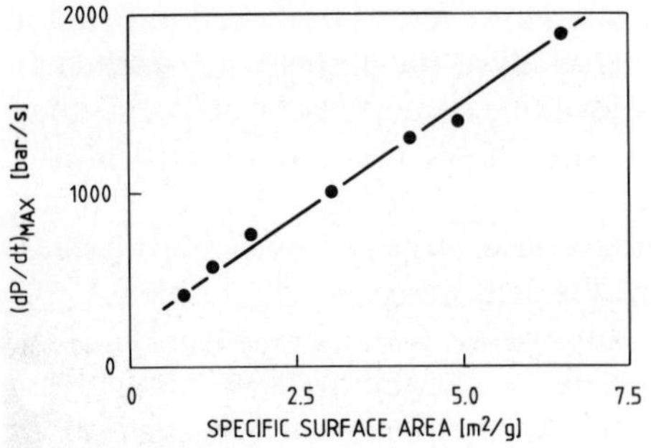

Figure 1.23 *Influence of specific surface area of aluminium dust on maximum rate of pressure rise in standard 1 m^3 ISO vessel (From Bartknecht, 1978)*

Figure 1.24 shows a comparatively coarse atomized aluminium powder of specific surface area only 0.045 m^2/g, whereas Figure 1.25 shows a fine aluminium flake powder. Note that the maximum rate of pressure rise of 2600 bar/s found for this powder in the 1.2 litre Hartmann bomb is not comparable to the values in Figure 1.23. This is due to different degrees of turbulence, degrees of dispersion into primary particles, and different vessel volumes.

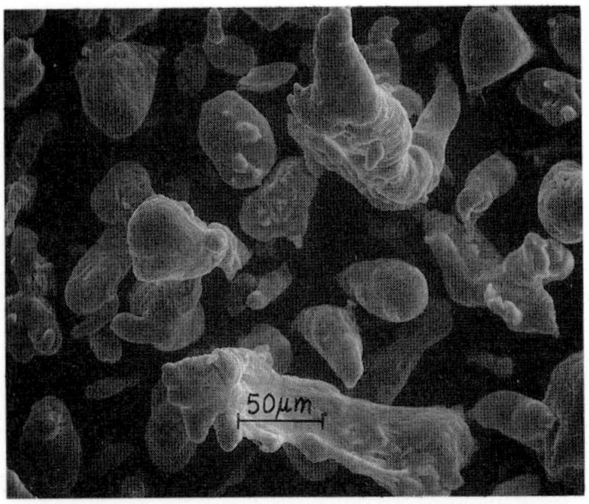

Figure 1.24 *Scanning electron microscope picture of atomized aluminium: typical particle size 50 μm: minimum ignition energy 3000 mJ (Courtesy of W. C. Wedberg)*

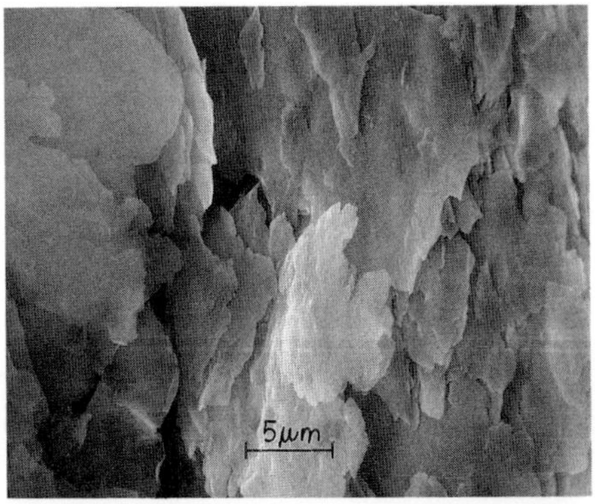

Figure 1.25 *Scanning electron microscope picture of aluminium flakes of thickness < 1 μm. Minimum ignition energy < 1–2 mJ: $(dP/dt)_{max}$ in Hartmann bomb 2600 bar/s (Courtesy of W. C. Wedberg)*

Figures 1.26 and 1.27 shows typical particle shapes in ground silicon in the comparatively coarse and the fine particle size region. The shapes are not very different for the two fractions. It should be noted that the size fraction 37–53 μm is not able to propagate a dust flame. It is necessary to add a tail of much finer particles. The influence of the detailed shape of the particle size distribution on the ignitability and explosibility of metal dust clouds needs further investigation.

Figure 1.28 summarizes some data for the maximum rate of pressure rise for various dusts as functions of median or average particle size.

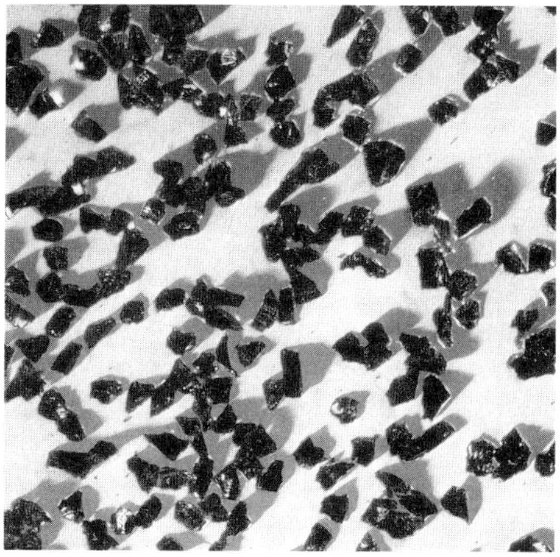

Figure 1.26 *Optical microscope picture of metal-shadowed (shadowing angle 25° with focal plane) 37–53 μm fraction of ground silicon.*

Figure 1.27 *Scanning electron microscope picture of fine fraction of ground silicon: typical particle size 2–3 μm (Courtesy of W. C. Wedberg)*

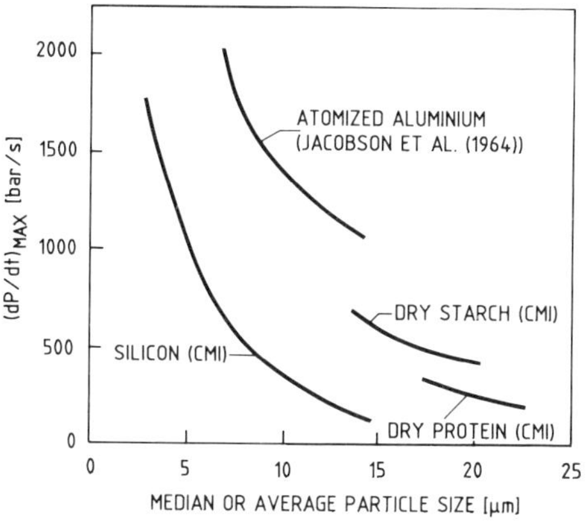

Figure 1.28 *$(dP/dt)_{max}$ in Hartmann bomb of clouds in air of silicon dust, aluminium dust, and dusts from natural organic materials, as functions of particle size (From Eckhoff* et al, *1986)*

Figure 1.29 illustrates how the minimum explosible dust concentration is influenced by the particle size. The particles used in these experiments were close to monodisperse, i.e. of narrow size distributions. In practice the distributions may be quite wide, and simple relationships for monosized dusts may not be valid. Hertzberg and Cashdollar (1987) interpreted the data in Figure 1.29 in terms of the existence of a critical particle size above which the devolatilization process becomes the critical factor in the flame propagation process. Below this size, devolatilization is so fast that the combustion is controlled by gas mixing and gas combustion only. It should be noted that the limiting particle size at the minimum explosible dust concentration is not necessarily the same as at higher concentrations where the explosions are more violent.

Figure 1.30 shows how particle size influences the minimum ignition energy for three different dusts. The vertical scale is logarithmic, and it is seen that the effect is very strong.

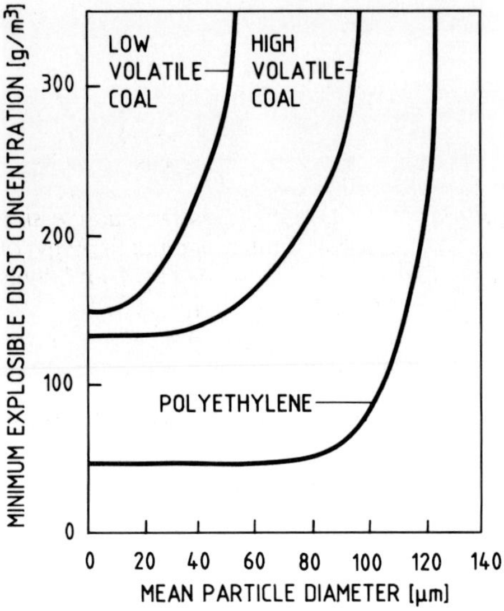

Figure 1.29 *Influence of mean particle diameter on minimum explosible concentration for three different dusts in 20 litre USBM vessel (From Hertzberg and Cashdollar, 1987)*

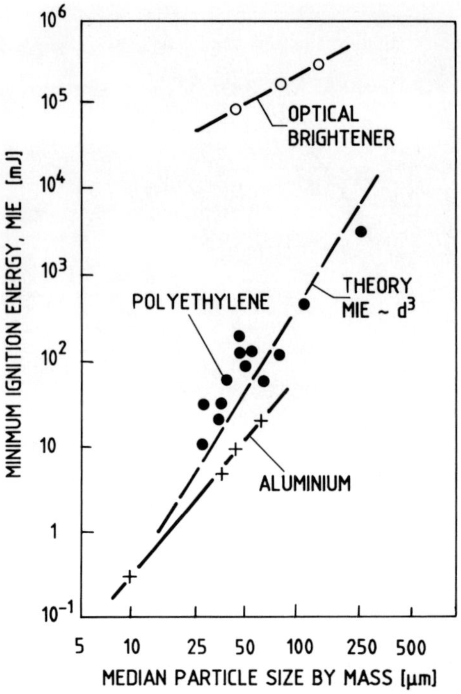

Figure 1.30 *Minimum electric spark ignition energy (MIE) of clouds in air of an optical brightener, polyethylene and aluminium, as functions of median particle size. From Bartknecht (1987):* Theoretical line for polyethylene *(From Kalkert and Schecker, 1979)*

Kalkert and Schecker (1979) developed a theory indicating that MIE is proportional to the cube of the particle diameter, as illustrated in Figure 1.30 by their theoretical prediction of the relationship for polyethylene.

Investigations at CMI showed that a 50–150 μm fraction of atomized aluminium powder could not be ignited as a cloud in air, even with a welding torch. This contradicts somewhat with the data in Figure 1.30. The discrepancy could perhaps be due to the presence of a fine particle size fraction in the powders used by Bartknecht (1978). This emphasizes the need for considering the entire size distribution rather than just a mean size.

Figure 1.31 gives some independent experimental results for MIE as a function of particle size for methyl cellulose, confirming the trends in Figure 1.30.

1.3.3
DEGREE OF DUST DISPERSION EFFECTIVE PARTICLE SIZE

In his experimental studies of burning times of pulverized fuels, Bryant (1973) found that persistent agglomeration was the reason for comparatively long burning times for

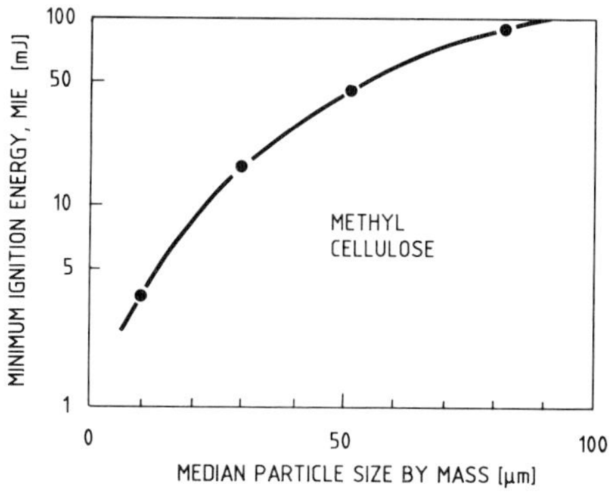

Figure 1.31 *Influence of median particle size by mass on the minimum ignition energy of clouds of methyl cellulose in air. Experiments performed at Chr. Michelsen Institute, Bergen*

apparently small particles. The situation is illustrated in Figure 1.32. A stable agglomerate behaves as a large single particle of the size of the agglomerate.

Eckhoff and Mathisen (1977/1978) investigated the influence of the degree of dispersion of maize starch grains on the rate of pressure rise during explosions in the 1.2 litre Hartmann bomb (see Chapter 7). As shown in Figure 1.22, maize starch consists of fairly monosized and close-to-spherical grains of typical diameters 10–15 μm. The degree of dispersion of the individual starch grains in the Hartmann bomb was studied by mounting a microscope slide with a double-sticky tape inside a specially made 1.5 litre dummy vessel that fitted to the dust dispersion cup of the Hartmann bomb, (see Figures 7.4 and 7.5). Microscopic analysis of the dust deposited on the tape revealed a considerable fraction of stable agglomerates, which were probably formed during production of the starch. It was found that various qualities of maize starch had different degrees of agglomeration. This was reflected in differences in combustion rate, in agreement with Figure 1.32. Figure 1.33 shows a scanning electron micrograph of typical stable maize starch agglomerates found in a commercial maize flour purchased in Norway. Figure 1.34 shows the results of Hartmann bomb experiments with this flour as purchased and after removal of the agglomerates retained by a 37 μm sieve, and a maize starch purchased in USA, all of which passed a 37 μm sieve. Figure 1.34 shows a consistent increase of $(dP/dt)_{max}$ as the effective particle size decreases.

The extent to which a certain powder or dust will appear in agglomerated form when dispersed in a cloud, very much depends on the intensity of the dispersion process. This is discussed in detail in Chapter 3. In general, the tendency of powders and dusts to form agglomerates increases with decreasing particle size, in particular in the range below 10 μm.

1.3.4 DUST CONCENTRATION

Figure 1.3 illustrates the comparatively narrow explosible range of dust concentrations in air. However, neither ignition sensitivity nor explosion rate is constant within the explosible range. Typical patterns of variation with dust concentration are illustrated in Figure 1.35. C_l is the minimum explosible concentration C_{stoich} the stoichiometric concentration, and C_u the maximum explosible concentration.

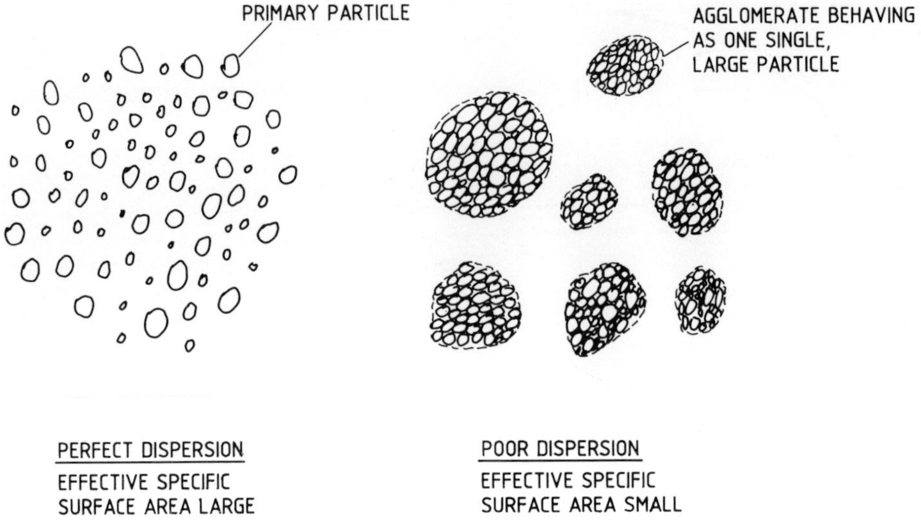

Figure 1.32 *Illustration of perfectly dispersed dust cloud and cloud consisting of agglomerates of much larger effective particle sizes than those of the primary particles*

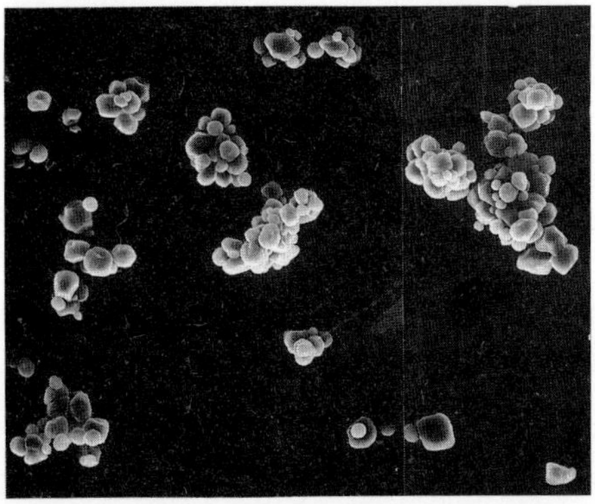

Figure 1.33 *Scanning electron microscope picture of stable agglomerates of primary maize starch grains. Diameters of primary grains typically 10–15 μm (Courtesy of W. C. Wedberg)*

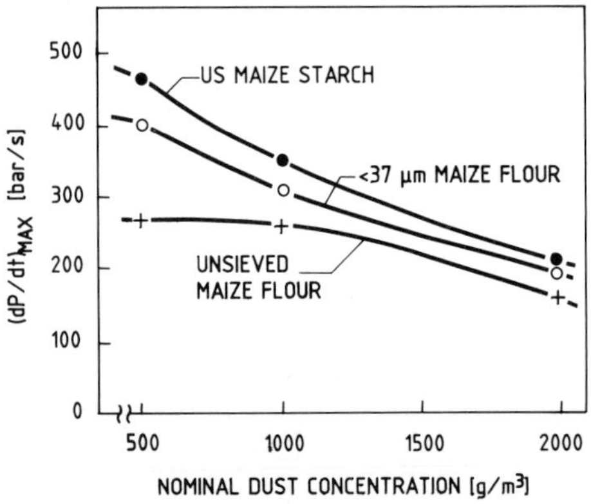

Figure 1.34 *Maximum rate of pressure rise in the 1.2 litre Hartmann bomb of maize starches containing different fractions of agglomerates (From Eckhoff and Mathisen, 1977/1978)*

For maize starch of low moisture content in air at normal pressure and temperature, the minimum explosible concentration equals about 70 g/m^3, the stoichiometric concentration 235 g/m^3, the worst-case concentration about 500 g/m^3, and the maximum explosible concentration probably somewhere in the range 1500–2500 g/m^3. (Note: Figure 4.16 in Chapter 4 suggests a worst-case concentration equal to the stoichiometric concentration, based on laminar flame speed measurements. However, peak values of $(dP/dt)_{max}$ in

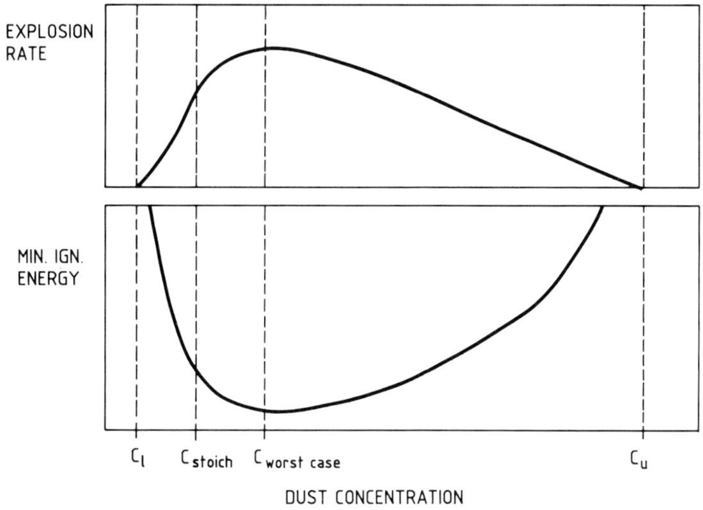

Figure 1.35 *Illustration of typical variation of explosion rate and minimum electric spark ignition energy (MIE) with dust concentration within the explosible range*

closed bomb experiments most often seem to occur at higher concentrations than stoichiometric.) For metal dusts the minimum explosible concentrations are normally considerably higher than for organic dusts and coals. For example for zinc dust it is about 500 g/m^3. The stoichiometric and worst-case concentrations will then also be correspondingly higher.

Figure 1.36 shows a set of results from experiments with maize starch (11% moisture) in the 1.2 litre closed Hartmann bomb. The maximum rate of pressure rise peaks at about 400–500 g/m^3, whereas the maximum pressure reaches a fairly constant peak level in the range from 500 g/m^3 and upwards. Figure 1.37 shows some results from large-scale experiments with the same starch in a 22 m high experimental silo of volume 236 m^3 and vented at the top. The results indicate a peak in the maximum vented explosion pressure at a concentration range not very different from the one that gave the highest $(dP/dt)_{max}$ in the Hartmann bomb experiments.

However, measuring the dust concentration distribution in the 236 m^3 silo was not straightforward and undue emphasis should not be put on this coincidence.

Figure 1.38 illustrates the influence of dust concentration on the ignition sensitivity by some experimental data from Bartknecht (1979).

1.3.5 TURBULENCE

In practical terms, turbulence in the present context is a state of rapid internal, more or less random movement of small elements of the dust cloud relative to each other in three dimensions. If the cloud is burning, turbulence will give rise to mixing of the hot

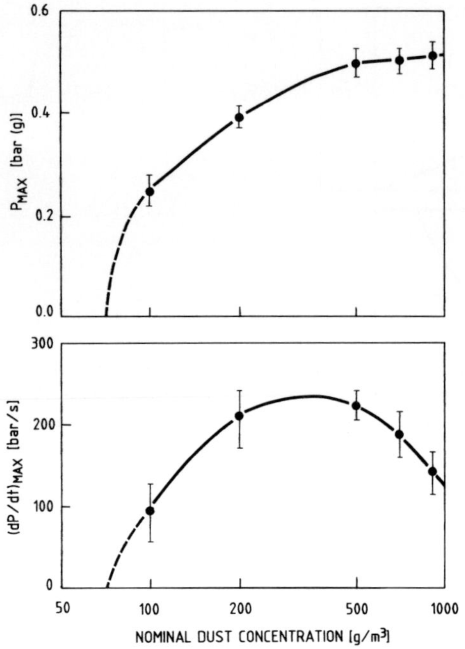

Figure 1.36 *Influence of nominal dust concentration in Hartmann bomb on maximum explosion pressure and maximum rate of pressure rise. Maize starch containing 11% moisture. The bars through the points show ± 1 standard deviation (From Eckhoff* et al., *1985)*

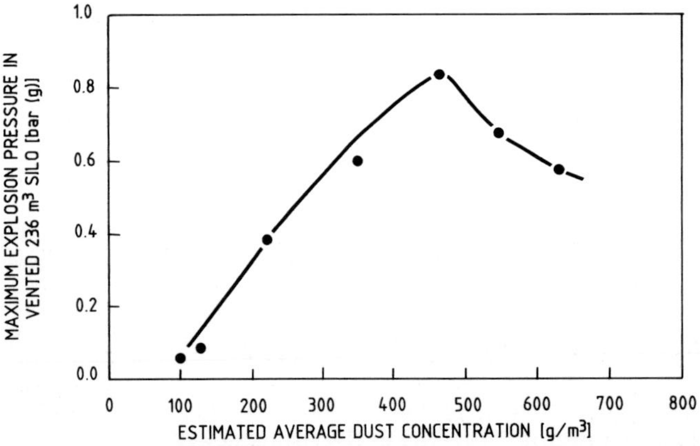

Figure 1.37 *Influence of estimated average dust concentration in exploding cloud in 236 m^3 silo of L/D = 6, on maximum explosion pressure in vented silo. Vent area at top of silo 5.7 m^2. Maize starch containing 11% moisture. Ignition close to bottom of silo (From Eckhoff* et al., *1985)*

burnt/burning parts of the cloud with the unburnt parts, and the cloud will become a kind of three-dimensional laminate of alternating hot burnt/burning and cold unburnt zones. Therefore a turbulent cloud will burn much faster than when a single plane flame sheet propagates through a quiescent cloud.

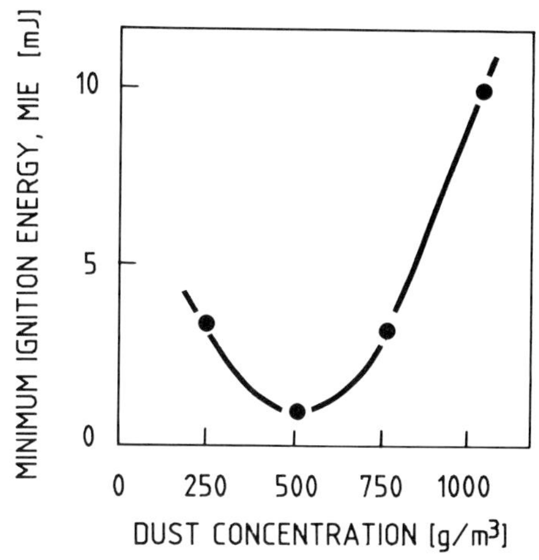

Figure 1.38 *Influence of average dust concentration in 1 m^3 ISO standard vessel on the minimum electric spark ignition energy of clouds of an antioxidant in air (From Bartknecht, 1979)*

In the case of ignition of the dust cloud, whether by an electric spark or a hot surface, the turbulence will disturb the heat transfer by removing heat from the ignition zone by rapid convection. Therefore, ignition of a turbulent dust cloud generally requires higher energies/temperatures than ignition of quiescent clouds.

In the context of dust explosions two kinds of turbulence, differing by their origin, have to be considered. The first is turbulence generated by the industrial process in which the dust cloud is formed, whether an air jet mill, a mixer, a cyclone, a bag filter, a pneumatic transport pipe, or a bucket elevator. This kind of turbulence is often called initial turbulence. The second kind is generated by the explosion itself by expansion-induced flow of unburnt dust cloud ahead of the propagating flame. The level of turbulence generated in this way depends on the speed of the flow and the geometry of the system. Obstacles, like the buckets in a bucket elevator leg, enhance the turbulence generation under such conditions.

In long ducts or galleries a positive feed-back loop can be established by which the flame can accelerate to very high speeds and even transit into a detonation. This is discussed in Chapter 4.

Figure 1.39 shows a characteristic example of the influence of initial turbulence on the rate of dust explosions in closed bombs. The dust cloud is generated in a closed vessel by dispersing a given mass of dust by a short blast of air.

In the early stages during dust dispersion, the dust cloud can be quite turbulent, but the turbulence fades away with time after the dispersion air has ceased to flow. Therefore, if explosion experiments with the same dust are performed in similar vessels at different delays between dust dispersion and ignition, they will have different initial turbulence. As Figure 1.39 shows, the explosion violence, in terms of the maximum gradient of the pressure rise versus time, decreased markedly, by at least an order of magnitude, as the initial turbulence faded away. However, Figure 1.39 also shows that the maximum explosion pressure remained fairly constant up to about 200 ms. This reflects the fact that the maximum pressure is essentially a thermodynamic property, as opposed to the rate of pressure rise, which contains a strong kinetic component.

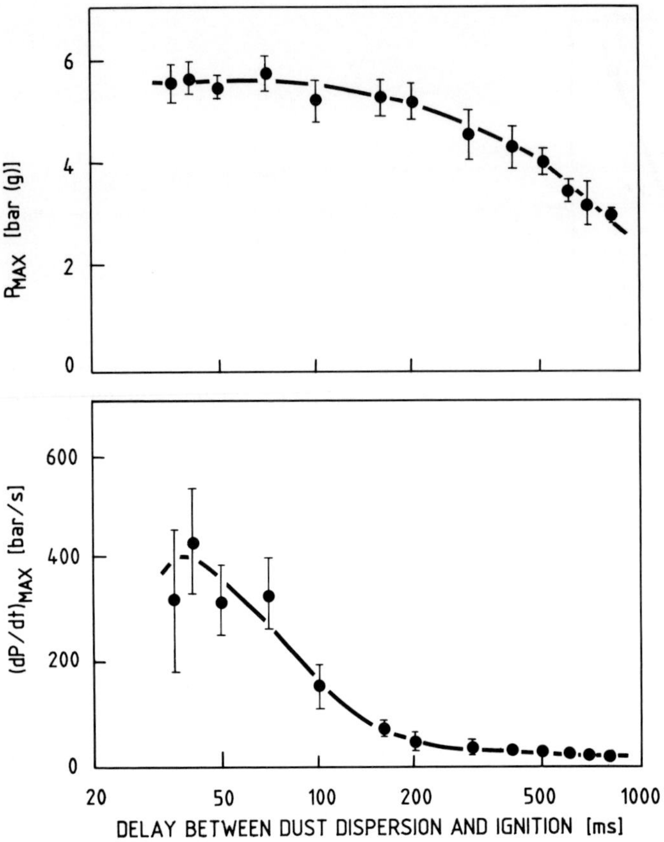

Figure 1.39 *Influence of initial turbulence on explosion rate of a dust cloud. Experiments with 420 g/m³ of lycopodium in air in the 1.2 litre Hartmann bomb. Five experiments per delay. Bars indicate ± 1 standard deviation (From Eckhoff, 1977)*

Christill *et al.* (1989), having developed a comprehensive model for predicting flame propagation and pressure development in gas explosions, implying the k-ϵ model of turbulence (see Section 4.4.1 in Chapter 4), suggested that similar models might be developed even for turbulent dust explosions. Other work along similar lines is discussed in Section 4.4.8 in Chapter 4.

Figure 1.40 shows the strong influence of initial turbulence on the minimum electric spark ignition energies of dust clouds. In this case turbulence acts in the direction of safety, making it much more difficult to ignite the dust cloud compared with the quiescent state. The effect is quite dramatic, the minimum ignition energy increasing by several orders of magnitude. This is fortunate in the context of the possible generation of electrostatic discharges in the presence of explosible dust clouds, because such discharges are normally generated when the cloud is in turbulent motion. Section 5.3.4 in Chapter 5 gives some further information.

Further analysis of the role of turbulence on propagation of dust flames is given in Chapter 4.

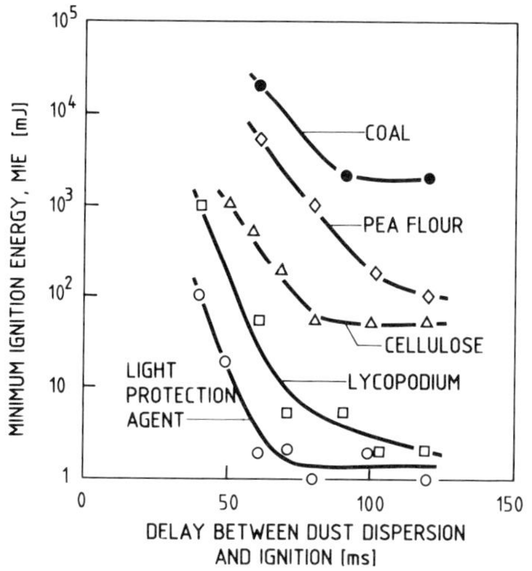

Figure 1.40 *Influence of initial turbulence on the minimum electric spark ignition energy of dust clouds. Experiments with various dusts in a 20 litre spherical explosion bomb (From Glarner, 1984)*

1.3.6 OXYGEN CONTENT OF OXIDIZER GAS

As one would intuitively expect, both explosion violence and ignition sensitivity of dust clouds decrease with decreasing oxygen content of the gas in which the dust is suspended. Wiemann (1984) investigated the influence of the oxygen content of the gas (air + nitrogen) on the maximum pressure and maximum rate of pressure rise of coal dust explosions in a 1 m^3 closed vessel. The results, shown in Figure 1.41, show that both the explosion pressure and the rate of pressure rise decreased with decreasing oxygen content. Furthermore, the explosible dust concentration range was narrowed, in particular on the fuel-rich side. It is worth noting that a reduction of the oxygen content from that of air to 11. 5% caused a reduction of the maximum rate of pressure rise by a factor of ten or more, whereas the maximum pressure was reduced by less than a factor of two. This illustrates the strong influence of the oxygen content on the kinetics of the combustion process. The reduction of the maximum pressure is approximately proportional to the reduction of the oxygen content, as would be expected from thermodynamical considerations.

Figure 1.42 shows some earlier results from the work of Hartmann (1948). The trend is similar to that of Wiemann's results in Figure 1.41. The maximum explosion pressure is approximately proportional to the oxygen content down to 16–17%, whereas the maximum rate of pressure rise falls much more sharply. For example at 15% oxygen, i.e. 71% of that in air, $(dP/dt)_{max}$ is only 13% of the value in air.

The influence of the oxygen content in the oxidizing gas on the minimum explosible dust concentration was studied in detail by Hertzberg and Cashdollar (1987). Some results for a high-volatile-content coal dust are shown in Figure 1.43. For particles smaller than about 10 μm a reduction of the oxygen content from that of air to 15.5% caused only a moderate increase, from 130 g/m^3 to 160 g/m^3, of the minimum explosible concentration. However, as the particle size increased, the influence of reducing the oxygen content became pronounced. At a mean particle size of 50 μm, 15.5% oxygen was sufficiently low to prevent flame propagation. It seems probable that the particle size fractions used by

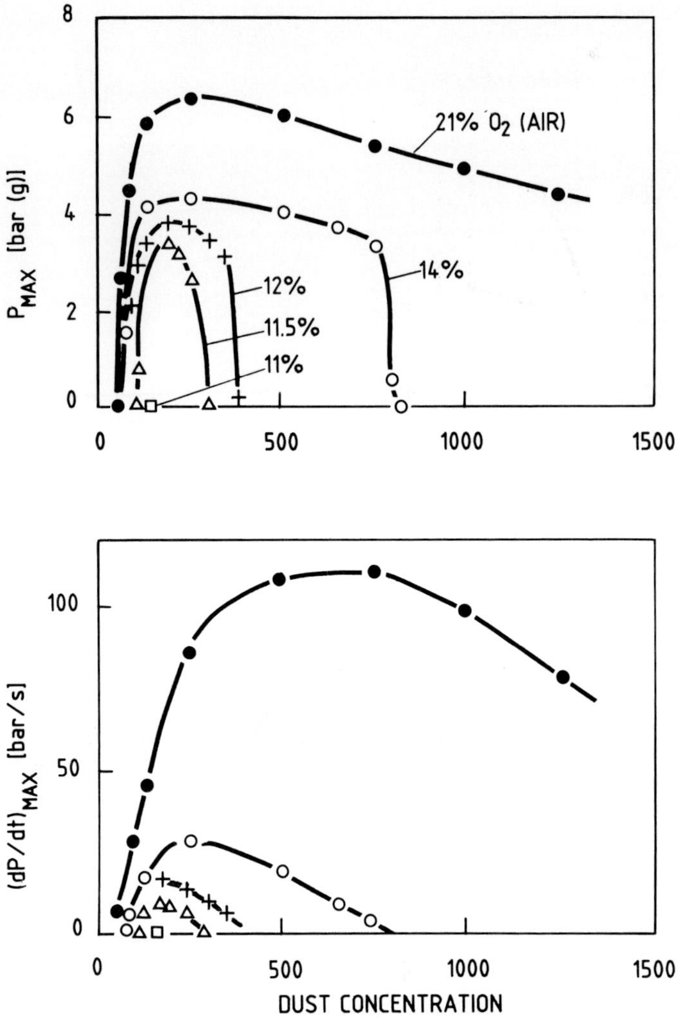

Figure 1.41 *Influence of oxygen content in the gas on the maximum explosion pressure and maximum rate of pressure rise of brown coal dust for various dust concentrations. Nitrogen as inert gas. 1 m³ ISO standard explosion vessel (From Wiemann, 1984)*

Hertzberg and Cashdollar (1987) were quite narrow. This can explain why particles of larger mean diameters than 100 μm did not give explosions in air at all, irrespective of dust concentration. In practice, most powders/dusts involved in dust explosions have comparatively wide particle size distributions, and characterizing their fineness by only a mean particle size can be misleading in the context of dust explosibility assessment. It would be expected that many coal dusts of larger mean particle diameter than 100 μm would be explosible in air if they contain a significant 'tail' of fine particles.

Sweiss and Sinclair (1985) investigated the influence of particle size of the dust on the limiting oxygen concentration in the gas for flame propagation through dust clouds.

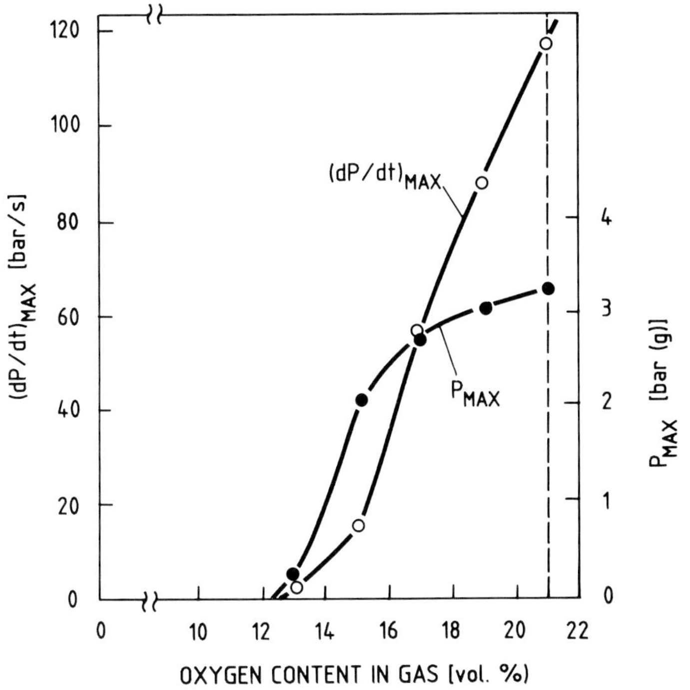

Figure 1.42 *Influence of oxygen content in gas on maximum pressure and maximum rate of pressure rise in explosions of 100 g/m*^{3} *of < 74 μm ethyl cellulose molding powder in the 1.2 litre Hartmann bomb (From Hartmann, 1948)*

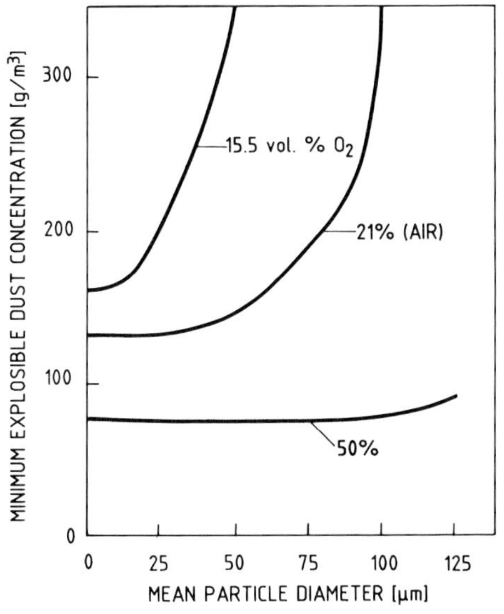

Figure 1.43 *Influence of oxygen content in gas on the minimum explosible concentration of coal dust (high volatile content) versus particle size (From Hertzberg and Cashdollar, 1987)*

Natural and synthetic organic dusts were studied. The results from experiments with narrow size fractions indicated that the limiting oxygen concentration decreased with decreasing particle size down to 100 μm. Below 100 μm the limiting oxygen concentration was practically independent of particle size. However, addition of only 5% by mass of a fine dust (≈ 60 μm) to a coarse main dust (200–1000 μm) reduced the limiting oxygen concentration by at least 60% of the difference between the values of the coarse dust only and the fine dust only.

Wiemann (1984) found that for brown coal, dust particle size had a comparatively small influence on the limiting oxygen concentration for inerting. Thus, at an initial temperature of 50°C and nitrogen as inert gas, the values were 11.8 vol% for a median particle size of 19 μm, and 12.4 vol% for 52 μm.

The results in Figure 1.44, produced by Walther and Schacke (1986), show that the maximum permissible oxygen concentration for inerting clouds of a polymer powder was independent of the initial pressure over the range 1–4 bar (abs.). For oxygen concentrations above this limit, the data in Figure 1.44 can be represented by the simple approximate relationship

$$P_{max}[bar(g)] = 0.35 \times P_o[bar(abs)] \times (vol\% O_2) \tag{1.12}$$

where P_o is the initial pressure.

Figure 1.45 illustrates the influence of the oxygen content of the gas on the minimum ignition temperature of a dust cloud. For < 74 μm Pittsburgh coal dust there is a

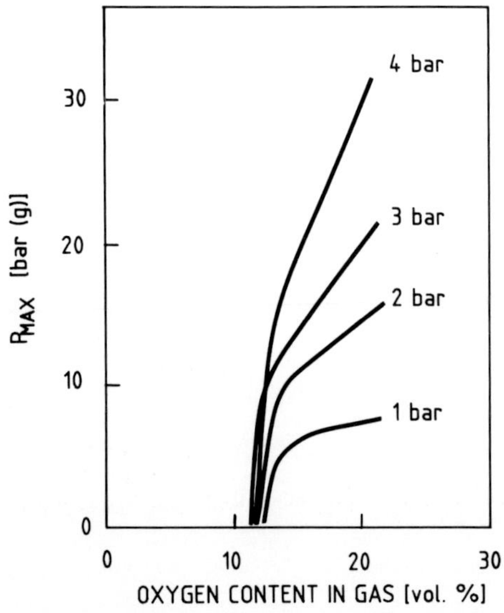

Figure 1.44 *Influence of oxygen content in gas on the maximum explosion pressure for a polymer powder for various initial pressures. 1 m³ closed ISO vessel (From Walther and Schacke, 1986)*

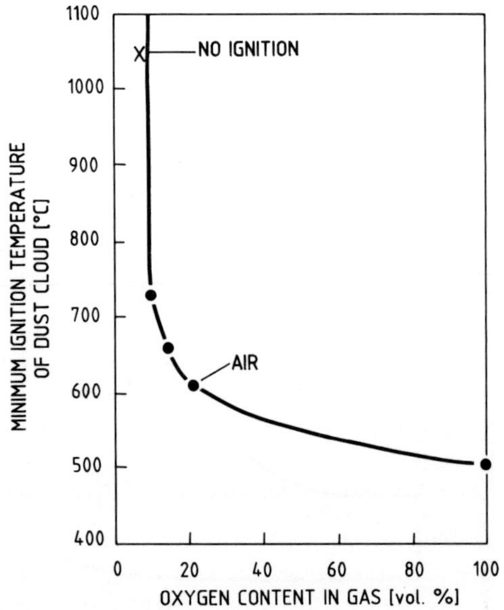

Figure 1.45 *Influence of oxygen content in gas on minimum ignition temperature of < 74 μm Pittsburgh coal dust in the Godbert-Greenwald furnace (From Hartmann, 1948)*

systematic increase from 500°C in pure oxygen via 600°C in air to 730°C in 10 vol% oxygen.

The influence of the oxygen content in the gas on the minimum electric spark ignition energy of dust clouds is illustrated by the data in Figure 1.46 for a sub-atmospheric pressure of 0.2 bar (abs). A reduction from 21 vol% to 10 vol% increased the minimum ignition energy by a factor of about 2. This is of the same order as the relative increase found by Hartmann (1948) for atomized aluminium, namely a factor of 1.4 from 21 vol% to 15 vol% oxygen, and a factor of 2.0 from 21 vol% to 8.5 vol% oxygen. However, as the oxygen content approached the limit for flame propagation, a much steeper rise of the minimum ignition energy would be expected. This is illustrated by Glarner's (1984) data for some organic dusts in Figure 1.47.

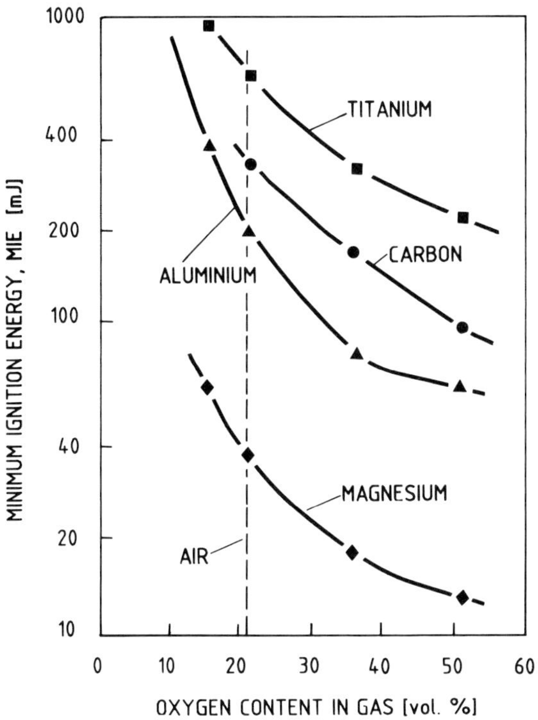

Figure 1.46 *Influence of oxygen content in atmosphere on minimum electric spark ignition energy of dust clouds of various materials. Initial pressure 0.2 bar (abs): mean particle diameter 40 μm. Equivalence ratio 0.65, i.e. excess oxygen for combustion: MIE defined for 80% probability of ignition (From Ballal, 1979)*

It should finally be mentioned that Wiemann (1984) found that the maximum oxygen concentration for inerting clouds of a brown coal dust of median particle diameter 52 μm varied somewhat with the type of inert gas. For an initial temperature of 150°C, the values were 10.9 vol% for nitrogen, 12.3 vol% for water vapour and 13.0 vol% for carbon dioxide. The influence of initial temperature was moderate in the range 50–200°C. Thus,

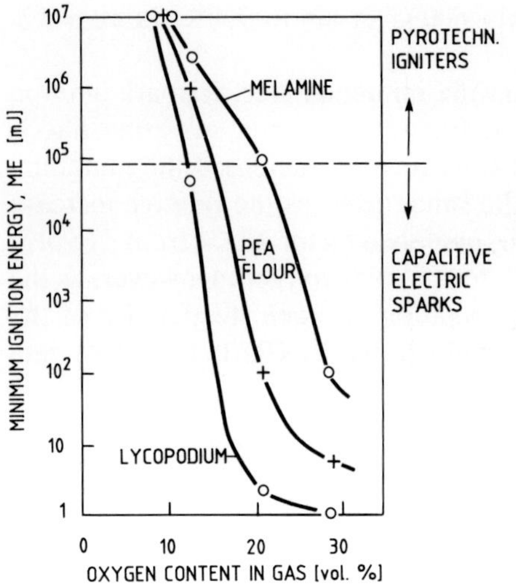

Figure 1.47 *Influence of oxygen content in gas on minimum ignition energy of dust clouds (From Glarner, 1984)*

the value for nitrogen dropped from 12.4 vol% at 50°C to 10.4 vol% at 200°C. For carbon dioxide the corresponding values were 14.0 and 12.5 vol% respectively.

1.3.7 INITIAL TEMPERATURE OF DUST CLOUD

Figure 1.48 summarizes results obtained by Wiemann (1987) and Glarner (1983) for various coals and organic dusts, indicating a consistent pattern of decreasing minimum explosible dust concentrations with increasing initial temperature. Furthermore, as the minimum explosible concentration decreases towards zero with increasing temperature, the data seem to converge towards a common point on the temperature axis. For gaseous hydrocarbons in air, Zabetakis (1965) proposed linear relationships between the minimum explosible concentration and the initial temperature, converging towards the point 1300°C for zero concentration. For methane/air and butane/propane/air, Hustad and Sönju (1988) found a slightly lower point of convergence, i.e. 1200°C. However, linear plots of the data in Figure 1.48 yield points of convergence for zero minimum explosible concentration in the range 300–500°C, i.e. much lower than the 1200–1300°C found for hydrocarbon gases. This indicates that the underlying physics and chemistry is more complex for organic dusts than for hydrocarbon gases.

The influence of the initial temperature of the dust cloud on the minimum electric spark ignition energy is illustrated in Figure 1.49 by the data of Glarner (1984). For the organic materials tested a common point of convergence for the straight regression lines at 1000°C and 0.088 mJ is indicated. This means that MIE values for organic dusts at elevated temperatures can be estimated by linear interpolation between this common point and the measured MIE value at ambient temperature.

Figure 1.50 shows how the initial temperature influences the maximum explosion pressure and rate of pressure rise. The reduction of P_{max} with increasing initial temperature is due to the reduction of the oxygen concentration per unit volume of dust cloud at a

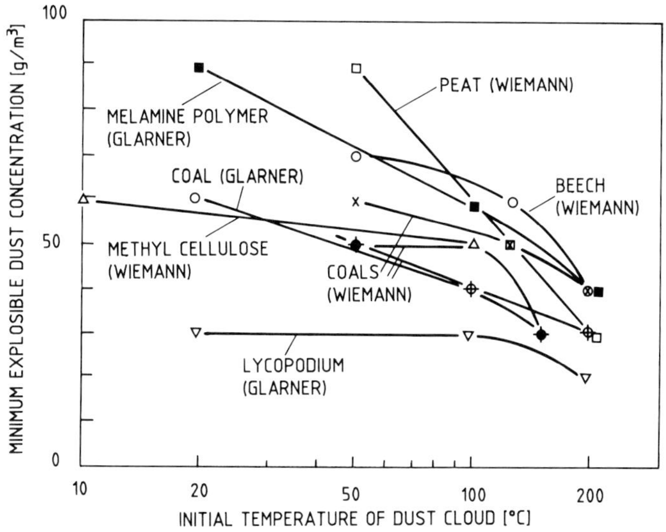

Figure 1.48 *Influence of initial temperature of dust clouds on minimum explosible dust concentration in air at 1 bar (abs.) (Data from Wiemann, 1987 determined in a 1 m³ closed vessel with 10 kJ chemical igniter and from Glarner, 1983 determined in a 20 litre closed vessel with 10 kJ chemical igniter)*

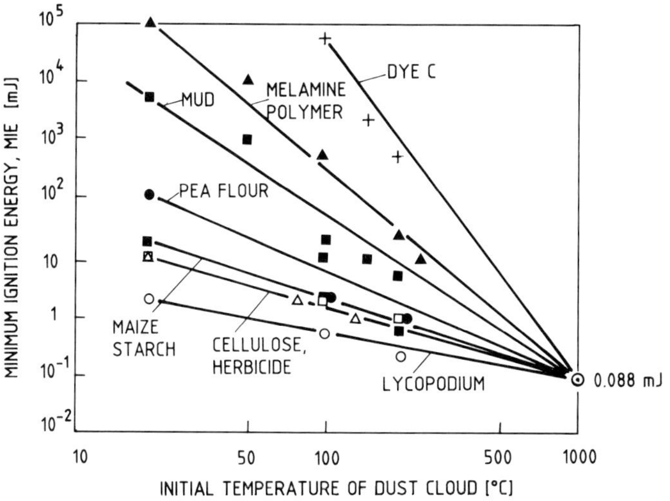

Figure 1.49 *Influence of initial temperature of dust cloud on minimum electric spark ignition energy (From Glarner, 1984)*

given initial pressure, with increasing initial temperature. The trend for $(dP/dt)_{max}$ in Figure 1.50 is less clear and reflects the complex kinetic relationships involved.

Figure 1.51 indicates an approximately linear relationship between the reciprocal of the normalized initial temperature and the normalized maximum explosion pressure for some organic materials and coals.

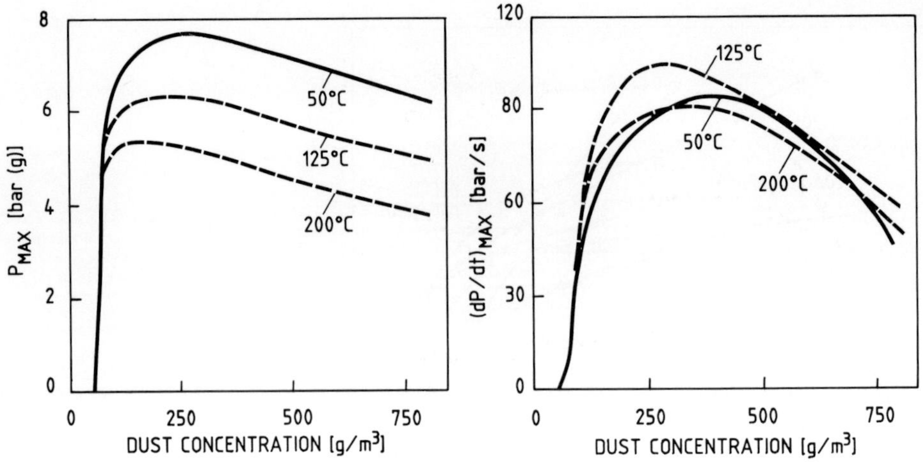

Figure 1.50 *Influence of initial temperature of dust cloud on explosion development in 1 m^3 closed vessel. Bituminous coal dust in air (From Wiemann, 1987)*

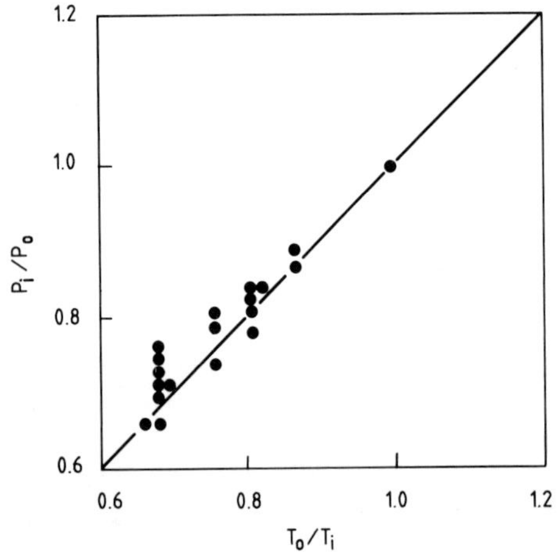

Figure 1.51 *Influence of normalized initial temperature (K) of dust clouds on normalized maximum explosion pressure (absolute). T_o is the lowest initial temperature investigated for a given dust, mostly 323 K. T_i is the actual initial temperature. P_o and P_i are the maximum explosion pressures for initial temperatures T_o and T_i respectively. Data for coals, beech, peat, jelly agent, milk powder, methyl cellulose and naphthalic acid anhydride (From Wiemann, 1987)*

1.3.8
INITIAL PRESSURE OF DUST CLOUD

Wiemann's (1987) data for brown coal dust in air in Figure 1.52 illustrate the characteristic pattern of the influence of initial pressure on the maximum explosion pressure in closed vessels (constant volume). Two features are apparent. First the peak maximum pressure (abs.) is close to proportional to the initial pressure (abs.). Secondly, the dust concentration that gives the peak maximum pressure is also approximately proportional to the initial pressure, as indicated by the straight line through the origin and the apexes of the pressure-versus-concentration curves. This would indicate that there is a given ratio of mass dust to mass air that gives the most efficient combustion, independently of initial pressure.

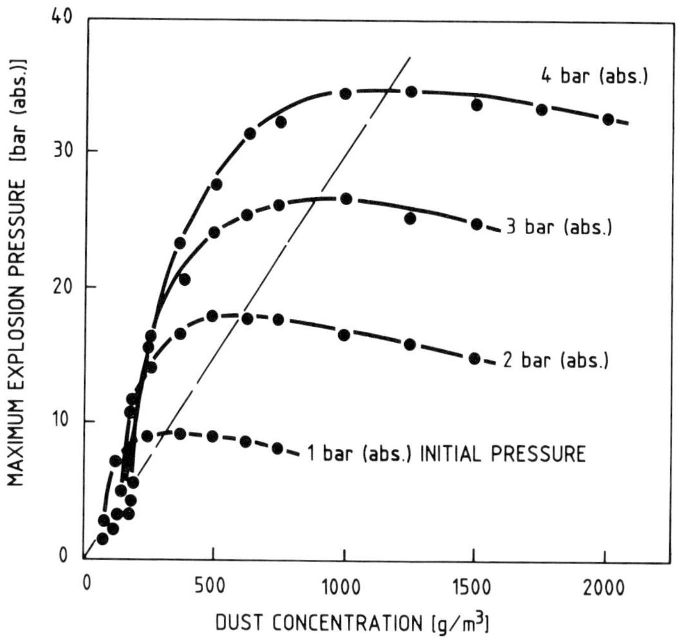

Figure 1.52 *Maximum explosion pressure in 1 m^3 closed vessel as function of dust concentration for different initial pressures. Brown coal dust in air. Straight line through the origin passes through the apexes of the curves (From Wiemann, 1987)*

Walther and Schacke (1986) presented results for polymer powder/air explosions in a 20 litre closed vessel, revealing the same trends as Figure 1.52, from an initial pressure of 3 bar (abs.) and down to 0.2 bar (abs.) These results are also in complete agreement with the earlier results for starch presented by Bartknecht (1978), covering the initial pressure range 0.2–2.0 bar (abs.). Figure 1.53 summarizes the results from the three investigations.

The results in Figure 1.54, obtained by Pedersen and Wilkins for higher initial pressures, indicate that the trend of Figure 1.53 extends at least to 12 bar (abs.). This is in agreement with corresponding linear correlations found for methane/air up to 12 bar (abs.) initial pressure, as shown by Nagy and Verakis (1983). For clouds of fuel mists in air, Borisov and Gelfand (1989) found a linear correlation between initial pressure and maximum explosion pressure up to very high initial pressures, approaching 100 bar.

Figure 1.54 also gives the maximum rate of pressure rise as a function of initial pressure. The excellent linear correlation is the result of somewhat arbitrary adjustment of the dust dispersion conditions with increasing quantities of dust to be dispersed.

The more arbitrary nature of the rate of pressure rise is reflected by the data in Figure 1.55, which show that in Wiemann's experiments $(dP/dt)_{max}$ started to level out and depart from the linear relationship for initial pressures exceeding 2 bar (abs.).

Figure 1.56 illustrates how the minimum explosible concentration of dusts increases systematically with increasing initial pressure. Hertzberg and Cashdollar (1988) attributed the close agreement between polyethyene and methane to fast and complete devolatilization of polyethylene in the region of the minimum explosible concentration. In the case of coals, only the volatiles contribute significantly to flame propagation in this concentration range. A more detailed discussion of these aspects is given in Chapter 4.

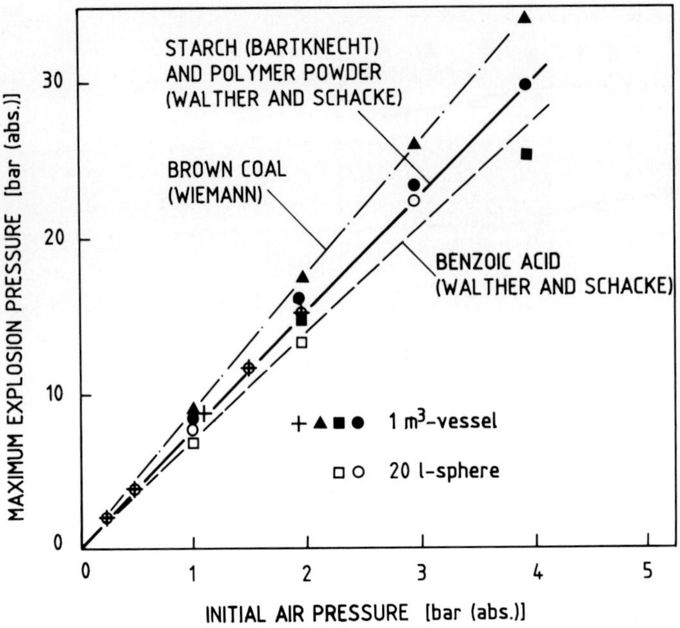

Figure 1.53 *Maximum explosion pressure at constant volume as a function of initial air pressure (From Bartknecht 1978, Walther and Schacke 1986 and Wiemann 1987)*

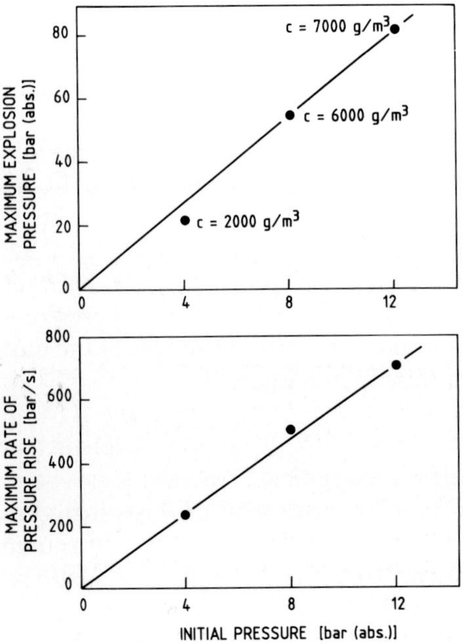

Figure 1.54 *Influence of initial pressure on maximum pressure and maximum rate of pressure rise in explosions of clouds of sub-bituminous coal dust in air in a 15 litre closed bomb: median particle size by mass 100 μm (From Pedersen and Wilkins, 1988)*

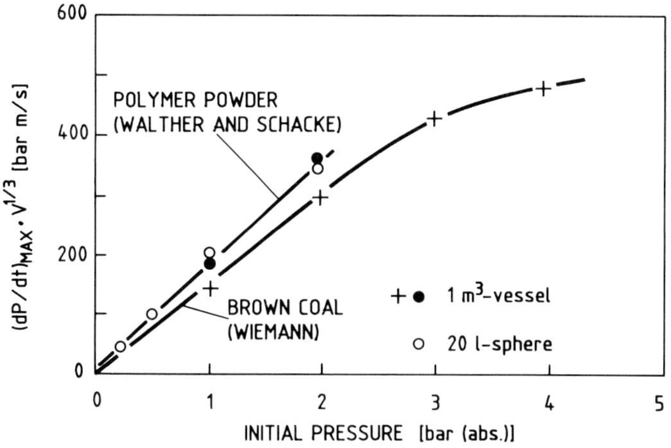

Figure 1.55 *Normalized highest* (dP/dt)$_{max}$ *as a function of initial pressure for explosions of Polymer and brown coal dusts in closed compatible 1 m^3 and 20 litre vessels (From Walther and Schacke, 1986 (Polymer) and Wiemann, 1987 (brown coal)*

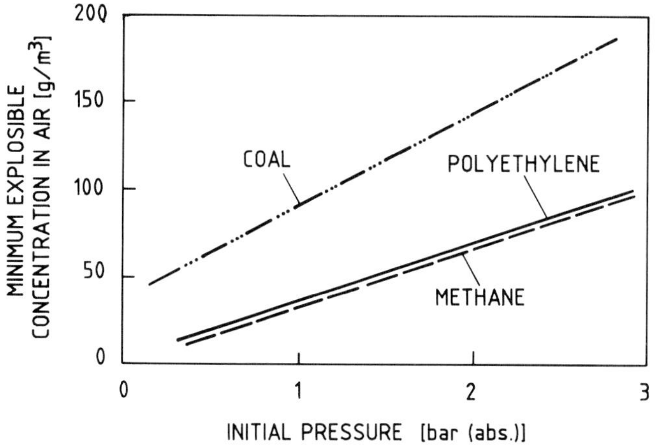

Figure 1.56 *Influence of initial pressure on the minimum explosible concentration of two dusts and methane in air (From Hertzberg and Cashdollar, 1988)*

1.3.9 COMBUSTIBLE GAS OR VAPOUR MIXED WITH DUST CLOUD ('HYBRID' MIXTURES)

It is not clear who was the first researcher to study the influence of comparatively small amounts of combustible gas or vapour on the ignitability and explosibility of dust clouds. However, more than a century ago Engler (1885) conducted experiments in a wooden explosion box of 0.25 m square cross section and 0.5 m height, and essentially open at the bottom. The box was filled with a mixture of air and marsh gas (methane) of the desired concentration, and a cloud of fine charcoal dust, which was unable to give dust explosions in pure air, was introduced at the container top by a vibratory feeder. Engler made the interesting observation that methane concentrations as low as 2.5 vol% made clouds of the charcoal dust explosible, whereas the methane/air alone without the dust, did not burn unless the gas content was raised to 5.5–6 vol%. One generation later, Engler (1907) described a simple laboratory-scale experiment by which the hybrid effect could be demonstrated. The original sketch of the apparatus is shown in Figure 1.57.

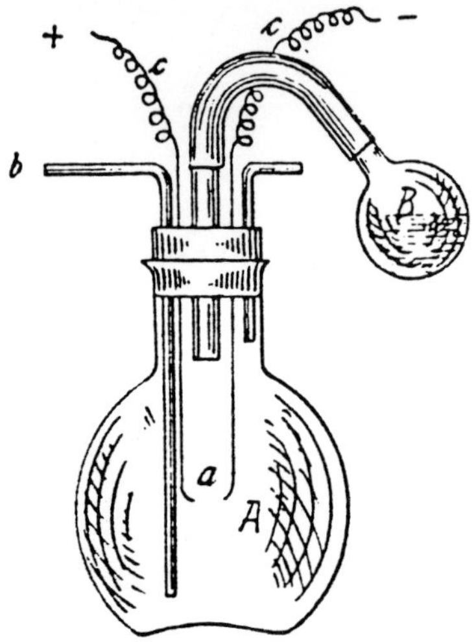

Figure 1.57 *Apparatus for demonstrating the hybrid interaction of combustible dust and gas:* ***A*** *is a glass explosion vessel of volume 250–500 cm^3,* ***B*** *is a glass dust reservoir connected to* ***A*** *via a flexible hose,* ***b*** *is the inlet tube for the dispersing air and* ***a*** *the gap for the spark ignition source (From Engler, 1907)*

The experimental procedure was first to raise reservoir **B** to allow an appropriate quantity of dust (unable to propagate a flame in pure air) to drop into vessel **A**. A continous train of strong inductive sparks was then passed across the spark gap **a**, whilst a short blast of air was injected via **b** by pressing a rubber bulb, to generate a dust cloud in the region of the spark gap. With air only as the gaseous phase no ignition took place. The entire vessel **A** was then replaced by another one of the same size and shape, but filled with a mixture of air and the desired quantity of combustible gas, and the experiment was repeated. Engler advised the experimenter to protect himself against the flying fragments of glass that could result in the case of a strong hybrid explosion!

Adding small percentages of combustible gas to the air, influences the minimum explosible dust concentration, depending on the type of dust. This is illustrated by the data of Foniok (1985) for coals of various volatile contents, shown in Figure 1.58. The effect is particularly pronounced for dusts that have low ignition sensitivity and low combustion rate in pure air. A similar relationship for another combination of dust and gas is shown in Figure 1.59.

Nindelt *et al.* (1981) investigated the limiting concentrations for flame propagation in various hybrid mixtures of dusts and combustible gases in air. The dusts and combustible gases were typical of those represented in the flue gases from coal powder plants.

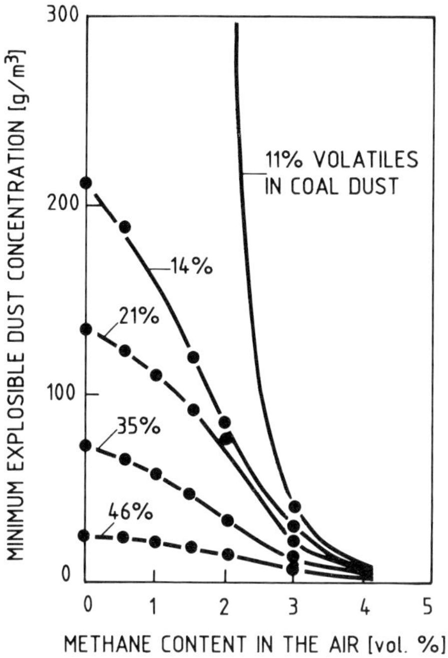

Figure 1.58 *Influence of methane content in the air on the minimum explosible concentration of coal dusts of different volatile contents. Average particle size 40 μm with 100% < 71 μm: 4.5 kJ ignition energy (From Foniok, 1985)*

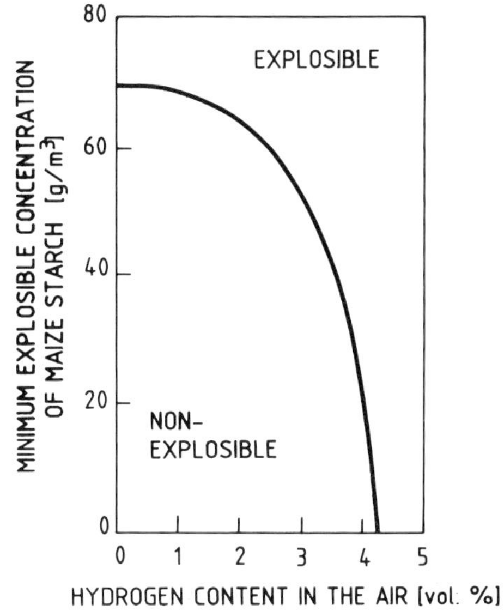

Figure 1.59 *Influence of small percentages of hydrogen in the air on the minimum explosible concentration of maize starch at normal ambient conditions (From Hertzberg and Cashdollar, 1987)*

Reeh (1979) determined the critical minimum contents of volatiles in coals and methane in the air, for self-sustained flame propagation in clouds of coal dust in a 200 m experimental mine gallery. With no methane in the air, flame propagation was possible only for volatile contents above 14%. With 1 vol% methane in the air, the critical value was 13%; for 2% methane, about 12% and for 3% methane, about 9% volatiles.

Cardillo and Anthony (1978) determined empirical correlations between the content of combustible gas (propane) in the air and the minimum explosible concentration of polypropylene, polyethylene and iron. It is interesting to note that iron responded to the propane addition in the same systematic way as the organic dusts. For no propane in the air the minimum explosible iron dust concentration was found to be 200 g/m^3, whereas for 1 vol% propane it was 100 g/m^3.

The influence of small fractions of methane in the air on the minimum electric spark energy for igniting clouds of coal dusts was investigated systematically by Franke (1978). He found appreciable reductions in MIE, by factors of the order of 100, when the methane content was increased from zero to 3 vol%.

Pellmont (1979) also investigated the influence of combustible gas in the air on the minimum ignition energy of dust clouds. A set of results, demonstrating a quite dramatic effect for some dusts, is given in Figure 1.60. Pellmont found that the most ignition

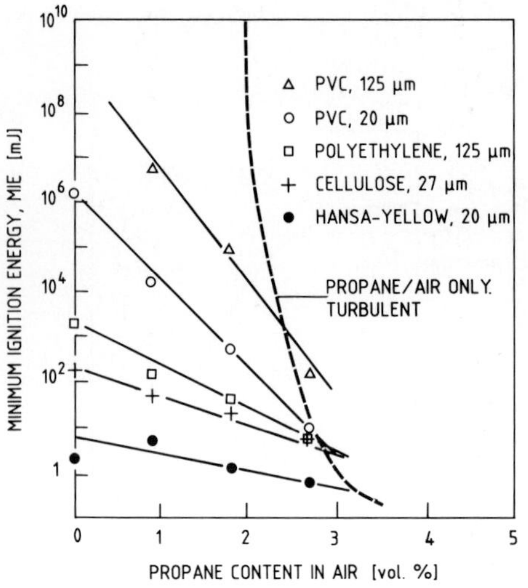

Figure 1.60 *Influence of small fractions of propane in the air on the minimum electric spark ignition energy of clouds of various organic dusts at normal ambient conditions (From Pellmont, 1979)*

sensitive concentration of the various dusts decreased almost linearly with increasing content of propane in the air. For example for 20 μm PVC in pure air the most sensitive concentration was 500 g/m^3 , whereas with 2 vol% propane in the air, it was 250 g/m^3. Figures 1.61 and 1.62 give some results presented by Foniok (1985). In agreement with the findings of Pellmont, Foniok also observed that the dust concentration that was most sensitive to ignition, and at which the reported MIE values were determined, decreased systematically with increasing combustible gas content in the air. For example for the 31% volatile dust, for which data are given in Figure 1.61, the most sensitive concentration was 750 g/m^3 with no methane in the air, whereas with 3.5% methane in the air, it dropped to 200 g/m^3.

Torrent and Fuchs (1989), probably using more incendiary electric sparks of longer discharge times than those used by Foniok (1985), found little influence of methane content in the air on MIE for coal dusts up to 2 vol% methane. For all the coal dusts tested but one, MIE in pure air was $<$ 100 mJ. For one exceptional coal dust containing 18% moisture and 12% ash, MIE dropped from 300 mJ for no methane to about 30–50 mJ for 2% methane.

It has been suggested that hybrid mixtures involving dusts that are very easy to ignite even without combustible gas in the air (MIE $<$ 10 mJ) may be ignited by electrostatic brush discharges, but definite proof of this has not been traced.

Figure 1.63 illustrates how the content of combustible gas in the air influences the percentage of inert dust required for inerting coal dust clouds.

One of the first systematic investigations of the influence of combustible gas in the air on the explosion violence of dust clouds was conducted by Nagy and Portman (1961). Their results are shown in Figure 1.64. The dust dispersion pressure is a combined arbitrary measure of the extent to which the dust is raised into suspension and dispersed, and of the turbulence in the dust cloud at the moment of ignition. As can be seen, the maximum explosion pressure, with and without methane in the air, first rose, as the dust dispersion was intensified. However, as the dust dispersion pressure was increased further, the dust without methane started to burn less efficiently, probably due to quenching by intense

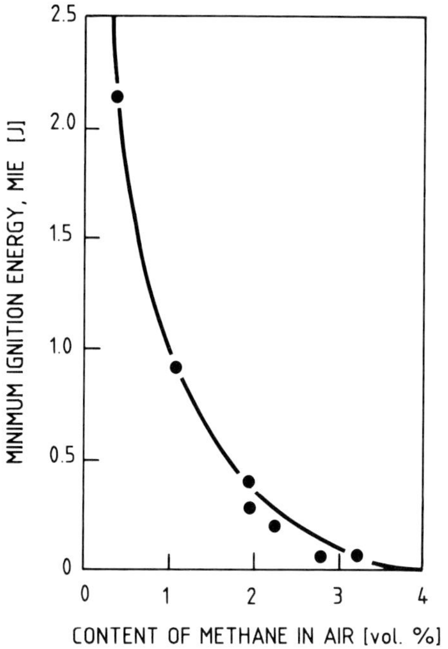

Figure 1.61 *Influence of methane content in the air on the minimum electric spark ignition energy of a coal dust of 31% volatile content. Average particle size 40 μm. NOTE: Presumably short-duration sparks from low-inductance, low-resistance capacitive discharge circuit (From Foniok, 1985)*

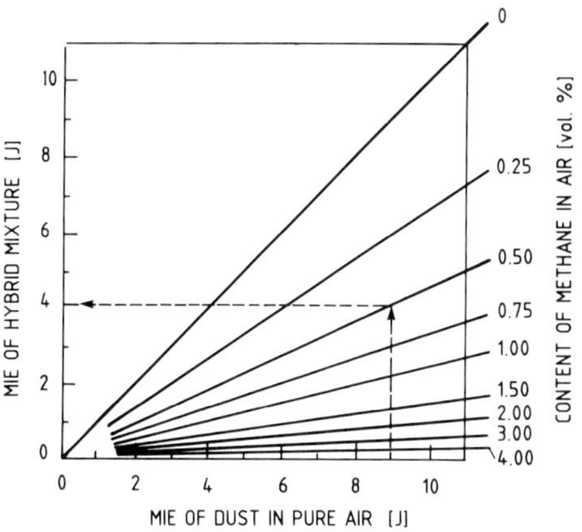

Figure 1.62 *Nomograph for minimum ignition energy of hybrid mixtures of dust and methane in air as a function of the methane content in the air and the minimum ignition energy of the dust in air only. NOTE: Presumably short-duration sparks from low-inductance, low-resistance capacitive circuit (From Foniok, 1985)*

turbulence. In the presence of methane this effect did not appear, presumably due to faster combustion kinetics. The influence of the methane was even more apparent for the maximum rate of pressure rise, which, for a dust dispersion pressure of 30 arbitrary units, had dropped to less than 100 bar/s without methane, whereas with 2% methane it had increased further up to 500 bar/s. This comparatively simple experiment revealed important features of the kinetics of combustion of turbulent clouds of organic dusts. It should also be mentioned that Ryzhik and Makhin (1978) investigated the systematic decrease of the induction time for ignition of hybrid mixtures of coal dust/methane/air, in the methane concentration range 0–5 vol%.

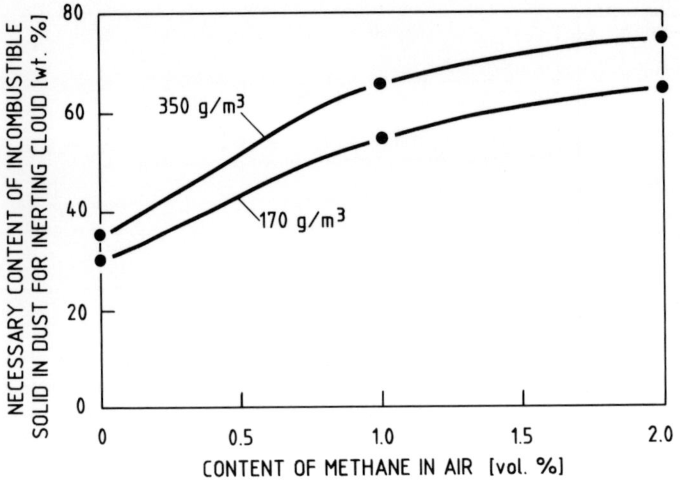

Figure 1.63 *Necessary mass percentage of incombustible solid material for inerting clouds of dry coal dust of 38% volatiles and 10% ash in air containing various low percentages of methane (From Torrent and Fuchs, 1989)*

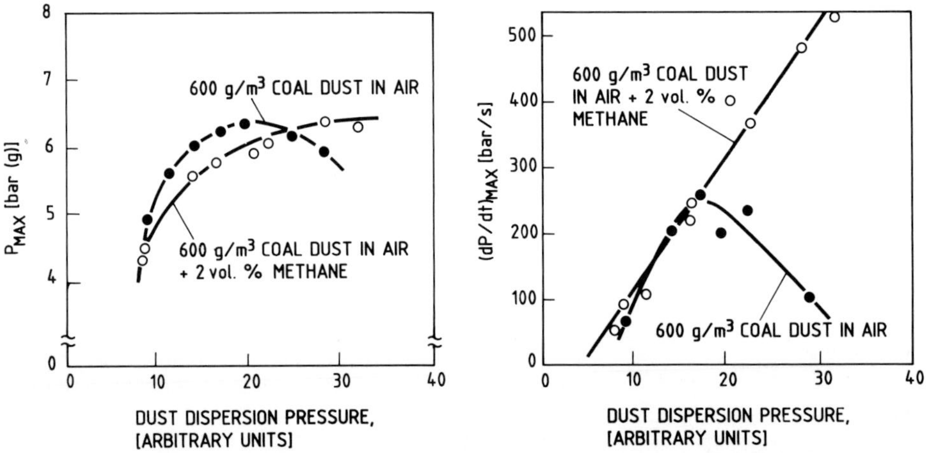

Figure 1.64 *Influence of 2 vol% methane in the air on maximum explosion pressure and maximum rate of pressure rise of coal dust in a 28 litre closed vessel at various levels of initial turbulence (From Nagy and Portman, 1961)*

Reeh (1978) conducted a comprehensive investigation of the influence of methane in the air on the violence of coal dust explosions. He concluded that the influence was strongest in the initial phase of the explosion. In the fully developed large-scale high-turbulence explosion it made little difference whether gas or coal dust was the fuel.

Further illustrations of the influence of combustible gas or vapour in the air on the explosion violence are given in Figures 1.65 from Bartknecht (1978) and 1.66 from Dahn

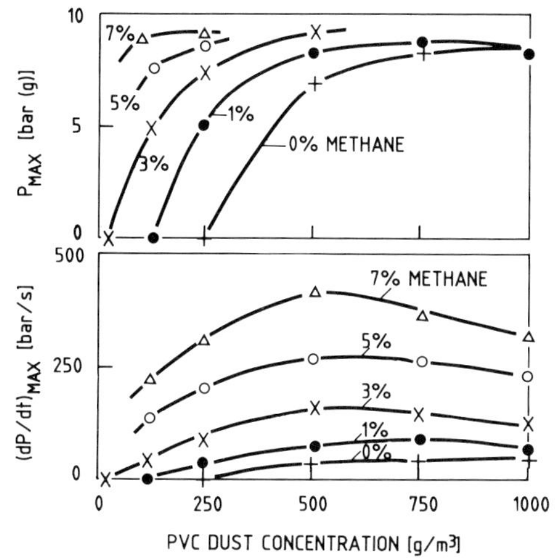

Figure 1.65 *Influence of small fractions of methane in the air on maximum explosion pressure and maximum rate of pressure rise in a 1 m³ closed vessel. 10 kJ pyrotechnical igniter (From Bartknecht, 1978)*

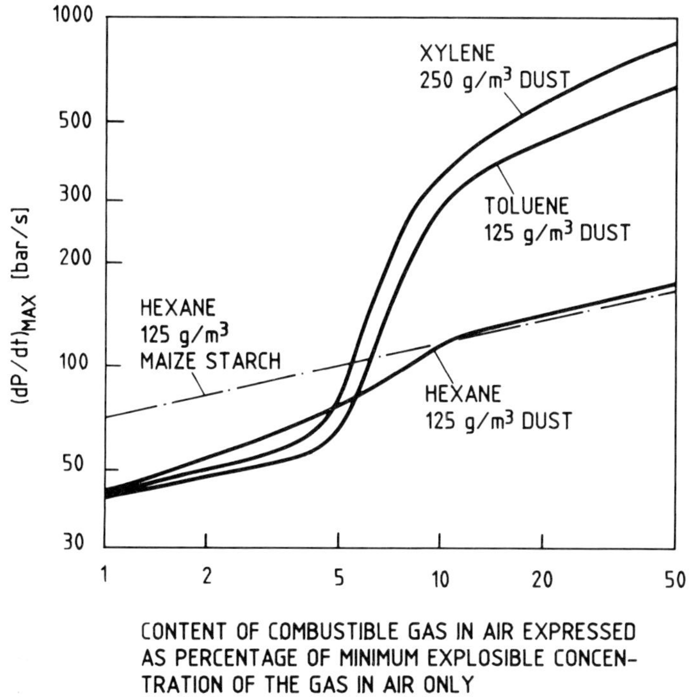

Figure 1.66 *Influence of low concentrations of various organic solvent vapours in the air on the maximum rate of pressure rise during explosions of organic dusts in a 20 litre closed vessel (From Dahn, 1986)*

(1986). Dahn studied the influence of small fractions of xylene, toluene or hexane in the air, on the maximum rate of pressure rise of explosions of a combustible waste dust in a 20 litre closed bomb. The waste dust originated from shredded materials including paper and plastics. Its moisture content was 20% and the particle size < 74 μm. Results for maize starch of 4–5% moisture content in hexane/air are also shown in Figure 1.66.

Torrent and Fuchs (1989) found that both maximum explosion pressure and maximum rate of pressure rise of a dry coal dust of 38% volatiles and 10% ash in a closed 20 litre vessel, increased by 30% when 3 vol% methane was added to the air. There was a significant decrease of the dust concentrations that gave the most violent explosions, with increasing methane content, from 600–700 g/m^3 without methane to about 300 g/m^3 with 3 vol% methane. This agrees with the trend found by Foniok (1985) for the minimum ignition energy.

1.3.10 INERTING BY MIXING INERT DUST WITH COMBUSTIBLE DUST

This principle of inerting the dust cloud is of little practical interest apart from in mining. In coal mines, stone dust has been used extensively for this purpose for a long time. Comprehensive information concerning that specific problem was provided by Cybulski (1975). Michelis (1984) indicated that satisfactory protection against propagation of coal dust explosions in mine galleries cannot be obtained unless the total content of combustible material in the mixture coal dust/lime stone is less than 20 wt%. This is not always easy to achieve in practice, and supplementary means of protection (water barriers, etc.) must be employed.

A useful, more general analysis of the problem of inerting combustible dust clouds by adding inert dust was given by Bowes, Burgoyne and Rasbash (1948).

Table A3 in the Appendix gives some experimental data for the percentages of inert dusts required for inerting clouds in air of various organic dusts and coals.

1.3.11 CONCLUDING REMARK

Section 1.3 has been included primarily to bring into focus the various important parameters that influence ignitability and explosibility of dust clouds, and to indicate main trends of their influence.

The extent to which the reader will find quantitative data that satisfy specific needs, is bound to be limited. In particular, size distributions and specific surface areas of dusts of a given chemistry can vary considerably in practice. However, the quantitative information provided can help in identifying the type of more specific information that is needed in each particular case. In many cases the required data will have to be acquired by tailor-made experiments.

1.4
MEANS FOR PREVENTING AND MITIGATING DUST EXPLOSIONS

1.4.1
THE MEANS AVAILABLE: AN OVERVIEW

The literature on the subject is substantial. Many authors have published short, general surveys on means of preventing and mitigating dust explosions in the process industry. A few fairly recent examples are Gibson (1978), Scholl, Fischer and Donat (1979), Kühnen and Zehr (1980), Field (1982a), Woodcock and Reed (1983), Siwek (1986, 1987), Field (1987), Swift (1987, 1987a) and Bartknecht (1988). The books mentioned in Section 1.1.1.5 also contain valuable information.

Table 1.9 gives an overview of the various means that are presently known and in use. They can be divided in two main groups, namely means for preventing explosions and means for their mitigation. The preventive means can again be split in the two categories prevention of ignition sources and prevention of explosible/combustible cloud. One central issue is whether only preventing ignition sources can give sufficient safety, or whether it is also necessary in general to employ additional means of prevention and/or mitigation. In the following sections the means listed in Table 1. 9 will be discussed separately.

Table 1.9 Means of preventing and mitigating dust explosions: a schematic overview

PREVENTION				MITIGATION	
Preventing ignition sources		Preventing explosible dust cloud			
a	Smouldering combustion in dust, dust flames	f	Inerting by N_2, CO_2 and rare gases	j	Partial inerting by inert gas
b	Other types of open flames (e.g. hot work)	g	Intrinsic inerting	k	Isolation (Sectioning)
c	Hot surfaces	h	Inerting by adding inert dust	l	Venting
d	Electric sparks and arcs Electrostatic discharges	i	Dust concentration outside explosible range	m	Pressure resistant construction
e	Heat from mechanical impact (metal sparks and hot-spots)			n	Automatic suppression
				o	Good housekeeping (dust removal/cleaning)

1.4.2
PREVENTING IGNITION SOURCES

1.4.2.1
Introduction

The characteristics of various ignition sources are discussed in 1.1.4, and some special aspects are elucidated more extensively in Chapter 5. Test methods used for assessing the ignitability of dust clouds and layers, when exposed to various ignition sources are discussed in Chapter 7.

Several authors have published survey papers on the prevention of ignition sources in process plant. Kühnen (1978) discussed the important question of whether preventing ignition sources can be relied upon as the only means of protection against dust explosions. His conclusion was that this may be possible in certain cases, but not in general. Adequate knowledge about the ignition sensitivity of the dust, both in cloud and layer form, under the actual process conditions, and proper understanding of the process, are definite pre-conditions. Schäfer (1978) concluded that relying on preventing ignition sources is impossible if the minimum electric spark ignition energy of the dust is in the region of vapours and gases (< 10 mJ). However, for dusts of higher MIE he specified several types of process plants that he considered could be satisfactorily protected against dust explosions solely by eliminating ignition sources.

In a more recent survey, Scholl (1989) concluded that the increased knowledge about ignition of dust layers and clouds permits the use of prevention of ignition sources as the sole means of protection against dust explosions, provided adequate ignition sensitivity tests have shown that the required ignition potential, as identified in standardized ignition sensitivity tests, is unlikely to occur in the process of concern. Scholl distinguished between organizational and operational ignition sources. The first group, which can largely be prevented by enforcing adequate working routines, includes:

- Smoking.
- Open flames.
- Open light (bulbs).
- Welding (gas/electric).
- Cutting (gas/rotating disc).
- Grinding.

The second group arises within the process itself and includes:

- Open flames.
- Hot surfaces.
- Self-heating and smouldering nests.
- Exothermic decomposition.
- Heat from mechanical impact between solid bodies (metal sparks/hot-spots).
- Exothermic decomposition of dust via mechanical impact.
- Electric sparks/arcs, electrostatic discharges.

1.4.2.2
Self-heating, smouldering and burning of large dust deposits

The tendency to self-heating in powder/dust deposits is dependent on the properties of the material. Therefore, the potential of self-heating should be known or assessed for any material before admitting it to storage silos or other part of the plant where conditions are favourable for self-heating and subsequent further temperature rise up to smouldering and burning.

Possible means of preventing self-heating include:

- Control of temperature, moisture content and other important powder/ dust properties before admitting powder/dust to e.g. storage silos.

- Adjustment of powder/dust properties to acceptable levels by cooling, drying etc. , whenever required.
- Ensuring that heated solid bodies (e.g. a steel bolt heated and loosened by repeated impacts) do not become embedded in the powder/dust mass.
- Continuous monitoring of temperature in powder mass at several points by thermometer chains.
- Monitoring of possible development of gaseous decomposition/oxidation products for early detection of self-heating.
- Rolling of bulk material from one silo to another, whenever onset of self-heating is detected, or as a routine after certain periods of storage, depending on the dust type.
- Inerting of bulk material in silo by suitable inert gas, e. g. nitrogen.

Thermometer chains in large silos can be unreliable because self-heating and smouldering may occur outside the limited regions covered by the thermometers.

Inerting by adding nitrogen or other inert gas may offer an effective solution to the self-heating problem. However, it introduces a risk of personnel being suffocated when entering areas that have been made inert. In the case of nitrogen inerting, humans get serious respiratory problems when the oxygen content in the atmosphere drops to 17–18 vol% (air 21 vol%).

If inerting is adopted, it is important to take into account that the maximum permissible oxygen concentration for ensuring inert conditions in the dust deposit may be considerably lower than the maximum concentration for preventing explosions in clouds of the same dust. Walther (1989) conducted a comparative study with three different dusts, using a 20 litre closed spherical bomb for the dust cloud experiments and the Grewer furnace (see Chapter 7) for the experiments with dust deposits. In the case of the dust clouds, oxidizability was quantified in terms of the maximum explosion pressure at constant volume, whereas for the dust deposits it was expressed in terms of the maximum temperature difference between the test sample and a reference sample of inert dust, exposed to the same heating procedure. The results are shown in Figure 1. 67. In the case of the pea flour it is seen that self-heating took place in the dust deposit right down to 5 vol% oxygen or even less, whereas propagation of flames in dust clouds was practically impossible below 15 vol% oxygen. Also for the coals there were appreciable differences.

Extinction of smouldering combustion inside large dust deposits e.g. in silos is a dual problem. The first part is to stop the exothermic reaction. The second, and perhaps most difficult part, is to cool down the dust mass. In general the use of water should be avoided in large volumes. Limited amounts of water may enhance the self-heating process rather than quench it. Excessive quantities may increase the stress exerted by the powder/dust mass on the walls of the structure in which it is contained, and failure may result. Generally, addition of water to a powder mass will, up to the point of saturation, reduce the flowability of the powder and make discharge more difficult (see Chapter 3).

Particular care must be taken in the case of metal dust fires where the use of water should be definitely excluded. Possible development of toxic combustion products must also be taken into account.

The use of inert gases such as nitrogen and carbon dioxide has proven to be successful both for quenching of the oxidation reaction and the subsequent cooling of smouldering combustion in silos. However, large quantities of inert gas are required, of the order of

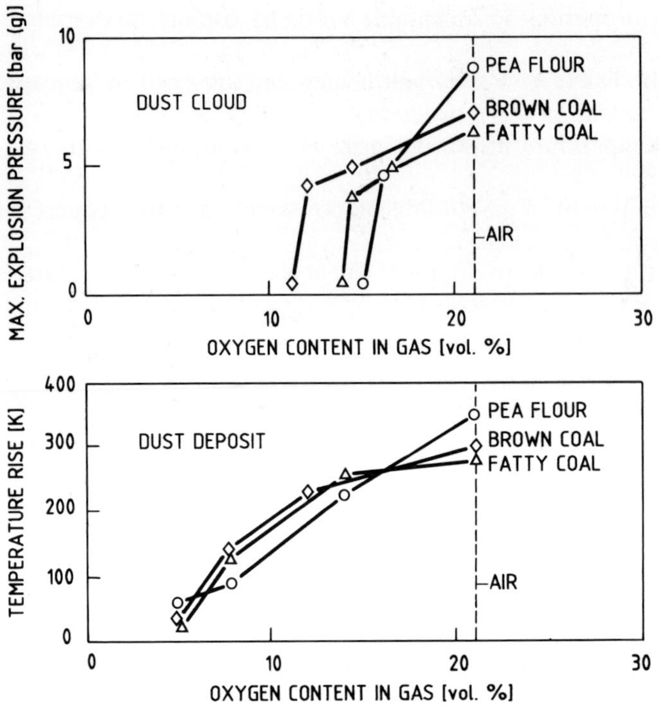

Figure 1.67 *Comparison of the influence of oxygen content in the gas on the oxidizability of dust clouds and dust deposits (From Walther, 1989)*

10 tonnes or more for a fair size silo. In the case of fine-grained products as wheat flour or maize starch, the permeability of the inert gas may be too low for efficient inerting of large bulk volumes.

Further details concerning extinction of powder and dust fires are given by Palmer (1973) and Verein deutscher Ingenieure (1986). The use of inert gas for extinction of smouldering fires in silos was specifically discussed by Dinglinger (1981) and Zockoll and Nobis (1981). Chapter 2 gives some examples of extinction of smouldering fires in practice.

Some synthetic organic chemicals, in particular cyclic compounds, can decompose exothermally and become ignited by a hot surface, a smouldering nest, frictional heat or other ignition source. Such decomposition does not require oxygen, and therefore inerting has no effect. Zwahlen (1989) gave an excellent account of this special problem. He pointed out that this type of exothermic decomposition can only be avoided by eliminating all potential ignition sources. However, by taking other processing routes one can eliminate or reduce the problem. Zwahlen suggested the following possibilities:

- The hazardous powder is processed in the wet state, as a slurry or suspension.
- If wet processing is impossible, one should avoid processes involving internal moving mechanical parts that can give rise to ignition.
- If this is not possible, strict control to prevent foreign bodies from entering the process must be exercised. Furthermore, detectors for observing early temperature and

pressure rise, and sprinkler systems must be provided. Adiabatic exothermal decomposition of bulk powder at constant volume can, due to the very high powder concentration, generate much higher pressures than a dust explosion in air.

- Generally the processed batches of the powder should be kept as small as feasible.
- Use of additives that suppress the decomposition tendency may be helpful in some cases.

1.4.2.3
Open flames/hot gases

Most potential ignition sources of the open flame type can be avoided by enforcing adequate organizational procedures and routines. This in particular applies to prohibition of smoking and other use of lighters and matches, and to enforcement of strict rules for performing hot work. Hot work must not be carried out unless the entire area that can come in contact with the heat from the work, indirectly as well as directly, is free of dust, and hazardous connections through which the explosion may transmit to other areas, have been blocked.

Gas cutting torches are particularly hazardous because they work with excess oxygen. This gives rise to ignition and primary explosion development where explosions in air would be unlikely.

In certain situations in the process industry, hot gaseous reaction products may entrain combustible dust and initiate dust explosions. Each such case has to be investigated separately and the required set of precautions tailored to serve the purpose in question.

Factory inspectorates in most industrialized countries have issued detailed regulations for hot work in factories containing combustible powders or dusts.

1.4.2.4
Hot surfaces

As pointed out by Verein deutscher Ingenieure (1986), hot surfaces may occur in industrial plants both intentionally and unintentionally. The first category includes external surfaces of hot process equipment, heaters, dryers, steam pipes and electrical equipment. The equipment where hot surfaces may be generated unintentionally include engines, blowers and fans, mechanical conveyors, mills, mixers, bearings and unprotected light bulbs.

A further category of hot surfaces arises from hot work. One possibility is illustrated in Figure 1.10. During grinding and disc-cutting, glowing hot surfaces are often generated, which may be even more effective as initiators of dust explosions than the luminous spark showers typical of these operations. This aspect has been discussed by Müller (1989).

A hot surface may ignite an explosible dust cloud directly, or via ignition of a dust layer that subsequently ignites the dust cloud. Parts of glowing or burning dust layers may loosen and be conveyed to other parts of the process where they may initiate explosions.

It is important to realize that the hot surface temperature in the presence of a dust layer can, due to thermal insulation by the dust, be significantly higher than it would normally be without dust. This both increases the ignition hazard and may cause failure of equipment due to increased working temperature. The measures taken to prevent ignition by hot surfaces must cover both modes of ignition. The measures include:

- Removal of all combustible dust before performing hot work.
- Prevention/removal of dust accumulations on hot surfaces.
- Isolation or shielding of hot surfaces.
- Use of electrical apparatus approved for use in the presence of combustible dust.
- Use of equipment with minimal risk of overheating.
- Inspection and maintenance procedures that minimize the risk of overheating.

1.4.2.5
Smouldering nests

Pinkwasser (1985, 1986) studied the possibility of dust explosions being initiated by smouldering lumps ('nests') of powdered material that is conveyed through a process system. The object of the first investigation (1985) was to disclose the conditions under which smouldering material that had entered a pneumatic conveying line would be extinguished, i.e. cooled to a temperature range in which the risk of ignition in the downstream equipment was no longer present. In the case of > 1 kg/m^3 pneumatic transport of screenings, low-grade flour and C3 patent flour, it was impossible to transmit a 10 g smouldering nest through the conveying line any significant distance. After only a few metres, the temperature of the smouldering lump had dropped to a safe level. In the case of lower dust concentrations, between 0.1 and 0.9 kg/m^3, i.e. within the most explosible range, the smouldering nest could be conveyed for an appreciable distance as shown in Figure 1.68, but no ignition was ever observed in the conveying line.

In the second investigation Pinkwasser (1986) allowed smouldering nests of 700°C to fall freely through a 1 m tall column containing dust clouds of 100–1000 g/m^3 of wheat flour or wheat starch in air. Ignition was never observed during free fall. However, in some tests

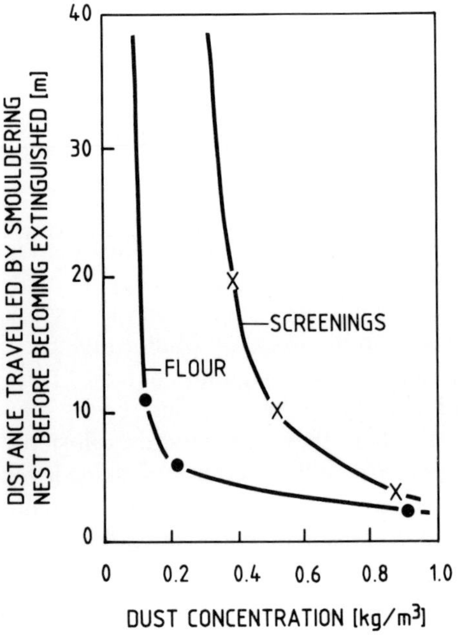

Figure 1.68 *Distance travelled in pneumatic transport pipe by smouldering nest before becoming extinguished, as a function of dust concentration in the pipe. Air velocity in pipe 20 m/s (From Pinkwasser, 1985)*

with nests of at least 25 mm diameter and weight at least 15 g, ignition occurred immediately after the nest had come to rest at the bottom of the test column. This may indicate the possibility that a smouldering nest falling freely through a dust cloud in a silo without disintegrating during the fall, has a higher probability of igniting the dust cloud at the bottom of the silo than during the fall.

Jaeger (1989) conducted a comprehensive laboratory-scale investigation on formation of smouldering nests and their capability of igniting dust clouds. He found that only materials of flammability class larger than 3 (see the Appendix) were able to generate smouldering nests. Under the experimental conditions adopted it was found that a minimum smouldering nest surface area of about 75 cm^2 and a minimum surface temperature of 900°C was required for igniting dust clouds of minimum ignition temperatures $\leqslant$ 600°C.

Zockoll (1989) studied the incendivity of smouldering nests of milk powder, and concluded that such nests would not necessarily ignite clouds of milk powder in air. One condition for ignition by a moving smouldering nest was that the hottest parts of the surface of the nest were at least 1200°C. However, if the nest was at rest, and a milk powder dust cloud was settling on to it, inflammation of the cloud occurred even at nest surface temperatures of about 850°C.

Zockoll suggested that in the case of milk powder, the minimum size of the smouldering nest required for igniting a dust cloud is so large that carbon monoxide generation in the plant would be adequate for detecting formation of smouldering nests before the nests have reached hazardous sizes.

Alfert, Eckhoff and Fuhre (1989) studied the ignition of dust clouds by falling smouldering nests in a 22 m tall silo of diameter 3.7 m. It was found that nests of low mechanical strength disintegrated during the fall and generated a large fire ball that ignited the dust cloud. Such mechanically weak nests cannot be transported any significant distance in e.g. pneumatic transport pipes before disintegrating. It was further found that mechanically stable nests ignited the dust cloud either some time after having come to rest at the silo bottom, or when being broken during the impact with the silo bottom. However, as soon as the nest had come to rest at the silo bottom, it could also become covered with dust before ignition of the dust cloud got under way.

Infrared radiation detection and subsequent extinction of smouldering nests and their fragments during pneumatic transport, e.g. in dust extraction ducts, has proven to be an effective means of preventing fire and explosions in downstream equipment, for example dust filters. One such system, described by Kleinschmidt (1983), is illustrated in Figure 1.69. Normally the transport velocity in the duct is known, and this allows effective extinction by precise injection of a small amount of extinguishing agent at a convenient distance just when the smouldering/burning nest or fragment passes the nozzles. Water is the most commonly used extinguishing agent, and it is applied as a fine mist. Such systems are mostly used in the wood industries, but also to some extent in the food and feed and some other industries. The field of application is not only smouldering nests, but also glowing or burning fragments from e.g. sawing machines and mills.

1.4.2.6
Heat from accidental mechanical impact

Mechanical impacts produce two different kinds of potential ignition sources, namely small flying fragments of solid material and a pair of hot-spots where the impacting bodies

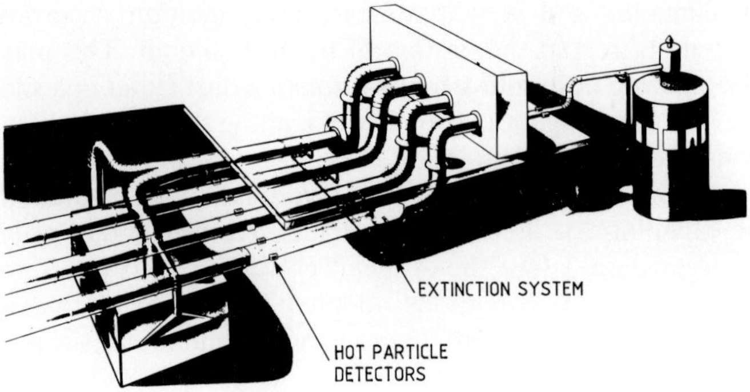

Figure 1.69 *Illustration of automatic system for detection and extinction of smouldering nests and their fragments, applied to a multiduct dust filter system (From Kleinschmidt, 1983)*

touch. Sometimes, e.g. in rotating machinery, impacts may occur repeatedly at the same points on one or both of the impacting bodies, and this may give rise to hot-spots of appreciable size and temperature. The hazardous source of ignition will then be a hot surface, and what has been said in 1.4.2.4 applies.

When it comes to single accidental impacts, there has been considerable confusion. However, research during the last decade has revealed that in general the ignition hazard associated with single accidental impacts is considerably smaller than often believed by many in the past. This in particular applies to dusts of natural organic materials such as grain and feedstuffs, when exposed to accidental sparking from impacts between steel hand tools like spades or scrapers, and other steel objects or concrete. In such cases the ignition hazard is probably non-existent, as indicated by Pedersen and Eckhoff (1987). The undue significance that has often been assigned to 'friction sparks' as initiators of dust explosions in the past, was also stressed by Ritter (1984) and Müller (1989).

However, if more sophisticated metals are involved, such as titanium or some aluminium alloys, energetic spark showers can be generated, and in the presence of rust, luminous, incendiary thermite flashes can result. Thermite flashes may also result if a rusty steel surface covered with aluminium paint or a thin smear of aluminium, is struck with a hammer or another hard object. However, impact of ordinary soft unalloyed aluminium on rust seldom results in thermite flashes, but just in a smear of aluminium on the rust. For a given combination of impacting materials, the incendivity of the resulting sparks or flash depend on the sliding velocity and contact pressure between the colliding bodies. See Chapter 5.

Although the risk of initiation of dust explosions by accidental single impacts is probably smaller than believed by many in the past, there are special situations where the ignition hazard is real. It would in any case seem to be good engineering practice to:

- Remove foreign objects from the process stream as early as possible.
- Avoid construction materials that can give incendiary metal sparks or thermite flashes.
- Inspect process and remove cause of impact immediately in a safe way whenever unusual noise indicating accidental impact(s) in process stream is observed.

Figures 1.70 and 1.71 show two examples of how various categories of foreign objects can be removed from the process stream before they reach the mills.

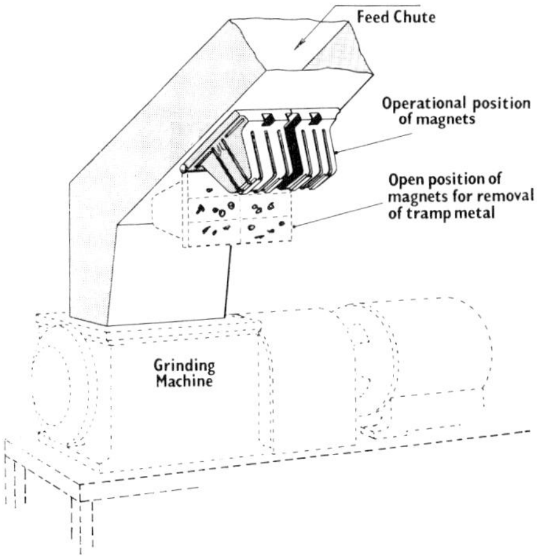

Figure 1.70 *A permanent magnetic separator fitted in the feed chute of a grinding mill to remove magnetic tramp metal (From DEP, 1970)*

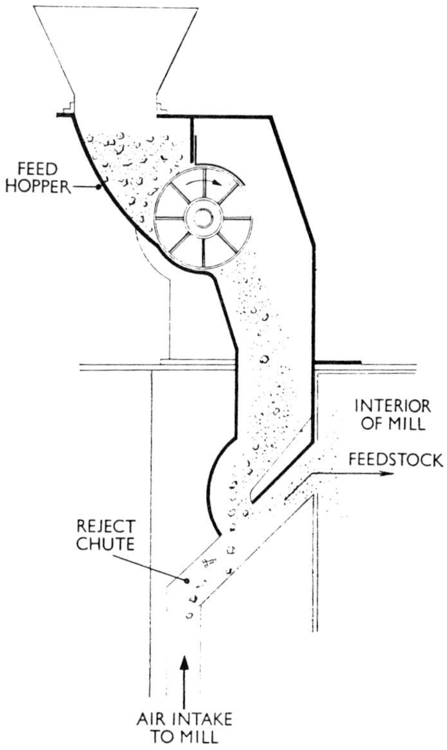

Figure 1.71 *A pneumatic separator can be used to remove most foreign bodies from the feed stock: the air current induced by the mill is adjusted to convey the feed stock and to reject heavier foreign bodies (From DEP, 1970)*

1.4.2.7
Electric sparks and arcs: electrostatic discharges

The various types of electric sparks and arcs and electrostatic discharges are described in Section 1.1.4.6. Sparks between two conducting electrodes are discussed in more detail in Chapter 5. Sparks or arcs due to breakage of live circuits can occur when fuses blow, in rotating electric machinery and when live leads are accidentally broken. The main rule for minimizing the risk of dust explosions due to such sparks and arcs is to

- Obey regulations for electrical installations in areas containing combustible dust (see Section 1.5.11).

The electrostatic hazard is more complex and it has not always been straightforward to specify clearly defined design guidelines. However, Glor (1988) has contributed substantially to developing a unified approach. As a general guideline he recommends the following measures:

- Use of conductive materials or materials of low dielectric strength, including coatings, (breakdown voltage across dielectric layer or wall < 4 kV) for all plant items that may accumulate very high charge densities (pneumatic transport pipes, dust deflector plates, and walls of large containers that may become charged due to ionization during gravitational compaction of powders). This prevents propagating brush discharges.
- Earth all conductive parts of equipment that may become charged. This prevents capacitive spark discharges from equipment.
- Earth personnel if powders of minimum ignition energies (MIE) < 100 mJ are handled. This prevents capacitive spark discharges from humans.
- Earth electrically conductive powders (metals etc.) by using earthed conductive equipment without non-conductive coatings. This prevents capacitive discharges from conductive powder.
- If highly insulating material (resistivity of powder in bulk > 10^{10} Ωm) in the form of coarse particles (particle diameter > 1 mm) is accumulated in large volumes in silos, containers, hoppers, etc., electrostatic discharges from the material in bulk may occur. These discharges can be hazardous when a fine combustible dust fraction of minimum ignition energy < 10–100 mJ is present simultaneously. So far, no reliable measure is known to avoid this type of discharge in all cases, but an earthed metallic rod introduced into the bulk powder will most probably drain away the charges safely. It is, however, not yet clear whether this measure will always be successful. Therefore the use of explosion venting, suppression or inerting should be considered under these circumstances.
- If highly insulating, fine powders (resistivity of powder in bulk > 10^{10} Ωm) with a minimum ignition energy ≤ 10 mJ as determined with a low-inductance capacitive discharge circuit, is accumulated in large volumes in silos, containers, hoppers, etc., measures of explosion protection should be considered. There is no experimental evidence that fine powders without any coarse particles will generate discharges from powder heaps, but several explosions have been reported with such powders in situations where all possible ignition sources, with the exception of electrostatics have been effectively eliminated.
- If combustible powders are handled or processed in the presence of a flammable gas or vapour (hybrid mixtures), the use of electrically conductive and earthed equipment is absolutely essential. Insulating coatings on earthed metallic surfaces may be tolerated provided that the thickness is less than 2 mm, the breakdown voltage less than 4 kV at locations where high surface charge densities have to be expected, and conductive powder cannot become isolated from earth by the coating. If the powder is non-conducting (resistivity of the powder in bulk > 10^6 Ωm), measures of explosion prevention (e.g. inert gas blanketing) are strongly recommended. If the resistivity of the powder in bulk is less than 10^6 Ωm, brush discharges, which would be incendiary for flammable gases or vapours, can also be excluded.

Glor pointed out, however, that experience has shown that even in the case of powders of resistivities in bulk < 10^6 Ωm it is very difficult in practice to exclude all kinds of effective ignition sources when flammable gases or vapours are present. In such cases large amounts of powders should therefore only be handled and processed in closed systems blanketed with an inert gas.

Further details, including a systematic step-by-step approach for eliminating the electrostatic discharge ignition hazard, were provided by Glor (1988). He also considered the specific hazards and preventive measures for different categories of process equipment and operations, such as mechanical and pneumatic conveying systems, sieving operations, and grinding, mixing and dust collecting systems.

1.4.3 PREVENTING EXPLOSIBLE DUST CLOUDS

1.4.3.1 Inerting by adding inert gas to the air

The influence of the oxygen content of the gas on the ignitability and explosibility of dust clouds was discussed in Section 1.3.6. For a given dust and type of added inert gas there is a certain limiting oxygen content below which the dust cloud is unable to propagate a self-sustained flame. By keeping the oxygen content below this limit throughout the process system, dust explosions are excluded. As the oxygen content in the gas is gradually reduced from that of air, ignitability and explosibility of the dust cloud is also gradually reduced, until ultimately flame propagation becomes impossible. Figure 1.72 shows some of the results from the experiments by Palmer and Tonkin (1973) in an industrial-scale experimental facility. The solid lines are drawn between the experiments that gave no

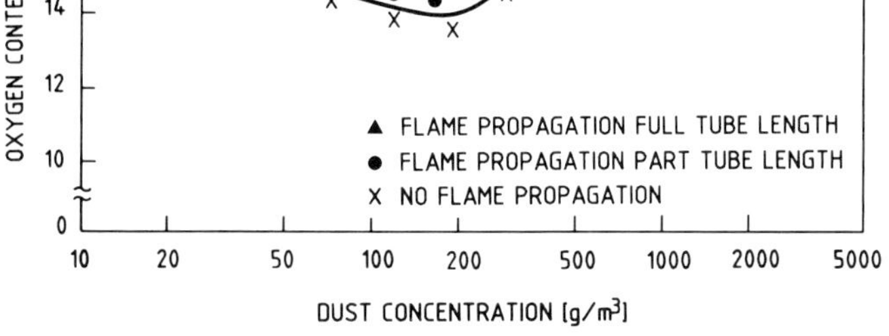

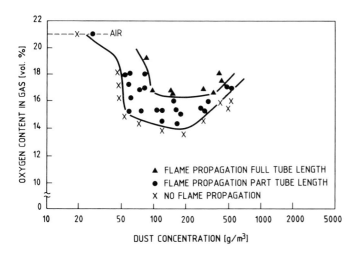

Figure 1.72 *Concentration range of flammability of clouds of phenol formaldehyde (15 μm mean particle diameter) as a function of the oxygen content in the gas. Inert gas added to air: CO_2. Experiments in vertical tube of diameter 0.25 m and length 5 m. Upwards flame propagation (From Palmer and Tonkin, 1973)*

flame propagation at all and flame propagation in part of the tube and between the experiments in which the flame propagated the entire length of the tube and only part of the tube length.

Schofield and Abbott (1988) and Wiemann (1989) have given useful overviews of the possibilities and limitations for implementing gas inerting in industrial practice. Five types of inert gases are in common use for this purpose:

- Carbon dioxide.
- Water vapour.
- Flue gases.
- Nitrogen.
- Rare gases.

Fischer (1978) also included halogenated hydrocarbons (halons) in his list of possible gases for inerting. However, due to the environmental problems caused by these substances, they may no longer be permitted for protecting against explosions and fires.

The choice of inert gas depends on several considerations, such as availability and cost, possible contaminating effects on products, and effectiveness. In the case of dusts of light metals, such as aluminium and magnesium, exothermic reactions with CO_2 and also in some situations with N_2 are known, and the use of rare gases may have to be considered in certain cases.

The Appendix gives some data for the maximum permissible oxygen concentration in the gas for inerting clouds of various dusts.

The design of gas inerting systems depends on whether the process is continuous or of the batch type, the strength of the process equipment and type and source of inert gas. Two main principles are used for establishing the desired atmosphere in the process:

- Pressure variation method.
- Flushing method.

The pressure variation method either operates above or below atmospheric pressure. In the former case, the process equipment, initially filled with air at atmospheric pressure, is pressurized to a given overpressure by inert gas. When good mixing of air and inert gas has been obtained, the process equipment is vented to the atmosphere and the cycle repeated until a sufficiently low oxygen content has been reached. The alternative is to first evacuate the process equipment to a certain underpressure, and fill up with inert gas to atmospheric pressure, and repeat the cycle the required number of times. By assuming ideal gases, there is, as shown by Wiemann (1989), a simple relationship between the oxygen content c_2 (vol%) at the end of a cycle and the content c_1 at the beginning, as a function of the ratio of the highest and lowest absolute pressures of the cycle.

$$c_2 = c_1(P_{max}/P_{min})^{1/n} \tag{1.13}$$

where $n = 1$ for isothermal and $n = C_p/C_v$ for adiabatic conditions.

The flushing method is used if the process equipment has not been designed for the significant pressure increase or vacuum demanded by the pressure variation method. It is useful to distinguish between two extreme cases of the flushing method, namely the replacement method (plug flow) and the through-mixing method (stirred tank). In order to maintain plug flow, the flow velocity of inert gas into the system must be low (< 1 m/s) and the geometry must be favourable for avoiding mixing. In practice this is very difficult to achieve, and the stirred tank method, using high gas velocities and turbulent mixing, is normally employed. It is essential that the instantaneous through-mixing is complete over the entire volume, otherwise pockets of unacceptably high, hazardous oxygen concentrations may form. Wiemann (1989) referred to the following equation relating thc oxygen content c_2 (vol%) in the gas after flushing and the oxygen content c_1 before flushing:

$$c_2 = (c_1 - c_r)e^{-v} + c_r \quad (1.14)$$

where c_r is the content of oxygen, if any, in the inert gas used, and v is the ratio of the volume of inert gas used in the flushing process, and the process volume flushed. Leaks in the process equipment may cause air to enter the inerted zone. Air may also be introduced when powders are charged into the process. It is important therefore to control the oxygen content in the inerted region, at given intervals or sporadically, depending on the size and complexity of the plant. The supply of inert gas must also be controlled.

Oxygen sensors must be located in regions where the probability of hitting the highest oxygen concentrations in the system is high. A sensor located close to the inert gas inlet is unable to detect hazardous oxygen levels in regions where they are likely to occur. Wiemann (1989) recommended that the maximum permissible oxygen content in practice be 2–3 vol% lower than the values determined in standard laboratory tests. (See Chapter 7 and the Appendix).

Various types of oxygen detectors are in use. The fuel cell types are accurate and fast. However, their lifetime is comparatively short, of the order of 1/2−1 year, and they only operate within a comparatively narrow temperature range. Zirconium dioxide detectors are very sensitive to oxygen and cover a wide concentration range with high accuracy and fast response. They measure the partial pressure of oxygen irrespective of temperature and water vapour. However, if combustible gases or vapours are present in the gas, they can react with oxygen in the measurement zone and cause systematically lower readings than the actual overall oxygen content, which can be dangerous. There are also oxygen detectors that utilize the paramagnetic or thermomagnetic properties of oxygen. Even these detectors are sufficiently fast and accurate for monitoring inerting systems for industrial process plants. However, nitrogen oxides can cause erratic results.

Wiemann emphasized two limitations of the gas inerting method when applied to dust clouds. First, as already illustrated by Figure 1.67, inerting to prevent dust explosions does not necessarily inert against self-heating and smouldering combustion. Secondly, as also mentioned earlier, the use of inert gas in an industrial plant inevitably generates a risk of accidental suffocation. The critical limit where severe breathing problems arise is 17–18 vol% oxygen. If flue gases are used, there may also be toxic effects.

Fischer (1978) also mentioned several technical details worth considering when designing systems for inerting of process plant to prevent dust explosions. He discussed specific examples of protection of industrial plant against dust explosions by gas inerting. Heiner (1986) was specifically concerned with the use of carbon dioxide for inerting silos in the food and feed industry.

The actual design of gas inerting systems can take many forms. Combinations with other means of prevention and mitigation of dust explosions are often used. Figure 1.73 illustrates nitrogen inerting of a grinding plant.

In Table 1.9 partial inerting, as opposed to complete inerting discussed so far, has been included as a possible means of mitigating dust explosions. This concept implies the

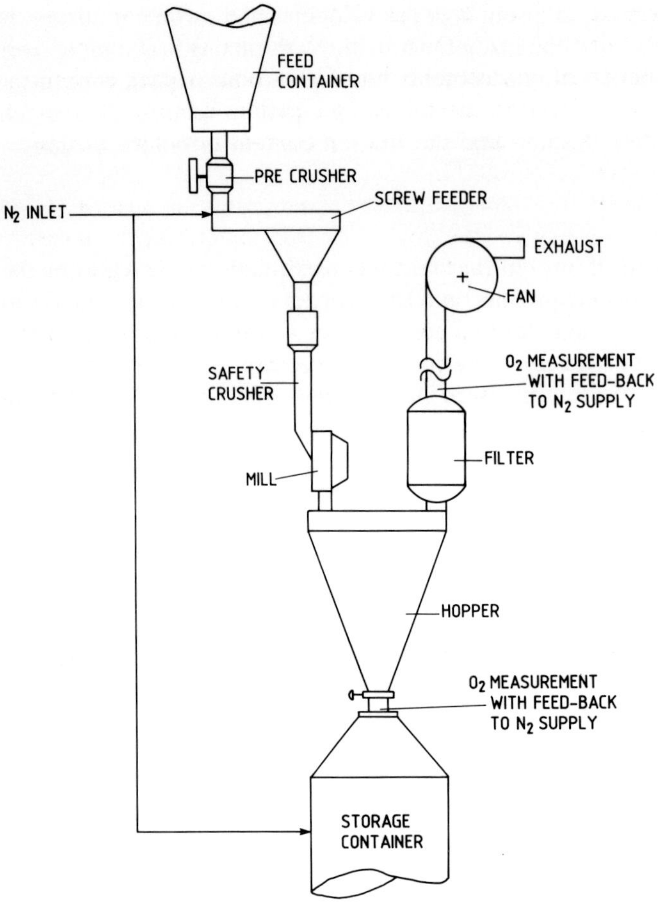

Figure 1.73 *Grinding plant inerted by nitrogen. Inerting combined with water spraying and explosion venting (simplified version of illustration from Bartknecht, 1978)*

addition of a smaller fraction of inert gas to the air than required for complete inerting. In this way both the ignition sensitivity, the explosion violence and the maximum constant-volume explosion pressure can be reduced appreciably, which means a corresponding reduction of the explosion risk. Partial inerting may be worth considering in combination with other means of prevention/mitigation when complete inerting is financially unacceptable.

1.4.3.2
Dust concentration outside explosible range

In principle one could avoid dust explosions by running the process in such a way that explosible dust concentrations were avoided (see Section 1.3.4). In practice, however, this is difficult in most cases, because the dust concentration inside process equipment most often varies in unpredictable and uncontrollable ways.

On the other hand, maintaining the powder/dust in the settled state by avoiding entrainment or fluidization in the air is one way of ensuring that the dust concentration is either zero or well above the upper explosible concentration. Good process design can significantly reduce the regions in which explosible dust concentrations occur, as well as the frequencies of their occurrence. One example is the use of mass flow silos instead of the traditional funnel flow type (see Perry and Green, 1984).

There are some special cases where it may be possible to avoid explosible dust clouds by actively keeping the dust concentration below the lower explosible limit. One such case is dust extraction ducts, another is cabinets for electrostatic powder coating, and the third is dryers. The latter case will be discussed in Section 1.5.3.5.

Ritter (1978) indicated that the measure of keeping the dust concentration below the minimum explosible concentration can also be applied to spray dryers, and Table 1.13 in Section 1.5.2 shows that Noha (1989) considered this a means of protection for several types of dryers. Noha also included dust concentration control when discussing explosion protection of crushers and mills (Table 1.12), mixers (Table 1.14) and conveyors and dust removal equipment (Table 1.15). However, in these contexts the dust concentration is below the minimum explosible limit due to the inherent nature of the process, rather than because of active control.

One essential requirement for controlling dust concentration is that the concentration can be adequately measured. Nedin *et al.* (1971) reviewed various methods used in the metallurgical industry in the USSR, mostly based on direct gravimetrical determination of the dust mass in isokinetically sampled gas volumes. Stockham and Rajendran (1984) and Rajendran and Stockham (1985) reviewed a number of dust concentration measurement methods with a view to dust control in the grain, feed and flour industry. In-situ methods based on light attenuation or backscattering of light were found to be most suitable.

Ariessohn and Wang (1985) developed a real-time system for measurement of dust concentrations up to about 5 g/m^3 under high-temperature conditions (970°C). Midttveit (1988) investigated an electrical capacitance transducer for measuring the particle mass concentration of particle/gas flows. However, such transducers are unlikely to be sufficiently sensitive to allow dust concentration measurements in the range below the minimum explosible limit.

Figure 1.74 shows a light attenuation dust concentration measurement station developed by Eckhoff and Fuhre (1975) and installed in the 6 inch diameter duct extracting dust from the boot of a bucket elevator in a grain storage plant. The long-lifetime light source was a conventional 12 V car lamp run at 4 V. A photoresistor and a bridge circuit was used for measuring the transmitted light intensity at the opposite end of the duct diameter.

The light source and photoresistor were protected from the dust by two glass windows flush with the duct wall. The windows were kept free from dust deposits by continuous air jets (the two inclined tubes just below the lamp and photoresistor in Figure 1.74).

Figure 1.75 shows the calibration data for clouds of wheat grain dust (10% moisture) in air. The straight line indicates that Lambert-Beer's simple concentration law for molecular species in fact applies to the system used.

Figure 1.76 illustrates a type of light attenuation dust concentration measurement probe developed more recently, using a light emitting diode (LED) as light source and a photodiode for detecting transmitted light. This concept was probably first introduced by Liebman, Conti and Cashdollar (1977), with subsequent improvement by Conti, Cashdollar and Liebman (1982). The particular probe design in Figure 1.76 was used successfully by Eckhoff, Fuhre and Pedersen (1985) for measuring concentration distributions of maize

Figure 1.74 *Light attenuation dust concentration measurement station mounted in the dust extraction duct on a bucket elevator boot in a grain storage facility in Stavanger, Norway (From Eckhoff and Fuhre, 1975)*

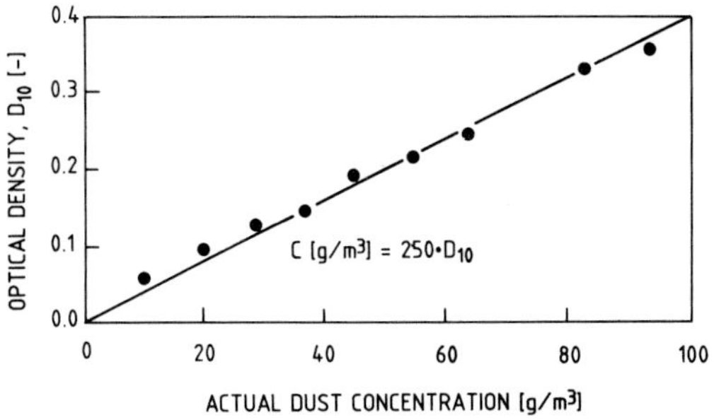

Figure 1.75 *Optical density of clouds in air of wheat grain dust containing 10% moisture: length of light path 150 mm: optical density D_{10} defined as*

$$\log_{10}\left(\frac{\text{Incident light intensity}}{\text{Light intensity after 150 mm}}\right)$$

(From Eckhoff and Fuhre, 1975)

starch in a large-scale (236 m^3) silo. The compressed air for flushing the glass windows of the probe was introduced via the metal tubing constituting the main probe structure.

However, in the case of dust explosions in the silo, the heat from the main explosion and from afterburns, required extensive thermal insulation of the probes in order to prevent damage.

The light path length of 30 mm was chosen to cover the explosible range of maize starch in air. The calibration data are shown in Figure 1.77. If this kind of probe is to be used for continuous monitoring of dust concentrations below the minimum explosive limit, e.g. in the range of 10 g/m^3, considerably longer paths than 30 mm will be required to make the

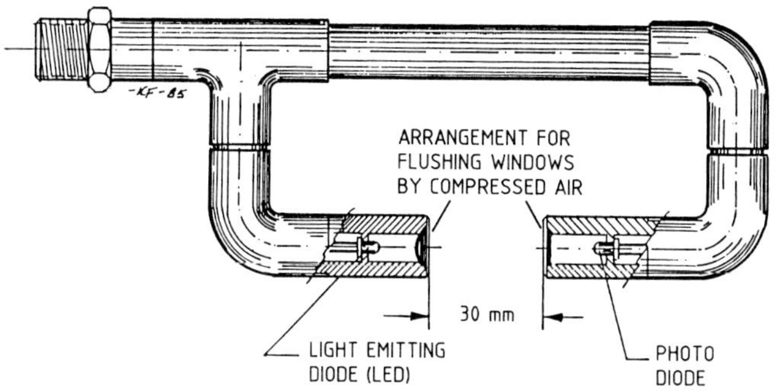

Figure 1.76 *Light attenuation probe for measurement of concentration of dust clouds, used by Eckhoff, Fuhre and Pedersen (1985) for measurement of concentration of maize starch in air in large-scale dust explosion experiments.*

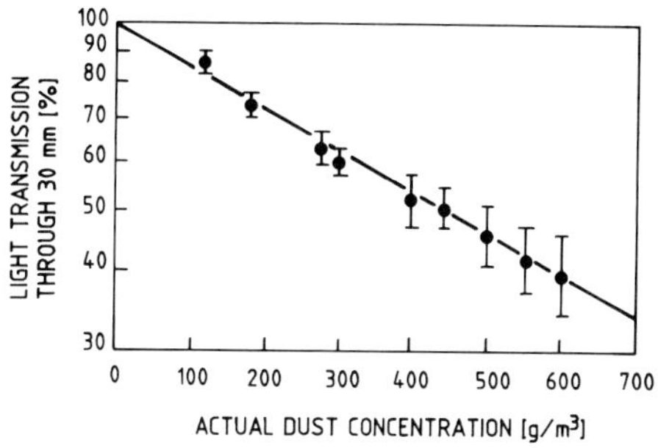

Figure 1.77 *Calibration data for light attenuation dust concentration probe in Figure 1.76, for native maize starch in air (From Eckhoff, Fuhre and Pedersen, 1985)*

instrument sufficiently sensitive. Other dust materials and particle sizes and shapes may also require other path lengths. In general it is necessary to calibrate light attenuation probes for each particulate dust and concentration range to be monitored.

The use of dust control in dust extraction systems is most likely to be successful if a small dust fraction is to be removed from a coarse main product, e.g. grain dust from grain, or plastic dust from pellets. By monitoring dust concentrations and controlling air flows the desired level of dust concentration can be maintained. However, if the air velocities are too low to prevent dust deposition on the internal walls of the ducting over time, dust explosions may nevertheless be able to propagate through the ducts (see Section 1.3.4 and also Chapter 4).

Possible dust entrainment and formation of explosible dust clouds by the air blast preceding a propagating dust explosion, may also occur in mixers, conveyors, etc. where sufficient quantities of fine dust are present as deposits. This means that in many cases dust concentration control is only feasible for preventing the primary explosion initiation, but not propagation of secondary explosions.

1.4.3.3 Adding inert dust

This principle is used in coal mines, by providing sufficient quantities of stone dust either as a layer on the mine gallery floor, or on shelves, etc. The blast that will always precede the flame in a dust explosion will then entrain the stone dust and coal dust simultaneously and form a mixture that is incombustible in air, and the flame, when arriving, will become quenched.

In other industries than mining, adding inert dust is seldom applicable due to contamination and other problems. It is nevertheless interesting to note the special war-time application for protecting flour mills against dust explosions initiated by high-explosive bombs, suggested by Burgoyne and Rashbash (1948). The Appendix contains some data for the percentage inert dust required for producing inert dust clouds with various combustible materials.

1.4.4 PREVENTING EXPLOSION TRANSFER BETWEEN PROCESS UNITS VIA PIPES AND DUCTS: EXPLOSION ISOLATION

1.4.4.1 Background

There are three main reasons for trying to prevent a dust explosion in one process unit from spreading to others via pipes and ducts.

Firstly, there is always a desire to limit the extent of the explosion as far as possible.

Secondly, a dust flame propagating in a duct between two process units tends to accelerate due to flow-induced turbulence in the dust cloud ahead of the flame. For a sufficiently long duct this may result in a vigorous flame jet entering the process unit at the down-stream end of the duct. The resulting extreme combustion rates can generate very high explosion pressures even if the process unit is generously vented. This effect was demonstrated in a dramatic way for flame-jet-initiated explosions of propane/air in a generously vented 50 m^3 vessel, by Eckhoff *et al.* (1980, 1984), as shown in Figure 1.78. There is no reason for not expecting very similar effects for dust explosions.

The third main reason for preventing flame propagation between process units is pressure piling. This implies that the pressure in the unburnt dust cloud in the downstream process unit(s) increases above atmospheric pressure due to compression caused by the expansion of the hot combustion gases in the unit where the explosion starts, and in the connecting duct(s). As shown in Section 1.3.8, the final explosion pressure in a closed vessel is proportional to the initial pressure. Therefore, in a coupled system, higher explosion pressures than would be expected from atmospheric initial pressure can occur transiently due to pressure piling. This was demonstrated in a laboratory-scale gas explosion experiment by Heinrich (1989) as shown in Figure 1.79.

In spite of the marked cooling by the walls in this comparatively small experiment, the transient peak pressure in V_2 significantly exceeded the adiabatic constant volume pressure of about 7.5 bar(g) for atmospheric initial pressure. Extremely serious situations can arise if flame jet ignition and pressure piling occurs simultaneously.

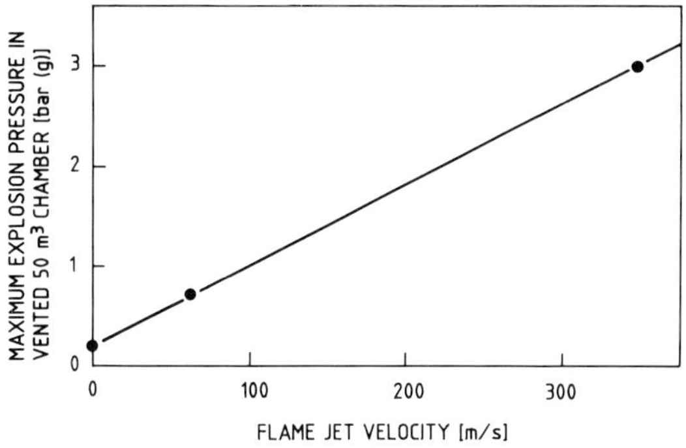

Figure 1.78 *Influence of flame jet ignition on the maximum explosion pressure for stoichiometric propane/air in a 50 m^3 vented chamber: vent orifice diameter 300 mm: vent area 4.7 m^2, no vent cover (From Eckhoff* et al.*, 1980)*

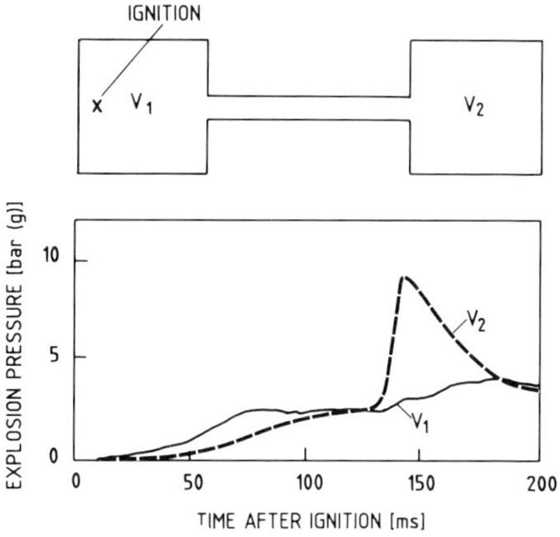

Figure 1.79 *Pressure development in two closed vessels of 12 litre each, filled with 10 vol% methane in air at atmospheric initial pressure and connected with a 0.5 m long duct, following ignition at location indicated (From Heinrich, 1989)*

1.4.4.2
Published overviews of methods for isolation

Basically there are two categories of methods, namely the passive ones being activated directly by the propagating explosion itself, and the active ones, which require a separate flame/pressure sensor system that triggers a separately powered system for operating the isolation mechanism. For obvious reasons, the passive systems are generally preferable if they function as intended and are otherwise suitable for the actual purpose.

Several authors have discussed the different technical solutions that have been used for interrupting dust explosions in the transfer system between process equipment. Walter (1978) concentrated on methods for stopping or quenching explosions in ducts. The methods included automatic, very rapid injection of extinguishing agent in the duct ahead of the flame front, and various kinds of fast response mechanical valves. Scholl, Fischer and Donat (1979) also included the concept of passive flame propagation interruption in ducts by providing a vented 180° bend system (see Figure 1.82). Furthermore, they discussed the use of rotary locks for preventing explosion transfer between process units or a process unit and a duct.

Czajor (1984) and Faber (1989) discussed the same methods as covered by Scholl, Fischer and Donat, and added a few more.

1.4.4.3
Screw conveyors and rotary locks

One of the first systematic investigations described in the literature is probably that by Wheeler (1935). Two of his screw conveyor designs are shown in Figure 1.80.

The removal of part of the screw ensures that a plug of bulk powder/dust will always remain as a choke. Wheeler conducted a series of experiments in which rice meal explosions in a 3.5 m^3 steel vessel were vented through the choked screw conveyors and through a safety vent at the other end of the vessel. Dust clouds were ejected at the downstream end of the conveyors, but no flame.

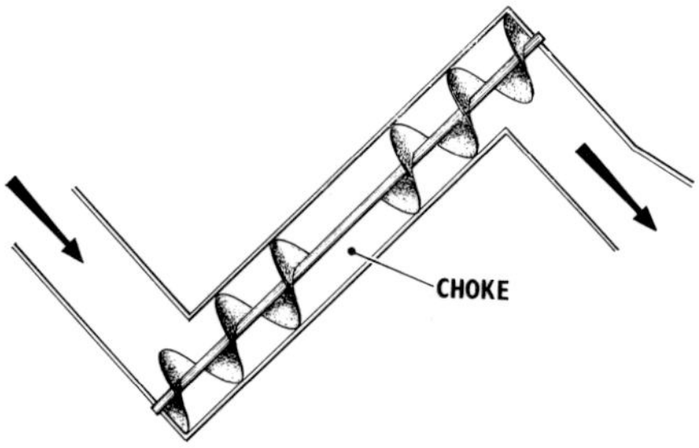

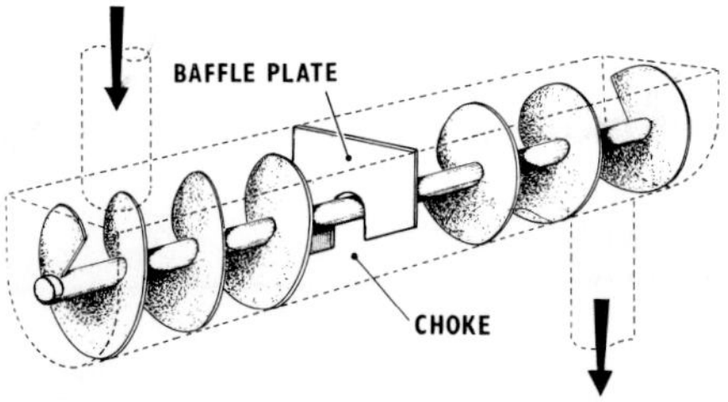

Figure 1.80 *Screw conveyors designed to prevent transmission of dust explosions (From Wheeler, 1935)*

Wheeler also conducted similar experiments with rotary locks. A hopper section mounted on top of the rotary lock was connected to the 3.5 m^3 explosion vessel. Even when the hopper was empty of rice meal, there was no flame transmission through the rotary lock. When the hopper contained rice meal and the rotary lock was rotating, there was not even transmission of pressure, and the rice meal remained intact in the hopper.

In recent years Schuber (1989) and Siwek (1989) conducted extensive studies of the conditions under which a rotary lock is capable of preventing transmission of dust explosions. Schuber provided a nomograph by which critical design parameters for explosion-transmission-resistant rotary locks can be determined. The minimum ignition energy and minimum ignition temperature of the dust must be known. However, the nomograph does not apply to metal dust explosions. Explosions of fine aluminium are difficult to stop by rotary locks. Schuber's work is described in detail in Chapter 4 in in the context of the maximum experimental safe-gap (MESG) for dust clouds.

Figure 1.81 illustrates how a rotary lock may be used to prevent transmission of a dust explosion from one room in a factory to the next.

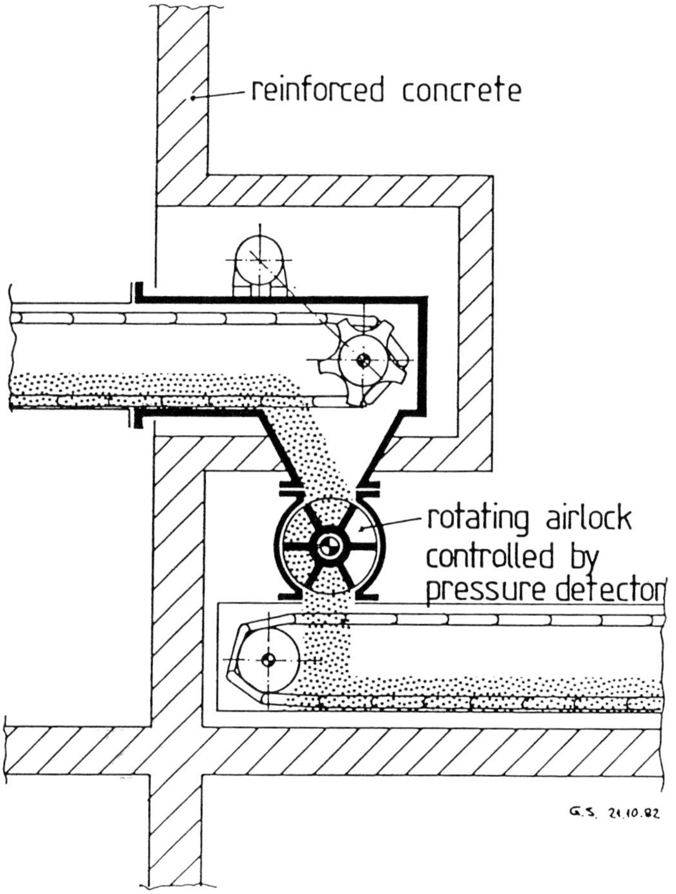

Figure 1.81 *Explosion isolation of two rooms using a rotary lock (Courtesy of Th. Pinkwasser and G. Schuber, Bühler, Switzerland)*

1.4.4.4
Passive devices for interrupting dust explosions in ducts

The device illustrated in Figure 1.82 was described relatively early by Scholl, Fischer and Donat (1979) and subsequently by others.

The basic principle is that the explosion is vented at a point where the flow direction is changed by 180°. Due to the inertia of the fast flow caused by the explosion, the flow will tend to maintain its direction rather than making a 180° turn. However, the boundaries for the applicability of the principle have not been fully explored. Parameters that may influence performance include explosion properties of dusts, velocity of flame entering the device, direction of flame propagation, and direction, velocity and pressure of initial flow in duct. Faber (1989) proposed a simplified theoretical analysis of the system shown in Figure 1.82, as a means of identifying proper dimensions. Figure 1.83 shows a commercial unit.

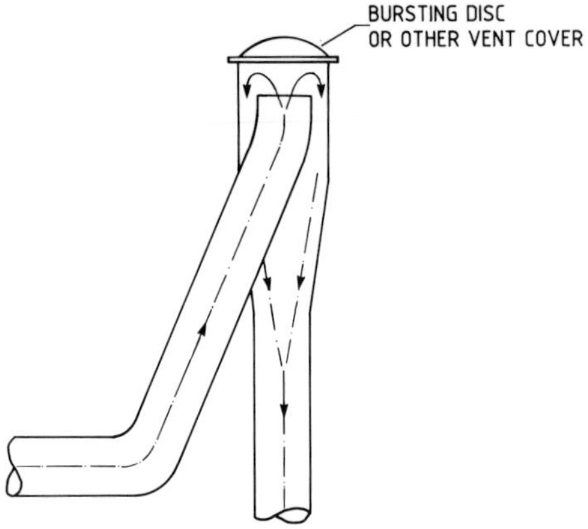

Figure 1.82 *Section through device for interrupting dust explosions in ducts by combining change of flow direction and venting. Flow direction may also be opposite to that indicated by arrows*

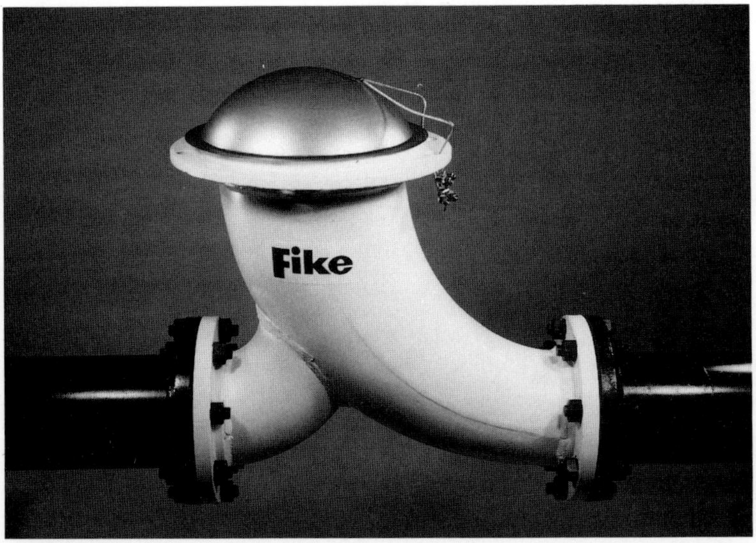

Figure 1.83 *Device for interrupting dust (and gas) explosions in ducts by combining change of flow direction and venting (Courtesy of Fike Corporation, USA)*

Figure 1.84 illustrates how the same basic principle may be applied to 90° bends at corners of buildings.

Another passive device for interrupting dust (and gas) explosions in ducts is the Ventex valve described by Rickenbach (1983) and illustrated in Figure 1.85.

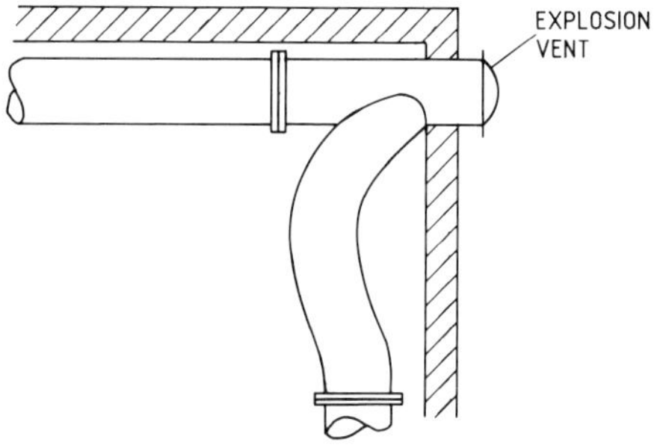

Figure 1.84 *Arrangement for interrupting/mitigating dust explosions in ducts by venting at 90° bends in corners of buildings*

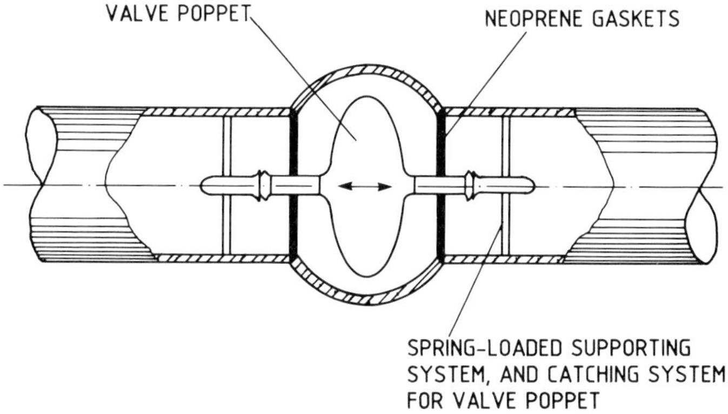

Figure 1.85 *Ventex valve for passive interruption of dust explosions in ducts (From Rickenbach, 1983)*

In normal operation the dust cloud being conveyed in the duct, flows around the valve poppet without causing any significant off-set as long as the flow velocity is less than about 20 m/s. However, in case of an explosion in the duct, the preceding blast pushes the valve poppet in the axial direction until it hits the neoprene gasket, where it is held in position by a mechanical catch lock, which can be released from the outside. Because of the inserts, the Ventex valve is perhaps more suitable when the dust concentration is low than for clouds of higher concentrations.

Active Ventex valves are also being used. In this case a remote pressure or flame sensor activates a separately powered system that closes the valve in the desired direction prior to arrival of the flame.

1.4.4.5
Active devices for interrupting dust explosions in ducts

Bartknecht (1980, 1982), Ebert (1983), Brennecke (1987) and Chatrathi and DeGood (1988) discussed the ability of various types of fast-closing slide valves to interrupt dust explosions in ducts. The required closing time depends on the distance between the remote pressure or flame sensor, and the valve, and on the type of dust. Often closing times as short as 50 ms, or even shorter, are required. This most often is obtained by using an electrically triggered explosive charge for releasing the compressed air or nitrogen that operates the valve. The slide valve must be sufficiently strong to resist the high pressures of 5–10 bar(g) that can occur on the explosion side after valve closure (in the case of pressure piling effects and detonation, the pressures may transiently be even higher than this).

Figure 1.86 shows a typical valve/compressed gas reservoir unit. Figure 1.87 shows a special valve that is triggered by a fast-acting solenoid instead of by an explosive charge. This permits non-destructive checks of valve performance.

Bartknecht (1978) described successful performance of a fast-closing (30 ms) compressed-gas-operated flap valve, illustrated in Figure 1.88.

Figure 1.89 illustrates an active (pressure sensor) fast-closing compressed-gas-driven valve that blocks the duct at the entrance rather than further downstream.

The last active isolation method of dust explosions in ducts and pipes to be mentioned is interruption by fast automatic injection of extinguishing chemicals ahead of the flame. The system is illustrated in Figure 1.90.

This is a special application of the automatic explosion suppression technique, which will be described in Section 1.4.7. Bartknecht (1978, 1987) and Gillis (1987) discussed this special application and gave some data for design of adequate performance of such systems. Important parameters are type of dust, initial turbulence in primary explosion, duct diameter, distance from vessel where primary explosion occurs, method used for detecting onset of primary explosion, and type, quantity and rate of release of extinguishing agent.

1.4.5
EXPLOSION-PRESSURE-RESISTANT EQUIPMENT

1.4.5.1
Background

If a dust cloud becomes ignited somewhere in the plant, a local primary dust explosion will occur. As will be discussed in Sections 1.4.6 and 1.4.7, there are effective means of reducing the maximum explosion pressure in such a primary explosion to tolerable levels. However, in some cases it is preferred to make the process apparatus in which the primary explosion occurs so strong that it can withstand the full maximum explosion pressure under adiabatic, constant volume conditions. Such pressures are typically in the range 5–12 bar(g) (see the Appendix, Table A1).

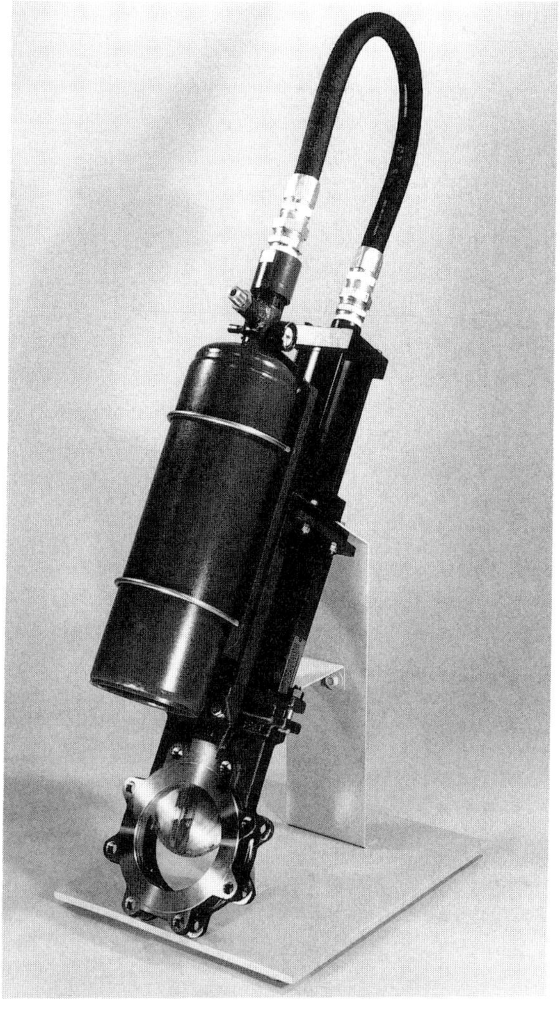

Figure 1.86 *Compressed-gas-driven fast-closing slide valve actuated by an explosive charge (Courtesy of Fike Corporation, USA)*

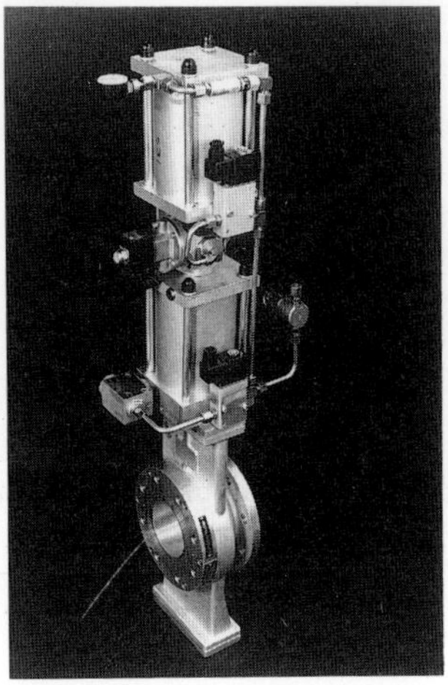

Figure 1.87 *Compressed-gas-driven fast-closing slide valve actuated by a fast solenoid (Courtesy of IRS, Germany)*

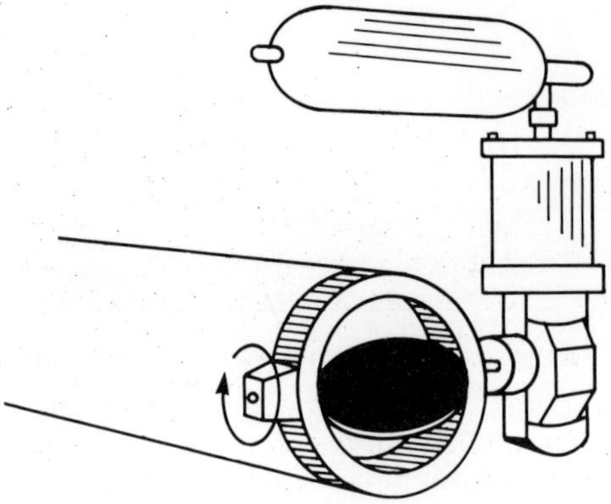

Figure 1.88 *Sketch of a compressed-gas-driven fast-closing flap valve*

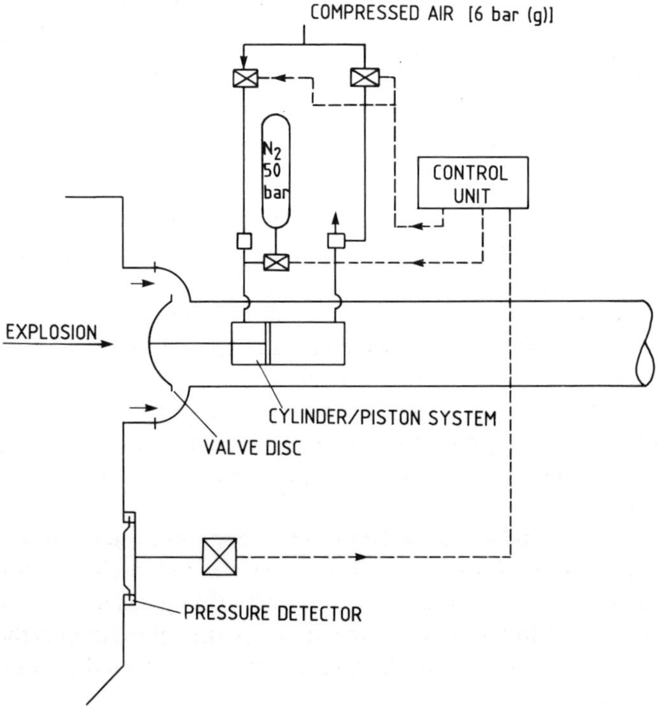

Figure 1.89 *Active fast-closing compressed-gas-driven valve system for blocking opening between process unit where primary explosion occurs, and duct/pipe (From Faber, 1989)*

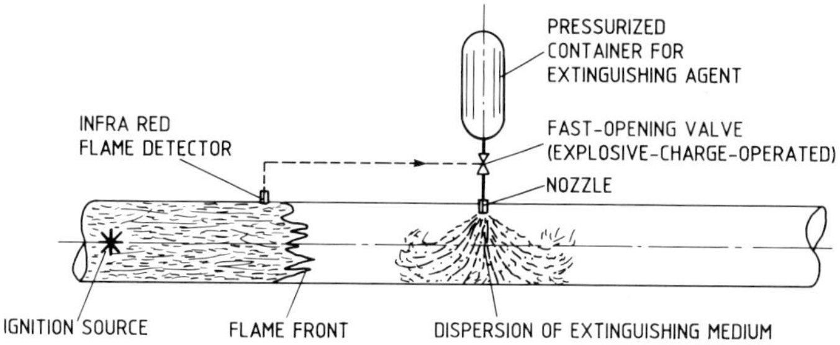

Figure 1.90 *Illustration of system for interrupting dust explosions in ducts by fast automatic injection of extinguishing agent ahead of the flame*

1.4.5.2
The 'explosion strength' of a process unit

The development of a stringent philosophy for the design of process equipment that has to withstand dust explosions is to a large extent due to the work of Donat (1978, 1984). More recent summaries of the subject were given by Kirby and Siwek (1986), Pasman and van Wingerden (1988) and Margraf and Donat (1989).

Donat (1978) introduced the useful distinction between pressure-resistant design and pressure-shock-resistant design. The first applies to pressure vessels, which must be capable of withstanding the maximum permissible pressure for long periods without becoming permanently deformed. In principle this concept could be used for designing explosion resistant equipment, by requiring that the process unit be designed as a pressure vessel for a maximum permissible working pressure equal to the maximum explosion pressure to be expected. However, experience has shown that this is a very conservative and expensive design. Pressure-shock-resistant design means that the explosion is permitted to cause slight permanent deformation of the process unit, as long as the unit does not rupture. This means that, for a given expected maximum explosion pressure, a considerably less heavy construction than would be required for pressure vessels is sufficient. The difference is illustrated in Figure 1.91, which applies to enclosures made of ferritic steels (plate steels). The pressure vessel approach would require that the apparatus be constructed so heavy that the maximum deformation during an explosion inside the vessel would not exceed two-thirds of the yield strength, or one-quarter of the tensile strength. The pressure-shock-resistant approach allows the explosion pressure to stress the construction right up to the yield point.

For austenitic (stainless) steels the stress-versus-strain curve does not show such a distinct yield point as in Figure 1.91. In such cases the pressure vessel approach specifies the maximum permissible working stress as two-thirds of the stress that gives a strain of 1%, whereas for the pressure-shock-resistant design the maximum permissible stress is the one that gives a strain of 2%. However, in the latter case, repair of deformed process equipment must be foreseen, should an explosion occur.

If dust explosions in the plant of concern were fairly frequent events, one would perhaps consider the use of the pressure vessel design approach, because the deformations that will often result with the pressure-shock-resistant design, would be avoided. This is a matter of

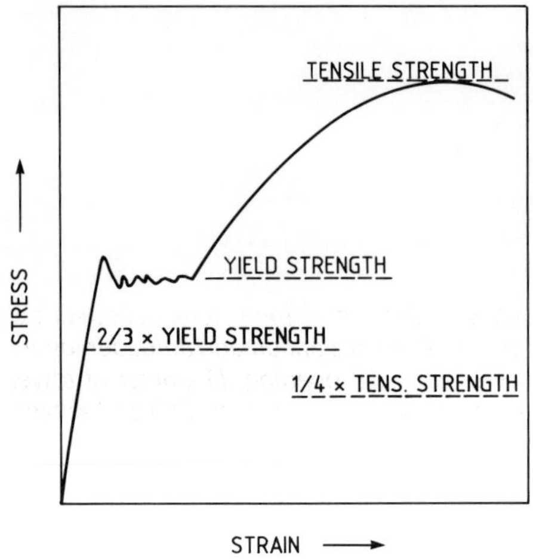

Figure 1.91 *Schematic stress-versus-strain curve for ferritic steel (From Kirby and Siwek, 1986)*

analysing cost versus benefit. From the point of view of safety, the main concern is to protect personnel, i.e. to avoid rupture of process equipment.

The field of structural response analysis has undergone substantial development over the past decades. Finite element techniques are now available for calculating stress and strain distributions on geometrically complex enclosure shapes, resulting from any given internal overpressure. Two examples are shown in Figures 1.92 and 1.93.

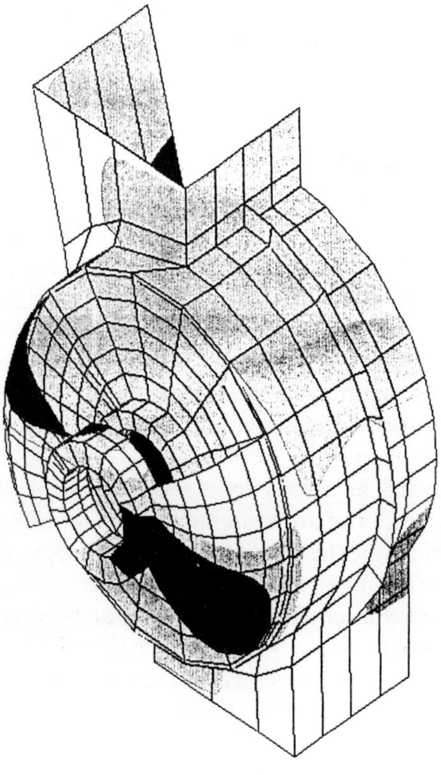

Figure 1.92 *Finite element design of rotary lock housing capable of withstanding 10 bar(g) internal pressure (Courtesy of Th. Pinkwasser, Bühler, Switzerland)*

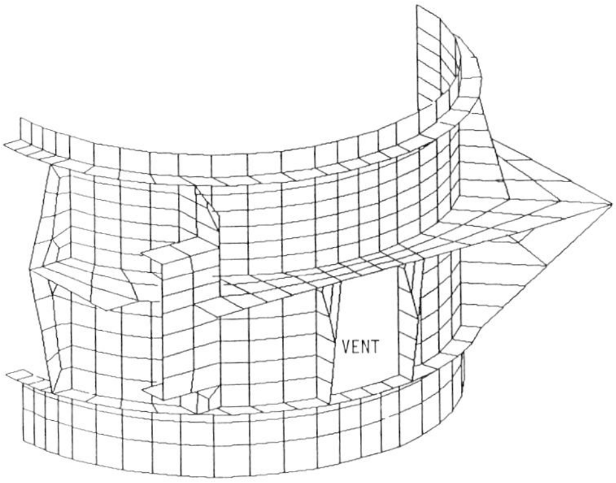

Figure 1.93 *Section of finite element network of cylindrical casing of a pneumatic unloader tower, with explosion vent opening. Diameter of tower 2 m (Courtesy of Th. Pinkwasser, Bühler, Switzerland)*

1.4.5.3
Influence of dynamics of explosion load

Pasman and van Wingerden (1988) discussed the influence of the dynamic characteristics of the explosion load on the structural response. Typical dust explosion pressure pulses in industrial equipment have durations in the range 0.1–1.0 s. In general, the shorter the load pulse, the stiffer and stronger the equipment will behave. Some quantitative data illustrating this were given by Kirby and Siwek (1986). However, the energy transfer from the dust cloud to the enclosure walls can be enhanced if the load pulse frequency equals the characteristic resonance frequency of the enclosure system. In this case acceleration and inertial forces can become important, and the load will exceed the value that would result if the maximum explosion pressure was applied as a static load.

Pasman and van Wingerden conducted a series of propane/air and acetylene/air explosions in various equipment typical of the powder production and handling industry. These included bins, ducts, an elevator head, eight cyclones, and a fan housing. The observed structural response (deformation, etc.) was correlated with the maximum explosion pressure and details of the construction of the equipment (number and dimensions of bolts in flanges, plate thicknesses, etc.). In spite of the complexity of the problem, it was possible to indicate some quantitative design criteria.

It nevertheless seems that direct explosion testing of full-scale process equipment prototypes will remain a necessity for some time. But, as illustrated in Figures 1.92 and 1.93, finite element techniques for structural response calculations are developing rapidly, and if these can be coupled to realistic dynamic explosion loads, the computer may replace full-scale explosion tests in a not too distant future.

Valuable further information concerning the response of mechanical structures to various types of explosion load was provided by Baker *et al.* (1983) and Harris (1983).

1.4.6 EXPLOSION VENTING

1.4.6.1 What is explosion venting?

The basic principle is illustrated in Figure 1.94. The maximum explosion pressure in the vented explosion, P_{red}, is a result of two competing processes:

- Burning of the dust cloud, which develops heat and increases the pressure.
- Flow of unburnt, burning and burnt dust cloud through the vent, which relieves the pressure.

The two processes can be coupled via flow-induced turbulence that can increase the burning rate.

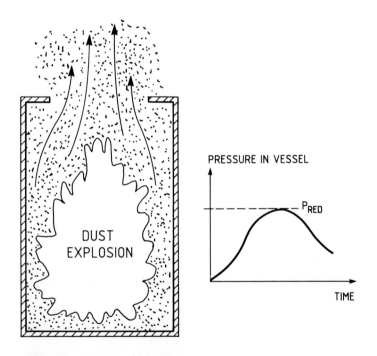

Figure 1.94 *Illustration of the basic principle of dust explosion venting: provision of an opening for controlled discharge of unburnt, burning and burnt dust cloud to keep the maximum pressure inside the vessel below a predetermined limit, P_{red}.*

The maximum permissible pressure, P_{red}, depends on the construction of the enclosure, and on whether pressure vessel design or pressure-shock-resistant design is adopted, as discussed in Section 1.4.5. Constructions of comparatively thin steel plate may require reinforcement for obtaining the P_{red} required. An example is shown in Figure 1.95.

Figure 1.95 *Reinforced vented 6m³ bag filter enclosure. P_{red} = 0.4 bar(g). Pressure-shock-resistant construction. The vent cover is a 0.85 m² three-layer bursting panel (Rembe, Germany) (Courtesy of Infastaub, Germany)*

1.4.6.2 Vent area sizing

Through the systematic work by Bartknecht (1978) and others it has become generally accepted that the required area of the vent opening depends on:

- Enclosure volume.
- Enclosure strength (P_{red}).
- Strength of vent cover (P_{stat}).
- Burning rate of dust cloud.

For some time it was thought by many that the burning rate of the dust cloud was a specific property of a given dust, which could be determined once and for all in a standard 1 m³ closed vessel test (K_{St}-value, see Chapter 4).

However, some researchers, including Eckhoff (1982), emphasized the practical significance of the fact that a given dust cloud at worst-case concentration can have widely different combustion rates, depending on the turbulence and degree of dust dispersion in the actual industrial situation. The influence of the dust cloud combustion rate on the maximum vented explosion pressure is illustrated in Figure 1.96.

During the 1980s new experimental evidence in support of the differentiated view on dust explosion venting has been produced, as discussed in detail in Chapter 6. In the course of the last decade the differentiated nature of the problem has also become gradually accepted as a necessary and adequate basis for vent sizing. This also applies to the latest revision of the VDI-3673 dust explosion venting code being currently drafted by Verein deutscher Ingenieure (1991) in Germany.

As discussed in both Chapter 4 and Chapter 7, a measure of the combustion rate of a dust cloud in air can be obtained by explosion tests in a standardized closed vessel. In these tests the maximum rate of rise of the explosion pressure is determined as a function of dust concentration, and the highest value is normally used for characterizing the combustion rate. Eckhoff, Alfert and Fuhre (1989) found that in practice it is difficult to discriminate between dusts of fairly close maximum rates of pressure rise, and it seems

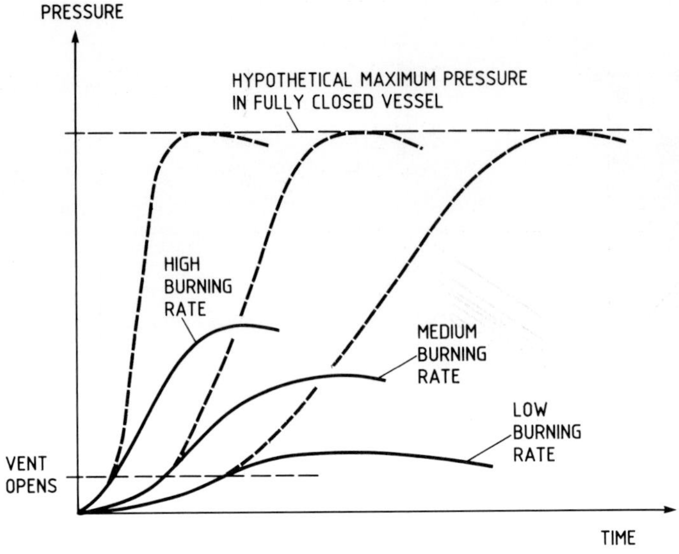

Figure 1.96 *Explosion pressure versus time in vented dust explosions with a given dust at worst-case concentration in a given enclosure with a given vent, for three different dust cloud burning rates (different turbulence intensities and degrees of dust dispersion)*

reasonable to work with a few, rather wide hazard classes of dusts. The classification used in the past in F. R. Germany comprises three classes. The first, St1, covers dusts that generate up to 200 bar/s in the 1 m^3 closed vessel test adopted by the International Standardization Organization (1985). The second class, St2, covers the range 200–300 bar/s, whereas the most severe class, St3, comprises dusts of > 300 bar/s. Pinkwasser (1989) suggested that the large St1 class be split in two at 100 bar/s, which may be worth while considering.

Various vent area sizing methods used in different countries are discussed in Chapter 6. Figure 1.97 summarizes what presently seems to be a reasonable compromise for dusts in the St1 class. The example shown is a 4.5 m^3 enclosure designed to withstand an internal pressure of 0.4 bar(g). If the process unit is a mill or other equipment containing highly turbulent and well-dispersed dust clouds, the vent area requirement is 0.48 m^2. If, however, the equipment is a silo, a cyclone or a bag filter, the required vent area is smaller, in the range 0.1–0.25 m^3.

Further details concerning vent area sizing, e.g. for enclosures of large length-to-diameter ratios, are given in Chapter 6. Scaling of vent areas may be accomplished using approximate formulae, as also discussed in Chapter 6.

1.4.6.3 Vent covers

A wide range of vent cover designs are in use, as shown in the comprehensive overview by Schofield (1984). Some designs are based on systematic research and testing, whereas others are more arbitrary. Beigler and Laufke (1981) carried out a critical inventory of

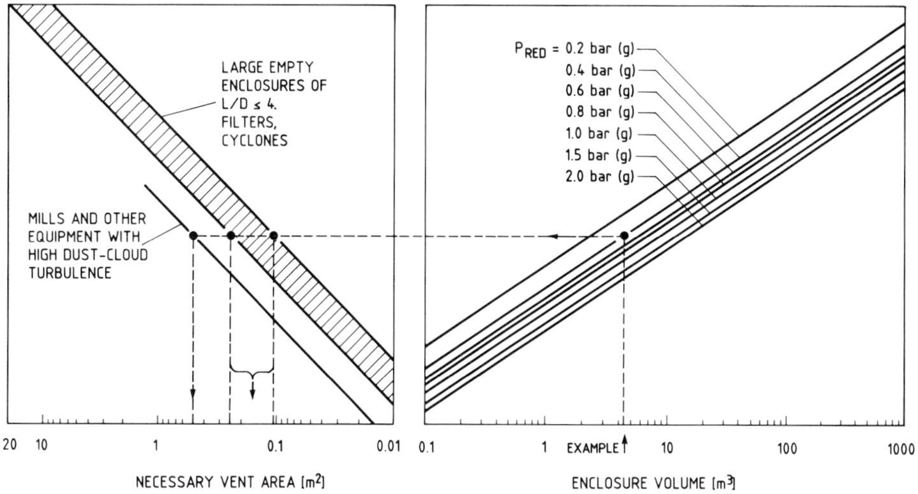

Figure 1.97 *Modified nomograph from VDI 3673 (1979) for St1 dusts ($0 < K_{St} < 200$ bar × m/s) and static vent cover opening pressures P_{stat} of ≤ 0.1 bar(g). Length of diameter ratio of enclosure ≤ 4. The example shown is an enclosure of volume 4.5 m^3 and strength P_{red} of 0.4 bar(g)*

vent covers used in the Swedish process industries for venting of process equipment as well as work rooms. Their conclusion was that a number of the vent covers inspected would not have performed adequately in the event of an explosion. They emphasized the need for ensuring that the static opening pressure of the vent cover is sufficiently low, and remains so over time, and that the mass of the cover is sufficiently small to permit rapid acceleration once released. Beigler (1983) subsequently developed an approximate theory for the acceleration of a vent cover away from the vent opening.

One quite simple type of vent cover is a light but rigid panel, e.g. an aluminium plate, held in position by a rubber clamping profile as used for mounting windows in cars. The principle is illustrated in Figure 1.98.

Other methods for keeping the vent cover in place include various types of clips. When choosing a method for securing the panel, it is important to make sure that the pressure, P_{stat}, needed to release the vent panel is small compared with the maximum tolerable explosion pressure, P_{red}. It is further important to anchor the vent panel to the enclosure to be vented, e.g. by means of a wire or a chain. Otherwise the panel may become a hazardous projectile in the event of an explosion. Finally, it is also important to make sure that rust formation or other processes do not increase the static opening pressure of the vent cover over time.

Bursting panels constitute a second type of vent covers. In the past, such panels were often 'home made', and adequate data for the performance of the panels were lacking. A primary requirement is that P_{stat}, the static bursting pressure of the panel, is considerably lower than the maximum permissible explosion pressure, P_{red}. Figure 1.99 shows a classic example of what happens if P_{stat} is larger than P_{red}. The enclosure bursts, whereas the explosion panel remains intact.

Today high-quality bursting panels are manufactured by several companies throughout the world. Such panels burst reliably at the P_{stat} values for which they are certified. An example of such a panel is shown in Figure 1.100 (see also Figure 1.95).

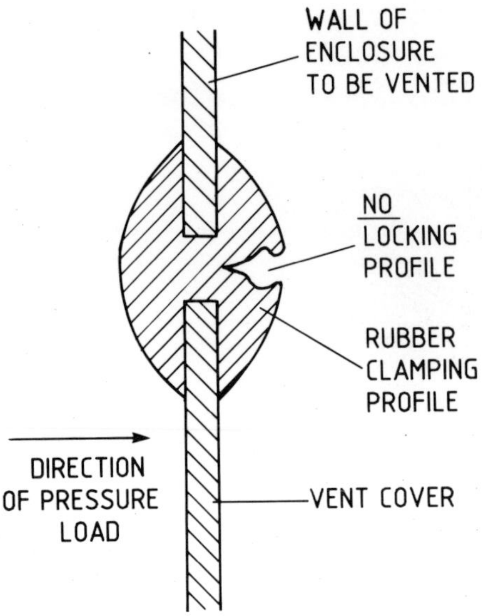

Figure 1.98 *Vent cover plate held in position by means of a rubber clamping profile*

Figure 1.99 *Damaged cyclone after a dust explosion. The vent cover was too strong to open before the cyclone itself ruptured (From DEP, 1970)*

Such panels are manufactured in a wide range of sizes and shapes, and coatings may be provided that allow permanent contact with various types of chemically aggressive atmospheres. Often a backing film of Teflon is used as an environmental protection to prevent the vent panel from contaminating the product inside the enclosure that is equipped with the vent. However, the upper working temperature limit of teflon is about 230°C. Brazier (1988) described special panels designed for service temperatures up to 450°C.

Figure 1.101 shows a bursting panel design that was originally developed for bucket elevators, but which may have wider applications. It consists of the bursting panel itself, which is a 0.04 mm thick aluminium foil of P_{stat} 0.1 bar(g), supported by a 0.5 mm metal gauze, and a second 0.5 mm metal gauze for further cooling of the combustion gases ('flame arrester'). Additional layers of metal gauze may be added as required for adequate cooling. The combustion gases should be cooled to the extent that unburnt discharged dust

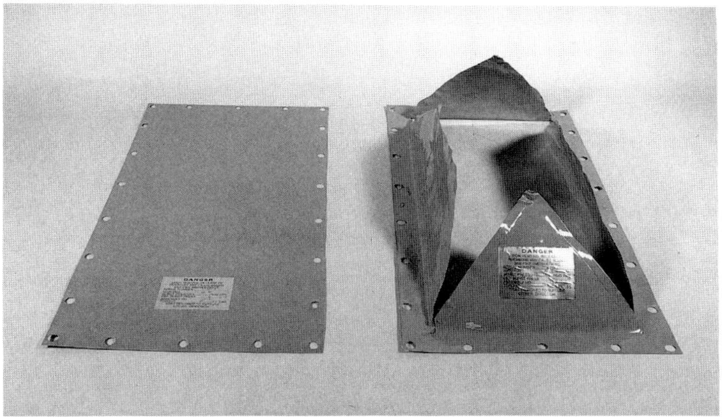

Figure 1.100 *Epoxy-coated explosion vent panel. Left: as mounted on vent. Right: after having relieved an explosion (Courtesy of Fike Corporation, USA)*

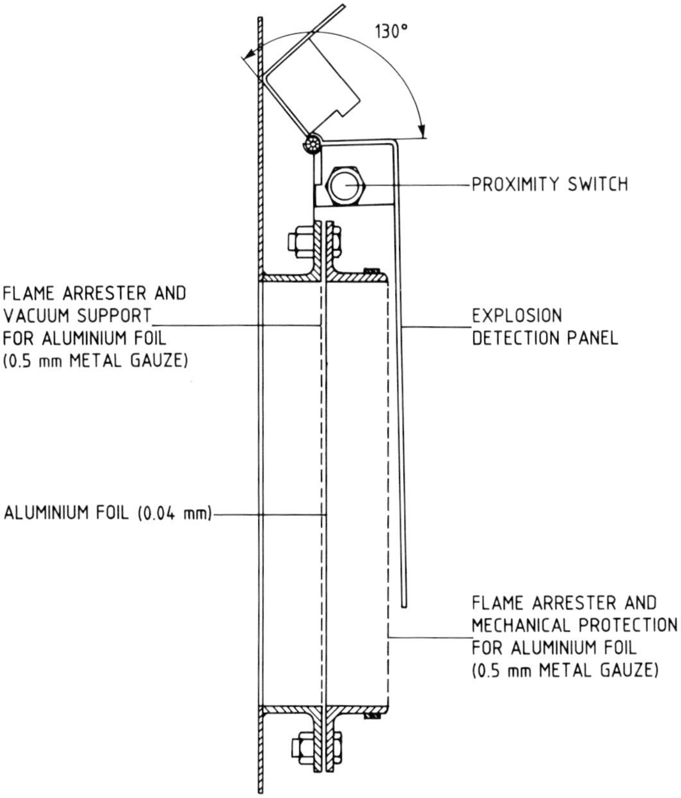

Figure 1.101 *Bursting panel combined with metal gauze for mechanical protection of vent panel and cooling of combustion gases. Displacement of explosion detection panel operates proximity switch (Courtesy of Th. Pinkwasser, Bühler, Switzerland)*

and dust that may be whirled up in the building are not ignited. In the case of an explosion, the explosion detection panel will be blown out and operate a proximity switch that triggers whatever pre-programmed automatic actions that should be taken (close-down of plant, closing of valves, automatic suppression, etc.).

He Kuangguo *et al.* (1987) investigated the dynamic strength and venting characteristics of bursting discs of various materials. Reasonable agreement was found between experimental results and theoretical predictions.

Hinged explosion doors constitute a third category of vent covers. Such doors may take a variety of different forms, depending on the equipment to be vented and other circumstances. Various kinds of calibrated locking mechanisms to ensure release at the predetermined P_{stat} have been developed. Hinged doors may be preferable if explosions are relatively frequent. Figure 1.102 shows an example of the use of hinged doors as vent covers.

Figure 1.102 *Four hinged explosion doors of 0. 8 m^2 each, with energy dissipation buffers, mounted on inlet hopper to a twin-rotor hammer mill for grinding household waste and bulky refuse: P_{red} = 1.0 bar(g) (Courtesy of Th. Pinkwasser, Bühler, Switzerland)*

Figure 1.103 shows the opened explosion doors on a milling plant similar to that in Figure 1.102, just after a dust explosion.

Donat (1973) discussed various advantages and disadvantages of bursting panels and hinged doors. Siwek and Skov (1989) analysed the performance of hinged explosion doors during venting with and without vent ducts (see Section 1.4.6.5). Both theoretical and experimental studies were carried out and a computer model of the venting process was developed.

Figure 1.103 *Hinged vent doors on a mill similar to that in Figure 1.102, just after a dust explosion. Damaged shock absorbers are replaced by new ones after each explosion before the doors are closed (Courtesy of Th. Pinkwasser, Bühler, Switzerland)*

The final category of vent covers to be mentioned are the reversible ones, i.e. covers that close as soon as the pressure has been relieved. The purpose of such covers is to prevent secondary air being sucked into the enclosure after the primary explosion has terminated, and giving rise to secondary explosions and fires. The reversible vent covers include counter-balanced hinged doors and spring-loaded, axially traversing vent covers.

One type of reversible hinged explosion vent cover is shown schematically in Figure 1.104.

The baffle plate is spring-loaded and acts as a shock absorber when being hit by the vent cover. Additional shock absorption is provided by the air cushion formed between the vent cover and baffle plate during impact. The adjustable pre-stressing device is used to set the static opening pressure, P_{stat}, of the vent cover to the desired level. Figure 1.105 shows the type of dust explosion vent illustrated in Figure 1.104 installed in the roof of a silo.

Käppler (1978) discussed the successful use of reversible hinged explosion doors on dust filter enclosures.

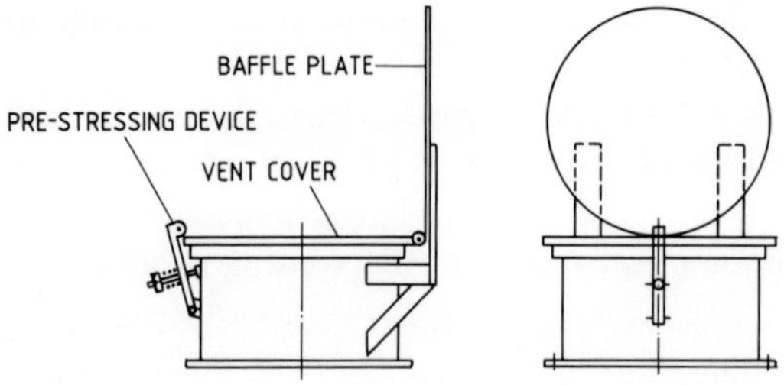

Figure 1.104 *Reversible low-mass hinged explosion door, which closes again by gravity once the explosion has been relieved (Courtesy of Silo-Thorwestern, Germany)*

Figure 1.105 *Reversible low-mass hinged explosion doors (see Figure 1.104) installed in the roof of a silo (Courtesy of Th. Pinkwasser, Bühler, Switzerland)*

One problem that can arise when using reversible explosion covers, is implosion due to the internal underpressure that follows the cooling of the gases inside the enclosure, when sealed just after the explosion. Wiemann, Bauer and Möller (1989) showed experimentally and theoretically that the internal underpressure can be limited to a desired non-damaging level by providing a small opening through which the small quantity of air required for preventing implosion is allowed to enter the enclosure in a controlled manner. Wiemann, Bauer and Möller presented a nomograph from which the necessary leak opening cross section can be determined from the vessel volume and the maximum permissible underpressure.

1.4.6.4
Potential hazards caused by venting

Venting of dust explosions prevents rupture of the enclosure in which the explosion takes place. However, significant hazards still remain. These include:

- Ejection of strong flame jets from the vent opening.
- Emission of blast waves from the vent opening.
- Reaction forces on the equipment, induced by the venting process.
- Emission of solid objects (vent panels and parts that can be torn off by the venting process).
- Emission of toxic combustion products.

In general, flame ejection will be more hazardous the larger the vent and lower the static opening pressure of the vent cover. This is because with a large vent and a weak cover, efficient venting will start at an early stage of the combustion process inside the enclosure. Therefore, in the early stages of venting, large clouds of unburnt dust will be pushed out through the vent and subsequently ignited when the flame passes through the vent. The resulting, secondary fire ball outside the vent opening can present a substantial hazard. If, on the other hand, the enclosure is strong, allowing the use of a small vent and a high P_{red}, only combustion products are vented, and the flame outside the vent is considerably smaller.

If a dust explosion is vented indoors, the blast waves and flame jet may generate serious secondary explosions in the work rooms (see Section 1.1.3). Some methods for preventing this are discussed in the following.

1.4.6.5
Vent ducts

One traditional solution to the flame jet problem is the use of vent ducts. This implies that a duct of cross sectional area at least equal to the vent area is mounted between the vent and a place where a strong flame jet will not present any hazard. The principle is illustrated in Figure 1.106.

Vent ducts will generally increase the flow resistance, and therefore also the pressure drop to the atmosphere. Consequently, adding a vent duct increases the maximum explosion pressure in the vented vessel. Furthermore, the pressure increases with increasing duct length, increasing number of sharp bends and decreasing duct diameter. These trends are confirmed by experiments.

Figures 1.107 and 1.108 give some results from small-scale experiments. The comparatively high pressures in Figure 1.107 for dextrin are due to the use of smaller vent and duct diameters than those employed for acquiring the data in Figure 1.108.

The same trend as exhibited by Figures 1.107 and 1.108 is found in larger scale, as shown by the data from TNO (1979) in Figure 1.109.

Walker (1982) analysed available data at that time and proposed the general relationship for the maximum explosion pressure in a vented vessel with duct as shown in Figure 1.110. It is felt that this correlation still holds good as a first approximation. For example, the data in Figure 1.109 are reasonably well accounted for in Figure 1.110.

Aellig and Gramlich (1984) studied the influence of various geometrical features of the vent duct design, in particular the details of the coupling between vessel, vent and duct, and geometry of bends. They proposed an overall correlation that looks similar to that of Walker in Figure 1.110, but the ratio of duct volume to vessel volume was used as parameter instead of the duct length.

Pineau (1984) conducted comprehensive series of experiments with explosions of wheat flour and wood dust in vented vessels of 0.1 m^3 and 1.0 m^3 volumes connected to vent

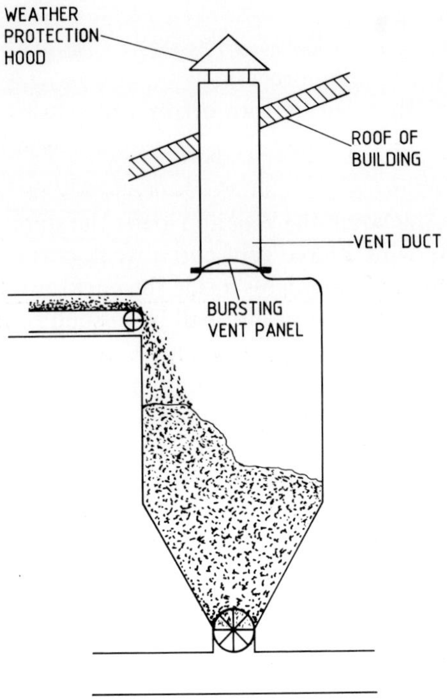

Figure 1.106 *Illustration of the use of vent duct for guiding discharged unburnt dust cloud and flames to a safe place*

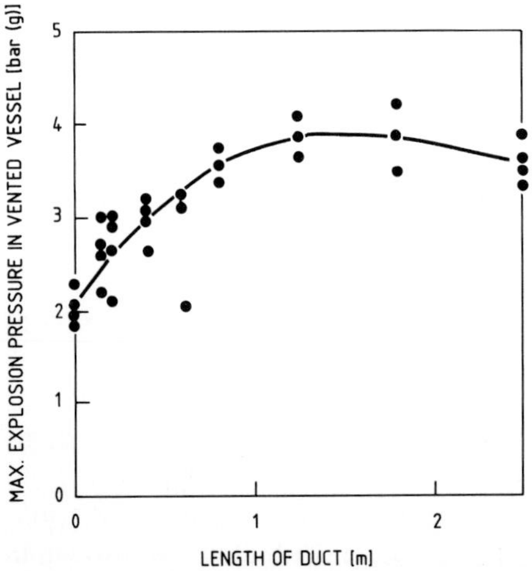

Figure 1.107 *Influence of the length of a straight vent duct of internal diameter 25 mm, on the maximum pressure during explosions of dextrin/air in a 20 litre spherical vessel. No vent cover (From Kordylewski, Wach and Wójcik, 1985).*

ducts of various diameters and lengths, with and without bends. Some experiments were also conducted with larger vessels of volumes 2.5–100 m^3 vented through ducts. In general the main trends observed in small-scale experiments were confirmed for the large scales, and it was recommended that vent ducts be as short as possible and have a minimum of sharp bends.

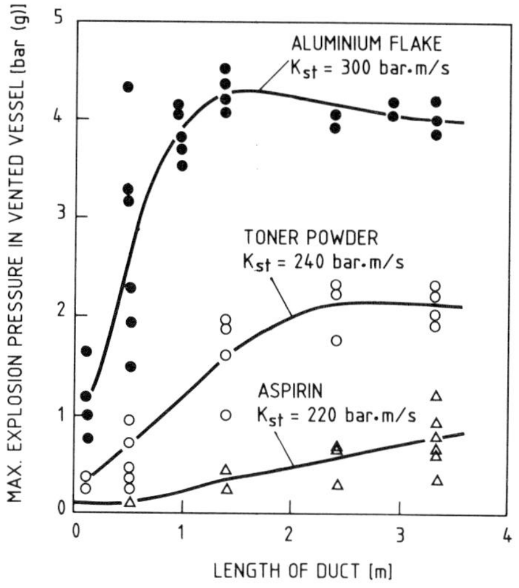

Figure 1.108 *Influence of the length of a straight vent duct of internal diameter 130 mm, on the maximum pressure during explosions of three different dusts in air in a 20 litre spherical vessel. Vent cover of diameter 130 mm and bursting strength 0.1 bar(g) between vessel and duct (From Crowhurst, 1988)*

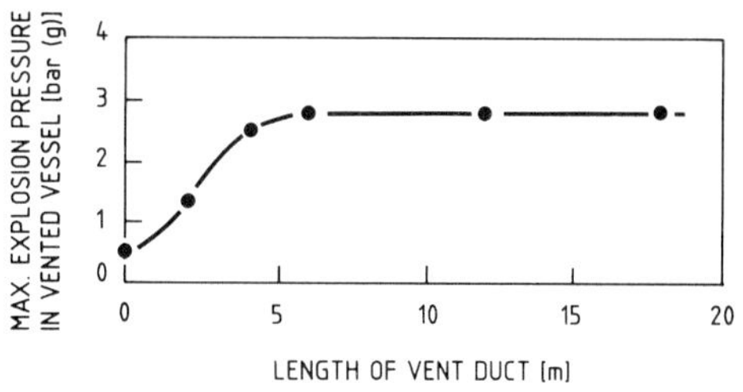

Figure 1.109 *Influence of the length of a straight vent duct of internal diameter 0.35 m, on the maximum dust explosion pressure in a 1 m³ vessel vented into the duct via a 0.35 m diameter bursting disc of bursting pressure 0.47 bar(g) (From TNO, 1979)*

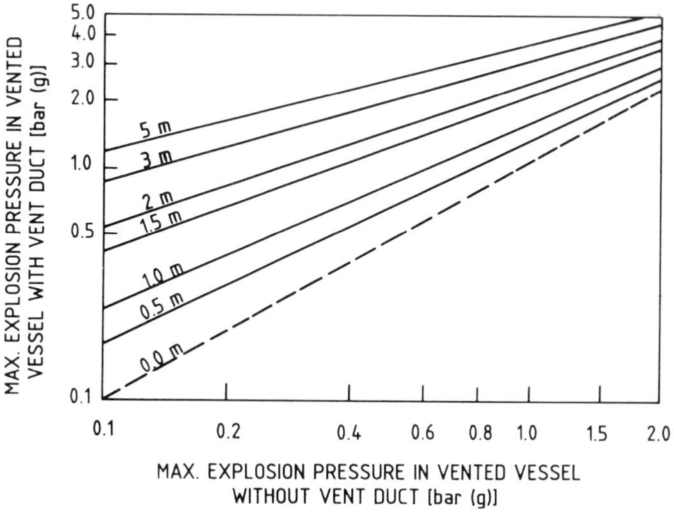

Figure 1.110 *Maximum pressure in vented vessel with vent duct as a function of maximum explosion pressure without duct, for various duct lengths. Diameter of duct equal to diameter of vent. No sharp bends (From Walker, 1982)*

More recently, Lunn, Crowhurst and Hey (1988) conducted a comprehensive theoretical and experimental study of the effect of vent ducts on the maximum explosion pressure in vented vessels. Experiments were performed in a 20 litre vessel (same experiments as Crowhurst, 1988), and in a large-scale 18.5 m^3 vessel. Figure 1.111 shows the 18.5 m^3 vessel fitted with a straight duct, whereas Figure 1.112 shows the same vessel during a coal dust explosion with a 90° bend at the end of the duct.

Figure 1.111 *18.5 m^3 vented explosion vessel connected to a straight vent duct (Courtesy of Health & Safety Executive, UK). For a much clearer picture see colour plate 1*

Figure 1.112 *Coal dust explosion in 18.5 m^3 vessel vented through a duct with a 90° bend at the end (Courtesy of Health & Safety Executive, UK). For a much clearer picture see colour plate 2*

In general the trends of the experimental data for the five dusts coal, aspirin, toner, polyethylene and aluminium used by Lunn, Crowhurst and Hey were similar to that in Figure 1.109. The maximum explosion pressure in the vessel increased systematically with duct length, or with the length-to-diameter ratio of the duct. The theoretical analysis generally confirmed this trend and yielded predictions in reasonable agreement with the experimental data, although some discrepancies were found. The theory developed by Lunn, Crowhurst and Hey may serve as a useful tool for estimating the influence of various types of vent ducts on P_{red}. The K_{St}-value, which is numerically identical with the maximum rate of pressure rise in the standard 1 m^3 ISO test, was used as a measure of the inherent explosibility of the dusts. The K_{St} values ranged from 144 bar × m/s for the coal to 630 bar × m/s for the aluminium.

1.4.6.6 The quenching tube

This promising new concept was developed by Alfert and Fuhre (1989) in cooperation with Rembe GmbH, F. R. Germany. (See also Anonymous, 1989.) The main principle is illustrated in Figure 1.113.

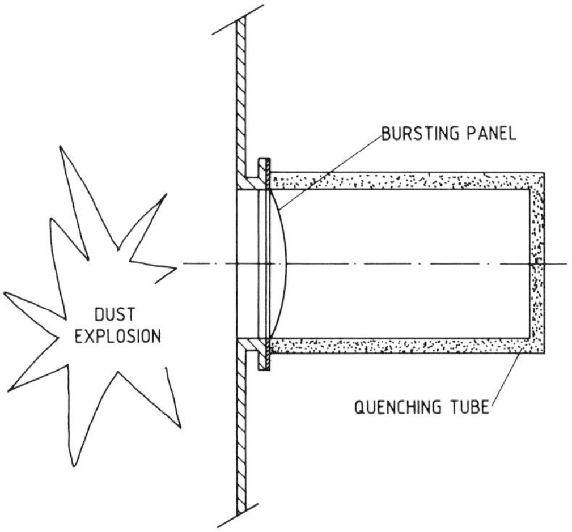

Figure 1.113 *Schematic illustration of a quenching tube for dust and flame-free venting of dust explosions*

If a dust explosion occurs in the enclosure to be vented, the bursting panel, which constitutes an integral part of the quenching tube assembly, bursts, and the explosion is vented through the comparatively large wall area of the quenching tube. The wall is designed to yield low pressure drop, but high retention efficiency for dust particles and efficient cooling of combustion gases. This means that flame ejection from the vent is effectively prevented and the blast effects significantly reduced.

Furthermore, burning lumps of powder and other smaller objects that could be ejected through an open vent, are retained inside the quenching tube. However, any toxic gaseous combustion products, e. g. carbon monoxide, will escape to the atmosphere.

The increase of the maximum explosion pressure in the vented enclosure due to the flow resistance through the quenching tube wall is mostly moderate, and can normally be compensated for by a moderate increase of the vent area.

It seems reasonable to expect that the quenching tube concept will be developed further in the coming years and gain wide application in areas where vent ducts have traditionally been the only solution, or where one has been forced to use open venting indoors in spite of the potential hazard associated with this. Bucket elevator legs and silos in congested areas, where normal venting is prohibited, are likely areas of application. The very high temperatures of burning light metal dust clouds (magnesium, aluminium, silicon, etc.) makes heavy demands on the design of the quenching tube wall, but there is no a priori reason for not assuming that even this problem will be solved.

Whether the remaining problem of possible toxic gas emission can be tolerated, depends on the actual circumstances, and must be considered specifically in each particular case.

Figure 1.114 shows a commercial prototype of a quenching tube.

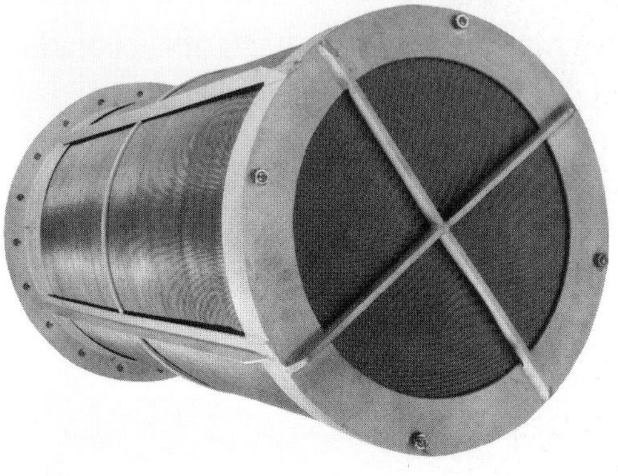

Figure 1.114 *Commercial prototype of a quenching tube (Courtesy of Rembe GmbH, Germany)*

Figure 1.115 shows venting of a 5.8 m^3 bag filter unit without and with the quenching tube. The white smoke in the lower picture is mostly condensed water vapour.

1.4.6.7
Reaction forces and blast effects

Experience has shown that the reaction forces from dust explosion venting can increase significantly both the material damage and the extent of the explosion. Equipment can tilt and ducts be torn off, and secondary dust clouds can be formed and ignited. Whenever installing a vent it is therefore important to make an assessment of whether the equipment to be vented is able to withstand the reaction forces from the venting, should an explosion occur. A very simple, static consideration says that the maximum reaction force equals the maximum pressure difference between the interior of the vessel being vented and the atmosphere, times the vent area. Careful experiments by Hattwig and Faber (1984) revealed that in actual explosion venting, the reaction force is about 20% higher than the value resulting from the simplified static consideration. The experimental relationship found by Hattwig and Faber is

$$F_{max}[MN] = 0.12 \times A[m^2] \times P_{max}[bar(g)] \qquad (1.15)$$

This equation can be used for estimating maximum reaction forces to be expected in practice. P_{max} is then the maximum permissible pressure P_{red} for which the vent is designed. Brunner (1983) found that the experimental reaction force was reduced by about 6% by vent ducts.

Figure 1.115 *Venting of a polypropylene/air explosion in a 5.8 m^3 bag filter unit without (top) and with quenching tube (bottom) (Courtesy of F. Alfert and K. Fuhre, Chr. Michelsen Institute, Norway). For a much clearer picture see colour plate 3*

As already discussed in Section 1.4.5.3, a given pressure pulse will interact with the mechanical structure that is exposed to it. This is also a relevant aspect in the present context. As pointed out by Pritchard (1989), the strength of some materials, including structural steels, is highly strain rate sensitive. This means that the stress at which plastic deformation starts, depends on the rate of loading. On the other hand, the damage to a structure also depends on how quickly the structure responds to the pressure loading. The natural period of vibration of the mechanical structure is normally used as a measure of the response time. If the duration of the pressure peak is long compared with the natural period of vibration, the loading can be considered as being essentially a static load. If, on the other hand, the pressure pulse is short compared with the response time of the structure, the damage is determined by the impulse, i.e. the time integral of pressure. Pritchard (1989) provided a qualitative illustration of these relationships, shown in Figure 1.116.

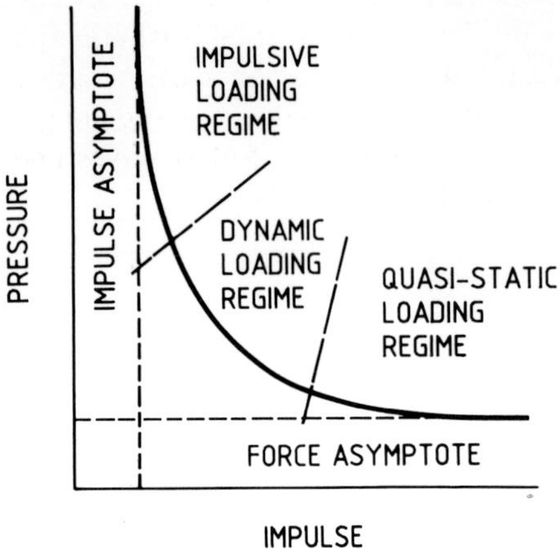

Figure 1.116 *Pressure-impulse diagram for single-degree-of-freedom elastic structure loaded with an ideal blast wave (From Pritchard, 1989)*

Brunner (1983, 1984) performed a detailed theoretical and experimental study of the structural response of supports and buildings due to the reaction forces from dust explosion venting of a vessel. An analysis of experimental explosion pressure-versus-time revealed two different regimes, namely $P_{max} < 0.9$ bar(g) and $P_{max} > 0.9$ bar(g). In the first regime, the pressure pulses generally had several peaks, whereas in the high pressure regime there was normally only one main peak. Theoretical equations for pressure-versus-time were developed for both regimes.

Brunner considered both linear and nonlinear model systems for the structures subjected to the reaction forces and developed response spectra for both fully elastic and ductile systems. The theoretical strain predictions were in good agreement with experimental results. Some practical guidelines for safe design of structures subjected to reaction forces from explosion venting were proposed.

Hattwig (1980) investigated the blast peak pressure, P_{blast}, outside a vented dust and gas explosion as a function of the distance D from the explosion and found that

$$P_{blast} = \frac{A \times P_{max} \times 1m}{D[m]} \tag{1.16}$$

where the dimensionless parameter A is given by

$$\log_{10} A = -\frac{0.26}{F[m^2]} + 0.49 \tag{1.17}$$

and P_{max} is the maximum explosion pressure inside the vented enclosure. Both P_{blast} and P_{max} are gauge pressures.

According to Kuchta (1985) the static, or 'side-on', gauge pressure of a blast wave front is

$$\Delta P_s = P_o \frac{2\gamma(M_o^2 - 1)}{(\gamma - 1)} \tag{1.18}$$

where P_0 is the ambient absolute pressure, γ the specific heat ratio of air, and M_0 the ratio of the actual wave front velocity to the velocity of sound. However, the total blast pressure that is sensed by an object exposed to a blast wave is the sum of the static gauge pressure and the dynamic pressure $1/2\ \rho\ V^2$ due to the gas flow (V is the gas velocity and ρ the gas density). Strehlow (1980) gave an instructive overview of the nature of blast waves and their damaging potential. A useful review was also given by Pritchard (1989).

1.4.7 AUTOMATIC SUPPRESSION OF DUST EXPLOSIONS

1.4.7.1 General concept

According to Dorn (1983), the first patent for a fast fire suppression system, a 'rapid dry powder extinguisher', was allotted to a German company as early as 1912. The Second World War accelerated the development. The British Royal Air Force found that 80% of the total losses of aircraft in combat were due to fire. Based on this evidence a military requirement was issued specifying a light-weight high-efficiency fire extinguishing system for protecting air craft engines and their fuel systems. A similar situation arose in Germany. As a result new, fast acting fire extinguishers were developed based on three main principles:

- Extinguishing agent permanently pressurized
- Large diameter discharge orifice
- Very fast opening of valve for immediate release of extinguishing agent by means of an explosive charge

These principles, combined with a fast-response flame or pressure rise detection system, form the basis even of today's automatic explosion suppression systems. Figure 1.117 illustrates the operation of a dust explosion suppression system.

The suppressor contains a suitable extinguishing agent (suppressant) and a driving gas, normally nitrogen at 60–120 bar. The onset of pressure rise in the vessel due to the growing dust flame is detected and an electric signal triggers the explosive charge that opens the suppressor valve. A special nozzle design ensures that the suppressant is distributed evenly throughout the vessel volume. In principle the pressure sensor can be made sensitive enough to detect even a very small initial flame. However, if the pressure rise for triggering the opening of the suppressor valve is chosen so small that similar pressure variations may occur due to normal plant operation, false activation of the suppression system becomes likely. This is not desirable and therefore the triggering pressure is normally chosen sufficiently high to avoid false alarms. The use of two pressure

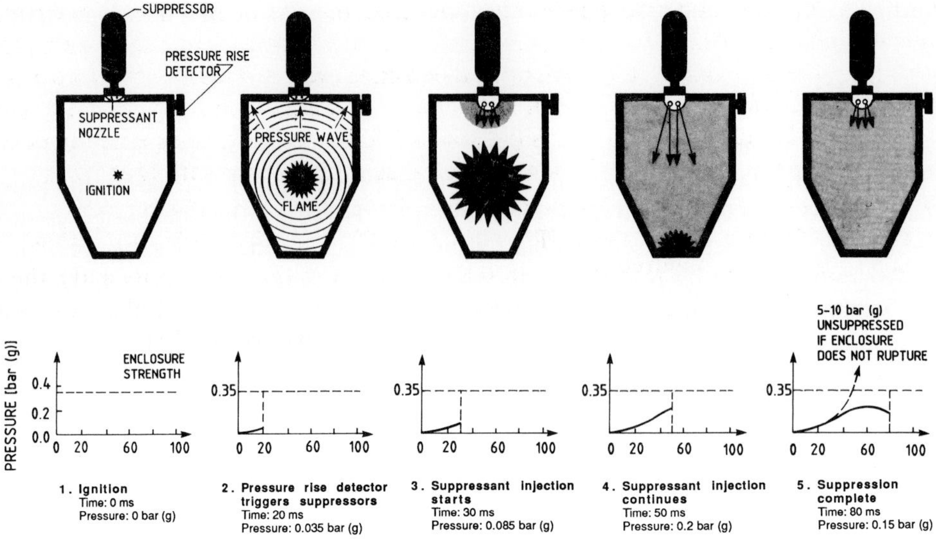

Figure 1.117 *Illustration of sequence of events and typical time scale of automatic suppression of dust explosions in process equipment. Actual figures apply to a starch explosion in a 1 m^3 vessel (Courtesy of Graviner Ltd., UK)*

detectors oriented at 90° to each other can make it easier to discriminate between pressure rise due to explosions and other disturbances.

Figure 1.118 shows a pressure detector of the membrane type, which is the most common type used in automatic dust explosion suppression systems.

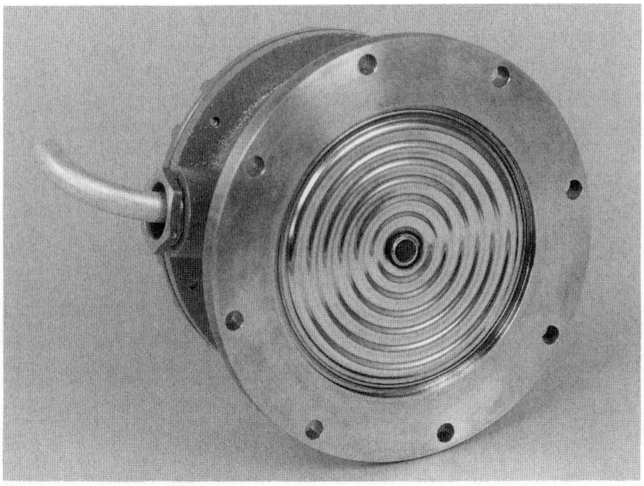

Figure 1.118 *Pressure sensor of the membrane type used for activating automatic dust explosion suppression systems. Diameter of membrane about 100 mm (Courtesy of Graviner Ltd., UK)*

UV or infrared optical flame sensors may be used instead of pressure sensors for detecting the initial explosion. However, careful consideration is required before doing so, because explosible dust clouds have high optical densities even at distances of only 0.1 m. This can make it difficult to sense a small initial flame in a large cloud. Optical detectors may be used in advance inerting systems (see below) for detecting flames entering ducts between process units. Figure 1.119 shows a typical suppressor unit with pressure gauge for controlling the driving gas pressure, and suppressant dispersion nozzle.

Figure 1.120 shows a very large suppressor developed for suppressing explosions in large volumes of several hundred m^3.

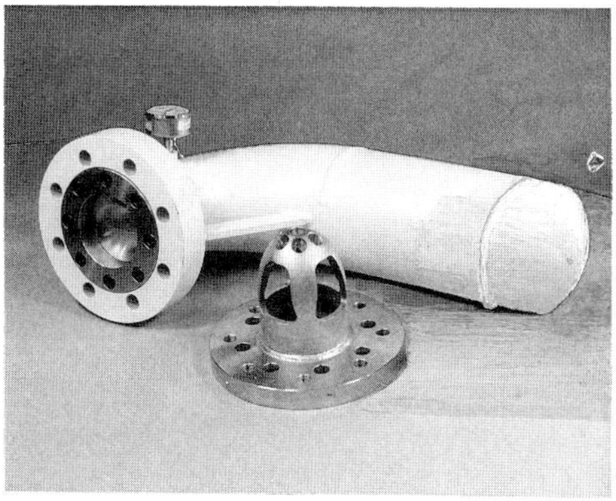

Figure 1.119 *Typical suppressor and nozzle system (Courtesy of Fike Corp., USA)*

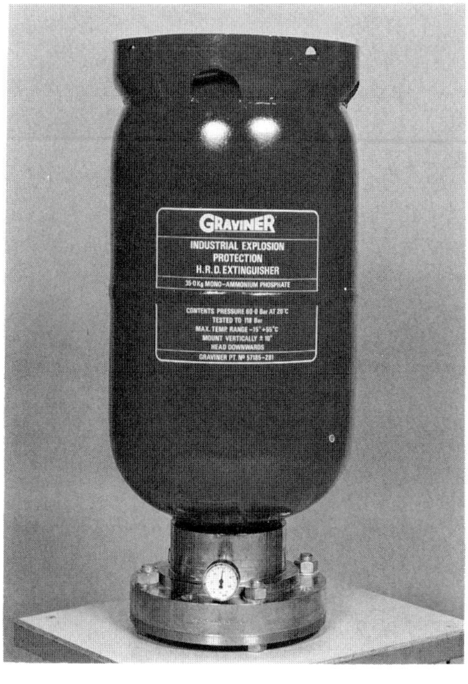

Figure 1.120 *Large 45 litre high-rate suppressor for very fast discharge of 35 kg of $NH_4H_2PO_4$ powder. Diameter of explosive-charge-operated valve 127 mm (5 inches). Driving gas: nitrogen at 60 bar (Courtesy of Graviner Ltd., UK)*

Figure 1.121 shows a special explosion suppression unit that is completely self-contained even with respect to power supply. This gives great flexibility with respect to mounting the unit at any desired location. However, regular inspection and testing of power supply, etc. is required. This unit was originally designed for using halon as suppressant, but transfer to powder suppressants is probably not too difficult.

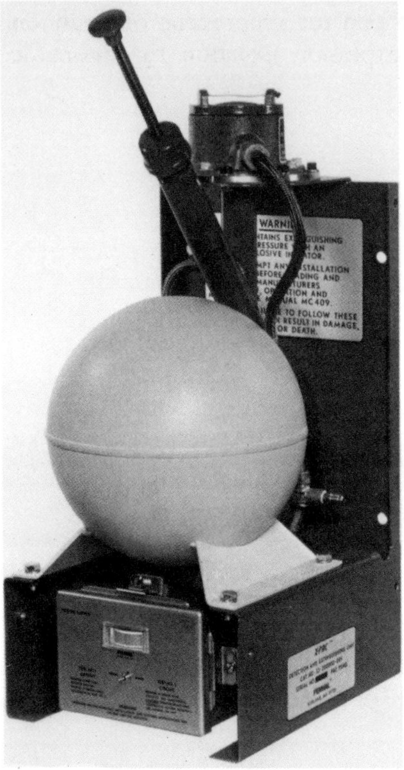

Figure 1.121 *Self-contained automatic explosion suppression unit (X-PAC) consisting of a pressurized spherical suppressant container with an explosive-charge-operated valve, a pressure detector and a long-life lithium battery power unit (Courtesy of Fenwal Inc., USA)*

The status on explosion suppression technology has been reviewed repeatedly in the literature. A fairly early paper discussing large-scale experimental research in France in the late 1960s was presented by Winter (1970). Bartknecht (1978) gave a comprehensive discussion of extensive research in F. R. Germany and Switzerland in the 1970s. A summary covering similar evidence was given by Scholl (1978). Singh (1979) summarized theoretical and experimental work from various countries including UK, USA, F. R. Germany and Switzerland. Moore (1981) discussed the results of his own comprehensive experimental and theoretical research, which resulted in a basis for systematic design of industrial suppression systems. He introduced the concept of critical mass M_t of suppressant that is just sufficient for suppressing the flame when being evenly distributed throughout the flame volume. He assumed that there exists a critical minimum mass concentration of any given suppressant for suppressing a flame of a given dust, and that a suppressant cloud of this concentration or higher must occupy at least the flame volume for successful suppression. It then follows that the critical mass M_t increases with time because the flame volume increases with time.

A similar line of thought was applied to the mass of suppressant actually delivered at any time after onset of flame development. Successful suppression would result if $M_{t,delivered} > M_{t,required}$. This is illustrated in Figures 1.122 and 1.123.

Moore, Watkins and Vellenoweth (1984) reviewed the status in the early part of the 1980s, including industrial experience with a number of automatic dust explosion suppression installations. More recently Hürlimann (1989) presented a detailed, comprehensive review of dust explosion suppression in general, and the research conducted by Ciba Geigy, Switzerland, in particular. Siwek (1989a) discussed recent research on explosion suppression in large vessels as well as explosion isolation by automatic suppression systems.

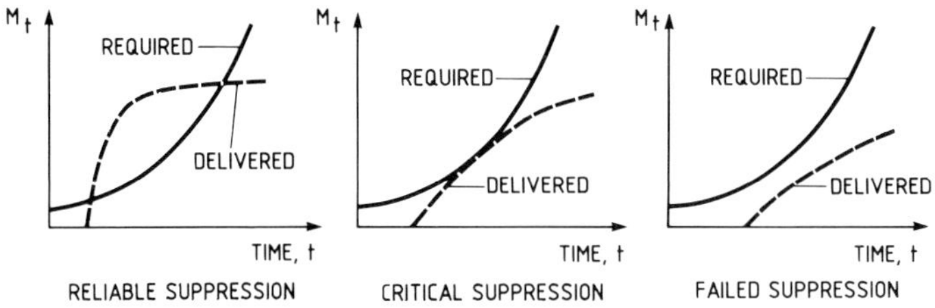

Figure 1.122 *Illustration of mass of suppressant required and delivered as functions of time, for reliable suppression, critical suppression and failed suppression (From Moore 1981, 1987)*

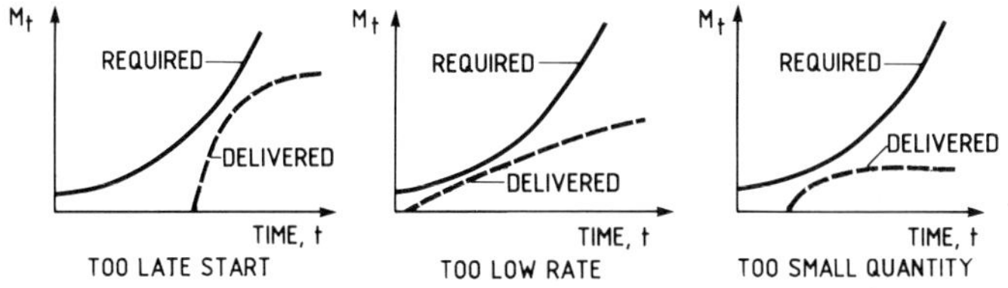

Figure 1.123 *Illustration of how failed suppression can result from too late start of suppressant injection, too low injection rate, and too small quantity of suppressant injected (From Moore, 1981, 1987)*

Automatic dust explosion suppression had proved to be feasible for organic dusts of maximum rate of pressure rise in the standard 1 m^3 closed ISO-vessel of up to 300 bar/s (i.e. K_{St} = 300 bar m/s, see Chapter 4). It remained uncertain, however, whether the method could also be used for aluminium dusts of K_{St} in the range 300–600 bar m/s. Moore and Cooke (1988) investigated this experimentally in a 18.5 m^3 vessel, using aluminium flake of K_{St} = 600 bar m/s. A special powdered suppressant consisting essentially of $NaHCO_3$ (ICI Dessicarb™) had proved to be the most effective for suppressing aluminium dust explosions, and was therefore used in all experiments.

However, it was found that for aluminium flake of K_{St} = 320 bar m/s, even under optimum conditions for suppression, it was difficult to ensure lower suppressed explosion pressures than about 2 bar(g). In the case of dusts of natural organic materials and plastics of K_{St} up to 300 bar m/s, the corresponding suppressed explosion pressures would have been 0.2–0.4 bar(g).

Moore and Cooke (1988) concluded that reliable suppression of aluminium flake explosions is difficult. However, they showed that a combination of explosion suppression and venting can reduce the maximum explosion pressure to a level significantly lower than the level for venting only. For an aluminium flake cloud of K_{St} = 600 bar m/s and a static opening pressure of the vent cover of 0.5 bar(g), venting only (about 1 m^2 vent area) gave 8.2 bar(g). When combined with optimal suppression, the maximum pressure was 3.8 bar(g). However, although this is considerably lower than 8.2 bar, it is still a high pressure.

It should be mentioned that Senecal (1989), over the range 240 to 340 bar m/s investigated, found that the correlation between K_{St} and reduced explosion pressure in similar suppression experiments, was rather poor.

1.4.7.2
Design of dust explosion suppression systems

As discussed by Moore (1983), one distinguishes between three different suppression strategies:

- *Advance inerting*: Detect explosion, identify its location, activate appropriate suppressors and establish suppressant barriers to prevent explosion spread to other process units.
- *Local suppression*: Detect initial explosion, identify its location, activate appropriate suppressors for ensuring no flame propagation beyond explosion kernel.
- *Total Suppression*: Detect explosion and deluge entire system with suppressant to ensure that explosion is totally suppressed.

The design of any particular industrial suppression system depends on the suppression strategy chosen, the type of suppressant, the chemical and explosibility properties of the dust, the nature of the process/enclosure to be protected (mill, cyclone, silo, etc.), the volume and shape of the enclosure, and on other actions taken to prevent or mitigate dust explosions in the plant. Moore and Bartknecht (1987) conducted dust explosion suppression experiments in large vessels of volumes up to 250 m^3 and were able to show that successful suppression of explosions in clouds of organic dusts is possible even in such large volumes. However, as the vessel volume increases, more suppressant and faster injection are required for successful suppression. The actual design of suppression systems depends very much on the specific design of the suppressors, and other details which vary somewhat from supplier to supplier. Therefore it is difficult to specify generally applicable quantitative design criteria. Figure 1.124 gives an example of a design guide developed by one specific equipment supplier, based on the experiments with organic dusts by Moore and Bartknecht (1987).

As can be seen, three standardized types of suppressors were employed. The smallest type, of volume 5.4 litres, was used for vessel volumes up to 5 m^3, whereas 20 litre suppressors were used in the range 5–30 m^3, and the very large 45 litre type for the larger volumes. The large-volume range was only verified experimentally up to 250 m^3, for which ten of the 45 litre suppressors were required for successful suppression of St 2 dust explosions (organic dusts). For St 1 dusts, seven such suppressors were sufficient.

Moore (1989) compared venting and suppression, referring to Figure 1.124, and showed that the two explosion protection methods are to a great extent complementary. In

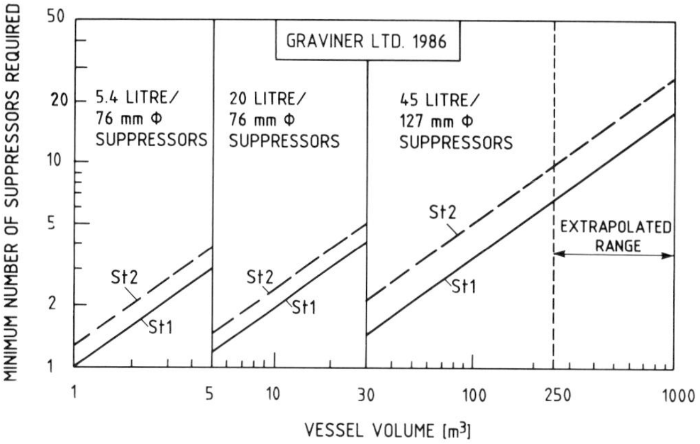

Figure 1.124 *Guide for designing dust explosion suppression systems for vessels of various volumes, based on the Graviner suppressor system.*

Pressure resistance of vessel: $\geqslant$ *1.0 bar(g)*
Explosion pressure detection level: $\leqslant$ *0.1 bar(g)*
Suppressant: $NH_4H_2PO_4$
Driving gas pressure: *60 bar*

St 1 means dusts that give maximum rates of pressure rise in the standard closed 1 m^3 *ISO vessel of up to 200 bar/s (*K_{St} *= 200 bar m/s). St 2 means dusts that give 200–300 bar/s (*K_{St} *= 200–300 bar m/s). (From Moore and Bartknecht, 1987)*

practice, cost effective safety is achieved by using either one of the two methods, or a combination of both.

Moore, Watkins and Vellenoweth (1984) provided a number of specific examples of automatic dust explosion suppression systems in industrial practice. One of these is shown in Figure 1.125.

Kossebau (1982) discussed the particular problem of suppressing dust explosions in bucket elevators, as illustrated in Figure 1.126, whereas Schneider (1984) was concerned with applying the suppression method to dust explosions in milling and grinding plants.

1.4.7.3
Influence of type of suppressant (extinguishing agent)

Traditionally halogenated hydrocarbons (halons) were used as suppressants in automatic dust explosion suppression systems. However, long before the environmental problems caused by these chemicals became a major issue, Bartknecht (1978) showed that powder suppressants, such as $NH_4H_2PO_4$, were in general much more effective for suppressing dust explosions than halons. Therefore powder suppressants have been used for suppressing dust explosions for many years. But powders differ in their suppressive power, and efforts have been made to identify the most effective ones.

Figure 1.127 shows that addition of only 30 weight% of $NH_4H_2PO_4$ powder is required to prevent flame propagation in dust clouds in air of Pittsburgh bituminous coal, whereas with $CaCO_3$ dust (limestone) 70 weight% is needed.

Similar systematic investigations were undertaken by Szkred (1983). He used a coal dust of 38% volatiles, 7% moisture and 38 μm mean particle size as fuel and found that 25

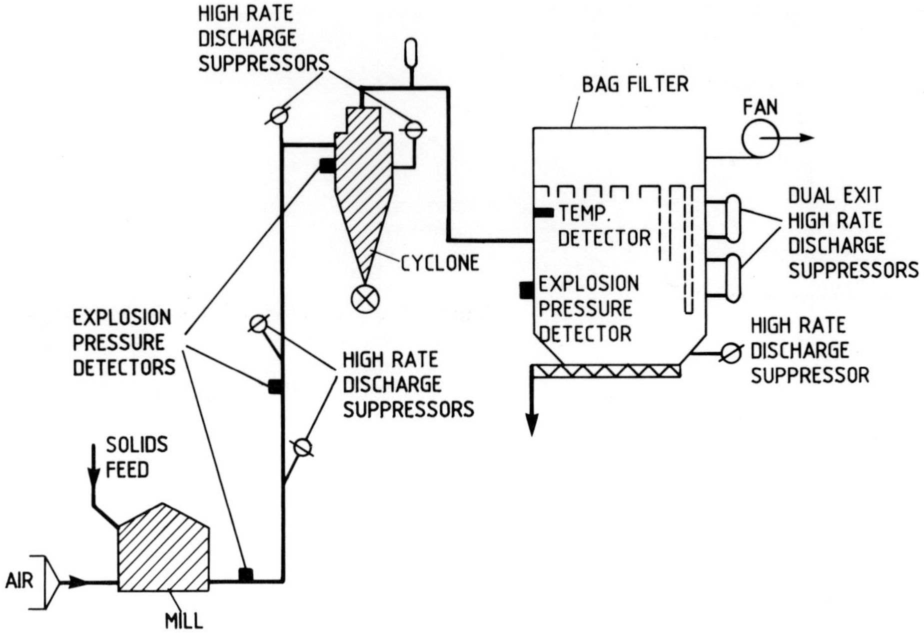

Figure 1.125 *Dust explosion protection of a grinding plant consisting of a mill, a cyclone and a bag filter, using a comprehensive automatic explosion suppression system. (From Moore, Watkins and Vellenoweth, 1984)*

Figure 1.126 *Application of automatic dust explosion suppression to bucket elevators (Courtesy of Th. Pinkwasser, Bühler, Switzerland). For a much clearer picture see colour plate 4*

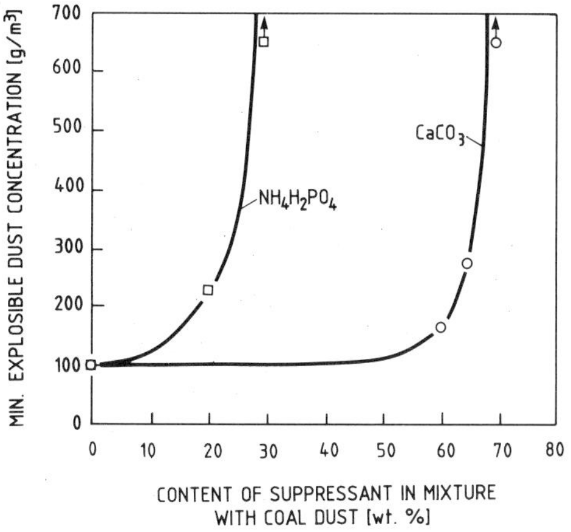

Figure 1.127 *Influence of chemistry of inorganic powder suppressant on mass percentage required for suppressing coal dust explosions. 400 J ignition source. 20 litre closed explosion vessel. (From Hertzberg* et al., *1984)*

weight% $NH_4H_2PO_4$, 35 weight% NaCl and 80 weight% $CaCO_3$ were required for suppressing flame propagation.

As already mentioned, the new powdered suppressant Dessicarb (> 98.5% $NaHCO_3$) has so far proved to be the most effective agent for suppressing aluminium dust explosions. Furthermore, this material is suitable for being used even in the food industry. It is soluble in water and can therefore be removed effectively by water only.

1.4.8 CONTROL AND INTERLOCKING SYSTEMS FOR PREVENTING AND MITIGATING DUST EXPLOSIONS IN INTEGRATED PROCESS PLANTS

1.4.8.1 Overview

The subject has been discussed in two papers by Faber (1985, 1989a). A wide range of sensors for automatic measurement of a number of physical and chemical process variables are in use. Micro processor technology has made it simple to utilize the signals from the sensors for control and interlocking purposes in a variety of ways.

The variety of process variables measured/detected includes:

- Rotational speed, position and translatory motion of mechanical objects, level of dusts and powders in silos, filter hoppers, etc.
- Temperature in powder and dust deposits, in bearings and electrical motors, and in gas flows.
- Gas pressure in process equipment and connecting ducts.
- Concentration of specific components in gases, e.g. oxygen in inert atmospheres and carbon monoxide in the case of self-heating.
- Presence of flames and hot gases.
- Concentration of dust suspended in a gas.

- Simple, digital quantities, e.g. whether an explosion vent door has opened or remains closed.

A comprehensive account of physical and chemical principles used for measuring such quantities, and of instrumentation using these principles, has been given by Bentley (1988).

Faber (1989a) mentioned three different objectives for monitoring process variables:

- Normal process control.
- Warning in case of abnormal process conditions.
- Triggering and control of measures for mitigating hazardous process conditions, e.g. dust explosions.

1.4.8.2 A practical example

Faber (1989a) used the plant for grinding and drying of coal shown in Figures 1.128 and 1.129 as an example. Such plants are used for producing the fuel for pulverized-coal-fired power plants. The basic process is simple. Lump coal is fed via a belt conveyor and a rotary lock to a rotary mill which is flushed with gas for drying of the coal and pneumatic transport of ground material to a gas classifier. The classifier separates the conveyed ground coal in a coarse fraction that is returned to the mill, and a fine product fraction that is removed from the gas in a cyclone and a subsequent filter. The coal dust collected in the cyclone and filter is conveyed to a coal dust silo.

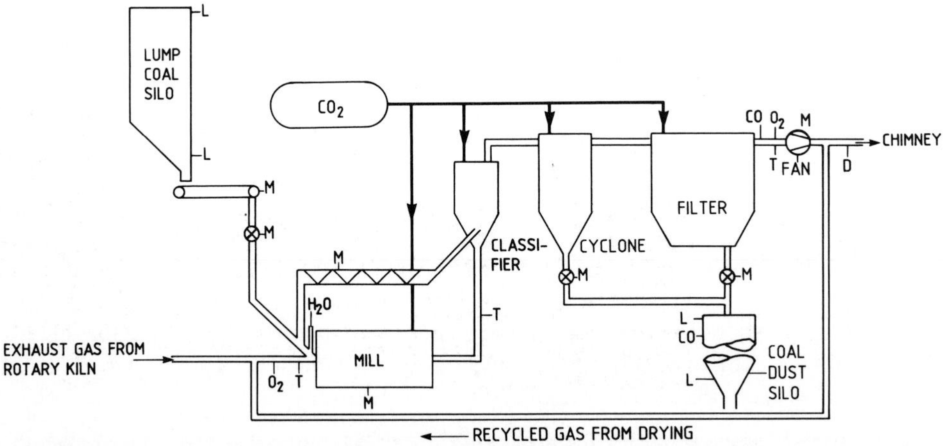

Figure 1.128 *Comprehensive sensor system for monitoring, controlling and interlocking of a process for milling and drying of coal. Explosion protection based on inerting with CO_2:*
CO – Carbon monoxide concentration sensors
D – Dust concentration sensor
L – Level sensors for coal and coal dust in silos
M – Movement sensors for mechanical components
O_2 – Oxygen concentration sensors
T – Temperature sensors
(From Faber, 1989a with minor adjustments)

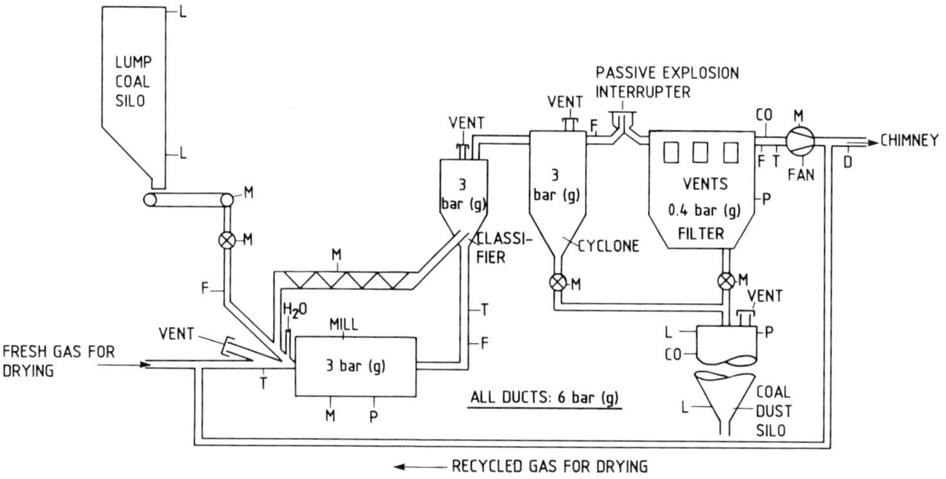

Figure 1.129 *Comprehensive sensor system for monitoring, controlling and interlocking of a process for milling and drying of coal. Explosion protection based on venting and explosion shock resistant design:*
CO – Carbon monoxide concentration sensors
D – Dust concentration sensor
F – Flame sensor
L – Level sensors for coal and coal dust in silos
M – Movement sensors for mechanical components
P – Pressure sensors
T – Temperature sensors
(From Faber, 1989a with minor adjustments)

Figures 1.128 and 1.129 show how the plant can be protected against damaging dust explosions utilizing two alternative measures, namely, inerting and venting. The instrumentation for monitoring, controlling and interlocking the process varies somewhat with the protective measure chosen.

In the case of inerting, one relies on keeping the plant inerted. To achieve this, as Figure 1.128 shows, continuous monitoring and control of a range of process variables is recommended. Automatic alarms can be activated as soon as a variable attains an unacceptable value. Interlocking by automatically turning off the power to the mill, the fan, the conveyor and the rotary locks, should the oxygen level become too high, adds to the safety. In case the temperature in the gas from the mill should become unacceptably high, water can be added automatically to the mill feed, as indicated. Development of carbon monoxide in the coal dust silo or filter indicates that smouldering combustion has developed, and this should also lead to automatic close-down of the plant.

If venting and explosion-shock-resistant design is the basic measure against damaging dust explosions (Figure 1.129), the probability of ignition is higher than in the case of inerting. The plant is therefore designed to be able to withstand dust explosions without becoming damaged, but such events are clearly undesirable. Therefore, continual monitoring and control of a series of process variables is again recommended. In addition to the sensors in Figure 1.128, Figure 1.129 indicates sensors for detection of abnormal pressure rise in the mill, in the filter and in the coal dust silos, and detectors for flames in the ducts from the mill. On the other hand, measurement of oxygen concentration is of less interest in this case, because one has to accept that the oxygen content can be as high as in air (21 vol%).

Besides stopping the mill, the fan, the rotary locks and the conveyors, the pressure and flame sensors can be used to activate various kinds of active isolation devices in the ducting between the various process units which are not already isolated by the rotary locks, screw conveyor or passive explosion interrupter (see Section 1.4.4).

It may be argued that the instrumentation suggested in Figures 1.128 and 1.129 is excessive. This is a matter of discussion in each specific case. The main purpose has been to indicate the possibilities that exist and from which one should select the appropriate measures to suit each specific application.

1.4.9 PREVENTION AND REMOVAL OF DUST ACCUMULATIONS OUTSIDE PROCESS EQUIPMENT: GOOD HOUSE-KEEPING

1.4.9.1 General outline

The main prerequisite for disastrous secondary explosions in factories is that sufficient quantities of combustible dust have accumulated outside the process equipment to permit development of large secondary dust clouds (see Section 1.1.3). In other words, the possibility of extensive secondary explosions can be eliminated if the outside of process equipment, and shelves, beams, walls and floors of work rooms are kept free of dust.

Significant quantities of dust may accumulate accidentally outside process equipment due to discrete events such as bursting of sacks or bags or erratic discharge from silos or filters. In such cases it is important that the spilt dust be removed immediately. In case of large dust quantities the main bulk may be sacked by hand using spades or shovels, whereas industrial, explosion-proof vacuum cleaners should be used for the final cleaning. In the case of moderate spills, dust removal may be accomplished by vacuum cleaning only.

Effective dust extraction should be provided in areas where dusting occurs as part of normal operation, e.g. at bagging machines.

Considerable quantities of dust can accumulate outside process equipment over a long time due to minor but steady leaks from process equipment. The risk of such leaks is comparatively large if the working pressure inside the process equipment is higher than ambient pressure, whereas running the process at slightly lower than ambient pressure reduces the leaks.

It is important that process equipment be inspected regularly for discovery and sealing of obvious accidental leak points as early as possible. However, often one has to accept a certain unavoidable level of dust leaks from process equipment. It is then important to enforce good housekeeping routines by which accumulations of explosible dust outside process equipment are removed at regular intervals, preferably by explosion-proof vacuum cleaning.

Use of compressed air for blowing spilled dust away should be prohibited. By this method dust is not removed, but only transferred to another location in the same room.

Besides, dust explosions can result if the dust concentration in the cloud that is generated is in the explosible range and an ignition source exists in the same location.

1.4.9.2
Industrial explosion-proof vacuum cleaners

The subject was discussed by Kühnen (1978a), Wibbelhoff (1984) and Beck and Jeske (1989). Beck and Jeske listed the requirements to mobile type 1 vacuum cleaners recommended in F. R. Germany for removal of combustible dusts:

- The fan must be on the clean side and protected against impacts by foreign bodies.
- The electric motor and other electric components must satisfy the general requirements to such components that are to be used in areas containing explosible dusts. Motors must be protected against short-circuits and overheating.
- The exhaust from the vacuum cleaner must be guided in such a way that it does not hit dust deposits and generate dust clouds.
- All electrically conducting parts of the equipment, including hose and mouthpiece, must be earthed with a resistance to earth of less than 1 MΩ.
- Vacuum cleaner housings must be constructed of materials that are practically non-flammable. Aluminium and aluminium paints must not be used.
- A clearly visible sign saying 'No suction of ignition sources' should be fitted to the housing of the vacuum cleaner.

Figure 1.130 shows an example of a large mobile vacuum cleaner for combustible dusts in industry.

Figure 1.130 *Large mobile vacuum cleaner for explosible dusts in industry. Vessels and connecting ducts are designed to withstand internal explosion pressures of 9 bar(g). Power requirement 45–55 kW (Courtesy of Edelhoff Polytechnik GmbH, Germany)*

Sometimes it can be useful to install stationary vacuum cleaning systems rather than having mobile ones. Figure 1.131 is a schematic illustration of the main principle. A central dust collecting station with suction fan is connected to a permanent tube system with a number of plug-in points for the vacuum cleaning hose at strategic locations.

The importance of good housekeeping is sometimes overlooked. It should always be remembered that clean work rooms excludes the possibility of extensive secondary explosions. Besides, cleanliness improves the quality of the working environment in general.

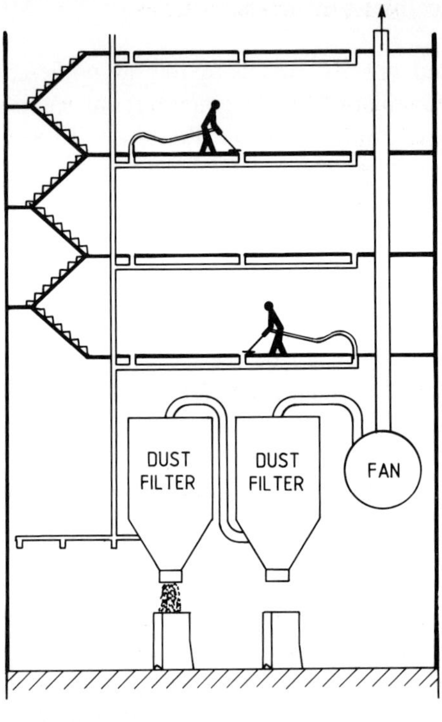

Figure 1.131 *Schematic illustration of a central stationary vacuum cleaning system with a number of alternative plug-in points for vacuum cleaning hoses*

1.4.10 DUST CONTROL BY ADDITION OF LIQUID

As discussed in Chapter 3, adding liquids to dusts can give rise to particle enlargement by formation of agglomerates held together by liquid bridges or capillary forces. Furthermore, if the main product is coarse, such as grain of wheat, oats, etc., adding a suitable liquid may soften the grain surface and reduce dust formation by rubbing and abrasion during handling and transport.

Adding liquids for controlling dust formation and dusting has primarily been used in the grain and feedstuffs industries. However, it is not unlikely that the method may also find other applications.

In grain handling and storage plants, addition of small quantities of refined mineral oil, vegetable oils or lecithin to the grain has turned out to be effective for suppressing dust cloud generation. The method was investigated by Lai *et al.* (1981) and Lai *et al.* (1986). One type of system used in practice is illustrated in Figure 1.132.

The oil may be sprayed on to the grain stream by means of conventional spraying equipment used in agriculture. The drop size should be sufficiently small to ensure even distribution of oil across the entire grain stream, but not so small that the oil becomes air-borne (aerosol). This would indicate an optimal drop size diameter somewhere in the range 0.1–1.0 mm. The oil may wet and penetrate into the surface of the grain. This counteracts formation of new fine dust by rubbing and impact. The oil layer on the grain surface may also act as an adhesive for fine dust particles. The oil further causes agglomeration of the fine primary dust particles to larger effective particles.

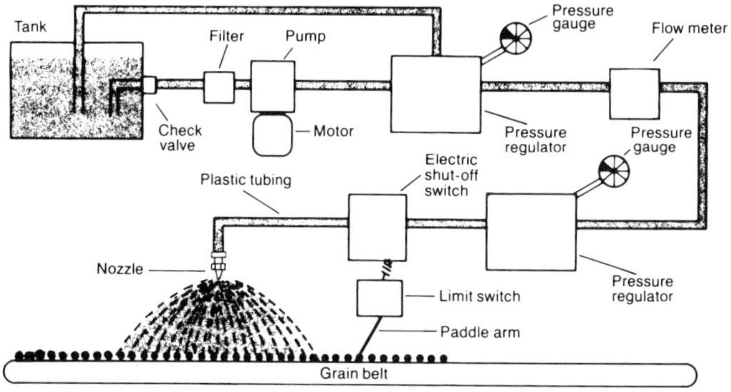

Figure 1.132 *Illustration of system for spraying small quantities of oil on to grain on a belt conveyor (Courtesy of American Soybean Association)*

The relative significance of the two mechanisms, namely grain surface wetting and adhesion of dust, and dust agglomeration, to some extent depends on the way in which the oil is applied. If oil spraying occurs while the grain is resting on a belt conveyor, the grain wetting mechanism may play a main role. If, however, the oil spray is applied inside the inclined feed duct to a bucket elevator bottom, where the high turbulence of the air flow causes most of the dust to be in suspension, direct agglomeration of dust particles is the more likely main result of adding the oil.

The latter configuration was used successfully by Johansen (1989). When handling various kinds of grain (wheat and barley, containing 700–800 g dust per tonne of grain), application of only 100 g soybean oil per tonne of grain was sufficient to reduce the dust level outside the process equipment substantially. The amount of dust, per tonne of grain, collected in the dust filters for the process stream was nearly the same as with no oil added. This was because of high air flow rates in the dust extraction system, which ensured collection of practically all the dust. However, the content of fine, unagglomerated particles in the collected dust, was considerably reduced by adding oil, as shown in Table 1.10.

The oil treatment method does not eliminate the dust explosion hazard. However, it reduces the hazard significantly in two ways: First, the quantity of the airborne fine dust that normally escapes from the process equipment and accumulates in workrooms, galleries, etc., is substantially reduced. Secondly, the clouds of agglomerated dust inside the process equipment have a lower ignition sensitivity and explosibility than the clouds of unagglomerated dust that would have existed inside the equipment in the absence of oil treatment. Some figures for dust collected in grain handling plants with and without oil treatment are given in Table 1.10. The independence of the minimum ignition temperature on oil treatment is in accordance with this parameter being rather insensitive to changes in particle size for organic dusts.

Table 1.10 Influence of treatment of wheat grain with soybean oil on the effective particle size, ignitability and explosibility of the grain dust resulting from handling of the grain: 115–230 g of oil per tonne of grain (From Johansen, 1989, 1990)

	Weight % of particles				Min. electric spark ign. energy [mJ]	Min. ignition temp.dust cloud [°C]	K_{St} [bar·m/s]	P_{max} [bar(g)]
	<125 μm	<63 μm	<32 μm	<10 μm				
Without oil	75	60	50	25	10 - 100	430	115	7
With oil	50	40	30	10	100 -1000	430	80	7

According to Johansen (1989) the oil spraying dust control method, when applied to a grain storage and handling plant, in fact reduced the running cost of the plant, in addition to reducing the dust explosion hazard.

In the case of products which are fine in themselves, such as wheat flour and tapioca, oil addition for suppressing dust is less suitable than in the case of a coarse main product containing a small dust fraction. However, in some cases circumstances permit addition of larger amounts of oil, up to several per cent, which can give a significant reduction of dust emission even for such fine products.

1.4.11 CONSTRUCTION AND LAYOUT OF BUILDINGS

It is important to distinguish between ideal requirements and realistic possibilities. In all circumstances it is strongly recommended that the dust explosion problem be taken into account as early as possible in the planning process, whether a completely new plant is to be constructed, or an existing plant rebuilt.

Ideally any factory in which dust explosions may occur should be located at a safe distance from other buildings. Furthermore, the various parts of the factory should be separate to enable effective isolation of the explosion to the section of the factory where it starts.

Buildings should be one-storey whenever otherwise suitable. If multi-storey buildings have to be used, the parts of the plant representing the greatest explosion hazard should be located as high up as possible, preferably on the roof. Alternatively, the hazardous plant items can be located in special, isolated, well vented niches as illustrated in Figure 1.133. Depending on the location, the floor and roof of the niche may also have to be explosion proof.

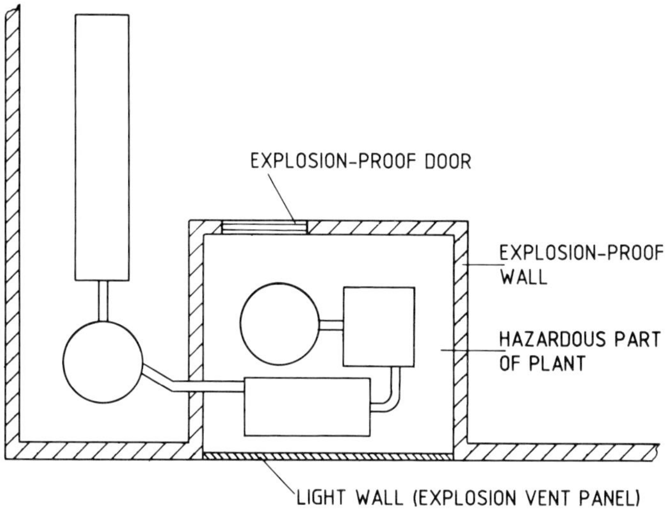

Figure 1.133 *Schematic illustration (top view) of location of hazardous part of plant in an isolated, well vented niche within the factory building*

In modern facilities for grain, feed and flour the bucket elevator legs are sometimes mounted on the outside of the buildings rather than inside. Venting of the elevator legs can then be performed outdoors.

In the past, floors and roofs of factory buildings were often supported by recesses in comparatively weak walls with no reinforcement as illustrated in Figure 1.134(a). In the case of an explosion, the walls were displaced outwards even at very modest overpressures, and the floors and roof fell down into the building as illustrated in Figure 1.134(b). Clearly, under such circumstances the consequences of even minor dust explosions in the building could be catastrophic.

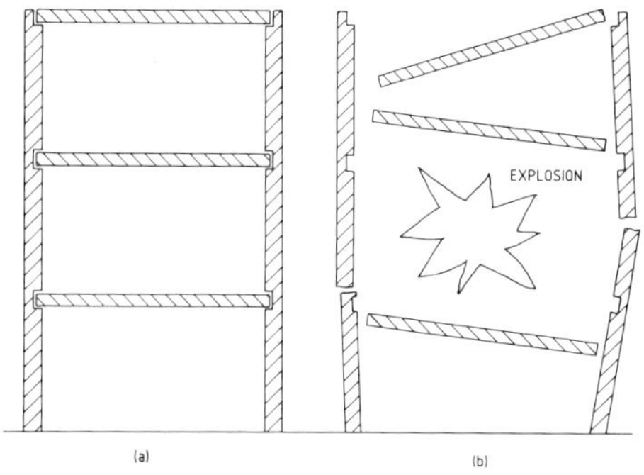

Figure 1.134 *Schematic illustration of hazardous construction of factory buildings. Minor overpressures due to an internal dust explosion will displace and break the weak walls and cause the roof and floors to fall down*

However, if the building is constructed of reinforced concrete, it can be made sufficiently strong to enable the windows to serve as vents. Figure 1.135 shows an actual example of successful venting of an explosion inside a building through the windows. It is

Figure 1.135 *Result of malted barley dust explosion in grain silo facility in Oslo, Norway in 1987. The windows provided sufficient venting to prevent destruction of building, which is of reinforced concrete (Courtesy of A. Fr. Johansen, Oslo Port Silo, Norway)*

important, however, to ensure that flying pieces of glass do not present a hazard to humans. To avoid this hazard it may be necessary to replace glass panes by anchored, transparent plastic panes.

As long as there are no special reasons for choosing other solutions, it is recommended that factory buildings in which dust explosions may occur, be constructed as indicated in Figure 1.136. The basic principle is that the roof and intermediate floors be supported by a strong frame structure. The walls are light-weight panel sections that function as vent

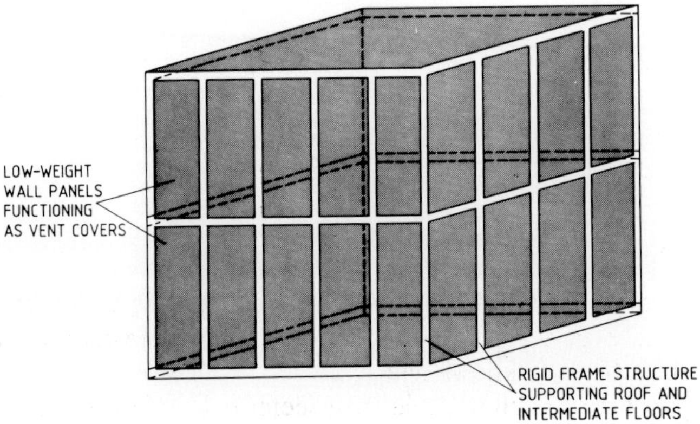

Figure 1.136 *Recommended construction of factory buildings to prevent collapse of building in the event of a dust explosion*

covers, should an explosion occur. If required, the panels may be anchored to the frame structure (see Figure 2.7).

Some final points to be taken into account when planning the layout and construction of factory buildings to reduce the explosion and fire hazard, include:

- Safe escape routes in case of explosion and fire.
- Fire-resistant construction materials.
- Fire-resistant doors.
- Electrical installations according to latest regulations/recommendations.

1.4.12 THE 'HUMAN FACTORS'

Proper build-up and maintenance of an integrated system for preventing and mitigating dust explosions very much depends on human relations and human behaviour.

A number of different personnel categories may be involved, including:

- Workers in the plant.
- Foremen in the plant.
- Workers from the maintenance department.
- Plant engineers.
- Safety engineers.
- Purchasing department officers.
- Safety manager.
- Middle management.
- Top management.
- Suppliers of equipment.
- Dust explosion experts/consultants.

Adequate prevention and mitigation of dust explosions cannot be realized unless there is meaningful communication between the various categories of personnel involved. If such communication is lacking, the result can easily become both unsatisfactory and confusing, as illustrated in Figure 1.137.

In general terms meaningful communication may be defined as conveyance and proper receipt and appreciation of adequate information whenever required. However, in order to receive, appreciate and use the information in a proper way, one must have

- Adequate knowledge.
- Adequate motivation.
- Adequate resources and deciding power.

Knowledge about dust explosions can be acquired by reading, listening to lectures, talking to experts etc., although experience from actual explosion prevention and mitigation work is perhaps the best form of knowledge.

Genuine motivation is perhaps more difficult to achieve. It seems to be a law of life that people who have themselves experienced serious explosion accidents possess the highest level of motivation, in particular if the accident caused injuries and perhaps even loss of life. This applies to workers as well as top management. However, high levels of

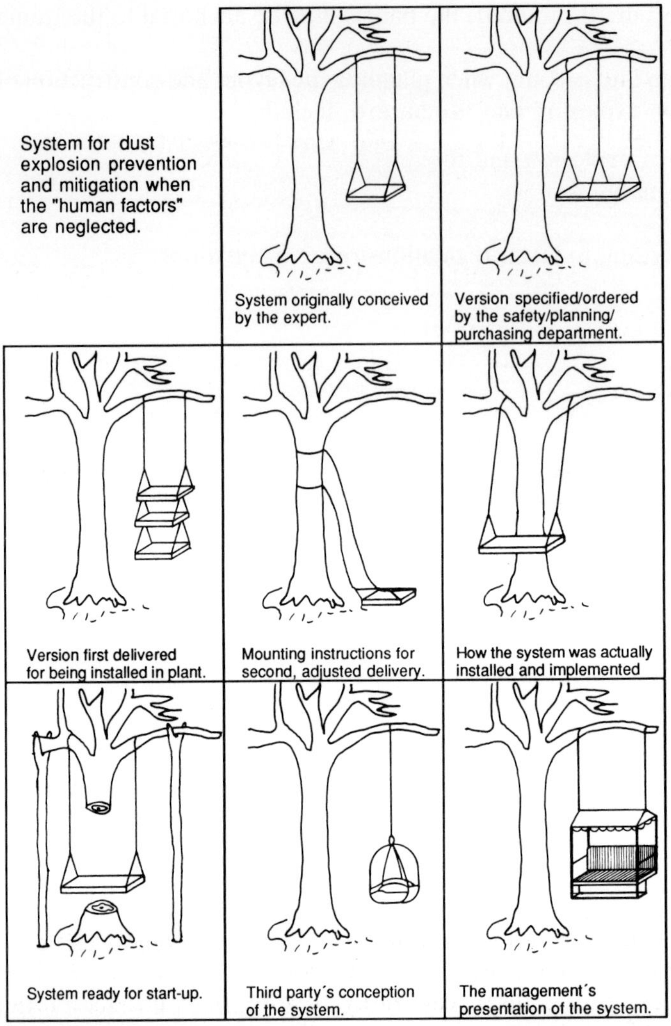

Figure 1.137 *Implementation of system for dust explosion prevention and mitigation when 'human factors' are neglected (Adapted version of original by Klapp, 1977)*

motivation can also result from good demonstrations of real explosions, including their initiation by various ignition sources, as well as their propagation and damaging effects. Video and film can also be a good help, if used properly.

The final element, adequate resources and the authority to put the good plans into practice, is in reality controlled by the top management. Verhaegen (1989) concluded from this that the real responsibility for establishing and running a proper safety assurance system must always lie on the top management. Summarizing the experience of a large, multi-national chemical company, Verhaegen suggested that the following ten essential elements be involved to ensure proper safety management:

- Top management responsibility.
- Safety statement (explicit commitment from top management).
- Objectives and goals (specification of long and short term expectations).
- Stated standards (written guidelines and rules).
- Safety committees (a dedicated organization for handling safety issues at all levels).
- Safety audits (regular re-examination of work practices).
- Accident records (written analyses of accidents. Why did they happen? How can similar future accidents be prevented?).
- Safety personnel (qualified specialists essential as advisers, but responsibility remains with top management).
- Motivation (by information and involvement, etc.).
- Training (a continual process; courses essential; the message must get through!).

Verhaegen emphasized the problem that a good safety organization is in reality often kept active by one or two dedicated individuals. If they change position within the company, or even leave, the safety organization may suffer. Management should foresee this problem and provide a workable solution.

Burkhardt (1989) gave an informative, more theoretical psychological analysis of the role of human factors in accident prevention in general. Atkinson (1988) and Proctor (1988) discussed various aspects of the training of safety personnel.

1.5 SELECTING APPROPRIATE MEANS FOR PREVENTING AND MITIGATING DUST EXPLOSIONS

1.5.1 BASIC PHILOSOPHY, COST ESTIMATION AND RISK ANALYSIS

1.5.1.1 The optimal solution, or striking the balance

The extensive menu of means of preventing and mitigating dust explosions, summarized in Table 1.9, has been discussed in Section 1.4.

Noha (1989) emphasized that the concepts of 'primary' and 'secondary' means of protection against dust explosions, used in F. R. Germany in the past, can be misleading by indicating that mitigation is of secondary importance as compared to prevention. The rational approach is to seek an optimal combination of means of both categories, for each specific application. In doing so, Noha suggested the need for breaking the problem down and evaluating specifically:

- The efficacy of the protective means.
- The technical feasibility.
- The environmental acceptability.
- The financial acceptability.

Figure 1.138 illustrates the situation.

One pitfall related to assessing the efficacy of the protective means is the selection of the dust sample on which the assessment is to be based. Noha (1989) mentioned as an example a comparatively coarse polypropylene powder to which < 1% of fine calcium stearate had been added to increase flowability. Such additives have large specific surface areas and correspondingly low minimum ignition energies. But as long as they are homogeneously mixed with the polypropylene, the small fraction of additive has little influence on the ignitability and explosibility of the polypropylene powder as a whole. However, if segregation occurs, the fine, reactive additive may accumulate in certain areas of the process, for example in a filter. This can create a much more hazardous situation than would have been anticipated on the basis of the properties of the polypropylene powder. In such cases it may be wise to base the assessment of the efficacy of the protective means on the properties of the additive, rather than on the characteristics of the main product. This not least applies to the incendivity of the dusts in terms of their minimum ignition energies.

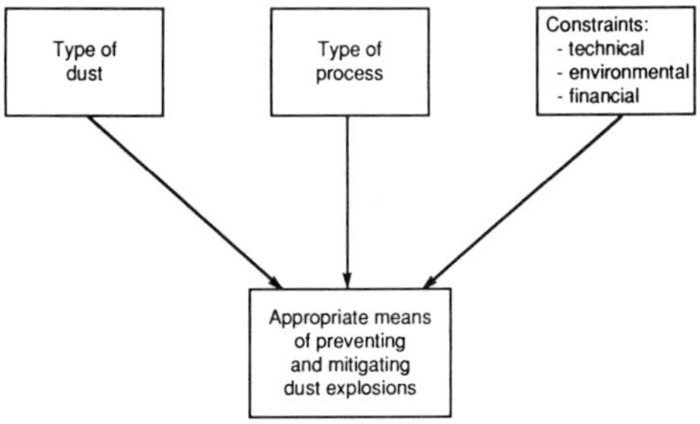

Figure 1.138 *The appropriate set of means for preventing and mitigating dust explosions depends on type of dust, type of process, and on the boundary conditions in terms of plant layout, type of building (s), environmental constraints, financial constraints, etc*

The protective means to be used must be technically and financially feasible. For example, there is no point in installing vents on an enclosure that is so weak that it will be unable to withstand the maximum pressure to be expected, even with the largest vent area that can be provided.

Traditional venting may sometimes be unacceptable due to the inevitable emission of unburnt/burning/burnt dust. This is particularly so in congested urban areas and for some special synthetic organic powders like pesticides, pharmaceuticals and dye stuffs. However, the further development of the quenching tube for dust and flame free venting of dust explosions (see Section 1.4.6.6) may alter this situation, and make venting a feasible means of mitigating dust explosions even in some of these situations.

In the case of very reactive dusts, of K_{St} values $\gg$ 300 bar m/s, automatic explosion suppression must most often be excluded because the injection of the suppressant is too slow to produce any significant mitigating effect on the explosion development. Inerting is only feasible if sufficient inert gas is available at an acceptable cost, whereas reinforcement of process equipment to explosion-shock resistant standard my often be both technically and financially unacceptable.

Figure 1.139 outlines a general approach to fighting the dust explosion hazard in industry. Sometimes the required ignitability and explosibility data for the dust(s) in question are available from earlier test work or from the literature. However, most often specific laboratory testing is needed.

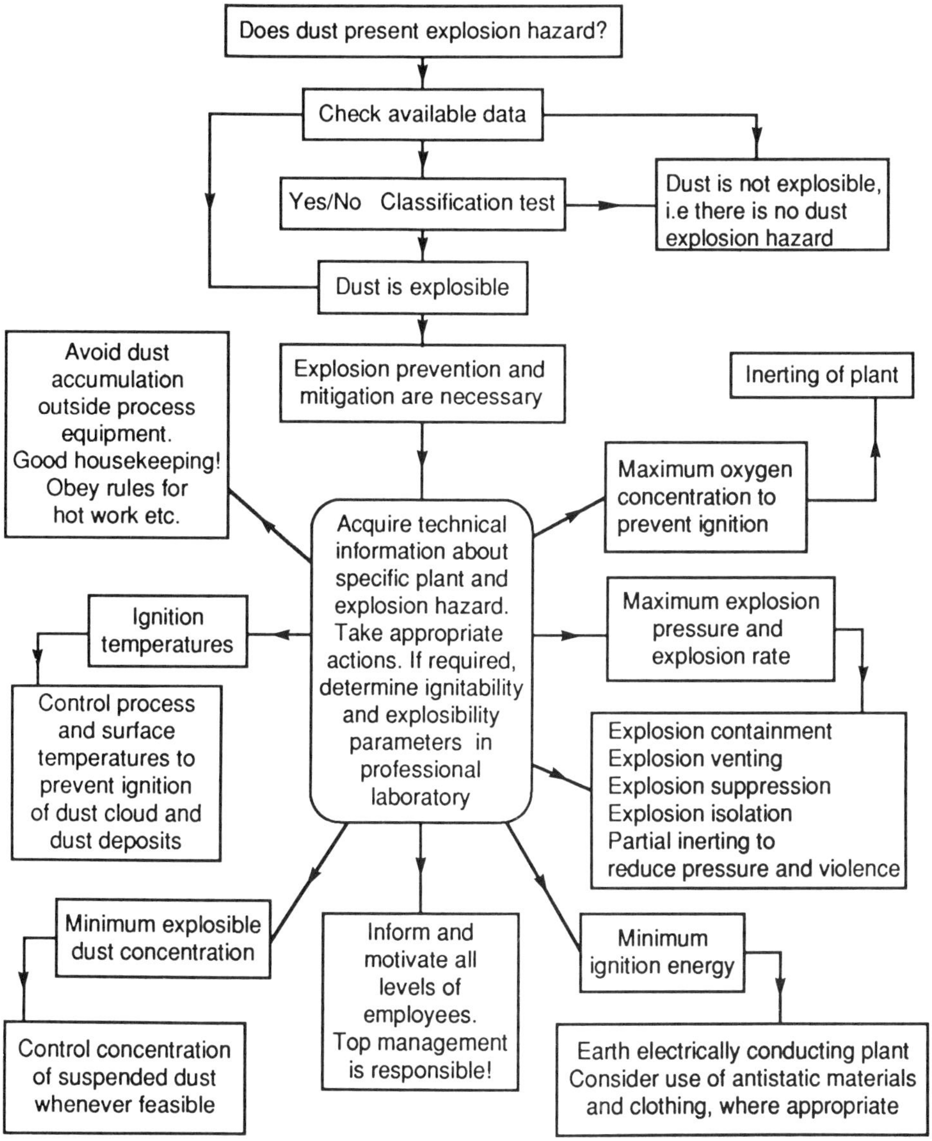

Figure 1.139 *Outline of general approach to practical dust explosion protection. (Modified and extended version of scheme suggested by Field, 1982)*

1.5.1.2
Cost considerations

Ritter (1978) compared the cost of the various means of dust explosion prevention and mitigation by means of cost indices. The index for proper elimination of ignition sources was per definition set equal to unity and used as a basis for cost comparison for all types of plant units except milling plants. Ignition source elimination included use of approved electrical equipment in all areas, earthing of all conducting equipment to avoid static electricity, avoidance of overheating by friction, safe maximum temperatures on all heated surfaces, and avoidance/elimination of smouldering nests.

A summary of Ritter's figures is given in Table 1.11.

Table 1.11 Relative costs for various means of preventing and mitigating dust explosions using the costs of eliminating ignition sources as a basis for comparison (From Ritter, 1978)

Plant type	Elimination of ignition sources	Inerting	Pressure resistant at 7 bar(g)	Venting. Pressure resistant at 2 bar(g)	Automatic suppression, pressure resistant 2 bar(g)	Dust concentration < C_{min}
Silo plant	1.0	1.3-1.5	3.0	1.5-1.7	2.1-2.6	Not possible
Spray dryer	1.0	1.7	3.1	1.7	2.0	1.1
Milling plant*	Not possible	1.3	1.8	1.4	1.5	Not possible
Bag filter	1.0	1.5	2.3	1.8	1.7	Not possible
Fluidized bed dryer	1.0	1.4	2.8	1.5	1.5	Not possible

*Ignition sources cannot be eliminated in mills, and therefore the cost of the plant itself is used as the cost basis of index 1.0.

Table 1.11 indicates that the use of pressure resistant equipment is generally comparatively expensive. However, the cost of pressure shock resistant equipment would be significantly lower. The relative costs in Table 1.11 of inerting, venting and automatic suppression are fairly equal, perhaps with a slight increase in the order mentioned. However, technology has changed somewhat since 1978, and the more liberal vent area requirements justified by more recent research (see Chapter 6), would suggest that venting may be somewhat cheaper than indicated by Table 1.11. Table 1.11 should rather serve as an illustration of the usefulness of systematic cost comparison, than as a final, generally valid ranking of costs.

1.5.1.3
Hazard analysis

This is a large subject in itself, covered by a substantial quantity of published information. The term 'hazard analysis' comprises a number of different systematic methods for identifying, and sometimes also quantifying the hazards to be associated with a given process or plant. In principle such analyses can also be used as a basis for optimizing the selection of means for preventing and mitigating dust explosions.

Cox (1986, 1987) gave an informative summary of the various techniques in use for hazard analysis, which is quoted more or less literally, under the following five headings:

1. *Hazard Surveys or Inventories* These methods are essential preliminaries to many safety studies. The survey consists of making an inventory of all stocks of hazardous material or energy and noting relevant details of storage conditions. When carried out at the conceptual stage of a project, such a survey can contribute to layout optimization and may suggest process changes to reduce stored quantities. It generates information that can be used in a preliminary risk assessment, but the hazard survey itself is little more than a 'screening' exercise designed to identify problem areas.

2. *Hazard and Operability Studies (HAZOP) and Failure Modes and Effects Analysis (FMEA)* These two techniques have very similar objectives and methods of approach. The purpose is to identify systematically all of the possible ways in which the system investigated could fail, and to evaluate these and formulate recommendations for preventive and mitigating measures.

FMEA is the simpler of the two techniques. The procedure is to take each plant item and component in turn, list all possible failure modes and consider the consequences of each. The results are recorded in a standard format in which recommendations for action can be included. The weakness of FMEA is that there is no specified method for identifying the failure modes and their effects. The engineer is expected to do this from first principles or past experience and the only discipline imposed on him or her is that of the reporting format itself.

HAZOP overcomes this difficulty by introducing a systematic method for identifying failure modes. This involves scrutiny of a large number of possible deviations from normal operating conditions, which are generated by applying guide words such as more, less, reverse, etc., to each of the parameters describing process conditions in each component, plant, item, or line in the plant. However, HAZOP in its original form has disadvantages, and some industrial companies have modified the way in which the results of the study are handled. Instead of 'recommendations', the output is 'identified problems', which leaves more room for a coordinated rational design revision, which is not only cheaper, but probably safer also.

3. *Analysis of Systems Reliability by Fault Tree Analysis* This method is applied to complex systems, whether the complexity is due to the nature of the process itself, or to the instrumentation required for running the process. In the basic technique, the 'Fault Tree Analysis', the failure modes must first be identified, e.g. by HAZOP. These failure modes are named 'top events'. An example of a 'top event' could be a dust explosion in a milling plant.

For each 'top event', the analyst must then identify all those events or combinations of events that could lead directly to the failure. The precise logical relationship between cause and effect is expressed by AND or OR gates and is usually presented in diagrammatic form. The immediate causes of the top event have their own contributory causes, and these can be presented in a similar way, so that a complete fault tree is built up. This process ceases when all of the causative factors at the bottom of the tree are of a simple kind for which frequencies of occurrence or probabilities can be estimated.

The synthesis of fault tree is a job that is best done by an engineer with good experience of the type of system under consideration; it is much easier to teach such a person how to construct a fault tree than to teach a reliability specialist everything about the system. However, the quantitative analysis of a fault tree is a separate activity in which the reliability specialist will play the dominant role.

An illustrative example of a quite comprehensive fault tree for a grain dust explosion in a grain storage facility was given by National Materials Advisory Board (1982).

4. Risk Analysis by Event Tree Analysis Risk analysis consists of four major steps: identification of a representative set of failure cases, calculation of consequences, estimation of failure probabilities and assessment of overall impact.

Failure cases are identified first by establishing the location of the main inventories of hazardous material and then by detailed scrutiny of the process flow and instrumentation diagrams using checklist methods or HAZOP.

Once the failure cases have been identified, the consequences of the failure must be calculated. Event tree analysis is a useful method in this process. An event tree is the reverse of a fault tree, starting with the initial or 'bottom events' and exploring all possible 'top events' that can result from it. Each outcome has further outcomes and all of these can be related by means of decision gates. At each gate the conditional probabilities must be estimated for each of the alternative branches. On this basis the probabilities of the final hazard, or 'top event', can be calculated.

Criteria have been suggested whereby calculated risks can be judged. Almost all criteria proposed so far are based on the concept of comparability with the existing general risk background. Cost/benefit and 'risk perception' arguments have been advanced, but they have not yet been developed to a practical and accepted form for being used in risk analysis.

Risk analysis has been criticized by pointing at

- Inaccurate mathematical models.
- Incomplete analysis of actual practical problem.
- Inaccurate primary failure probability data
- Inadequate acceptability criteria.
- Difficulty of checking final result.
- Complexity and labouriousness of method.

Hawksley (1989) discussed the conditions under which the various elements of quantitative risk analysis are useful in the assessment of risks in practice.

5. Safety Audits Once a plant enters operation, hardware and procedures will start to change from those originally established by the commissioning team. Usually, there are good reasons for this: the plant engineers and operators may find simpler or more economic procedures, and the operational requirements themselves may change. However, it is also quite possible that safety standards fall off with time because experience of satisfactory operation leads to overconfidence and a false sense of security.

For these reasons, Safety Audits are used in many operating companies. These may vary from a half-day tour by the works manager to a review lasting several weeks carried out by a team of engineers covering different disciplines and independent of the regular operational management of the plant. For the most penetrating audits, the study should not be announced in advance.

In practice the assessment of dust explosion hazards is bound to be subjective, because the problem is too complex for quantitative analytical methods to yield an indisputable answer. In Figure 1.140, four different scenarios for a given industrial plant are indicated.

Scenario 'A', which was assumed by Pinkwasser and Häberli (1987) for the grain, feed and flour industry, suggests that most of the dust explosion hazard can be eliminated by 'soft' means such as training, motivation, improving the organization, good house-keeping and proper maintenance. The alternative Scenario 'B' suggests that concentrating on preventing ignition sources gives the greatest benefit. Scenarios 'C' and 'D' focus on keeping the dust cloud non-explosible, and using mitigating measures respectively. Other scenarios can easily be envisaged.

However, experience suggests that some scenarios, depending on the type of powder/dust and plant, are more credible than others. For example, it can be argued that a plant producing or handling fine aluminium flake is well represented by Scenario 'C' in Figure 1.140, because inerting by nitrogen will probably reduce the dust explosion risk from high to acceptably low.

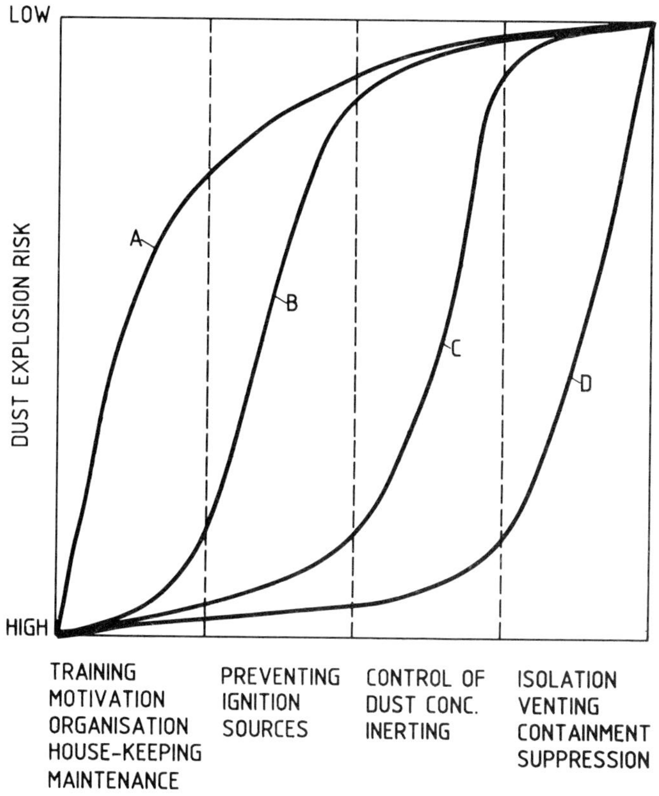

Figure 1.140 *Effect of various means of preventing and mitigating dust explosions. Four different scenarios for a given plant. Extended and generalized presentation based on original (Scenario 'A') by Pinkwasser and Häberli (1987)*

In spite of the objections that can be raised against the use of the various hazard analysis methods, several authors have suggested that risk analysis and other methods be adopted specifically for reducing the dust explosion hazard in the process industry, including grain, feed and flour storage plants. These include Beck (1974, 1985), Sorgdrager (1985), Kameyama *et al.* (1982), Lai *et al.* (1985) and Piotrowski *et al.* (1988).

1.5.2 SELECTION SCHEME SUGGESTED BY NOHA FOR THE CHEMICAL PROCESS INDUSTRY

1.5.2.1 General background

Noha (1989) restricted his analysis to four groups of process equipment, namely:

- Crushing and milling equipment.
- Dryers.
- Mixers.
- Conveyors and dust removal equipment.

Informative descriptions and illustrations of the various kinds of process equipment typical of powder producing and handling plant are provided by Perry and Green (1984).

Before deciding how a given plant should be protected against dust explosions, it is necessary to evaluate the extent to which protection is really necessary. This requires knowledge of ignitability and explosibility parameters of representative dust samples, as well as information about the plant design, layout and operation. It may be necessary to investigate the possibility of generation of hybrid mixtures (dust + explosible gas or vapour).

Tables 1.12, 1.13, 1.14 and 1.15 give Noha's suggestions for selecting appropriate means for preventing and mitigating dust explosions in four categories of process equipment in the chemical process industry. The symbol X indicates the most appropriate means of protection, whereas (X) implies that the use of the means indicated is possible, but that it is not implemented very often in practice.

Noha emphasized that a given plant item, for example a specific dryer, should not be considered in isolation. It is always necessary also to consider the entire plant or the part of it that is likely to become involved in a dust explosion in the system. Identification of probable ignition sources and ignition points is an important part of the analysis.

1.5.2.2 Crushing and milling equipment

The justification for some of the suggestions in Table 1.12 for crushing and milling equipment are as follows:

In crushers and roll mills, the concentration of fine dust that can give explosions is mostly below the minimum explosible concentration just because of the nature of the process itself. In the case of screen mills and air jet mills the probability of ignition sources

Table 1.12 Appropriate means for preventing and mitigating dust explosions in chemical process plant (From Noha, 1989)

Means of explosion prevention/mitigation — Crushing and milling equipment	Dust concentr. < min. expl. concentr.	Inerting by adding inert gas	Intrinsic inerting	Evacuation of process equipment	Addition of inert solids	Elimination of ignition sources	Explosion resistant equipment	Explosion venting	Automatic explosion suppression	Explosion isolation
Ball mills		X		(X)	(X)		X	(X)		
Vibratory mills		(X)			(X)		X	(X)		
Crushers	X					(X)	X	(X)	(X)	
Roll mills	X					(X)				
Screen mills		(X)				X	(X)			
Air jet mills		(X)				X	(X)			
Pin mills		(X)			(X)		X	(X)	(X)	
Impact mills					(X)		X	(X)	(X)	
Rotary knife cutters	(X)						X	(X)	(X)	
Hammer mills		(X)			(X)		X	(X)	(X)	

Table 1.13 Appropriate means for preventing and mitigating dust explosions in chemical process plant (From Noha, 1989)

Means of explosion prevention/mitigation — Powder Dryers	Dust concentr. < min. expl. concentr.	Inerting by adding inert gas	Intrinsic inerting	Evacuation of process equipment	Addition of inert solids	Elimination of ignition sources	Explosion resistant equipment	Explosion venting	Automatic explosion suppression	Explosion isolation
Spray dryers (nozzle)	X	(X)	X			X		(X)	(X)	
Spray dryers (disc)	X	(X)	X					(X)	(X)	
Fluidized bed dryers		(X)	(X)			X	(X)	(X)	(X)	
Stream dryers	(X)	(X)				X	(X)	(X)		
Spin-flash dryers						(X)	X	(X)	(X)	
Belt dryers	X					(X)				
Plate dryers	X					(X)				
Paddle dryers	X	X		(X)	(X)	X	(X)	(X)	(X)	

Table 1.14 Appropriate means for preventing and mitigating dust explosions in chemical process plant (From Noha, 1989)

Powder Mixers \ Means of explosion prevention/mitigation	Dust concentr. < min. expl. concentr.	Inerting by adding inert gas	Intrinsic inerting	Evacuation of process equipment	Addition of inert solids	Elimination of ignition sources	Explosion resistant equipment	Explosion venting	Automatic explosion suppression	Explosion isolation
With mixing tools:										
- High-speed	(X)	X		(X)			X	(X)	(X)	
- Low-speed	(X)	(X)		(X)		X	(X)	(X)	(X)	
Without mixing tools:										
- Drum mixers		(X)		(X)		X	(X)			
- Tumbling mixers		(X)		(X)		X	(X)			
- Double cone mixers		(X)		(X)		X	(X)			
Air flow mixers:										
- Fluidized bed mixers						X	(X)	(X)		
- Air mixers						X	(X)	(X)		

Table 1.15 Appropriate means for preventing and mitigating dust explosions in chemical process plant (From Noha, 1989)

Powder/dust conveyors and dust removal equipment \ Means of explosion prevention/mitigation	Dust concentr. < min. expl. concentr.	Inerting by adding inert gas	Intrinsic inerting	Evacuation of process equipment	Addition of inert solids	Elimination of ignition sources	Explosion resistant equipment	Explosion venting	Automatic explosion suppression	Explosion isolation
Screw conveyors	(X)	(X)				(X)	X	(X)		
Chain conveyors	(X)						X	(X)	(X)	
Bucket elevators		(X)				(X)	X	X	(X)	
Conveyor belts	X									
Shaker loaders						X		(X)		
Rotary locks		(X)				X	X	(X)		
Pneumatic transport equipment	(X)	(X)				X	(X)			(X)
Dust filters and cyclones		(X)				X	X	(X)	(X)	
Industrial vacuum cleaning installations	X					X				

can be regarded as low. Inerting is most common in the case of batch mills, whereas other mill types are often made sufficiently strong to be able to withstand an internal dust explosion.

Mayerhauser (1978) considered dust explosion protection of mills and air classifier specifically. He concluded that pressure resistant/pressure shock resistant construction and inerting were the two most suitable methods for these kinds of equipment.

Ruttmann (1989) described the systematic design of one specific plant, in which inerting was used for protecting a combined milling and mixing system against dust explosions.

Whenever possible one should use mill types that minimize dust cloud formation and generation of ignition sources by high-speed impact. Figure 1.141 shows a type of open slow-speed screw shredder that to an increasing extent replaces enclosed high-speed hammer mills. Because of the slow motion, both dust cloud formation and the probability of ignition source generation are minimized. Furthermore, the open construction provides generous venting should an explosion nevertheless occur.

In Table 1.12 Noha also indicated that adding inert dust to the explosible dust is a means of preventing dust explosions in crushers and mills in principle. However, as pointed out in Section 1.4.3.3, most often this method is not feasible in practice, due to contamination of the product by the inert additive.

Figure 1.141 *Open slow-speed screw shredder for size reduction of combustible waste materials. Top: Complete installation in operation. Bottom: The shredding screw section in the lower part of the installation (Courtesy of Th. Pinkwasser, Bühler, Switzerland)*

1.5.2.3
Dryers for powders and granular materials

In the case of dryers (Table 1.13) the special protective method 'intrinsic inerting' can be a good solution in some cases, particularly for spray dryers. This method implies that the required quantity of inert gas is generated in the plant itself by controlled combustion in the hot-gas generator and recirculation of the gas. Such hot-gases mainly consist of nitrogen, carbon dioxide and water vapour. The residual concentration of oxygen is kept at a a sufficiently low level to ensure inert conditions. 'Intrinsic inerting' of dryers was discussed specifically by Hammer (1978) and Klais (1989).

Spray dryers normally operate at dust concentrations significantly below the lower explosible limit, which clearly adds to safety. However, dust deposits are often generated on walls, etc. and smouldering nests may develop, depending on the local temperature and oxygen concentration. Klais (1989) emphasized that oxygen concentrations as low as 4 vol%, which exclude dust explosions for most organic powders, may still be too high for prevention of certain auto-oxidation processes in the dried powder when deposited as a layer or a heap. (See also Figure 1.67 in Section 1.4.2.2.) If a smouldering nest loosens and gets carried away with the product stream, it can initiate a dust explosion in downstream cyclones and silos.

In the case of disc type spray dryers, one cannot fully exclude the possibility that a disc that flies away and impacts with the walls of the dryer generates a hot-spot of sufficient size and temperature to initiate a dust explosion.

The powder/dust in belt and plate dryers are mostly in deposited form. In paddle dryers the dust concentration would normally be expected to exceed the upper explosible limit in the areas where occurrence of an ignition source might be envisaged.

Gibson, Harper and Rogers (1985) evaluated the fire and explosion risk in powder dryers with particular emphasis on detection of exothermic decomposition. Their conclusion was that existing methods for dust explosion risk evaluation, combined with an adequate method for characterization of the exothermic decomposition properties of the powder provide a means of specifying safe drying conditions. Most often control of the atmosphere to keep the dust cloud non-explosible, or the use of venting, automatic suppression, or explosion (shock) resistant equipment is required. However, in certain cases safety can be based solely on avoidance of decomposition and ignition.

1.5.2.4
Powder/dust mixers

In mixing equipment (Table 1.14) ignition sources may be avoided as long as there are no fast-moving mixing tools. Inerting by adding, for example, nitrogen or other inert gas is feasible in batch mixers, whereas continuous mixers may preferably be designed to withstand the pressure rise caused by a possible dust explosion.

Radandt (1969) discussed dust explosion protection of mixing silos and containers and concluded that the choice of means of preventing and mitigating dust explosions depends on the specific equipment and process design.

1.5.2.5
Powder/dust conveyors and dust removal equipment

As Table 1.15 for conveyors and dust removal equipment shows, Noha recommended the use of explosion resistant construction whenever fast-moving mechanical elements constitute part of the system. This for example applies to bucket elevators, which can preferably be equipped with cylindrical, pressure resistant legs, vented to a safe place via the elevator top. In some cases elevator legs can preferably be mounted along the wall outside the building and vented directly to the atmosphere.

Some materials collected in filters may form consolidated plugs in the dust discharge hopper at the filter bottom. This may give rise to frictional heating and self-ignition, which can result in dust explosions. Provided that the main enclosure as well as the filter bag supports are properly earthed, incendiary electrostatic discharges would not normally be expected in filters. (In the case of hybrid mixtures, the situation may be different.) However, if the equipment upstream of the filter, such as mills and spray dryers, can generate ignition sources that may be conveyed to the filter, the filter must be protected against possible explosions. By adopting cylindrical/conical body shape, the use of pressure shock resistant filter enclosures is a feasible possibility.

According to Noha (1989), industrial vacuum cleaners mostly operate at dust concentrations below the lower explosible limit. The fan is normally located on the clean side of the filter, and does therefore not constitute an ignition hazard. However, the possibility of internal dust explosions in industrial vacuum cleaners cannot be fully excluded in general, and therefore such equipment is often designed to be explosion (shock) resistant. Various aspects of preventing and mitigating dust explosions in pneumatic and other systems for conveying powders and dusts were discussed by Palmer (1973a) and Eckhoff (1982a). Pinkwasser (1985) described the extinction of smouldering powder nests in a dust cloud during pneumatic transport, whereas Göpfert (1981) discussed means of dust explosion prevention and mitigation in continuous conveying equipment in general. Palmer (1975) paid specific attention to dust explosions in dust collecting plant including cyclones and filters.

1.5.2.6
Conclusion

Noha (1989) emphasized that Tables 1.12–1.15 should be regarded as a starting point for discussion rather than as a final answer. The solution ultimately adopted must be the result of detailed analysis of the relevant factors in each specific case. General guidelines are useful as a point of departure, but the end result will always be a product of tailor-making!

1.5.3 SPECIAL ASPECTS FOR SOME SPECIFIC GROUPS OF POWDERS/DUSTS: A BRIEF LITERATURE SURVEY

1.5.3.1 Grain and feed dusts, and flour

The literature on preventing and mitigating dust explosions in the grain, feed and flour industry is substantial. This is easy to understand in view of the large losses of life and property that have been caused by dust explosions in this branch of industry. One of the earliest contributions, by Weber (1878), is still relevant in many respects. In particular he emphasized the marked influence of the moisture content of the dust/flour on the explosion hazard. Almost a century later, Theimer (1972) gave his well-known summary of causes, and means of prevention, of dust explosions in grain storage facilities and flour mills.

A useful summary of existing knowledge and experience at that time was collected during an international symposium arranged by National Materials Advisory Board, USA (1978). Aldis and Lai (1979) reviewed literature related to the engineering aspects of grain dust explosions. National Materials Advisory Board (1982) produced a set of well-structured recommendations for selection of adequate means for preventing and mitigating dust explosions in grain storage facilities and flour mills. The reason why the use of soybean oil for reducing dust formation in grain storage facilities (see Section 1.4.10) was not mentioned, is simple. The potential of this very promising method of dust control was just not known at that time. This illustrates that knowledge and technology is changing continuously, necessitating regular updating of sources of information.

A most informative document was issued by the Committee on Agriculture (1982), of the US House of Representatives. A number of witnesses, including both people from industry and scientists, were asked to express their views on how to reduce the probability of dust explosions in the US grain industry. Several witnesses emphasized the need for improved dust control.

Solymos (1985) discussed various 'dry' methods of dust control as a means of preventing dust explosions in grain storage facilities. Erling (1984) outlined the very comprehensive system for preventing and mitigating fires and dust explosions in the rebuilt Roland flour mill in F. R. Germany. The mill suffered a catastrophic explosion and fire in 1979.

Radandt (1987) reviewed the prevention and control of dust explosions in the grain, feed and flour industry in F. R. Germany in general, whereas Zhang Fenfen and Zhang Chunxiao (1987) considered grain dust explosions and their prevention in grain storage facilities in P. R. China.

Tesler and Semyonov (1988, 1989) discussed new concepts for reducing the dust explosion hazard in grain storage facilities, with particular reference to the situation in the USSR. The latter paper included schemes of explosion protection systems for integrated grain storage facilities, and quantitative methods for design of equipment, structures and buildings.

Venting of large silos in the grain, feed and flour industry has been a topic for discussion for a long time. Experimental full-scale work conducted in Norway and discussed in

Chapter 6 has provided new evidence. Pinkwasser and Häberli (1987) described specific designs of relief panels in the roof of large silo cells.

It is well known that bucket elevators are often the site of the primary dust explosion. Wilcoxen (1981) reported on an actual dust explosion in a grain storage facility in which the bucket elevators were in part located outdoors, and fitted with explosion vents. Due to the vents in the elevator legs, the extent of the explosion and resulting damage was comparatively minor. It was concluded that the design adopted had proved to be successful.

The French organization for standardization, Afnor (1986), issued a recommendation for mitigating dust explosions in the grain, feed and flour industry by venting. However, in view of the fast development in the field, one may wish to revise the recommendation at some points by including recent experimental evidence.

The ignitability and explosibility characteristics of dusts influence the choice and design of means of dust explosion prevention and mitigation. Ignitability and explosibility in turn depend on basic dust chemistry, effective particle size (see Section 1. 3. 3) and moisture content. Contributions elucidating various relationships for grain, feed and flour dusts were published by Hartmann *et al.* (1950), Jacobson *et al.* (1961), Eckhoff (1977/78), Eckhoff and Mathisen (1977/78), Enright and Bullock (1983, 1983a), Chiotti and Yoshizaki (1983) and Ambroziak (1985)

1.5.3.2
Milk powder, fish meal and sugar

According to Beever (1984) the number of reported fires and explosions in operations involving spray drying of milk increased during the early 1980s. Spray drying of milk is known to be particularly apt to generate self-heating and charring of the dried product. In milk spray dryers there will always be some regions containing explosible dust clouds. The question is only whether an ignition source is also present. Self-heating/charring in deposited dried milk powder can generate effective ignition sources, and Beever (1984) concluded that glowing/burning powder deposits was the most likely source of ignition in milk spray dryers. She estimated the minimum thicknesses for self-ignition of deposits of various types of milk powders at 7–14 mm for 200°C ambient temperature and 100–320 mm for 100°C.

Following an extensive explosion in a milk spray dryer in France, Pineau (1984a, 1985) conducted a comprehensive study of the self-ignition properties of milk powders, and of their ignitability and explosibility properties in cloud form.

Fish meals constitute another product group that can give rise to dust explosions. One example will be given in Chapter 2. Self-heating properties of fish meals as functions of meal chemistry, moisture content, etc. were discussed by Dreosti (1980). Eckhoff (1980) gave some further data for the ignitability and explosibility of fish meals.

In a series of full-scale sugar dust explosion experiments in the dust removal plant of a disused sugar factory Scholl (1973) demonstrated the potential of this material to give rise to serious dust explosions. Meek and Dallavalle (1954) tried to correlate explosibility properties of various types of sugar (C_{18}, C_{12} and C_6) with molecular structure and particle size. However, possibly due to a very weak ignition source and non-homogeneous dust clouds this was only partly successful.

1.5.3.3 Wood, cellulose and peat dusts

The fire and explosion hazards in the production of chip board, hard board and wood powder have been recognized for a long time. As new insight and technology have become available, the methods of preventing and mitigating the hazards have been improved.

Thelning and Laufke (1970) mainly focused on mitigation, in particular by venting and automatic suppression of explosions, and fire extinction by carbon dioxide and water vapour. Schmid (1972) gave detailed recommendations for both fire, explosion and environmental protection of chip board producing plants. He included prevention of ignition sources by recommending removal of foreign stone and metal objects before admitting the raw material to the process, and avoidance of overheating. The specific processes of chip preparation, pressing and cutting of the board, and the final finishing of the board surface were considered separately.

Arvidsson *et al.* (1977) conducted a very comprehensive investigation of the explosion and fire hazards in the production of chip board. They summarized their recommendations for preventive and mitigating measures in a list of 59 specific points, paying particular attention to removal of foreign objects in the plant feed, transport, storage and further size reduction of undried wood chip, drying of the chip, storage of the dried chip, milling of the dried chip, finishing of the board, and general housekeeping. Special attention was paid to the chip drying process. Continuous control of the oxygen concentration in the drying gas, and maintaining it as low as possible, was strongly recommended.

The work of Arvidsson *et al.* (1977) was presumably not known to Drossel (1984), who suggested a similar list of means for preventing and mitigating dust explosions and fires in chip board production. However, Drossel included automatic extinction of potential ignition sources in the form of small glowing wood or board fragments in gas and dust extraction ducts as an additional element. This method, which results from new technological development, has proved particularly useful in the wood industry, and was described in greater detail by Schröder (1984) (see also Section 1.4.4).

Scholl (1975) investigated the flame development following ignition inside mobile vacuum collectors for wood dust and wood chips. He found that only smouldering or open fires occurred, but no dust explosions. Furthermore, fire was only initiated if the ignition source was comparatively energetic, and the dust/chip mixture contained an appreciable fraction of fine dust ($< 100\ \mu m$). May *et al.* (1981) concentrated on the prevention of fire and explosion in wood chip dryers. They found that considerable overall improvement could be achieved by adequate process control and energy economization. Pfeiffer *et al.* (1985) investigated particle size distributions of airborne dusts from wood sawing and finishing operations. The particle size at which 30% of the dust mass was finer varied between 22 μm and 10 μm depending on operation and wood type. The mass fractions of very fine dust ($< 7\ \mu m$) varied between 20 and 2%.

Natural cellulosic dusts that can give rise to explosions are also generated in the cotton and linen textile industry. A catastrophic linen dust explosion in Harbin, P. R. China, will be discussed in Chapter 2. Kuczynski (1987) suggested a comprehensive scheme for protecting cotton processing plants against dust explosions. Particular attention should be paid to dust collecting systems and systems for storage of raw materials. Early detection of self-heating/self-ignition in deposits of dust and raw material, and adequate systems for extinction, was recommended. It was found that automatic injection of $NH_4H_2PO_4$ in connecting ducts to other plant sections provided effective isolation of the primary explosion (see Section 1.4.4).

As a result of the country's large peat resources, the use of peat as fuel has become a major concern in Finland. The ignitability and explosibility of peat dust depend on the origin and decomposition of the peat, and on its moisture content and particle size distribution. Weckman *et al.* (1981) investigated the possibilities for reducing the fire and explosion hazard in Finnish peat handling plants, with particular reference to peat power plants. They concluded that every effort should be made to prevent ignition, but it would also be necessary to take mitigating actions, should explosions nevertheless be initiated. The recommended means of mitigation were use of pressure (shock) resistant equipment, explosion venting, automatic explosion suppression, and isolation.

1.5.3.4
Coal dust/pulverized coal (excl. mines)

The literature on the ignitability and explosibility of coal dust is extensive. Originally the objective was limited to reducing the dust explosion hazard in coal mines and associated activities, as discussed in detail by Cybulski (1975). However, as pulverized coal has become an increasingly important fuel both for general heat production in power plants and for other purposes (cement furnaces, etc.), the coal dust explosion problem has also become an important issue in these areas.

The influences of the chemical composition, particle size and moisture content of coal dust on its ignitability and explosibility has been studied systematically since early in this century. Much information was collected by Nagy *et al.* (1965) and Cybulski (1975). Ignitability and explosibility properties of coal dust have also been investigated by Carpenter and Davies (1958), Scholl (1981), Bracke (1984), Enright (1985), Nettleton (1986), Wall *et al.* (1988) and Woskoboenko (1988). Torrent *et al.* (1988) found a good statistical correlation between two canonical variables representing the explosibility properties and the chemical composition of coal dusts respectively. Some further data related to the combustion of coal dust are given in Chapter 4 and in the Appendix.

Self-heating leading to self-ignition can be a significant problem when storing coal powder/dust in bulk. This problem and means of prevention and mitigation have been discussed by Korotov and Polferov (1978), Heinrich (1981), Thatcher (1982), Chauvin *et al.* (1985), Wiemann and Scholl (1985), and Braun (1987). Schlieper (1984) was particularly concerned with self-ignition of pulverized coal during transport by rail and road.

The extinction of coal dust explosion flames by various gaseous and pulverized solids additives was studied by Rahimian *et al.* (1982) in a laboratory-scale jet-stirred reactor. Most additives tested were just thermal heat sinks, whereas NaCl and $NH_4H_2PO_4$ also caused chemical reaction chain termination. Rae and Thompson (1979) investigated the effectiveness of various halogenated hydrocarbons as inerting agents and suppressants for coal dust explosions. However, due to the negative environmental effects of such substances they are currently being replaced by other extinguishing agents. Scherrer (1984) and Wehland (1984) discussed prevention of self-ignition in dust deposits and of explosions in dispersed dust in plants for production and storage of pulverized coal, by inerting with combustion gases, nitrogen or carbon dioxide.

The overall dust explosion protection of coal pulverizing plants was discussed by Birolini and Sammartin (1979), Wibbelhoff (1981), Diliberto (1983), Carini and Hules (1987) and Dansk Brandvaerns-Komité (1987). Fire and explosion protection of systems for conveyance and storage of pulverized coal was treated by Körner (1984) and Chauvin *et al.* (1987), whereas dust removal from pulverized coal plants was considered by Parpart (1979). Mullinger (1987) was concerned with fire and explosion protection of pulverized firing systems, whereas Egesoe (1978) discussed dust explosion prevention in systems for preparing and burning of coal dust in cement kilns. Patzke (1984) considered venting of dust explosions in plants for milling and drying of coal.

Finally, Ruygrok *et al.* (1983) were concerned with prevention and mitigation of coal dust explosions in surface facilities for transport, storage and handling of coal. The possibility of gas explosions due to release of methane from the coal, in particular from anthracites, was also investigated.

1.5.3.5
Polyester/epoxy powders for electrostatic powder coating

Electrostatic powder coating is to an increasing extent replacing traditional liquid paint spraying systems for industrial painting of metal products. The basic principle is that the metal object is first covered with an even layer of electrostatically bound epoxy/polyester powder. By subsequent treatment in an oven, the powder melts and hardens to an even, strong protective/decorative coating.

In the actual process the powder is transported pneumatically from a powder hopper to an electrostatic spraying gun. As the powder particles flow through the spraying gun, they become electrostatically charged by passing a strong electrostatic field of the order of tens of kilovolts. The charged particles are then attracted to and deposited on the earthed workpiece. The powder will continue to be deposited on the earthed workpiece until, at a certain powder layer thickness, the layer acts as an insulator and prevents further deposition of powder. Powder that is not deposited on the workpiece is normally collected in a powder recovery unit by a dust extraction system.

As technology developed and knowledge increased, the overall concepts of preventing and mitigating dust explosions in electrostatic powder coating systems were revised periodically. An early summary was given by Eckhoff and Enstad (1975). One of the preventive measures recommended was to keep the dust concentration in the spraying boot lower than the minimum explosible concentration. In a later paper Liere (1983) omitted this possibility, concentrating instead on inerting, automatic flame extinction, and isolation. Bartknecht (1986) and Liere (1989) conducted realistic full-scale explosion experiments in a powder spraying cabin and showed that dust flames in clouds of concentrations just above the minimum explosible concentrations are weak and slow. Bartknecht and Liere also determined ignitability and explosibility properties of typical polyester and epoxy powders used for electrostatic powder coating. Eckhoff, Pedersen and Arvidsson (1988) were not able, in a subsequent investigation, to reproduce the lowest minimum explosible dust concentrations of 15 g/m^3 reported by Bartknecht. In view of the fact that the minimum explosible concentration of typical hydrocarbon gases in air is about 35 g of gas per m^3 of air, and gas phase combustion is the basic flame propagation process for organic dusts, the value of 15 g/m^3 seems unrealistically low. Eckhoff *et al.* further found that up to 50 wt% of non-combustibles the minimum explosible dust concentration increased systematically with increasing non-combustibles in the powder, in such a way that the minimum explosible concentration of the combustible fraction was constant, in the range of 32–35 g/m^3. A dust containing 50 wt% non-combustibles, therefore, had a minimum explosible concentration of 65–70 g/m^3.

Both Bartknecht (1986) and Eckhoff *et al.* (1988) observed that some coating powders had exceptionally low minimum electric spark ignition energies, of < 3 mJ.

CENELEC (1989) issued a comprehensive European standard for electrostatic powder coating, where keeping the dust concentration in the spraying cabinet and dust extraction system below the minimum explosible concentration, was re-introduced as a central preventive measure. Another preventive measure was use of antistatic materials to avoid accumulation of electrostatic charge. Mitigating measures included interlocking systems, and use of non-combustible construction materials.

1.5.3.6
Aluminium and magnesium powder/dust

The fire and explosion hazards associated with production and handling of aluminium and magnesium powders has been the subject of extensive research for many years. As for metal powders in general, the hazard increases with decreasing particle size right down into the range below 1 μm. Dust clouds in air of very fine aluminium and magnesium powders have exceptionally low electric spark minimum ignition energies and give exceptionally violent explosions (see Appendix A). On the other hand, coarser aluminium powders, e.g. of particle diameters 100 μm, only present a moderate explosion hazard. However, if a comparatively coarse aluminium powder contains a fine dust fraction, even if it represents only a few per cent by mass, the explosion hazard can be considerably increased. For metal dusts like aluminium it is particularly true that keeping a watch on the explosion hazard to a large extent means keeping a watch on particle size.

Beck *et al.* (1984) discussed prevention and mitigation of dust explosions in aluminium grinding plants. By using wet grinding (e.g. water), the aluminium particles can be collected as a slurry and the dust explosion problem can be eliminated altogether. Alternatively the grinding operation itself can be dry, whereas the fine metal dust is collected in a liquid either immediately after the grinding point or in a separate wet collector further downstream. In general the need for measures for preventing and mitigating dust explosions depends on the extent to which the process is dry.

Beck *et al.* (1984) recommended several types of measures, adapted to the nature of the actual process. The list included interlocking systems to prevent grinding without dust extraction or insufficient liquid (water) supply, location of fans in dust-free areas, prevention of mechanical/electric sparks and hot surfaces, no smoking, and good housekeeping (cleanliness) in work rooms.

Reinke (1987) described the safety measures taken in a plant for production of comparatively coarse atomized aluminium powder (63–1200 μm). The fine fraction < 63 μm, representing the most severe explosion hazard, was separated out in an air jet filter. A high-speed automatic isolation valve was installed in the duct between the filter and the other part of the process, and the filter enclosure was equipped with a vent. Detectors for air flow and pressure were integrated in the interlocking system.

In plants producing very fine aluminium and magnesium powders, extensive gas inerting is necessary. For aluminium, nitrogen is normally suitable as inert gas, whereas a rare gas (helium or argon) is required for magnesium. However, in order to enable the particle surface to become oxidized, and thus avoid extreme reactivity when the powder/dust is later exposed to air, a certain fraction of oxygen, normally between 3–5 vol%, should remain in the inerting gas. NFPA (1987) discussed inerting and other necessary measures more extensively.

Eckhoff and Alfert (1988) reviewed the influence of particle size on the ignitability and explosibility properties of aluminium powders.

1.5.3.7
Silicon, silicon alloys and other metals

As indicated by Table 1.1 in Section 1.1.2, silicon dust has the potential of generating nearly the same explosion strength as aluminium dust of the same particle size. This has also been confirmed in practice. Fine silicon dust has given rise to catastrophic explosions in production and handling plants (see Chapter 2). As in the case of magnesium and aluminium dust clouds, clouds of silicon in air burn with a very high temperature, and thermal radiation from the burning cloud represents a severe threat to personnel.

If silicon is alloyed with iron, ignitability and explosibility is generally reduced as the iron content increases. On the other hand, the presence of magnesium in silicon alloys significantly increases the explosion hazard. In particular the minimum electric spark ignition energy drops significantly if the magnesium content approaches 5–10 wt% or more. In general the understanding of the influence of various alloy compounds on the ignitability and explosibility of silicon alloys is incomplete, and specific investigation is often required.

Eckhoff *et al.* (1986) investigated the ignitability and explosibility of silicon dust clouds in air and confirmed that the minimum electric spark ignition energy decreases and the explosion violence increased systematically with decreasing particle size. However, very fine powders/dusts of particle sizes in the range of 1 μm and even smaller, may be difficult to disperse completely into primary particles, and may therefore behave as if they were coarser. This can complicate correlation of primary particle size with ignitability and explosibility data (see Chapter 3 for further details on dust dispersion).

In the case of manganese and ferromanganese, flashes that can initiate flame propagation in dust clouds are easily produced by mechanical impacts of lumps of the material, or in crushing operations. (This particular feature has also been observed with ferro-silicon-magnesium.) Clouds of fine manganese dust in air can have very low minimum electric spark ignition energies, of the order of 1 mJ. On the other hand, flame propagation in clouds in air of dusts of manganese and manganese alloys is comparatively slow, and the flame temperature comparatively low. Qian Qiyong *et al.* (1987) studied how dust explosions and fires in the cyclone separator of a ferromanganese milling plant could be prevented, in spite of unavoidable flashes in the crushing and milling units. As part of the work they also studied ignition of layers of ferromanganese dusts on a hot-plate. Even for a layer thickness of only 2 mm, the minimum ignition temperature was as low as 320°C.

Allenbach (1984) proposed a special system for classifying the fire and explosion hazards to be associated with dusts of various metals and ferro alloys in industrial plants. He introduced three combustibility classes based on observation of the flame development in clouds of freshly ground < 44 μm dust fractions in a laboratory-scale explosion vessel.

- Class 1 – Very active: Very violent flame propagation
- Class 2 – Active: Quite fast flame propagation
- Class 3 – Combustible: Slow propagation of weak flame

The hazard of a particular powder/dust was evaluated by combining the flammability class of the ground < 44 μm dust sample and the actual particle size of the product. Allenbach provided a list of the combustibility class ratings of a wide range of ferro alloys and other metals. All listed calcium alloys and most alloys containing aluminium and magnesium were of Class 1. The other metals and alloys tested, including boron alloys, chromium and chromium alloys, manganese and its alloys, and silicon and silicon alloys were of Class 2 or Class 3.

Allenbach did not provide sufficient information about the experimental apparatus and procedure, to permit further evaluation of his proposed classification system.

Ma *et al.* (1987) and Xiao *et al.* (1987) produced kinetic data and mathematical models for the oxidation of calcium silicon alloys, which may prove useful in future modelling of dust explosions involving these materials.

1.5.3.8
Miscellaneous powders/dusts

Baklygin and Nikitina (1978) investigated the minimum explosible dust concentration and minimum ignition temperature (dust layer) of various dust mixtures generated in the mixing plant of the Moscow Tyre Works.

Gehring *et al.* (1978) studied the explosiveness of clouds in air of dust of a propellant containing 84% nitrocellulose, 10% dinitrotoluene, 5% dibutylphthalate and 1% dephenylamine. For a < 75 μm fraction of this particular propellant the minimum explosible dust concentration in air was 100–200 g/m^3, whereas the minimum electric spark ignition energy of dust clouds was about 150 mJ. This means that when dispersed as clouds in air, such materials exhibit ignitability and explosibility properties that are similar to, or even less severe than those of normal organic solid fuels like starch and proteins of the same particle size. However, the pressure and temperature waves generated by the initial dust explosion may in some cases initiate more hazardous secondary exothermal reactions in adjacent condensed propellant deposits.

The fire and dust explosion hazard connected with mine blasting of oil shale has been considered by several authors, including Cashdollar *et al.* (1984), Richmond and Beitel (1984), Weiss *et al.* (1985), Miron and Lazzara (1985), Weiss *et al.* (1986), Sapko *et al.* (1986) and Hertzberg and Cashdollar (1988). Karim *et al.* (1979), in a more basic investigation, studied the combustion of oil sand fragments in hot, flowing, oxidizing gas.

Dust explosions can also result from mining of sulphide ores, containing substances like pyrite, pyrrhotite, arsenopyrite, sphalerite and galena. The hazards of sulphide dust explosions also include the toxic effects of the combustion products. Various aspects of sulphide ore dust explosions were discussed by Polikarpov (1984), Enright (1984, 1984a) and Amaratunga (1988).

Finally, a quite special dust explosion hazard should be mentioned. It arises when burnt-out fuel rods in nuclear power plants are cut in reprocessing plants and fine zircaloy dust is generated. Zircaloy is essentially zirconium with small percentages of antimony,

antimony, iron and nickel. It is used as cladding for nuclear fuel rods. Because of the hazardous radioactivity of the zircaloy dust, very special precautions must be taken when assessing the ignitability and explosibility properties of the dust. Andriessen *et al.* (1987), Hensel (1988) and Hattwig (1988) discussed the methods used and the results obtained, and suggested possible means of preventing and mitigating zircalloy dust explosions in reprocessing plants.

1.5.4 STANDARDS, RECOMMENDATIONS, GUIDELINES

Most industrialized countries have their own official codes of practice for preventing and mitigating dust explosions in industry. This, for example, is the case in Germany, UK, France, Holland, Sweden, Norway, USA and USSR. Normally the official national factory inspectorate, or health-and-safety inspectorate, is the responsible authority issuing the codes and controlling whether they are practised.

In addition independent bodies in many countries issue their own regulations, some of which are in reality considered authoritative. Examples of such bodies are National Fire Protection Association (NFPA) in USA and Verein deutscher Ingenieure (VDI) in Germany. Sometimes various industrial branches in a country, such as the grain, feed and flour industry, or the ferro alloy industry, issue their own set of specific guidelines. It is important then that these comply with the general authoritative codes of the country.

Codes and standards are also issued on an international level, through cooperation between many countries. Examples of international organizations set up for such work are the International Standardization Organization (ISO), the International Electrotechnical Commission (IEC), and European Community bodies (CEN, CENELEC).

All the various codes, standards, regulations and guidelines are, or should be, periodically revised to keep pace with the development of knowledge and technology. One should therefore always make sure that the document at hand is the latest, valid version.

REFERENCES

Abbot, J. A. (1988) Survey of Dust Fires and Explosions in UK 1979–1984: Dust Explosion Prevention and Protection – Latest Development. In *Proceeding of Internat. meeting organ. by IBC/BMHB,* British Material Handling Board, UK

Aellig, A., and Gramlich, R. (1984) Einfluss von Ausblasrohren auf die Explosionsdruckentlastung. *VDI-Berichte* **494.** pp. 175–183, VDI-Verlag GmbH, Düsseldorf

Afnor (1986) Sécurité des silos. Atténuation des effets des explosions par les évents de décharge. Calcul des surfaces d'évents. (December, 1986) Afnor 86820, U54–540

Aldis, D. F. and Lai, F. S. (1979) Review of Literature Related to Engineering Aspects of Grain Dust Explosions. US Dept. Agriculture. *Miscell. Publ.* No. 1375

Alfert, F., Eckhoff, R. K., and Fuhre, K. (1989) Zündwirksamkeit von Glimmnestern und heissen Gegenständen in industriellen Anlagen. *VDI-Berichte* **701.** pp. 303–319, VDI-Verlag GmbH, Düsseldorf

Alfert, F., and Fuhre, K. (1989) Flame and Dust Free Venting of Dust Explosions by Means of a Quenching Pipe, Report No. 89/25820-1, January, Chr. Michelsen Institute, Bergen, Norway

Allenbach, C. R. (1984) Combustibility Characteristics of Fine Sized Ferro-Alloys and Metals. In *Proc. of 42, Electric Furnace Conference*, (December) Toronto, Canada

Amaratunga, L. M. (1988) Sulphide Dust Explosion. Laboratory Ore Dust Ignition and Degree of Oxidation Test. *Symp. Ser. Austr. Inst. Min. Metall.* **60** pp. 483–489

Ambroziak, Z. K. (1985) Investigation of Dust Explosion Hazards in Baking Industry. *Internat. Symp., Control of the Risks in Handling and Storage of Granular Foods* (April), APRIA, Paris

Andriessen, H., Kroebel, R., Bereznai, T., *et al.* (1987) Untersuchungen zum Brand- und Explosionsverhalten von Zircaloy-Feinteilchen. *Bericht EUR 11120 des Amts für amtliche Veröffentlichhungen des Europäischen Gemeinschaften*

Anonym, (1989) Neue Wege im Staubexplosionsschutz. Druckentlastung ohne Flammenausbreitung und Immissionen. *Chemie-Technik*, **18** pp. 56–57

Ariessohn, P. C., and Wang, J. C. F. (1985) Recent Development of a Real-Time Particulate Mass Sampling System for High Temperature Applications. *Combustion and Flame* **59** pp. 81–91

Arvidsson, T., Back, E., and Östman, B. (1977) Brand- och explosionsrisker i spånskivefabriker. Part 4. Final Report (April), *STFI-meddelande.* Series B No. 442 (FSB:57), Svenska Träforskningsinstitutet, Stockholm, Sweden

Atkinson, N. (1988) Everyone Makes Mistakes, but Can You Always Live with the Result? *Process Engineering*, October, pp. 35–37

Baker, W. E., Cox, P. A., Westine, P. S., *et al.* (1983) Explosion Hazards and Evaluation. In *Fundamental Studies in Engineering Series*, Vol. 5 Elsevier Sci. Publ. Co.

Baklygin, V. N., and Nikitina, Z. K. (1978) Assessment of the Fire and Explosion Hazards of Dust Mixtures of Ingredients in Preparation. *Internat. Polym. Science and Technol.*, **5** No. 5

Ballal, D. R. (1979) Ignition and Flame Quenching of Quiescent Dust Clouds of Solid Fuels. In *Proc. Roy. Soc. Lond. Series A*, (July) London

Bartknecht, W. (1978) *Explosionen – Ablauf und Schutzmassnahmen.* Springer-Verlag

Bartknecht, W. (1979) Forschung in der Sicherheitstechnik. *Chemie-Technik* **8** pp. 493–503

Bartknecht, W. (1980) Der Staubexplosionsgefahr is mit Schnellschluss-Schiebern wirksam zu begegnen. *Maschinenmarkt* **86** pp. 1133–1136

Bartknecht, W. (1982) How Valve Action can Combat the Transmission Effect (of Dust Explosions). *Bulk-Storage, Movement, Control.* (July/August) pp. 37–40

Bartknecht, W. (1986) Latest Findings concerning Explosion and Ignition Behaviours of Coating Powders with Special Consideration of their Inflammation Behaviour in Coating Cabins. In *Proc. of 1. Br.-Ger. Conference on Powder Coating.* Techn. + Kommun. Verlag GmbH, Berlin 54 pages

Bartknecht, W. (1987) Staubexplosionen – Ablauf und Schutzmassnahmen. Springer-Verlag

Bartknecht, W. (1988) Brennbare Stäube und hybride Gemische: Teil 2. Schutzmassnahmen gegen das Entstehen und die Auswirkungen von Staubexplosionen. *Staub-Reinhalt. Luft* **48** pp. 417–425

Beck, G. (1974) Ein Beitrag zum systematischen Staubexplosionsschutz. *Chem. Techn.* **26** pp. 426–427

Beck, G. (1985) Risikoanalysen – Voraussetzung für einen optimalen Staub-explosionsschutz in der Land- und Nahrungsgüterwirtschaft der DDR. *Agrartechnik* **35** pp. 68–70

Beck, H. (1982) Schadenanalyse von Staubexplosionen. *Staub-Reinhalt. Luft* **42** pp. 118–123

Beck, H., and Jeske, A. (1982) Dokumentation Staubexplosionen. Analyse und Einzelfalldarstellung. Berufsgenossenschaftliches Institut für Arbeitssicherheit. Report No. 4/82, St. Augustin, F. R. Germany

Beck, H., Foerster, H., and Faber, M. (1984) Staubexplosionen in Aluminiumschleifereien und Massnahmen zu ihrer Verhütung. *Proc. 9. Int. Symp. Prev. Occ. Accid. & Diseas. Chem. Ind.*, Luzern, Switzerland, (June) pp. 981–1011

Beck, H., and Jeske, A. (1988) Staubexplosionen. Gefahren – Dokumentation – Auswertung. *Staub-Reinhalt. Luft* **48** pp. 35–39

Beck, H., and Jeske, A. (1989) Prüfung fahrbarer Industriestaubsauger zum Einsatz in durch Staub explosionsgefährdeten Bereichen. (Bauart-1-Staubsauger). *VDI-Berichte* **701,** VDI-Verlag

GmbH, Dusseldorf pp. 881–897

Beever, P. F. (1984) Fire and Explosion Hazards in the Spray Drying of Milk. *Proc. of Seminar*, (December) Portlaoise, Ireland

Beigler, S. E., and Laufke, H. (1981) Inventering av tryckavlastere. Ingeniörvetenskapsakademien, Report 200, Stockholm, Sweden

Beigler, S. E. (1983) Utforming och dimensionering av tryckavlastare. Report to Arbetarskyddsstyrelsen, Kemisektion 3, (June) Solna, Sweden

Bentley, J. P. (1988) *Principles of Measurement Systems,* 2. ed. Longman Scientific and Techn., Harlow, UK

Beyersdorfer, P. (1925) *Staub-Explosionen.* Verlag von Theodor Steinkopff, Dresden und Leipzig

Birolini, P., and Sammartin, L. (1979) Explosionseigenschaften von Kohlenstaub und ihre Berücksichtigung beim Bau von Kohlenstaubmahlanlagen. *Zement-Kalk-Gips*, **32** pp. 613–616

Bodurtha, F. T. (1980) *Industrial Explosion Prevention and Protection,* McGraw-Hill, New York

Borisov, A., and Gelfand, B. (1989) Personal Communication with R. K. Eckhoff in Bergen, Norway (December)

Bowes, P. C., Burgoyne, J. H., Rasbash, D. J. (1948) The Inflammability in Suspension of Mixtures of Combustible and Incombustible Dusts. *Journ. Soc. Chem. Ind.* **67** pp. 125–130

Bracke, J. (1984) Ontvlambaarheid van Steenkoolstof. *Annales des Mines de Belgique*, pp. 291–301

Braun, E. (1987) Self-Heating Properties of Coal. Report NBSIR 87-23554 US Dept. Commerce, Center for Fire Research, Gaithersburg, MD 20899, USA

Brazier, G. (1988) Temperature and Vent Design. *Solids Handling* **10** pp. 84–85

Brennecke, H. (1987) A New Digital Pressure Measurement System for Fast and Reliable Detection of Dust Explosions in their Initial Phase. *Shenyang International Symposium on Dust Explosions*, NEUT, Shenyang, P. R. China. pp. 487–499

Brown, K. C., and James, G. J. (1962) Dust Explosions in Factories: A Review of the Literature. SMRE Res. Rep. No. 201 Safety in Mines Research Establishment, Sheffield, UK

Brunner, M. Y. (1983) Bauwerkbeanspruchungen durch die Rückstosskräfte druckentlasteter Staubexplosionen in Behältern. Doctorate Thesis, ETH, Zürich, Switzerland

Brunner, M. Y. (1984) Bauwerkbeanspruchungen durch die Rückstosskräfte druckentlasteter Staubexplosionen *VDI-Berichte* **494,** VDI-Verlag GmbH, Düsseldorf pp. 227–232

Bryant, J. T. (1973) Powdered Fuel Combustion: Mechanism of Particle Size. *Combustion and Flame* **20** pp. 138–139

Burgoyne, J. H., and Rasbash, D. H. (1948) The Prevention of Dust Explosions Caused by High-Explosive Bombs in flour Mills. *J. Soc. Chem. Ind.* **67** pp. 130–139

Burkardt, F. (1989) Human Factors in Accident Prevention. In *Proceedings of 6. Internat. Symp. Loss Prevention and Safety Promotion in the Process Industries* (June), Norwegian Society of Chartered Engineers, Oslo, Norway

Cardillo, P., and Anthony, E. J. (1978) The Flammability Limits of Hybrid Gas und Dust Systems. *La Rivista dei Combustibili*, **XXXII** pp. 390–395

Carini, R. C., and Hules, K. R. (1987) Coal Pulverizer Explosions. In *Industrial Dust Explosions*, ASTM Special Techn. Publ. 958, (eds K. L. Cashdollar and M. Hertzberg) pp. 202–216 ASTM, Philadelphia, USA

Carpenter, D. L., and Davies, D. R. (1958) The Variation with Temperature of the Explosibility Characteristics of Coal Dust Clouds Using Electric Spark Ignition. *Combustion and Flame*, **2** pp. 35–53

Cashdollar, K. L., Hertzberg, M., and Conti, R. S. (1984) Explosion Hazards of Oil Shale Dusts: Limits, Pressures, and Ignitability. *Proc. 17, Oil Shale Symp.* (Aug.), Colorado School of Mines Press, pp. 243–254

CENELEC, (1989) Requirements for the Selection, Installation and Use of Electrostatic Spraying Equipment for Flammable Materials, Part 2: Hand-Hold Electrostatic Powder Spray Guns (March) CENELEC, Brussels

Chatrathi, K., and De Good, R. Explosion Isolation Systems used in Conjunction with Explosion Vents. Unpublished manuscript, Fike Corporation, Blue Springs, MO, USA
Chauvin, R., Lodel, R., and Philippe, J. L. (1985) Spontaneous Combustion of Coal. CERCHAR, 60550 Verneuil-en-Halatte, France
Chauvin, R., Lodel, R., Nomine, M., *et al.* (1987) Safety in Storage and Transport of Pulverized Coal. In *Proc. of 1st Europ. Dry Fine Coal Conf.*, Harrogate, UK, p. 111/2–1
Chiotti, P., and Yoshizaki, S. (1983) Adsorption of Moisture by Grain Dust and Control of Dust Hazards. *Proc. of Conference on Particulate Systems: Technology and Fundamentals* , Ames Lab., Ames, IA 50011, USA
Christill, M., Nastoll, W., Leuckel, W., *et al.* (1989) Der Einfluss von Strömungsturbulenz auf den Explosionsablauf in Staub/Luft-Gemischen. VDI-Berichte 701, pp. 123–141 VDI-Verlag GmbH, Düsseldorf
Committee on Agriculture (1982) *Review of Grain Elevator Safety*. Hearing before the Subcomm. Wheat, Soybeans and Feed Grains, Comm. Agric. House of Repr. 97 Congr., 2. Sess., July 21 (1982). US Government Printing Office, Serial No. 97-YYY, Washington DC
Conti, R. S., Cashdollar, K. L., and Liebman, I. (1982) Improved Optical Probe for Monitoring Dust Explosions. *Rev. Sci. Instrum.* **53** pp. 311–313
Cox, R. A. (1986) Appraisal of the Utility of Risk Analysis in the Process Industry. *Proc. of 5th Internat. Symp., Loss Prev. Safety Prom. Process Ind.*, (Sept.) Paper No. 14. Cannes, France
Cox, R. A. (1987) An Overview of Hazard Analysis. *Proceedings of Internat. Symp. Prev. Major Chem. Accidents*, (February) (Ed. J. L. Woodward) Center for Chemical Process Safety, Washington DC
Crowhurst, D. (1988) Small-scale Dust Explosions Vented Through Ducts. *Proc. of 3rd Internat. Colloquium on Dust Explosions*, (October) Szxyrk, Poland
Cross, J., and Farrer, D. (1982) Dust Explosions, Plenum Press, New York/London
Cybulski, W. (1975) *Coal Dust Explosions and their Suppression*. (English translation from Polish), published by the Foreign Scientific Publications Department of the National Center for Scientific, Technical and Economical Information, Warsaw, Poland
Czajor, W. (1984) Explosionstechnische Entkopplung von Apparaturen. *VDI-Berichte* **494,** pp. 233–238. VDI-Verlag GmbH, Düsseldorf
Dahn, C. J. (1986) Contribution of Low-Level Flammable Vapour Concentrations to Dust Explosion Output. *Proc. of 2nd Internat. Coll. Dust Explosions*, (Nov.), Warsaw, Poland
Dansk Brandvaernskomité (1987) Brandmaessige forhold forbundet med kulformaling og den dertil hoerende opplagring og transport av kulmel. Dansk Brandvaernskomité, Copenhagen, Denmark
DEP (1970) *Dust Explosions in Factories*, Health and Safety at Work, Booklet No. 22. Department of Employment and Productivity. Her Majesty's Stationery Office, London
Diliberto, M. C. (1983) Coal Dust Explosion Hazards and Prevention, *Proc. of 3rd Conf. on Coal Technology Europe*, Ind. Pres. Group **4** pp. 7–30
Dinglinger, G. (1981) Inertisierung von Silos durch Stickstoff im Schadensfalle. *Getreide, Mehl und Brot* No. 12
Donat, C. (1973) Einsatz von Berstsicherungen bei langsamen und schnellen Druckanstieg. *Chemie-Ingenieur-Technik* **45** pp. 790–796
Donat, C. (1978) Apparatefestigkeit bei Beanspruchung durch Staubexplosionen. *VDI-Berichte* **304,** pp. 139–149, VDI-Verlag GmbH, Düsseldorf
Donat, C. (1984) Explosionsfeste Bauweise von Apparaturen. *VDI-Berichte* **494,** pp. 161–167, VDI-Verlag GmbH, Düsseldorf
Dorn, J. (1983) Total Gas Explosion Suppression Systems. *Proceedings of an informal seminar on suppression of gas explosions on offshore gas and oil installations*. Report No. 833402-2, (April) pp. 70–82, Chr. Michelsen Institute, Bergen, Norway
Dreosti, G. M. (1980) Spontaneous Heating of Fish Meal, News Summary – No. 49 Special issue on processing, (Oct.), Internat. Assoc. Fish Meal Manuf., Orchard Parade, Potters Bar, Herts., UK.

Drossel, K. (1984) Brand- und Explosionsschutzmassnahmen in einem Sponplattenwerk. *VDI-Berichte* 494, pp. 287–292, VDI-Verlag GmbH, Düsseldorf

Ebert, F. (1983) Explosionsschutz mit Schnellschluss-Schieber, *Staub-Reinhalt. Luft* **43** pp. 14–17

Eckhoff, R. K., and Enstad, G. G. (1975) Motvirkning av skader ved støveksplosjoner i anlegg for elektrostatisk pulversprøyting. En vurdering av aktuelle tiltak. Report No. 74173-1, (December), Chr. Michelsen Institute, Bergen, Norway

Eckhoff, R. K., and Fuhre, K. (1975) Investigations Related to the Explosibility of Agricultural Dusts in Air. Part 3. Report No. 72001/RKE/KF, (May), Chr. Michelsen Institute, Bergen, Norway

Eckhoff, R. K. (1977) The Use of the Hartmann Bomb for Determining K_{St} Values of Explosible Dust Clouds. *Staub-Reinhalt. Luft* **37** pp. 110–112

Eckhoff, R. K. (1977/1978) Pressure Development During Explosions in Clouds of Dusts from Grain, Feedstuffs and Other Natural Organic Materials. *Fire Research* **1** pp. 71–85

Eckhoff, R. K., (1977/1978) and Mathisen, K. P. A Critical Examination of the Effect of Dust Moisture on the Rate of Pressure Rise in Hartmann Bomb Tests. *Fire Research* **1** pp. 273–280

Eckhoff, R. K. (1980) Powder Technology and Dust Explosions in Relation to Fish Meal. News Summary – No. 49 Special issue on processing, (October), Internat. Assoc. Fish Meal Manuf., Orchard Parade, Potters Bar, Herts., UK. pp. 63–112

Eckhoff, R. K., Fuhre, K., Krest, O., *et al.* (1980) Some Recent Large-Scale Gas Explosion Experiments in Norway. Report No. 790750–1, (January), Chr. Michelsen Institute, Bergen, Norway

Eckhoff, R. K. (1982) Current Dust Explosion Research at the CMI. In *Fuel-Air Explosions*, (Eds J. H. S. Lee and C. M. Guirao) pp. 657–678, University of Waterloo Press, Canada

Eckhoff, R. K. (1982a) Støveksplosjonsfaren i pneumatiske pulvertransportanlegg. *Proceedings of Seminar on Pneumatic Powder Transport*, (October), Norwegian Society of Chartered Engineers

Eckhoff, R. K., Fuhre, K., Guirao, C. M., *et al.* (1984) Venting of Turbulent Gas Explosions in a 50 m^3 Chamber. *Fire Safety Journal* **7** pp. 191–197

Eckhoff, R. K., Fuhre, K., and Pedersen, G. H. (1985) Vented Maize Starch Explosions in a 236 m^3 Experimental Silo. Report No. 843307–2, (December), Chr. Michelsen Institute, Bergen, Norway

Eckhoff, R. K., Parker, S. J., Gruvin, B., *et al.* (1986) Ignitability and Explosibility of Silicon Dust Clouds. *J. Electrochem. Soc.* **133** pp. 2631–2637

Eckhoff, R. K., and Alfert, F. (1988) Fire and Explosion Hazards in the Production and Handling of Aluminium Powder. Report No. 88/02103–1, (June), Chr. Michelsen Institute, Bergen, Norway

Eckhoff, R. K., Pedersen, G. H., and Arvidsson, T. (1988) Ignitability and Explosibility of Polyester/Epoxy Resins for Electrostatic Powder Coating. *J. Hazard. Materials* **19** pp. 1–16

Eckhoff, R. K., Alfert, F., and Fuhre, K. (1989) Venting of Dust Explosions in a 5.8 m^3 Bag Filter under Realistic Conditions of Dust Cloud Generation. *VDI-Berichte* **701,** pp. 695–722. VDI-Verlag GmbH, Düsseldorf

Egesoe, V. (1978) Opberedning af og fyring med kulstoev i cementindustrien, gjennomgaaet specielt med henblik på forebyggelse af kulstoevseksplosioner. *Proc. of Meeting arranged by the Danish Association of Engineers*, Copenhagen (April)

Engler, C. (1885) Beitrage zur Kenntniss der Staubexplosionen. *Chemische Industrie*, (June), pp. 171–173

Engler, C. (1907) Einfacher Versuch zur Demonstration der gemischten Kohlenstaub- und Gasexplosionen. *Chemiker-Zeitung* No. **28** pp. 358–359

Enright, R. J., and Bullock, M. H. (1983) Explosibility of Australian Wheat Dusts. School of Civil and Mining Engineering, (April), The University of Sydney, Australia

Enright, R. J., and Bullock, M. H. (1983a) Changes in the Characteristics of Dust Produced During the Transportation and Storage of Wheat. School of Civil and Mining Engineering, (April), The University of Sydney, Australia

Enright, R. J. (1984) Sulphide Dust Explosions in Metalliferous Mines. Res. Rep. No. R466, (February), School of Civil and Mining Engineering, The University of Sydney, Australia
Enright, R. J. (1984a) Sulphide Dust Explosions in Metalliferous Mines. *Proc. Australas. Inst. Min. Metall.* No. 289, pp. 253–257, Austr. Inst. Min. Metall., Parkville, Australia
Enright, R. J. (1985) Effect of Moisture on Explosion Parameters of Coal Dust. *Proc. of 21st Int. Conf. Saf. Min. Res. Inst.*, (Oct.) Sydney, Australia
Erling, H. P. (1984) Brand- und Explosionsschutzmassnahmen in einer Getreidemühle. *VDI-Berichte* **494,** pp. 281–285 VDI-Verlag GmbH, Düsseldorf
Faber, M. (1985) Sensoren in der Sicherheitstechnik von Explosionsgeschützten Industrieanlagen. *Technisches Messen* tm **52** pp. 273–276
EuropEx (1990) Partial Print-Out of EuropEx Data Base. European Information Centre for Explosion Protection, Hove-Antwerp, Belgium
Faber, M. (1989) Explosionstechnische Entkopplung. *VDI-Berichte* **701,** pp. 659–680, VDI-Verlag GmbH, Düsseldorf
Faber, M. (1989a) Steuerungs- und Verriegelungseinrichtungen bei Anwendung von Explosionsschutzmassnahmen in komplexen Systemen. *VDI-Berichte* **701,** pp. 899–916, VDI-Verlag GmbH, Düsseldorf
Faraday, M., and Lyell, C. (1845) Report on the Explosion at the Haswell Collieries, and on the Means of Preventing Similar Accidents. *Philosophical Magazine* **26** pp. 16–35
Field, P. (1982) *Dust Explosions*, Elsevier Sci. Publ. Co., Oxford, UK
Field, P. (1982a) Industrial Dust Explosion Hazards Assessment, Prevention and Protection. Solidex March 1982. Health and Safety in the Bulk Solids Handling Industry, Harrogate
Field, P. (1987) Basic Philosophy and Practical Approach to Dust Explosion Protection in the UK. *Proc. of Shenyang International Symposium on Dust Explosions*, pp. 1–44, NEUT, Shenyang, P. R. China
Fischer, P. (1978) Primärer Explosionsschutz durch Inertisierung. *VDI-Berichte* **304,** pp. 85–90, VDI-Verlag GmbH, Düsseldorf
Foniok, R. (1985) Hybrid Dispersive Mixtures and Inertized Mixtures of Coal Dust. Explosiveness and Ignitability. *Staub Reinhalt. Luft* **45** pp. 151–154
Franke, H. (1978) Bestimmung der Mindestzündenergie von Kohlenstaub/Methan/Luft-Gemischen (Hybride Gemische). *VDI-Berichte* 304, pp. 69–72, VDI-Verlag GmbH, Düsseldorf
Freitag, H. H. (1965) *Handbuch der Raumexplosionen.* Verlag Chemie GmbH, Weinheim
Gehring, J. W., Friesenhahn, G., and Rindner, R. M. (1978) Exploratory Study of M-1 Propellant Dust Explosibility. (Report ARLCD-CR-78022, September) US Army Armament Res. Devel. Command, Dover, NJ 07801, USA
Gibson, N., and Lloyd, F. C. (1965) Incendivity of Discharges from Electrostatically Charged Plastics. *Brit. J. Appl. Phys.* **16** pp. 1619–1631
Gibson, N. (1978) Design of Explosion Protection for Dust Control Equipment. *Proc. of Symp. Dust Control*, (Manchester, March 1978), Inst. Chem. Engrs., London
Gibson, N., Harper, D. J., and Rogers, R. L. (1985) Evaluation of the Fire and Explosion Risk in Drying Powders. *Plant/Operations Progress* **4** pp. 181–189
Gillis, J. P. (1987) Prevention of Dust Explosion Propagation through Ducting. *Proceedings of Powder and Bulk Solids Conference*, Cahners Expos. Group, pp. 335–343
Glarner, Th. (1983) Temperatureinfluss auf das Explosions- und Zündverhalten brennbarer Stäube. Doctorate Thesis No. 7350, ETH, Zürich
Glarner, Th. (1984) Mindestzündenergie – Einfluss der Temperatur. *VDI-Berichte* **494,** pp. 109–118, VDI-Verlag GmbH, Düsseldorf
Glor, M. (1985) Hazards due to Electrostatic Charging of Powders. *Journal of Electrostatics* **16** pp. 175–191
Glor, M. (1988) *Electrostatic Hazards in Powder Handling,* Research Studies Press Ltd., John Wiley & Sons Inc.
Göpfert, H. (1981) Staubexplosionen in Verbindung mit Stetigfördern. Proceedings of 'Transmatic

81', pp. 562–569. Institut für Fördertechnik, Karlsruhe
Haase, H. (1972) *Statische Elektrizität als Gefahr.* Verlag Chemie GmbH, 2. Ed., Weinheim, F. R. Germany
Hammer, P. R. (1978) Eigeninertisierung an Sprühtrocknern, Probleme aus der Licht des Herstellers. *VDI-Berichte* **304,** pp. 91–95, VDI-Verlag GmbH, Düsseldorf
Hammond, E., and Kaye, B. H. (1963) The Growing Interest in Powders. *New Scientist*, **7**. (Nov.), pp. 324–326
Harmanny, A. (1990) Flame Jet Hazards. *EuropEx Newsletter*, Ed. 13, (Sept.) pp. 9–16
Harris, R. J. (1983) *The Investigation and Control of Gas Explosions in Buildings and Heating Plant.* British Gas Corporation. E & FN Spon, London
Hartmann, I. (1948) Recent Research on the Explosibility of Dust Dispersions. *Industrial and Engineering Chemistry* **40** pp. 752–758
Hartmann, I., Cooper, A. R., and Jacobson, M. (1950) Recent Studies on the Explosibility of Cornstarch. Rep. Inv. 4725, US Bureau of Mines, Washington
Hattwig, M. (1980) Auswirkung von Druckentlastungsvorgängen auf die Umgebung. Report ISSN 0172–7613, (June), BAM, Berlin
Hattwig, M., and Faber, M. (1984) Rückstosskräfte bei Explosionsdruckentlastung. *VDI-Berichte* 494, pp. 219–226. VDI-Verlag GmbH, Düsseldorf
Hattwig, M., Hensel, W., and Osswald, R. (1983) The Prevention of Dust Explosions and Fires in a Nuclear Reprocessing Plant. *Proc. of 3rd Internat. Coll. Dust Explosions*, (Oct.), Szczyrk, Poland
Hawksley, J. L. (1989) The Selective Use of the Elements of Quantitative Risk Assessment. *Proceedings of 6th Internat. Symp. Loss Prevention and Safety Promotion in the Process Industries*. (June), Norwegian Society of Chartered Engineers, Oslo, Norway
Heiner, H. (1986) Inertisierung und Explosionsschutz in Silos durch CO_2. *Die Ernährungsindustrie* No. 6 pp. 32–35
Heinrich, H.-J. (1981) Grundlagen für die Einstufung von Kohlestäuben in die Gefahr-klasse 4.2 der Beförderungsvorschriften. Armts- und Mitteilungsblatt der BAM **11** pp. 326–330
Heinrich, H.-J. (1989) Ablauf von Gas- und Staubexplosionen – Gemeinsamkeiten und Unterschiede. *VDI-Berichte* **701,** pp. 93–112, VDI-Verlag GmbH, Düsseldorf
Hensel, W. (1988) Staubexplosionsprobleme bei der Zerkleinerung abgebrannter. Brennelemente in Wiederaufbereitungsanlagen. Paper prepared for the *Amts- und Mitteilungsblatt* BAM, Berlin
Hertzberg, M., Cashdollar, K. L., Zlochower, I., *et al.* (1984) Inhibition and Extinction of Explosions in Heterogeneous Mixtures. *Proceedings of 20th Symp. (Internat.) on Combustion*, pp. 1691–1700 The Combustion Institute, Pittsburgh, USA
Hertzberg, M., and Cashdollar, K. L. (1987) Introduction to Dust Explosions. *Industrial Dust Explosions,* ASTM Special Techn. Publ. 958, (eds K. L. Cashdollar and M. Hertzberg), pp. 5–32, ASTM, Philadelphia, USA
Hertzberg, M., and Cashdollar, K. L. (1988) Prevention of Oil Shale Dust Explosions. *Proc. Internat. Conf. 'Oil Shale and Shale Oil', Beijing,* (May) pp. 575–583. Published by Chemical Industry Press, Beijing, P. R. China
Hürlimann, H. (1989) Explosionsunterdrückung von Staubexplosionen. VDI-Berichte 701, pp. 617–657, VDI-Verlag GmbH, Düsseldorf
Hustad, J. E., and Sönju, O. K. (1988) Experimental Studies of Lower Flammability Limits of Gases and Mixtures of Gases at Elevated Temperatures. *Combustion and Flame* **71** pp. 283–294
International Standardization Organization (1985) Explosion Protection systems. Part 1: Determination of Explosion Indices of Combustible Dusts in Air. ISO 6184/1, ISO, Geneva
Jacobson, M., Nagy, J., Cooper, A. R., *et al.* (1961) Explosibility of Agricultural Dusts. Rep. Inv. 5753, US Bureau of Mines, Washington
Jacobson, M., Cooper, A. R., and Nagy, J. (1964) Explosibility of Metal Powders. Rep. Inv. 6516, US Bureau of Mines, Washington
Jaeger, N. (1989) Zündwirksamkeit von Glimmnestern in Staub/Luft-Gemischen. *VDI-Berichte* 701, pp. 263–294. VDI-Verlag GmbH, Düsseldorf

Jeske, A., and Beck, H. (1987) Dokumentation Staubexplosionen. Analyse und Einzelfalldarstellung. Report No. 2/87 Berufsgenossenschaftliches Institut für Arbeitssicherheit, St. Augustin, F. R. Germany

Jeske, A., and Beck, H. (1989) Evaluations of Dust Explosions in the Federal Republic of Germany. *EuropEx Newsletter*, Edition 9, (July), pp. 2–4

Johansen, A. H. (1989) Economical and Safety Aspects of New Dust Suppression System. *Proc. of Grain Dust Suppression Seminar*, Oslo 9–10 Oct. Norwegian Grain Corporation

Johansen, A. H. (1990) Private Communication to R. K. Eckhoff from A. H. Johansen, Norwegian Grain Corporation

Kalkert, N., and Schecker, H.-G (1979) Theoretische Überlegungen zum Einfluss der Teilchengrösse auf die Mindestzündenergie von Stäuben. *Chem.-Ing.-Tech.* **51** pp. 1248–1249

Kameyama, Y., Lai, F. S., Sayama, H., *et al.* (1982) The Risk of Dust Explosions in Grain Processing and Handling Facilities. *J. Agric. Engng. Res.* **27** pp. 253–259

Käppeler, G. (1978) Konstruktive Möglichkeiten zur Druckentlastung von Filterapparaten mit Explosionsklappen. *VDI-Berichte* **304,** pp. 135–138, VDI-Verlag GmbH, Düsseldorf

Karim, G. A., Bardon, M., and Hanafi, A. (1979) Combustion of Oil Sand Fragments in Hot Oxidizing Streams. *Combustion and Flame* **36** pp. 291–303

Kirby, G. N., and Siwek, R. (1986) Preventing Failures of Equipment Subject to Explosions. *Chemical Engineering*, (June) pp. 125–128

Klais, O. (1989) Erfahrungen mit selbstinertisierten Trocknungsanlagen. *VDI-Berichte* **701,** pp. 849–860, VDI-Verlag GmbH, Düsseldorf

Klapp, E. V. (1977) Sicherheitsbegriff, Schädigung und Schadensereignis in Chemieanlagen – und Apparatebau. *Chem.-Ing.-Tech.* **49** pp. 535–541

Kleinschmidt, H.-P. (1983) Funkenlöschanlagen lösen ein brennendes Problem. WLB *Wasser, Luft und Betrieb* No. 11 p. 36

Kordylewski, W., Wach, J., Wójcik, J. (1985) Role of Ducts in Sonic Venting of Explosions. Report from Instytut Techniki Cieplnej i Mechaniki Plynow, Technical University of Wroclaw, Poland

Korol'chenko, A. Ja. (1986) Pozarovzrivoopasnost' prom i slennoi pili. Fire and Explosion Hazards of Industrial Dusts. *Chimija*, Moskva

Körner, H. (1984) Sicherheitsmassnahmen bei der Lagerung und pneumatischen Förderung von Kohlenstaub. *Giesserei* **71** pp. 902–905

Korotov, E. I., and Polferov, K. Ya. (1978) Some Causes of the Ignition of Pulverized Fuel in Boiler Bins and Ways of Preventing it. *Combustion* **50** pp. 27–29

Kossebau, F. (1982) Explosion-Protected Bucket Elevators. *Proc. of Oyez/IBC Symp. on Control and Prevention of Dust Explosions*, (November), Basle, Switzerland

Kuchta, J. M. (1985) *Investigation of Fire and Explosion Accidents in the Chemical, Mining and Fuel-Related Industries – A Manual.* Bulletin 680, US Bureau of Mines, US Dept. Interior. Washington

Kuczynski, R. (1987) Brand- und Explosionsschutzmassnahmen für Baumwollverarbeitungsbetrieve. *Staub-Reinhalt. Luft* **47** pp. 157–160

Kühnen, G. (1978) Möglichkeiten zur Vermeidung von Zündquellen. *VDI-Berichte* **304,** pp. 97–102, VDI-Verlag GmbH, Düsseldorf

Kühnen, G. (1978a) Staubexplosionsschutz bei technischen Arbeitsmitteln, speziell bei Industrie-Staubsaugern. *Die Berufsgenossenschaft* No. 4, April, Erich Schmidt Verlag, Bielefeld, F. R. Germany

Kühnen, G. and Zehr, J. (1980) Schutz von Staubexplosionen – Theorie und Praxis. *Staub-Reinhalt. Luft* **40** pp. 374–379

Laar, G. F. M. van (1981) De ontstekingsmogelijkheid van explosieve stof-lucht-mengels door sigaren, sigaretten en glimnesten. Report PML 1980–118, Prins Maurits Labor. TNO

Laar, G. F. M. van, and Zeeuwen, J. P. (1985) On the Minimum Ignition Energy of Dust-Air Mixtures. *Archivum Combustionis* **5** pp. 145–159

Lai, F. S., Miller, B. S., Martin, C. R., *et al.* (1981) Reducing Grain Dust with Oil Additives.

Transactions of the ASAE **24** pp. 1626–1631

Lai, F. S., Shenoi, S., and Fan, L. T. (1985) Fuzzy Fault-Tree Analysis of Grain Dust Explosions. Part. Multi-Phase Proc., Int. Workshop Fine Particle Soc., USA

Lai, F. S., Martin, C. R., Pomeranz, Y., *et al.* (1986) Oils and Lecithin as Dust Suppression Additives in Commercially Handled Corn, Soybeans and Wheat: Efficacy of Treatments and Effect on Grain Quality. Final Report to American Soybean Association and National Grain and Feed Association, USA

Liebman, I., Conti, R. S., and Cashdollar, K. L. (1977) Dust Cloud Concentration Probe. *Rev. Sci. Instrum.* **48** pp. 1314–1316

Liere, H. (1983) Massnahmen zum Schutz vor Brand und Staubexplosionen bei der industriellen Pulverbeschichtung. *Staub-Reinhalt. Luft* **43** pp. 398–402

Liere, H. (1989) Explosions- und Zündverhalten von Beschichtungspulvern in Beschichtungskabinen. *VDI-Berichte* **701,** pp. 321–350 VDI-Verlag GmbH, Düsseldorf

Lunn, G. A. (1984) *Venting of Gas and Dust Explosions – A Review.* The Institution of Chemical Engineers, UK

Lunn, G. A. (1988) *Guide to Dust Explosion Prevention and Protection. Part 3: Venting of Weak Explosions and the Effect of Vent Ducts.* The Institution of Chemical Engineers, UK

Lunn, G., Crowhurst, D., and Hey, M. (1988) The Effect of Vent Ducts on the Reduced Explosion Pressures of Vented Dust Explosions. *Journ. Loss Prevention in the Process Industries* **1** pp. 182–196

Lüttgens, G., and Glor, M. (1989) *Understanding and Controlling Static Electricity.* Expert Verlag, Ehningen bei Böblingen

Lüttgens, G. (1985) Collection of Accidents Caused by Static Electricity. *Journal of Electrostatics* **16** pp. 247–255

Ma Zhi, Xiao Xingguo, Deng Xufan *et al.* (1987) Oxidation Kinetics of Ca-Si Alloy Powder. *Proc. of Shenyang International Symposium on Dust Explosions*, NEUT, Shenyang, P. R. China. pp. 137–148.

Margraf, D., and Donat, C. (1989) Explosionsfeste Bauweise für den maximalen Explosionsdruck. *VDI-Berichte* **701,** pp. 511–527, VDI-Verlag GmbH, Düsseldorf

May, H. A., Melhorn, L., and Marutzky, R. (1981) Vermeidung der Brand- und Explosionsgefahren und Verminderung der Emissionen bei der Spänetrocknung. *Staub-Reinhalt. Luft* **41** pp. 416–420

Mayerhauser, D. (1978) Mahlen und Sichten von explosionsfähigen Stäuben. *VDI-Berichte* **304,** pp. 169–172, VDI-Verlag GmbH, Düsseldorf

Meek, R. L., and Dallavalle, J. M. (1954) Explosive Properties of Sugar Dusts. *Ind. & Eng. Chem.* **46** pp. 763–766

Michelis, J. (1984) Massnahmen zur Bekämpfung von Explosionen im Steinkohlenbergbau unter Tage. *VDI-Berichte* **494,** pp. 259–264, VDI-Verlag GmbH, Düsseldorf

Midttveit, Ø. (1988) Measurement of Phase Velocities and Concentrations in Flowing Gas/Solids Suspensions. Report No. 88/03880–1, Chr. Michelsen Institute, Bergen, Norway

Miron, Y., and Lazzara, C. P. (1985) Fire Hazards of Oil Shale Dust Layers on Hot Surfaces. *Proc. 18th Oil Shale Symp.*, (August), pp. 83–100. Colorado School of Mines Press

Moore, P. E. (1983) Graviner and Deugra Gas Explosion Suppression Systems. *Proceedings of an Informal Seminar on Suppression of Gas Explosions on Offshore Gas and Oil Installations.* Report No. 833402–2, (April), Chr. Michelsen Institute, Bergen, Norway, pp. 31–69

Moore, P. E., Watkins, G. K. P., and Vellenoweth, A. C. (1984) Explosion Suppression – Its Effectiveness and Limits of Applicability. *VDI-Berichte* **494,** pp. 247–257, VDI-Verlag GmbH, Düsseldorf

Moore, P. E. (1987) Suppression of Maize Dust Explosions. In *Industrial Dust Explosions*, ASTM Special Techn. Publ. 958, (eds K. L. Cashdollar and M. Hertzberg), pp. 281–293, ASTM, Philadelphia, USA

Moore, P. E., and Bartknecht, W. (1987) Extending the Limits of Explosion Suppression Systems.

Staub-Reinhalt. Luft **47** pp. 209–213
Moore, P. E., and Cooke, P. L. (1988) Suppression of Metal Dust Explosions. Report No. 88/49, (November) British Materials Handling Board
Moore, P. E. (1989) Industrial Explosion Protection – Venting or Suppression? *Proc. of I. Chem. E. Symp.* Series No. 115 pp. 257–279
Morozzo, Count (1795) Account of a Violent Explosion which Happened in a Flour-Warehouse, at Turin, December the 14th, 1785, to which are Added some Observations on Spontaneous Inflammations. *The Repertory of Arts and Manufactures* **2** pp. 416–432
Müller, R. (1989) Zündfähigkeit von mechanisch erzeugten Funken und heissen Oberflächen in Staub/Luft-Gemischen. *VDI-Berichte* **701,** pp. 421–466, VDI-Verlag GmbH, Düsseldorf
Mullinger, P. J. (1987) Fire and Explosion Protection for Pulverized Firing Systems. *Proc. of 1st Europ. Dry Fine Coal Conf.*, Harrogate, UK pp. III/1/-1 to 18
Nagy, J., and Portman, W. M. (1961) Explosibility of Coal Dust in an Atmosphere Containing a Low Percentage of Methane. Rep. Inv. 5815 US Bureau of Mines, Washington
Nagy, J., Dorsett, H. G., and Cooper, A. R. (1965) Explosibility of Carbonaceous Dusts. Rep. Inv. 6597, US Bureau of Mines, Washington
Nagy, J., and Verakis, H. C. (1983) *Development and Control of Dust Explosions,* Marcel Dekker, Inc., New York
National Materials Advisory Board (1978) Publications NMAB-352–1 and NMAB-352–2 *International Symposium on Grain Elevator Explosions*, National Academy of Sciences, Washington DC.
National Materials Advisory Board (1982) *Prevention of Grain Elevator and Mill Explosions* Publication NMAB-367–2, National Academy Press, Washington DC.
Nedin, V. V. (1971) *Preduprezdenie vnezapnich vzrivov gazodispnich sistem.* Naukova Pumka, Kiev, USSR
Nedin, V. V., Nejkov, O. D., Alekseev, A. G., *et al.* (1971) Explosibility of Metal Powders. Naukova Pumka Kiev, (in Russian)
Nettleton, M. A. (1986) The Effects of the Properties of Ignition Source and Particles on the Ignition of Coal Dust Flames. *Archivum Combustionis* **6** pp. 125–138
NFPA (1957) *Report of Important Dust Explosions.* National Fire Protection Association, USA
NFPA (1987) *Manufacture of Aluminium and Magnesium Powder.* (1987 Edition). National Fire Protection Association, Quincy, MA 02269, USA
Nindelt, G., Lukas, W., and Junghans, R. (1981) Untersuchungsergebnisse zur Explosionsneigung hybrider Gemische beim Elektroabscheiderbetrieb. *Staub-Reinhalt. Luft* **41** pp. 184–189
Noha, K. (1989) Auswahlkriterien für Explosionsschutzmassnahmen. *VDI-Berichte* **701,** pp. 681–693. VDI-Verlag GmbH, Düsseldorf
North East University of Technology (1987) *Proceedings Shenyang International Symposium on Dust Explosions*, (September) NEUT, Shenyang, P. R. China
Palmer, K. N. (1973) *Dust Explosions and Fires.* (New edition in preparation.) Chapman and Hall, London
Palmer, K. N. (1973a) Dust Explosion Hazards in Pneumatic Transport. Fire Research Note No. 992, Fire Research Station, Borehamwood, Herts, UK
Palmer, K. N., and Tonkin, P. S. (1973) Use of Inert Gas to Prevent Dust Explosions. *Proceedings of European Symposium*, (Ed. F. J. Weinberg), Combustion Institute, London
Palmer, K. N. (1975) Explosions in Dust Collection Plant. *The Chemical Engineer* pp. 136–142
Parpart, J. (1979) Entstaubung von Kohlenstaubanlagen. *Zement-Kalk-Gips* **32** pp. 265–269
Pasman, H. J., and Wingerden, C. J. M. van (1988) Explosion Resistance of Process Equipment. *Proc. of Conference on Flammable Dust Explosions*, (November) St. Louis, USA
Patzke, J. (1984) Explosionsdruckentlastung von Kohlenmahltrocknungsanlagen. VDI-Berichte 494, pp. 271–276, VDI-Verlag GmbH, Düsseldorf
Pedersen, G. H., and Eckhoff, R. K. (1987) Initiation of Grain Dust Explosions by Heat Generated During Single Impact Between solid Bodies. *Fire Safety Journal* **12** pp. 153–164
Pedersen, G. H., and Wilkins, B. A. (1988) Explosibility of Coal Dusts and Coal Dust/Limestone

Mixtures at Elevated Initial Pressures. Report No. 88/02101, Chr. Michelsen Institute, Bergen, Norway

Pellmont, G. (1979) Explosions- und Zündverhalten von hybriden Gemischen aus brennbaren Stäuben und Brenngasen. Doctorate Thesis No. 6498, ETH, Zürich

Perry, R. H., and Green, D. (1984) *Perry's Chemical Engineers' Handbook.* (6th Edition), McGraw-Hill, USA

Pfeiffer, W., Kühnen, G., and Armbruster, L. (1985) Feinheit von Holzstäuben. *Staub-Reinhalt. Luft* **45** pp. 515–518

Pineau, J. P. (1984) Dust Explosions in Vessels Connected to Ducts. *VDI-Berichte* **494,** pp. 67–80. VDI-Verlag GmbH, Düsseldorf

Pineau, J. P. (1984a) Sécurité incendie et explosion des installations de fabrication de poudre de lait. *Annales des Mines de Belgique* **7–8** pp. 302–318

Pineau, J. P. (1985) Sécurité incendie et explosion des installations de fabrication de poudre de lait. *Proc. of Internat. Symp. Control of Risks in Handling and Storage of Granular Foods*, (April) APRIA, Paris

Pinkwasser, Th. (1985) On the Extinction of Smouldering Fires in Pneumatic Conveyors. *Proc. of Internat. Symp. Control of Risks in Handling and Storage of Granular Foods*, (April) APRIA, Paris

Pinkwasser, Th. (1986) On the Ignition Capacity of Free-Falling Smouldering Fires. *Euromech Colloquium 208, Explosions in Industry*, (April)

Pinkwasser, Th., and Häberli, P. (1987) Explosion Pressure Relief in Large-Volume Storage Bins. *Bulk Solids Handling* **7** pp. 83–85

Pinkwasser, Th. (1989) Private Communication to R. K. Eckhoff from Th. Pinkwasser, Bühler, Switzerland

Piotrowski, T., Mrzewinski, T., and Proskurmicka, H. (1988) New Classification System of Dust Hazards in Industrial Technological Processes. Research Report from Institute of Organic Industry, Warsaw, Poland

Polikarpov, A. D. (1984) Formation of an Air Shock Wave During Explosion of Sulphide Dust. (Translation from Fiziko-Tekhnicheskie Probl. Raz. Pol. Isk.) Plenum Publishing Corp. pp. 212–216

Pritchard, D. K. (1989) A Review of Methods for Predicting Blast Damage from Vapour Cloud Explosions. *Journ. Loss Prevention Process Ind.* **2** pp. 187–193

Proctor, A. (1988) Tailored Training. *Process Engineering*, (October), pp. 41–45

Proust, Ch., and Pineau, J. P. (1989) Dust Explosions: Risk Assessment. *Proc. of Internat. Symp., Dust Explosion Protection*, 11–13 September, EuropEx, Antwerp

Qian Qiyong, Wang Taisheng and Xiao Hechai (1987) The Explosion Prevention of Ferromanganese Powder in the 1.2 m^3 Cyclone Separator. *Proc. of Shenyang International Symposium on Dust Explosions*, NEUT, Shenyang, P. R. China, pp. 500–505

Radandt, S. (1969) Betriebs- und Unfallsicherheit pneumatischer Mischsilos und Mischbehälter. *Die Mühle + Mischfuttertechnik* **106** pp. 851–852

Radandt, S. (1987) Prevention and Control of Dust Explosions in the Grain, Feed and Flour Industry in the Federal Republic of Germany. Influence of Results from Recent Research. *Proc. of Shenyang International Symposium on Dust Explosions*, NEUT, Shenyang, P. R. China. pp. 450–468

Rahimian, S., Choi, T., and Essenhigh, R. H. (1982) Extinction of Coal Dust Explosion Flames by Additives, Powder and Bulk Solids Handling and Processing, Cahners Exposition Group, pp. 126–129

Rajendran, N., and Stockham, J. D. (1985) Grain Dust Measurement Techniques: An Evaluation. *Transactions of the American Society of Agricultural Engineers*, **28** pp. 2030–2036

Reeh, D. (1978) Das Explosionsverhalten von Staub/Gas/Luft-Gemischen (Hybride Gemische). *VDI-Berichte* 304, pp. 73–79. VDI-Verlag GmbH, Düsseldorf

Reeh, D. (1979) Das Explosionsverhalten von Staub/Gas/Luft-Gemischen (Hybride Gemische).

Erdöl und Kohle-Erdgas-Petrochemie vereinigt mit Brennstoff-Chemie **32** p. 38
Reinke, W. (1987) Safety Measures in Aluminium Powder Production. Aluminium Powder Safety Workshop, (June), Pocono Manor, PA, USA
Richmond, J. K., and Beitel, F. P. (1984) Dust Explosion Hazards Due to Blasting of Oil Shale, *Proc. 17. Oil Shale Symp.* (August), Colorado School of Mines Press
Rickenbach, H. (1983) Explosion Barriers in Conveyor Installations for Dust-Air Mixtures. *Swiss Chem.* **5** No. 9a
Ritter, K. (1989) Beispiele des Anlagenschutzes mit Kostenbetrachtungen. *VDI-Berichte* **304,** pp. 157–168, VDI-Verlag GmbH, Düsseldorf
Ritter, K. (1984) Mechanisch erzeugte Funken als Zündquellen. *VDI-Berichte* **494,** pp. 129–144. VDI-Verlag GmbH, Düsseldorf
Ruttmann, G. (1989) Inertisierung einer Mahl- und Mischanlage. *VDI-Berichte* **701,** pp. 861–874. VDI-Verlag GmbH, Düsseldorf
Ruygrok, J. P. J., Laar, G. F. M. van, and Zeeuwen, J. P. (1983) Coal Explosion Hazards. *Bulk Systems International* pp. 52–55
Ryzhik, A. B., and Makhin, V. S. (1978) Ignition of Methane/Air Suspensions of Coal Dust. *Combustion, Explosion and Shock Waves* **14** pp. 517–519
Sapko, M. J., Weiss, E. S., and Cashdollar, K. L. (1986) Methane Released During Blasting at the White River Shale Project. *Proc. 19. Oil Shale Symp.*, (August), Colorado School of Mines Press pp. 59–68
Schäfer, H. K. (1978) Die Vermeidung von Zündquellen für Staubexplosionen. *VDI-Berichte* **304,** pp. 103–106. VDI-Verlag GmbH, Düsseldorf
Schenk, E. (1984) Explosions- und Zündverhalten von Flockmaterial. *VDI-Berichte* **494,** pp. 53–57. VDI-Verlag GmbH, Düsseldorf
Scherrer, E. (1984) Inertisierungsmassnahmen an Braunkohlenstaub-Silos. *VDI-Berichte* **494,** pp. 99–103. VDI-Verlag GmbH, Düsseldorf
Schlieper, H. (1984) Transport von Kohlenstaub mit Silofahrzeugen. *VDI-Berichte* **494,** pp. 265–270. VDI-Verlag GmbH, Düsseldorf
Schmid, W. (1972) Brand-, Explosions- und Umweltschutz bei der Spanplattenfertigung. *Arbeitsschutz* No. **12** pp. 409–422
Schneider, D. (1984) Erfahrungen mit Staubunterdrückungssystemen in Mahlanlagen. *VDI-Berichte* **494,** pp. 293–307. VDI-Verlag GmbH, Düsseldorf
Schoeff, R. W. (1989) News Release and Summary for 1988. Dust Explosion Statistics – USA. *EuropEx Newsletter*, Edition 9, July, pp. 8–9
Schofield, C. (1984) *Guide to Dust Explosion Prevention and Protection. Part 1: Venting*, The Institution of Chemical Engineers, UK
Schofield, C., and Abbott, J. A. (1988) *Guide to Dust Explosion Prevention and Protection. Part 2: Ignition, Prevention, Containment, Inerting, Suppression and Isolation*, The Institution of Chemical Engineers, UK
Scholl, E. W. (1973) Explosionsversuche mit Zuckerstaub in Entstaubungsanlagen einer stillgelegten Zuckerfabrik. Forschungsbericht Nr. 109, Bergewerkschaftliche Versuchsstrecke, Dortmund
Scholl, E. W. (1975) Brand- und Explosionsgefahren an fahrbaren Einzel-Absaug-geräten für Hobelspäne und Holzstaub. Die Berufsgenossenschaft, (April), pp. 129–135
Scholl, E. W. (1978) Explosionsunterdrückung. *VDI-Berichte* **304,** pp. 151–156. VDI-Verlag GmbH, Düsseldorf
Scholl, E. W., Fischer, P., and Donat, C. (1979) Vorbeugende und konstruktive Schutzmassnahmen gegen Gas- und Staubexplosionen. *Chem.-Ing.-Tech.* **51** pp. 8–14
Scholl, E. W. (1981) Brenn- und Explosionsverhalten von Kohlenstaub. *Zement-Kalk-Gips* **34** pp. 227–233
Scholl, E. W. (1989) Vorbeugender Explosionsschutz durch Vermeiden von wirksamen Zündquellen. *VDI-Berichte* **701,** pp. 477–489. VDI-Verlag GmbH, Düsseldorf
Schröder, H. (1984) Funkenerkennung und-Løschung in Anlagen der Holzindustrie. *VDI-Berichte*

494, pp. 239–245. VDI-Verlag GmbH, Düsseldorf

Schuber, G. (1989) Ignition Breakthrough Behaviour of Dust/Air and Hybrid Mixtures Through Narrow Gaps, *Proceedings of 6th Internat. Symp. Loss Prevention and Safety Promotion in the Process Industries*. (June), Norwegian Society of Chartered Engineers, Oslo, Norway

Senecal, J. A. (1989) Deflagration Suppression of High K_{St} Dusts. *Plant/Operations Progress* **8** pp. 147–151

Singh, J. (1979) Suppression of Internal Explosions in Process Plant. *Journ. Occup. Accidents* **2** pp. 113–123

Siwek, R. (1986) Preventive Protection Against Explosions and Constructive Explosion Protection to Prevent their Effects, Euromech. Colloquium 208, Explosions in Industry, (April), Göttingen, F. R. Germany

Siwek, R. (1987) Explosionsschutz in Apparaturen der chemisch-pharmazeutischen Industrie. *Pharm. Ind.* **49** pp. 1165–1175

Siwek, R. (1989) New Knowledge about Rotary Air Locks in Preventing Dust Ignition Breakthrough. *Plant/Operations Progress*, AIChE **8** pp. 165–176

Siwek, R. (1989a) Explosion Suppression in Large Volumes. Dust Explosion Protection. *EuropEx Symposium*, (September), Antwerp, Belgium

Siwek, E., and Skov, O. (1989) Modellberechnung zur Dimensionierung von Explosionsklappen auf der Basis von praxisnahen Explosionsversuchen. *VDI-Berichte* **701,** pp. 569–616. VDI-Verlag GmbH, Düsseldorf

Solymos, L. (1985) Dust Explosion Prevention in Grain Elevators by Cutting the Indoor Dust Quantity. *Proc. of Internat. Symp. Control of Risks in Handling and Storage of Granular Foods*, (April), APRIA, Paris

Sorgdrager, W. (1985) Risk Analysis of Grain Dust Explosions, *Proc. of Part. Multi-Phase Proc., Int. Workshop*. Fine Particle Soc.

Stockham, J. D., and Rajendran, N. (1984) Establishing a Reliable Grain Dust Measurement Technique for the Bucket Elevator. Fire and Explosion Research Report DCE–84–080 National Grain and Feed Association, USA,

Strehlow, R. A. (1980) Accidental Explosions. *American Scientist* **68** pp. 420–428

Sweis, F. K., and Sinclair, C. G. (1985) The Effect of Particle Size on the Maximum Permissible Oxygen Concentration to Prevent Dust Explosions. *Journal of Hazardous Materials* **10** pp. 59–71

Swift, I. (1987) Protection Methods Against Dust Explosions. *Powder and Bulk Engineering*, (September), pp. 22–29

Swift, I. (1987a) Explosion Protection Methods Against Dust Explosions. *Proc. of Powder and Bulk Solids Conference*, (May), Chicago

Szkred, T. (1983) Inertisierung von Staub/Luft-Gemischen mit Inhibitoren. *Staub-Reinhalt. Luft* **43** pp. 392–397

Tesler, L. A., and Semyonov, L. I. (1988) Peculiarities of Initiation and Propagation of Explosions in Grain Silos, Flour Mills, Feed Mills, and Related Security Problems. *Archivum Combustionis* **8** pp. 33–41

Tesler, L. A., and Semyonov, L. I. (1989) New Solutions to Help Reduce the Explosion Hazard in Grain Elevators. Paper presented at 12th ICDERS, (23–28 July), University of Michigan, USA

Thatcher, J. J. (1982) Coal Dust Ignitions and Spontaneous Combustion Fires: How to Minimize them. Powder and Bulk Storage Handling and Processing, Int. Progr., Cahners Exposition Group pp. 123–125

Thelning, L., and Laufke, H. (1970) Brand- och explosionsrisker vid board-tilverkning enligt torra metoden. Report from Svenska Brandförsvarsföreningen, (September) Stockholm, Sweden

Thorpe, D. G. L., Sampuran Singh, Cartwright, P., *et al.* (1985) Electrostatic Hazards in Sugar Dust in Storage Silos. *Journal of Electrostatics* **16** pp. 193–207

TNO (1979) Onderzoek naar stofexplosie-eigenschappen van een aantal stoffen dat voorkomt in bed ijren die bloem, mangvoerders, zetmeel en zuivelprodukten verwerken en opslaan. TNO Report G6175, TNO, The Netherlands

Torrent, J. G., Armada, I. S., and Pedreira, R. A. (1988) A Correlation Between Composition and Explosibility Index for Coal Dust. *Fuel* **67** pp. 1629–1632
Torrent, J. G., and Fuchs, J. C. (1989) Flammability and Explosion Propagation of Methane/Coal Dust Hybrid Mixtures. *Proc. of 23rd Int. Conf. Saf. Min. Res. Inst.*, (September), Washington DC
Verein deutscher Ingenieure (1957) Brennbarer Industriestäube, Forschungsergebnisse. VDI-Berichte 19. VDI-Verlag GmbH, Düsseldorf
Verein deutscher Ingenieure (1978) Sichere Handhabung brennbarer Stäube. *VDI-Berichte* **304.** VDI-Verlag GmbH, Düsseldorf
Verein deutscher Ingenieure (1984) Sichere Handhabung brennbarer Stäube. *VDI-Berichte* **494.** VDI-Verlag GmbH, Düsseldorf
Verein deutscher Ingenieure (1986) *Staubbrände und Staubexplosionen.* VDI-Richtlinie 2263. VDI-Verlag GmbH, Düsseldorf
Verein deutscher Ingenieure: (1989) Sichere Handhabung brennbarer Stäube. *VDI-Berichte* **701.** VDI-Verlag GmbH, Düsseldorf
Verein deutscher Ingenieure (1991) *Druckentlastung von Staubexplosionen.* VDI-Richtlinie 3673, (Draft). VDI-Verlag GmbH, Düsseldorf
Verhaegen, H. (1989) Safety, a Management Task. *Proceedings of 6th Internat. Symp. Loss Prevention and Safety Promotion in the Process Industries.* (June), Norwegian Society of Chartered Engineers, Oslo, Norway
Walker, W. J. (1982) Venting: A Gap in Safety Practice. *Process Engineering*, (December), pp. 35–37
Wall, T. F. Phong-Anant, D., Gururajan, V. S., *et al.* (1988) Indicators of Ignition for Clouds of Pulverized Coal. *Combustion and Flame* **72** pp. 111–118
Walter, W. (1978) Explosionssperren für staubführende Rohrleitungen. *VDI-Berichte* **304,** pp. 173–178. VDI-Verlag GmbH, Düsseldorf
Walther, C.-D., and Schacke, H. (1986) Evaluation of Dust Explosion Characteristics at Reduced and Elevated Initial Pressures. (Poster summary), Bayer AG, Leverkusen, F. R. Germany
Walther, C.-D. (1989) Einfluss der Sauerstoffkonzentration auf Staubexplosionen und Staubbrände. *VDI-Berichte* 701, pp. 195–214, VDI-Verlag GmbH, Düsseldorf
Weber, R. (1878) Preisgekrönte Abhandlung über die Ursachen von Explosionen und Bränden in Mühlen, sowie über die Sicherheitsmassregeln zur Verhütung derselben. *Verh. Ver. Gew. Fliess.*, Berlin pp. 83–103
Weckman, H., Hyvärinen, P., Olin, J., *et al.* (1981) Reduction of Fire and Explosion Hazards at Peat Handling Plants. Research Report 2/1981 Technical Research Centre of Finland, Espoo
Wehland, P. (1984) Inertisierung von Kohlenmahltrocknungsanlagen. *VDI-Berichte* **494,** pp. 277–280. VDI-Verlag GmbH, Düsseldorf
Weiss, E. S., Cashdollar, K. L., and Sapko, M. J. (1985) Dust and Pressure Generated During Commercial Oil Shale Mine Blasting. Part I., *Proc. 18th Oil Shale Symp.*, (August), Colorado School of Mines Press pp. 68–76
Weiss, E. S., Cashdollar, K. L., and Sapko, M. J. (1986) Dust and Pressure Generated During Oil Shale Mine Blasting. Part II., *Proc. 19th Oil Shale Symp.*, (August), Colorado School of Mines Press pp. 47–58
Wheeler, R. V. (1935) *Report on Experiments into the Means of Preventing the Spread of Explosions of Carbonaceous Dust.* His Majesty's Stationery Office, London
Wibbelhoff, H. (1981) Derzeitige sicherheitstechnische Anforderungen an Kohlen-Mahl-Trocknungs-Anlagen. *Steine und Erden* pp. 61–66
Wibbelhoff, H. (1984) Fahrbare Gross-Staubsauger in explosionsgeschützter Ausführung. *Steine und Erden* pp. 168–169
Wiemann, W. (1984) Einfluss der Temperatur auf Explosionskenngrössen und Sauerstoffgrenzkonzentrationen. *VDI-Berichte* **494** pp. 89–97. VDI-Verlag GmbH, Düsseldorf
Wiemann, W., and Scholl, E.-W. (1985) Selbstentzündung von Braunkohlenstaub bei Verminder-

ten Sauerstoffgehalt. *Staub-Reinhalt. Luft* **45** pp. 147–150
Wiemann, W. (1987) Influence of Temperature and Pressure on the Explosion Characteristics of Dust/Air and Dust/Air/Inert Gas Mixtures. In *Industrial Dust Explosions*, ASTM Special Techn. Publ. 958, (eds. K. L. Cashdollar and M. Hertzberg) pp. 202–216. ASTM, Philadelphia, USA
Wiemann, W. (1989) Vermeidung von Staubexplosionen durch Inertisieren. *VDI-Berichte* **701,** pp. 491–510. VDI-Verlag GmbH, Düsseldorf
Wiemann, W. Bauer, R., and Möller, F. (1989) Unterdruck-Sicherung von Silos nach Staubexplosionen bei Anwendung von Explosionsklappen. *VDI-Berichte* **701,** pp. 775–800. VDI-Verlag GmbH, Düsseldorf
Wilcoxen, J. B. (1981) Design for Grain Dust Explosion Works! Unpublished manuscript, Borton Inc., Hutchinson, Kansas, USA
Winter, J. (1970) Etude d'un arrêtbarrage déclenché. *Revue de l'Industrie Minerale* **52**, pp. 549–558
Woodcock, C. R., and Reed, A. R. (1983) Designing to Minimize the Explosion Hazard in Bulk Solids Handling Systems. Powder and Bulk Solids Handling and Processing. Tech. Progr. Ind. Powder Inst.
Woskoboenko, F. (1988) Explosibility of Victorian Brown Coal Dust. *Fuel* **67** pp. 1062–1068
Xiao Xingguo, Ma Zhi, Wang Wenxiu, *et al.* (1987) Oxidation and Combustion of Ca-Si Alloy Dust Clouds and its Mathematical Model. *Proc. of Shenyang International Symposium on Dust Explosions*, NEUT, Shenyang, P. R. China, pp. 149–160
Yong Fan Yu (1985) On the Electrostatic Charging of some Finely Divided Materials in Modern Agricultural Pneumatic Transport Systems. *Journal of Electrostatics* **16** pp. 209–217
Zabetakis, M. G. (1965) *Flammability Characteristics of Combustible Gases and Vapours*. Bulletin 627, US Bureau of Mines, Washington
Zeeuwen, J. P., and Laar, G. F. M. van, (1985) Ignition Sensitivity of Flammable Dust/Air Mixtures. *Proc. of Internat. Symp. Control of Risks in Handling and Storage of Granular Foods*, (April), APRIA, Paris
Zehr, J. (1957) Anleitung zu den Berechnungen über die Zündgrenzwerte und die maximalen Explosionsdrücke. *VDI-Berichte* **19** pp. 63–68
Zehr, J. (1965) Eigenschaften brennbarer Stäube und Nebel in Luft. *Handbuch der Raumexplosionen*, (ed. by H. H. Freytag), pp. 164–186, Verlag Chemie GmbH, Weinheim, F. R. Germany
Zhang Fenfen and Zhang Chunxiao (1987) Grain Dust Explosion and its Prevention in China Grain Silos. *Proc. of Shenyang International Symposium on Dust Explosions*, NEUT, Shenyang, P. R. China pp. 557–567
Zockoll, C., and Nobis, P. (1981) Untersuchungen zur Siloinertizierung im Schadensfall. ASI-Informationen 8.41, Berufsgenossenschaft Nahrungsmittel und Gaststätten, Mannheim
Zockoll, C. (1989) Zündwirksamkeit von Glimmnestern in Staub/Luft-Gemischen. VDI-Berichte 701, pp. 295–301. VDI-Verlag GmbH, Düsseldorf
Zwahlen, G. (1989) Deflagrationsfähigkeit – ein gefährliches Zersetzungsverhalten abgelagerter Stäube. *VDI-Berichte* **701,** pp. 167–186. VDI-Verlag GmbH, Düsseldorf

Chapter 2

Case histories

2.1 INTRODUCTION

Experience has shown that 'learning by doing' is an effective way of acquiring new knowledge. Unfortunately this also applies to learning about dust explosions. Those who have experienced a dust explosion in their own plant, whether workers or management, have a much more profound appreciation of the realism of this hazard than those who have only heard or read about dust explosions in general terms. Real understanding in turn produces the proper motivation for minimizing the probability of occurrence of such events in the future.

Clearly, accidental dust explosions are highly undesirable in any plant, and one therefore seeks less dramatic means of transferring knowledge and motivation. One way is the use of case histories, i.e. fairly detailed accounts of dust explosions that have actually occurred elsewhere.

The number of well documented dust explosions worldwide is considerable and only a small fraction can be covered in the present text. Because of his close cooperation with Norwegian industry in investigating accidental dust explosions for nearly 20 years, the author has had access to detailed information on many explosions that have occurred in Norway through the years. It was natural, therefore, to include some of this information in the present book.

On the other hand, it was considered appropriate also to include accidents in other countries than in Norway. However, some well-known explosions that have been described extensively elsewhere in the open literature, have not been included. This for example applies to the catastrophic wheat flour explosion in the Roland Mill in Bremen, F. R. Germany, which has been discussed in detail by the Fire and Police Authorities of Bremen (1979). Also, many of the large dust explosions in USA after 1975 have been discussed in detail by Kauffman (1982, 1987) and Kauffman and Hubbard (1984). A few of these will nevertheless be included in the present account.

2.2 THE EXPLOSION IN A FLOUR WAREHOUSE IN TURIN ON 14TH DECEMBER, 1785

This is probably the most frequently quoted of all dust explosions that have occurred. However, only very rarely are details of Count Morozzo's (1795) fascinating account mentioned. It is therefore considered appropriate to start this sequence of case histories with the full original account of the wheat flour explosion in Mr. Giacomelli's bakery in Turin. The explosion was a comparatively minor one, but there is still much to learn from Count Morozzo's analysis. The considerations related to the low moisture content of the flour due to dry weather are important and still relevant. The same applies to the primary explosion causing a secondary explosion by entrainment of dust deposits.

LIV. Account of a violent Explosion which happened in a Flour-Warehouse, at Turin, December the 14th, 1785; to which are added some Observations on spontaneous Inflammations; by Count Morozzo.

From the Memoirs of the Academy of Science of Turin.

The Academy having expressed a desire to have a particular account of the explosion which I mentioned to them a few days after it happened, I have made all possible haste to fulfil their desires, by ascertaining, with the utmost attention, all the circumstances of the fact, so as to be able to relate it with the greatest exactness.

I shall take the liberty to add to it a short account of several spontaneous inflammations, which have happened to different substances, and which have been the cause of very great misfortunes. Although the greater number of these phenomena is already well known to philosophers, I trust the collecting them together in this place will not be displeasing, as it is impossible to render too well known facts which so strongly interest the public utility.

On the 14th of December, 1785, about six o'clock in the evening, there took place in the house of Mr. Giacomelli, baker in this city, an explosion which threw down the windows and window-frames of his shop, which looked into the street; the noise was as loud as that of a large cracker, and was heard at a considerable distance. At the moment of the explosion, a very bright flame, which lasted only a few seconds, was seen in the shop; and it was immediately observed, that the inflammation proceeded from the flour-warehouse, which was situated over the back shop, and where a boy was employed in stirring some flour by the light of a lamp. The boy had his face and arms scorched by the explosion; his hair was burnt, and it was more than a fortnight before his burns were healed. He was not the only victim of this event; another boy, who happened to be upon a scaffold, in a little room on the other side of the warehouse, seeing the flame, which had made its passage that way, and thinking the house was on fire, jumped down from the scaffold, and broke his leg.

In order to ascertain in what manner this event took place, I examined, very narrowly, the warehouse and its appendages; and, from that examination, and from the accounts of the witnesses, I have endeavoured to collect all the circumstances of the event, which I shall now describe.

The flour-warehouse, which is situated above the back shop, is six feet high, six feet wide, and about eight feet long. It is divided into two parts, by a wall; and arched ceiling extends over both, but the pavement of one part is raised about two feet higher than that of the other. In the middle of the wall is an opening of communication, two feet and a half wide, and three feet high; through it the flour is conveyed from the upper chamber into the lower one.

The boy, who was employed, in the lower chamber, in collecting flour to supply the bolter below, dug about the sides of the opening, in order to make the flour fall from the upper chamber into that in which he was; and, as he was digging, rather deeply, a sudden fall of a great quantity took place, followed by a thick cloud, which immediately caught fire, from the lamp hanging to the wall, and caused the violent explosion here treated of.

The flame shewed itself in two directions; it penetrated, by a little opening, from the upper chamber of the warehouse, into a very small room above it, where, the door and windowframes being well closed and very strong, it produced no explosion; here the poor boy, already mentioned, broke his leg. The greatest inflammation, on the contrary, took place in the smaller chamber, and, taking the direction of a small staircase, which leads into the back shop, caused a violent explosion, which threw down the frames of the windows which looked into the street. The baker himself, who happened then to be in his shop, saw the room all on fire some moments before he felt the shock of the explosion.

The warehouse, at the time of the accident, contained about three hundred sacks of flour. Suspecting that this flour might have been laid up in the warehouse in a damp state, I thought it right to enquire into that circumstance. I found, upon examination, that it was perfectly dry; there was no appearance of fermentation in it, nor was there any sensible heat.

The baker told me that he had never had flour so dry as in that year (1785), during which the weather had been remarkably dry, there having been no rain in Piedmont for the space of five or six months: indeed, he attributed the accident which had happened in his warehouse to the extraordinary dryness of the corn.

The phænomenon, however striking at the time it happened, was not entirely new to the baker, who told me that he had, when he was a boy, witnessed a similar inflammation; it took place in a flour-warehouse, where they were pouring flour through a long wooden trough, into a bolter, while there was a light on one side; but, in this case, the inflammation was not followed by an explosion.

He mentioned to me several other instances, which I thought it my duty to enquiry into; amongst them, one which had happened to the widow Ricciardi, baker in this city, where (there being, on the other side of the wall of the flour-warehouse, a lock-smith's forge) the flour was heated to such a degree, that a boy who went into the warehouse could not remain there, so much were his feet scorched by the heat; this flour was of a dark brown colour, and whilst the people were examining it, sparks began to appear, and fire spread itself around, without producing any flame, like a true *pyrophorus** .

He also informed me, that an inflammation like that above-mentioned had happened at the house of a baker in this city, called Joseph Lambert; it was occasioned by shaking some large sacks, which had been filled with flour, near a lighted lamp, but the flame, though pretty brisk, did not do any mischief.

According to the foregoing accounts, it appears to me, that it is not difficult to explain the phænomenon in question. The following is the idea I have conceived of it: as the flour fell down, a great quantity of inflammable air, which had been confined in its interstices, was set free; this, rising up, was inflamed by the contact of the light; and, mixing immediately with a sufficient quantity of atmospheric air, the explosion took place on that side where there was the least resistance. As to the burning of the hair, and the skin, of the boy who was in the warehouse, the cause of it must be attributed to the fire of the fine particles of the flour, which, floating in the atmosphere, were kindled by the inflammable air, in the same manner as the powder from the stamina of certain vegetables, (particularly of the pine, and of some mosses,) when thrown in the air, takes fire if any light is applied to it.

But it may be objected, that as the flour was not at all damp, and had not any sensible degree of heat, there should not be any fermentation in it, and consequently no inflammable air should be produced: to this I answer,

First. That flour is never entirely free from humidity, as is evidently shewn by distillation.

Secondly. That although the degree of heat was not so great as to set free inflammable air by fermentation, a sufficient quantity was set free, by what may be called a mechanical mean, to inflame upon the contact of light; and to disengage, at the same time, all that which communicated with the atmospheric air.

Thirdly. We must recollect that flour also furnishes alkaline inflammable air, which is produced from the glutinous vegeto-animal part of the corn; and we know that this kind of inflammable air is of a very active nature.

After having described this singular event, I shall beg leave to collect together, in this place, all the known facts respecting spontaneous inflammations produced by different substances. A circumstantial account of these phænomena cannot but be very interesting to those concerned in government; not only as it may tend to prevent the unhappy accidents which result from them, but also as it may sometimes hinder the suspicion and persecution of innocent persons, on account of events which are produced merely by natural causes.

*I was very anxious to ascertain by experiments, whether it were possible to bring flour alone into the state of pyrophorus, but it was in vain; for though I calcined flour with a strong heat, in a small retort, with the same precautions as used in making other pyrophori, I never could succeed in making it take fire by exposure to the air. By joining alum with it I obtained a true pyrophorus, as Lemery had already done.

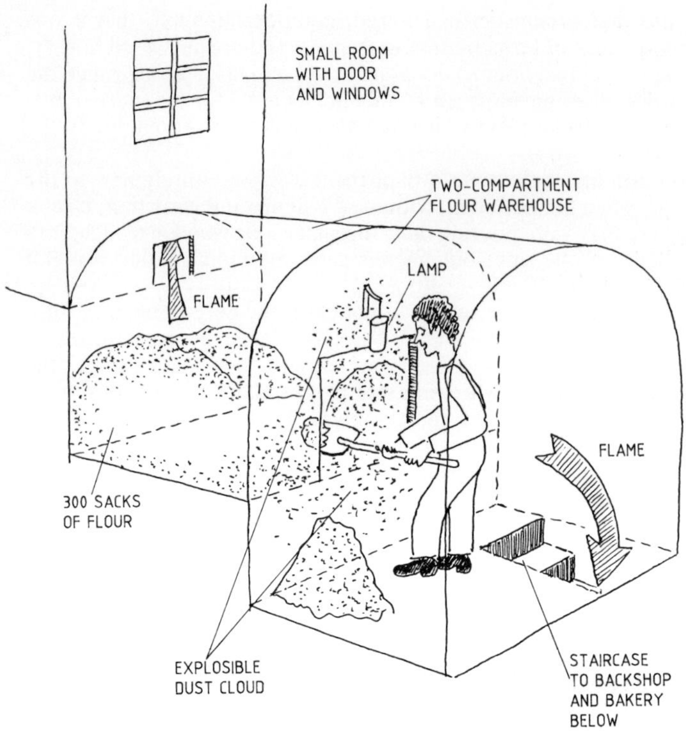

Figure 2.1 *Reconstruction of possible scene of wheat flour explosion in Mr. Giacomelli's bakery on 14th December, 1785, as described by Count Morozzo (1795)*

2.3 GRAIN DUST EXPLOSIONS IN NORWAY

2.3.1 WHEAT GRAIN DUST, STAVANGER PORT SILO, JUNE 1970

The explosion, which was discussed by Astad (1970) and Mo (1970), occurred in Norway's largest and newly built import grain silo in Stavanger on a hot and dry summer day. Fortunately, no persons were killed, but some workers suffered first-degree burns. Although the extent of flame propagation was considerable, the material damage was moderate due to the comparatively strong reinforced concrete structure of the buildings and the venting through existing openings.

The entire event lasted for a period of about 25–30 seconds, during which a sequence of six or seven distinct, major explosions were heard. In the middle of this sequence there was an interval of 10–12 seconds. The flame propagated a total distance of about 1500 meters, through a number of bucket-elevators, horizontal conveyors, ducting, filters and rooms in the building. Dust explosions occurred in six of the large, cylindrical storage silos of total volume 2000 m^3 each, in one large, slightly smaller silo, in seven of the slimmer, intermediate silos of capacities 400 or 1000 m^3, in one 150 m^3 silo, and in seven loading-out silos with capacities of 50 m^3 each. The six largest silos had no venting, whereas the explosions in the large silo of slightly smaller volume, and in all the intermediate and loading-out silos, were vented through 0.4 m^2 manholes, which had their covers flung open.

It is of interest to note that only one silo was damaged in the incident, namely one of the six unvented, large storage silos, which had its roof blown up, as shown in Figure 2.2. It is thus clear that the maximum explosion pressures in all the other 21 silos, vented and unvented, were lower than about 0.2 bar(g), which would be required to blow up the actual type of silo roof.

Figure 2.2 *Damaged silo roof after the wheat grain dust explosion in Stavanger in June 1970 (Courtesy of Egil Eriksson)*

Almost all the windows, except those in the office department, were blown out, as was a large provisional light wall at the top of the head house, as shown in Figure 2.3. The legs of all of the five bucket elevators (0.65 m × 0.44 m cross section) were torn open from bottom to top. The dust extraction ducts were also in part torn open.

Figure 2.3 *Provisional light-weight wall acting as vent during the wheat grain dust explosion in Stavanger in June 1970 (Courtesy of Egil Eriksson)*

The source and site of initiation of the explosion was never fully identified. However, two hypotheses were put forward. The first was self-ignition of dust deposited in the boot of the elevator in which the explosion was supposed to start. The self-ignition process was thought to have been initiated by a bucket that had been heated by repeated impacts until it finally loosened and fell into the dust deposit in the elevator boot. The second hypothesis was that the chain of events leading to ignition started with welding on the outside of the grain feed duct to one of the elevator boots. Due to efficient heat transfer through the duct wall, self-heating could then have been initiated in a possible dust deposit on the inside of the duct wall. (See Figure 1.10 in Chapter 1).

Lumps of the smouldering deposit could then have loosened and been conveyed into the elevator boot and initiated an explosion in the dust cloud there.

2.3.2 WHEAT GRAIN DUST, NEW PART OF STAVANGER PORT SILO, OCTOBER 1988

The explosion was described by Olsen (1989). Because of effective mitigation by explosion suppression and venting, both the extent of and damage caused by the explosion were minor. There were neither fatalities nor injuries. The incident deserves attention, however, because the chain of events leading to explosion initiation was identified, and because the incident illustrates that proper measures for explosion mitigation are effective.

The explosion occurred in a bucket elevator head immediately after termination of transfer of Norwegian wheat grain between two silo cells. At the moment of explosion the transport system was free of grain. In this new part of Stavanger Port Silo, the bucket elevator legs are cylindrical and mounted outdoors, along the wall of the head house. A number of vents are located along the length of the legs. The vent covers on the elevator legs involved were blown out, which undoubtedly contributed to reducing the extent of the explosion. There was no significant material damage, either by pressure or by heat.

Figure 2.4 illustrates the head of the bucket elevator in which the explosion occurred. Because of a slight offset, the steel cover plate for the felt dust seal for the pulley shaft touched the shaft and became heated by friction during operation of the elevator. The hot steel plate in turn ignited the felt seal, from which one or more glowing fragments dropped into the wheat grain dust deposit on the inclined surface below, and initiated smouldering combustion in the deposit. Figure 2.5 shows the burnt/charred felt seal when investigated just after the explosion. Just after the elevator had stopped, there was presumably still enough dust in the air to be ignitable by the smouldering dust, and to be able to propagate a flame. Alternatively, some of the smouldering dust may have slid down the inclined surface and become dispersed into an exploding dust cloud. Just after the explosion, some smouldering dust was still left on the inclined plate below the elevator pulley.

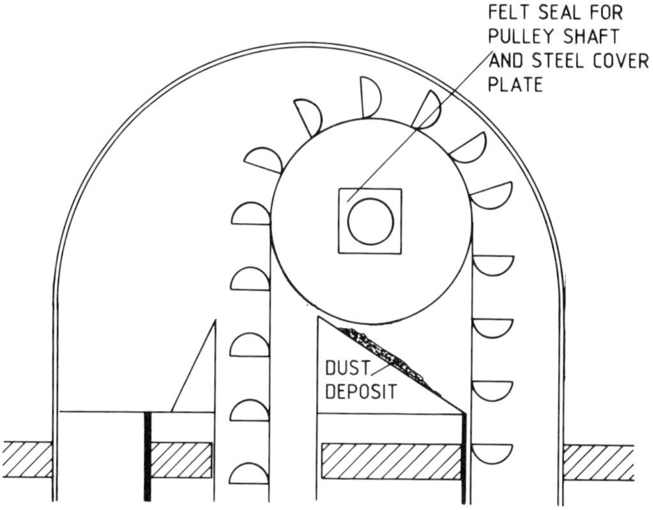

Figure 2.4 *Schematic illustration of head of bucket elevator at new part of Stavanger Port Silo, where the minor 1988 wheat grain dust explosion was initiated (Courtesy of O. Olsen, Stavanger Port Silo, Norway)*

2.3.3 GRAIN DUST (BARLEY/OATS), HEAD HOUSE OF SILO PLANT AT KAMBO, JUNE (1976)

This explosion, described by Storli (1976), caused considerable material damage, but due to fortunate circumstances there were neither fatalities nor significant injuries. The dust involved was from Norwegian barley or oats.

The explosion probably started in a bucket elevator, initiated by burning/glowing material from an overheated hammer mill. The primary explosion developed into a secondary explosion in the head house itself, which pushed out most of the front wall of the head house, as shown in Figure 2.6.

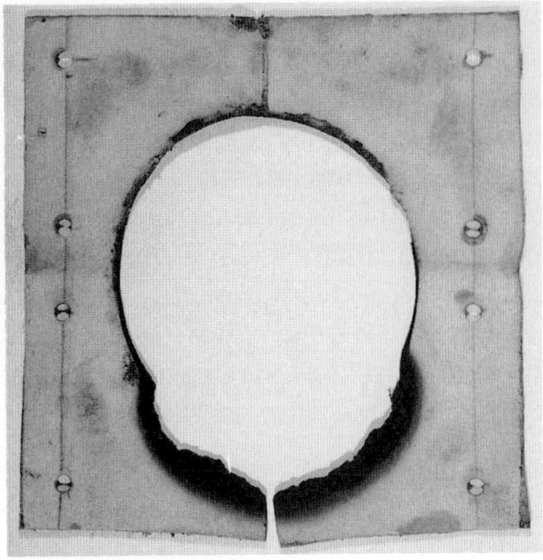

Figure 2.5 *Burnt/charred felt seal for elevator pulley shaft at Stavanger Port Silo (Courtesy of O. Olsen, Stavanger Port Silo, Norway)*

Figure 2.6 *Damaged silo head house after a grain dust explosion at Kambo, Norway, in June 1976 (Courtesy of Scan Foto, Oslo, Norway)*

Two of the bucket elevators had bulged out along the entire length and the dust extraction ducting had become torn apart, and this gave rise to the secondary explosion. Because the floors were supported by the wall, and the connections between wall and floors were weak, the entire wall sheet was pushed out at a quite low explosion pressure, leaving the floors unsupported at the front. (See Figure 1.134 in Chapter 1.)

After the explosion, the head house was reconstructed utilizing the principle illustrated schematically in Figure 1.136 in Chapter 1. The floors were supported by a rigid framework, and the light-weight wall elements can serve as vent covers, should an explosion occur again. The reconstructed head house is shown in Figure 2.7.

Figure 2.7 *Reconstructed head house (1977) after grain dust explosion at Kambo, Norway, in June 1976. The entire head house front is covered with light-weight wall elements that can serve as explosion vent panels (Courtesy of J. Kosanovic, Kambo, Norway)*

2.3.4 MALTED BARLEY DUST, OSLO PORT SILO, JULY 1976

The explosion, described by Johansen (1976), occurred in an old silo building in the central harbour area of Oslo at about 07.30 on a dry summer morning. The material damage was extensive, and much debris was thrown into the surroundings. However, due to several fortunate circumstances, there were neither loss of life nor severe injuries.

The dust involved was from malted barley, of only 5–6% moisture content. The ignition source was not identified, but the explosion probably started in a silo cell and propagated to other cells through the common dust extraction system. The primary explosions in the silo cells blew up the cell roofs, which were part of the floor of the silo loft, and gave rise to an extensive secondary explosion in the loft, blowing up the entire silo roof. The result is shown in Figure 2.8. The damage was so extensive that the entire building had to be demolished.

Figure 2.8 *Damaged silo building after malted barley dust explosion at Oslo Port Silo in July 1976 (Courtesy of A. Fr. Johansen, Oslo Port Silo, Norway)*

2.3.5
MALTED BARLEY DUST, OSLO PORT SILO, JUNE 1987

The explosion, described by Johansen, Johansen and Mo (1987), occurred on a warm and dry summer day during unloading of malted barley from a ship. There were neither fatalities nor injuries, and no damage was made to the building apart from broken window panes and a broken silo cell roof.

As in the 1976 explosion (Section 2.3.4), the malted barley was quite dry, containing only a few per cent moisture. The explosion probably started in a main dust filter in which smouldering combustion had developed due to frictional heating caused by packing of dust in the unloading screw at the filter bottom. Due to build-up of explosion pressure in the filter, the air flow in the dust extraction duct to the filter was reversed, and the explosion propagated upstream to the silo cell to which the duct was connected. The resulting explosion in the silo cell blew up the part of the concrete floor of the loft that was also the roof of that particular silo cell, and a fairly strong explosion occurred in the loft. The explosion also propagated from the filter to a bucket elevator that was torn open, which gave rise to a secondary explosion in the room. Furthermore, the explosion propagated to the lorry loading station of the silo plant.

As Figure 1.135 in Chapter 1 shows, the windows of the main building served as vents and probably prevented damage of the main structure of the building.

2.4 FOUR GRAIN DUST EXPLOSIONS IN USA, 1980–1981 (Source: Kauffman and Hubbard, 1984)

2.4.1 INLAND GRAIN TERMINAL AT ST. JOSEPH, MISSOURI, APRIL 1980

The explosion, which occurred in the middle of the day, killed one person and injured four. Material damage was estimated at US$ 2 mill.

The explosion probably started in the dust cloud in a silo cell that was used for receipt and delivery of grain. The probable ignition source was an electric arc between the electric wires of the lower level indicator in the silo. Repeated filling and discharge of grain had pulled the level indicator from the wall and the electric arc occurred between the bare wires that had subsequently been pulled out of their conduit.

There was severe structural damage to almost all of the silos in the head house and moderate damage to most of the head house structure. Most of the head house silo roofs were blown up, destroying the spout floor and the top of the cleaner floor. Rupture of the silos around the edge of the head house caused failures in the outside wall. The casings of all bucket elevators, steel as well as concrete, had opened up in many places. A silo complex comprising 18 cells suffered severe explosion damage to the gangway connecting it to the head house, to the gallery, to the far end of the tunnel, and to a small group of silos centred around an air-shaft approximately one-third of the way along the gallery. At the location of the air-shaft, the gallery wall and roof had been completely destroyed. Beyond this point the explosion damage to the gallery was still significant, but not as severe. The exterior concrete silo walls had been extensively shattered, leaving in many places only the reinforcing rods. A detailed view down the air-shaft made after the grain was removed is shown in Figure 2.9. As can be seen, the concrete fragments were quite small and much concrete had been removed from the steel reinforcement. Concrete fragments from this area of the plant had been thrown about hundred meters into the adjacent railway yard.

2.4.2 RIVER GRAIN TERMINAL AT ST. PAUL, MINNESOTA, JUNE 10, 1980

The explosion occurred just before lunch time. There were no fatalities, but 13 persons were injured. The material loss was estimated at about US$ 0.3 mill.

The probable cause of the explosion was that an electrician was repairing live electrical equipment in a truck-receiving cross tunnel, while the elevator was unloading grain trucks.

Figure 2.9 *Air-shaft along the damaged walls of reinforced concrete silo cells of the grain terminal at St. Joseph, Missouri, USA, 1980 (Courtesy of C. W. Kauffman, University of Michigan, USA)*

The ignition source probably was electric arcing in an open electric junction box located within an explosible dust cloud.

The blast and flame front moved in one direction along the tunnel into the head house basement. There were open spouts to the bucket elevators and with the secondary explosion in the basement initiated by the cross tunnel explosion, the explosion was carried into all the bucket elevators and the dust extraction systems. The building was of structural steel with non-supporting metal clad walls, and this allowed rapid pressure relief by the wall panels being blown out. (See Figure 1.136 in Chapter 1.) Therefore, the blast that went out of the head house and up one of the bucket elevators did not do much damage to the galleries. This was fortunate because, as Figure 2.10 shows, the level of housekeeping in the gallery at the moment of the explosion was rather poor. With a stronger blast entering the gallery and a flame following, a serious secondary gallery explosion could have resulted.

Figure 2.11 shows another example of unacceptably large quantities of accumulated dust. Kauffman (1982) used this photograph as a reminder when emphasizing that even a dust layer of only 0.5 mm thickness may be able to propagate a dust flame when being entrained by a blast wave preceding a propagating dust flame. This experience has been transformed into a simple rule of thumb saying that if footprints are visible, the dust layer is unacceptably thick.

During the explosion at St. Paul, the flame front and pressure wave from the primary explosion cross tunnel explosion also travelled into the three tunnels under the grain storage tanks. However, these tunnels were clean and the blast was unable to pick up

Figure 2.10 *Accumulation of dust in gallery of river grain terminal at St. Paul, Minnesota, USA, 1980 (Courtesy of C. W. Kauffman, University of Michigan, USA)*

Figure 2.11 *Dust accumulation on the floor in the head house of a grain silo plant in USA (Courtesy of C. W. Kauffman, University of Michigan, USA)*

sufficient dust to sustain the flame propagation, and the explosion dissipated. However, the pressure wave continued down the three tunnels, sweeping away objects in its path, and finally damaging the aeration fans before venting itself to the atmosphere at the tunnel ends.

2.4.3 TRAIN-LOADING COUNTRY GRAIN TERMINAL AT FONDA, IOWA, USA, JULY 15, 1980

In this explosion, which occurred in the early afternoon, there were neither fatalities nor injuries. The material loss was modest, estimated at US$ 0.03 mill.

The probable cause of the explosion was electrical welding on a bucket elevator. However, the ignition source was not the welding spot itself, but probably a hot-spot in the casing of the elevator boot caused by poor electrical contact between the earthing clamp and the earthed elevator casing. The hot-spot either ignited the maize grain in the elevator boot, which in turn ignited the maize dust cloud, or the dust cloud was ignited directly by the hot-spot.

The explosion was transmitted to a second bucket elevator and blast waves and flames propagated upwards in the legs of both elevators, bursting the casings. Figure 2.12 shows one of the elevator legs after the explosion.

Kauffman (1982) emphasized the essential role played by bucket elevators in fourteen carefully investigated grain dust explosions in USA. In five of the fourteen accidents, the explosion originated in the bucket elevator. In six other accidents, bucket elevators were able to effectively amplify and propagate the explosion, although the combustion process did not originate there. Only in three of the fourteen cases, the bucket elevators were not involved. Kauffman (1982) also discussed why the bucket elevator is so much involved in the explosions. When in operation, the elevator contains an explosible dust cloud that is confined. Potential ignition sources can result from friction, foreign objects, hot bearings, belt and bucket slip, etc. As Kauffman stated 'Once a combustion process enters a bucket elevator, things can only get worse'.

Fortunately, in the Fonda explosion in 1980, the bucket elevator explosions did not result in secondary explosions in the head house. Proper housekeeping could be one reason for this.

2.4.4 LARGE EXPORT GRAIN SILO PLANT AT CORPUS CHRISTI, TEXAS, USA, APRIL 1981

In this catastrophic explosion, nine persons lost their lives and 30 were injured. The material loss was also substantial, estimated at US$ 30 mill.

The probable cause of ignition was smouldering lumps of milo grain that entered a bucket elevator together with the grain and ignited the dust cloud in the elevator. The milo was being unloaded from hopper-bottom railway carriages. The grain had been stored in these carriages for 30 days and the weather had been quite warm. A fine screen had been

Figure 2.12 *Damaged bucket elevator casing following a maize dust explosion in a grain terminal at Fonda, Iowa, USA, 1980 (Courtesy of C. W. Kauffman, University of Michigan, USA)*

put over the rail dump to prevent the larger lumps of the milo from entering the elevator. However, smaller lumps of smouldering milo nevertheless probably entered one of the operating bucket elevators and ignited the dust cloud there.

From this elevator the explosion propagated into the other elevators and eventually broke out into the head house basement either through the dust control system, spout mixers, and/or the head house silos. It then travelled out of the basement into a tunnel to the basement of a large concrete silo complex, where the combustion process entered the hooded conveyors where it found more than sufficient dust to sustain the combustion process. As it travelled within this enclosure, the flame accelerated and generated a pressure wave moving ahead of it. Approximately halfway down the basement of the silo complex the conveyor hoods blew up, throwing a large cloud of dust throughout the basement. The trailing flame front then arrived at this dust cloud and a very rapid combustion process developed. This explosion then vented itself in four different directions. It blew out the north basement wall, it went upwards through the grain silo cells, westwards through the dog house, and eastwards back into the head house, which eventually exploded. The explosion then propagated further through the dust extraction system and into the hooded conveyors in the middle of the basement of the second large concrete silo complex, through which it was channelled to the railway dump area on the north and the shipping gallery on the south. The explosion in the basement of the second silo complex was vented through the basement windows.

Figure 2.13 shows the silo plant just after the explosion. The entire gallery of the nearest large silo complex was totally demolished, and some of the silo cells had blown out along the entire length. The head house was also badly damaged.

Figure 2.13 *Demolished Corpus Christi grain silo plant after major grain dust explosion in 1981 (Courtesy of C. W. Kauffman, University of Michigan, USA)*

The extensive destruction of the railway dump area is shown in Figure 2.14. The wall cover sheets of the shelter have been shattered and blown away from the frame structure.

According to Kauffman and Hubbard (1984) the housekeeping in the Corpus Christi plant was excellent. Therefore the only explanation for the extensive flame propagation is accumulation of large dust quantities inside the process and dust extraction equipment, including ducting.

Figure 2.14 *Destroyed railway dump area of the Corpus Christi grain silo plant after major grain dust explosion in 1981 (Courtesy of C. W. Kauffman, University of Michigan, USA)*

2.5
A DUST EXPLOSION IN A FISH MEAL FACTORY IN NORWAY IN 1975

The explosion, described by Eckhoff (1980), took place at the end of a hot and dry day in August 1975, in one of the many fish meal factories located along the Norwegian west coast. A young worker lost his life due to severe burns, and another was injured. At the time of the explosion, the Norwegian factory inspectorate had just about released its very first set of rules for fighting industrial dust explosions. Hence, the general appreciation of the dust explosion hazard in Norwegian industry was still meagre.

The part of the factory involved in the explosion was the fish meal grinding plant, illustrated schematically in Figure 2.15. This plant was located in a 30 m tall building that also contained several fairly large storage and mixing silos. A photograph of the building, taken just after the explosion and showing the damaged roof, is given in Figure 2.16.

The three silos indicated on Figure 2.15, which played a key role in the development of the explosion, were 12 m high and had diameters of about 3 m. The wooden floor of the loft of the building also served as the common roof of the three silos. Close to the top of the silos there were 0.1m × 1 m open slots in the common wall between silos No. 1 and No. 2, and No. 2 and No. 3. The original purpose of the three silos was to store the production of fish meal accumulated during the night shift, allowing the screening operation to be limited to day shifts. However, it soon turned out that the hopper parts of

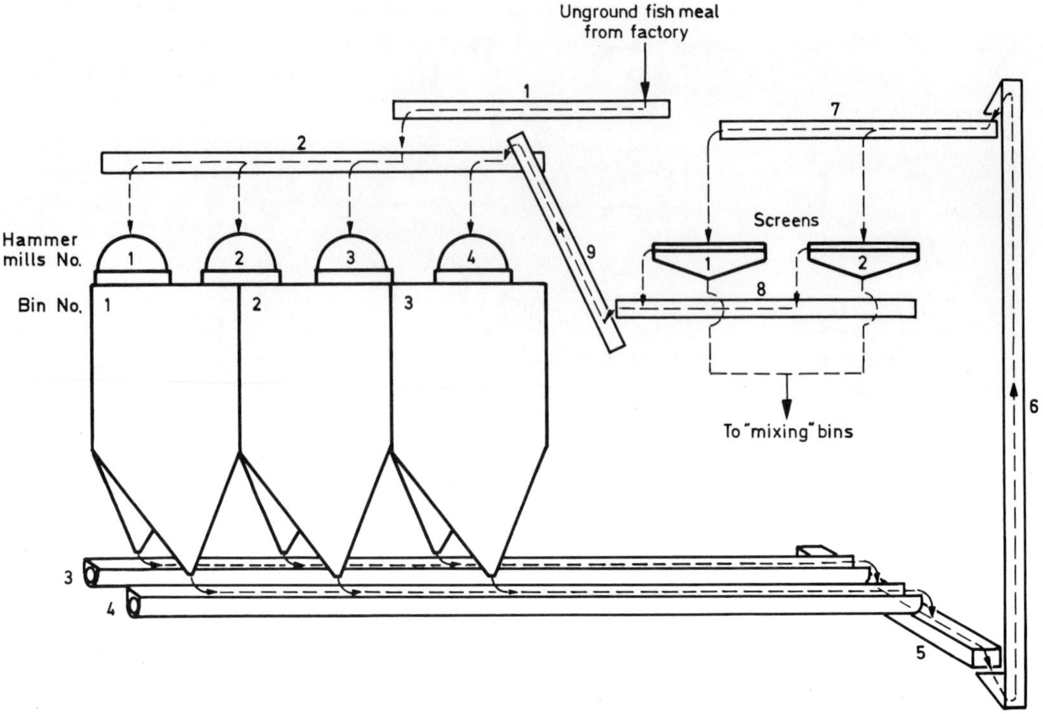

Figure 2.15 *Diagram of the fish meal grinding plant that was afflicted with a dust explosion in 1975*

Figure 2.16 *The building of the fish meal factory in Norway that was afflicted with a dust explosion in 1975*

the silos were not properly designed, and severe flow problems were encountered when attempting to discharge the fish meal by means of the screw conveyors at the hopper outlets. Therefore the use of the silos as buffer stores had to be abandoned. But instead of feeding the output from the hammer mills directly to the screens, the long transport loop via the large silos was maintained, the silos being mostly empty because of the large capacity of the screws at the hopper outlets. Nevertheless, arching problems still occurred across the hopper outlet just above the screw conveyors, and breaking such arches became part of the regular duties of the staff operating the plant.

Although the grinding of the fish meal in the hammer mills in the loft produced large quantities of fine dust, no dust extraction system had been installed. As a consequence, the interior of the three large, empty silos acted as dust collectors, and considerable quantities of dust accumulated on the internal walls. Furthermore, appreciable amounts of dust escaped to most other parts of the building. Besides having a much larger specific surface area than the main fish meal product, this fine dust would also in periods of hot and dry weather, as on the day of the explosion, become quite dry. Because of the heat liberated by the production process itself, the temperature in the loft of the silo building would frequently be in the range 25–30°C. On the exceptionally hot day of the explosion, the temperature in the loft in the middle of the day was 45°C.

One particular feature of the screw conveyors of this plant was that the bolts fixing the screw blades to the shaft (bolts of lengths 110–120 mm and diameters 12–16 mm) broke fairly regularly, presumably as a result of material fatigue. Figure 2.17 shows part of one of the screws with three bolt heads.

Figure 2.17 *Part of screw conveyor in the exploded fish meal plant showing fixing bolts*

In spite of frequent bolt failures, the plant had no provisions for trapping tramp metal, such as broken bolts, before it reached the hammer mills. Neither were there any instructions for controlling the screws to replace any defective bolts in advance. As a consequence, entrance of broken bolts and other tramp metal into the hammer mills was a fairly frequent event. The presence of bolts in the mills created a most unpleasant noise, which warned the operators of the plant. The normal procedure for removal of bolts from the mill was to open the 250 mm × 180 mm door in the mill chute, shown by the arrow in Figure 2.18, and wait until the foreign metal object eventually found its way out of the opening.

Just prior to the explosion, as part of the routine during start-up of the night shift, one of the workers went up to the loft where he at once heard, by the sound from the mills, that foreign objects had entered several of them. By means of the usual procedure, bolts were first removed from mills Two and Four. However, there was still noise of foreign objects

Figure 2.18 *View of hammer mill No. 1, in which the fish meal explosion started, the door in the mill chute and the man hole in the floor. The latter acted as vent for the explosion in silo No. 1*

and the source was mill No. 1. However, as soon as the door in the mill chute had been opened, a rapidly growing cloud of 'sparks' (probably burning fish meal particles) was discovered. At the same time flames just below mill No. 1 were observed through a narrow slot close to the mill. The main explosion occurred immediately after these observations had been made, blowing the hatch off the man-hole and ejecting a strong flame through the loft room and against the roof of the building. According to the observer, the flame was bluish in colour, similar to that of a brazing lamp. This first blast was then followed by a kind of whistling or howling that moved in the direction from mill No. 1 to mill No. 4. This may have been flame propagation from silo No. 1 via silo No. 2 to silo No. 3 through the 0.1 m × 1 m slots at the top of the common wall between two neighbouring silos. At this moment the witness found his way out and escaped from the loft, which was now on fire.

The explosion was also observed from the outside by two persons who just happened to pass by. One distinct and fairly strong explosion could be heard. This was followed by a large pyramidal flame lasting for 30–45 seconds and extending 4–5 m above the roof of the building. The explosion was sufficiently strong to blow out windows in the building even in other parts than the loft.

It seems highly probable that the ignition took place in hammer mill No. 1 and that the ignition process was closely related to the presence of tramp metal in the mill. It seems unlikely that sparks struck between two steel objects would be able to ignite clouds of the fish meal in the mill. However, a metal object can be heated, even to glowing, by repeated impact/friction, and thus act as a hot surface for direct initiation of dust explosion in a cloud. It is not unlikely that this latter process has been in operation in the actual case, because after the explosion it was discovered that a 14 × 7 mm strip of steel was wedged into one of the 3 mm slots of the bottom screen plate of Mill No. 1. In view of the high rotation speed of the mill (25 rev/s) such an object could easily have been heated to appreciable temperatures by repeatedly being struck by one of the mill hammers.

This accident shows that normal fish meals can, under unfavourable circumstances, give rise to quite severe dust explosions. This is so in spite of the fact that the explosions produced by such materials in standard laboratory tests are relatively weak compared with those produced by many other dusts. Because the pressure needed to blow up very weak structures like the wooden floor of the loft, is low, even a modest dust explosion will be hazardous under the actual circumstances. The housekeeping was very poor. Neither was there any dust extraction system nor any routine for frequent and regular removal of the considerable amounts of fine dust accumulated throughout the plant and building. This dust certainly was the main source of the extensive secondary explosion/fire sweeping through the entire loft. Because of the dry weather, the dust moisture content was probably low and the dust easy to entrain and disperse.

The process design was inadequate in that the large silos below the hammer mills did not serve their purpose and merely acted as large potential dust explosion bombs.

Also prevention of potential ignition sources was inadequate. In the grinding plant no provision was made for removing foreign metal objects before they entered the hammer mills. By allowing tramp metal into the hammer mills at all, a considerable risk of creating potential ignition sources was introduced. Furthermore, the procedure for removing broken conveyor screw bolts from the hammer mills by opening the door in the mill chute was indeed questionable.

Hence, three key ingredients needed for generating a serious dust explosion were present: large enclosures that were empty apart from explosible dust clouds, large quantities of dust throughout the entire building, and an ignition source.

2.6 SMOULDERING GAS EXPLOSION IN A SILO PLANT IN STAVANGER, NORWAY, IN NOVEMBER 1985

This accident, described by Braaten (1985), was not primarily a dust explosion, but an explosion of combustible gases released from a solid organic material during self-heating in a silo cell. At the first glance such an event may seem out of place in the context of dust explosions. However, smouldering combustion is most often related to powders and dusts, and therefore the initial smouldering gas explosion will in most cases entrain combustible dust and the explosion can easily develop into a normal dust explosion.

The cause of events was in accordance with Figure 1.9 in Chapter 1. The explosion occurred in a fairly modern reinforced concrete silo complex used for storage of various feed stuffs. Pellets of Canadian rape seed flour had been stored in one of the silos for some time when it was discovered that the material in the bottom part of the silo had become packed to a solid mass and could not be discharged through the silo exit. Some time later, one week before the explosion, flames were observed in the silo. The fire brigade was called and covered the pellets in the silo with foam from above. Various unsuccessful attempts were then made at discharging the pellets mass at the silo bottom. During this phase there was considerable development of smoke, which mixed with the air not only in the silo cell in question, but also in the silo loft above the cells. It is probable that the

Figure 2.19 *View of damaged loft of silo plant in Stavanger, Norway, after smouldering gas explosion in November 1985 (Courtesy of Øyvind Ellingsen, Stavanger Aftenblad, Norway)*

smoke contained combustible gases, e.g. CO, and that the strong explosion that occurred just after the top of the pellets had been covered with foam once more, was mainly a gas explosion. However, any dust deposits in the loft may also have become involved. The entire roof of the building was blown up, and debris was thrown into the surrounding area. Because the explosion occurred in the middle of the night (0300), and just after the fire brigade had left, nobody was killed or hurt.

2.7 SMOULDERING GAS EXPLOSIONS IN A LARGE STORAGE FACILITY FOR GRAIN AND FEEDSTUFFS IN TOMYLOVO IN THE KNIBYSHEV REGION OF USSR

This extensive series of explosions were of the same nature as the smouldering gas explosion discussed in Section 2.6. The report of the event was provided by Borisov and Gelfand (1989).

The large storage facility for grain and feedstuffs consisted of four sections of 60 silo cells each, i.e. 240 silo cells altogether. As indicated in Figures 2.20 and 2.21, each cell was of 3 m × 3 m square cross section and 30 m height. The first explosion occurred in December 1987 in a silo cell containing moist sunflower seed, which was not supposed to be stored in such silos due to the risk of self-heating. However, this had nevertheless been done, and the resulting self-heating developed into extensive smouldering decomposition during which methane and carbon monoxide was produced and mixed with the air in the empty top part of the silo, above the powder bed surface. It is reasonable to believe that the primary explosion was in this mixture of explosive gas and air, and that the ignition source was the smouldering combustion when it penetrated to the powder bed top surface, as illustrated in Figure 1.9 in Chapter 1. However, dust deposits on the internal silo walls and roof may well have become entrained by the initial blast and involved in the explosion.

This was only the first of a large series of 20–30 subsequent explosions that took place in the same facility, in one silo cell after the other, during 1988 and 1989.

There are two main reasons for this continued explosion activity in the silo complex. The most important is the heat transfer from a silo cell in which smouldering combustion is taking place, to the neighbouring cells. Such heat transfer was facilitated by the large contact surface area between the cells provided by the square cross-section. Furthermore, the pre-fabricated construction elements used throughout the entire facility, as shown in Figures 2.20 and 2.21, may have been comparatively poor heat insulators.

The second main reason for the repeated explosions was that sunflower seed was not the only material in the facility that was not supposed to be stored there. Some of the silo cells contained buckwheat and wheat grain of higher moisture contents than the maximum permissible limits for storage in such facilities.

During the period of repeated explosions a series of attempts were made at breaking the unfortunate chain of events. Cells were opened at the top for inspection. However, this admitted fresh air to the smouldering mass and enhanced the combustion process.

Figure 2.20 *Corner of silo complex in Tomylovo, USSR, damaged by smouldering gas explosions, 1987–1989 (Courtesy A. Borisov and B. Gelfand, USSR Academy of Science, Moscow)*

Attempts were made at quenching and cooling the powder mass by liquid nitrogen, but this was only partly successful.

It was agreed that the use of water for extinguishing the smouldering combustion in the silo cells was not feasible. Limited quantities of water would probably enhance the self-heating process rather than quench it, whereas use of extensive quantities would increase the load on the silo walls and cause collapse of the structure.

At one stage it was discussed whether the whole facility could be blown up to put an end to the problem. However, this was considered too hazardous. The final solution chosen was to just leave the entire facility to itself and await a natural termination of the problem over time.

Besides obeying the rules specifying which materials can be stored in silos, systematic use of portable gas analysers for early detection of hydrogen, methane and carbon monoxide in the silo cells was suggested as the best means for preventing similar accidents in the future.

Figure 2.21 *Central part of silo complex in Tomylovo, USSR, damaged by smouldering gas explosions, 1987–1989 (Courtesy of A. Borisov and B. Gelfand, USSR Academy of Science, Moscow)*

2.8 SMOULDERING GAS EXPLOSION AND SUBSEQUENT SUCCESSFUL EXTINCTION OF SMOULDERING COMBUSTION IN PELLETIZED WHEAT BRAN IN A SILO CELL AT NORD MILLS, MALMÖ, SWEDEN, IN 1989

A cross-section of the silo is shown in Figure 2.22.

The course of events, as recorded by Templin (1990), was as follows:

Saturday 28th January, 0700: The night shift stopped the production for the weekend according to schedule, and all activity in the grain silo plant terminated.

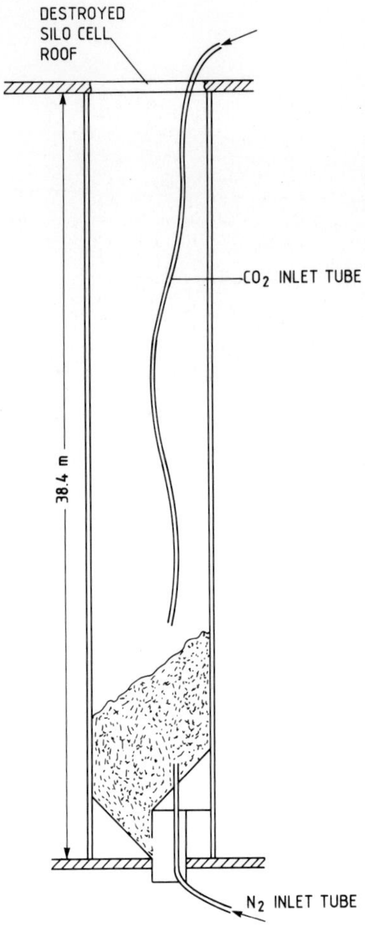

Figure 2.22 *Cross-section of silo in Malmö, Sweden, in 1989, with smouldering wheat bran pellets, showing inlets for carbon dioxide and nitrogen for extinction and cooling (Courtesy of G. Templin and B. Persson, Nord Mills, Sweden)*

Saturday 28th January, 1000: According to Nord Mills' safety procedures, the safety guard team made its inspection round through the entire plant. Nothing special was observed. No persons were encountered.

Saturday 28th January, 2300: A bang, muffled by the noise of strong winds, was heard in the neighbourhood, but no action was taken.

Sunday 29th January, 0930: During its scheduled inspection round, the safety guard team discovered fragments of shattered window panes spread over the entire yard. Inspection of the roof of the silo building revealed that the roof of an intermediate star cell had blown up, as shown in Figure 2.23, and that dense smoke was emerging from the open cell top. The height of the cell involved was about 36 m and its cross sectional area about 20 m^2. Most of the silo was empty, the pelletized wheat bran occupying only the seven first metres above the cell bottom.

Figure 2.23 *Damaged silo roof after smouldering gas explosion resulting from smouldering wheat bran pellets in a 38 m tall concrete silo in Malmö in 1989 (Courtesy of G. Templin and B. Persson, Nord Mills, Malmö, Sweden)*

Sunday 29th January, rest of day: Fire brigade and other personnel were called, and the entire plant area was cordoned off. About 2000 kg of gaseous carbon dioxide was pumped into the burning silo from above through a long vertical pipe extending right down to the surface of the smouldering pellets.

Monday 30th January, early morning: The discharge valve at the cell bottom was removed, and discharge of the pellets mass, using a mobile suction unit, was started. This gave rise to increased smoke production, and at 03.30 more carbon dioxide was loaded into the silo cell from above.

Tuesday 31st January: The discharge operation was interrupted. Carbon dioxide was emerging through the bottom silo exit, and more was loaded into the silo at the top.

Wednesday 1st February: More carbon dioxide was loaded into the silo at the top. From 03.00 to 12.50 the smoke development was enhanced by vibrations due to operation of another silo cell. The smoke temperature just above the pellets was 96°C, and just above the silo top 45°C.

Thursday 2nd February–Wednesday 8th February: Smoke development and temperature rise was suppressed temporarily by loading several tonnes of carbon dioxide into the silo from the top, but there was only slow permanent progress. Temperature rise was observed in the material stored in the four larger adjacent silo cells.

Thursday 9th February–Saturday 11th February: Holes were drilled through the silo bottom and at intervals a total of several tonnes of nitrogen were pumped into the pellets from below, while carbon dioxide was charged from above.

Monday 13th February–Wednesday 15th February: Some 6000 kg of carbon dioxide and 3000 kg of N_2 was injected into the burning pellets. Temperatures in the burning and adjoining cells and contents of oxygen, CO and CO_2 in the gas above the pellets, were monitored regularly.

Monday 20th February: The smouldering combustion in the wheat bran pellets had finally been brought to an end.

This case history illustrates that fighting smouldering combustion in large silo complexes is not only a matter of quenching, or terminating the oxidation reaction, but also indeed a matter of cooling massive bulks of poor heat conductors to a temperature level at which the combustion process will not start again once air is re-admitted to the system.

2.9 LINEN FLAX DUST EXPLOSION IN HARBIN LINEN TEXTILE PLANT, P. R. CHINA, MARCH 1987

2.9.1 GENERAL OUTLINE

In the middle of the night (2.39 AM) on March 15, 1987, the spinning section of the large linen textile plant in Harbin, P. R. China, was afflicted with a catastrophic dust explosion. The losses were substantial. Out of the 327 women and men working night shift in the spinning section when the explosion occurred, 58 lost their lives and 177 were injured. 13 000 m^2 of factory area was demolished.

This explosion accident has been discussed in detail by Xu Bowen (1988) and Zhu Hailin (1988). Xu Bowen *et al.* (1988) reconstructed a possible course of the explosion development on the basis of a seismic recording of the explosion by the State Station of Seismology, located only 17 km from the Harbin Linen Textile Plant.

2.9.2 EXPLOSION INITIATION AND DEVELOPMENT, SCENARIO 1

Figure 2.24 illustrates the 13 000 m^2 spinning section through which the explosion swept, and the possible locations and sequence of the nine successive explosions that comprised the event according to Xu Bowen (1988) and Xu Bowen *et al.* (1988). These workers based their reconstruction of the explosion on three independent elements of evidence. First, they identified the location of the various explosion sites throughout the damaged plant. Secondly, they ranked the relative strengths of the local explosions by studying the extent and nature of the damage. Thirdly, they arranged the various local explosions in time by means of the relative strengths of the nine successive explosions, identified by decoding the seismic recording of the event.

Figure 2.25(A) shows a direct tracing of the amplitude-modulated seismic signal actually recorded 17 km from the explosion site. Figure 2.25(B) shows the sequence of nine energy pulse impacts on the earth at the location of Harbin Linen Textile Plant, deduced from the signal in Figure 2.25(A). Figure 2.25(C) finally shows the theoretical prediction of the seismic signal to be expected from the sequence of explosions in Figure 2.25(B). The agreement between the (A) and (C) signals is striking, which supports the validity of the energy impact pulse train (B).

Table 2.1 summarizes the findings of Xu Bowen *et al.* (1988) that led to the suggestion of the explosion development indicated in Figure 2.24. According to this scenario the explosion was initiated in one of the nine units in the central dust collector system. All nine units were connected by ducting. The ignition sources were not identified, but an electrostatic spark was considered as one possibility, a local fire or glow as another. The initial flame was transmitted immediately to the next dust collecting unit and both units (1) exploded almost simultaneously, giving rise to the first major impact pulse in Figure 2.25(B). The explosion then propagated through the other seven dust collecting units in the central collecting plant (2), and into the pre-carding area, where the blast wave preceding the flame had generated an explosible dust cloud in the room, which was ignited by the flame jet from the dust collectors (3). The room explosion propagated further to the carding and pre-spinning shops (4), and right up to the eastern dust collectors, where another distinct explosion (5) occurred. The final four explosion pulses were generated as the explosion propagated further into the underground linen flax stores, where it finally terminated after having travelled a total distance of about 300 m. The chain of nine explosions lasted for about 8 seconds.

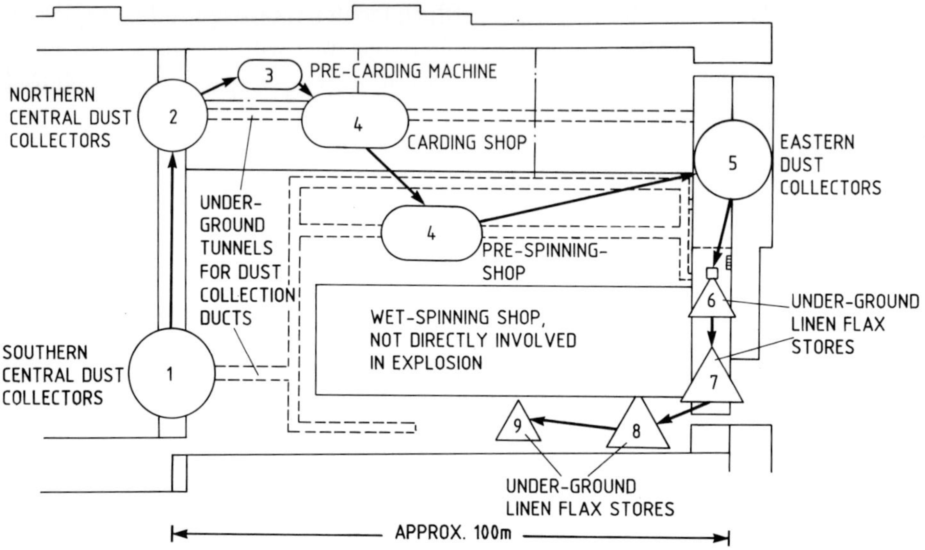

Figure 2.24 *Schematic illustration of the 13 000 m^2 spinning section of the Harbin Linen Textile Plant, P. R. China, that was afflicted with a catastrophic dust explosion on 15th March 1987. Numbered circles, ovals and triangles indicate location and sequence of a postulated series of nine successive explosions (From Xu Bowen* et al., *1988)*

Table 2.1 Sequence, relative strengths and locations of nine successive dust explosions in the Harbin Linen Textile Plant, Harbin, P. R. China, 15th March 1987, postulated on the basis of damage analysis in the plant and a seismic recording of the explosion event (From Xu Bowen *et al.*, 1988)

Explosion No.	Onset of explosion [s]	Seismic energy [T erg]	Location of explosion in plant
1	0.0	50.7	Southern central dust collector
2	0.6	5.4	Northern central dust collector
3	1.2	2.5	Pre-carding machine
4	1.6	7.6	Carding and pre-spinning shops
5	3.0	6.8	Eastern dust collectors
6	4.8	1.4	
7	6.0	3.9	Under-ground linen flax stores
8	7.3	2.2	
9	8.2	0.45	

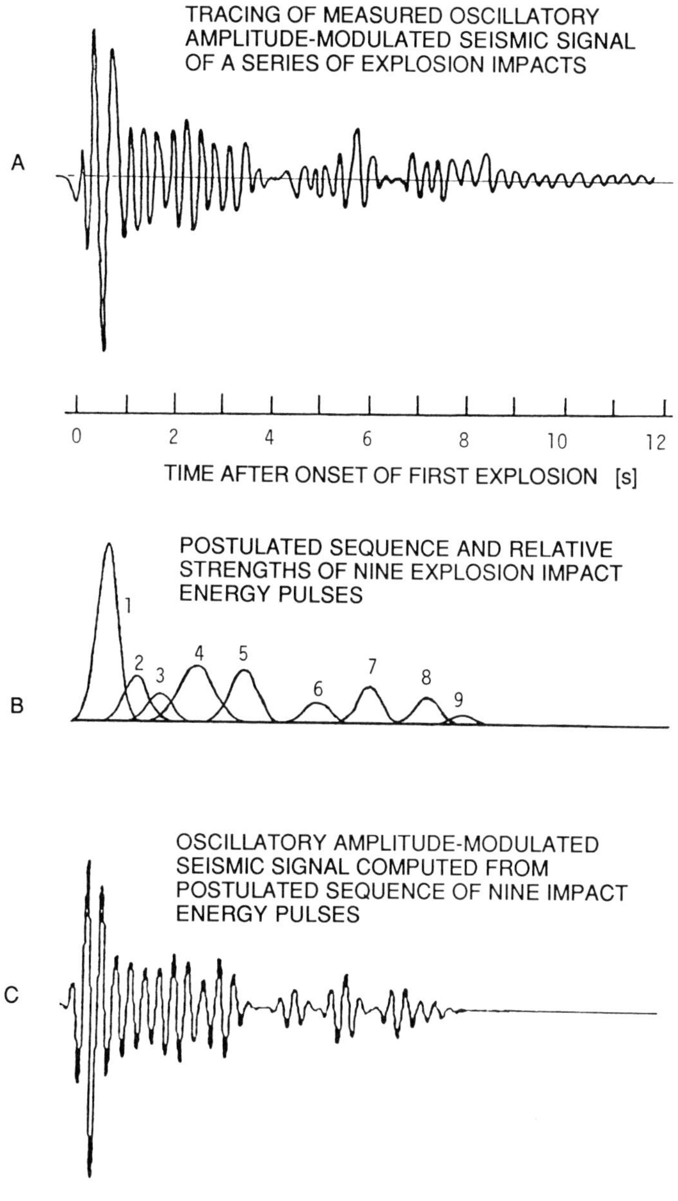

Figure 2.25 *Sequence of nine impact energy pulses from nine successive explosions in the Harbin Linen Textile Plant, Harbin, P. R. China, 15th March 1987, postulated on the basis of a seismic record of the event (From Xu Bowen* et al., *1988)*

2.9.3 EXPLOSION INITIATION AND DEVELOPMENT, SCENARIO 2

This alternative scenario originates from the investigation of Zhu Hailin (1988), who found evidence of an initial smouldering dust fire caused by a live 40 W electrical portable light lamp lying in a flax dust layer of 6–8 cm thickness in a ventilation room. He also found evidence of flame propagation through the underground tunnels for the dust collection ducting. On the basis of his analysis, Zhu suggested that the explosion was initiated in the eastern dust collectors (5 in Figure 2.24) from which it transmitted to nine units of the central dust collecting plant (1 and 2 in Figure 2. 24) via the ducting in the underground tunnels. Severe room explosions were initiated when the ducting in the tunnel ruptured, and the resulting blast dispersed large quantities of dust in the workrooms into explosible clouds that were subsequently ignited. From the eastern dust collectors the explosion also propagated into the underground flax stores.

It is not unlikely that even this scenario could be developed further in such a way as to agree with the evidence from the seismic recording.

2.9.4 ADDITIONAL REMARK

The investigation of the Harbin disaster exposed the great difficulties in identifying the exact course of events of major explosions creating massive damage. In addition to causing pain and grief, loss of life also means loss of eye witnesses. Besides, the immediate need for fire fighting and rescue operations, changes the scene before the investigators can make their observations. Also, the explosion itself often erases evidence, e.g. of the ignition source.

This problem was also shared by the experts who investigated the Harbin explosion, and it seems doubtful that the exact course of events will ever be fully resolved.

However, the Harbin disaster unambiguously demonstrated the dramatic consequences of inadequate housekeeping in industrial plants where fine dust that can give dust explosions, is generated.

2.10 FIRES AND EXPLOSIONS IN COAL DUST PLANTS

2.10.1 METHANE EXPLOSION IN 17000 m^3 COAL SILO AT ELKFORD, BRITISH COLUMBIA, CANADA, IN 1982

As mentioned in Section 1.5, handling and storage of coal can, in addition to the dust explosion hazard, also present a gas explosion risk, due to release of methane from some types of coal. An account of such an explosion was given by Stokes (1986).

The silo of height 48 m and diameter 21 m that exploded, was used for storage and load-out of cleaned, dried metallurgical coal. The capacity of the silo was 15000 tonnes.

Prior to the explosion accident, a methane detector had been installed in the roof of the silo. The detector activated a warning light in the silo control room when a methane concentration of 1% was detected, and an alarm light was activated when detecting 2% methane. A wet scrubber was located in the silo head house to remove dust from the dust-laden air in the silo during silo loading. A natural ventilation methane stack was also located in the silo roof to vent any build-up of methane gas from the silo.

The explosion occurred early in the morning on 1st May, 1982, devastating the silo roof, head house, and conveyor handling system. Witnesses stated that a flash was noticed in the vicinity of the head house, followed seconds later by an explosion which displaced the silo top structures. This was followed by an orange-coloured fire ball that rolled down the silo walls and extinguished prior to reaching the base of the silo. Fortunately, neither injury nor death resulted, and damage to surrounding structures was minimal, although large blocks of concrete and reinforcing steel had been thrown several hundred metres from the silo. However, the plant itself had suffered substantial damage.

The silo was full of coal 24 hours prior to the explosion. During the evening before the explosion, 10 000 tonnes of coal were discharged. At the same time, conveying of deep-seam coal into the silo commenced and continued until the explosion occurred. At the time of the explosion, there were approximately 12 300 tonnes of coal in the silo, of which 7600 tonnes were deep-seam coal. Testing had shown that this quality of coal has a high methane emission rate and produced a low volatile coal dust. Clouds in air of this dust could not be ignited unless the air was mixed with methane.

The ignition source was not identified, but the following three possible sources were considered:

- Spontaneous combustion of the stored coal.
- An electrical or mechanical source.
- Hot coal from the thermal dryer.

During ten years of operation, with coal being stored in different environments for varying lengths of time, spontaneous combustion had never presented a problem, and consequently was not considered to be a probable source of ignition. During demolition of the damaged silo, all electrical and mechanical components were recovered and inspected and did not show any evidence of being the ignition source. Stokes (1986) did not exclude the remaining possibility that hot coal from the thermal dryer was the source of ignition.

2.10.2 METHANE/COAL DUST EXPLOSION IN A COAL STORAGE SILO AT A CEMENT WORKS AT SAN BERNARDINO COUNTY, CALIFORNIA, USA

This incident was reported by Alameddin and Foster (1984). A fire followed by an explosion occurred inside a coal silo of 900 tonnes capacity while the silo was nearly empty, and the remaining 85 tonnes of coal were being discharged. Prior to the explosion, a hot-spot of 0.6 m × 1.0 m had been detected on the lower part of the silo wall by means of an infrared heat detector. The hot-spot originated from smouldering combustion in the coal in the silo. This process liberated methane, carbon monoxide and other combustible gases from the coal. The explosion probably resulted from ignition of a mixture of combustible gas and airborne coal dust in the space above the bulk coal by the smouldering fire or glow when it reached the surface of the coal deposit. (See Figure 1.9 in Chapter 1.)

It was concluded that the supply of carbon dioxide from the top, which was used for suppressing the fire and preventing explosion, was insufficient to prevent the development of an explosible atmosphere in the space above the bulk coal.

In order to prevent similar accidents in the future, it was recommended that a carbon dioxide system be installed in both the top and bottom of the coal silo. Sufficient inerting gas should be added for development of a slight positive pressure inside the silo. The inerting gas must be of sufficient quantity to insure a nonexplosible atmosphere above the coal and sufficient pressure to prevent a sudden inrush of fresh air into the silo.

2.10.3 GAS AND DUST EXPLOSION IN A PULVERIZED COAL PRODUCTION/ COMBUSTION PLANT IN A CEMENT FACTORY IN LÄGERDORF IN F. R. GERMANY, IN OCTOBER 1980

According to Patzke (1981), who described this explosion accident, the explosion occurred while coal of about 30% volatiles was milled at a rate of 55 tonnes per hour. The start-up of the cement burner plant followed a compulsory break of at least 20 minutes of the milling operation to allow all airborne dust to settle out. A few seconds after the main gas valve had been opened, there was a violent explosion. The probable reason was a failure in the system for electric ignition of the gas. Within the period of six seconds before the gas valve was reclosed automatically, about 1 m^3 of gas had been discharged to the atmosphere of the hot combustion chamber and become mixed with the air to an explosible gas cloud. The temperature of the walls of the chamber was sufficiently high to ignite the gas, and a gas explosion resulted. The blast and flame jet from this comparatively mild initial explosion was vented into the milling system where a large, turbulent dust cloud was generated and ignited, resulting in a violent secondary dust explosion.

Various parts of the milling plant, some unvented and some vented, had all been designed to withstand the pressure generated in an extensive dust explosion. Furthermore, a passive device for explosion isolation of the type shown in Figure 1.82 in Chapter 1 had been installed upstream of an electrostatic dust filter.

Apart from deformation of some explosion vent doors, the dip tubes of two cyclones, and the coal feeder upstream of the mill, the plant had been able to withstand the explosion without being damaged. The passive explosion isolation device effectively protected the electrostatic filter from becoming involved in the system.

2.10.4 FURTHER EXPLOSION/FIRE INCIDENTS INVOLVING COAL

Andersson (1988) gave a step-by-step account of the process of extinction of a smouldering fire in a 50 m^3 coal dust silo in Arvika in Sweden, in August 1988. It was necessary to pay attention to the risk of explosion of combustible gases driven out of the coal by the heat from the fire.

First gaseous carbon dioxide was loaded into the silo at the top to build up a lid of inert atmosphere immediately above the coal deposit. Then all the coal was discharged carefully through the exit at the silo bottom. In this particular case, supply of carbon dioxide at the silo bottom was considered superfluous.

Wibbelhoff (1981) described a dust explosion in a coal dust burner plant of a cement works in F. R. Germany, in March 1981. Prior to the explosion, an electrical fault had caused failure of an air blower. The explosion occurred just after restart of the repaired blower. During the period in which the blower was out of operation, dust had accumulated on the hot surfaces inside the furnace and ignited, and as soon as the blower was restarted, the glowing/burning dust deposits were dispersed into a dust cloud that exploded immediately.

Pfäffle (1987) gave a report of a dust explosion in the silo storage system of a pulverized coal powder plant in Düsseldorf, F. R. Germany, in July 1985. The explosion occurred early in the morning in a 72 m^3 coal dust silo. The silo ruptured and burning material that was thrown into the surroundings initiated a major fire, which was extinguished by means of water. Fortunately no persons were killed or injured in this primary accident. However, during the subsequent cleaning-up process, a worker was asked to free the damaged silo of ashes by hosing it down with water. It then appeared that a glowing fire had developed in the dust deposit that was covered by the ashes. The worker had been warned against applying the water jet directly to the smouldering fire, but for some reason he nevertheless did this. The result was an intense dust flame that afflicted him with serious third degree burns. The smouldering fire was subsequently extinguished by covering its surface with mineral wool mats, and subsequently soaking the whole system with water containing surface-active agent.

2.11 DUST EXPLOSION IN A SILICON POWDER GRINDING PLANT AT BREMANGER, NORWAY, IN 1972

In this serious explosion accident, five workers lost their lives and four were severely injured. The explosion that occurred in the milling section of the plant, was extensive, rupturing or buckling most of the process equipment and blowing out practically all the wall panels of the factory building. Figure 2.26 gives a flow chart of the plant. Figure 2.27 shows the total damage of the entire grinding plant building, whereas Figure 2.28 gives a detailed view of the extensive damage.

Eye-witnesses reported that the flame was very bright, almost white. This is in accordance with the fact that the temperature of silicon dust flames, as of flames of aluminium and magnesium dust, is very high due to the large amounts of heat released in the combustion process per mole of oxygen consumed. (See Table 1.1 in Chapter 1.). Because of the high temperature, the thermal radiation from the flame is intense, which was a main reason for the very severe burns that the nine workers suffered.

The investigation after the accident disclosed a small hole in a steel pipe for conveying Si-powder from one of the mechanical sieves to a silo below. An oxygen/acetylene cutting torch with both valves open was found lying on the floor about 1 m from the pipe with the

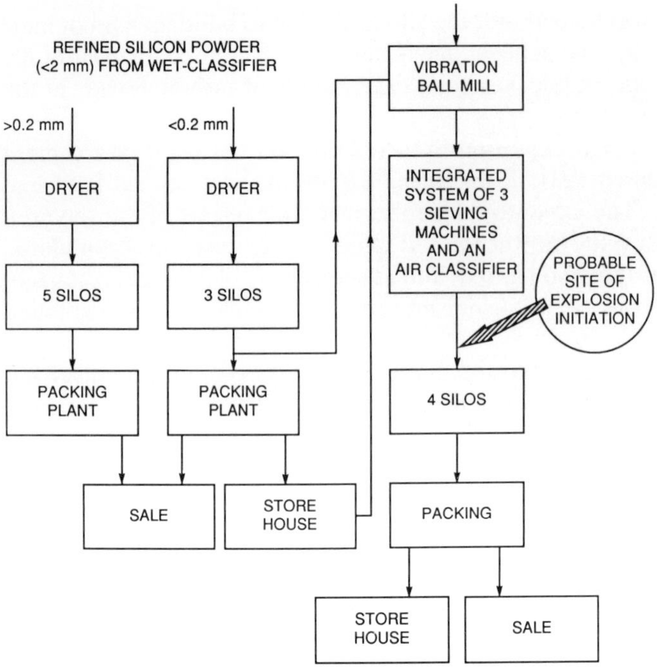

Figure 2.26 *Flow chart of dry part of plant for production of refined silicon products at Bremanger, Norway. The grinding plant that was totally damaged in the explosion in 1972 is shown to the right in the chart*

Figure 2.27 *Totally destroyed milling section of silicon powder production plant at Bremanger, Norway, after the dust explosion in October 1972*

Figure 2.28 *Detailed view of the extensive material damage caused by the silicon dust explosion at Bremanger, Norway, October 1972*

hole. According to Kjerpeseth (1990) there was strong evidence of the small hole having been made by means of the cutting torch just at the time when the explosion occurred. At the moment of the explosion, part of the plant was closed down due to various repair work. However, the dust extraction system was operating, and this may in part explain the rapid spread of the explosion throughout the entire plant. The interior of the pipe that was perforated had probably not been cleaned prior to the perforation. In view of the high temperature and excessive thermal power of the cutting torch, and not least the fact that it supplied oxygen to the working zone, a layer of fine dust on the internal pipe wall may well have become dispersed and ignited as soon as the gas flame had burnt its way through the pipe wall. The blast from the resulting primary silicon dust explosion then raised dust deposits in other parts of the plant into suspension and allowed the explosion to propagate further until it eventually involved the entire silicon grinding building.

The grinding plant was not rebuilt after the explosion.

2.12 TWO DEVASTATING ALUMINIUM DUST EXPLOSIONS

2.12.1 MIXING SECTION OF PREMIX PLANT OF SLURRY EXPLOSIVE FACTORY AT GULLAUG, NORWAY, IN 1973

The main source of information concerning the original investigation of the accident is Berg (1989). The explosion occurred during the working hours, just before lunch, while ten workers were in the same building. Five of these lost their lives, two were seriously injured, two suffered minor injuries, whereas only one escaped unhurt. A substantial part of the plant was totally demolished, as illustrated by Figure 2.29.

Figure 2.29 *Scene of total demolition after aluminium dust explosion in the premix plant of a slurry explosive factory at Gullaug, Norway, in August 1973 (Courtesy of E. Berg, Dyno Industries, Gullaug, Norway)*

The premix preparation plant building was completely destroyed. Debris was found up to 75 m from the explosion site. The explosion was followed by a violent fire in the powders left in the ruins of the plant and in an adjacent storehouse for raw materials.

The explosion occurred when charging the 5.2 m^3 batch mixer, illustrated in Figure 2.30. It appeared that about 200 kg of very fine aluminium flake, sulphur and some other ingredients had been charged at the moment of the explosion. The total charge of the formulation in question was 1200 kg.

The upper part of the closed vertical mixing vessel was cylindrical, and the lower part had the form of an inverted cone. The feed chute was at the bottom of the vessel. The mixing device in the vessel consisted of a vertical rubber-lined screw surrounded by a rubber-lined earthed steel tube. The powders to be mixed were transported upwards by the screw, and when emerging from the top outlet of the tube, they dropped to the surface of the powder heap in the lower part of the vessel, where they became mixed with other powder elements and eventually re-transported to the top.

The construction materials of the mixer had been selected so as to eliminate the formation of mechanical sparks. This was probably why both the screw and the internal wall of the surrounding earthed steel tube were lined with rubber.

During operation the 5.2 m^3 vessel was flushed with nitrogen, the concentration of oxygen in the vessel being controlled by a direct reading oxygen analyser. According to the foreman's statement, the oxygen content at the moment of explosion was within the specified limit.

After the explosion, the central screw part of the mixer, with the mixer top, was retrieved, as shown in Figure 2.31, about 12 m away from the location that the mixer had prior to the explosion. More detailed investigation of the part of the screw that was shielded by the steel tube, revealed, as shown in Figure 2.32, that the screw wings had been deformed bi-directionally as if an explosion in the central part had expanded violently in both directions. This evidence was considered as a strong indication of the explosion having been initiated inside the steel tube surrounding the screw. The blast and

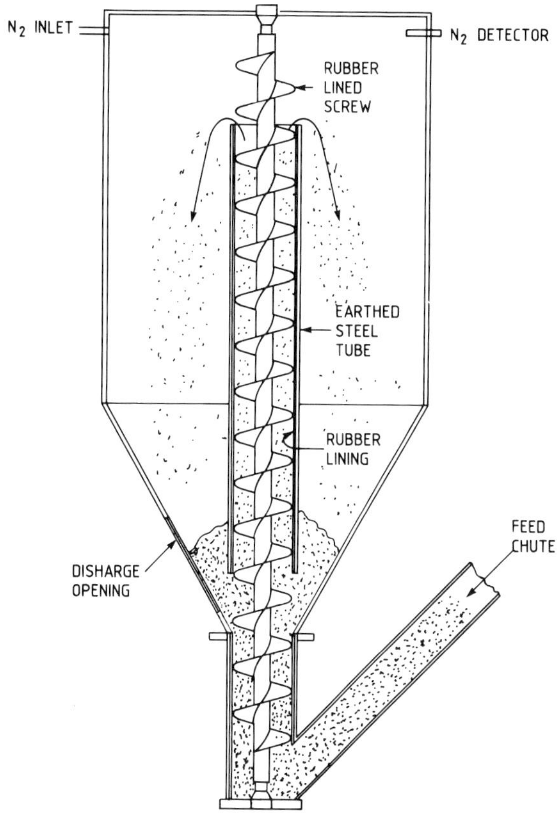

Figure 2.30 *Cross section of mixer for producing dry premix for slurry explosives at Gullaug, Norway, in 1973 (Courtesy of E. Berg, Dyno Industries, Gullaug, Norway)*

Figure 2.31 *Top of 5.2 m³ premix mixer, and 3.3 m long mixing screw with surrounding steel tube (see Figure 2.30), as found after the explosion 12 m away from location of the mixer prior to the explosion (Courtesy of E. Berg, Dyno Industries, Gullaug, Norway)*

Figure 2.32 *Section of screw after splitting and removal of surrounding steel tube, showing bi-directional deformation of the screw wings from the explosion centre. Part of rubber lining of steel tube removed from upper half of tube (Courtesy of E. Berg, Dyno Industries, Gullaug, Norway)*

flame from this primary explosion in turn generated and ignited a larger dust cloud in the main space inside the mixer, and finally the main bulk of the powder in the mixer was thrown into suspension and ignited when the mixer ruptured, giving rise to a major explosion in the workrooms.

Subsequent investigations at Chr. Michelsen Institute, Bergen, Norway, revealed that clouds in air of the fine aluminium flake powder was both extremely sensitive to ignition and exploded extremely violently. The minimum electric spark ignition energy was of the order of 1 mJ, and the maximum rate of pressure rise in the Hartmann bomb 2600 bar/s. Both these values are extreme. The thickness of the aluminium flakes was about 0.1 μm, which corresponds to a specific surface area of about 7.5 m^2/g. (See Section 1.1.1.3 in Chapter 1.)

The investigation further disclosed that the design of the nitrogen inerting system of the mixer was inadequate. First the nitrogen flow was insufficient to enable reduction of the average oxygen concentration to the specified maximum level of 10 vol% within the time allocated.

Secondly, even if the flow had been adequate, both the nitrogen inlet and the oxygen concentration probe were located in the upper part of the vessel, which rendered the measured oxygen concentration unreliable as an indicator of the general oxygen level in the mixer. It is highly probable that the oxygen concentration in the lower part of the mixer, and in particular in the space inside the tube surrounding the screw, was considerably higher than the measured value. This explains why a dust explosion could occur in spite of the use of a nitrogen inerting system.

The final central concern of the investigators was identification of the probable ignition source. In the reports from 1973 it was concluded that the primary explosion in the tube surrounding the screw was probably initiated by an electrostatic discharge. However, this conclusion was not qualified in any detail. In more recent years the knowledge about various kinds of electrostatic discharges has increased considerably (see Section 1.1.4.6). It now seems highly probable that the ignition source in the 1973 Gullaug explosion was a propagating brush discharge, brought about by the high charge density that could be accumulated on the internal rubber lining of the steel tube surrounding the screw, because of the earthed electrically conducting backing provided by the steel tube itself. The discharge could then have occurred through a hole in the lining (see Figure 1.14).

2.12.2 ATOMIZED ALUMINIUM POWDER PRODUCTION PLANT AT ANGLESEY, UK, IN 1983

This accident has been discussed in detail by Lunn (1984), and the following brief summary is based on Lunn's account.

The explosion occurred on a Saturday evening in July 1983. Only three employees were working on the site at the time of the explosion. Two of these were injured whereas the third escaped unhurt. The plant was substantially damaged.

Figure 2.33 shows the basic layout of the plant.

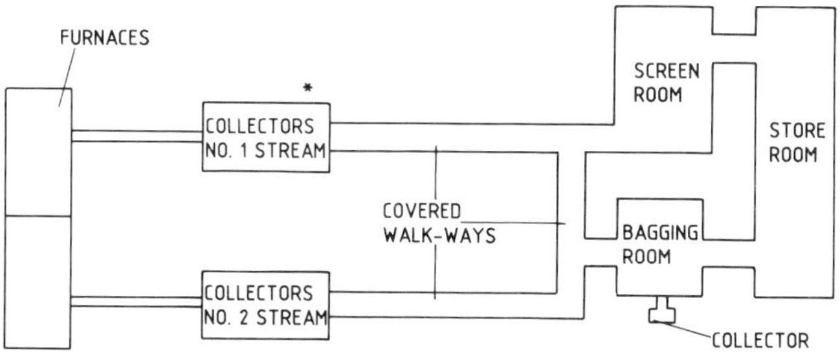

Figure 2.33 *Layout of plant for atomized aluminium powder production, in Anglesey, UK, which was damaged by an extensive dust explosion in July 1983. Ignition probably occurred in the No. 1 stream collector system* (From G. Lunn, 1984)*

Molten aluminium from the furnaces was broken up into small droplets by a jet of air. The aluminium powder so formed was carried by a current of air along sections of horizontal ducting at ground level before entering a riser which delivered it to a two-stage collecting system. There were two parallel collector streams, as shown in Figure 2.33. After the powder had been separated out in the collectors, the air passed through a fan and out to the atmosphere via a vertical stack. The powder dropped through rotary valves into a 'Euro-bin', one for each stream. When full, the bins were transported along a covered walk-way from beneath the collector to the screen-room where the aluminium powder was separated into particle-size fractions. The fractions were bagged in the bagging-room, and the bagged powder was taken through a short corridor to the store room.

The explosion swept through almost the entire plant. Examples of the extensive damage are given in Figures 2.34 and 2.35. Figure 2.34 shows the No. 2 stream collector plant and Figure 2.35 the screen room.

Figure 2.34 *Damaged No. 2 stream collectors after a dust explosion in an aluminium powder production plant at Anglesey, UK, in 1983 (Courtesy of G. Lunn, Health and Safety Executive, UK)*

Figure 2.35 *Damaged screen room after a dust explosion in an aluminium powder production plant at Anglesey, UK, in 1983 (Courtesy of G. Lunn, Health and Safety Executive, UK)*

According to Lunn (1984), neither the ignition source nor the location of the point of ignition was identified conclusively, but the fact that only No. 1 stream was in operation at the moment of the explosion would indicate that the explosion started there. The damage picture suggested that ignition could have occurred either before or within the first stage of the No. 1 stream collectors. Air blasts from the initial explosions then stirred up dust deposits in the walk-ways and screen room, allowing the flame to propagate into these areas.

The combination of a turbulent aluminium dust cloud ejected at a relatively high pressure from the No. 1 stream collectors, and a large, energetic and turbulent ignition source provided by the flames ejecting from the open vents generated ideal conditions for a dust explosion in the space between the No. 1 and No. 2 stream collectors capable of generating a significant blast overpressure. In fact, the damage to the No. 2 stream collectors (Figure 2.34) suggested that an overpressure had been exerted downwards, collapsing the structure. However, the evidence also suggested that a relatively violent explosion inside the No. 2 stream collectors had taken place. Air movement from an external explosion, and collapse of the structure could be sufficient to disperse dust inside the collectors. Ingress of flame from the external explosion into the collectors through tears in the bodywork caused by the collapse would provide multiple ignition sources.

An external explosion occurring some distance from the ground between the two collectors would also explain the damage to the cladding on the furnace room and the covered walk-way beneath the No. 2 stream collectors. The cladding on the furnace room had not been blown out by an internal explosion, but must have been pulled away from its fastenings by suction. This could have been caused by air movement generated by an explosion in the open air between the collectors. Similarly, cladding on the walk-way has been pulled away rather than blown out.

REFERENCES

Alameddin, A. N., and Foster, R. K. (1984) Evaluation of a Coal Bin Explosion Accident in Cushenbury Cement Plant. Report D4839-S497, (August) Industrial and Electrical Safety Division, Safety and Health Technology Center, Denver, Colorado, USA

Andersson, B. (1988) Kolpulversilo hotade explodera. Såhär löstes problemet. Sirenen, *Räddningsverkets Tidning* Nr. **3**, October p. 4

Astad, A. (1970) Private communication to R. K. Eckhoff from director A. Astad. Stavanger Port Silo, Norway

Berg, E. (1989) Private communication to R. K. Eckhoff from E. Berg, Dyno Industries, Gullaug, Norway

Borisov, A., and Gelfand, B. (1989) Private communication to R. K. Eckhoff from A. Borisov and B. Gelfand, USSR Academy of Science, Moscow

Braaten, T. S. (1985) Investigation of Silo Plant at Kvalaberget, Stavanger, Norway, after Explosion on 22nd November 1985. Norwegian Factory Inspectorate, Internal Report, 27 (November)

Eckhoff, R. K. (1980) Powder Technology and Dust Explosions in Relation to Fish Meals. Paper given at Internat. Symp. Processing of Fish Meal and Oil, Athens, October 6, 1980. Report No. 803301–2 (June) Chr. Michelsen Institute, Bergen, Norway

Fire and Police Authorities of Bremen (1979) Brand- und Explosionsschaden Bremer Rolandmühle am 6. Februar 1979. Eine Dokumentation. Issued by the Fire and Police Authorities of Bremen

Johansen, A. Fr. (1976) Så smalt det på Vippetangen. *Kornmagasinet* No. **3** p. 11

Johansen, A. Fr., Johansen, A. H., Mo, A. (1987) Rapport over støveksplosjonen ved Oslo Havnesilo – Vippetangen, 29. Juni (1987) Internal Report 20th August, Norwegian Grain Corporation

Kauffman, C. W. (1982) Agricultural Dust Explosions in Grain Handling Facilities. In *Fuel-Air Explosions*, ed. by J. H. S. Lee and C. M. Guirao, University of Waterloo Press, Canada pp. 305–347

Kauffman, C. W., and Hubbard, R. F. (1984) *An Investigation of Fourteen Grain Elevator Explosions Occurring Between January 1979 and April 1981*. Occupational Safety and Health Administration (OSHA) (May) Washington DC

Kauffman, C. W. (1989) Recent Dust Explosion Experiences in the US Grain Industry. In *Industrial Dust Explosions*, ASTM Special Techn. Publ. 958, (ed. K. L. Cashdollar and M. Hertzberg), pp. 243–264, ASTM, Philadelphia, USA

Kjerpeseth, E. (1990) Private communication to R. K. Eckhoff from E. Kjerpeseth, Elkem-Bremanger, Svelgen, Norway

Lunn, G. A. (1984) Aluminium Powder Explosion at ALPOCO, Anglesey, UK. Report No. SMR 346/235/0171, (September), Health and Safety Executive, Explosion and Flame Laboratory

Mo, A. (1970) Private communication to R. K. Eckhoff from A. Mo, Norwegian Grain Corporation Norway

Morozzo, Count (1795) Account of a Violent Explosion which Happened in a Flour-Warehouse, at Turin, December the 14th, 1785, to which are Added some Observations on Spontaneous Inflammations. *The Repertory of Arts and Manufactures* **2** pp. 416–432

Olsen, O. (1989) Private communication to R. K. Eckhoff from O. Olsen, Stavanger Port Silo, Norway

Patzke, J. (1981) Sicherheitstechnische Betriebserfahrungen bei der Kohlen-mahlung in Zementwerk Lägerdorf. *Zement-Kalk-Gips* **34** pp. 238–242

Pfäffle, H. (1987) Braunkohlenstaubverpuffung – Ursache, Verlauf und Folgerungen im Kraftwerk Lausward. *VGB Kraftwerkstechnik* **67** pp. 1163–1167

Stokes, D. A. (1986) Fording Coal Limited Silo Explosion. *CIM Bulletin* **79** No. 891, pp. 56–60

Storli, K.: (1976) Private communication to R. K. Eckhoff from K. Storli, Norwegian Factory Inspectorate

Templin, G. (1990) Private communication to R. K. Eckhoff from G. Templin, Nord Mills, Malmö, Sweden

Wibbelhoff, H. (1981) Explosion in Braunkohlenstaub-Feuerungsanlage. *Steine und Erden* No. **3** pp. 112–113

Xu Bowen, (1988) The Explosion Accident in the Harbin Linen Textile Plant. *EuropEx Newsletter*, Edition **6**, January pp. 2–3

Xu Bowen *et al.*, (1988) The Model of Explosion Accident Determined by the Seismic Record. Unpublished English manuscript concerning the Harbin Linen Textile Plant explosion, given by Xu Bowen to R. K. Eckhoff, (November)

Zhu Hailin, (1988) Investigation of the Dust Explosion in Harbin Linen Factory. Unpublished English manuscript given to R. K. Eckhoff by Zhu Hailin (November)

Chapter 3

Generation of explosible dust clouds by re-entrainment and re-dispersion of deposited dust in air

3.1 BACKGROUND

The dust concentration range, within which flames can propagate through a cloud of combustible dust in air, spans from the order of 50 g/m^3 to a few kg/m^3. The bulk density of powders and dusts, when settled in a layer or a heap, range from a few hundred kg/m^3 and upwards. Therefore, there is a gap of a factor of at least 100 between the maximum explosible dust concentration and the bulk density in the settled state. Consequently, in order for an explosible dust cloud to be formed, the dust must be suspended in the air to the extent that the concentration of dust per unit volume of cloud drops into the explosible range.

In dust explosion research, the important role played by this resuspension process has often been overlooked, or underestimated. It is realized that particle size plays a key role both with respect to the ignition sensitivity and the explosibility of dust clouds. However, it has not always been realized that fine, cohesive powders cannot be dispersed in a gas as individual primary particles unless particle agglomerates are exposed to very high shear or tensile stresses. This means that the effective particle size in a dust cloud can be much larger than the size of the primary particles.

It is interesting to note that Professor Weber, one of the pioneers of dust explosion research, stressed the importance of dust cohesion and dispersibility more than 100 years ago. In his excellent paper on the ignitability and explosibility of flour (Weber 1878) emphasizes that 'cohesion of the flour, which is caused by inter-particle adhesion, plays an important role with respect to the ability of the flour to disperse into explosible dust clouds.' Weber suggested that two large dust explosion disasters, one in Szczecin (Stettin) and one in München, were mainly due to the high dispersibility of the flours. He also demonstrated, using simple but convincing laboratory experiments, that the dispersibility or dustability of a given flour increased with decreasing moisture content in the flour.

In some special situations such as in air jet mills, explosible dust clouds may be generated *in situ*, i.e. the dust particles become suspended in the air as they are produced. However, in most cases explosible dust clouds are generated by re-entrainment and re-dispersion of powders and dusts that have been produced at an earlier stage and allowed to accumulate as layers or heaps. Such accumulation may either be intentional, as collection of powders and dusts in silos, hoppers and bag filters, or unintentional as deposition of dust on beams, external surfaces of process equipment or walls and floors of work and storage rooms.

Re-suspension and re-dispersion of dust may either occur intentionally, e.g. by handling and transport in various process equipment (powder mixers, bucket elevators, pneumatic transport etc.), or unintentionally by bursting of sacks and bags that contain powder, leaks of dust from process equipment, or by sudden blasts of air generated by an explosion that has started elsewhere in the plant.

The characterization of the 'state' of a dust cloud is far more complicated than characterizing the 'state' of a premixed quiescent gas mixture. For a quiescent gas the thermodynamic state is completely defined by the chemical composition, the pressure and the temperature. For a dust cloud, however, the state of equilibrium will be complete separation, with all the particles settled out at the bottom of the system.

In the context of dust explosions, the relevant 'state' will therefore always be dynamic. In various industrial environments as well as in experiments with dust clouds, gravity and other inertia forces act on the dust particles, giving rise to a complex dynamic picture. In the ideal static dust cloud, all the particles would be located in fixed positions, either ordered or at random. The closest approximation to the ideal dust cloud that can be encountered in practice is probably a cloud in which the particles are settling in quiescent gas under the influence of gravity alone.

3.2 STRUCTURE OF PROBLEM

Formation of explosible dust clouds from powder deposits implies that particles originally in contact in the powder deposit must be separated and distributed in the air to give concentrations within the explosible range. There are two aspects to consider. The first is the spectrum of forces originally acting on and between the particles in the deposit, resisting the separation of the particles. The second aspect is the forces and energy required for the separation process under various conditions.

Eckhoff (1976) suggested that a global dispersibility parameter for a powder deposit may be defined by considering these two aspects. A given mass of powder at equilibrium with the ambient atmosphere, contains a finite number of inter-particle bonds, each of which requires a specific amount of work to be broken. The total minimum work W_{min} needed to break all these bonds in one unit mass of powder, could in principle be calculated by integrating the work required for breaking all the individual inter-particle bonds. The influence of gravity would depend on whether the particles would have to be raised into suspension or whether dispersion would be downwards. One could then define a theoretical upper limit value of the dispersibility for that specific powder deposit by:

$$D_{max} = \frac{1}{W_{min}} \tag{3.1}$$

When defined in this way, the 'dispersibility' has the dimension mass per unit of energy or work, and is thus a measure of the quantity of powder that can be completely dispersed by spending one unit of energy from external sources in the process. However, no realistic dispersion process can be one-hundred per cent efficient. This can be accounted for by incorporating an efficiency factor, K:

$$D_{real} = \frac{K}{W_{min}}, \; 0 < K < 1 \tag{3.2}$$

The particle size distribution of the powder has a great influence on W_{min} at a given powder bulk density. It also is well known that powders consisting of small particles are compressible. The reason is that the various inter-particle forces other than gravity are stronger than the gravity forces and therefore permit the formation of loosely packed particle arrangements that would have collapsed had gravity been the only force in operation. This means that the number of inter-particle bonds per unit mass of cohesive powder can be increased by compacting the powder, i.e. by increasing the bulk density of the powder deposit. Therefore W_{min} also increases with the degree of compaction. Moisture influences W_{min} by influencing the strength of certain types of inter-particle bonds.

The logical link between W_{min} and nature and number-density of the inter-particle bonds in a powder has given rise to detailed studies of various types of inter-particle bonds. Attempts have further been made at predicting aggregated powder-mechanical strength properties from microscopic inter-particle structure and forces. This kind of work is concerned with the quantity D_{max} (Equation (3.1.)

However, the efficiency factor $0 < K < 1$ in Equation (3.2) allows D_{real} to have any value between zero and D_{max}, depending on the way in which the work W_{min} is applied to the powder to be dispersed. This in turn depends on the geometrical arrangement of the powder and the form of the mechanical energy available for the dispersion process. If a comparatively coarse non-cohesive powder is for example charged into a silo from a hopper at the silo top, the potential energy of the powder, when being transformed to kinetic energy in the gravity field, may be sufficient to generate well dispersed explosible dust cloud in the silo. The same applies if deposits of this powder are falling down from shelves and beams in a factory workroom.

However, very energetic air flows may be required to raise deposits of such a powder on the factory floor into explosible suspensions.

When considering the other end of the scale, cohesive powders composed of very small particles, inter-particle forces play a minor role and inter-particle bonds may not be broken unless the particle agglomerates are exposed to large shear forces. This means that complete dispersion into primary particles is only possible in high velocity flow fields, or if the particles are exposed to high-velocity impacts.

Consequently, the understanding of how explosible dust clouds can be generated, requires knowledge both of the nature of the powder (W_{min}) and of the actual dispersion process (K). The dispersion process in turn depends very much on the actual industrial situation, which will be different in bucket elevators, pneumatic transport systems, fluidized beds, various kinds of mills, driers, mixers, cyclones, filters and silos. Therefore, intimate knowledge of the nature of the industrial environment is required.

It has not been possible to trace any comprehensive theory of the generation of dust clouds, leading from the properties of the powder deposit, via the nature of the energy available for dispersion, to the structure of the dust cloud. However, in view of the wide variation in possible boundary conditions in industrial practice, one would not expect to find one single, unified theory covering all possible situations. On the contrary, each specific situation needs to be analysed separately. Much work has been conducted on various limited elements inherent in the total problem complex. Some of this will be reviewed in the following in sufficient detail for the genuine nature of the various problems to become visible. This is considered important in a new text on dust explosions because in the past, dust explosion research has often been conducted without paying appropriate attention to the central role played by powder mechanics/particle technology.

3.3 ATTRACTION FORCES BETWEEN PARTICLES IN POWDER OR DUST DEPOSITS

Two categories of inter-particle forces exist, one that operates even in dry powders, and one that is due to the presence of a viscous liquid. Useful summaries have been given by Green and Lane (1964), Corn (1966), Rumpf (1974), Schubert (1979) and Enstad (1980).

3.3.1 VAN DER WAALS' FORCES

The van der Waals' force F_w between two spherical particles has been estimated theoretically by integrating London–van der Waals' forces over all interacting pairs of molecules. The resulting expression is:

$$F_w = \frac{A}{a^2}\frac{x_2 x_2}{(x_1 + x_2)} \tag{3.3}$$

where A is a constant, a the smallest distance between the sphere surfaces and x_1 and x_2 the diameters of the two spheres.

Van der Waals' forces between particles are of significance as long as $x < 100$ nm. If $x_1 \gg x_2$, the force is only determined by the size of the smallest particle, and equation (3.3) reduces to

$$F_w = \frac{A}{a^2} x_2 \tag{3.4}$$

Most particles in real life are not smooth spheres, but of irregular shape and surface topography. Schubert (1979) showed that F_w between a plane surface and a point on an irregular particle of diameter x, having a small elevation of radius r that touches the plane surface, is:

$$F_w = A \times \left(\frac{x}{(r + a_0)^2} + \frac{2r}{a_0^2}\right) \tag{3.5}$$

The distance, a_0, is the smallest distance that can exist between two bodies in touch, and it is estimated at 0.4 nm.

3.3.2 ELECTROSTATIC FORCES

When considering electrostatic forces, one distinguishes between electrically conducting and non-conducting particles. In the case of conducting particles, electrostatic inter-particulate attraction between touching particles may occur even if the particles did not initially carry any net excess charge, provided their electron work functions are different. Electrons will then be transferred from one particle to the other. Different electron work functions can occur in particle systems of apparently identical materials, due to differences in impurities, oxide layers etc. Provided the smallest distance a between the two surfaces is shorter than 100 nm, i.e. the particles are in electric contact, the electrostatic contact attraction force between the two conducting particles is:

$$F_{e,c} = \pi\epsilon_1\epsilon_0 \times \frac{U^2}{a} \times \frac{x_1 x_2}{(x_1 + x_2)} \tag{3.6}$$

Here ϵ_0 is the permittivity of vacuum and ϵ_1 the dielectric constant of the gas surrounding the particles. U is the contact potential between the two particle surfaces.

For electrically non-conducting particles, such as plastics, the electrostatic contact force is negligible. In this case, electrostatic attraction between particles is caused by excess charges on the particle surfaces, acquired tribo-electrically during preceding production and handling. The attraction force between two non-conducting particles having total excess opposite charges on the surfaces of q_1 and q_2, equals:

$$F_{e,n} = \frac{q_1 q_2}{4\pi\epsilon_1\epsilon_0} \times \frac{1}{\left(\frac{x_1 + x_2}{2} + a\right)^2} \tag{3.7}$$

For $a \gg (x_1 + x_2)$, equation (3.7) reduces to Coulomb's equation for attraction between two opposite point charges. If a is much smaller than the diameter of the largest particle, $F_{e,n}$ will essentially be independent of a.

Equations (3.3)–(3.7) are all concerned with the attraction between two single particles under idealized conditions. It is clear, therefore, that these equations are of limited value for predicting inter-particle attraction forces in real powders and dusts where many particles are interacting and particle shape and surface properties may be complex. In the case of electrostatic forces, realistic assessment of the particle charges q_1 and q_2 is also difficult, even for idealized particle geometries.

In industrial practice the relative humidity of the air will have different values, and this will influence the strength of the electrostatic attraction forces between particles in powders. This influence was investigated by Nguyen and Nieh (1989). They proposed a general mechanism of charge elimination in flowing powders in humid air by 'hydrated ion clusters' $(H_2O)_nH^+$ and $(H_2O)_nOH^-$ and their polymers.

Ross (1988), working with clouds of lycopodium in air, was able to significantly reduce electrostatic agglomeration of particles, as well as electrostatic adhesion to the wall of an experimental flame tube, when the air was ionized by means of an alpha emitter mounted on the flame tube wall.

3.3.3 INTER-PARTICLE FORCES DUE TO LIQUIDS

It is a common experience from practice in industry that dry dusts are usually easier to disperse than moist dusts (one exception can be heavily electrostatically charged dry plastic powders). Even small quantities of adsorbed moisture can in some cases increase the attraction forces between particles considerably. Adsorbed layers of up to 3 nm thickness can adhere firmly to the particle surface and make it more smooth. This can reduce the effective distance between two touching particles appreciably. Even for a spherical particle as small as 1 μm diameter the volume of a 3 nm layer of liquid water constitutes only 2% of the particle volume. (The situation is different if the moisture is also absorbed by the interior of the particle, rather than being just adsorbed on its surface.)

The next stage is reached when the moisture content in the powder has become so high that excess water starts to form liquid bridges between particles, as illustrated in Figure 3.1(a). If the moisture content increases further, a transition range is reached which is characterized by some of the space between particles being completely filled with water (Figure 3.1(b)). Figure 3.1(c) illustrates the capillary range where the capillary under-pressure is the main source of the cohesion. If the water content is increased beyond this point, the system is transformed from a cohesive powder to a suspension of particles in a liquid (Figure 3.1(d)).

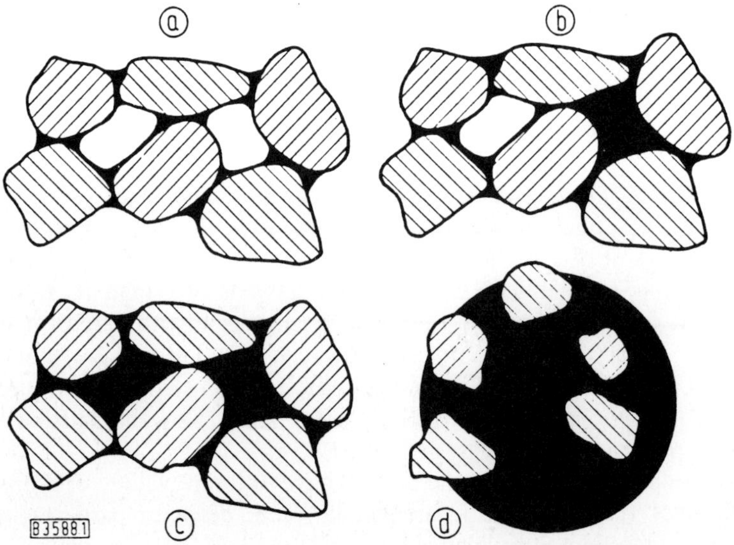

Figure 3.1 *Distribution of a liquid in a powder (From Schubert, 1973)*

In order to assess the strength of liquid bridges between particles in a powder (Figure 3.1 (a)), Schubert (1973) used the approximate relationship derived by Rumpf (1970) for the tensile strength σ_T of a bed of monosized spheres (see 3.4.1):

$$\sigma_T = \frac{1-\epsilon}{\epsilon} \times \frac{F(\epsilon)}{x^2} \tag{3.8}$$

Here ϵ is the porosity of the bed, $F(\epsilon)$ the mean inter-particle force (dependent on ϵ) and x the particle diameter. Equation (3.8) is derived from Equation (3.10) via the relationship $\epsilon \times k(\epsilon) \simeq 3.1 \simeq \pi$ found experimentally for spherical particles.

Schubert's equation for the tensile strength of a powder due to inter-particle liquid bridges is as follows:

$$\sigma_T = \frac{\gamma}{x} \times \frac{1-\epsilon}{\epsilon} \times F_F\left(\epsilon, S, \delta, \frac{a}{x}\right) \tag{3.9}$$

Here γ is the surface tension of the liquid. $F_F(\epsilon, S, \delta, a/x)$ is the dimensionless liquid-bridge inter-particle attraction force, where S is the fraction of the total pore volume between the particles that is filled with liquid, and a and δ as shown in Figure 3.2. Equation (3.9) cannot be solved analytically, but Schubert (1973) arrived at a graphical solution.

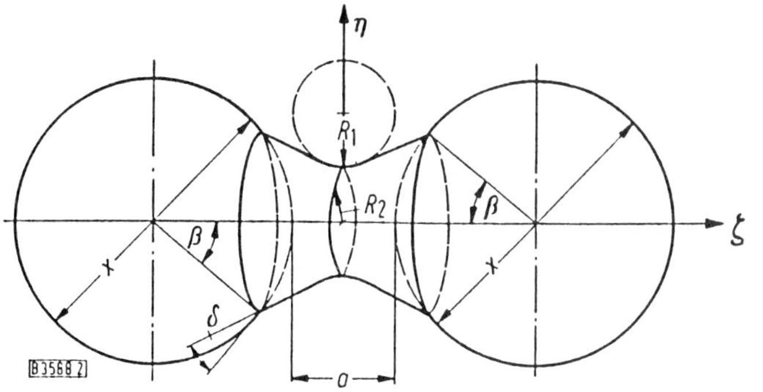

Figure 3.2 *Liquid bridge between two identical spherical particles (From Schubert, 1973)*

The liquid bridge regime extends up to about $S = 0.25$ (Schubert's experiments with 70 μm limestone particles). This regime is the most relevant one with a view to transformation of dust deposits into explosible dust clouds. For a powder of specific density of 1 g/cm^3 packed to a porosity ϵ of 0.4, $S = 0.25$ represents a moisture content of 14% (neglecting moisture absorbed by the interior of the particles). The transition regime in which the liquid partly forms bridges between particles and partly fills the voids completely, spans from $S = 0.25$ to $S = 0.8$. When the voids between the particles are just filled up with liquid, the tensile strength of the bulk powder is determined solely by the internal underpressure caused by capillary forces. In practice this will be the case for $0.8 < S < 1.0$.

Figure 3.3. summarizes some of Schubert's (1973) experimental and theoretical results. He found that equation (3.9), using $a/x = 0.05$, gave excellent agreement with the experiments in the liquid bridge regime, for which there is a strong increase of σ_T as S increases from zero to 0.1.

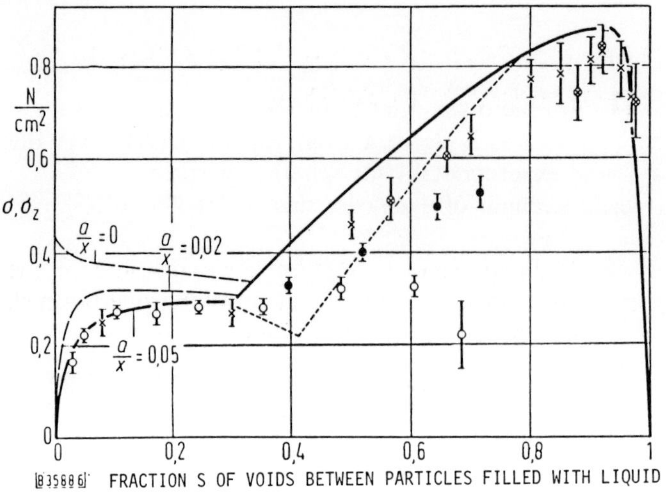

Figure 3.3 *Tensile strength σ_T of a powder bed as a function of the fractions of the voids between the particles that are filled with liquid. Experiments with limestone of 70 μm particle diameter. $\epsilon = 0.415$. ———, — — — and ---- are theoretical calculations using different assumptions (From Schubert, 1973)*

For particles of density 1 g/cm^3 packed to a porosity of 0.4, $S = 0.1$ corresponds to a moisture content of 6.25%. It is therefore to be expected that the influence of the moisture content on the dispersibility of the powder would be particularly strong in the range of a few per cent moisture. However, this does not apply if a significant fraction of the moisture is absorbed by the interior of the particles rather than adhering to the particle surfaces.

As S increases and moves into the capillary pressure region, the tensile strength of the powder bed increases further. As Figure 3.3 shows, the tensile strength of the powder bed in the region just before complete saturation is three times the maximum tensile strength in the liquid bridge region.

However, as pointed out by Enstad (1980), the tensile strength of the powder bed in the capillary under-pressure regime can never exceed a pressure difference of one atmosphere. In the liquid bridge regime there is no such limitation, and for small particle diameters $\ll$ 70 μm equation (3.9) can easily give tensile strengths corresponding to pressure differences of several atmospheres. In this range of particle sizes the shape of the curve of σ_T (S) will differ from that in Figure 3.3, by having its maximum in the liquid bridge range of $S < 0.25$.

Adding liquids to dusts is sometimes used intentionally in industry for reducting dust dispersibility. One application of this method is addition of soya bean oil to grain for preventing generation of grain dust clouds in grain storage plants. See Section 1.4.10 in Chapter 1.

3.4 RELATIONSHIP BETWEEN INTER-PARTICLE ATTRACTION FORCES AND STRENGTH OF BULK POWDER

3.4.1 THEORIES

The question arises whether it would be possible to deduce some measure of the inter-particle forces in powder deposits from measurement of bulk powder properties such as shear strength and tensile strength. As already mentioned, Rumpf (1970) developed the following equation for the relationship between the bulk strength σ of a powder bed of monosized particles and the mean inter-particle force $F(\epsilon)$, the coordination number $k(\epsilon)$ (average number of neighbouring particles with which a given particle is in contact), particle diameter x and porosity of the powder bed ϵ:

$$\sigma = \frac{1-\epsilon}{\pi} k(\epsilon) \frac{F(\epsilon)}{x^2} \tag{3.10}$$

Equation (3.10) shows that for geometrically similar powder beds, differing only in particle size x, and assuming that the mean attraction force per inter-particle contact is independent of particle size, the strength of the bulk powder is inversally proportional to x^2, i.e. the powder strength increases strongly as the particle size decreases.

Rumpf (1970) was able to show that equation (3.10) is valid not only for spherical particles, but also for irregular ones provided certain statistical conditions concerning the arrangement of the particles in the bed and the particle shape are fulfilled. By extending his treatment to beds containing a variety of particle sizes, he arrived at the equation:

$$\sigma = \frac{1-\epsilon}{f_0 M_{30}} \times \int_{x=0}^{\infty} [k(x) \times F(x, n(x)) \times n(x)]dx \tag{3.11}$$

Here f_0 is a particle shape factor and M_{30} the 'third moment' of the particle size distribution (distribution of particle volume).

For integration of equation (3.11) the coordination number $k(x)$ as a function of particle size, and the inter-particle force $F(x, n(x))$ as a function of particle size and particle size distribution must be known. The practical usefulness of equation (3.11) is therefore limited, but it establishes a formal logical link between the bulk strength of a powder, and the mean microscopic inter-particle attraction force.

Molerus (1978) also studied the link between inter-particle forces and bulk powder strength. He made use of the following empirical relationship between the adhesive force F between a limestone particle and a plane metal surface, and the external force N used initially for pressing the particle against the surface:

$$F = F_0 + \kappa N \tag{3.12}$$

F_0 is the attraction force between particles that are just touching without having been pressed against the plate by an external force. On the basis of theoretical considerations of the inter-particle forces in a cohesive bulk powder. Molerus developed a relationship of the same form as equation (3.12), where F_0 and κ where expressed in terms of the Hamaker constant, the plastic yield pressure of the particle material, a characteristic distance of adhesion (about 0.9 nm) and the size of the spot where the particles are touching. Encouraging agreement with experiment was obtained for limestone. Molerus then developed a theoretical model for the connection between such inter-particle forces and the cohesive properties of the bulk material by assuming that

1. Van der Waals' forces and deformation of the contact areas where the particles are touching each other, are responsible for the inter-particle adhesion.
2. The particles are monosized spheres.
3. the coordination number $k(\epsilon)$ is a unique function of the porosity of the particle bed.
4. Equation (3.10) is generally applicable for relating the macroscopic tensile and shear strength of the bulk powder to the corresponding microscopic inter-particle forces.
5. Breakdown of inter-particle adhesion occurs at a critical ratio between shear force and compressive force defining the internal angle of friction of the powder bed.

The theory predicts yield loci (see 3.4.2.1) for a bulk powder, with the corresponding cohesion and tensile strength values, as a function of the degree of compaction (or porosity ϵ). Encouraging agreement between experiments and theoretical prediction was found for a cohesive baryte powder.

3.4.2 MEASUREMENT OF THE MECHANICAL STRENGTH OF COHESIVE BULK POWDERS AND DUSTS

3.4.2.1 Basic concepts

If a sample of dry sand is subjected to a compressive force, the volume reduction, or reduction in the porosity ϵ, will be very small. Furthermore, as soon as the compressive force is released, the sand will flow freely again. Such behaviour is characteristic of non-cohesive powders, in which inter-particle forces of the nature discussed in Section 3.3 play little or no role compared with gravity. If, however, a sample of finer dust or powder, such as an organic pigment, is subjected to compression, the powder sample will shrink and the porosity ϵ be reduced. Removal of the compressive force will not cause the powder sample to return to its original state of loose packing, but it will maintain a lower porosity and stick together as a lump. The larger the compressive force, the lower the resulting ϵ, and the stronger the powder sample will become.

The science of powder mechanics, which deals with these relationships in a systematic way, was established by the pioneering work of Jenike (1964). Jenike used Sokolovski's (1960) theory of the statics of soils as his starting point. Schwedes (1976) has given a concise summary of the basic concepts in Jenike's theory. The powder mechanical state of one specific cohesive powder sample of a given porosity ϵ is characterized by the so-called yield locus, as illustrated in Figure 3.4. The yield locus is an envelope curve for all the Mohr circles describing stress combinations causing yield, referred to a specific powder sample for which σ_1 was the maximum principal consolidation stress during preparation of the sample. The porosity (and bulk density) of the specific powder in question is a unique function of σ_1. S is the tensile force, N the normal force and A the area of the powder specimen in the shearing plane. The quantity f_c is the maximum principal stress at failure when the powder sample is in a situation where the minor principal stress is zero. σ_T is the tensile strength of the powder sample and c is the cohesion, defined as the shear strength of the powder sample at zero normal load.

For a given type of cohesive powder, there exists a continuous range of yield loci, each locus being characterized by a given porosity $\epsilon(\sigma_1)$. Both f_c, the cohesion c and the tensile strength σ_T increase systematically with decreasing ϵ, or increasing σ_1. The straight line $\tau = \sigma_N \times \tan \phi_e$ is called the effective yield locus. The angle ϕ_e is a measure of the internal friction in the powder during steady flow (plastic deformation).

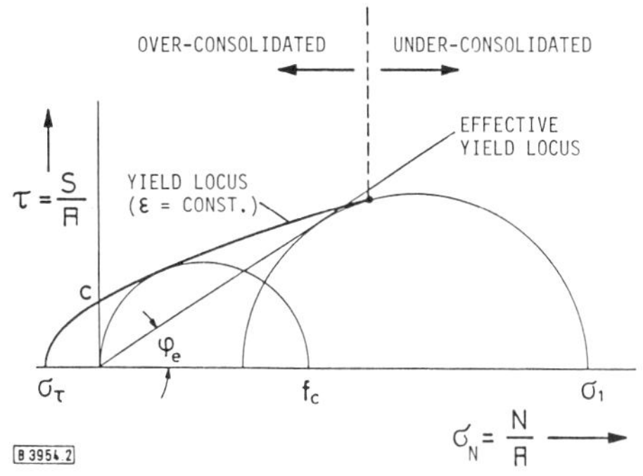

Figure 3.4 *Yield locus and effective yield locus of a given powder at a given porosity ϵ (From Schwedes, 1976)*

For a non-cohesive, free-flowing powder, the yield locus and the effective yield locus will coincide and pass through the origin, and both σ_T and c will be zero.

3.4.2.2
Shear cells

Yield loci as illustrated in Figure 3.4 are determined by means of shear cells. A cross section of the well-known Jenike cell is shown in Figure 3.5.

This cylindrical cell of 95 mm diameter is split, and the upper ring can be pushed horizontally in relation to the lower, fixed part. The test procedure for obtaining a point on a yield locus (Figure 3.4) consists of two steps. First the powder is consolidated during plastic flow to a given porosity ϵ under the action of a major principal stress σ_1. In the second step the sample is shear strained at a constant strain rate, while being compressed by a constant normal stress $\sigma_N = N/A$, where N is the normal force and A is the cross section of the cell (71 cm^2). The shear force S, which is recorded continuously during the process, will increase with the strain to a maximum value, at which the powder sample fails, and S drops suddenly. This maximum value of S defines the $\tau = S/A$ value that together with the corresponding $\sigma_N = N/A$ gives a point on the yield locus. By shearing identical powder samples (the same $\epsilon(\sigma_1)$), at different σ_N, the entire yield locus is determined.

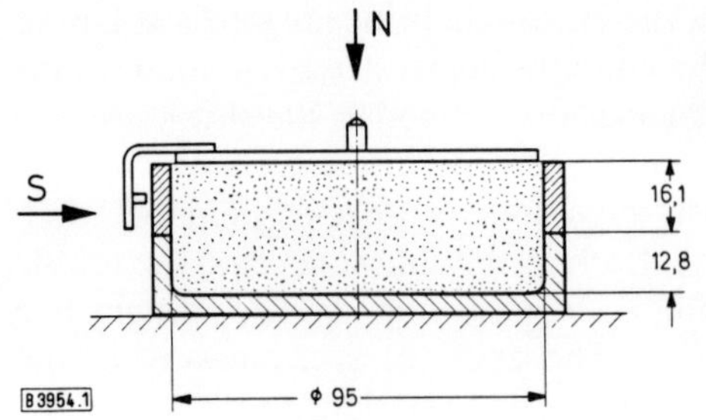

Figure 3.5 *Vertical cross section of the Jenike shear cell for measuring the mechanical strength of powders. All dimensions in mm (From Schwedes, 1976)*

In the context of dust dispersibility, the mechanical 'strength' of a given powder, consolidated to a given porosity ϵ by a major principal stress σ_1, can be characterized either by $f_c(\epsilon)$, $c(\epsilon)$ or $\sigma_T(\epsilon)$ (Figure 3.4). The Jenike shear cell gives a measure of $f_c(\epsilon)$. $c(\epsilon)$ can only be estimated by extrapolating Jenike cell failure loci to $\sigma_N = 0$, which may be uncertain, whereas $\sigma_T(\epsilon)$ cannot be determined by the Jenike shear cell. Recently a detailed standardized procedure for conducting Jenike shear cell tests has been worked out via international cooperation (EFCE Working Party Mech. Part. Solids (1989)).

The validity of $f_c(\epsilon)$ from the Jenike shear cell in absolute terms has been questioned. Arthur, Dunstan and Enstad (1985) have developed a new, biaxial test apparatus that enables a more direct measurement of $f_c(\epsilon)$ right down to very low consolidation stresses where $f_c \simeq \sigma_1$.

3.4.2.3
Tensile strength testers

Figure 3.6 illustrates the traditional split-plate tilting-table tensile strength tester. Schubert and Wibowo (1970) also used a more sophisticated cell by which the capillary underpressure during tensile strain of powder saturated with liquid, could be measured.

By slowly increasing the tilting angle α shown in Figure 3.6, a point is reached where the powder sample ruptures. When the mass of the system that travels down the inclined plane after rupture, is known, the tensile force is also known, assuming that frictional losses can be neglected. This is a reasonable assumption when the cell is supported by steel balls as indicated in Figure 3.6.

The ratio of the estimated tensile force at the point of rupture, and the cross sectional area of the powder sample in the plane of rupture has traditionally been taken as a measure of the tensile strength of the powder. However, Schubert and Wibowo (1970) investigated the influence of the depth of the powder bed on the measured tensile strength. Although the maximum tensile force just before rupture increased somewhat with the bed thickness, the ratio of the two decreased as the thickness increased. This is because it is impossible to apply the tensile force evenly over the entire rupture plane. Instead the tensile stress in the powder will be concentrated in the region close to the bottom of the cell, where the movable and stagnant bottom plates separate. When rupture occurs, it will propagate from the bottom and upwards in the powder bed. Therefore, tensile strength values of powders determined from just one bed thickness, are bound to be arbitrary numbers, although relative comparison of different powders may be possible. Schubert and Wibowo (1970) suggested that this problem can be overcome by determining the nominal tensile strength (tensile force just before rupture divided by rupture area) for various powder bed thicknesses and extrapolating to zero thickness. A typical set of results are given in Figure 3.7.

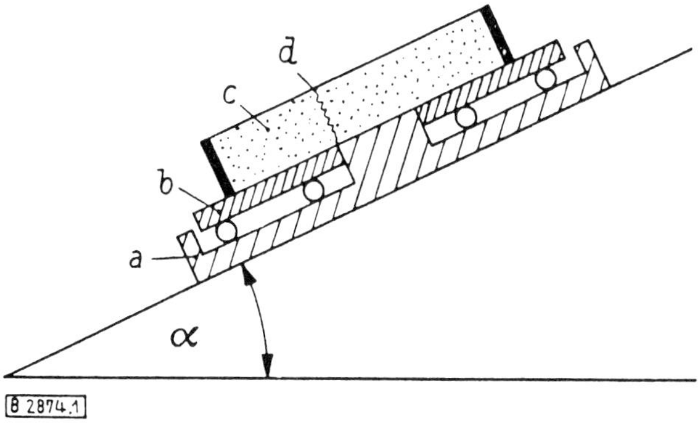

Figure 3.6 *Split-plate tilting-table tensile strength tester for powders. a – base plate; b – moveable plate; c – powder/dust sample; d – rupture plane; (From Schubert and Wibowo, 1970)*

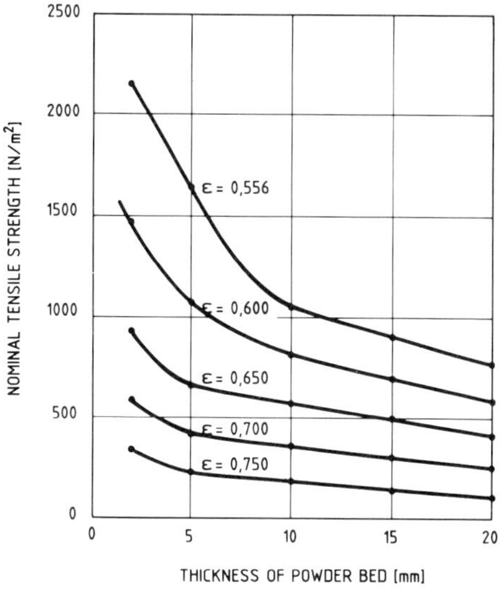

Figure 3.7 *Influence of powder bed thickness and powder porosity* ϵ *on nominal tensile strength of a fine limestone powder of mean particle diameter 3* μm *(From Schubert and Wibowo, 1970)*

The question is now if the tensile strength σ_T for a powder, determined by Schubert and Wibowo's extrapolation method, fits together with the yield loci from shear cell measurements, as would be expected from Figure 3.4. Eckhoff, Leversen and Schubert (1978) investigated this using a fine SiC powder. The results from the tensile strength measurements are shown in Figure 3.8, whereas Figure 3.9 shows that σ_T values from extrapolation for the various major principal stresses σ_1 (i.e. various porosities ϵ) could be joined to the yield loci by approximately straight lines, assuming isostatic conditions in the tensile tests. However, if uni-axial conditions are assumed, the deviations between the extrapolated yield loci and the experimental shear cell data in the low stress regime become pronounced.

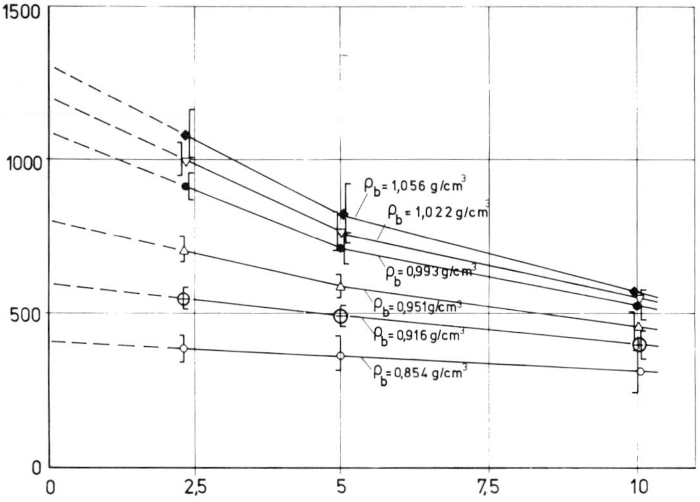

Figure 3.8 *Nominal tensile strength of a fine SiC powder as a function of the powder bed thickness and bulk density (or porosity ϵ) (From Eckhoff, Leversen and Schubert, 1978)*

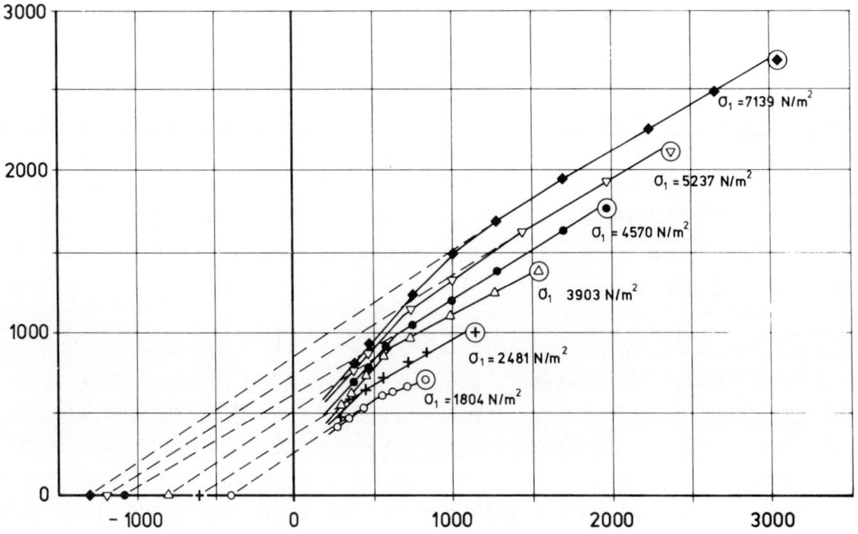

Figure 3.9 *Combination of shear test data and tensile test data assuming isostatic conditions in tensile tests (From Eckhoff, Leversen and Schubert, 1978)*

The results indicate that the Jenike shear cell underestimates the shear strength at low normal stresses. When performing the necessary extrapolation of yield loci data for estimating f_c and c by the Jenike cell, results for $\sigma_N < 0.3\,\sigma_1$ should definitely be discarded. Even σ_N data in the range $0.3\,\sigma_1 < \sigma_N < 0.5\,\sigma_1$ should be treated with caution.

This emphasizes the need for improved methods for measuring basic properties of powders, as proposed by Arthur, Dunstan and Enstad (1985).

An interesting experimental study of the correlation between the tensile strength of a bulk powder and its dispersibility in a gas was performed by Yamamoto (1990).

3.5 DYNAMICS OF PARTICLES SUSPENDED IN A GAS

3.5.1 TERMINAL SETTLING VELOCITY OF A PARTICLE IN THE GRAVITATIONAL FIELD

Terminal settling velocities of particles in air have been determined experimentally in numerous investigations. An early example is the work of Zeleny and McKeehan (1910), who conducted careful measurements of the terminal velocities of spherical drops and particles of paraffin, black wax and mercury in air at atmospheric pressure and room temperature. The measurements were in excellent agreement with Stokes' theory for the laminar flow regime.

Some pollens and spores were also included in this investigation, but for these particles the experimental terminal settling velocities were generally somewhat lower than the theoretical Stokes' velocity. This also applied to lycopodium, the spore of club moss, which has been widely used all over the world in dust explosion research (Eckhoff, 1970). Lycopodium particles are close to monosized with an arithmetic mean diameter of about 30 μm. The particle density is about 1.18 g/cm^3. According to Figure 3.10 this corresponds to a Stokes' terminal velocity of 0.035 m/s, whereas the experimental value was only 0.017 m/s. The difference by a factor of two was attributed to the formation of eddies in the wake of the spore and rotational settling, due to assymetric particle shape and a very rough surface texture (see Figures 3.11 and 3.12). If, on the other hand, a lower particle density based on the hydrodynamic envelope volume is used, agreement with Stokes law might be found. Geldart (1986) gives a simple method for measurement of appropriate particle densities of porous particles.

Figure 3.10 gives the terminal settling velocity in air in the gravitational field for smooth spherical particles of various diameters and densities. The straight parts of the lines in Figure 3.10 essentially represent the Stokes' law regime for the terminal settling velocity, v_t, of smooth spherical solid particles in a quiescent gas:

$$v_t = \rho_p x^2 g/(18\,\mu) \tag{3.13}$$

As smooth, spherical particles get smaller than a few μm diameter, they will attain somewhat higher terminal settling velocities than predicted by Stokes' law (Cunningham slip correction). For comparatively large particles, the viscous drag becomes greater than

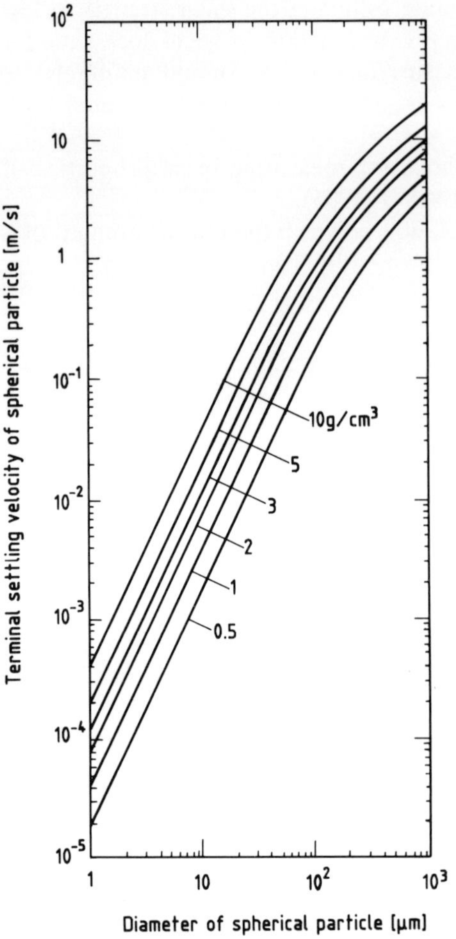

Figure 3.10 *Terminal settling velocities for spherical particles of various diameters and densities in air at atmospheric pressure and 20°C (From Perry & Chilton, 1973)*

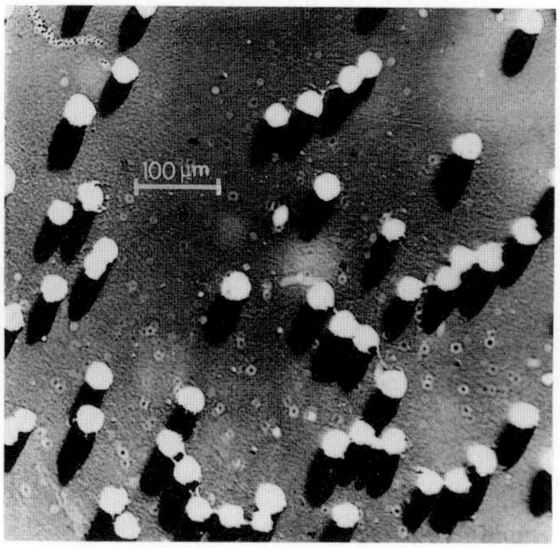

Figure 3.11 *Optical micrograph of a metal shadowed sample of lycopodium. Shadowing angle 20°*

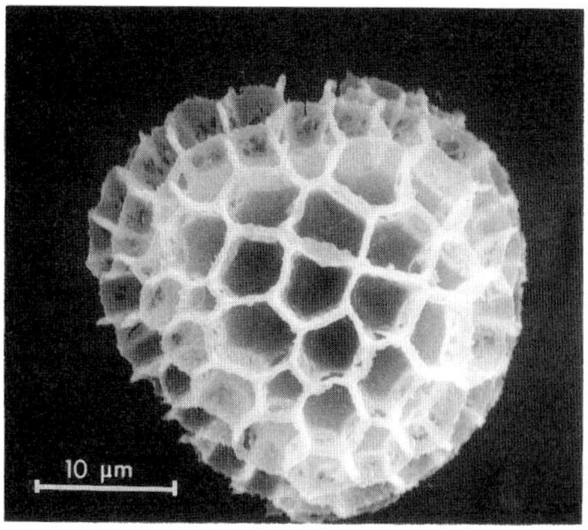

Figure 3.12 *Scanning electron micrograph of a single lycopodium particle showing the rough surface topography*

assumed in Stokes' law, and the terminal settling velocities will be lower than predicted. This is the reason for the curving of the lines in Figure 3.10 in the range of large particles.

The settling velocities indicated in Figure 3.10 apply even to particles in a dust cloud, provided the particle concentration is not too high and particle agglomeration can be neglected. For solids volume fractions below 0.001, the hindered settling effect causes less than 1% reduction of the settling velocities given in Figure 3.10 (Perry & Chilton, 1973). For a dust of particle density 1 g/cm^3, a volume fraction of 0.001 corresponds to a dust concentration of 1 kg/m^3, which would be in the upper part of the explosible range. Therefore Figure 3.10 is also adequate for a rough evaluation of the gravitional settling velocities of particles in explosible dust clouds.

3.5.2 DRAG ON A PARTICLE IN GENERAL

Figure 3.10 covers the terminal settling velocities of the particle sizes of primary interest in relation to dust explosion problems, and as shown, Stokes' laminar theory applies over most of the range. However, in many situations in industry, and particularly during dust explosions, general inertia forces may dominate over the gravity force, and other flow regimes may be of primary interest. The Reynolds' number of the particle is an important indicator of the flow regime. Reynolds' number for a particle of diameter x travelling in a gas is defined as:

$$Re = \frac{\rho_g v_{rel} x}{\mu} \tag{3.14}$$

where ρ_g is the density of the gas, v_{rel} the relative velocity between the particle and the gas, and μ the viscosity of the gas. The drag coefficient C_D is another important parameter. It is the ratio between the drag force acting on the particle, and the product of the cross sectional area of the particle and the dynamic pressure acting on that area. For laminar flow conditions (Stokes' range).

$$C_D = \frac{24}{Re}, \tag{3.15}$$

The change of the drag coefficient C_D as Reynolds' number increases, is shown in Figure 3.13 for three different particle shapes.

According to Haider and Levenspiel (1989) one can find more than 30 equations in the literature, which relate the drag coefficient C_D to the Reynolds' number for spherical particles falling at their terminal velocities. They also give more recent experimental data confirming that Figure 3.13 is adequate for isometric particles of sphericities Φ of 0.7–1.0, where Φ is defined as the ratio of the surface area of a sphere having the same volume as the particle, and the actual surface area of the particle. For discs of lower Φ values, in the range 0.2–0.02, C_D at a given Re are higher, by a factor of the order of 10, than shown by the curve in Figure 3.13.

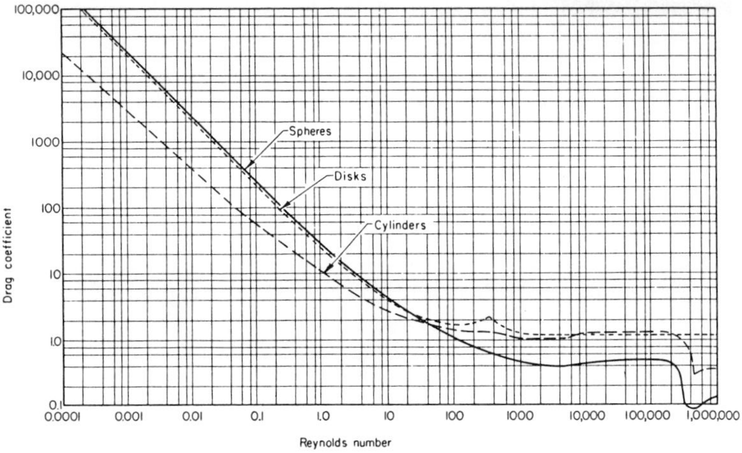

Figure 3.13 *Drag coefficient C_D for particles of various shapes, moving in a fluid at various Reynolds' numbers (From Perry and Chilton, 1973)*

Haider and Levenspiel also presented a series of graphs, corresponding to Figure 3.10, for the terminal settling velocities of non-spherical particles of various sphericities Φ.

The general expression for the terminal gravitational settling velocity of a particle in a gas is:

$$v_t = \left[\frac{V_p \times \rho_p \times 2g}{A \times \rho_g \times C_D}\right]^{1/2} \tag{3.16}$$

where V_p is the particle volume, A the projected particle area in a plane perpendicular to the gas flow direction, and ρ_p the particle density.

Rumpf's (1975) discussion of the various regimes of Re for smooth spherical particles is summarized in Table 3.1.

In the context of a dust particle in a gas, $Re = 10^5$ is an extremely high number. As an example a 100 μm diameter particle in air at atmospheric pressure and room temperature will have a relative velocity with respect to the gas of 17 km/s, which is far beyond even detonation front velocities.

Table 3.1 Ranges of drag forces on smooth spherical particles moving in a quiescent, non-compressible viscous medium (From Rumpf, 1975)

$Re \leq 0.25$	Range of Stoke's drag, i.e. C_D equals 24/Re.
$0.25 < Re < 10^3$	Significant deviation from perfect streamline flow round the particle and eddy formation in its wake starts at about Re = 25. The regime of eddy formation is fully developed at Re = 10^3. Navier-Stokes equations are applicable up to Re = 100.
$10^3 < Re < Re_c$ ($Re_c \approx 3 \cdot 10^5$)	The size of the eddy liberation zone in the wake of the particle remains approximately constant, and C_D is also approximately constant, and equal to 0.4-0.5.
$Re \approx Re_c$	At this point the laminar boundary layer round the upstream part of the particle breaks down, and the boundary region becomes fully turbulent, and C_D suddenly drops to the order of 0.1.
$Re > Re_c$	In this supercritical range, C_D again starts to increase with Re.

Considerations based on assuming non-compressible conditions only hold at low Mach-numbers (the Mach number is defined as the ratio between the relative velocity between the particle and the gas, and the speed of sound in the gas). Figure 3.14 shows the variation of *Re* for the particle with the relative velocity for particles of various diameters, travelling in air at atmospheric pressure and 20°C. For transformation to higher gas temperatures, Sutherland's formula for the influence of temperature (absolute) on the viscosity of gases is useful (Smithsonian, 1959):

$$\mu(T) = \mu_0\left(\frac{273 + C}{T + C}\right)\left(\frac{T}{273}\right)^{3/2} \tag{3.17}$$

For air, μ_0 (the viscosity at 0°C) is 1.7×10^{-5} kg/sm, whereas the temperature constant C equals 118 K.

According to Rumpf (1975) the assumption of non-compressible conditions holds with reasonable accuracy up to Mach number 0.6, provided $Re > 100$. For smaller *Re* the situation at such large Mach number becomes very complicated because the gas cannot any longer be regarded as a continuum.

Figure 3.14 shows that at $v_{rel} = 200$ m/s, i.e. a Mach number of 0.6, *Re* is 13 for a 1 μm particle, 130 for a 10 μm particle and 1300 for a 100 μm particle. Therefore, the condition of Mach number < 0.6 and Re > 100 means that the particles must be larger than about 8 μm.

If the particle shape differs appreciably from sphericity, as illustrated in Figure 3.15, Stokes' law for the terminal velocity of a sphere cannot be applied unless some equivalent particle diameter is used, as indicated in Figure 3.13. This is often done by regarding an arbitrary particle as having a nominal 'Stokes' diameter equal to that of a sphere of the same density, which has the same terminal velocity as the arbitrary particle.

According to Herdan (1960), calculations have been made of the drag on ellipsoids and infinitely long cylinders, flat blades and infinitely thin discs. The theoretical drag depends on the particle orientation with respect to the direction of motion. Thus the viscous drag for a disc moving edge on is equal to that on a sphere with a diameter 16/9π times that of the disc, compared with 24/9π times when the disc is moving broadside on. As a rough approximation it has been suggested that the viscous drag on a particle of any shape, taking an averaged orientation, is equivalent to the drag on a sphere having the same surface area as the particle. Rumpf (1975) also discusses the influence of the particle shape on the drag acting on the particle.

The particle density may not be known in some cases, as discussed by Rudinger (1980). One may then define an 'aerodynamic' or 'kinetic' diameter as the diameter of a spherical particle of density 1 g/cm^3 that has the same terminal setting velocity as the particle.

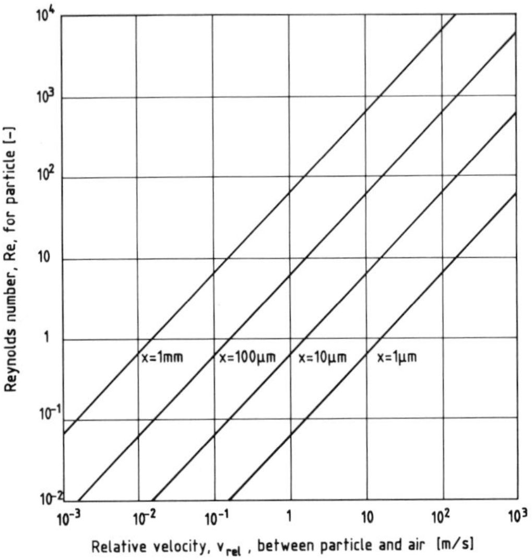

Figure 3.14 *Reynolds' number for a spherical particle of diameter x moving relative to air of 20°C and atmospheric pressure, at velocity v_{rel}*

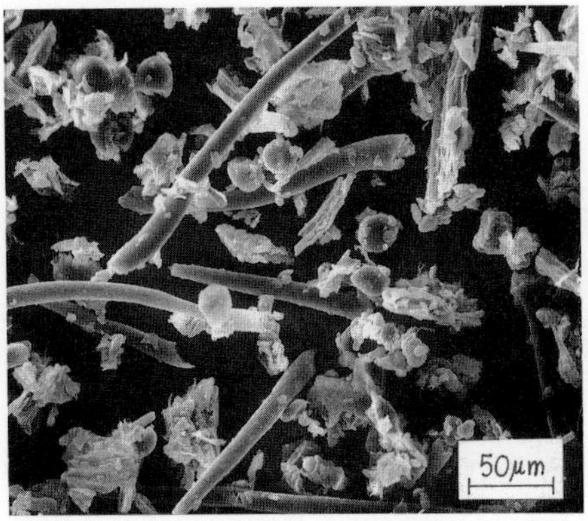

Figure 3.15 *Particles in sample of dust from Australian wheat grain. Elongated fibrous particles (hairs) are typical of wheat grain dusts*

3.5.3
MOVEMENT OF A PARTICLE IN AN ARBITRARY FLOW

In an arbitrary, non-steady flow the influence of gravity can be neglected whenever the drag force exerted on the particle by the motion of the gas is considerably greater than the weight of the particle. As an illustration, Rudinger (1980) has discussed the case where a particle is introduced into a gas flow of velocity:

$$v(t) = v_o + bt \tag{3.18}$$

at time $t = 0$. The initial velocity of the particle is zero. The constant b can either be positive or negative. Then the velocity $v_p(t)$ of the particle at time t equals:

$$v_p(t) = v(t) - b\tau_v + v_0(b\tau_v/v_0 - 1)\exp(-t/\tau_v) \tag{3.19}$$

τ_v is called 'the velocity relaxation time of the motion', and is a characteristic time constant for the particle to reach its terminal velocity.

Rudinger differentiates between three cases of equation (3.19). In the first case the flow is stationary, i.e. b is zero, and $v_p(t)$ approaches v_0 asymptotically. If b has a finite, positive value, $v_p(t)$ approaches $v(t) - b\tau_v$ asymptotically. For negative b, $v_p(t)$ catches up with and starts to exceed $v(t)$ at the time

$$\tilde{t} = \tau_v ln(1 - v_0/b\tau_v) \tag{3.20}$$

whereafter it approaches $v(t) - b\tau_v$ asymptotically.

The three different cases are illustrated in Figure 3.16.

In a turbulent dust cloud, b will vary with time and space. The flow will change continuously both in direction and magnitude, the particles will be moving in all directions, and never attain the same velocity as the gas element in which it is at any instant. The fact that real particles will not only be in translatory motion, but also rotate, adds to the complexity of the problem. The irregular movement of particles will cause the local dust concentration to vary irregularly with time.

A number of experimental and theoretical studies have been published on various aspects of the interaction of dust particles and gas in turbulent flows. Some of these will be discussed in Section 3.8.

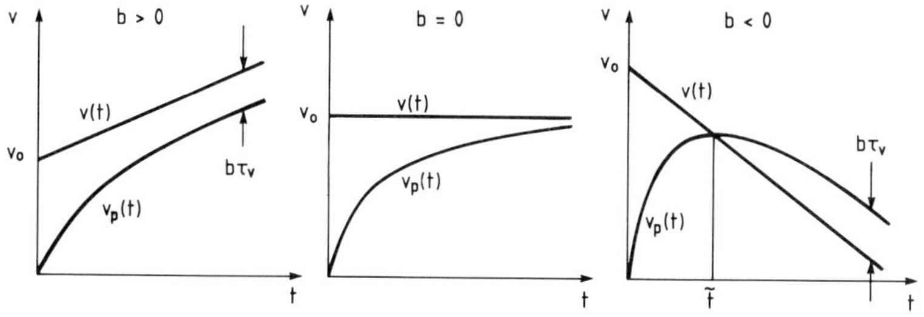

Figure 3.16 *Velocity $v_p(t)$ of a particle introduced in a gas flow of velocity $v_0 + bt$ at $t = 0$ (From Rudinger, 1980)*

3.5.4
SPEED OF SOUND IN A DUST CLOUD

The speed of sound plays an important role in all compressible flow phenomena, including dust explosions. Rudinger (1980) distinguishes between two extreme cases. In the first case the particles are considered to be in equilibrium with the gas at all times, i.e. the particles follow the gas movement exactly and have the same temperature as the gas. Provided that the volume fraction of the particles in the cloud is small, as it will be in an explosible dust cloud, the equilibrium speed of sound, a_e, is given by the expression:

$$\left(\frac{a_e}{a}\right)^2 = \frac{(1-\phi)(1-\phi+\delta\phi)}{(1-\phi+\gamma\delta\phi)} \qquad (3.21)$$

where a is the sound speed in the particle free gas, ϕ is the mass fraction of particles in the dust cloud, δ is the ratio between the specific heat of the particle material and the specific heat at constant pressure of the gas, and γ is the specific heat ratio c_p/c_v for the gas.

Values for the specific heat of various solids as a function of temperature, partly based on interpolation, are given in Table 3.2.

Table 3.2 Specific heats of various solids [kJ/°C × kg]. (Handbook of Chemistry and Physics, 1963)

Material	Temperature [°C]				
	0	100	300	600	900
Al	0.87	0.94	-	1.16	-
C (graphite)	0.45	0.80	-	-	1.90
Fe	0.44	0.48	0.58	-	-
Mg	0.97	1.08	1.17	1.30	-
Si	0.7	-	-	-	-
Wood	1.30	-	-	-	-
Polyethylene	2.3	-	-	-	-
Polymethyl Methacrylate (PMMA)	1.5	-	-	-	-
Polypropylene	2.1	-	-	-	-
Epoxy resin	1.1-1.7	-	-	-	-
Phenol-formaldehyde	1.5-1.7	-	-	-	-

For air at atmospheric pressure and room temperature, the specific heat at constant pressure is 1.0 kJ/°C kg. Most of the values in Table 3.2 are within a factor of two upwards and downwards of the air value. A variation spectrum of δ of 0.5–2 has only modest influence on a_e. For a dust cloud of $\phi = 0.5$, which is in the rich or central part of the explosible dust concentration range, $\delta = 0.5$ gives $a_e = 0.66$ a, whereas $\delta = 2.0$ gives $a_e = 0.63$ a. For a cloud of $\phi = 0.1$, i.e. in the lean concentration range, $\delta = 0.5$ gives $a_e = 0.88$ a, whereas $\delta = 2.0$ gives $a_e = 0.84$ a.

These examples also show that the 'equilibrium' sound speed in explosible dust clouds may be up to 40–50% lower than in the dust free gas.

The other extreme value of the sound speed in a dust cloud considered by Rudinger (1980) is the so-called 'frozen-flow' speed of sound. In this case it is assumed that the changes of the gas flow are so fast that the particles cannot respond at all and remains fixed in space. The 'frozen' sound speed will be somewhat higher than the sound speed, a, in the dust free gas. However, if the particle volume fraction is negligible, as it will be in an explosible dust cloud, the 'frozen' sound speed becomes practically identical to the sound speed in the particle free gas.

In practice, the sound speed in a dust cloud will have a value somewhere between the 'equilibrium' and 'frozen' values, depending on the frequency of the sound wave, which in the context of dust explosions is determined by the characteristic dimension of the enclosure in which the explosion takes place.

3.5.5 PROPAGATION OF LARGE-AMPLITUDE PRESSURE WAVES IN DUST CLOUDS

Rudinger (1980), also discusses the propagation of shock waves and large-amplitude waves of arbitrary form in dust clouds.

Shock waves are of primary importance in the propagation of dust cloud detonations, but are also generated in fast, high-turbulence dust cloud deflagrations. Because the volume fraction of the particles in an explosible dust cloud at atmospheric pressure is very small, it can be neglected in the theoretical treatment.

The speed of a shock wave is at least of the order of the speed of sound. This means that even for a particle of only 0.1 μm diameter the velocity and thermal relaxation times τ_v and τ_T are about 10^3 times longer than the period during which a shock passes the particle. Therefore, the dynamic and thermal conditions of particles are the same immediately after the shock front has passed, as just before it passes, and particle movement can be omitted from the equations describing conservation of mass momentum and energy of the gas across the shock front itself.

However, immediately after a shock has passed, the dust cloud is in a state of non-equilibrium and the particles will start to move in relation to the gas. The distance behind the shock required to reach velocity equilibrium between particles and gas is of the order of 0.5 m (0.3 m for 10 μm glass spheres in air at a particle mass fraction $\phi = 0.17$ according to Rudinger (1980)). Temperature equilibrium is established at a similar distance behind the shock. However, these estimates are somewhat uncertain because they depend on a number of assumptions.

The theoretical analysis or arbitrary non-steady large-amplitude pressure waves through dust clouds is even more complicated than the shock wave analysis. As pointed out by Rudinger (1980) it is necessary to solve a complete set of partial differential equations, using the method of characteristics. An analysis of this kind was undertaken by Rudinger and Chang (1964).

3.6 DISLODGEMENT OF DUST PARTICLES FROM A DUST OR POWDER DEPOSIT BY INTERACTION WITH AN AIRFLOW

3.6.1 AIR FLOW PARALLEL WITH A MONOLAYER OF PARTICLES ON A PLANE, SMOOTH SURFACE

A simple configuration for investigating particle dislodgement is a monolayer of particles adhering to a plane, smooth surface. This well defined geometry enables systematic comparison between the drag force exerted on the particle by the gas, and the adhesion force between the particle and the substrate. Corn and Stein (1965) carried out particle monolayer dislodgement studies in a small laboratory-scale wind tunnel of cross section only 1 mm × 25 mm. In such systems the gas velocity profile is well defined and thus also the gas velocity past the particles, and the drag forces acting on them can be estimated fairly accurately. Figure 3.17 shows the velocity profile calculated by Corn and Stein (1965) for the air flow in their tiny 1 mm × 25 mm wind tunnel. The thickness of the laminar sub-layer close to the substrate was calculated to be 40 μm, but in reality the transition between the laminar sub-layer and the buffer layer is not sharp. Figure 3.18 shows some results from Corn and Stein's re-entrainment experiments in one of their small high-velocity wind tunnels. 430 glass spheres were initially placed on the wall of the test chamber and exposed to air streams with successively increasing mean velocities, and the number and size distributions of the remaining particles after each run were determined by microscopy.

As can be seen, the size distribution on a number basis was systematically shifted towards smaller particles with increasing air velocity, showing that in a given air stream and with particle diameters of the order of 1 to 10 μm, a small particle is more difficult to re-entrain than a larger one. Comparison was also made between the force needed to separate a particle from a substrate by centrifugation, and the calculated drag force required for separation in an air stream, and fair agreement, mostly within one order of magnitude, was found between theory and experiment.

Singer, Greninger and Grumer (1967) carried out experiments in a wind tunnel of somewhat larger cross section. Figure 3.19 shows the same effect as exhibited by Figure 3.18.

As the average air velocity in the wind tunnel is increased, the particle size for which 75% entrainment is obtained, is shifted systematically towards smaller particles for all the three types of particles. Singer, Greninger and Grumer (1967) compared their results with those of Corn and Stein (1965) and concluded that the two studies agreed within a factor of five.

Figures 3.18 and 3.19 illustrate the fact that in the cohesive size range, small particles are more difficult to dislodge and entrain in an air flow than larger particles. This has important implications for the understanding of dispersion of cohesive powders and dusts in air in practical industrial situations.

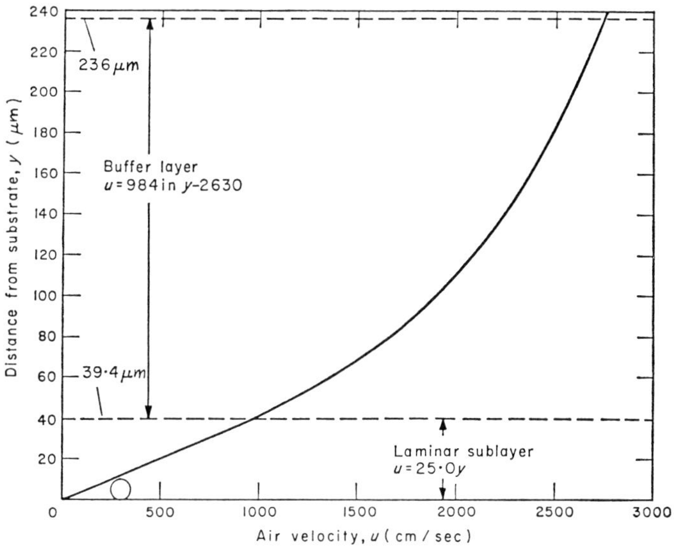

Figure 3.17 *Example of calculated air velocity profile in the boundary layer near the wall in a shallow wind tunnel of 1 × 25 mm cross section (From Corn and Stein, 1965)*

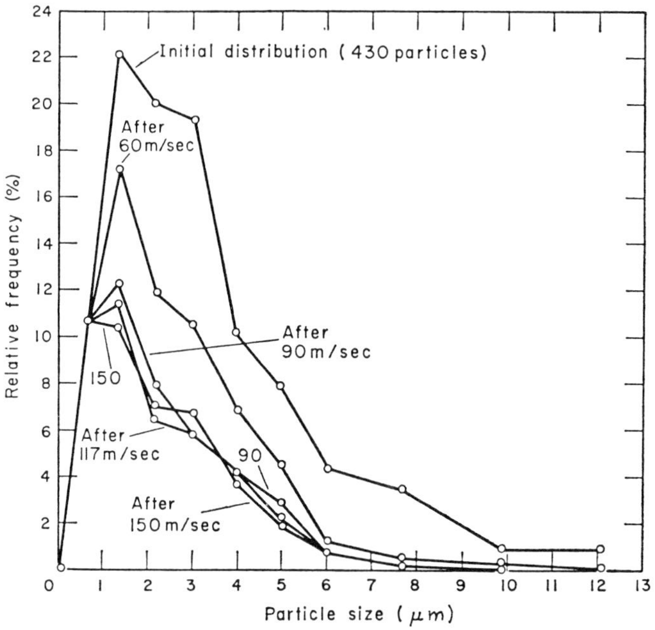

Figure 3.18 *Dislodgement of fly ash particles adhering to a glass substrate in high velocity air flows in a 2 mm × 6.4 mm wind tunnel (From Corn and Stein, 1965)*

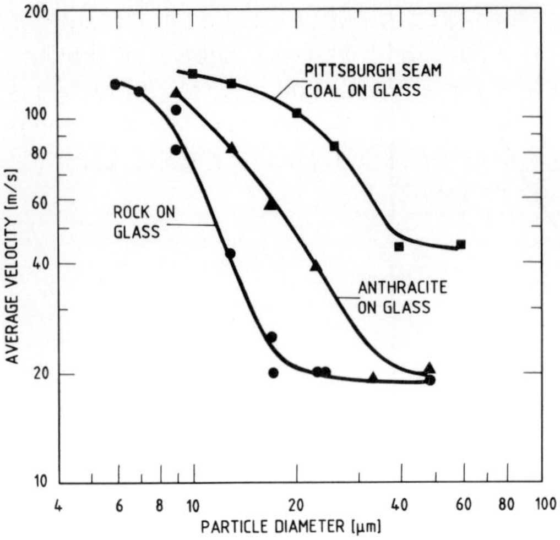

Figure 3.19 *Average air velocity for 75% dislodgement of different sizes of rock and coal particles from a smooth glass surface in a 51 mm × 76 mm wind tunnel (From Singer, Greninger and Grumer, 1967)*

3.6.2 AIR FLOW PARALLEL WITH THE SURFACE OF A POWDER OR DUST DEPOSIT

Several investigations have also been carried out on the entrainment of particles from powder beds by a gas flowing past the bed. Under steady conditions of turbulent gas flow parallel with the surface of a powder bed of uniform roughness, the Prandtl-Karman relation for rough boundaries applies (Bagnold, 1960)

$$v = 5.75 \times \left(\frac{\tau_0}{\rho}\right)^{1/2} \times \log_{10} \frac{30\,z}{x} \tag{3.22}$$

Here v is the mean gas velocity parallel with the powder surface, measured at a distance z from the surface, x is the characteristic surface roughness dimension (characteristic particle size), τ_0 the shear stress at the interface between gas flow and powder surface, and ρ the density of the gas. The term $(\tau_0/\rho)^{1/2} = v_*$ is called the 'drag velocity' and has the dimensions of a velocity. It characterizes a specific gas flow.

Bagnold (1960) suggested a two stage mechanism for the re-entrainment process. In the first stage the horizontal gas flow fluidizes a relatively thin layer of the powder surface, whereby the inter-particle bonds are broken. In the second stage, the detached particles are moved upwards against gravity by eddies in the turbulent gas. This requires that at least some of the eddies have upwards vertical gas velocities exceeding the gravitational settling velocity of the particle in the gas. Bagnold reported experiments showing that in the case of deposits of particles of uniform size, the gas flow required to generate such conditions is much higher than that needed to produce the initial fluidization of the powder surface layer. His experimental values for v_* for initial fluidization of the surface of beds of monosized silica sand are shown in Figure 3.20. Bagnold suggested that the measured increase of v_* as the particle diameter becomes smaller in the range 80–40 μm is not primarily caused by inter-particle adhesion, but rather by the way in which the viscous gas interacts with the particle surface. (Inter-particle forces will, however, dominate when the particles become considerably smaller than 40 μm.)

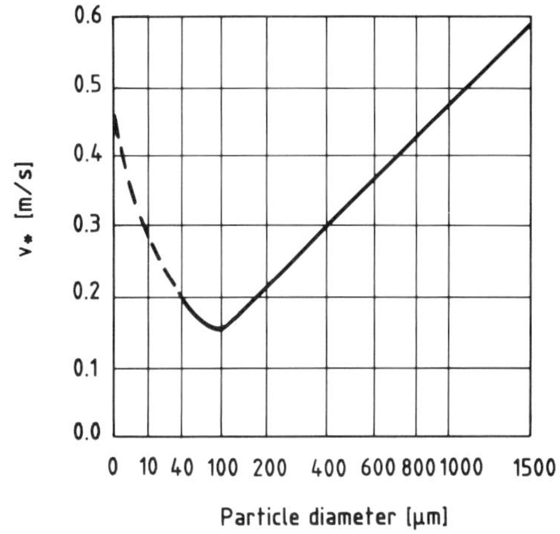

Figure 3.20 *Critical drag velocity v_* for initial fluidization of the surface of a bed of monosized silica sand as a function of particle size of the sand (From Bagnold, 1960)*

Even if the gas flow passing over the powder bed is turbulent, there will be a thin laminar boundary layer of thickness of the order of $\mu/(\rho v_*)$, where μ is the viscosity of the gas and ρ its density. If the particles are of the same order or smaller than the thickness of the laminar layer, they cannot be caught by the turbulent eddies and entrained in the gas flow. Furthermore, reducing the particle size also reduces the effect of the disturbance of one particle in the surface layer by the impact of others.

According to Gutterman and Ranz (1959), the thickness δ of the laminar boundary layer of gas in contact with a smooth powder surface is given by

$$\delta \frac{(\tau_0 \rho)^{1/2}}{\upsilon} \simeq 11.4 \tag{3.23}$$

where τ_0 is the shear stress at the interface between the flowing gas and the powder surface, υ is the kinematic viscosity of the gas and ρ the gas density. The total boundary layer is then the sum of the laminar sublayer and the buffer layer, as illustrated in Figure 3.17. The simple approximate expression (an alternative to the Prandtl-Karman equation (3.22)) for the gas velocity gradient in the laminar layer near the powder surface, adopted by Gutterman and Ranz (1959), is

$$v = \left(\frac{\tau_0}{\upsilon \rho}\right) z \tag{3.24}$$

For the experimental conditions employed by Gutterman and Ranz, δ was at least 250 μm. Therefore, in the case of a smooth dust surface, most particle sizes associated with dust explosions would be submerged in the boundary layer and subjected to velocities according to equation (3.24).

In order to estimate the aerodynamic force acting on a particle of diameter x in the powder layer surface, Gutterman and Ranz assumed spherical particles, no interparticle forces except gravity, the effective velocity of the laminar flow acting on the particle is given by equation (3.24) for $z = x/2$, the aerodynamic drag force on the particle is the same as if the particle had been suspended in an infinite gas volume, and the aerodynamic drag is resisted by the particles having to roll over neighbouring particles against gravity.

Gutterman and Ranz then arrived at the following set of equations for the critical shear stress at the particle bed surface for initiation of particle movement

$$C_D \times Re^2 = 0.65 \times 10^{12} \times x^3 \times \rho_p \times \phi \tag{3.25}$$

$$\tau_0 = 5.9 \times 10^{-10}\, Re/x^2 \; [N/m^2] \tag{3.26}$$

where C_D is the viscous drag coefficient as discussed in Section 3.5.2, Re the Reynolds number, x the particle diameter, ρ_p the particle density, ϕ the internal friction factor of the bulk powder ($0 < \phi < 1$). ϕ was measured in a shear box similar to the Jenike shear cell (Section 3.4.2). The powder was charged gently into the shear box by means of a funnel, whereafter the box was rapped sharply three times to obtain a standard degree of consolidation of the powder. The shear force required for causing powder samples prepared in this way to fail was measured as a function of the vertical force acting on the sample (similar to the determination of failure loci as discussed in Section 3.4.2). The plot of shear stress at failure versus vertical force usually gave an approximately straight line, and the tangent of the angle between this line and the vertical force axis was defined as the internal friction factor ϕ. When comparing this approach with the comprehensive approach described in Section 3.4.2, it seems that a measure of the degree of consolidation of the powder sample, either in terms of the porosity ϵ, the bulk density, or the major principal consolidation stress σ_1, was lacking in this early work of Gutterman and Ranz (1959).

In order to determine τ_0 from equation (3.26), $C_D \times Re^2$ was first calculated from equation (3.25), whereafter Re was found by trial and error from the universal $C_D(Re)$ graph (Figure 3.13).

Gutterman and Ranz also conducted wind tunnel experiments with different powder types, and found reasonable agreement between the critical experimental τ_0 for onset of particle movement on the powder surface and the theoretical values from equations (3.25) and (3.26). Reasonable agreement was also found between the corresponding theoretical and experimental critical gas velocities for initial particle movement. Initial bulk movement (fluidization) of the powder surface was the result of a cascade process starting with a particle upstream being lifted into the air flow. When this particle impinged on the bed surface, one or more new particles were ejected from the bed, and their return to the bed surface ejected further particles and so on.

Bagnold (1960) largely limited his studies to silica sand in the non-cohesive range of particle diameters > 40 μm. He was fully aware of the strong influence of cohesion in the range of smaller particles, but found that the knowledge of the nature of inter-particle forces was insufficient to allow him to conduct any systematic studies. He nevertheless carried out an entrainment experiment with a smooth layer of fine, uncompressed cement in a wind tunnel. Even at a wind speed of 36 m/s, measured 10 cm above the powder layer, there was no continuing disruption of the powder surface. However, as also implied by Bagnold, deposits of fine, cohesive powders can be easily disrupted if the characteristic surface roughness is considerably larger than the particle size and the laminar boundary layer. This is particularly so if the surface topography of the bed is characterized by sharp edges rather than by rounded contours.

Figure 3.21 illustrates how agglomerates of fine cohesive particles can be entrained by and carried along with the air flow as apparent single 'particles'. As long as the agglomerate is not exposed to shear or tensile stresses that exceed its cohesive strength, it will not be broken down further.

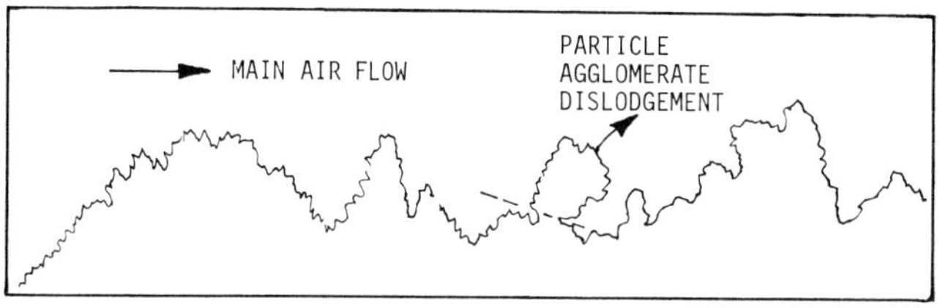

Figure 3.21 *Rough surface topography of a deposit of fine, cohesive particles*

In the case of powders having very wide particle size distributions, the entrainment of the large particles can include mechanical disturbance of the fine ones and facilitate their deagglomeration (breaking of cohesive inter-particle bonds) and entrainment. This process is called 'saltation'.

Fairchild *et al.* (1985) studied the re-entrainment of fine, cohesive aluminium particles of < 10 μm diameter in a wind tunnel, without and with large 'saltation' particles in the air flow sweeping over a fine-particle bed. The saltation particles were monosized spheres of 100, 240 and 500 μm diameter, and they were introduced into the air stream upstream of the bed of fine particles after stationary flow conditions had been established. Measurement of dust concentration as a function of distance above bed surface was conducted between 10 and 150 mm. It was concluded that, within the experimental range, resuspension of particles from a bed of loosely packed aluminium particles increased monotonically with increasing gas velocity and size of saltation particles.

Singer *et al.* (1967) studied the entrainment of coal and rock dust in an air stream passing over a loosely packed dust ridge placed on the floor of a laboratory scale wind tunnel, as illustrated in Figure 3.22.

The properties of the three dusts tested are given in Table 3.3

Photographic studies disclosed various mechanisms of dust dispersion. These included erosion from a dust surface, and denudation from a dust surface under the influence of a pulsating air stream. In erosion the dust is dispersed particle by particle from the deposit surface. In denudation, the entire dust layer leaves the surface suddenly without the particles beig separated at the instant of lifting. Denudation was considerably faster than erosion for similar deposit geometries.

Table 3.3 Properties of dust used in dust ridge entrainment experiments (Singer, Greninger and Grumer, 1967)

Dust type	Pittsburgh coal	Anthracite	Rock
Bulk density, [g/cm^3]	0.56	0.47	1.20
Solid density, [g/cm^3]	1.37	1.61	2.76
Porosity, ε	0.59	0.71	0.57
Median particle size, [μm]	34	10	27

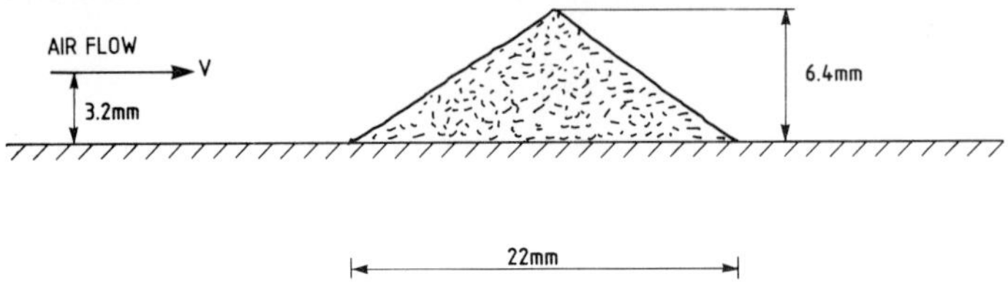

Figure 3.22 *Cross section of typical dust ridge used in wind tunnel dust entrainment experiments. Length of ridge 25 mm. (From Singer, Greninger and Grumer, 1967)*

Even at air velocities only slightly higher than the minimum air velocity for particle entrainment, the ridge dispersion was relatively rapid, having a characteristic time constant of less than 0.1 s. Minimum air velocities for dust dispersion at half ridge height above the wind tunnel floor, were calculated to be 10–20 m/s, using classical boundary layer theory. There was no clear difference between the minimum velocities for Pittsburgh coal and the finer anthracite. However, as Table 3.3 shows, the finer anthracite had a considerably higher porosity than the coal, and this probably compensated for finer particles being more difficult to entrain than larger ones at the same bed porosity. As would be expected, the bulk density of the dust ridge had significant influence on the minimum air velocity for dispersion of the ridge. It was further suggested that the tensile strength, σ_T, of the powder deposit (Section 3.4.2) was a significant factor.

Based on resolution of velocity vectors, Singer *et al.* (1967) proposed a simple empirical model for estimation of lift and drag coefficients on particles in deposits exposed to an air flow. The model neglected both the pressure difference between the windward and the leeward sides of the dust ridge, and the surface roughness. It took the following form:

$$C_D = k(Re)^m \cos \beta$$
$$C_L = k(Re)^m \sin \beta \qquad (3.27)$$

where C_D and C_L are the drag and lift coefficients, *Re* is a special Reynolds number based on the upstream air velocity at midheight of the dust ridge, the ridge height, and the density and viscosity of air. β is the angle between the base and the windward side of the ridge, and *m* and *k* are empirical constants.

Singer *et al.* (1967) also found that large-amplitude air stream pulsations of up to 33 Hz superimposed on the main air flow by a rotating vane in a vent duct, broke dust ridges into lumps. The lumps were lifted almost vertically into the turbulent pulsating air stream where they were eventually dispersed as individual particles into the turbulent core.

Iversen (1985) determined re-entrainment rates of fine powders of Al, Cu, Mo and W of average particle size 5 μm in a wind tunnel of width 0.50 m and height 0.71 m. The length of the powder layer was 1.8 m and its width 0.14 m. The bed was prepared by dispersing dust by air guns and allowing the dispersed dust to settle under gravity and form the bed.

The data for particle mass collected as a function of height above the suface, wind speed, and particle density were analysed using the following solution of the equations for diffusion from a two-dimensional source oriented laterally to the mean wind direction:

$$C = C_0 \exp(-B(z/z_1)^{(n+1)/n}) \quad (3.28)$$

Here C is the dust concentration at height z above the powder bed surface, and n is a velocity profile exponent defined by:

$$v = v_1(z/z_1)^{1/n} \quad (3.29)$$

and

$$B = v_1 z_1 (n/(n+1)^2)\lambda y \quad (3.30)$$

where λ is a diffusion coefficient and y the coordinate in the wind direction.

Equation (3.28) was used for calculating the average vertical flux, q_v, of particles from the bed surface (equal to horizontal flux divided by the area of the powder bed) for molybdenum particles. The following empirical equation was found to fit the experimental data for all four powders:

$$q_v = 2.3 \times 10^4 \times \rho_p \times v_* \times (v_*/v_{*,\min} - 1) \quad (3.31)$$

where ρ_p is the particle density, v_* the actual 'drag velocity' of the air (see paragraph following equation 3.22), and $v_{*,\min}$ the minimum 'drag velocity, for entrainment of particles (Figure 3.20). The $v_{*,\min}$ values for the four powders were 20, 23, 24 and 27 m/s, in order of increasing particle density.

Akiyama and Tanijiri (1989) used a wind tunnel of 3.6 m length and a rectangular cross section of 30 mm width and 150 mm height in their study of re-entrainment of dust particles from a powder bed having its plane surface flush with the wind tunnel floor. The particles studied included glass beads, talc, alumina and fly ash of volume-surface diameters ranging from 15 to 80 μm, solid densities in the range from 2.3 to 4.0 g/cm^3, and bulk porosities in the uncompressed state from 0.47 to 0.77.

The bed of particles to be tested in the wind tunnel was conditioned in a humidistat of relative humidity H for more than six hours before being exposed to the re-entrainment experiment. The humidity of the air in the wind tunnel was not controlled, but it was assumed that the short test period of about 60 s did not significantly influence the humidity inside the bed. In order to obtain $H = 0$, the particle bed was kept at 177°C for more than 10 hours.

With a powder bed of 220 mm length and 30 mm width, and an average air velocity of 15 m/s in the wind tunnel the entrained particle mass per unit time was independent of relative humidity up to 65%. For higher humidities there was a drop of the entrainment rate with increasing humidity, increasing with decreasing particle size. However, at the given conditions, some of the particle systems tested could not be entrained at all, even at low air humidities. It should be pointed out that the particles investigated were non-hygroscopic in the sense that moisture did not penetrate into the bulk of the individual particles, but accumulated only on the particle surface. For some natural organic materials the influence of the relative humidity may therefore be more complex.

Akiyama and Tanijiri then investigated the relationships between the entrainment rate and the four powder mechanical properties, angle of repose, angle of spatula, compressibility, and cohesiveness or cohesion. All these parameters are somewhat arbitrary and not easy to relate to the more fundamental powder mechanical properties described in Section 3.4. They are determined in a set of somewhat arbitrary tests specified in terms of apparatuses and procedures. An overall dimensionless flowability coefficient F was defined as a function of the four measured parameters, and the rates of re-entrainment measured in the wind tunnel were correlated with F for the various powders. Reasonable monotonic correlations comprising all the seven powders were obtained for the three overall wind tunnel velocities 8, 12 and 15 m/s investigated.

Ural (1989, 1989a) postulated that the dispersibility of dusts can be characterized by two parameters, namely the minimum aerodynamic shear stress required for dust entrainment from a horizontal surface, and the settling velocity distribution of a dust cloud. This is an interesting approach which will be discussed in greater detail in Chapter 7 treating various test methods related to the dust explosion hazard.

It should finally be noted that Bagnold (1960) briefly mentioned the re-entrainment of a powder layer by a sudden blast of gas rather than by a steady flow. This clearly is an important case in the context of dust explosions. Even if the Mach number is considerably smaller than unity, and the static pressure gradient in the direction of air movement is negligible, the dynamic pressure gradient (gas velocity gradient) can be considerable.

3.6.3
ENTRAINMENT OF PARTICLES BY AN UPWARDS AIR FLOW THROUGH A PARTICLE BED

Entrainment of particles of equal shape in a fluidized bed configuration as illustrated schematically in Figure 3.23 was studied by Schofield *et al.* (1979).

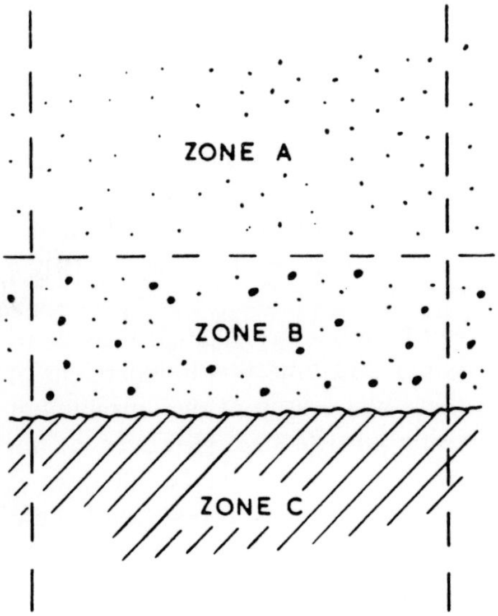

Figure 3.23 *Schematic representation of the emission of dust from a fluidized bed (From Schofield, Sutton and Waters, 1979)*

Let v_n be the minimum local air velocity inside zone C needed for lifting a particle of aerodynamic cross section a_n from zone C into zone B, and v_m the average vertical air velocity in zones B and A corresponding to v_n in zone C. v_m is often denoted the 'superficial' gas velocity through a fluidized bed. Because the effective cross section for vertical air flow in zone C is smaller than in zones A and B, $v_m < v_n$. Therefore the largest particles that are injected from zone C to zone B will drop back into zone C. Only particles of aerodynamic cross sections smaller than a maximum value a_m will be lifted further into zone A. At a given v_m, all the particles in zone C of smaller aerodynamic cross section than a_m will eventually be extracted from zone C and pass through zone B into zone A. Therefore the concentration of these particles in zone C can only be regarded as approximately constant during the initial phase of the fluidization process. This was accounted for in the investigation by Schofield *et al.* (1979), who used a fluidized bed of 46 cm^2 cross section in their experiments. All experimental data were acquired during the initial fluidization phase. Grade emission curves as illustrated in Figure 3.24 were determined for a chalk powder exposed to various values of v_m.

The grade emission curve expresses the mass per unit time at which particles smaller than a given size are emitted from the bed under a given set of experimental conditions. For example, with reference to Figure 3.24 and 0.205 m/s air velocity, particles smaller than 10 μm are emitted at a rate of 20 mg/min., whereas particles smaller than 2 μm are emitted at 4.5 mg/min. This means that particles between 10 and 2 μm are emitted at 15.5 mg/min.

The grade emission curve represents a useful empirical concept, which permits relative comparisons of 'dustability' of various powders and dusts. Schofield *et al.* (1979) gives results illustrating the effect on the grade emission curve of dust moisture content and particle size distribution of the initial dust bed.

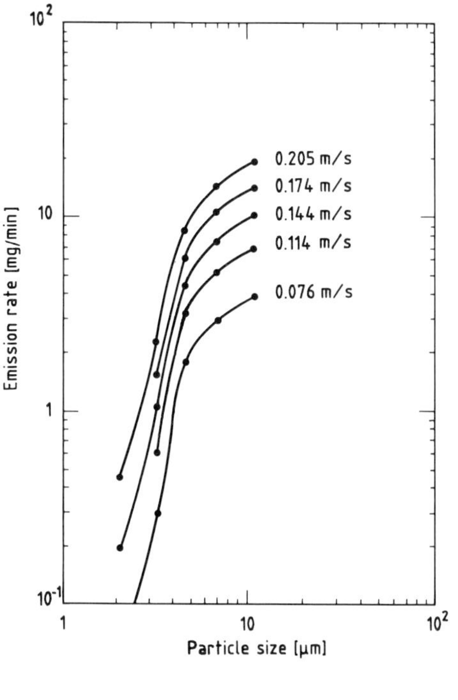

Figure 3.24 *Grade emission curves for a chalk powder at various fluidization velocities (From Schofield, Sutton and Waters, 1979)*

In the fluidized bed studies of cohesive powders by Geldart and Wong (1984), the expansion of the entire powder bed due to an air flow through the bed was used as an indication of the strength of the interparticle forces, or cohesive strength of the powder. The data were analysed using the Richardson-Zaki equation:

$$v_{m,\epsilon} = v_{m,\epsilon=1} \times \epsilon^n \tag{3.32}$$

where $v_{m,\epsilon}$ is the superficial gas velocity through the bed, and $v_{m,\epsilon=1}$ is the minimum superficial velocity needed for dispersing the entire bed. ϵ is the porosity of the bed and a direct measure of the bed expansion, and n an empirical constant. For laminar liquid/solid systems, n has been found to be equal to 4.65. For gas/solid systems n is generally higher than 4.65 and therefore $n/4.65 > 1$. Geldart and Wong (1984) correlated the ratio $n/4.65$ with the ratio between the tapped and loose bulk density of a range of cohesive powders and found

$$\frac{n}{4.65} = 0.65 \times \left(\frac{\rho_{\text{tapped}}}{\rho_{\text{loose}}}\right)^{4.16} \tag{3.33}$$

Both the loose and the tapped densities are sensitive to the methods of sample preparation, and Geldart and Wong (1984) specify detailed experimental procedures for determining the two densities.

Geldart and Wong (1984) also found a correlation similar to equation (3.33) using the superficial gas velocity v_m:

$$\frac{n}{4.65} = 1.65 \times \left(\frac{v_{m,\epsilon=1}}{v_{m,\epsilon}}\right)^{0.132} \tag{3.34}$$

Expansion of beds of cohesive powders is caused by proliferation and enlargement of horizontal and inclined cracks. Powders become more cohesive as the particle size is reduced. For any given superficial air velocity, Geldart and Wong (1984) found that the bed expansion ratio (or ratio of ϵ after and before expansion) increased with decreasing mean particle size down to about 12 μm. However, a further decrease of the particle size caused the bed expansion ratio to drop markedly. This was attributed to the generation of vertical cracks and channels in very cohesive powders.

3.7 DISPERSION OF AGGLOMERATES OF COHESIVE PARTICLES SUSPENDED IN A GAS, BY FLOW THROUGH A NARROW NOZZLE

The effective 'particles' in clouds of very fine, cohesive dusts will often be large agglomerates of primary particles rather than the small primary particles themselves. Depending on the actual degree of dispersion, or deagglomeration, the effective particle size distribution in the dust cloud can differ considerably for the same cohesive powder. This will give corresponding differences in both ignition sensitivity and explosibility of the dust cloud, because an agglomerate will behave as a single particle of the agglomerate size.

Bryant (1973) studied the degree of agglomeration of fine boron carbide particles of diameters 1 μm or less, dispersed as a cloud in a gas. He generated the cloud by blowing dust through a narrow nozzle and measured the mean effective 'particle' size (agglomerates) as a function of the injection pressure (injection velocity). At a pressure of 3.5 bar(g) the mean diameter of the particles in the cloud was 6.2 μm, whereas at 7 bar(g) it had been reduced to 3.5 μm.

This important phenomenon was investigated in greater detail by Yamamoto and Suganuma (1984), and their findings are of significance both to the actual industrial dust explosion hazard and to the design of experimental methods for assessment of ignitability and explosibility of clouds of cohesive dusts.

Figure 3.25 shows the dispersing nozzle used in the investigation by Yamamoto and Suganuma (1984).

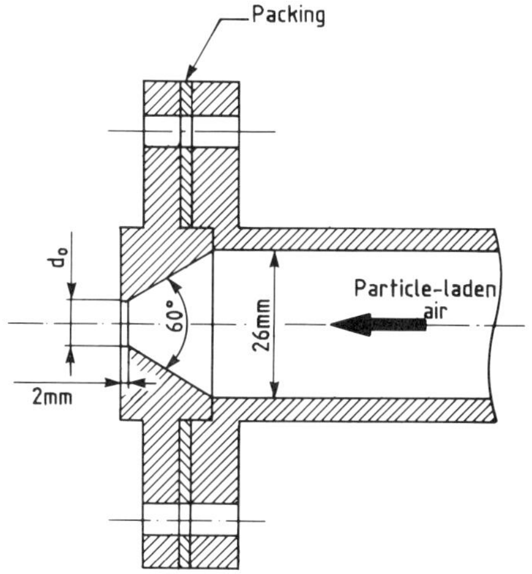

Figure 3.25 *Schematic view of nozzle for dispersing agglomerates of cohesive dust particles (After Yamamoto and Suganuma, 1984)*

The dust was first dispersed in the upstream air flow by simply feeding it into the 26 mm diameter air supply pipe from a vibration feeder via a funnel. For cohesive powders this gave comparatively poor dispersion and large effective particle size. The primary dust cloud was then forced through the narrow nozzle where the agglomerates were dispersed to varying extents depending on the flow conditions in the nozzle. A sample of the secondary, dispersed dust cloud was sucked through a five or six stage cascade impactor from which the effective, aerodynamic *in situ* size distribution in the secondary dust cloud was obtained. (Cascade impactors, sedimentation balances and other methods for determining particle size distributions are described by Herdan (1960), Green and Lane (1964), Allen (1981), Kaye (1981) and Bunville (1984).) Figure 3.26 gives a set of typical results. The distribution of effective particle sizes is shifted systematically towards smaller particles as the dispersion process in the nozzle gets more effective, i.e. as the average air velocity through the nozzle increases.

The effect is quite dramatic. For an air velocity in the nozzle of 10.5 m/s, the median effective particle size is somewhat larger than 10 μm, whereas for velocities in the range 100–150 m/s it is only 1 μm. For the primary dust cloud, which was generated in a way that would be typical in industry, the median particle size would probably be considerably larger than 10 μm. It can be observed from Figure 3.26 that at the highest air velocities in the nozzle, the distribution of the sizes of 'effective' particles in the secondary dust cloud approached the size distribution found in a sedimentation balance after having dispersed the powder in a liquid in a way that would be expected to produce close to perfect dispersion.

Yamamoto and Suganuma arrived at the following empirical relationship for the efficiency of the nozzle dispersion process:

$$\frac{x_a}{x_s} = 31.3 \times h^{-0.2} \tag{3.35}$$

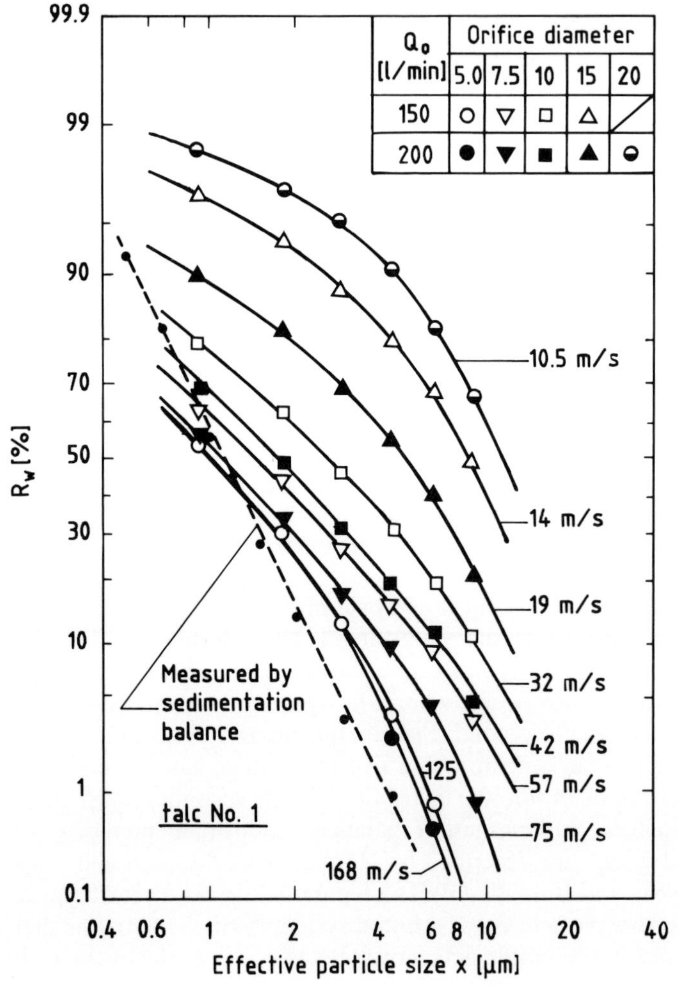

Figure 3.26 *Effective particle size distributions of an airborne talc dust after dispersal by different orifices and air velocities. R_w is the percentage by weight of the effective 'particles' that are larger than the size x (From Yamamoto and Suganuma, 1984)*

where x_a is the effective *in situ* median particle diameter determined by the cascade impactor for the actual secondary dust cloud, and x_s is the ultimate median particle size determined by the sedimentation balance. The parameter h is defined by:

$$h = 0.4 \times \Delta p_o \overline{v} / d_o \quad [J/m^3 \times s] \tag{3.36}$$

where Δp_o is the pressure drop across the dispersing nozzle, $\overline{v}$ is the mean air velocity through the nozzle and d_o is the orifice diameter.

It is interesting to compare the results in Figure 3.26 with those of Corn and Stein (1965) in Figure 3.18 from particle dislodgement experiments in a narrow wind tunnel of cross section comparable with those of the smallest nozzles in Figure 3.26. The order of air velocities required for dislodging particles in the size range 1–10 μm in Corn and Stein's experiments is the same as required for breaking up agglomerates of 1–10 μm in the Yamamoto and Suganuma's nozzle dispersion experiments. This indicates that the adhesive forces between particles in an agglomerate and a particle and a plane substrate are of the same nature (probably mostly van der Waal forces), and that viscous drag forces are dominant dislodging forces in both cases.

3.8 DIFFUSION OF DUST PARTICLES IN A TURBULENT GAS FLOW

Gutterman and Ranz (1959) determined the dust concentration gradient in a turbulent air flow, following the injection of a given quantity of dust in a closed-loop laboratory-scale wind tunnel system. The average solid volume concentration of dust was about 200 cm^3 per 1 m^3 of air, i.e. in the explosible concentration range for most combustible dusts.

Typical experimental dust concentration profiles are shown in Figure 3.27.

According to Gutterman and Ranz (1959), the general differential equation for the distribution of dust concentration in a dust cloud moving in a two-dimensional flow can be written as

$$\frac{\partial c}{\partial t} + v_y \frac{\partial c}{\partial y} + v_z \frac{\partial c}{\partial z} + v_{\text{term}} \frac{\partial c}{\partial z} = D_{\text{eff}} \left(\frac{\partial^2 c}{\partial y^2} + \frac{\partial^2 c}{\partial z^2} \right) \tag{3.37}$$

Here c is dust concentration, t time, v_y the average gas velocity in the horizontal flow direction y, v_z the average gas velocity in the vertical direction z, v_{term} the terminal particle settling velocity in the gravitional field, and D_{eff} is the effective diffusion coefficient for the particles. D_{eff} is a function of both y and z. The system studied by Gutterman and Ranz was stationary in both time and y-direction, and the average vertical gas velocity v_z was zero. Therefore

$$V_{\text{term}} \times c = D_{\text{eff}} \frac{\partial c}{\partial z} \tag{3.38}$$

Here c is a function of z only, and $\partial c / \partial z$ could be determined from experimental $c(z)$ correlations (Figure 3.27), v_{term} can be calculated, and therefore an 'experimental'

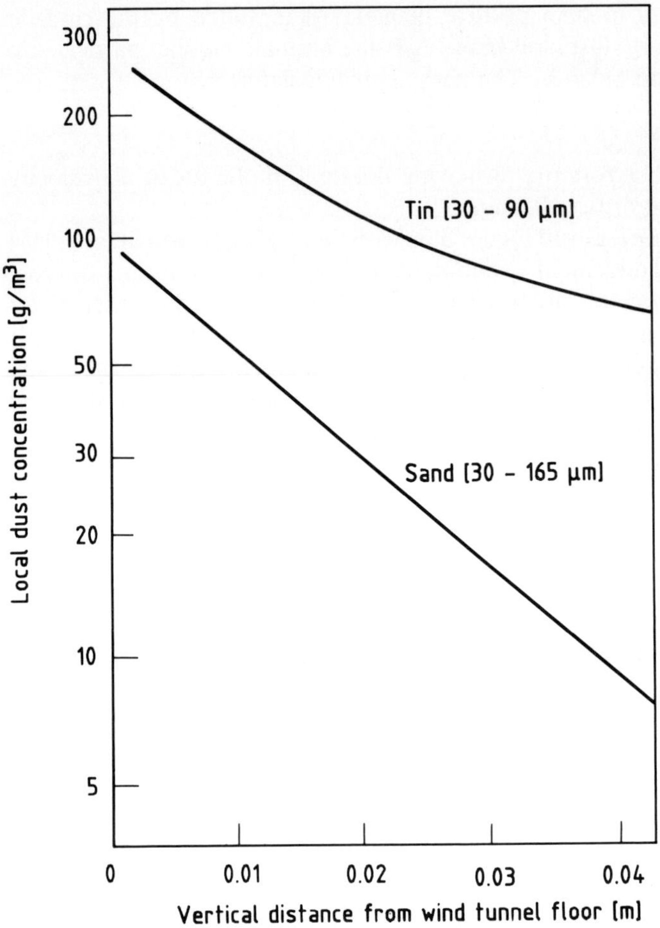

Figure 3.27 *Dust concentration gradients in horizontal turbulent air flow in a wind tunnel (From Gutterman and Ranz, 1959)*

diffusion coefficient, D_{exp} could be found. This was compared with the theoretical turbulent diffusion coefficient D_{turb} (related to the turbulent eddy viscosity) for the gas, and for particles that are so small that they follow the turbulent gas motion.

Some results are shown in the central column of Table 3.4.

Because of the small size of the glass beads, D_{exp} was very close to D_{turb}, i.e. the glass beads followed the gas motion fairly well, whereas the coarser sand particles and the high-density tin particles had considerably higher diffusion coefficients than the gas.

According to Gutterman and Ranz, turbulent gas diffusion behaviour of particles can be expected if the Weiss-Longwell criterion for diffusion of solid particles in an oscillatory gas velocity field

$$\frac{D_{eff}}{D_{turb}} = \frac{(18\ \mu/\rho_g x^2)^2}{\omega^2 + (18\ \mu/\rho_g x^2)^2} \qquad (3.39)$$

Plate 1 *18.5 m³ vented explosion vessel connected to a straight vent duct (Courtesy of Health & Safety Executive, UK)*

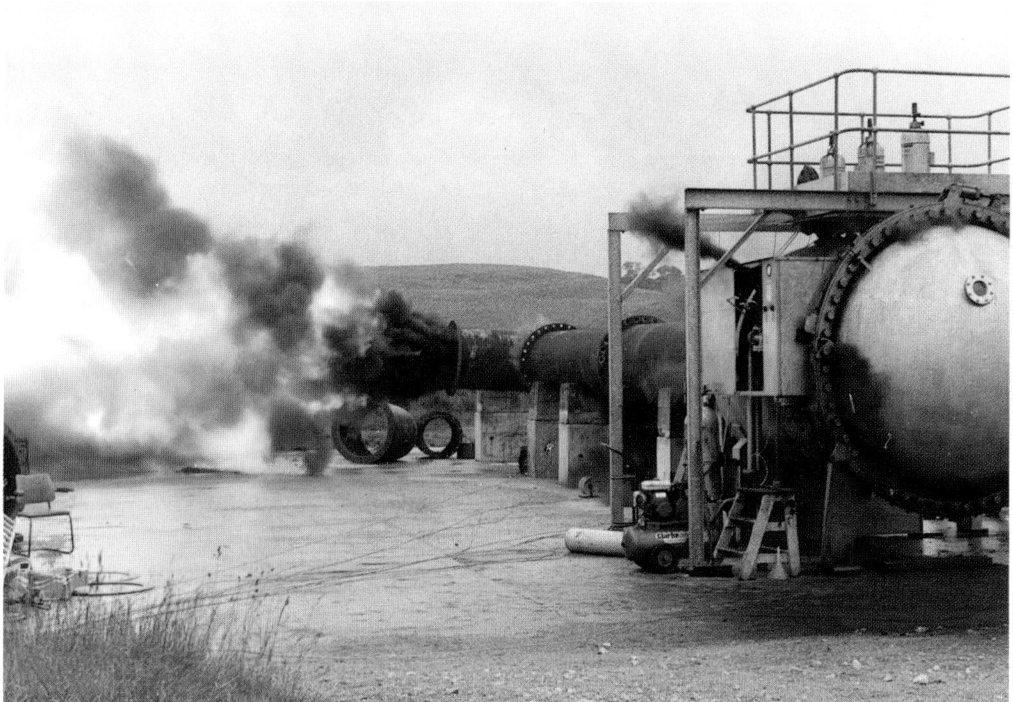

Plate 2 *Coal dust explosion in 18.5 m³ vessel vented through a duct with a 90° bend at the end (Courtesy of Health & Safety Executive, UK)*

Plate 3 *Venting of a polypropylene/air explosion in a 5.8 m³ bag filter unit without (top) and with quenching tube (bottom) (Courtesy of F. Alfert and K. Fuhre, Chr. Michelsen Institute, Norway)*

Plate 4 *Application of automatic dust explosion suppression to bucket elevators (Courtesy of Th. Pinkwasser, Bühler, Switzerland)*

is close to unity. This will be the case for small particle diameters x, for which $\omega^2 \ll 18\ \mu/\rho_g x^2$, ω being the average rotational frequency (radians per second) of the gas eddies.

Table 3.4 Average ratios between experimentally determined diffusion coefficients for various particle types and theoretical turbulent diffusion coefficients for the gas (From Gutterman and Ranz, 1959)

Particles	D_{exp}/D_{turb}	D_{exp}/D_{bounce}
Glass 3-40 μm 2.50 g/cm^3	1.1	-
Sand 30-165 μm 2.65 g/cm^3	5.6	1.8
Tin 30-90 μm 7.3 g/cm^3	4.3	-

Gutterman and Ranz also compared their experimental diffusion coefficients based on measured dust concentration gradients, with coefficients derived theoretically by assuming back-mixing of particles into the gas flow by irregular statistical bouncing when the particles hit the bottom and roof of the wind tunnel, was the governing diffusion mechanism. The third column of Table 3.4 gives the result for angular sand particles. This shows that for coarse particles in a narrow boundary zone of a few cm from the wall of the wind tunnel the theory of back-mixing by bouncing against the wall gives better agreement with experiments than the theory of turbulent gas diffusion.

Hwang, Singer and Hartz (1974) performed theoretical studies of the dispersion of dust in a turbulent gas flow in a duct, following the initial entrainment of the dust deposits from the duct wall. In particular they studied the entrainment of deposited dust by the non-stationary turbulent air blast ahead of a self-sustained dust explosion sweeping through a long duct. The objective was to predict the dust concentration in the gas flow as a function of time and location in the duct.

The dust flux leaving the duct walls was treated as originating from single or multiple stationary or moving sources. Formulas and sample computations for various types of dust sources in circular and rectangular channels were derived based on experimental dust entrainment rates. The theoretical results appeared to agree with the physical characteristics of explosion-driven dust dispersion in a 0.6 m diameter and 50 m long explosion tunnel.

In the theoretical analysis the process of turbulent mixing was treated as a diffusion process, using diffusion-type equations that had been successfully applied to the dispersion of dusts in pipes, open channels, and semi-infinite systems. The generalized form of the diffusion equation used was:

$$\frac{\partial c}{\partial t} + \bar{v} \times \text{grad } c = \text{div } (k \text{ grad } c) \tag{3.40}$$

where c is the dust concentration, k is the diffusion coefficient (assumed to be 25–100 cm^2/s) and $\overline{v}$ is the velocity with which the dust particles were convected in addition to being diffused. $\overline{v}$ differs from the gas velocity because of the inertia of the dust particles in the flow. It was assumed that the effect of gravity could be neglected during the initial period of the dispersion process, and that equation (3.40), employing an appropriate value of k, determined the gross behaviour of the dust cloud.

Figure 3.28 gives an example of the computational results obtained. Dust concentrations that would be in the middle of the explosible range for combustible dusts have developed at 2.5 m downstream of the dust source. However, at 3.5 m downstream the concentrations are still below the typical minimum explosible limit range.

Hinze (1975) discussed the Tchen-theory of diffusion of discrete solid particles in a fluid of homogeneous turbulence. In this theory the following assumptions are made:

1. The turbulence of the fluid is homogeneous and steady.
2. The domain of turbulence is infinite in extent.
3. The particle is spherical and so small that its motion relative to the ambient fluid follows Stokes' law of resistance.
4. The particle is small compared with the smallest structure present in the turbulence.

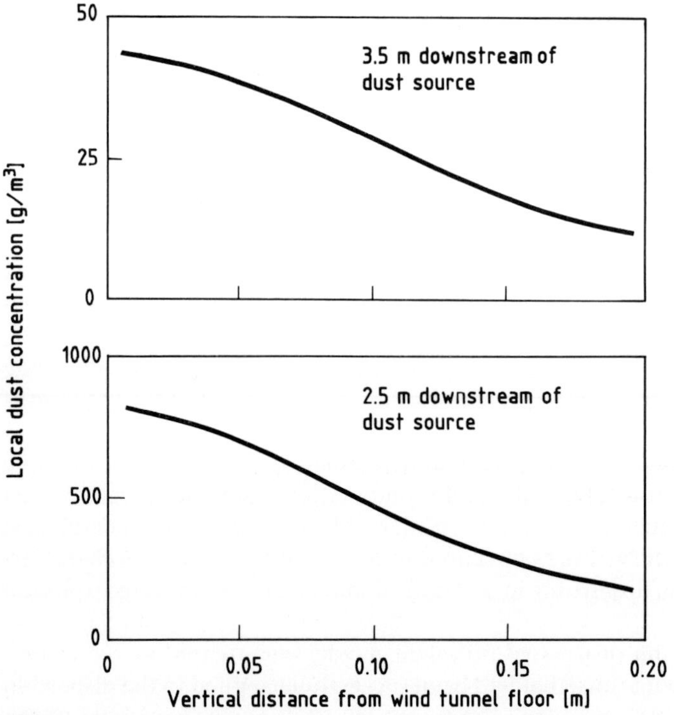

Figure 3.28 *Computed two-dimensional dust concentration distributions at two locations in a wind tunnel of square cross section 0.53 m × 0.53 m at 1.0 s after onset of dust dispersion. Dust source: 140 g of rock dust distributed as a 0.2 m long, even layer over the entire channel width. Average wind velocity 5 m/s, dust diffusion coefficient k = 50 cm²/s (From Hwang, Singer and Hartz, 1974)*

5. The particle will be embedded in the same fluid element during the motion.
6. Any external force acting on the particle originates from a potential field, such as the field of gravity.

All assumptions, except number 5, may in reality actually be satisfied. However, the mechanism of a real turbulence is such that it is hardly possible for assumption 5 to be satisfied. If the element of fluid containing a small discrete particle could be considered non-deformable, it could possibly satisfy this assumption, provided that its size was larger than the amplitude of the motion of the discrete particle relative to the fluid (no overshooting). However, in turbulent motion the fluid elements are distorted and stretched into long thin ribbons and it seems unreasonable that the fluid element should continue to contain the same discrete particles during this stretching process.

As part of an account on the use of laser-doppler anemometry for characterizing turbulence, Durst *et al.* (1981) also discussed various theories for the movement of small particles in a turbulent flow.

During the last decade a number of further experimental and theoretical studies on the interaction of dust particles and a gas in turbulent flows have been published. Some central papers are those by Alquier *et al.* (1979), Tomita *et al.* (1980), Genchev and Karpuzov (1980), Tadmor and Zur (1981), Ebert (1983), Elghobashi and Rizk (1983), Chen and Wood (1983), Beer *et al.* (1984), Lee (1984), Krol and Ebert (1985), Picart *et al.* (1986), Bachalo *et al.* (1987), Johansen (1987), Shrayber (1988) and Lee (1989). These and other similar investigations are important for the development of comprehensive computer codes for numerical simulation of combustion and explosion of dust clouds (see Chapter 4).

A number of different methods are now available for experimental investigation of the turbulence in gases and dust clouds. Some of those discussed by Smolyakov and Tkachenko (1983) are:

- Hot-wire and hot-film anemometer.
- Laser-doppler anemometer.
- Flow visualization by means of small particles (< 1 μm) as 'markers'.
- Flow analysis by thermal markers (rapid heating of a small gas volume by hot-wires. The movement of the heated gas volumes is followed by another set of hot-wires. Poor spatial resolution).
- Acoustic anemometer (poor spatial resolution).
- Electric discharge anemometer (corona- and glow-discharge).
- Cold-wire anemometer (for measurement of temperature fluctuations).

Durst *et al.* (1981) gave an in-depth discussion of one of the most versatile methods, the laser-doppler anemometer.

Beer, Chomiak and Smoot (1984) discussed the application of such methods in the study of turbulence effects in burning dust clouds. Laser anemometers may be used for local particle velocity measurements, particle sizing and concentration fluctuation measurements. Very accurate measurements of both mean and fluctuating particle temperatures are possible by other optical methods. As long as the flow is optically thin, which means low dust concentrations, flow visualization is not more difficult in dust clouds than in gas flows. Some techniques, like direct high-speed photography, are even simpler for two-phase combustion than for gas flames, due to the strong radiation of the flames.

However, the investigations are extremely time-consuming and difficult. Multipoint conditionally-sampled measurements have to be performed for flame structure studies. Advanced data-reduction techniques must be applied for evaluation and interpretation purposes and for extraction of information about individual events. The development of controlled excitation studies provides the possibility of investigating the details of the coherent structures through phase-lock on the induced perturbation.

Hatta *et al.* (1989) extended the theoretical equations for flow of dust/gas mixtures through nozzles to the complicated case of poly-sized particle systems. The equations covered both sub-sonic and supersonic gas flow. Some numerical solutions were discussed.

Fan *et al.* (1989) studied the flow of poly-sized particles/gas mixtures in a co-axial jet system, both theoretically and experimentally. Advanced instrumentation was used for experimental determination of particle movement. Numerical computations gave results in good agreement with the experiments.

Lockwood and Papadopoulos (1989) described a new powerful method for calculation of dispersion (not de-agglomeration) of solid particles in a turbulent flow. An equation, which correctly accounts for particle momentum conservation, was derived for the evolution of the probability of particle velocity and position. The method enabled determination of the position and velocity probability density functions for all cells within the computational domain at a fraction of the cost of conventional stochastic computations.

3.9 METHODS FOR GENERATING EXPERIMENTAL DUST CLOUDS FOR DUST EXPLOSION RESEARCH PURPOSES

3.9.1 BACKGROUND

Almost half a century ago Hartmann *et al.* (1943) found it necessary, when discussing research in the field of dust explosions, to make the following statement: 'Over the past 30 years, various investigators have worked on means of producing uniform dust clouds; comparison of results indicated that none of them has been wholly successful. The mechanisms to produce such a cloud, of sufficient volume to be usable for test work, remain to be perfected.'

Although a substantial amount of work has been carried out during the years since the statement of Hartmann was made, to overcome this basic problem in dust explosion research, his words are still valid.

It appears, however, that the problem does not merely arise from experimental difficulties. The basic question is perhaps not how to produce the 'perfect' experimental dust cloud, but rather whether a realistic definition of such a dust cloud can be given at all. The ideal static, fully dispersed and uniform dust cloud is impossible to realize in practice whether in the laboratory or in real life in industry. In any realistic dust cloud the particles and supporting gas will be in motion, the dust concentration is only to some extent uniform and the dispersion of agglomerates may not be complete.

Sophisticated means of overcoming the problems have been attempted. These include the use of free-falling explosion chambers by which the influence of gravity is eliminated (Ballal, 1983, Gieras *et al.*, 1986), and steady-state suspension of the dust cloud is a strong electrostatic field (Gardiner *et al.*, 1988). However, whilst such methods may provide useful insight into basic details of ignition and combustion of dust particles and clouds, they do not represent practical industrial conditions. Disregarding such highly sophisticated techniques, the methods used for the formation of experimental dust clouds for dust explosion research may be classified in the three main groups illustrated in Figure 3.29.

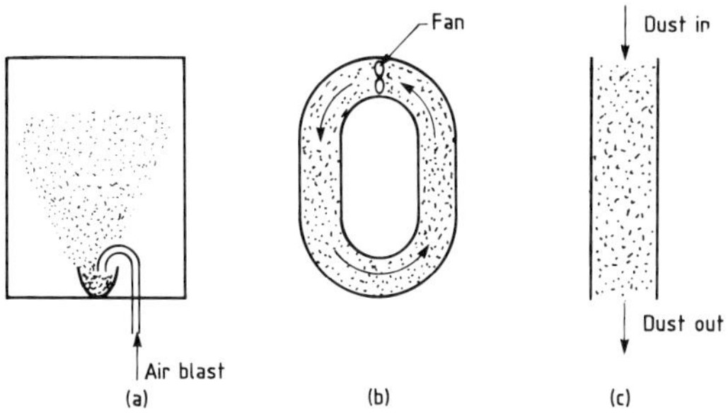

Figure 3.29 *Three basic principles used for generating dust clouds for dust explosion research. a, Transient clouds generated by dispersing a given quantity of dust by a short blast of air. b, Stationary circulation of a given quantity of dispersed dust in a closed system. c, Stationary formation of a dust cloud in an open system*

3.9.2
TRANSIENT DUST CLOUDS GENERATED BY A SHORT AIR BLAST

Due to the relatively simple equipment that is required and the minimal dust quantities needed, the transient dust cloud method has been adopted in the major part of published investigations, both in small and large scale.

According to Brown and James (1962), the transient air blast method was probably first introduced by Holtzwart and von Meyer (1891). Their very simple explosion apparatus consisted of a glass tube of 50 cm^3 capacity, fitted with a pair of platinum electrodes, across which an induction spark was passed. The dust was placed in a piece of a narrower tube attached to one end of the explosion tube, and the dust cloud was formed in the region of the spark by means of a short blast of compressed air. Engler (1907) used glass flasks of 250 to 500 cm^3 capacity as explosion vessels, (see Figure 1.57). In the spherical explosion bomb of 1.4 litre capacity which was used by Trostel and Frevert (1924), the dust was placed in a small cup near the bottom of the bomb, and the dispersing blast of air was introduced through a glass tube, entering the bomb through the bottom and, having a bend of 180°, facing the opening downwards towards the dust heap in the dispersion cup. In their explosion vessels of one litre capacity, Boyle and Llewellyn (1950) and Eckhoff (1970) used arrangements practically identical to that introduced by Trostel and Frevert.

The well-known Hartmann apparatus, which was first described by Hartmann *et al.* (1943), consists of a vertical cylinder having a volume of about 1.2 litre, supported by a metal bottom part shaped like a cup, in which the powder is placed (see Chapter 7). The dispersing air blast is introduced axially from below and deflected downwards towards the dust heap by means of a small conical 'hat' or 'mushroom'. As discussed by Dorsett *et al.* (1960), this apparatus, in the form of either an open tube or a closed bomb, has been used for the determination of the numerous values of minimum ignition energy, minimum explosible dust concentration, rates of pressure rise and maximum explosion pressure etc., which have through the years been published by the US Bureau of Mines.

Carpenter (1957) used a slightly modified form of the Hartmann apparatus, the main features, however, being identical. In a subsequent work Carpenter and Davies (1958) used a smaller, detached dust dispersion cup of 2 cm diameter fitted in the lower part of a cylindrical 275 cm^3 combustion chamber. Meek and Dallavalle (1954) employed a rather large explosion chamber of about 60 litres. The dust was dispersed from a polished funnel shaped cup, fitted with a special dispersing cone.

Various versions of the transient air blast method have also been used in a number of other investigations. Nagy *et al.* (1971) adapted this technique over a wide range of explosion vessel volumes ranging from 1 litre to 14 m^3. Moore (1979) employed the method in three different vessels of volumes from 1 to 43 litres, whereas Enright (1984) used it in three vessels of volumes from 1 to 20 litres.

It was realized that the simplest version of the transient air blast method, based on just directing a blast of air towards a dust heap, gave a rather poor dispersion of very fine, cohesive dusts. In order to improve dust dispersion, more refined versions of the air blast method have been developed, based on forcing the dust/air suspension through narrow nozzles (see Section 3.7).

This was for example done by Helwig (1965), who generated his dust clouds from a 100 cm^3 cylindrical 'whirling' chamber shown in Figure 3.30. The chamber was placed inside the 43 litre explosion bomb. By means of a blast of compressed air admitted through the bottom of the 'whirling' chamber, the dust was fluidized and the fluidized suspension forced through a number of holes in the chamber lid at the top. There is little doubt that the nozzle dispersion mechanism discussed in Section 3.7 played an essential role in this process.

In his 1 m^3 explosion vessel, Bartknecht (1971) used a dust dispersion system by which the dust was forced at high velocity by high pressure air through a number of 4–6 mm diameter holes in a U-shaped tube of 19 mm i.d. Bartknecht's 1 m^3 vessel and dust dispersion system has later been adopted as an ISO standard (International Organization for Standardization (1985)). From what has been said in Section 3.7 it is quite clear that this standard test method gives a high degree of dispersion even for very cohesive dusts.

Siwek (1977) developed a smaller spherical apparatus, which is capable of yielding approximately the same degree of dust dispersion and turbulence as in the 1 m^3 ISO vessel. However, obtaining an acceptable correlation with the 1 m^3 ISO vessel required a large experimental effort, starting with a 5 l bomb, continuing with one of 10 l volume, and ultimately finishing up with the final 20 l bomb. In particular it was necessary to investigate a range of different dispersion nozzle systems before finally arriving at one that produced turbulence and dust dispersion levels in acceptable agreement with those generated in the 1 m^3 standard chamber. It is not surprising that the dispersion system finally arrived at was very similar to the perforated U-tube system of the 1 m^3 vessel.

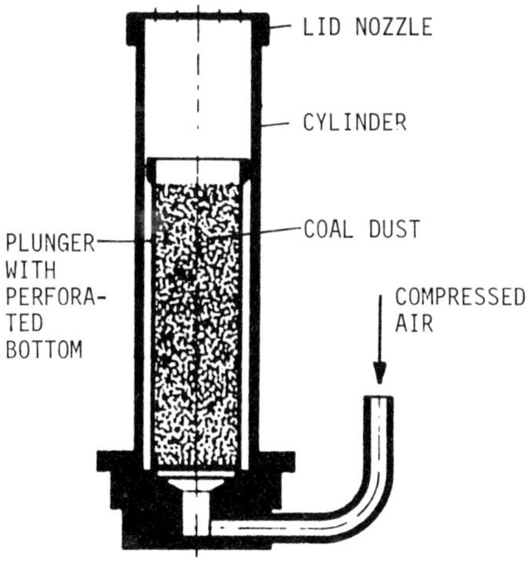

Figure 3.30 *'Whirling' chamber for fluidizing the dust sample and subsequently forcing the dense dust/air suspension through a series of parallel nozzle holes (From Helwig, 1965)*

However, recently Siwek (1988) introduced a new and quite different dispersal nozzle for the 20 litre sphere based on high-velocity impact of agglomerates on target plates. The new system was claimed to produce degrees of dust dispersion comparable to those generated by the original nozzle (see Figure 7.58).

Following the development of the 20 litre vessel by Siwek, an alternative 20 litre vessel has been proposed by Cashdollar and Hertzberg (1985). They mention the interesting possibility of inserting interchangeable dust dispersion units at the bottom of their vessel. This makes it possible to work with the intensity of dust dispersion that is relevant for the problem to be investigated.

The Institute of Iron and Steel in Kiev, USSR developed a dust dispersion unit which is particularly suitable for dispersing cohesive metal powders. The unit, which was mounted at the upper end of a 4 litre vertical cylindrical explosion vessel of internal diameter 110 mm, is shown in Figure 3.31. The basic philosophy behind this design is the same as for several of the methods already discussed. The dust cloud, after having been initially dispersed by the air blast in the conventional way, is forced through a system of narrow nozzles at high speed, causing further break-up of particle agglomerates before the cloud is admitted to the explosion vessel (see Section 3.7). The concentration distribution of the resulting transient dust cloud in the vessel, as a function of time and position in space, was investigated by means of a special gravimetric concentration sampling probe. The dust clouds were also studied by means of high-speed photography by replacing the explosion vessel by a glass container. It was found that generally a reasonably homogeneous cloud, filling the entire vessel volume, was obtained at some stage during the dispersion process.

Nedin *et al.* (1975) compared the transient dust concentration distributions generated by the US Bureau of Mines Hartmann tube dispersion system (See Chapter 7), and two systems used in the USSR for generating experimental laboratory-scale clouds for dust explosion testing and research.

In spite of the extensive use of the transient air blast method, it has its clear limitations. The method inevitably yields both time and space dependent dust concentration. Schläpfer (1951) found that powders of different density, particles size and surface properties were dispersed in different ways if exposed to the same air blast conditions.

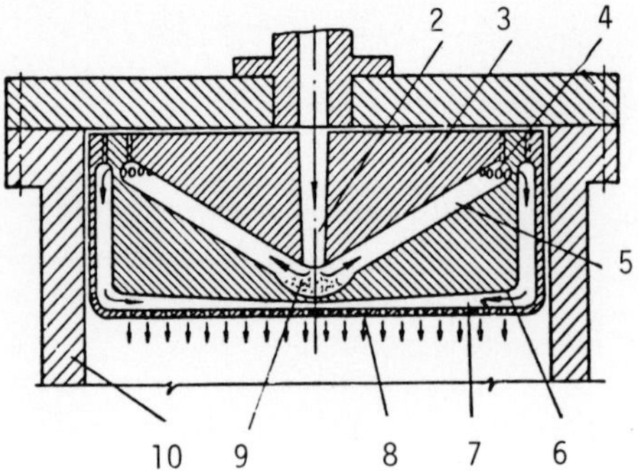

Figure 3.31 *Cross section of dust dispersion system developed at the Institute of Iron and Steel in Kiev, USSR (From Eckhoff, 1977)*

1. *Entrance for dispersing air blast.*
2. *Nozzle directing air towards dust heap.*
3. *Internal body of dispersion unit.*
4. *Narrow peripheral nozzles for agglomerate break-up.*
5. *Slot for flow of primary dust cloud to the narrow peripheral nozzles.*
6. *Intermediate body of dispersion unit.*
7. *Slot for flow of dispersed dust towards perforated bottom. Tapering ensures even distribution of dust concentration over explosion vessel cross section.*
8. *External body of dispersion unit, having the entire bottom penetrated by narrow nozzles for distribution of dispersed dust in the explosion vessel underneath.*
9. *Initial dust sample.*
10. *Wall of explosion vessel.*

Consequently it was difficult to make meaningful comparison of results obtained with different powders. Schläpfer concluded rather pessimistically that the best that can be expected from experiments based on the transient air blast technique is relative data for dusts of one material.

It should finally be mentioned that Proust and Veyssiere (1988) generated laminar transient dust clouds by gently fluidizing a given quantity of maize starch (6% moisture content) from a bed of about 600 g of starch, initially resting on a porous bed at the bottom of a vertical column of 0.2 m × 0.2 m cross section. The average velocity of the fluidizing air was of the order of 0.1 m/s. This set-up enabled the study of genuinely laminar flame propagation, but the method is probably limited to powders that are comparatively easy to disperse and have relatively narrow size distributions. (See also Chapter 4.)

3.9.3
STATIONARY DUST CLOUD IN A CLOSED CIRCULATION SYSTEM

Several workers have tried to experiment with systems in which a given quantity of dust, suspended in a given volume of air circulates in a closed loop. An early attempt was made by Bauer (1918), who simply used a fan located near the bottom of the vertical cylindrical explosion chamber, and rotating at 90° with respect to the axis of the chamber, for keeping the dust in suspension. However, the assumption that this arrangement was capable of producing a uniform dust concentration was not, according to Selle (1957) justified, because the dust could not be completely raised from the bottom of the vessel.

Eckhoff (1970) made an attempt to adopt a similar system as that used by Bauer, by using a co-axial fan at the bottom of the cylindrical 1 litre bomb and a 'beater' rotating at 90° to the cylinder axis. Howeever, it was impossible to generate a stationary dust cloud. The dust initially introduced in the bomb was raised into suspension, but the dust concentration decreased systematically and comparatively fast with time due to inevitable deposition of dust on parts of the vessel wall, etc.

Brown and Woodhead (1953) had arrived at the same conclusion for another version of the closed circuit apparatus. These workers studied the dust cloud formed in a closed loop of 7.5 cm diameter glass tubing, through which the dust dispersion was circulated at various rates of flow. They concluded that it was impossible to obtain a uniform cloud, and very high circulation velocities were required to prevent dust from depositing at the bends of the loop. Furthermore, at these high circulation speeds it was noted that significant comminution of the particles could take place.

A special version of the method based on the circulation of a constant quantity of dispersed powder in a closed apparatus was developed by Gliwitzky (1936). In order to keep the dust dispersed, two propellers, rotating with different speeds and in opposite directions, were situated co-axially inside a horizontal cylinder with open ends, which was placed inside the 43 litre explosion chamber. The closed circuit thus consisted of the inside of the internal cylinder and the annulus between this cylinder and the wall of the explosion chamber. However, this system was not found to be satisfactory, since even with easily dispersable aluminium flakes, the dust dispersion was incomplete.

It thus appears to be justified to conclude that none of the various versions of the closed circuit system which have been developed for the formation of experimental dust clouds has proved to offer an acceptable solution to the problem.

3.9.4 STATIONARY DUST CLOUD IN AN OPEN-CIRCUIT SYSTEM

Because of the limitations and shortcomings of the two other categories of methods, open-circuit dust cloud generators have been used by a number of workers, in spite of the comparatively large dust quantities and the more complicated apparatus required.

A simple version of the open-circuit principle for dust dispersion was in fact described by Weber (1878) more than a century ago as part of a comprehensive investigation of causes of dust explosions and fires in flour mills. The flour was placed on a 100 mm diameter sieve of a suitable mesh size, and by vibrating the sieve, a controlled column of falling dust was created in the region below the sieve. By means of an annulus, placed at the bottom of the sieve, the diameter of the dust column could be controlled. The measurement of the dust concentration in the column was carried out simply by inserting two parallel plates, separated by a fixed distance, into the falling dust column, the plates being perpendicular to the axis of the column. Since the diameter of the column could be measured, and the distance between the plates was known, the volume of the dust cylinder trapped between the plates was known, and consequently the dust concentration was given by the amount of flour which had settled on to the lower plate, divided by the volume of the dust cylinder initially trapped between the plates.

70 years later the idea of Weber was adopted by Jones and White (1948) who, by sieving the dust into a vertical cylinder, avoided the gradual distortion of the falling dust column, which in the case of Weber took place as the dust travelled away from the sieve. The sieving method for dispersing dust was also adopted by Craven and Foster (1967) as part of a more refined experimental set-up, and on a comparatively large scale by Palmer and Tonkin (1968).

Schläpfer (1951) used a conveyor screw for supplying the powder at constant rate at the bottom of the vertical dust explosion tube, in which air was flowing at constant rate upwards. The bottom part of the tube, where the dust was introduced, was narrow to ensure high turbulence during dispersion of the powder in the air stream. The upper part of the tube was considerably wider, and hence more laminar-like flow of the dust cloud could be obtained.

A similar arrangement was used by Cassel (1964). In this apparatus, in order to obtain a constant flow rate and dust concentration, the dust was dispersed by a jet of gas from a hypodermic needle which was directed vertically downwards towards a horizontal rotating dust layer of constant thickness. The arrangement also incorporated an electromagnetic vibrator, the purpose of which was to prevent the powder from depositing on the inner walls of the apparatus.

Line *et al.* (1957) used an apparatus where a turbulent dust cloud was initially formed by means of jets of oxidizer gas passing through a bed of the dust. This cloud was then directed downwards through a vertical cylindrical tube, under laminar conditions. The combustion chamber, having a considerably larger diameter, was attached to the tube, and consequently, on leaving the tube, the dust cloud formed a continuous wall-free column travelling downwards through the combustion chamber.

In order to investigate the burning velocity of laminar flames of lycopodium, Kaesche-Krischer and Zehr (1958) fed lycopodium into the lower end of a vertical 2 cm diameter tube where it was dispersed into a stationary dust cloud by an upwards moving stream of air. This arrangement made it possible to obtain stable flames in the concentration range of dust between 200 and 500 g/m^3. Mason and Wilson (1967), who also studied the burning velocity of stationary flames of lycopodium, described a dispersing arrangement where the lycopodium was elutriated from a fluidized bed. These workers were able to obtain stable dust flames in the concentration range 125 to 190 g/m^3.

Ballal (1983) also supplied the dust to be dispersed to a worm conveyor, from which it was fed at the desired rate at the top of the apparatus into the controlled downwards air flow. In the case of very fine powders, particle clusters or agglomerates tended to form at the output of the worm conveyor. The problem could be somewhat reduced by modifying the design of the screw and optimizing its speed of rotation. However, in order to ensure proper dust dispersion, the dust output from the worm conveyor was fed into a swirl chamber driven by a controlled air flow. A high level of turbulence shear created within this chamber by the colliding whirling jets produced a cluster-free, uniform dust dispersion. For high dust concentrations and especially for coal dust, a secondary dust dispersion source in the form of a high speed rotary disc was incorporated within the settling chamber. Thus a uniform and finely dispersed dust cloud could be produced. This was gently drawn into the explosion chamber by a small suction pump.

3.9.5 CONCLUSION

Provided the relatively large amounts of powder required and the relatively complicated experimental apparatus can be justified, the open circuit principle is the most satisfactory alternative for dust dispersion. However, as for the transient cloud systems, it is essential that the dust dispersion system is designed carefully, to ensure that the degree of agglomeration of the particles in the cloud and the cloud turbulence, corresponds with the state actually wanted. Quite often it is desirable or even necessary to keep the consumption of dust or powder at a minimum. In such cases the transient air blast technique may, provided the limitations of the technique are borne in mind, offer the best solution.

REFERENCES

Akiyama, T., and Tanijiri, Y. (1989) Criterion for Re-Entrainment of Particles. *Powder Technology*, **57** pp. 21–26

Allen, T. (1981) *Particle Size Measurement*, (3rd Edn.), Chapman and Hall, London

Alquier, M., Gruat, J., and Valentian, F. (1979) Influence of Large Eddies on the Suspension of Solid Particles. *Int. J. Multiphase Flow*, **5** pp. 427–436

Arthur, J. R. F., Dunstan, T., and Enstad, G. G. (1985) Determination of the Flow Function by Means of a Cubic Plane Strain Tester. *Int. J. Bulk Solids Storage in Silos*, **1** No. 2, pp. 7–10

Bachalao, W. D., Rudoff, R. C., and Houser, M. J. (1987) Particle Response in Turbulent Two-Phase Flows. *ASME (Am. Soc. Mech. Engrs.) FED*, **55** pp. 109–120

Bagnold, R. A. (1960) The Re-Entrainment of Settled Dusts. *Int. J. Air Poll.*, **2** pp. 357–363

Ballal, D. R. (1983) Flame Propagation through Dust Clouds of Carbon, Coal, Aluminium and Magnesium in an Environment of Zero Gravity. *Proc. R. Soc.* London. A385 pp. 21–51

Bartknecht, W. (1971) Brenngas- und Staubexplosionen. Forschungsbericht F45. Bundesinstitut für Arbeitsschutz, Koblenz

Bauer, G. (1918) On Dust Explosions, *Z. ges. Schiess.-u. Sprengstoffw.*, **13** pp. 272–273

Beer, J. M., Chomiak, J., and Smoot, L. D. (1984) Fluid Dynamics of Coal Combustion: A Review. *Prog. Energy Combust. Sci.*, **10** pp. 177–208

Boyle, A. R., and Llewellyn, F. J. (1950) The Electrostatic Ignitability of Dust Clouds and Powders. *J. Soc. Chem. Ind.*, London, **69** pp. 173–181

Brown, K. C., and Woodhead, D. W. (1953) Dust Explosion in Factories: Closed-Circuit Test Apparatus, Res. Rep. No. 86, Safety in Mines Research Establishment, Sheffield, UK

Brown, K. C., and James, G. J. (1962) *Dust Explosions in Factories: A Review of the Literature*, Res. Rep. No. 201, Safety in Mines Research Establishment, Sheffield, UK

Bryant, J. T. (1973) Powdered Fuel Combustion: Mechanism of Particle Size. *Combustion and Flame*, **20** pp. 138–139

Bunville, L. G. (1984) Commercial Instrumentation for Particle Size Analysis. In *Modern Methods of Particle Size Analysis, Chemical Analysis*, Vol. **73** pp. 1–42

Carpenter, D. L. (1957) The Explosibility Characteristics of Coal Dust Clouds Using Electric Spark Ignition, *Combustion and Flame*, **1** pp. 63–93

Carpenter, D. L., and Davies, D. R. (1958) The Variation with Temperature of the Explosibility Characteristics of Coal Dust Clouds Using Electric Spark Ignition, *Combustion and Flame*, **2** pp. 35–54

Cashdollar, K. L., and Hertzberg, M. (1985) 20-L Explosibility Test Chamber for Dusts and Gases, *Rev. Sci. Instr.* **56** pp. 596–602

Cassel, H. M. (1964) Some Fundamental Aspects of Dust Flames, Rep. Inv. 6551, US Bureau of Mines

Chen, C. P., and Wood, P. E. (1983) Turbulence Model Predictions of a Two-Phase Jet. *33rd Canadian Chem. Engng. Conference*, Vol. **1**, pp. 267–273

Corn, M., and Stein, F. (1965) Reentrainment of Particles from a Plane Surface. *Am. Ind. Hygiene Assoc. Journal* **26** pp. 325–336

Corn, M. (1966) Adhesion of Particles In *Aerosol Science* (ed C. N. Davies) Chapter **XI** pp. 359–392 Academic Press

Craven, A. D., and Foster, M. G. (1967) Dust Explosion Prevention – Determination of Critical Oxygen Concentration by a Vertical Tube Method. *Combustion and Flame*, **11** pp. 408–414

Durst, F., Melling, A., and Whitelaw, J. H. (1981) *Principles and Practice of Laser-Doppler Anemometry*. Second Edition, Academic Press

Dorsett, H. G., Jacobson, M., Nagy, J., *et al.* (1960) Laboratory Equipment and Test Procedures for Evaluating Explosibility of Dusts, Rep. Inv. No. 5624 US Bureau of Mines

Ebert, F. (1983) Zur Bewegung feiner Partikeln, die in turbulent strömenden Medien suspendiert sind. *Chem.-Ing. Techn.* **55** No. 12, pp. 931–939

Eckhoff, R. K. (1970) The Energy Required for the Initiation of Explosion in Dust Clouds by Electric Sparks. M.Phil. Thesis, University of London

Eckhoff, R. K. (1976) *Factors Affecting the Dispersibility of Powders*. Beretning **XXXIX**, 3. Chr. Michelsen Institute, John Grieg, Bergen, Norway

Eckhoff, R. K. (1977) Some Notes made during a Visit to the Research Institute of Materials Science Problems, Kiev, USSR, Rep. 77002-5 (Oct.), Chr. Michelsen Institute, Bergen, Norway

Eckhoff, R. K., Leversen, P. G., and Schubert, H. (1978) The Combination of Tensile Strength Data of Powders and Failure Loci from Jenike Shear Cell Tests. *Powder Technology* **19** pp. 115–118

EFCE–Working Party Mech. Part. Solids (1989) Standard Shear Testing Technique for Particulate Solids Using the Jenike Shear Cell. The Inst. Chem. Engrs./Europ. Fed. Chem. Engng. Rugby, UK

Elghobashi, S., and Rizk, M. A. (1983) The Effect of Solid Particles on the Turbulent Flow of a Round Gaseous Jet: A Mathematical and Experimental Study, Report on Contract No. DE–FG22–80PC–30303, US Dept. Energy

Engler, C. (1907) Einfacher Versuch zur Demonstration der gemischten Kohlenstaub- und Gasexplosionen. *Chemiker Zeitung* **31** pp. 358–359

Enstad, G. G. (1980) Inter-Particle Forces in Powders. A Literature Study. (In Norwegian). CMI–Report No. 803103–1 (March), Chr. Michelsen Institute, Bergen, Norway

Enright, R. J. (1984) Experimental Evaluation of the 1.2, 8 and 20 Litre Explosion Chambers. *Proc. 1st Int. Colloquium on Explosiveness of Industrial Dusts* (November), Baranow, Poland pp. 52–62

Fairchild, C. I., Tillery, M. I., and Wheat, L. D. (1985) Wind and Saltation Driven Particle Resuspension in a Wind Tunnel. *International Symposium/Workshop on Particulate and Multi-phase Processes*, Fine Particle Society, Harlow, UK

Fan Jianren, Zhao Hua, and Cen Kefa (1989) Numerical Modelling and Experimental Study of the Gas-Particle Coaxial Jets. *Chemical Reaction Engineering and Technology*, **5** (Sept.) pp. 26–35

Forsythe, W. E. (1959) *Smithsonian Physical Tables*. The Smithsonian Institution, Washington DC

Gardiner, D. P., Caird, S. G., and Bardon, M. F. (1988) An Apparatus for Studying Deflagration through Electrostatic Suspensions of Atomized Aluminium in Air. In *Proc. 13. Int. Pyrotech. Sem.* pp. 311–326

Geldart, D., and Wong, C. Y. (1984) Fluidization of Powders Showing Degrees of Cohesiveness. I. Bed Expansion. *Chem. Engng. Science* **39** pp. 1481–1488

Geldart, D. (1986) *Gas Fluidization Technology,* J. Wiley and Sons

Genchev, Zh. D., and Karpuzov, D. S. (1980) Effects of the Motion of Dust Particles on Turbulence Transport Equations. *J. Fluid Mech.* **101** Part 4, pp. 833–842
Gieras, M., Klemens, R., Wolanski, P., Wojcicki, S. (1986) Experimental and Theoretical Investigation into the Ignition and Combustion Processes of Single Coal Particles under Zero and Normal Gravity Conditions. *21st Int. Symp. Comb.*, The Comb. Inst., pp. 315–323
Gliwitzky, W. (1936) Pressure-Time Measurement in Aluminium Dust Explosions. *Z. Ver. Deutch. Ing.* **80** pp. 687–692
Green, H. L., and Lane, W. R. (1964) *Particulate Clouds: Dusts, Smokes and Mists*, 2nd edn., E. & F. N. Spon, London
Gutterman, A. M., and Ranz, W. E. (1959) Dust Properties and Dust Collection. *J. Sanitary Engng. Div. Proc. Am. Soc. Civ. Engrs.* SA**4**, (July), pp. 25–69
Haider, A., and Levenspiel, O. (1989) Drag Coefficient and Terminal Velocity of Spherical and Non-Spherical Particles. *Powder Technology* **58** pp. 63–70
Hartmann, I., Nagy, J., and Brown, H. R. (1943) Inflammability and Explosibility of Metal Powders. Rep. Inv. 3722, US Bureau of Mines
Hatta, N., Takuda, H., Ishii, R., *et al.* (1989) A Theoretical Study on Nozzle Design for Gas-Particle Mixture Flow. *ISIJ International* **29** pp. 605–613
Helwig, N. (1965) Untersuchungen über den Einfluss der Korngrösse auf den Ablauf von Kohlenstaubexplosionen. D.82 (Diss. T. H. Aachen). Mitteilungen der Westfählischen Berggewerkschaftskasse. Heft 24
Herdan, G. (1960) *Small Particle Statistics.* Butterworths, London
Hinze, J. O. (1975) *Turbulence,* 2nd Edition McGraw-Hill
Hodgman, C. D. (1963) *Handbook of Chemistry and Physics.* 44th Edition, The Chemical Rubber Publishing Co.
Holtzwart, R., and von Meyer, E. (1891) On the Causes of Explosions in Brown-Coal Briquette Works. *Dinglers J.* **280** pp. 185–190, 237–240
Hwang, C. C., Singer, J. M., and Hartz, T. N. (1974) Dispersion of Dust in a Channel by a Turbulent Gas Stream. Rep. Inv. 7854, US Bureau of Mines
International Standardization Organization (1985) Explosion Protection Systems: Part 1: Determination of Explosion Indices of Combustible Dusts in Air, Report No. ISO 6184/1, ISO, Geneva
Iversen, J. D. (1985) Particulate Entrainment by Wind. *International Symposium/Workshop on Particulate and Multiphase Processes*, Fine Particle Society
Jenike, A. W. (1964) Storage and Flow of Solids. Bulletin 123, Utah University, Utah, USA
Johansen, S. T. (1987) Numerical Modelling of the Acucut Air Classifier for Fine Powders, Report No. STF 34 F87058 (May) SINTEF, Trondheim, Norway
Jones, E., and White, A. G. (1948) Gas Explosions and Dust Explosions – A Comparison. *Proc. of Conference on Dust in Industry*, Society of Chemical Industry, (Sept.) pp. 129–139
Kaesche-Krischer, B., and Zehr, J. (1958) Untersuchungen an Staub/Luft-Flammen. *Z. Phys. Chemie (Neue Folge)* **14** pp. 384–387
Kaye, B. H. (1981) *Direct Characterization of Fine particles,* John Wiley and Sons, Inc., New York
Krol, M., (1985) Zur Partikelbewegung in turbulent strömenden hochkonzentrierten Feststoffsuspensionen. *Chem.-Ing. Techn.* **57** No. 3 pp. 254–255
Lee, L. W. (1984) The Convective Velocity of Heavy Particles in Turbulent Air Flows. *Chem. Eng. Commun.* **28** pp. 153–163
Lee, S. L. (1989) Particle Motion in a Turbulent Two-Phase Dilute Suspension Flow. *Particle and Particle Systems Characterization* **6** pp. 51–58
Line, L. E., Clark, W. J., and Rahman, J. C. (1957) An Apparatus for Studying the Burning of Dust Clouds. *6th Symp. (International) on Combustion.* Reinhold, New York pp. 779–786
Lockwood, F. C., and Papadopoulos, C. (1989) A New Method for the Computation of Particle Dispersion in Turbulent Two-Phase Flows. *Combustion and Flame* **76** pp. 403–413
Mason, W. E., and Wilson, M. J. G. (1967) Laminar Flames of Lycopodium Dust in Air.

Combustion and Flame **11** pp. 195–200
Meek, R. L., and Dallavalle, J. M. (1954) Explosive Properties of Sugar Dusts. *Industrial and Engineering Chemistry* **46** p. 763
Molerus, O. (1978) Effect of Interparticle Cohesive Forces on the Flow Behaviour of Powders. *Powder Technology* **20** pp. 161–175
Moore, P. E. (1979) Characterization of Dust Explosibility: Comparative Study of Test Methods. *Chemistry and Industry*, (July) pp. 430–434
Mulcahy, M. F. R., and Smith, I. W.: Kinetics of Combustion of Pulverized Fuel: A Review of Theory and Experiment. *Rev. Pure and Appl. Chem.*, **19** p. 81
Nagy, J., Seiler, E. C., Conn, J. W., *et al.* (1971) Explosion Development in Closed Vessels. Rep. Inv. 7507, US Bureau of Mines
Nedin, V. V., Nejkov, O. D., Alekseev, A. G., *et al.* (1975) A Comparative Investigation of Explosibility Characteristics of Dust Clouds. In *Prevention of Dust Ignition and Dust Explosions* (in Russian). Izdatel'stvo,Naukova Pumka, Kiev
Nguyen, T., and Nieh, S. (1989) The Role of Water Vapour in the Charge Elimination Process for Flowing Powders. *Journ. Electrostatics*, **22** pp. 213–227
Palmer, K. N., and Tonkin, P. S. (1968) The Explosibility of Dusts in Small-Scale Tests and Large-Scale Industrial Plant. *3rd Symp. Chem. Proc. Hazards with Special Ref. to Plant Design.* Inst. Chem. Engrs. Symp. Series No. 25, London
Perry, R. H., and Chilton, C. H. (1973) *Chemical Engineers' Handbook.* 5th edn, McGraw-Hill Book Co.
Picart, A., Berlemont, A., and Gouesbet, G. (1986) Modelling and Predicting Turbulence Fields and the Dispersion of Discrete Particles Transported by Turbulent Flows. *Int. J. Multiphase Flow* **12** No. 2 pp. 237–261
Proust, C., and Veyssiere, B. (1988) Fundamental Properties of Flame Propagation in Starch Dust/Air Mixtures. *Combust. Sci. Technol.*, **62** pp. 149–172
Ross, H. D. (1988) Reducing Adhesion and Agglomeration within a Cloud of Combustible Particles. NASA Technical Memorandum 100902, (July), Nat. Techn. Inform. Service, Springfield, VI, USA
Rudinger, G., and Chang, A. (1964) Analysis of non-steady Two-Phase Flow. *Phys. Fluids*, **7** pp. 1747–1754
Rudinger, G., (1980) *Fundamentals of Gas-Particle Flow*. Handbook of Powder Technology Series, Vol. 2, Elsevier
Rumpf, H. (1970) Zur Theorie der Zugfestigkeit von Agglomeraten bei Kraftübertragung an Kontaktpunkten. *Chem.-Ing.-Techn.*, **42** No. 8, pp. 538–540
Rumpf, H. (1974) Die Wissenschaft des Agglomerierens. *Chem.-Ing.-Techn.* **46** No. 1, pp. 1–11
Rumpf, H. (1975) *Mechanische Verfahrenstechnik,* Carl Hanser Verlag, München pp. 39–57
Schläpfer, P. (1951) Ueber Staubflammen und Staubexplosionen. Schweiz. Verein von Gas- und Wasserfachmännern. *Monatsbull.* No. 3, **31** pp. 69–82
Schofield, C., Sutton, H. M., and Waters, K. A. N. (1979) The Generation of Dust by Materials Handling Operations. *J. Powder & Bulk Solids Techn.*, **3** pp. 40–44
Schubert, H., and Wibowo, W. (1970) Zur experimentellen Bestimmung der Zugfestigkeit von gering verdichteten Schüttgütern. *Chem.-Ing.-Techn.*, **42** No. 8, pp. 541–545
Schubert, H. (1973) Kapillardruck und Zugfestigkeit von feuchten Haufwerken aus körnigen Stoffen. *Chem.-Ing.-Techn.*, **45** No. 6, pp. 396–401
Schubert, H. (1979) Grundlagen des Agglomerierens. *Chem.-Eng.-Techn.*, **51** No. 4, pp. 266–277
Schwedes, J. (1976) Fliessverhalten von Schüttgütern in Bunkern. *Chem.-Ing.-Techn.* **48** No. 4, pp. 294–300
Selle, H. (1957) Die Grundzüge der Experimentalverfahren zur Beurteilung brennbarer Industriestäube. *VDI-Berichte* **19** pp. 37–48
Shrayber, A. A. (1988) An Approximate Method for Determining the Parameters of Fluctuating

Motion of Particles in a Turbulent, Particle-Laden Gas Flow. *Fluid Mechanics – Soviet Research* **17** pp. 27–34
Singer, J. M., Greninger, N. B., and Grumer, J. (1967) Some Aspects of the Aerodynamics of the Formation of Float Coal Dust Clouds. *12th Int. Conf. Mine Safety Res. Establ.*, Dortmund
Siwek, R. (1977) 20-l-Laborapparatur für die Bestimmung der Explosionskenngrössen brennbarer Stäube. Diploma Thesis (Sept), Technical University of Winterthur, Switzerland
Siwek, R. (1988) Zuverlässige Bestimmung explosionstechnischer Kenngrössen in der 20-Liter Laborapparatur. *VDI-Berichte* **701** pp. 215–262
Smolyakov, A. V., and Tkachenko, V. M. (1983) *The Measurement of Turbulent Fluctuations.* (English Translation) Springer-Verlag
Sokolovski, V. V. (1960) *Statics of Soil Media,* (Translated to English from Russian by D. H. Jones and A. N. Schofield), Butterworths Scientific Publications, London
Tadmor, J., and Zur, I. (1981) Resuspension of Particles from a Horizontal Surface. *Atmospheric Environment*, **15** pp. 141–149
Tomita, Y., Tashiro, H., Deguchi, K., *et al.* (1980) Sudden Expansion of Gas-Solid Two-Phase Flow in a Pipe. *Phys. Fluids*, **23**(4) pp. 663–666
Trostel, L. J., and Frevert, H. W. (1924) The Lower Limits of Concentration for Explosion of Dusts in Air. *Chem. Metall. Engng.*, **30** pp. 141–146
Ural, E. A. (1989) Dispersibility of Dusts Pertaining to their Explosion Hazard, Factory Mutual Research Report J. I. OQ2E3.RK, (April), Norwood, Mass., USA
Ural, E. A. (1989a) Experimental Measurement of the Aerodynamic Entrainability of Dust Deposits. *12th Int. Coll. Dyn. Expl. React. Syst.* (July 24–28) Ann Arbor, Michigan, USA
Weber, R. (1878) Preisgekrönte Abhandlung über die Ursachen von Explosionen und Bränden in Mühlen, sowie über die Sicherheitsmassregeln zur Verhütung derselben. *Verh. Ver. Gew. Fliess., Berl.* pp. 83–103
Yamamoto, H., and Suganuma, A. (1984) Dispersion of Airborne Aggregated Dust by an Orifice. *International Chemical Engineering*, **24** pp. 338–345
Yamamoto, H. (1990) Relationship between adhesive force of fine particles and their dispersibility in gas. *Proc. 2. World Congress in Particle Technology,* Sept. 19–22, Kyoto, Japan, pp. 167–173
Zeleny, J., and McKeehan, L. W. (1910) Die Endgeschwindigkeit des Falles kleiner Kugeln in Luft. *Physik. Zeitschrift* **XI** pp. 78–93

Chapter 4

Propagation of flames in dust clouds

4.1 IGNITION AND COMBUSTION OF SINGLE PARTICLES

4.1.1 ALUMINIUM

Friedman and Macek (1962, 1963) studied the ignition and combustion of aluminium particles in hot gases of varying oxygen content. They concluded that ignition occurred only after melting of the oxide layer (melting point 2300 K) which coats the particle. The process of ignition did not appear to be affected by the moisture content of the hot ambient gas and was only slightly influenced by the oxygen content. At an oxygen content of only 2–3 mole per cent, ignition occurred at 2300 K, whereas at 35 mole per cent oxygen, it occurred at 2200 K. On the other hand, the concentrations of oxygen and water vapour had significant influence on the combustion of the metal. Oxygen promoted vigorous combustion, and, if its concentration was sufficiently high, there was fragmentation of particles. In the absence of moisture, diffusion and combustion took place freely in the gas phase, whereas in the presence of moisture, the process was impeded and confined to a small region, because the reactants had to diffuse through a condensed oxide layer on the surface of the molten particle.

Cassel (1964) injected single 60 μm diameter aluminium particles into the centre of a laminar aluminium dust flame of known spatial temperature distribution. Ignition of the particles occurred at 2570 K, but this was probably higher than the minimum temperature required for ignition, because the residence time of the particle in the hot environment was not more than 2 ms. This is shorter than the induction period required for self-heating of the particle from its minimum ignition temperature to the minimum temperature for self-sustained oxidation.

Cassel further observed that within 2 ms after ignition a concentric burning zone, of diameter about nine times the original particle diameter, developed around the particle. After 3 ms, a detached envelope appeared, which at first surrounded the particle concentrically, but then became elongated and gradually developed into a cylinder of length more than 10 times its diameter. This expanding oxide envelope, being in the liquid state, followed the relative motion of the ambient atmosphere.

Burning times of 60 μm aluminium particles located between the lobes of the aluminium-dust flame were found to be of the order of 10.5 ms (about 4.5 times longer than for magnesium particles burning under the same conditions). Cassel attributed this to the greater oxygen requirement for the oxidation of aluminium.

Prentice (1970) studied the ignition and combustion of single 300–500 μm aluminium particles in dry air, following initial heating and melting by a light flash from either a pulsed Nd-glass laser or a xenon-flash discharge lamp. In air (as opposed to in Ar/O_2)

oxide accumulated on the burning aluminium droplet. Because of this, the combustion process was terminated by fragmentation of the droplet (as shown by Nelson, 1965 for zirconium). The very fast flash-heating method generated fully developed metal droplets with practically no oxide on the surface. This presented initial conditions for studying the subsequent ignition and combustion processes, when the virgin droplets interacted with the surrounding air. Detailed SEM studies of the oxide layer build-up revealed a porous structure with a great number of fumaroles. Over the experimental range, the burning time to fragmentation increased linearly with the particle diameter from about 200 ms at 300 μm to 600 ms at 500 μm. Prentice studied the combustion of aluminium droplets in dry air over a range of pressures up to 4.5 bar (abs.). The particles were found to fragment in dry air at pressures up to about 2.4 bar (abs.). Fragmentation became quite weak and sporadic at this pressure and finally ceased as the pressure was raised to approximately 4.0 bar (abs.). The time to fragmentation was found to be inversely proportional to the air pressure, i.e. to the oxygen concentration.

Prentice also found that the nitrogen in the air played an active role in the combustion process, causing the oxide generated to adhere to the droplet surface and form an asymmetrical, spin-generating oxide layer that appeared to be a pre-condition for fragmentation. The driving gas causing particle fragmentation is in part aluminium vapour, but for combustion in air the major constituent is nitrogen from nitride.

Frolov *et al.* (1972) studied ignition and combustion of single aluminium particles in high-temperature oxidizing gases, as a function of particle size and state of the gas. Various theories were reviewed.

Grigorev and Grigoreva (1974) modified the theory of aluminium particle ignition by Khaikin *et al.* (1970), by including a fractional oxidation law accounting for possible changes of the structure of the oxide film during the pre-flame heating period. Experiments had revealed that the minimum ignition temperature of aluminium particles was independent of particle size, and Grigorev and Grigoreva attributed this to the oxidation rate depending very little on the thickness of the oxide layer.

Razdobreev *et al.* (1976) studied the ignition and combustion of individual 230–680 μm diameter aluminium particles in air, following exposure to stationary laser light fluxes. At incident fluxes approaching 150 W/cm^2 melting of the particle took place, but ignition occurred only at fluxes higher than 250 W/cm^2. Coefficients of reflection were not measured, but were assumed to be in the range 96 to 50%, which means that less than half of the incident light flux was absorbed by the particle. The time from onset of radiant heating to ignition increased with particle diameter from 100 ms for 230 μm, via 270 ms for 400 μm, to 330 ms for 680 μm.

Ermakov *et al.* (1982) measured the surface temperature of 400–1200 μm diameter aluminium particles at the moment of ignition. The heating was performed by a continuous laser of wavelength 10.6 μm at a constant flux incident on the particle in the range 1500–4500 W/cm^2, i.e. much higher than the experimental range of Razdobreev *et al.* (1976). The particle temperature was measured by a tungsten-rhenium thermocouple, whose junction of thickness 18–20 μm was located at the centre of the particle. Microscopic high-speed film records were made synchronously with the recording of the particle temperature at a rate up to 4500 frames/s. The simultaneous recording permitted detailed simultaneous comparison of the temperature of the particle with physical phenomena observed on the particle surface. The appearance of a flame in the form of a tongue on a limited section of the surface was noted at a particle temperature of

2070 ± 50 K. With further heating to 2170 K, the flame tongue propagated to the entire particle surface, and the particle temperature remained constant at 2170 K during the subsequent burning. This temperature is slightly lower than the melting point of the oxide, and Ermakov *et al.* challenged the oxide melting point hypothesis. They concluded that the ignition temperature obtained in their experiments showed that ignition is not caused by melting of the oxide film, but is a result of the destruction of the integrity of the film due to thermomechanical stresses arising during the heating process. This was indicated by photographs of the particle surface at the time that the flame tongue appeared. No influence of the incident heating flux density on the stationary combustion temperature of the particle was detected.

4.1.2 MAGNESIUM

Cassel and Liebman (1959) found that ignition temperatures of magnesium particles in air did not differ from those in pure oxygen. Therefore they excluded oxygen diffusion as the reaction rate controlling mechanism in the ignition process, and proposed a theory based on a simple chemical control Arrhenius term for describing the rate of heat generation per unit of particle surface area. An average value of the activation energy of 160 ± 13 J/mole was derived from the available experimental data.

Cassel and Liebman (1963) measured the ignition temperatures of single magnesium particles of 20 to 120 μm diameter by dropping the particles into a furnace containing hot air of known temperature. They found that the minimum air temperature for ignition decreased systematically with increasing particle size, being 1015 K for a 20 μm diameter particle, 950 K for 50 μm, and 910 K for 120 μm.

Cassel (1964) proposed a physical model for the combustion of individual magnesium particles, as illustrated in Figure 4.1. After ignition, the oxide layer that coats the particle prior to ignition, is preserved, only growing slightly in thickness. During combustion, the oxide shell encloses the evaporating metal drop, while superheated metal vapour diffuses through the semi-permeable shell to the outside and reacts with oxygen that diffuses toward the particle from the ambient atmosphere. The rate of burning of the particle is therefore governed by the rate of oxygen diffusion towards the reaction zone. In the initial stage of combustion the site of reaction is close to the outer surface of the oxide layer. However, owing to depletion of oxygen, this zone is detached from the oxide surface and shifted to a distance, L, from the particle shell. The rate of oxygen diffusion and the rate of combustion are determined by the gradient of oxygen partial pressure at $r_o + L$. This gradient remains approximately constant over the lifetime of the burning particle, except for the final stage, when the reaction zone withdraws to the oxide shell.

Cassel (1964) also suggested a theoretical model for the combustion of a magnesium particle. On the assumption that the location of the liquid drop inside the oxide shell is unimportant, and that the rate of oxygen diffusion is always slower than the rate of the chemical reaction, the burning rate of a magnesium particle is given by the quasi-stationary balance of the oxygen diffusion rate:

$$\overline{W}_{O_2} = 4\pi(r_0 + L)\frac{Dp}{RT}\,1n\,\frac{p - p_L}{p - p_\infty}, \tag{4.1}$$

and the rate of metal vaporization:

$$\overline{W}_{Mg} = -\frac{4\pi\rho r^2}{M\epsilon}\frac{dr}{dt} \tag{4.2}$$

Here D is the average oxygen diffusion coefficient at average temperature T, M is mole weight of magnesium, ρ is density of magnesium, ϵ is oxygen equivalent (=2 for oxidation of magnesium), p is absolute total pressure at distance r_0 (just outside of the oxide shell), and p_L and p_∞ are the partial pressures of oxygen at distances L and infinity.

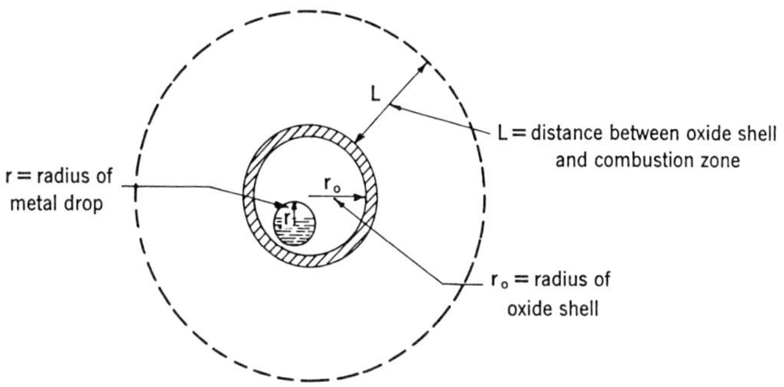

Figure 4.1 *Model of burning magnesium particle (From Cassel, 1964)*

The time τ required for complete combustion of a particle is obtained by combining equations (4.1) and (4.2) and integrating from the initial drop radius r_0 to zero. The resulting equation is:

$$\tau = \frac{\rho RT}{M\epsilon Dp}\frac{r_0^3}{3(r_0 + L)} \bigg/ 1n\left(\frac{p - p_L}{p - p_\infty}\right) \tag{4.3}$$

Equation (4.3) was used to derive values of (D/T) from observed τ values. It should be noted that ρ, p_∞, and D refer to different temperatures, namely the boiling point of the metal, the ambient gas temperature, and the temperature in the diffusion zone near the reaction front, T. The estimates of D assuming molecular diffusion, gave an unrealistically high T value of 4860 K for a magnesium particle burning in air. Cassel suggested therefore that the combustion of magnesium particles is governed predominantly by diffusion of atomic oxygen. He also suggested that the same must be true in any dust flame burning at 3000°K or more.

Liebman *et al.* (1972) studied experimentally the ignition of individual 28–120 μm diameter magnesium particles suspended in cold air, by an approximately square laser light pulse of 1.06 or 0.69 μm wavelength and 0.9 ms duration. The results suggest that during heating of a magnesium particle by a short flash of thermal radiation, the particle temperature first rises rapidly to the boiling point. Vaporized metal then expands rapidly from the particle surface, and vapour-phase ignition may occur near the end of the radiant pulse. In accordance with the model proposed by Cassel (Figure 4.1), ignition is assumed to occur at some distance from the particle surface where conditions (magnesium and oxygen concentrations, and temperature) are optimal. The onset of ignition was characterized by the rapid appearance of a large luminous zone. Radiant intensities required to ignite the particles were found to increase with particle size and the thermal conductivity of the ambient gas environment. In accordance with the results from hot gas ignition, there was little change in the radiant intensities required for ignition when replacing air by pure oxygen.

Florko *et al.* (1982) investigated the structure of the combustion zone of individual magnesium particles using various techniques of spectral analysis. They claimed that their results confirm the assumption that the oxide, after having been generated in the gas phase in the reaction zone, condenses between this zone and the surface of the burning particle. This observation is an interesting supplement to the observation made and the physical model proposed by Cassel (1964).

Florko *et al.* (1986) estimated the temperature in the reaction zone of burning magnesium particles as a function of the pressure of the ambient gas, by analysing the spectrum of the unresolved electron-vibration bands of the MgO molecules in the reaction zone. For large particles of 1.5–3 mm diameter, the reaction zone temperature was practically independent of the gas pressure and equal to 2700–2800 K in the range 0.3 to 1 bar (abs.). When the pressure was reduced to 0.05 bar (abs.) the reaction zone temperature dropped only slightly, to about 2600 K. The burning time of 1.5–3 mm diameter particles was proportional to the square of the particle diameter. For a 2 mm diameter particle at atmosphere pressure, the burning time was about 6 s. Extrapolation to 60 μm particle diameter gives a burning time of 5.4 ms, which is quite close to the times of a few ms found by Cassel (1964) for Mg particles of this size. When the pressure was reduced to 0.2 bar (abs.), Florko *et al.* found a slight reduction, by about 10%, of the burning time.

4.1.3 ZIRCONIUM

Nelson and Richardson (1964) and Nelson (1965) introduced the flash light heating technique for melting small square pieces of freely falling metal flakes to spherical droplets. They applied this method for generating droplets of zirconium, which were subsequently studied during free fall in mixtures of oxygen and nitrogen, and oxygen and argon. The duration of the light flash was only of the order of a few ms. A characteristic feature was the sparking or explosive fragmentation of the drop after some time of free fall. This was supposed to be due to the forcing out of solution of nitrogen, hydrogen, and carbon monoxide that had been chemically combined with the metal earlier in the combustion process. The experimental results for air at atmospheric pressure showed, as a first order approximation, that the time from droplet formation to explosive fragmentation was proportional to the initial particle diameter. The relative humidity of the air had only marginal influence on this time. The heat initially received by a given particle by the flash was not specified.

4.1.4 CARBON AND COAL

Research on explosibility of coal dust has long traditions. According to Essenhigh (1961), the possible role of coal dust in coal mine explosions was suggested as early as in 1630 by Edward Lloyd, when commenting on information received from Anthony Thomas concerning an explosion in England in about 1580. The role of coal dust in such explosions was certainly clear to Faraday and Lyell (1845), discussing the disastrous explosion in the Haswell collieries the year before. More systematic investigations into the ignitability and explosibility of coal dusts started at the end of the 19th and the beginning of the present century.

However, combustion of coal dust particles is not only related to the explosion problem. The increasing use of pulverized coal in burners for energy production has become an important area of research and development, and much information on the combustion of coal particles that is directly applicable to the coal dust explosion problem has been generated in that context. Furthermore, this use of pulverized coal in industry as well as in the public sector, has caused coal dust explosions to become a potential hazard not only in mines, but also in power generating plants utilizing powdered coal.

Coal normally contains both solid carbon and combustible volatiles. In addition there is usually some ash, and some moisture. The simplest system to study is the combustion of pure carbon or char. Nusselt (1924) proposed that the oxidation of pure carbon was essentially a direct conversion of solid carbon to CO_2 at the particle surface. However, later investigations have disclosed a more complex picture even for oxidation of pure carbon, as illustrated in Figure 4.2.

In zone I the concentration of O_2 is zero, whereas in Zone II the CO concentration is zero. At the carbon surface, S, CO_2 reacts with the solid carbon according to the endothermic scheme $CO_2 + C \rightarrow 2CO$. The required heat is supplied from the oxidation zone R, where the temperature is at maximum, and where the exothermic reaction $CO + \frac{1}{2} O_2 \rightarrow CO_2$ takes place. Using the theory of van der Held (1961), de Graaf (1965) found that the temperature in the oxidation zone R was about 2500 K for a coal surface temperature of 1800 K.

For low carbon surface temperatures of < 1400 K, a significant concentration of O_2 may exist right at the surface, and at very low surface temperatures of < 800 K, direct

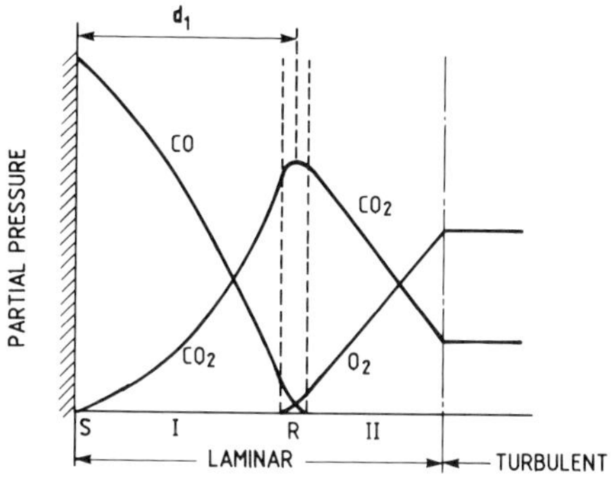

Figure 4.2 *Composition of laminar gas layer during combustion of solid carbon according to the theory of van der Held (1961) for surface temperatures > 1400 K. Nitrogen is not considered. S = carbon surface; R = reaction zone (From de Graaf, 1965)*

oxidation by oxygen according to the consecutive scheme $2C + O_2 \rightarrow 2CO$ and $2CO + O_2 \rightarrow 2CO_2$ takes place close to the surface. de Graaf carried out experiments that supported van der Held's theory.

However, conclusions from experiments with burning of comparatively large samples of carbon may not necessarily apply to the burning of very small particles. Ubhayakar and Williams (1976) studied the burning and extinction of single 50–200 μm diameter carbon particles in quiescent mixtures of oxygen and nitrogen, ignited by a light flash from a pulsed ruby laser. An initial objective of their study was to investigate whether a gas phase burning mechanism or a surface burning mechanism, possibly accompanied by pore diffusion, governs the combustion of sub-millimeter carbon particles. An additional objective was to obtain burning duration data for such small particles. The lowest mass fraction of oxygen used in the oxidizer gas was 0.5, which is considerably larger than in air. They concluded that in the temperature range of 2000–3500 K, the kinetics of the carbon oxidation could be represented by a surface reaction producing CO, and having an activation energy of 75 kJ/mole. As expected, the maximum temperature at the particle surface increased with increasing oxygen fraction in the oxidizer gas. At atmospheric pressure it was about 3000 K in pure oxygen and about 2200 K at an oxygen mass fraction of 0.6. Typical particle burning durations at atmospheric pressure were 60 ms for 100 μm diameter particles and 25 ms for 60 μm particles. For low oxygen mass fractions, extinction occurred before the particles had burnt away, and this explained why burning times for a given particle size were shorter in atmospheres of lower oxygen mass fractions than in pure oxygen.

In a purely theoretical investigation, Matalon (1982) considered the quasi-steady burning of a carbon particle which undergoes gasification at its surface by chemical reactions, followed by a homogeneous reaction in the gas phase. The burning rate M was determined as a function of the gas phase Damköhler number D_g (ratio of chemical and diffusion controlled reaction rates) for the whole range $0 < D_g < \infty$. The monotonic $M(D_g)$ curve, obtained for comparatively hot or cool particles, described the gradual transition from frozen flow to equilibrium. For moderate particle temperatures the transition was abrupt and the $M(D_g)$ curve was either S-shaped or Z-shaped, depending on the relative importance of the two competitive surface reactions $2C + O_2 \rightarrow 2CO$ and $C + CO_2 \rightarrow 2CO$.

Specht and Jeschar (1987) also investigated the governing mechanisms for combustion of solid carbon particles of various diameters. The chemical reactions considered were the same as discussed above, but it was found that their relative importance depends on particle size via its influence on the Damköhler number D_g.

On the basis of idealized considerations, Fernandez-Pello (1988) derived theoretical expressions for the instantaneous local mass burning rate and the overall regression rate (rate of reduction of the particle radius) for the combustion of a spherical condensed fuel (e.g. carbon) particle in a forced convective oxidizing gas flow. The model is illustrated schematically in Figure 4.3.

The equations derived are of the form:

$$\frac{dm}{dt} = -\frac{\lambda}{rC}(Re)^{1/2} f_1(B, G, \sigma) \quad (4.4)$$

$$\frac{dr^2}{dt} = \frac{\lambda}{\rho C}(Re)^{1/2} f_2(B, G, \sigma) \quad (4.5)$$

where

m is the remaining particle mass at time t
r is the particle radius at time t
λ is the thermal conductivity of the oxidizing gas
C is the mean specific heat of the reaction products
ρ is the density of the particle
Re is the particle Reynolds number referred to the velocity and viscosity of the oxidizing gas upstream of the particle
f_1 and f_2 are functions of a mass transfer number B, a normalized energy species function G, and the angular coordinate σ

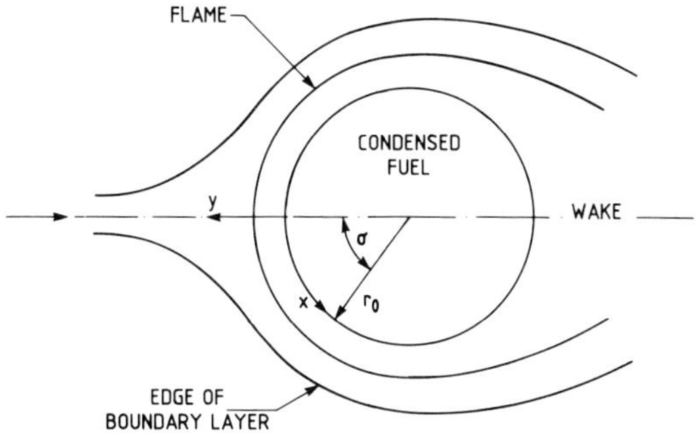

Figure 4.3 *Combustion of a condensed fuel particle in a forced convective oxidizing gas flow. Schematic illustration of theoretical model and coordinate system (From Fernandez-Pello, 1988)*

The predicted dependence of the overall particle regression rate, or the Nusselt number, on the Reynolds and mass transfer numbers was in qualitative agreement with semi-empirical correlations based on experiments with polymethyl methacrylate particles burning in mixtures of oxygen and nitrogen. Quantitative comparison between theory and experiments was difficult because of different definitions of the mass transfer number B and difference between theoretical and experimental environment conditions. However, it appeared that the theoretical analysis predicts higher (by a factor of approximately two) mass burning rates than those observed experimentally. The choice of the thermophysical properties of the fuel and oxidizer used in the theory, and the idealized assumptions implicit in the theoretical analysis could explain the quantitative disagreement with the experiments. The predicted variation of the particle radius with time is of the form $r_0^{3/2} - r^{3/2} \sim t$.

In the context of the dust explosion hazard, pure carbon is of limited interest, because clouds of pure carbon dust, e.g. graphite, in air at atmospheric pressure, are unlikely to represent a significant explosion hazard in practice. Therefore, coals containing volatiles are of greater practical interest. However, the volatiles complicate the ignition and combustion mechanisms, and the picture is less clear than for pure carbon combustion.

Gomez and Vastola (1985) compared the ignition and combustion of single coal and char particles in an isothermal flow reactor, by measuring the concentrations of CO and CO_2 in the downstream gas flow as functions of time. A sub-bituminous coal containing

22% moisture, 4.6% ash, 33.8% volatiles, and 39.6% fixed carbon was used in the study. For each run a single particle from a 850–1000 μm sieve fraction was injected into a reaction furnace swept with air. Experiments were performed at five temperatures: 928 K, 980 K, 1076 K, 1118 K and 1273 K. At each temperature two types of run were performed, namely coal combustion and char combustion. The char particles were prepared by injecting a coal particle into the reactor with a flowing nitrogen gas stream at the desired temperature. After pyrolysis was completed, the char was ignited by switching the carrier gas from nitrogen to air.

The main conclusion drawn by Gomez and Vastola from their experiment was that two chemical reactions compete for the oxygen surrounding the coal particle. The two reactions are quite different in nature, one involving the carbon surface (heterogeneous), and the other involving the volatiles (homogeneous). The gas concentration curves obtained for the heterogeneously oxidized char particles were considered typical for the heterogeneous reaction involving the carbon surface. Oxidation of coal particles could be heterogeneous, depending on the temperature. The gas concentration curves obtained for heterogeneous oxidation were similar to the curves for char combustion, except for an initial peak of carbon monoxide attributed to the combustion of volatiles on the surface or within the particle at low oxygen concentrations. However, when the coal particles ignited homogeneously, an initial pronounced peak of carbon dioxide was detected which was attributed to the gas phase combustion of the volatile matter at conditions of sufficient oxygen for burning most of the carbon in the volatiles to carbon dioxide. The initial peaks of carbon monoxide for heterogeneous coal ignition and carbon dioxide for homogeneous, can be used to measure the pyrolysis time during combustion.

Gomez and Vastola suggested that all the carbon in the volatiles is oxidized to carbon monoxide or carbon dioxide. This is because methane, the most difficult hydrocarbon to oxidize, which was detected in the volatiles of coal particles after pyrolysis in nitrogen, was not traced in the products from combustion in air.

If the particle burns under external diffusion control, the reaction proceeds on the external surface of the particle at a very low oxygen concentration. The particle diameter then reduces as the combustion advances, but the density of the remaining particle mass m at time t is the same as of the initial particle mass m_0. Integration of the reaction rate equation for this case, assuming spherical geometry, results in:

$$(m/m_0)^{2/3} = kt \tag{4.6}$$

where the global constant k embraces a number of constants and parameters. If this relationship describes the mechanism controlling the combustion process, a plot of the power two-thirds of the reduced mass m of the particle against time, determined experimentally, should result in a straight line. For char particles Gomez and Vastola's experiments gave straight lines at gas temperatures > 1100 K, whereas for coal particles straight lines were found for gas temperatures > 980 K.

The total combustion times, determined both by the method described above, and by independent light intensity measurements, varied from 5–10 s at a gas temperature of 1300 K, to 20 s at 930 K. These times are very long in the context of dust explosions, and are mainly due to the large particle diameter of about 1 mm, and partly to the comparatively low oxidizing gas temperatures in Gomez and Vastola's experiments.

Howard and Essenhigh (1965, 1966, 1967) discussing the results of their extensive research on coal particle combustion, first indicated that ignition of a bituminous coal particle generally occurs on the solid surface of the particle rather than in the volatile pyrolysis products. However, in their final conclusion (1966) they differentiated between various mechanisms on the basis of particle size. The classical view, of ignition taking place in the volatiles, still seemed to be valid for particle diameters larger than 65 μm. Smaller particles than this would, however, not be able to generate a sufficiently concentrated envelope of volatiles to prevent oxygen from diffusing to the solid carbon surface. For particle diameters smaller than 15 μm, the ignition reaction is more or less entirely heterogeneous oxidation at the particle surface.

The essential point in Howard and Essenhigh's argument is the assumption that for particles of smaller diameters than 100 μm, the total devolatilization time is independent of particle size. This implies that the average flow of volatiles per unit of particle surface area, increases with the particle size. For very small particles, the volatile flux is not sufficient for maintaining a volatile flame envelope round the particle.

In a more recent investigation of the devolatilization process by Johnson *et al.* (1988) Howard and Essenhigh's assumption of negligible influence of particle size on devolatilization rates (or total devolatilization times) was maintained for the range of particle sizes typical of most pulverized fuels and explosible dusts. These workers studied the devolatilization of monolayers of coal particles in an inert atmosphere, at heating rates from 100 to 1500 K/s. The results also indicated that for 10–1000 μm diameter particles of bituminous coals, resting on an electrically heated filament, the heating rate had little influence on the devolatilization yield, which was rather determined by the peak temperature. The maximum rate of devolatilization and maximum hydrocarbon yield occurred at peak temperatures between 700 and 1000 K.

Froelich *et al.* (1987) studied the combustion in air at 1400 K of single 80–100 μm diameter coal particles containing 30% volatile matter. They used the experimentally determined relationship between particle temperature (two colour pyrometer) and time in a furnace of known temperature to calculate the rate of gasification of the solid carbon of a coal particle. After about 5 ms in the furnace, the particle temperature reached a sharp peak of 2200 K, which was attributed to the devolatilization and ignition of the volatiles. A second, less sharp temperature rise, which started at about 10 ms and terminated at about 60 ms, had a peak value of about 1800 K and was associated with the gasification of the solid carbon.

In their theoretical analysis, Froelich *et al.* assumed that:

- The particle was a perfect and homogeneous sphere.
- The temperature of the particle was uniform.
- Either the diameter or the density of the particle remained constant (devolatilization or combustion of solid carbon).
- The furnace and the particle were black and grey bodies, respectively.
- The particle was in permanent thermal equilibrium with the gas and walls of the furnace.

The following equation was proposed:

$$\frac{xC_p\rho_c}{6}\frac{dT_p}{dt} = H_r + H_c + H_q \qquad (4.7)$$

where

H_r is radiative heat flux received by the particle per unit time
H_c is convective heat flux received by the particle per unit time
H_q is heat of reaction per unit time
C_p is specific heat capacity of the particle
T_p is temperature of the particle
ρ_c is density of the particle
x is diameter of the particle

H_r was determined from the Stefan-Boltzmann law by assuming that the particle is in radiative equilibrium with the furnace wall:

$$H_r = E\tau(T_f^4 - T_p^4) \tag{4.8}$$

where

E is total emissivity of the coal
τ is the Stefan-Boltzmann constant
T_f is the furnace wall temperature

The convective heat flux H_c was taken as:

$$H_c = h_c(T_g - T_p) \tag{4.9}$$

where

T_g is temperature of the gas around the particle
h_c is the convective heat transfer coefficient between the particle and the gas determined from the Nusselt number, assuming laminar flow around a spherical particle

The heat of reaction per unit time H_q was taken as:

$$H_q = AW\frac{\pi}{4}x^2 \tag{4.10}$$

where

W is the rate of devolatilization per unit of particle surface area, and
A is a constant.

W as a function of time was calculated from the experimentally determined particle temperature as a function time, by inserting equations (4.8), (4.9) and (4.10) in (4.7) and applying an iterative numerical method of solution. It was found that W had a peak of 4×10^{-2} kg/m^2 s at about 17 ms, and remained fairly constant at $3 \times 10^{-2} - 2 \times 10^{-2}$ kg/m^2 s from 20–40 ms to about 55 ms, whereafter it dropped rapidly to zero.

In their study of ignition and combustion of single coal particles, Gieras *et al.* (1985, 1986) eliminated the influence of gravity by performing the experiment during 1.4 s of free fall of the test chamber. In this way gravity driven convective heat transfer was avoided, and the exclusive roles of conductive and radiative heat transfer could be studied. The experiment was performed with one or more coal particles glued on to thin quartz needles. The smallest particle size that could be used without the needle and glue influencing the particle ignition and combustion significantly was about 300 μm. Therefore the most interesting particle sizes from a dust explosion point of view (diameters < 100 μm) could not be studied. However, the observed trends are nevertheless of interest.

In one series of experiments, pairs of equal-size particles separated by a fixed centre-to-centre distance D, were studied after one of the particles had been ignited by the flame from a burning 1 mm diameter drop of n-octane. For 700 μm diameter particles the maximum distance D_{max} for the second particle to become ignited by the first one, increased systematically with the volatile content of the coal and the oxygen content of the gas, as shown in Figure 4.4. It was also found that D_{max} was proportional to the particle diameter in the range 300–1200 μm investigated. For anthracite and coke in air, ignition of the second particle did not take place unless the particles were nearly touching, whereas particles of the coal of the highest volatile content in air could be separated by up to about two to three particle diameters.

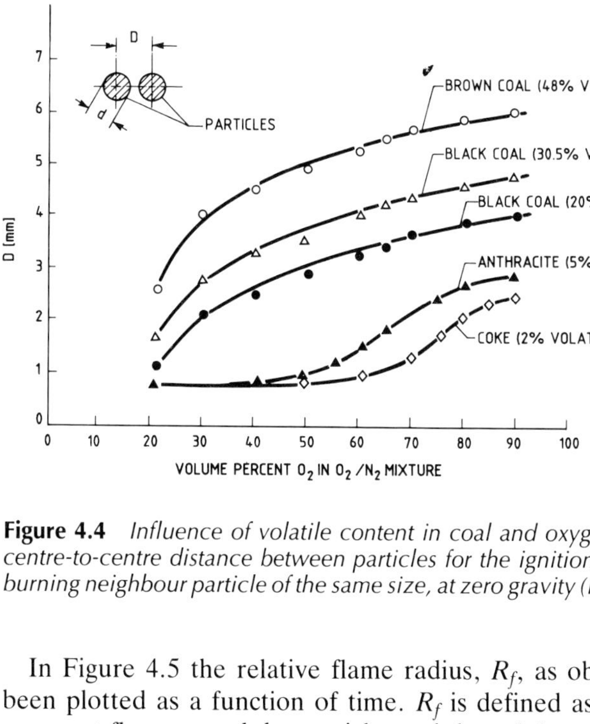

Figure 4.4 *Influence of volatile content in coal and oxygen concentration in gas on the maximum centre-to-centre distance between particles for the ignition of a 700 μm diameter coal particle by a burning neighbour particle of the same size, at zero gravity (From Gieras, Klemens and Wojcicki, 1985)*

In Figure 4.5 the relative flame radius, R_f, as observed on 48 fr/s movie photos, has been plotted as a function of time. R_f is defined as the ratio between the radius of the apparent flame round the particle, and that of the original particle. Figure 4.5 shows that the time required for reaching the maximum flame radius decreased and the maximum flame radius increased with increasing volatile content. This trend was interpreted in terms of the volatiles burning more rapidly than the char, in agreement with the general understanding of the combustion of coal particles.

In a further series of experiments, Gieras *et al.* (1985) studied the propagation of combustion through static linear chains of consecutive coal particles separated by a given optimal centre to centre distance D_{opt} depending on the volatile content. It was confirmed that the velocity of the 'one dimensional' flame propagation increased (approximately proportionally) with the volatile content of the coal.

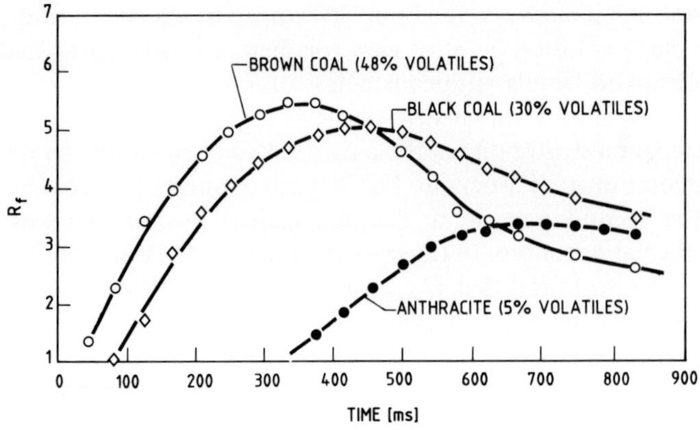

Figure 4.5 *Change of relative flame radius R_f with time during combustion of a 700 μm diameter coal particle at zero gravity (From Gieras, Klemens and Wojcicki, 1985)*

When similar inter-particle flame transfer experiments were conducted at normal gravity conditions, buoyancy played an important role (Gieras *et al.*, 1986). The maximum inter-particle distance for upwards flame transfer was then significantly larger than for horizontal transfer. This has important implications in dust explosions, e.g. for the definition of the concept of minimum explosible dust concentration. Under gravity conditions the limiting dust concentration for flame propagation will depend on whether the propagation occurs upwards, downwards or horizontally (see Section 4.2.6.2).

Wagner *et al.* (1987) studied the ignition and combustion of single coke and coal particles of diameters 63–125 μm in a vertical reactor containing hot oxidizing gas, through which the particles settled for predetermined periods (distances) before being captured and cooled rapidly. The initial volatile content for the materials investigated varied from 4.5% to 37%. The experimental data were compared with predictions by a numerical computer model, based on the earlier work by Field (1969) and Smith (1971). The model also treated the devolatilization process, by considering it as one single stage reaction of activation energy 228.5 kJ/mole. The combustion was considered as being controlled partly chemically and partly by diffusion processes. Both convective and radiative heat transfer was considered.

Figure 4.6 gives a set of experimental results for particles burning in air at atmospheric pressure and the corresponding predictions by the computer model. For all three coals and a gas temperature of 1170 K, devolatilization and combustion of volatiles is completed within about 0.5 s, whereas the burning-off time of the char increases markedly with decreasing content of volatiles.

Levendis *et al.* (1989) studied mechanisms and rates of oxidation of char particles in the size range from a few μm to several tens of μm. The specific surface area of the char particles varied with the origin of the char (polymers with pore-forming additives). When heated in an inert atmosphere, the char particles maintained their amorphous nature up to 1600 K. However, when oxidized at 1600 K, the carbon matrix underwent partial graphitization.

Vareide and Sönju (1987) developed approximate computer models for predicting burn-off of char particles. Two alternative assumptions concerning the particle size and

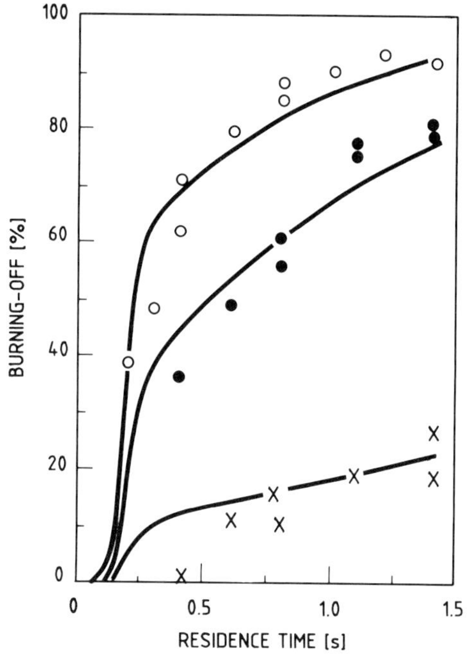

Figure 4.6 *Burning-off of 63–125 μm coal particles of various volatile contents as functions of residence time in hot gas (1170 K) in vertical reactor:*
○ = 37.1% volatiles
● = 20.7% volatiles
x = 7.3% volatiles
— = computer model predictions
(From Wagner et al.*, 1987)*

density were adopted, namely constant density and decreasing diameter, and constant diameter and decreasing density. The total burn-off time decreased with initial particle diameter. In the case of the shrinking particle model, the total burn-off time at 15 vol% O_2 and 1500 K was about 1 s for a 100 μm particle, and 0.1 s for a 10 μm particle. The corresponding burn-off times predicted by the constant-particle-diameter model were about 0.3 s and 0.04 s.

Essenhigh *et al.* (1989) gave a comprehensive survey of the status on coal particle ignition in the light of the historical development over the previous two decades. The possibility of extending the single-particle results to dust clouds was examined. Theories are available, but experimental verification is incomplete. The boundary between conditions that give heterogeneous ignition and those giving homogeneous ignition is not fully identified.

4.1.5 WOOD

Malte and Dorri (1981) developed a complete theory for the life of a single wood particle of diameter from 100 μm and upwards, in a wood waste furnace, of the grate type. The particle was followed from the movement of injection, via drying and pyrolysis to completion of combustion. A main objective was to study the extent to which small particles were entrained by the upwards air flow before combustion was completed.

Equation 3.16 in Chapter 3 was used for calculating the gravitational terminal settling velocity v_t of the particle. The drag coefficient C_D was determined experimentally for various particle sizes and shapes. One problem is that v_t depends on particle drying and devolatilization because these processes reduce the particle density.

The homogeneous particle temperature was calculated by integrating the following equations (4.11) to (4.15). The drying process was described by:

$$\dot{Q}/m_{DW} = (C + MC_l)\frac{dT}{dt} - h_v\frac{dM}{dt} \tag{4.11}$$

$$-\frac{dM}{dt} = \frac{\dot{Q}}{h_v m_{DW}}(1 - e^{-M/b}) \tag{4.12}$$

where

$\dot{Q}$ = rate of heat transfer to particle
C = specific heat of dry wood
C_l = specific heat of liquid water
m_{DW} = dry mass of wood particle
T = homogeneous particle temperature
h_v = latent heat of vaporization, including differential heat of wetting
M = fractional moisture content: mass H_2O/dry mass

and the parameter b (empirical correlation) equals:

$$b = \frac{1.08\,\dot{Q}}{h_v m_{DW}} + 0.14 \tag{4.13}$$

The pyrolysis process, neglecting particle swelling, was described by:

$$\dot{Q}/V_p = \rho C\frac{dT}{dt} - [C_v(T - T_0) - q]\frac{d\rho}{dt} \tag{4.14}$$

$$-\frac{d\rho}{dt} = (\rho - \rho_F)k \tag{4.15}$$

where

ρ = particle density at time t
ρ_F = final particle density
V_p = particle volume
C_v = specific heat of volatiles
T_0 = reference temperature
q = exothermic heat of pyrolysis at reference temperature
k = Arrhenius rate constant equal to A exp. $(-E/RT)$

The value of k varies with temperature, activation energy and the constant A. A and E in turn varies with details of the composition of the wood, the rate of heating etc. This aspect was investigated in some detail by Malte and Dorri (1981).

The computer model was used to simulate trajectories of wood particles of various sizes and shapes, in the waste furnace. It could be shown that particles of diameters smaller than 500 μm had a significant tendency to become entrained by the upwards air in the furnace and escape ignition and combustion at the hot grate at the furnace bottom.

4.2 LAMINAR DUST FLAMES

4.2.1 LAMINAR FLAME PROPAGATION IN PREMIXED, QUIESCENT GASES

The basic concepts of flame propagation in dust clouds are adopted from premixed gas propagation theory. It is appropriate, therefore, to briefly introduce some central aspects of the latter.

The linear rate at which a laminar combustion wave or reaction zone propagates relative to the unburnt gas of a flammable mixture is called the fundamental or laminar burning velocity, commonly denoted by S_u. As pointed out by Kuchta (1985), this velocity is a fundamental property of the mixture and depends primarily upon the thermal diffusivity $\lambda/\rho C_p$ of the unburnt gas, where λ is the thermal conductivity, ρ the density and C_p the specific heat at constant pressure of the unburnt gas, and on the chemical reaction rate and heat of combustion of the gas. The reaction zone in a premixed gas is normally quite thin, of the order of 1 mm. According to the classical Mallard-le Chatelier (1883) theory, the fundamental laminar burning velocity of a homogeneous gas mixture equals:

$$S_u = \frac{\lambda(T_b - T_i)}{\rho \times C_p \times l(T_i - T_u)} \tag{4.16}$$

where T_i is the ignition temperature of the gas mixture, and l the thickness of the reaction zone. One problem with this theory is that a relevant value of T_i is normally not known for a given gas mixture. The fundamental limitation of the theory is that it does not relate S_u to the heat release rate. Therefore more refined theories have been developed, as will be mentioned below.

Of great practical interest is the flame speed S_f, i.e. the speed of the flame front relative to an observer or fixed geometries. It may be defined as

$$S_f = S_u + S_g, \tag{4.17}$$

where S_g is the gas velocity component caused by the expansion and buoyancy of the combustion product gases. Figure 4.7 illustrates the experimental relationship between S_u, S_f and S_g for spherical flame propagation in CH_4 air as a function of equivalence ratio (fraction of stoichiometric fuel concentration). The maximum S_f and S_u values occur on the rich side of stoichiometric composition and the ratio S_f/S_u is about 6. Under ideal adiabatic conditions, the maximum S_f/S_u ratio is about 7.5, which is typical of the combustion product expansion ratio E for most organic fuels. The plane, one-dimensional flame speed may be calculated from the following expressions:

$$S_f = S_u E = S_u \rho_u/\rho_b \tag{4.18}$$

$$S_u \rho_u/\rho_b = S_u \frac{M_u T_b p_u}{M_b T_u p_b} \tag{4.19}$$

where M is molecular weight, T temperature (K), p pressure (absolute), ρ gas density, and the u and b subscripts refer to the unburnt and burnt states, respectively. In the case of spherical flame propagation the radial flame speed is given by equations (4.18) and (4.19) if the flame thickness is negligible compared with the radius of the spherical flame surface. For finite flame thicknesses methods for correcting for flame stretch have been developed, as shown by Kawakami *et al.* (1988)

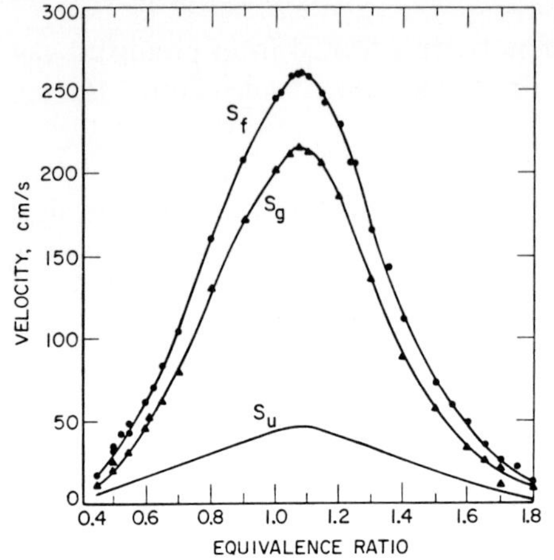

Figure 4.7 *Flame speed S_f, gas velocity S_g and burning velocity S_u versus equivalence ratio for spherical methane-air flame propagation and atmospheric pressure (From Kuchta, 1985. Originally from Andrews and Bradley, 1972)*

The burning velocity in air generally increases consistently with increasing initial temperature, whereas for many fuels it decreases somewhat with increasing pressure. When the ratio of O_2/N_2 in the oxidizing gas is either smaller or larger than in air, the burning velocity decreases or increases correspondingly. In pure oxygen, burning velocities are considerably higher than in air because of increased reaction rates and heats of reaction, particularly at stoichiometric fuel concentrations, which are much higher in oxygen than in air at the same total pressure.

Table 4.1 summarizes maximum S_u values for some gases mixed homogeneously with air, at atmospheric pressure and normal room temperature.

Table 4.1 Maximum fundamental burning velocities S_u for homogeneous mixtures of air and various combustible gases. Atmospheric pressure and normal room temperature (Data from Freytag, 1965; Zabetakis, 1965; and Kuchta, 1985)

Fuel gas	S_u [m/s]
Hydrogen	3.25
Acetylene	1.60
Ethylene	0.80
Methane to n-Heptane	0.42-0.47

Plate 5 *High-turbulence maize starch explosion in 500 m^3 bolted steel plate silo at Vaksdal, Norway, in April 1982 (Photographer: A. M. Fosse, Vaksdal)*

Plate 6 *Experimental site outside Bergen, Norway, with 236 m^3 steel silo, dust injection system and instrumentation cabins. Enclosed winding staircase along the silo wall to the left*

Plate 7 *Vented maize starch explosion in 236 m^3 steel silo in Norway*

Plate 8 *Maize starch explosion in 5.8 m^3 experimental bag filter unit in Norway. Vent area 0.16 m^2. Static opening pressure of vent cover 0.10 bar(g). Maximum explosion pressure 0.15 bar(g)*

Plate 9 *Silicon dust explosion in the welding torch ignition test apparatus used in Norway*

Plate 10 *Ignition of a dust cloud in the Godbert-Greenwald furnace*

In his book on combustion phenomena, Glassman (1977) reviewed various theories for the laminar burning velocity of gases. He showed the historical development from thermal diffusion theories, via 'particle' diffusion theories to comprehensive theories. The classical Mallard/le Chatelier theory (1883) (Equation (4.16) is a purely thermal diffusion theory, assuming the existence of a specific 'ignition' temperature for the combustible mixture. This theory was later improved by Zeldovich and Frank-Kamenetzkii, who included the diffusion of molecules. Their theoretical derivation was presented in detail by Semenov (1951), and also by Glassman (1977). Diffusion of free radicals and atoms was included at a later stage. Tanford and Pease (1947) in fact suggested that the flame propagation process in a gas mixture is essentially governed by the diffusion of free radicals, and not by the temperature gradient as assumed in thermal diffusion theories.

Glassman (1977) showed, however, that a modified form of the Mallard/le Chatelier equation (4.16) and the equation resulting from the more complex approach by Zeldovich, Frank-Kamenetzkii and Semenov, can both be expressed as

$$S_u \sim (\alpha G)^{1/2} \tag{4.20}$$

where α is the thermal diffusivity and G the chemical reaction rate.

4.2.2 DIFFERENCES BETWEEN FLAMES IN PREMIXED GAS AND IN DUST CLOUD

Leuschke (1965) pointed out some characteristic differences between a laminar premixed gas flame and a laminar dust flame. One important difference is that the reaction zone in the dust cloud is considerably thicker than in the gas, irrespective of the type of dust, and of the order of at least 10–100 mm. When discussing this feature of the dust flame, Cassel (1964) distinguished between two types of flames. The first, the Nusselt type, is controlled by diffusion of oxygen to the surface of individual, solid particles, where the heterogeneous chemical reaction takes place. In the second type, the volatile flame, the rate of gasification, pyrolysis, or devolatilization is the controlling process, and the chemical reaction takes place mainly in the homogeneous gas phase. In Nusselt type flames, the greater thickness of the combustion zone as compared with that of premixed gas flames, results from the slower rate of molecular diffusion, compared to diffusion in premixed homogeneous gases. In the case of the volatile flame type, the greater flame thickness is due to the pre-heating zone, where volatiles or pyrolysis gases are driven out of the particles ahead of the flame. When mixed with the air these gases and vapours burn almost as a premixed gas. The combustion of the remaining solid char particles occurs subsequently at a slower rate in the tail of the flame, and therefore the volatile flame in clouds of coals and organic dusts is also, in fact coupled to a Nusselt type flame.

In the case of metals, low-melting-point materials may oxidize in the vapour phase, but due to the oxide film round each particle this does not result in a homogeneous metal vapour/air flame. Because of the large heat of combustion per mole O_2 of for example aluminium and magnesium dust, compared with organic dusts, the temperature of the burning particles is very high and thermal radiation plays a central role in the transfer of heat in the combustion wave. Radiative heat transfer is also supposed to play a role in coal dust flames. However, because the thermal radiation is proportional to the fourth power of the temperature, the role of thermal radiation in coal dust flames is less important than in for example aluminium and magnesium dust flames. Radiative heat transfer in dust flames is a complex process, and it is of interest to note that Elsner *et al.* (1988) investigated the solid particle emissivity in dust clouds as a function of dust cloud thickness, specific surface area of the particles, dust concentration and absorption and scatter coefficients. Experiments were conducted with fluidized bed ash and quartz sand. Good agreement was found between experiments and a theoretical equation.

Leuschke (1965) conducted an illustrative series of experiments demonstrating the importance of radiative heat transfer in metal dust flames, using the experimental set-up illustrated in Figure 4.8.

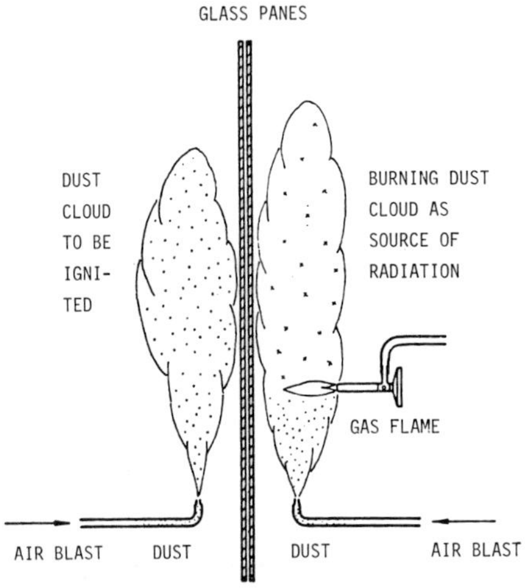

Figure 4.8 *Experiment demonstrating the ignition of a cloud of metal dust in air by radiation from a burning cloud of the same dust, through a double-glass window (From Leuschke, 1965)*

Two transient dust clouds were generated simultaneously on the two sides of a double glass window, one being ignited immediately by a gas flame. It was then observed whether the radiation from the burning cloud was able to ignite the other cloud.

Table 4.2, summarizing the results, shows that only the flames of Zr, Ti, Al and Mg were able to produce sufficient radiation to ignite the other cloud. Ignition of graphite was not accomplished at all, in agreement with the inability of graphite dust clouds to propagate a self-sustained flame in air at normal temperature and pressure. The reason why the gas flame coal could be ignited by the radiation from zirconium and titanium clouds, whereas the brown coal did not ignite, is not clear. Leuschke (1965) points out that clouds in air of iron and zinc powder, wood and cork dust, and lycopodium, ignited easily when exposed to light flashes of the type used for illumination in photography. As far as self sustained flame propagation in dust clouds is concerned, Table 4.2 confirms that radiative heat transfer is much more important in high temperature metal flames than in flames of organic materials and coal.

Table 4.2 Ignition of various dust clouds by radiation from various dust flames. Experiments according to Figure 4.8 (From Leuschke, 1965)

Cloud to be ignited by radiation \ Radiating cloud	Zirconium	Titanium	Aluminium (pyro)	Aluminium (greased)	Magnesium	Carbonyl iron	Iron	Graphite	Gas flame coal	Brown coal
Zirconium	+	+	+	+	+	-	-		-	-
Titanium	+	+	+	-	+					
Aluminium (pyro)	+	+	+	+	-					
Aluminium (greased)	+	+	+	-	+					
Magnesium	+	+	-	-	-					
Carbonyl iron	+	+	+	+	+	-				
Iron	+	+	-	-	-		-			
Graphite	-	-	-	-	-			-		
Gas flame coal	+	+	-	-	-				-	
Brown coal	-	-	-	-	-					-

+ = ignition, − = no ignition

With respect to the role of radiative heat transfer in dust flames, Cassel (1964) reasoned that losses from the heat generated in the combustion zone will necessarily make the maximum temperatures actually attained considerably lower than the temperatures predicted thermodynamically for adiabatic conditions. However, in the interior of sufficiently large dust clouds, temperatures will undoubtedly approach theoretical values. Therefore, as heat losses by radiation decrease with decreasing surface-to-volume ratio of the burning cloud, dust flames should show a positive correlation between flame size and burning velocity not encountered in combustible gas mixtures. Therefore, in the absence of other scale effects, larger high-temperature dust flames may be expected to burn faster than smaller ones.

Another difference between flame propagation in a premixed gas and dust cloud has been elucidated by Goral, Klemens and Wolanski (1988). They studied upwards propagation of flames in a lean methane/air mixture to which had been added inert particles (sand). It was found that the upwards flame velocity increased with increasing sand grain size, from 0.33 m/s for the 5.1% vol% methane/air with no sand particles, via 0.4 m/s for 40 μm particles, 0.65 m/s for 180 μm particles to 0.75 m/s for 360 μm particles. The effect was mainly attributed to the enhanced combustion due to the microturbulence generated in the wake of the falling particles. However, thermal radiation effects were also assumed to play a role.

4.2.3 EXPERIMENTAL BURNING VELOCITIES, FLAME THICKNESSES, QUENCHING DISTANCES, AND TEMPERATURES OF LAMINAR DUST FLAMES

In the case of premixed gases, the properties of laminar flames can be investigated in detail in special stationary burners. The same technique has been adopted in the study of laminar dust flames. However, as Lee (1987, 1988) pointed out, laminar dust flames are difficult to stabilize without causing significant cooling of the flame. Therefore such stabilized flames are non-adiabatic, and average burning velocities will be lower than for an adiabatic flame. Besides, the flame will not be uniform over its cross section, and burning velocities and flame thicknesses are not always easy to define. Nevertheless, much valuable information on the nature of laminar dust flames has been obtained from stationary burner flame studies.

4.2.3.1 Metal dusts

Cassel (1964) developed a special burner for studying stationary propagation of flat 'laminar' graphite and metal dust flames. Circular Mache-Hebra nozzles were used to ensure a reasonably uniform distribution of the upwards velocity of the dust cloud into the flame region. Once ignited, the flat dust flame floated approximately 20–30 mm above the burner port. The flame was stabilized by an enveloping divergent gas stream without using a pilot flame. Burning velocities were determined photographically both by measuring the minimum upwards vertical particle velocity in the preheating zone below the flame, and the particle velocity in the cold dust cloud further down.

Some results for dust clouds of 6 μm aluminium particles are given in Table 4.3. The results for argon/air mixtures show that both the burning velocity and the brightness temperature increase somewhat with nozzle diameter or flame area. This indicates that the values in Table 4.3 are minimum values in the dust explosion context. The brightness temperatures were measured by optical pyrometry. Because the burning dust cloud is not a black body, the true flame temperatures are higher than the brightness temperatures. Cassel, using the particle track method by Fristrom *et al.* (1954), estimated the true temperature of a 240 g/m^3 cloud of 6–7 μm diameter aluminium particles, burning in a mixture of 20 vol% O_2 and 80 vol% Ar at atmospheric pressure, to about 2850 K. If Ar was replaced by He, the temperature estimate rose to 3250 K. In both cases the ratio of the estimated true flame temperature and the brightness temperature is about 1.4.

If this factor is applied to the brightness temperatures in Table 4.3 of the flames in air, the flame temperature estimates will be 2500 K for 200 g/m^3, 2670 K for 250 g/m^3 and 2900 K for 300 g/m^3. Closed-bomb experiments with aluminium dust clouds in air give the highest peak pressures with dust concentrations above stoichiometric, typically in the range of 500 g/m^3. This could indicate that the temperature of a flame of 500 g/m^3 fine aluminium particles in air at atmospheric pressure would exceed 3000 K.

In the discussion published with Friedman and Macek's (1963) paper, Glassman asserted that the temperature of aluminium particle diffusion flames is not dependent on the concentration of oxygen in the atmosphere, except at very low concentrations. The flame temperature equals the boiling point of the oxide, i.e. 3800 K.

Cassel (1964) gives a photograph of a flat, laminar flame of 230 g/m^3 6 μm diameter aluminium particles in air at atmospheric pressure, which suggests a flame thickness of the order of 10 mm, i.e. at least ten times the characteristic flame thickness of laminar premixed gas flames.

Table 4.3 Burning velocities and brightness temperatures for flat, laminar flames of 6 μm aluminium particles in various oxidizer gases at atmospheric pressure (From Cassel, 1964)

Gas mixture	Dust concentration [g/m^3]	Nozzle diameter [cm]	Flame area [cm^2]	S_u [m/s]	Brightness temperature [K]
Air	200	0.95	1.13	0.21	1790
	250	0.95	1.33	0.30	1910
	300	0.95	1.54	0.35	2060
$O_2 + 4\,Ar$	200	0.45	0.21	0.21	1850
	250	0.45	0.26	0.28	1910
	300	0.45	0.31	0.32	1960
	200	0.95	0.87	0.23	1980
	250	0.95	1.08	0.32	2080
	300	0.95	1.17	0.38	2140
	200	1.30	1.30	0.27	2070
	250	1.30	1.42	0.36	2230
	300	1.30	1.48	0.41	2320
$O_2 + 4\,He$	200	0.95	0.87	0.70	2090
	250	0.95	1.08	1.00	2320
	300	0.95	1.23	1.15	2430

The burning velocity for the 6 μm aluminium particles in air varied, as seen from Table 4.3, with the dust concentration, being 0.21 m/s for 200 g/m^3 and 0.35 m/s for 300 g/m^3.

Other experiments by Cassel (1964) showed that the burning velocity of aluminium/air clouds also increased with decreasing particle size. At 200 g/m^3 it was roughly 0.2 m/s for a '<30 μm' atomized aluminium powder, and 0.4 m/s for a '<10 μm' quality. The latter value agrees favourably with the maximum value of 0.42 m/s determined by Ballal (1983) for aluminium of a volume surface mean diameter (D_{32}) of 10 μm. The maximum flame speed occurred close to the stoichiometric concentration 310 g/m^3. Ballal (1983) conducted his sophisticated experiments in a special vertical explosion tube during free fall (zero gravity conditions), and it is interesting to observe that for particle sizes of about 10 μm, gravitational effects did not seem to play a dominating role in the laminar flame propagation through aluminium dust clouds.

Gardiner *et al.* (1988) studied flame propagation in comparatively small, electrostatically suspended clouds of 20 μm volume surface mean diameter aluminium particles in air in a small semi-closed cylindrical vessel and found maximum flame speeds in excess of 2.0 m/s.

Alekseev and Sudakova (1983) measured radial flame speeds of spherical flames in essentially unconfined clouds of five different metal powders. The experimental dust clouds were generated by dispersing a given quantity of dust by means of a special atomizer during a period of 0.4 s. A glowing resistance wire coil or a pyrotechnical charge was used for igniting the dust cloud of about 10 litre volume at its centre. Flame propagation was recorded by high-speed photography. Dust concentration was assessed both from the volume of the dust cloud just prior to ignition, and by sampling of the cloud at various locations using a fast-response probe. Figure 4.9 gives some results for the five powders specified in Table 4.4. Particle size clearly plays a key role and explains for example why the magnesium powder (median particle size of about 45 μm) gave a considerably lower flame speed than the aluminium powder (median particle size of about 9 μm). As seen from Figure 4.9 the radial flame speed for the aluminium powder at 300 g/m^3 was about 1.5 m/s.

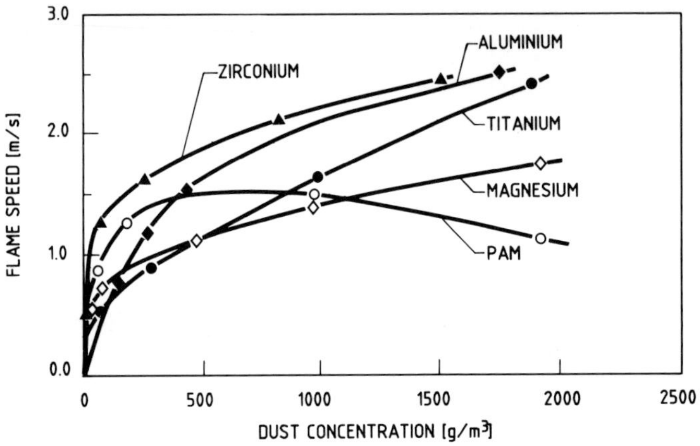

Figure 4.9 *Flame speed as a function of dust concentration in unconfined clouds of metal dusts. Spherical flame propagation (From Alekseev and Sudakova, 1983)*

Table 4.4 Size distributions of five metal powders used in flame propagation experiments (From Alekseev and Sudakova, 1983)

Powder	Cumulative mass fraction [%] finer than stated particle size					
Type	6 μm	10 μm	16 μm	25 μm	40 μm	50 μm
Zirconium	6	14	23	39	92	100
Aluminium	21	58	88	97	99	100
Titanium	1	1	4	8	22	100
Magnesium	2	5	10	20	75	100
PAM*	2	3	6	29	74	100

*Aluminium/magnesium alloy.

Experiments in closed bombs give pressure rise ratios up to 12.5 for explosions of aluminium dust in air (BIA/BVS/IES (1987)). For ideal adiabatic expansion and assuming a specific heat ratio of 1.4, this gives expansion ratios of up to 6.1, and according to equation (4.18), the radial flame speed is then 6.1 times the radial burning velocity. The burning velocity corresponding to a flame speed of 2.5 m/s is then about 0.4 m/s, i.e. close to the value found in laminar burner experiments for aluminium flames.

Jarosinski *et al.* (1987) determined the quenching distance for laminar flames in air of aluminium flakes of thickness 0.1 μm and average diameter 15 μm, and atomized aluminium particles of average diameter 8 μm. The smallest quenching distance found for both dusts was 10 mm. This occurred in the dust concentration range 700–1000 g/m^3.

4.2.3.2
Coal dusts

In a comprehensive survey of a number of investigations on the propagation of laminar pulverized coal dust/air flames, Smoot and Horton (1977) discuss factors influencing experimentally determined burning velocities, flame temperatures and flame thicknesses. Most experiments are performed by stabilizing dust flames in burners of various kinds. Due to heat losses by radiation from the hot dust particles, and conduction, typical stabilized burner flames will have temperatures that are lower than the adiabatic flame temperature. In principle heat losses can be avoided by using burners of very large diameters, or equipped with walls having temperature and emissivity profiles matching those of the flame. However, according to Smoot and Horton, the use of such devices had not been reported up to the time of their survey (1977).

Smoot and Horton found large differences in burning velocities observed by various investigators which could not be explained in terms of variations in dust properties or dust concentration. They considered incomplete dispersion of fine cohesive dusts as the main source of error. (See Chapter 3.) The data in Figure 4.10 illustrate how improved dispersion of a fine coal dust gives increased burning velocity, by 50% and even more. Some main conclusions from the survey of Smoot and Horton are given in Table 4.5.

Horton *et al.* (1977), investigating flat, laminar coal dust flames, found that the peak burning velocities for a 9 μm (mass average particle size) Pittsburgh coal dust in air was about 0.33 m/s, whereas a coarser fraction of the same coal (33 μm mass average fraction)

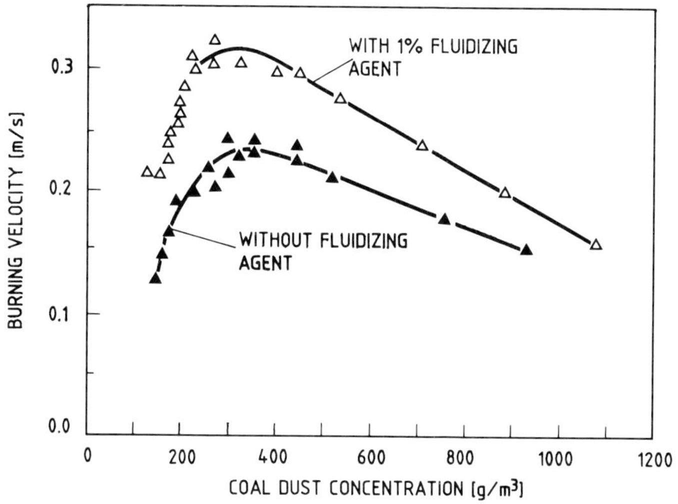

Figure 4.10 *Effect of very fine SiO_2 fluidizing agent (Acrosil) on the burning velocity of an air suspension of 10 μm, 28% volatile content Sewell coal dust (From Smoot and Horton, 1977)*

Table 4.5 Summary of some experimental observations for laminar coal dust flames in air at atmospheric pressure (From Smoot and Horton, 1977)

1.	Observed flame velocities depend upon the burner used.
2.	Peak burning velocities mostly range from 0.05 to 0.35 m/s, depending upon burner design, coal type and particle size. An exception was the high value of 0.86 m/s measured by Ghosh et al. (1957), which was attributed to the use of a furnace with preheated walls.
3.	Peak burning velocities occur at higher fuel concentrations than stoichiometric, somewhere in the neighbourhood of the stoichiometric concentration for combustion of the volatile matter. The peak flame velocity increases with the specific surface area of the coal dust.
4.	The rich flammability limit occurs at higher fuel concentrations relative to that giving the peak burning velocities, as compared to gaseous flames.
5.	Decreasing coal-dust particle size increases burning velocity on the lean side of the peak but may decrease it on the rich side. Also, smaller particles shift the peak and rich flammability limit to a leaner concentration.
6.	Increasing volatiles content increases the burning velocity and slightly shifts the peak to a leaner concentration.
7.	Oxygen enrichment beyond the 21 vol% in air increases burning velocity, as does addition of methane.
8.	Thicknesses of steady, laminar, coal dust flames are usually of the order of 5 mm, but larger thicknesses have been observed, especially for larger particles at high coal dust concentrations.
9.	Measured peak flame temperatures range from 1000 to 1500 K and may be correlated with coal dust concentration. These measured temperatures may be lower than the real temperatures due to inadequate measurement techniques.
10.	In the flame front, liberated volatile matter burns rapidly in the gas phase, whilst there is very little heterogeneous combustion of the char.
11.	In traversing the flame front, the irregularly shaped solid particles soften and become rounded and filled with blow holes, but remain of about the same size.
12.	A considerable amount of volatile matter remains in the char leaving the flame front, the amount being a strong function of coal dust concentration.
13.	The extent of coal devolatilization is especially related to coal dust concentration.
14.	The volatile material liberated during rapid pyrolysis in this type of flame has a higher C/H ratio than the volatile matter liberated during proximate analysis.
15.	Only small amounts of H_2 or CH_4 are observed in the flame.

gave peak velocities of about 0.22 m/s. A similar influence of particle size was found for a Pocahontas coal.

The question of what are the true laminar burning velocities for coal dust clouds to some extent remains unanswered. The true peak values are probably somewhat higher than 0.35 m/s, but certainly lower than the exceptional value of 0.86 m/s measured by Ghosh *et al.* (1957). (See Table 4.5 pt. 2.)

In a comprehensive investigation comprising several types of dusts, Ballal (1983) determined the laminar burning velocity in clouds of coal dust in air under zero gravity conditions, using a free-fall explosion tube. For a coal dust of 8 μm surface-volume diameter (D_{32}) and 13.8% volatile matter, the maximum burning velocity of 0.11 m/s was found for dust concentrations close to stoichiometric, i.e. 210 g/m^3. For coals of higher volatile contents, the maximum values were about 0.25 m/s (40% volatiles and

D_{32} = 12 μm), 0.17 m/s (27% volatiles and D_{32} = 11 μm) and 0.12 m/s (37% volatiles and D_{32} = 47 μm). The experimental concentration range did not extend beyond the stoichiometric concentration for which the maximum values were obtained. However, the trend of the experimental burning velocity-versus-dust concentration curves indicates that even higher burning velocities would have been found for dust concentrations somewhat higher than stoichiometric. It is interesting to note that the burning velocities measured by Ballal for coal/air under zero gravity conditions are close to those found under normal gravity conditions by Smoot and Horton (1977) and Horton *et al.* (1977).

Hertzberg *et al.* (1986) analysed experimental data from explosions of Pittsburgh seam bituminous coal dust in a closed bomb. When assuming that all the volatiles participated in the combustion reaction, and treating the char as an inert substance, they found that the theoretical adiabatic maximum explosion pressures and maximum flame temperatures were considerably higher than the experimental values. Maximum theoretical adiabatic flame temperatures were 2500 K for constant volume and 2200 K for constant pressure combustion. The experimental maximum value for constant volume was 1850 K. Details of the experimental method used for measuring coal dust flame temperatures are given by Cashdollar and Hertzberg (1983). Hertzberg *et al.* (1986) attribute the discrepancy between idealized theory and experiment to incomplete devolatilization. They found that the effective fraction β of volatiles that can take part in the combustion, is a function of the intrinsic devolatilization rate constant, the effective heating flux of the approaching flame, the decomposition chemistry and the time available for devolatilization. The experimental data for maximum constant-volume explosion pressures could be readily interpreted in terms of estimated β-factors. Figure 4.11 shows how the fraction of volatiles that is assumed to take part in the combustion of Pittsburgh seam bituminous coal dust, decreases with increasing dust concentration.

In a subsequent paper, Hertzberg *et al.* (1987) formulated a 3-stage model for the coal dust flame propagation:

1. Heating and devolatilization of dust particles.
2. Mixing of emitted volatiles with air in the space between the particles.
3. Gas phase combustion of premixed volatile/air.

Each stage is characterized by a time constant. For small particles and low dust concentrations, the combustion process is controlled by stage 3, whereas for large particles

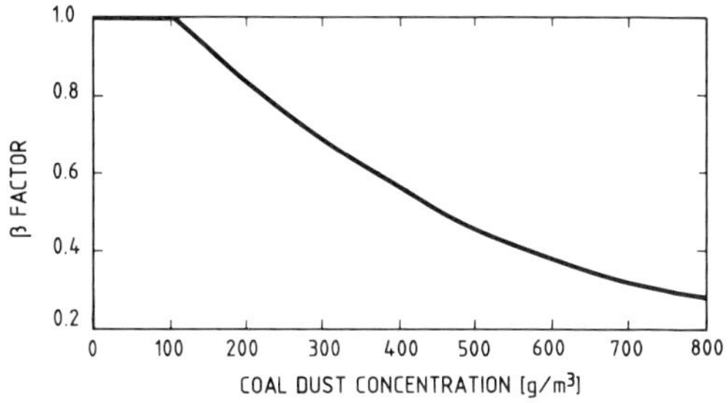

Figure 4.11 *Fraction of coal volatiles, β, assumed to contribute to flame propagation in order to obtain agreement between measured explosion pressures and calculated pressures for constant volume combustion (From Hertzberg* et al., *1986)*

and high dust concentrations, stage 1 is controlling the combustion rate. When discussing the influence of particle size on devolatilization in coal dust flames, Hertzberg *et al.* (1987) suggested that for particles smaller than 50–100 μm diameter, devolatilization is complete and not rate limiting for the combustion reaction, i.e. β in Figure 4.11 is equal to unity. On the basis of measurement of pyrolysis rates of single particles, and microscopic studies of particle morphology, they concluded that the pyrolysis wave preceding a coal dust flame is non-isothermal, with a velocity that is proportional to the net absorbed heat flux intensity and inversely proportional to the overall enthalpy change of the combustion reaction.

In view of Hertzberg *et al.*'s suggestion of a limiting particle diameter of 50–100 μm, it is interesting to consider the influence of particle size on maximum explosion pressure and maximum rate of pressure rise of lignite dust in air in a 1 m^3 vessel, as measured by Scholl (1981). As shown in Figure 4.12, there was no further systematic increase of the two parameters with decreasing particle size below 60–80 μm diameter, in accordance with what would be expected on the basis of the hypothesis of Hertzberg *et al.*

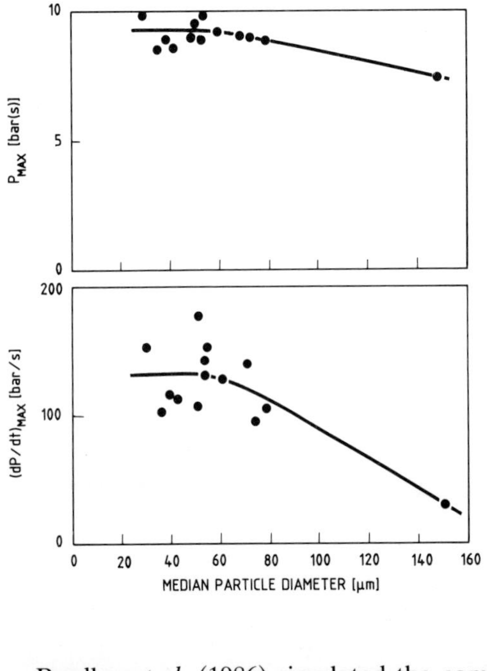

Figure 4.12 *Explosion characteristics of lignite dusts in a 1 m^3 closed vessel as a function of median particle size (From Scholl, 1981)*

Bradley *et al.* (1986) simulated the combustion of rapidly devolatilizing coal dusts by generating stabilized laminar flames of mixtures of < 10 μm diameter graphite dust and methane in air. The laminar burning velocities measured agreed well with theory of coal dust flame propagation, assuming rapid devolatilization and subsequent gas phase mixing, and no heat sink influence of the graphite particles. Apart from radiative losses from the particles, which were also accounted for in theory, the flames were in fact close to adiabatic. The theoretical prediction also agreed well with experimental burning velocities for coal dusts as long as the particle diameter did not exceed 10 μm and the volatile content of the coal was greater than about 25%.

In a subsequent study Bradley *et al.* (1989) investigated the burning velocities of CH_4/air/graphite dust flames near the minimum explosible concentration at sub-atmospheric pressure of 0.14 bar(abs.). On the basis of an indicated experimental peak flame temperature of 1550 K at the limit concentration for flame propagation, a theory was developed, which enabled computation of chemical species concentration profiles, gas temperatures and heat release rates for flames at atmospheric pressure. As an example it was found that the laminar burning velocity for a fuel concentration corresponding to an equivalence ratio of 0.72, decreased from 0.18 m/s for methane as the only fuel, to 0.06 m/s for a fuel mass ratio of CH_4/graphite of 0.2. The relevance of assuming that CH_4 /graphite mixtures can be used for simulating coal dust mass was investigated theoretically.

The lower experimentally determined limit of volatile content of the coal for a cloud of coal dust to be able to propagate a self sustained flame at normal atmospheric conditions is about 13% according to Cybulski (1975) and Ballal (1983), and 8–10% according to Scholl (1981).

It should be mentioned that Helwig (1965), who used a 43 litre closed bomb, found that the rate of explosions of coal dust containing 10–50% volatiles, did not increase monotonically with decreasing particle size. Instead the explosion rate for the finest fraction, of 0–10 μm particle diameter, was systematically lower than for the most explosible size range 20–30 μm. It is not clear whether incomplete dispersion of the finest particle fraction contributed to this effect.

Jarosinski *et al.* (1987) measured the quenching distance for flames in air of a < 74 μm bituminous coal dust of 32% volatile matter, and of the same dust ground to <5 μm particle diameter. The quenching distances were 190 mm for the < 74 μm dust and 25 mm for the < 5 μm one. The reason for these unexpectedly high values is not clear.

4.2.3.3
Organic materials

Laminar 20 mm diameter flames of lycopodium/air and polyvinyl alcohol/air were studied by Kaesche-Krischer and Zehr (1958) and Kaesche-Krischer (1959). The burning velocity, defined as the ratio of air flow and flame cone area, was determined photographically from the height of the flame cone. Some results are given in Figure 4.13. Lycopodium/air flames of dust concentrations lower than 180 g/m^3 and higher than 500 g/m^3 were difficult to stabilize (stoichiometric concentration $\simeq$ 125 g/m^3). The appearance of a stabilized lycopodium/air flame was very similar to that of a rich hydrocarbon/air flame, i.e. a blue flame front followed by a more or less luminous soot edge. Approximate thermocouple measurements of flame temperatures gave about 1800 K for a 180 g/m^3 flame, and 1100 K for a 500 g/m^3 flame. Figure 4.13 shows the measured burning velocities as a function of the dust concentration. In the range 180–300 g/m^3 the burning velocity of lycopodium flames has a maximum value of about 0.25 m/s. The corresponding concentration range for the PVA dust was 140–220 g/m^3. Figure 4.13 also shows that an increase of the oxygen percentage in the gas from 21 for air to 30, gave a significant increase of the measured burning velocities for both dusts, in accordance with expectations. The photographs provided by Kaesche-Krischer and Zehr (1958) indicate typical thicknesses of lycopodium flames of a few mm.

Kaesche-Krischer implied that the differences in the concentration ranges giving the highest burning velocities for the two dusts were due to a higher volatile content in the

PVA than in lycopodium, assuming that the flame essentially propagates through a homogenous mixture of volatiles and air. This is in accordance with the findings of Hertzberg *et al.* (1986) for coal dust and polyethylene.

Mason and Wilson (1967) investigated laminar flames of lycopodium in air in the dust concentration range 125 to 190 g/m^3. When accounting for wall cooling effects in their experiments, they arrived at maximum burning velocities similar to those found by Kaesche-Krischer and Zehr (1958), i.e. about 0.25 m/s. Mason and Wilson also conducted some temperature measurements in a 140 g/m^3 flame using a 25 μm thermocouple. Two mm below the flame front the temperature was 330–350 K, whereas 1.5 mm above the flame front it was about 1800 K. The latter figure is in complete agreement with the temperature measured by Kaesche-Krischer and Zehr (1958) in a 180 g/m^3 lycopodium/air flame. These measurements showed that the preheating zone was about 2 mm thick, and of the same order as for gases of similar burning velocities, and that the total thickness of a laminar lycopodium/air flame is of the order of a few mm.

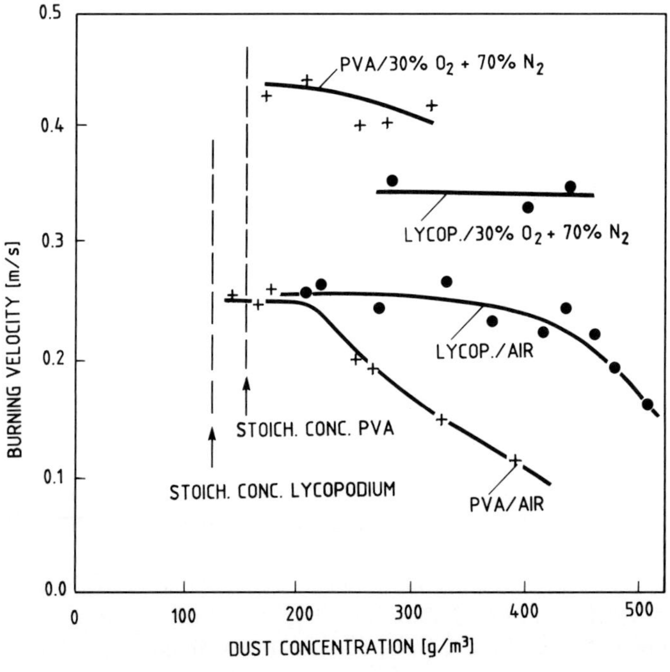

Figure 4.13 *Burning velocities of flames of lycopodium and polyvinylalcohol dust (< 60 μm particle diameter) flames as functions of dust concentration (Data from Kaesche-Krischer and Zehr, 1958 and Kaesche-Krischer, 1959)*

More recently Proust and Veyssiere (1988) studied the propagation of genuinely laminar dust flames in clouds of maize starch of 6% moisture content in air. They used the comparatively large apparatus illustrated in Figure 4.14. Dust clouds were generated in the vertical experimental glass duct of 0.2 m × 0.2 m cross section and 2 m height by

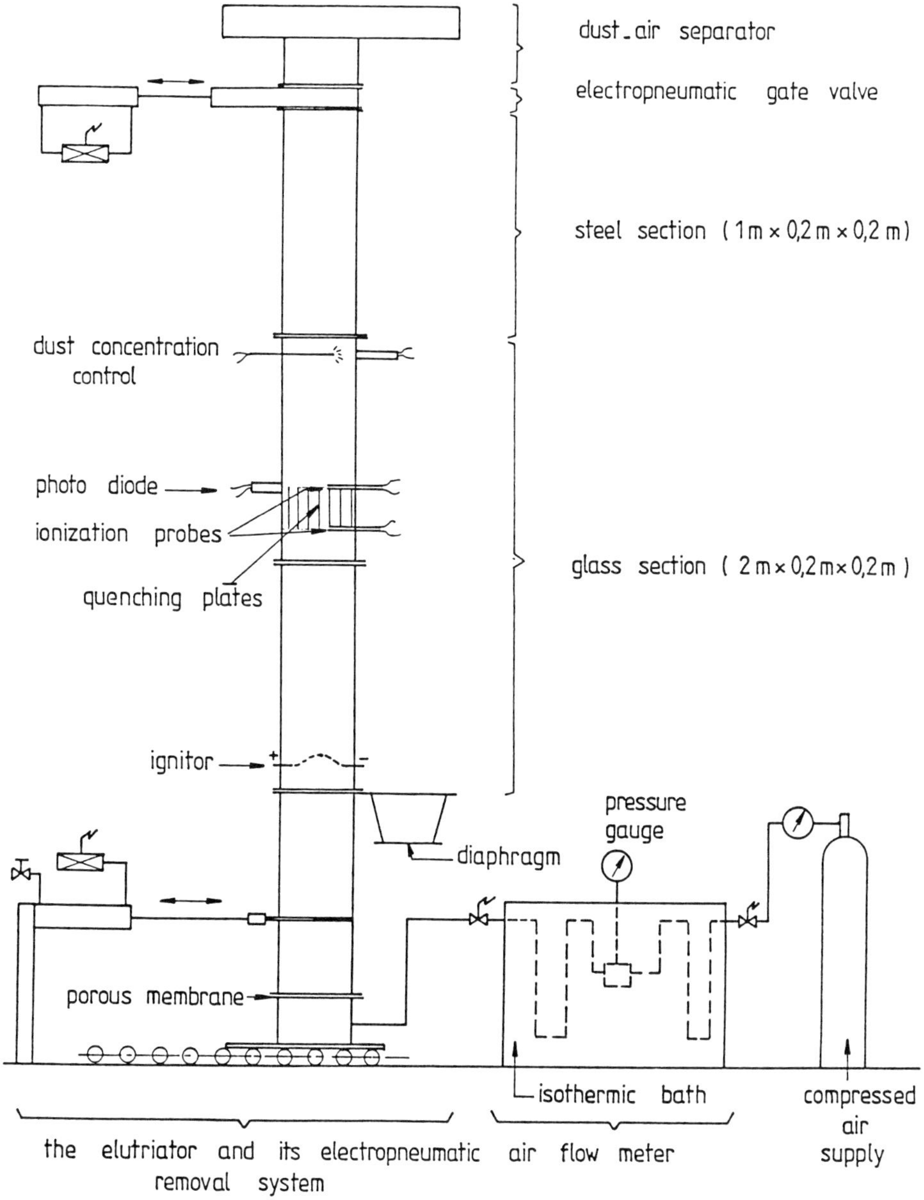

Figure 4.14 *Large vertical duct for studying flame propagation in dust clouds (From Proust and Veyssiere, 1988)*

low-velocity elutriation from a fluidized bed of 600 g of starch resting on a porous membrane at the bottom of the system. The average vertical air velocity was of the order of 0.1 m/s. A battery of vertical parallel 0.5 mm thick steel plates was inserted across the whole cross section of the duct when quenching distances were measured. Average dust concentrations were determined from the dust mass lost from the fluidized bed as a function of time and the air flow through the system. A laser tomography system was used for controlling the homogeneity of the dust cloud.

Laminar burning velocities were determined from the measured flame speeds and photographically estimated flame surface areas as in the case of Kaesche-Krischer (1959). However, in order to obtain proper laminar flame propagation it was necessary to avoid the build up of fundamental-mode standing acoustic wave motion in the duct. Such waves are easily generated by the gas expansion following the initial flame, and can subsequently interfere with the flame propagation. Proust and Veyssiere solved this problem by fitting a special damping diaphragm at the open bottom end of the duct (see Guenoche (1964)).

A series of photographs of the propagating laminar maize starch flame is shown in Figure 4.15. Figure 4.16 shows the upwards laminar flame front velocity (duct closed at upper end) as a function of the dust concentration. The velocity was measured by means of ionization probes. The maximum value of 0.63 m/s occurred close to the stoichiometric dust concentration 235 g/m^3. The corresponding laminar burning velocity was estimated to 0.27 m/s, i.e. very close to the maximum values found for lycopodium and polyvinyl-alcohol dust in air by other workers using the burner technique (see Figure 4.13).

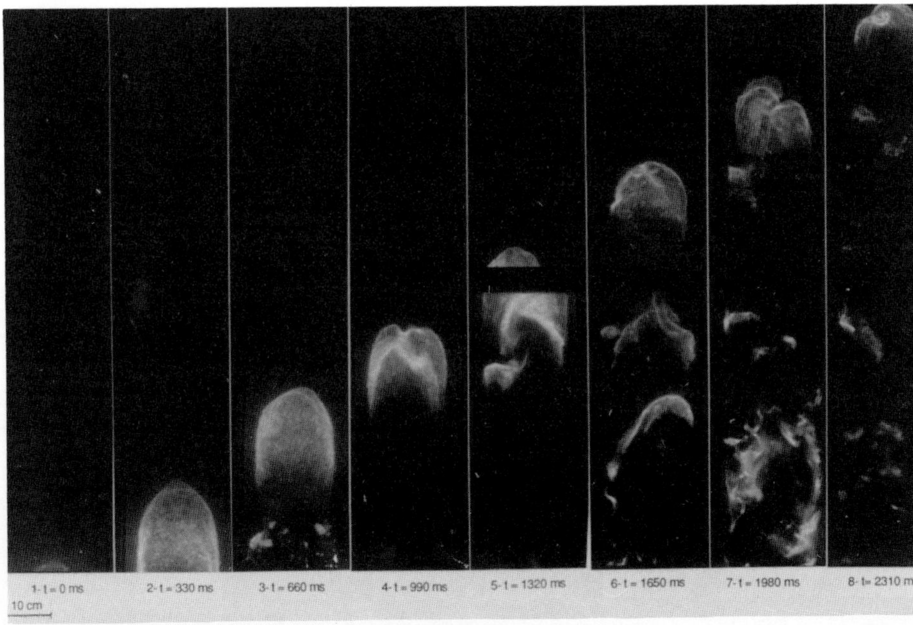

Figure 4.15 *Photographic records of an upward propagating laminar flame in a 120 g/m^3 cloud of maize starch in air (From Proust and Veyssiere, 1988)*

The flame temperature was measured by means of thermocouples of either 25 μm or 200 μm junction diameter. The results are shown in Figure 4.17.

The maximum value of about 1600 K was obtained close to the stoichiometric dust concentration of 235 g/m^3. This maximum is somewhat lower than the maximum

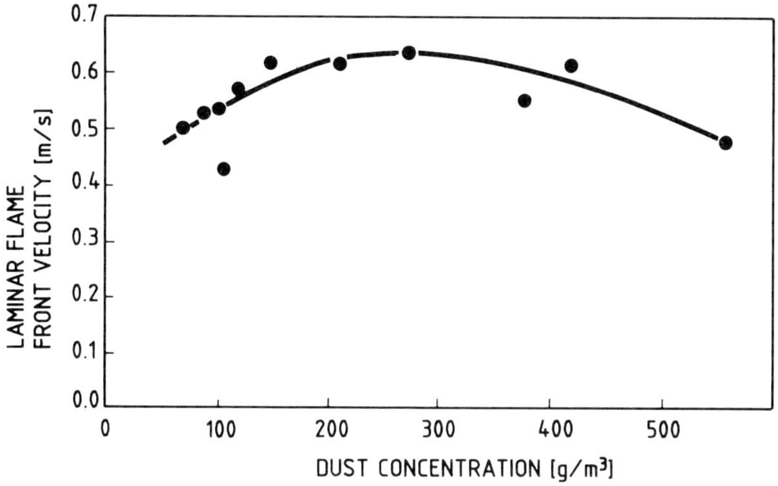

Figure 4.16 *Upwards laminar flame front velocity through a cloud of maize starch in air as a function of dust concentration (From Proust and Veyssiere, 1988)*

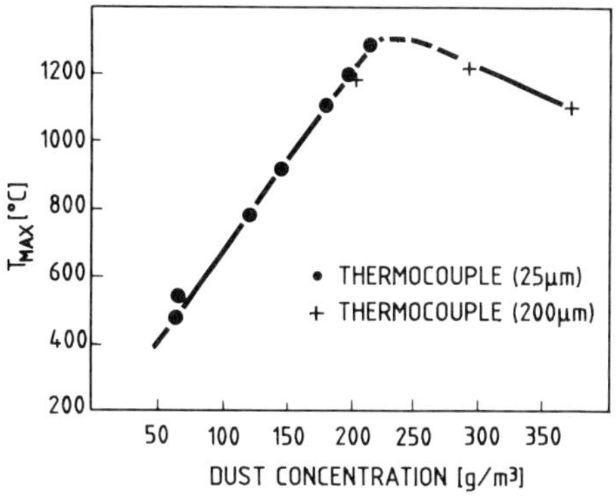

Figure 4.17 *Variation of the maximum temperature of maize starch flames with dust concentration (From Proust and Veyssiere, 1988)*

temperatures of about 1800 K measured in laminar burner flames of lycopodium and polyvinyl alcohol.

The results from measurement of quenching distances for laminar flames of maize starch in air are shown in Figure 4.18.

The quenching distance was defined as the maximum distance between the vertical parallel plates that prevented laminar flame propagation through the plate battery and further upwards in the test duct. As Figure 4.18 shows, the quenching distance depended on the dust concentration. Below about 80 g/m^3 flame propagation was impossible even with an inter-plate distance of 30 mm, and this therefore also was the minimum explosible concentration for upwards laminar flame propagation. With increasing dust concentration, the quenching distance decreased systematically and reached about 7 mm at about the stoichiometric concentration of 235 g/m^3. For higher dust concentrations, up to 550 g/m^3 the quenching distance remained unchanged at the minimum value of 7 mm.

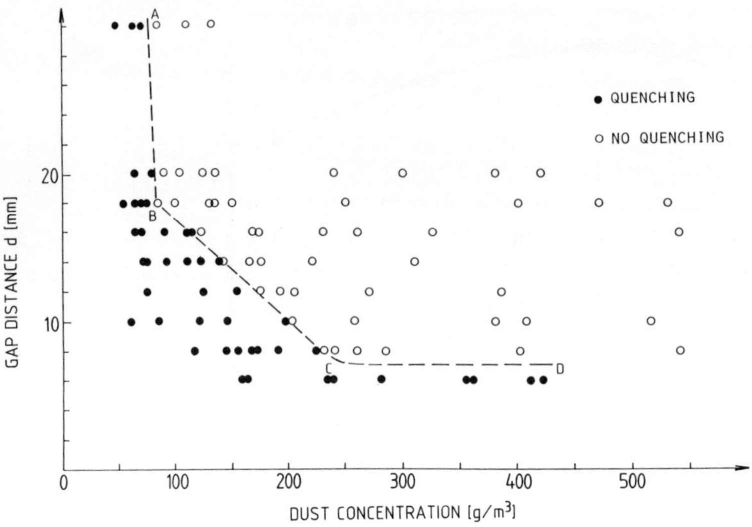

Figure 4.18 *The quenching distance of laminar flames of maize starch/air mixtures as a function of the dust concentration (From Proust and Veyssiere, 1988)*

The lowest value of about 7 mm for the quenching distance for maize starch/air mixture is in close agreement with the lowest value of about 6 mm found by Jarosinski *et al.* (1987) in a similar experimental configuration. However, these workers found their lowest value in the concentration range 500–1100 g/m^3, whereas the values in the range < 500 g/m^3 increased with decreasing concentration, being about 10 mm at 400 g/m^3.

Proust and Veyssiere (1988) also determined the thickness of the laminar starch dust/air flame, using the criterion for laminar gas flames proposed by Jarosinski (1984).

$$\text{Flame thickness} = 2(T_b - T_u)/(dT/dx)_{max} \tag{4.21}$$

In spite of some experimental difficulties they were able to identify a flame thickness of 3–4 mm at stoichiometric dust concentration. According to Jarosinski (1984) there is a factor of two between the quenching distance and the thickness of a laminar gas flame, which was also the result obtained by Proust and Veyssiere for their laminar flame of maize starch/air.

Comparing with properties of methane/air flames, Proust and Veyssiere summarized their results for 6% moisture content maize starch as shown in Table 4.6:

Table 4.6 Comparison of characteristic experimental parameters of laminar flames in air of maize starch (6% moisture) and methane at stoichiometric fuel concentrations (From Proust and Veyssiere, 1988)

Fuel	Maize starch	Methane
Laminar burning velocity [m/s]	27	45
Quenching distance [mm]	7	2
Flame thickness [mm]	3-4	1

4.2.3.4
Miscellaneous dust/gas mixtures

Characteristics of laminar flames of graphite in O_2/N_2 mixtures richer in O_2 than air have been determined by Cassel (1964), Chamberlain and Gray (1967), Bryant (1971) and Ballal (1983). Cassel (1964) and Ballal (1983) also give data for magnesium dust flames. For a given particle size, the burning velocities of magnesium dust clouds in air are somewhat higher than for aluminium dust clouds. Ballal (1983) further investigated the influence of higher oxygen concentration than in air, and of addition of hydrogen and methane to the gas phase (hybrid mixtures).

4.2.4
THEORIES OF ONE-DIMENSIONAL LAMINAR FLAME PROPAGATION IN DUST CLOUDS

4.2.4.1
Theory by Cassel *et al.*

In order to obtain an approximate equation for laminar burning in dust clouds, Cassel *et al.* (1949) modified the Mallard/le Chatelier (1883) theory for premixed gases by incorporating a term for thermal radiation effects due to the particles in a dust cloud. Their equation was:

$$S_u = \frac{\mu(T_b - T_i)/b + bw\sigma\alpha F(T_b^4 - T_u^4)/\rho_d r}{(c_p\rho + c_d w)(T_i - T_u)} \tag{4.22}$$

Here, S_u is the burning velocity and μ the heat conductivity. T_u, T_b, T_i are the temperatures of the unburnt and burnt masses, and of ignition. σ is the emissivity of the particle surfaces, whereas α is a correction factor, larger than 1, which accounts for the radiation of glowing combustion products (solids and gas). F is a geometrical view factor and b the thickness of the burning zone. c_p is the specific heat of the gas, ρ its density, whereas c_d is the specific heat of the dust, ρ_d its density and w its concentration, and r is the average particle radius.

Cassel *et al.* pointed out that the factor b, which is assumed to have the same value in both the conduction and the radiation terms, depends on both r, w and F. By introducing the burning time of a single particle, τ, and Equation (4.18), the factor b can be replaced by $\tau S_u\rho_u/\rho_b$.

Equation (4.22) then takes the form:

$$S_u^2 = \frac{K\rho_b}{\tau\rho_u} \frac{(T_b - T_i)}{\left(T_i - T_u - \dfrac{\tau w\sigma\alpha F\rho_u(T_b^4 - T_u^4)}{\rho_d\rho_b r(c_p\rho + c_d w)}\right)} \tag{4.23}$$

where K is the thermal diffusivity and equals $\mu/(c_p\rho + c_d w)$.

Assuming that oxygen diffusion governs the burning of individual particles, an upper limit for the burning velocity is obtained if τ is expressed in terms of the diffusion rate of oxygen:

$$\tau = \rho_d r^2 RT_u^{3/2}/(2MDpT_a^{1/2}) \qquad (4.24)$$

Here D is the diffusion coefficient at temperature T_u, R the gas constant, T_a the average ambient gas temperature around a particle as it passes through the reaction zone, p the average partial pressure of oxygen, M the oxygen equivalent of the fuel, expressed in grams of fuel per mol. of oxygen. Equation (4.23) thus takes the form:

$$S_u^2 = \frac{KDp}{r^2k} \frac{(T_b - T_i)}{\left(T_i - T_u - \dfrac{rwkF\sigma\alpha(T_b^4 - T_u^4)}{\rho_d Dp(c_p\rho + c_d w)}\right)} \qquad (4.25)$$

where $k = \rho d\rho_u RT_u^{3/2}/(2M\rho_b T_a^{1/2})$

Cassel *et al.* illustrated the implications of Equation (4.25) by first estimating the burning time of a representative dust particle from Equation (4.24). For instance, for a 25 μm diameter aluminium particle, a time τ of about 0.01s is obtained. Assuming a value of $S_u(\rho_u/\rho_b)$ of the order of 2.5 m/s from experimental data for S_u, the thickness of the burning zone in an aluminium dust flame is calculated to be of the order of 25 mm. This is 25–100 times greater than typical values for flames of premixed gases. This comparatively great thickness of the burning zone is a characteristic feature of laminar aluminium dust flames as is confirmed by experiments. (See Section 4.2.3.1.)

4.2.4.2
Ballal's theory for zero gravity conditions

Ballal (1983) postulated that the necessary and sufficient condition for self propagation of a laminar flame through a dust cloud is:

$$t_q = t_e + t_c \qquad (4.26)$$

where t_q is the quenching time, t_e the evaporation, pyrolysis or devolatilization time and t_c is the chemical reaction time. The criterion simply says that a flame can only propagate steadily if the quenching time just equals the sum of the time required to generate an explosible gas mixture and the time required for completion of the chemical gas phase reaction. Ballal claimed this approach to be universally applicable to dust clouds of any combustible material from metals to organic materials, and even liquid sprays. In the case of pure carbon in O_2/N_2, he considered the reaction $2C + O_2 \rightarrow 2CO$ as the 'evaporation' stage associated with t_e.

Evidence from flame propagation experiments under zero gravity conditions (Ballal (1983)) suggested that the laminar burning velocity of dust clouds in air is influenced by particle size, dust concentration, volatile matter content (for coal), heat loss by radiation from burning dust particles, and a mass transfer number B of the particles. B has the dimensions of dust concentration and equals the stoichiometric dust concentration for particles that react directly with oxygen in the solid state. If the main chemical oxidation reaction takes place in the gaseous phase, B is a complex function of boiling point, gas temperature, surface temperature, heat of combustion etc.

By considering the theoretical influence of these variables, Ballal (1983) arrived at the following expressions for t_q, t_e and t_c in Equation (4.26):

$$t_q = [\alpha_g/\delta_r^2 + (9q/c_{p,g}\rho_f)(C_1^2/C_3^3)(fD_{32})^{-1}\,\epsilon\sigma T_p^4/\Delta T_r]^{-1} \tag{4.27}$$

$$t_e = \frac{C_3^3\rho_f D_{32}^2}{8C_1(k/c_p)_g \ln(1 + B)} \quad \text{(truly evaporating particles)} \tag{4.28}$$

$$t_e = \frac{C_3^3\rho_f D_{32}^2}{8f^2C_1(k/c_p)_g\phi \ln(1 + B)} \quad \text{(carbon, coal)} \tag{4.29}$$

$$t_c = \delta_r/S_u \tag{4.30}$$

where the thickness of the reaction zone, δ_r, is defined as:

$$\delta_r = \frac{k_g}{c_{p,g}\rho_u S_u}\,\frac{\Delta T_r}{\Delta T_{pr}} \tag{4.31}$$

The notation for Equations (4.27) to (4.31) is as follows:

B Mass transfer number [–]
C_1 Ratio of mean particle diameters D_{20}/D_{32} [–]
C_3 Ratio of mean particle diameters D_{30}/D_{32} [–]
T_p Particle temperature [K]
T_{pr} Pre-heat zone temperature [K]
T_r Reaction zone temperature [K]
S_u Laminar burning velocity [m/s]
q Dust concentration [g/m^3]
k thermal conductivity [J/msK]
ϕ Equivalence ratio (=1 for stoichiometric conc.)
δ_r Thickness of reaction zone [m]
c_p Specific heat of gas at constant pressure [J/kgK]
g Subscript for gas
ρ_u Density of unburnt gas [kg/m^3]
ρ_f Density of particle [kg/m^3]
f Swelling factor for particle [–]
ϵ Emissivity of fuel particles [–]
σ Stefan-Boltzmann constant ($= 5.66 \times 10^{-8}$ J/sm^{2}K^4)
α Thermal diffusivity $k/c_p\rho$ [m^2/s]

By substituting Equations (4.27), (4.28) or (4.29), and (4.30) and (4.31) into Equation (4.26), a complex expression for the flame thickness δ_r results. The equation is composed of three main terms, namely a diffusion term, a chemical kinetics term and a radiative heat loss term. Once δ_r has been calculated, the corresponding S_u can be obtained from Equation (4.31).

Figure 4.19 shows that the theoretical prediction of S_u (solid lines) agrees well with the experimental data. Figure 4.19 also shows the predicted relative influence of the factors t_c, Q_r (radiative loss from particles) and f.

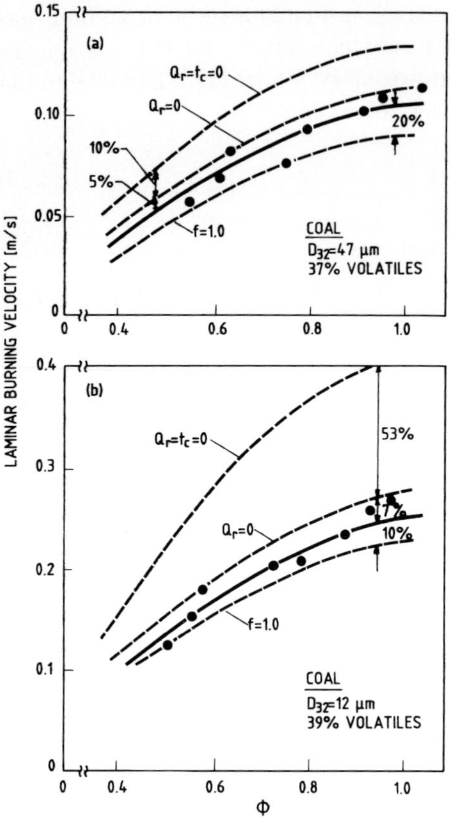

Figure 4.19 *Burning velocities of clouds of two coal dusts in air at zero gravity as functions of equivalence ratio (= 1 for stoichiometric mixtures). Data points are experimental values. Solid line is comprehensive theory. Dotted lines are simplified theory neglecting either radiative loses only, radiative losses and chemical reaction time, or particle swelling. Percentages indicate the roles of the respective factors (From Ballal, 1983)*

Figure 4.20 gives the theoretically predicted dimensionless flame thickness (real flame thickness divided by average surface/volume particle diameter D_{32}) as functions of the equivalence ratio (dimensionless dust concentration).

The 37% volatiles coal in Figure 4.19(a) has a burning velocity of about 0.11 m/s at stoichiometric concentration. According to Figure 4.20 the corresponding δ_r/D_{32} value is about 25, which for D_{32} = 0.047 mm gives δ_r = 1.18 mm. This is somewhat smaller than the experimental values in Section 4.2.3.2, and illustrates the limitations of the theory. Ballal (1983) pointed out that his theory is not applicable if:

1. The equivalence ratio $\phi \gg 1$, in the case of which radiation contributes positively to flame propagation.
2. Radiative heat transfer from shielding walls or pilot flames is significant.
3. The combustion is or becomes turbulent.
4. Gravitational effects play a significant role (particle diameter > 5 μm).

4.2.4.3
Theory by Ogle *et al.*

Ogle *et al.* (1984) presented a simplified thermal diffusion theory for plane, laminar flames in dust clouds, neglecting the velocity slip and temperature lag between the particle and gas phases. They first developed a model considering both radiation, convection and conduction (RCC). The governing equations were the continuity and thermal energy equations for the steady, one-dimensional laminar flow of a compressible, gray absorbing fluid of arbitrary optical thickness and constant physical properties:

$$G = \rho V = \rho_0 V_0 = \text{constant} \tag{4.32}$$

$$\rho c_p V \frac{dT}{dx} = k \frac{d^2T}{dx^2} + 2\sigma a T_f^4 E_2(ax) \tag{4.33}$$

In these equations ρ is the density, V is the velocity, x is the coordinate in the direction of flame propagation, G is the mass flux, c_p is the specific heat at constant pressure, k is the thermal conductivity, σ is the Stefan-Boltzmann constant, a is the absorption coefficient, T is the temperature, the subscript f denotes the flame position, E_2 is the exponential integral of order two. The exponential integral term represents the radiative absorption of energy emitted from the flame sheet at temperature T_f. The subscript zero denotes the initial ambient conditions.

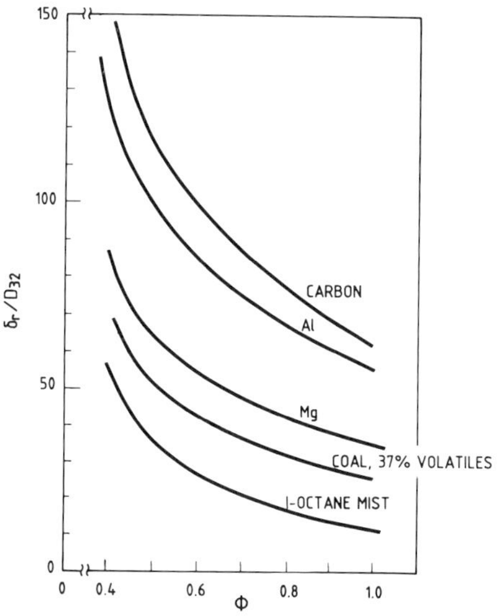

Figure 4.20 *Computed values of the flame thickness for dust clouds and iso-octane mist in air at atmospheric pressure, 290 K and zero gravity, as functions of the equivalence ratio (From Ballal, 1983)*

The boundary conditions were:

$$T(x = 0) = T_0$$

$$\frac{dT}{dx}(x \to +\infty) = 0 \tag{4.34}$$

$$T(x \to +\infty) = T_f$$

The first two boundary conditions allow determination of the temperature profile, and the third one specifies the burning velocity. By making certain assumptions, Equations (4.32)–(4.34) were solved to yield the temperature profile and the burning velocity:

$$(T - T_0)/(T_f - T_0) = 1 - e^{-1.5ax} \tag{4.35}$$

$$S_u = \frac{1}{\rho_0 c_p (T_f - T_0)} \left[\sigma T_f^4 - \frac{3}{2} ka(T_f - T_0) \right] \tag{4.36}$$

This is the RCC model. To evaluate the relative importance of conduction, the thermal conductivity can be set equal to zero yielding the radiation, convection model (RC). This results in the same temperature profile, but a different expression for the burning velocity:

$$S_u = \rho T_f^4/[\rho_0 c_p (T_f - T_0)] \tag{4.37}$$

It was found that the difference between burning velocities predicted by the RCC and RC models was negligible. Hence, conduction was negligible as compared to convection and radiation. The predicted burning velocity was 0.27 m/s for a flame temperature of 1750 K and increased almost linearly with flame temperature to 0.37 m/s for the adiabatic flame temperature 1950 K. Predicted burning velocities in the range 0.27–0.37 m/s for flame temperatures in the range 1750–1950 K are in reasonable agreement with experimental values.

Weber (1989) proposed a modification of the approach by Ogle *et al.* He used the mathematical condition for an inflection point (second derivative equal to zero) to obtain the burning velocity S_u as an eigenvalue from the two-point boundary value problem for a linear, second-order differential equation with arbitrary forcing. The flame was divided into a pre-heating zone from T_0 to T_i where T_i was the inflection point of the temperature-versus-distance profile, and a reaction zone from T_i to T_f. The application to dust flames, with thermal radiation, was considered.

4.2.4.4
Theory by Nomura and Tanaka for monosized particles

In the theory for plane flames developed by Nomura and Tanaka (1978) for monosized particles, it is assumed that the particles are initially arranged in a cubical pattern with centre-to-centre distance L in all three main directions. The relationship between L and the dust concentration C_d is given by:

$$L = \left[\frac{\pi}{6} D_p^3 \rho_p / C_d \right]^{1/3} \tag{4.38}$$

where D_p is the particle diameter and ρ_p the particle density. The flame propagation is assumed to occur as a one-dimensional wave composed of identical parallel elements of cross sectional area L^2, starting from a plane wall, as indicated in Figure 4.21.

Each particle is assumed to be located at the centre of a cubical air element of volume L^3, indicated by the dotted lines in Figure 4.21. When particle Number One burns, the surrounding gas element is heated adiabatically at constant pressure and expands in the x-direction, whilst the cross section L^2 normal to the x-axis is maintained constant. During this plug flow expansion the whole chain of subsequent gas elements are pushed to the right along the x-axis. The unburnt particles are assumed to follow their respective gas elements completely during this process.

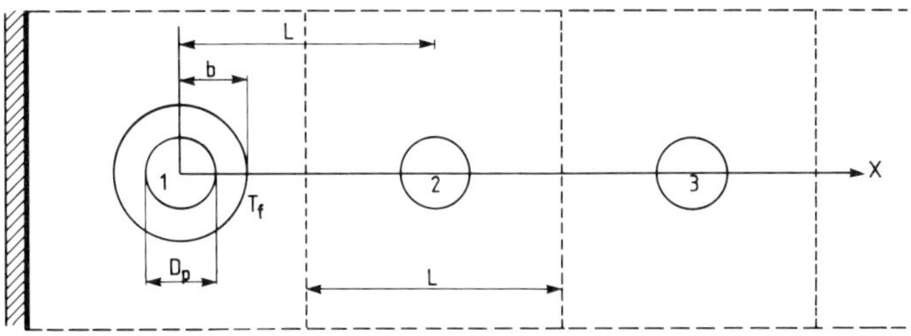

Figure 4.21 *Physical model forming the basis of the one-dimensional Nomura and Tanaka (1978) theory for laminar flame propagation through dust clouds of monosized particles*

When calculating the temperature profile due to combustion of particle No. 1, a one-dimensional model was used, corresponding to the particle being a plane of size L^2, normal to the x-axis rather than a sphere. The corresponding thermal diffusion equation is:

$$\frac{\partial T}{\partial t} = \alpha \frac{\partial^2 T}{\partial x^2} \tag{4.39}$$

where T is the gas temperature at distance x at time t and α is the thermal diffusivity. If the boundary condition at $x = 0$ is $T = T_f$, i.e. a constant flame temperature, and $T = T_0$ at $x = \infty$, the solution of Equation (4.39) is:

$$T(x, t) = (T_f - T_0) erf\left(\frac{x}{\sqrt{\alpha t}}\right) + T_0 \tag{4.40}$$

A dynamic heat balance for each particle was obtained by considering the heat transfer from the burning particle No. n, to the unburnt particle No. (n + 1) as given in Equation (4.41):

$$\frac{\pi}{6} D_p^3 \rho_p c_p \frac{dT_{dL}}{dt} = h\pi D_p^2 (T_{gL} - T_{dL}) + \frac{\pi D_p^2}{2} (a_p \epsilon_f F \sigma T_f^4 + a_p \epsilon_G \sigma T_G^4) - \pi D_p^2 \epsilon_p \sigma T_{dL}^4 \tag{4.41}$$

The notation not already explained is as follows:

c_p Specific heat of particle [J/gK]
T_{dL} Temperature of particle No. (n + 1) [K]
T_{gL} Temperature of gas surrounding particle No. (n + 1) [K]
T_G Temperature of hot gas sphere surrounding particle after burning [K]
h Heat transfer coefficient [J/(cm^{2}sK)]
a_p Absorptivity of particle [–]
ϵ_f Emissivity of flame [–]
ϵ_G Emissivity of hot gas surrounding particle after burning [–]

ϵ_p Emissivity of particle [–]
F Particle shape factor [–]
σ Stefan-Boltzmann constant ($= 5.66 \times 10^{-8}$J/(sm^{2}K^4))

The left-hand side of Equation (4.41) is the net heat input to particle No. (n + 1), whereas the three terms on the right-hand side are the convective heat flow to particle No. (n + 1) from the surrounding gas, the radiative heat flows to this particle from the flame front and the hot gas sphere round the burnt particle No. n, and the radiative heat loss from particle No. (n + 1).

Nomura and Tanaka analysed the various parameters in Equation (4.41) in detail and concluded that the radiative heat loss from particle No. (n + 1) was only about 10% of the radiative heat input to this particle from particle and gas element No. n. The simplified Equation (4.41), deleting the last term on the right-hand side, was then integrated from $t = 0$ to $t = \tau$, the total burning time of a particle, to identify the unknown time Δt_i when particle n reached its ignition temperature T_{ig}. T_{ig} was assumed to be known from experiments or other theory. The calculations started with n = 1 and were repeated for n = 2, 3 . . . up to n = 500. The time Δt_n for ignition of particle No. (n + 1) decreases with increasing n if the burning time τ is considerably larger than Δt_n. This is because more particles will then burn simultaneously and produce a greater heat flow to the next unburnt particle than if only one particle burns. In the examples shown by Nomura and Tanaka, Δt_n reached a constant value Δt_∞ for n > 100.

Nomura and Tanaka introduced the following expression for the burning time of a particle:

$$\tau = K_D \times D_p^2 \tag{4.42}$$

The burning constant K_D was assumed to be of the order of 1000 s/cm^2 for solid particles in general, and about 2000 s/cm^2 for coal particles specifically.

By using the corresponding τ from Equation (4.42), Δt_∞ was calculated, and the laminar burning velocity is then given by the simple relationship:

$$S_u = L/\Delta t_\infty \tag{4.43}$$

Calculated S_u values for coal dust in air at a dust concentration of 600 g/m^3 were 0.70 m/s for 20 μm diameter particles and 0.36 m/s for 40 μm diameter particles.

By requiring an experimental 'ignition temperature' of a particle, the Nomura-Tanaka theory suffers from the same basic weakness as the classical Mallard-le Chatelier (1883) theory for gases. The problem is that the 'ignition temperature' is not a true physical property of the particle, but depends on the actual circumstances under which the particle is ignited.

4.2.4.5
Specific theories for coal dust in air

Smoot and Horton (1977) gave a comprehensive review of the theoretical work on laminar coal dust/air flames up to the time of their paper, starting with the pioneering contributions on carbon/air flames by Nusselt (1924), and concluding with the unified theory for coal/air by Krazinski *et al.* (1977). The latter theory did not consider the devolatilization process and assumed that the particles had the same velocity as the surrounding gas.

However, both thermal radiation and conduction was accounted for, as well as char oxidation. The treatment of thermal radiation also included scattering effects. However, the theory was limited to low-volatile coals and was not confirmed by experiments. The predicted influence of particle size on the burning velocity was small.

In another paper, Smoot *et al.* (1977) presented their own, improved theory for laminar coal/air flame propagation, assuming particle/gas dynamic equilibrium and constant pressure. The general transformation method for computerized calculations of laminar burning velocities developed by Spalding *et al.* (1971), was adopted. Effects of gaseous diffusion, coal pyrolysis, char oxidation, and gaseous reaction were considered, whereas effects of gravity, viscous dissipation, forced diffusion, thermal diffusion and temperature gradients within particles were neglected. The unsteady state equations were solved numerically using finite difference techniques. The theory suggested that in a laminar coal-dust flame, gas phase diffusion and conduction, gas particle conduction, and coal pyrolysis are important rate-determining steps, while hydrocarbon and char oxidation may not be rate-limiting. The importance of gas phase diffusion processes in such flames was suggested.

The theory comprised six basic one-dimensional differential equations for:

1. Conservation of gas species.
2. Conservation of particle species.
3. Particle mass consumption rate.
4. Gas phase thermal energy balance.
5. Particle thermal energy balance including radiation.
6. Particle number balance.

Computed laminar burning velocities for coal dust in air, neglecting radiative effects, generally differed from experimental values by less than 25%.

Although not directly related to the theory of laminar flames it should be mentioned that Wolanski (1977) developed a comparatively simple one-dimensional theoretical model of coal dust combustion in a constant-pressure combustion chamber with recirculation of some of the exhaust gases. The model comprised five basic differential equations for:

1. Energy balance for the gas including heat conduction and convection.
2. Energy balance for the solid residue including conduction and radiation.
3. Mass balance for the released volatiles.
4. Mass balance for the solid residue.
5. Mass balance for oxygen.

The set of equations is similar to that used by Smoot *et al.* (1977).

Wolanski calculated gas and particle temperature-versus-time profiles with and without recirculation, and for various particle sizes and dust concentrations. For a coal of 35% volatiles, primary and secondary air temperatures of 360 and 600 K and wall temperature of 650 K, the calculated peak temperatures were about 1500 K for the gas and 3600 K and 2300 K for 10 μm and 80 μm diameter particles respectively.

Laminar burning of clouds of graphite dust in methane/air, and coal dust in air was investigated theoretically by Bradley *et al.* (1986). They calculated laminar burning velocities from the profile of net heat release rate Q versus dimensionless gas temperature τ, using Spalding's (1957) analytical approach. Their equation was:

$$Q = f_1(\tau)f_2(\tau)h \tag{4.44}$$

Here $f_1(\tau)$ is the ratio between the thermal gas conductivities at actual and unburnt gas temperatures, expressed as a function of gas temperature, $f_2(\tau)$ is the volumetric reaction rate, expressed as a function of gas temperature, and h is the heat of reaction. The use of this equation implies the calculation of the eigenvalue using the centroid of area expression given by Spalding (1957).

Bradley *et al.* (1986) assumed that the fuel was essentially premixed gas generated by rapid devolatilization of the coal particles and subsequent rapid mixing of the volatiles with the air. Furthermore, they assumed that the methane was the essential component of the volatiles, and that the presence of the char particles in the gas phase did not change the gas composition or chemical kinetics. The radiative loss from the char particles as they moved through the flame was computed. For a chemical heat release rate q per unit surface area of a smooth spherical particle, the total energy equation for a particle was taken as:

$$\alpha(T_p - T_g) = q - \epsilon\sigma T_p^4 - \frac{1}{3r^2}\frac{\partial r^3\rho_p h}{\partial t} \tag{4.45}$$

Here α is the convective heat transfer coefficient, T_p and T_g the particle and gas temperatures, ϵ the particle emissivity, assumed equal to unity throughout, σ the Stefan-Boltzmann constant, r, ρ_p and h the particle radius, density and enthalpy, and t the time. The equation neglects radiative absorption from the walls, gas and other particles.

The net heat release rate versus gas temperature profile was calculated using the comprehensive chemical kinetic model for methane/air combustion developed by Dixon-Lewis and Islam (1982) correcting for the rate of net energy supply from the particles due to their heating by oxidation of the char or graphite. The correction, which was generally found to be small compared with the heat release rate from the gas combustion, is given by:

$$H = 4\pi r^2 \alpha n(T_p - T_g) \tag{4.46}$$

where n is the number density of particles in the cloud, and the other notations as for equation (4.45).

Figure 4.22 shows a comparison of burning velocities predicted theoretically by Bradley *et al.* and experimental data from Smoot *et al.* In general, Bradley *et al.* found that their theory agreed well with experiments as long as devolatilization and gas phase mixing was sufficiently fast, and the char did not create a significant heat sink. This was found to be satisfied if the particle diameter was $< 10\ \mu m$ and the volatile content $> 25\%$.

The basic approach suggested by Hertzberg *et al.* (1982) and Hertzberg *et al.* (1987) is similar to that of Ballal (1983). It was assumed that three sequential processes are involved in the propagation of flame through a dust/air mixture:

1. Heating and devolatilization of dust particles.
2. Mixing of the volatiles with air.
3. Gas phase combustion of the premixed volatiles.

The characteristic time constants for the three consecutive processes are τ_{dv}, τ_{mx}, and τ_{pm}. It was realized that the process of particle heating and devolatilization is a complex combination of conductive, convective, and radiative heat exchange between the burnt

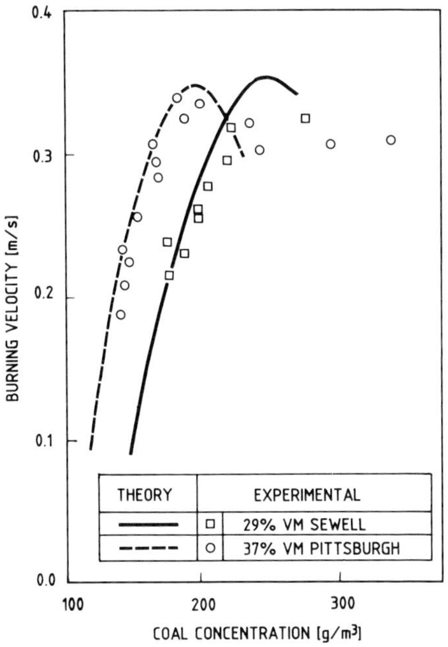

Figure 4.22 *Curves of theoretical burning velocities for clouds in air of coal dusts of 29 and 37% volatile matter and particle diameter 10 μm. Experimental points from Smoot* et al., *(1977) (From Bradley* et al., *1986)*

products and the unburnt reactants. However, the problem was simplified by handling those processes implicitly in the laminar burning velocity, S_u, which characterizes the overall rate of flame propagation. A laminar flame propagating at S_u has an overall reaction zone thickness of $\delta = \alpha/S_u$, where α is the effective diffusivity across the flame front. The overall reaction time for species passing through the reaction zone is $\tau = \delta/S_u$, and therefore:

$$S_u = (\alpha/\tau)^{1/2} \tag{4.47}$$

Per definition

$$\tau = \tau_{dv} + \tau_{mx} + \tau_{pm} \tag{4.48}$$

According to Hertzberg *et al.* (1982), the mixing process is normally comparatively rapid and τ_{mx} is shorter than both τ_{dv} and τ_{pm}. Furthermore, for small particles $\tau_{dv} \ll \tau_{pm}$, and the process is essentially controlled by premixed gas combustion. For larger particles it was assumed that the fraction of a particle that is devolatilized at a time t after the particle has entered the reaction zone equals:

$$\beta = 1 - (1 - 2\dot{x}_0 t/D_0)^3 \tag{4.49}$$

where $\dot{x}_0$ is the constant rate with which the pyrolysis/devolatilization wave progresses into the spherical particle of initial diameter D_0. It is further assumed that:

$$\dot{x}_0 = kS_u c\rho(T_b - T_u) \tag{4.50}$$

where k is the rate constant for the pyrolysis/devolatilization process, c is the heat capacity, ρ is the density of the unburnt mixture and T_b and T_u the gas temperatures of the burnt and unburnt mixture. As the dust particles become coarser and the dust concentration higher, the heating and devolatilization processes will begin to control the combustion rate, i.e. $\tau_{dv} > \tau_{pm}$. At conditions that give the highest burning velocities, approaching 0.40 m/s, the overall time constant τ is of the order of only 1 ms. Hertzberg *et al.* (1987) suggested that for such rapidly propagating dust flames, only the surface regions of the dust particles can contribute volatiles to the flame. The flame 'rides the crest' of a near-stoichiometric concentration of volatiles regardless of the dust concentration. This was considered the reason why Hertzberg *et al.* were unable to detect a sharp upper explosible concentration limit for dusts.

Although excess volatiles may continue to be emitted in the burnt gases at high dust concentrations, they are emitted too late to dilute the flame front with excess fuel vapour.

Krazinski *et al.* (1978) developed a theory for flame propagation in mixtures of monosized particles of low volatile coal dust and air, neglecting the role of the volatiles, but accounting for radiative heat transfer from the burning to the unburnt particles. For a stoichiometric mixture of air and 30 μm particles, an adiabatic burning velocity of 0.72 m/s was predicted. The flame thickness was of the order of several m, and this may in part explain why clouds of pure carbon in air are unable to propagate a flame in laboratory-scale apparatus.

Greenberg and Goldman (1989) developed a simplified theory for coal dust/air combustion for investigating the characteristics of a counter flow pulverized coal combustor. The model should be applicable even to laminar flames. It is related to the microscopic behaviour of the coal particles only, whereas the velocity, temperature and composition of the gas has to be obtained independently from experiments and/or other theories. The model includes drag between particle and gas, particle devolatilization and combustion, and heat transfer to and from the particles due to convection, radiation and chemical reactions.

4.2.5 THEORIES OF LAMINAR FLAME PROPAGATION IN CLOSED VESSELS

4.2.5.1 Theories by Nagy *et al.*

In sections 4.2.3 and 4.2.4 it was shown that both experiment and theory confirms that the concept of laminar burning is applicable to combustible dust clouds as well as to combustible premixed gases. Therefore the characteristic features of laminar dust explosions in closed vessels should be similar to those of laminar gas explosions in closed vessels. The explosion development in a closed spherical vessel was studied theoretically by Nagy *et al.* (1969). This treatment is also included in the book by Nagy and Verakis (1983). The following simplifying assumptions were made:

1. The equation of state for ideal gases is applicable.
2. Point ignition at the sphere centre by negligible energy supply.
3. Viscosity and heat capacities are constant.
4. Burning velocity is low compared to the velocity of sound, i.e. the pressure is spatially uniform throughout the vessel at any instant.
5. The thickness of the propagating reaction zone is negligible compared to the vessel radius.

The overall flame speed S_f with reference to the vessel was considered as the sum of three additive velocities, namely the laminar burning velocity S_u, the gas expansion or contraction velocity S_n due to the chemical change of number of molecules, and the gas expansion velocity S_e due to the heating of the gas.

The dependence of S_u on pressure P and temperature T_u in the unburnt mixture was taken as:

$$S_u = S_{u,r}(T_u/T_r)^2 \times (P_r/P)^{\beta} \tag{4.51}$$

where the index r refers to the reference state of 300 K and atmospheric pressure. β is an empirical constant that equals 0.5 or less for gases.

The problem was first simplified by treating the flame propagation as an 'isothermal' process, considering T_u as a constant equal to the mixture temperature T_0 before ignition, and T_b in the combustion products as a constant equal to the overall temperature T_m when all the mixture has burnt and the flame reaches the vessel wall.

The resulting analytical equation for the rate of pressure rise was:

$$\frac{dP}{dt} = \frac{3S_{u,r}T_0^2 P_r \beta P_m^{2/3}}{RT_r^2 P_0} (P_m - P_0)^{1/3} \left(1 - \frac{P_0}{P}\right)^{2/3} P^{(1-\beta)} \tag{4.52}$$

where R is the vessel radius and P_m the pressure when the flame reaches the vessel wall. This equation can be integrated analytically for $\beta = 0$. If $T_0 = T_r$, $P_0 = P_r$, $S_{u,o} = S_{u,r}$ and $\beta = 0$, the equation reduces to:

$$\frac{dP}{dt} = \frac{3S_{u,o}}{R} (P_m - P_0)^{1/3} P_m^{1/3} \left(1 - \frac{P_0}{P}\right)^{2/3} \left(\frac{P}{P_0}\right) \tag{4.53}$$

The maximum $(dP/dt)_{max}$ occurs when $P = P_m$, i.e.

$$(dP/dt)_{max} = \frac{3S_{u,o}}{R} (P_m - P_0)\,(P_m/P_0) \tag{4.54}$$

Equation (4.54) shows that this idealized isothermal treatment predicts that $(dP/dt)_{max}$ is inversely proportional to R, i.e. to the cube root of the vessel volume, in agreement with the frequently quoted 'cube root law'. However, this treatment also shows the strict conditions under which the 'cube root law' is valid. These conditions were explicitly pointed out by Eckhoff (1984/1985 and 1986) in a simplified analysis. First the thickness of the reaction zone or flame must be negligible compared to R. Secondly $S_u(T_u, P)$ must be independent of R. Under conditions of significant and unspecified turbulence, which are typical of dust explosion experiments in closed vessels, neither of these requirements may be fulfilled (see Section 4.4.3.3 for further discussion).

Nagy *et al.* (1969) extended the isothermal treatment to the more realistic adiabatic conditions for which T_u and T_b are not constants, but given by:

$$T_u = T_0 \left(\frac{P}{P_0}\right)^{\frac{\gamma_u - 1}{\gamma_u}} \tag{4.55}$$

$$T_b = T_m \left(\frac{P}{P_m}\right)^{\frac{\gamma_b - 1}{\gamma_b}} \tag{4.56}$$

Here γ_u and γ_b are the specific heat ratios for the unburnt and burnt mixture. Nagy *et al.* simplified the calculation by assuming an average value for γ, neglecting the difference between γ_u and γ_b. The resulting Equation (4.57) for (dP/dt), assuming that the initial conditions $S_{u,0}$, T_0 and P_0 equals the reference conditions $S_{u,r}$, T_r and P_r, is similar to Equation (4.53), but contains γ as a complicating parameter, and must be integrated numerically.

$$\frac{dP}{dt} = \frac{3S_{u,o}}{R} \times \frac{\gamma P_m^{2/3\gamma}}{P_0^{(2-1/\gamma-\beta)}} \times (P_m^{1/\gamma} - P_0^{1/\gamma})^{1/3} \times \left[1 - \left(\frac{P_0}{P}\right)^{1/\gamma}\right]^{2/3} P^{(3-2/\gamma-\beta)} \tag{4.57}$$

Values of both β and $S_{u,0}$ can be determined from Equation (4.57) and experimental data for $P(t)$, by plotting the experimental $(dP/dt)/(1 - (P_0/P)^{1/\gamma})^{2/3}$ as a function of P in a double logarithmic diagram. Then β is determined from the slope, and $S_{u,0}$ from the intercept with the ordinate axis (log P = 0). This theoretical treatment yielded a reasonable burning velocity for 7.7 Vol% acetylene in air, namely $S_{u,0}$ = 1.1 m/s, which is close to values from direct measurements.

However, when applying this approach to data from maize starch explosions in a 3 m^3 rectilinear closed vessel, β was found to be 0.36, which appears reasonable, but $S_{u,0}$ was found to be 3.15 m/s, which is about ten times the experimental laminar burning velocities for maize starch in air. Nagy *et al.* pointed out the fact that this high apparent value was most probably due to the turbulent conditions in the explosion. It is therefore necessary, when trying to determine laminar burning velocities from closed-bomb dust explosion experiments, to correct for the inevitable initial turbulence in such experiments. Nagy and Verakis (1983) attempted to do this and derived laminar burning velocities for clouds in air of various dusts by applying a modified form of Equation (4.53) to experimental dust explosion data from the elongated 1.2 litre Hartmann bomb. Corrections were made for the increase in the initial pressure due to the dispersing air. The first modification made in Equation (4.53) was that the ratio $3/R$ for a spherical vessel was replaced by the general ratio A/V for any arbitrary vessel shape, where A is the internal surface area of the vessel and V the vessel volume. Secondly, the initial laminar burning velocity, $S_{u,o}$, at atmospheric pressure and 300 K, was replaced by the corresponding turbulent burning velocity $S_{u,o}\,\alpha$, where α is a turbulence enhancement factor > 1. Furthermore, β was taken as equal to zero. The generalization of the theory to non-spherical vessels was justified by referring to the work on premixed gases by Ellis (1928) and Ellis and Wheeler (1928), and later work at US Bureau of Mines. These investigations indicated that in non-spherical vessels the initially spherical flame front gradually assumes the vessel shape.

The modified version of Equation (4.53) suggests that a straight line should result if (dP/dt) is plotted as a function of $(1 - (P_0/P))^{2/3} \times (P/P_0)$. The slope of this line determines the apparent turbulent burning velocity $S_{u,o}\,\alpha$. It was then simply assumed that α = 3.0 could be used as a representative average value for all the Hartmann bomb experiments. The resulting $S_{u,o}$ values are given in Table 4.7. No information on particle size is given explicitly and therefore the possibilities of detailed interpretation are limited.

The values are generally of the same order as laminar burning velocities determined by other methods, but it is clearly unsatisfactory to have to rely on somewhat arbitrary estimates of the factor α.

Table 4.7 Laminar burning velocities at atmospheric pressure and 300 K for combustible dust/air clouds estimated from experiments in the closed 1.2 litre Hartmann bomb at a dust concentration of 500 g/m^3 (From Nagy and Verakis, 1983)

Dust	Max. pressure, P_m [bar(g)]	Max. rate of pressure rise, $(dP/dt)_{max}$ [bar/s]	Estimated burning velocity, $S_{u,o}$ at atm. pressure and 300 K [m/s]
Alfalfa	4.55	76	0.20
Aluminium, atmomized	6.20	480	0.75
Aluminium, flaked	6.50	690	0.95
Antimony	0.55	7	0.49
Cellulose acetate	5.40	152	0.29
Cellulose acetate	5.85	248	0.43
Cellulose acetate	6.85	414	0.52
Chromium	3.50	255	1.08
Cinnamon	7.85	270	0.26
Citrus peel	3.50	76	0.33
Corn (maize) starch	7.95	620	0.59
Cotton linters	3.30	10	0.06
Gilsonite	6.15	262	0.43
Hemp hurd	7.10	690	0.82
Hydroxyethyl cellulose	7.30	180	0.20
Hydroxypropyl cellulose	5.80	200	0.33
Lignite, brown	5.70	172	0.30
Magnesium	6.75	760	0.98
Onion	1.25	7	0.16
Pittsburgh coal	5.70	160	0.26
Polyethylene	5.70	172	0.30
Polypropylene	4.75	193	0.46
Shellac	5.05	248	0.56
Sorbic acid	5.50	345	0.66
Stearic acid	6.00	290	0.46
Sulphur, 100% 44 μm	3.85	213	0.75
Titanium	6.20	760	1.15

4.2.5.2
Three-zone model by Bradley and Mitcheson

Bradley and Mitcheson (1976) carried the theoretical analysis further by first giving further support to the useful relation

$$\frac{P - P_0}{P_m - P_0} = \frac{m_b}{m_0} = n \tag{4.58}$$

suggested by Lewis and von Elbe (1961). Equation (4.58) simply says that the fractional pressure rise equals the fractional mass burnt, and rests on a number of assumptions of chemical and physical nature. Simplified analytical solutions of pressure versus time obtained by using this equation agreed fairly well with comprehensive computer solutions. Equation (4.58) replaces assumptions concerning the density and specific heat ratio of the burnt fraction. Bradley and Mitcheson further emphasized the importance of knowing the dependence of S_u on pressure and temperature, and they referred to a number of suggested relationships, including Equation (4.51) proposed by Nagy *et al.* (1969).

In the complete three-zone computer model of Bradley and Mitcheson (1976), Equation (4.58) was superfluous because most basic relationships were accounted for directly. Flame propagation was considered as consumption of unburnt combustible mixture in small mass decrements dm_u. However, in reality this mass does not become burnt instantaneously, but passes through a reaction zone of finite thickness and this was accounted for. The overall model therefore comprises three zones, namely the volumes of unburnt, reacting and burnt mixture, the sum of which equals the known vessel volume. The inclusion of a finite reaction zone is of particular interest in the context of dust explosions, where reaction zone thicknesses are generally much larger than in laminar premixed gases.

The flame was in turn considered as consisting of two zones, namely a pre-heat zone extending from unburnt mixture temperature T_u to its ignition temperature, T_{ig}, and a reaction zone in which the temperature increased from T_{ig} to the ideal equilibrium temperature T_f. This picture is in fact a agreement with the classical model by Mallard and le Chatelier (1983). T_{ig} is not a fundamental constant for a given mixture, but depends on the method of determination.

The unburnt gas was assumed isotropic, but each burnt gas element arising from each mass decrement dm_u was treated independently in order to estimate its temperature after isotropic compression. Any energy exchange between mixture elements by conduction, convection or radiation, was neglected.

The comprehensive computer model gave good agreement with experimental data for pressure-versus-time in laminar closed-bomb explosions of methane/air mixtures. However, no comparisons with dust explosions were made.

4.2.5.3 Theory by Nomura and Tanaka

Nomura and Tanaka (1980) extended their theory for plane laminar burning of dust clouds at constant pressure (Nomura and Tanaka (1978)) to laminar burning in closed vessels. By making certain assumptions, they derived the general equation:

$$\frac{P^{1/\gamma} - P_0^{1/\gamma}}{P_m^{1/\gamma} - P_0^{1/\gamma}} = \frac{m_b}{m_0} \tag{4.59}$$

which is slightly different from Equation (4.58) by having all three pressures raised to the power of $1/\gamma$, where γ is the average specific heat ratio for burnt and unburnt mixture.

As before (Nomura and Tanaka, 1978) it was assumed that the dust cloud consisted of monosized particles arranged in a regular, static pattern. However, in the present case ignition occurred at a point, as opposed to at an infinite plane, and the flame propagation was spherical, as opposed to the plane, one-dimensional propagation considered earlier. Consequently, the particle centres were considered as being located at concentric spherical shells, rather than in the regular cubical grid structure applicable to plane flames.

In the spherical geometry, the relationship between the average interparticle distance L, the particle density ρ_p, the particle diameter D_p, and the dust concentration C_d was defined as:

$$L = \left(\frac{\rho_p}{C_d}\right)^{1/3} \times D_p \tag{4.60}$$

which differs from Equation (4.38) by the factor $(\pi/6)^{1/3}$.

Equation (4.41) was used in a simplified form by neglecting all thermal radiation except that from the flame front to the next particle shell. The resulting equation for the maximum rate of pressure rise in a spherical vessel with central point ignition was:

$$(dP/dt)_{\max} = \frac{3\gamma D_p}{R\Delta t} \times \left(\frac{\rho_p}{a}\right)^{1/3} \times P_m \left[1 - \left(\frac{P_0}{P_m}\right)^{1/\gamma}\right] \tag{4.61}$$

which conforms with the 'cube-root' law as long as all constants at the right-hand side are independent of the vessel radius R. It is implicitly assumed during the derivation of this equation that the thickness of the flame zone is negligible compared to the vessel radius R. The constant a in Equation (4.61) has the dimensions of mass per unit volume and equals the effective dust concentration that can burn completely consuming the oxygen available. For dust concentrations C_d up to stoichiometric the parameter $a = C_d$, whereas for higher concentrations, it maintains the stoichiometric value.

The Δt is the time required for the flame to propagate from the $(n - 1)$th to the nth particle shell. For starch dusts of $D_p < 50$ μm, Δt was found to be independent of n for $n > 30$. Therefore the burning velocity equals $S_u = L/\Delta t_\infty$, as defined by Equation (4.43). Nomura and Tanaka derived Δt_∞ as a complex function of particle and combustion properties.

Nomura and Tanaka (1980) also extended their theoretical treatment to non-spherical vessel shapes. This was done by maintaining spherical flame propagation for any part of the flame that had not reached the vessel wall. As soon as a part of the flame reached the wall, flame propagation stopped for that part. Heat loss to the vessel wall was not considered. Under these conditions the theoretical analysis showed that the 'cube root' relationship was valid even for elongated, cylindrical vessels, as long as they were geometrically similar.

Figure 4.23 illustrates the theoretical development of pressure with time in an elongated cylinder. At time t_1 the spherical flame reaches the cylinder wall and at time t_0 the entire dust cloud has burnt.

Nomura and Tanaka tried to correlate their theoretical results for laminar flame propagation with experimental data from dust explosions in closed vessels. However, inevitable and unknown turbulence in the experimental dust clouds could not be accounted for, and the value of the correlation therefore seems limited.

4.2.5.4
Simplified theory by Ogle *et al.*

Ogle *et al.* (1983) proposed a simplified three-element theory for the development of a dust explosion in a closed vessel. The first element was a model for the burning time τ of a particle:

$$\tau = \frac{r_0}{k_c Y_{O_2}} \tag{4.62}$$

where r_0 is the characteristic size of the particle [m] obtained from morphological Fourier analysis, k_c a first order rate constant [m/s] and Y_{O_2} the initial mass fraction of oxygen in the gas phase.

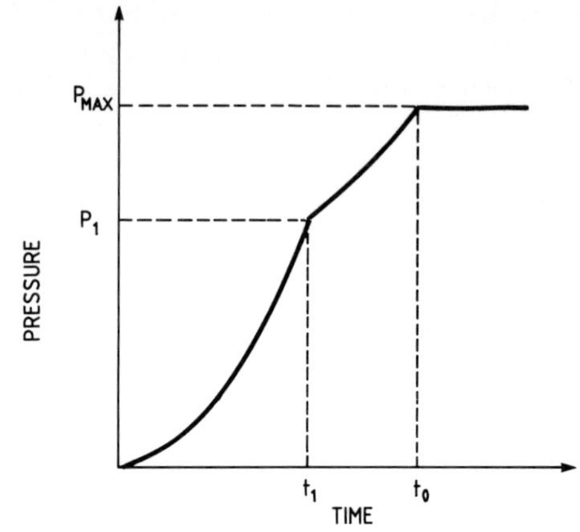

Figure 4.23 *Theoretical relation between pressure and time in dust explosions in a closed elongated cylindrical vessel (From Nomura and Tanaka, 1980)*

The second element was a model for the laminar burning velocity of the dust cloud, based essentially on the classical Mallard-le Chatelier (1883) model for premixed gases, with an additional term for thermal radiation. The resulting equation for the burning velocity is:

$$S_u = B + (B^2 + A)^{1/2} \tag{4.63}$$

where

$$B = 1/2 \left(\frac{\epsilon\sigma}{\rho_u c_p}\right)\left(\frac{T_f^4 - T_0^4}{T_i - T_0}\right)$$

and

$$A = \frac{\lambda}{\rho_u c_p \tau}\left(\frac{T_f - T_i}{T_i - T_0}\right)$$

The various symbols denote:

ρ Initial density of the gas phase [kg/m^3]
σ Stefan-Boltzmann constant (= 5.66×10^{-8} J/s m^{2}K^4)
ϵ Emissivity [–]
c_p Heat capacity of gas at constant pressure [J/kg K]
T_f Flame temperature [K]
T_i Ignition temperature [K]
T_0 Initial temperature [K]

λ Thermal conductivity [J/s m K]
τ Burning time of a dust particle [s]

Equation (4.63) differs somewhat from Equation (4.22) derived by Cassel *et al.* (1949), but rests on a similar basic philosophy.

The third model element was the equation for the rate of pressure rise:

$$(dP/dt)_{\max} = \frac{3\gamma r^2 P}{(R^3 - r^3)} \times \left(\frac{\rho_u}{\rho_b} - 1\right) S_u \tag{4.64}$$

where γ is the specific heat ratio, r the radius of the spherical flame, and ρ_u and ρ_b the densities of the unburnt gas and the combustion gases. Equation (4.64) is based on the approximation $dr/dt = (\rho_u/\rho_b)S_u$.

Estimates for S_u for aluminium dust clouds using the theory by Ogle *et al.*, gave considerably higher burning velocities, by a factor of four, than experimental values from laminar burners.

4.2.5.5
Computer model by Continello

The laminar flame propagation through a coal dust/air suspension in a spherical enclosure was studied by Continello (1988) by means of a one-dimensional, spherically-symmetric mathematical model. An Eulerian formulation was adopted for the gas phase mass continuity, species and energy balance equations, while a Lagrangian formulation was employed for the mass, energy and momentum balance equations for the particles.

For the 'gas phase' the following assumptions were made: The flow is laminar and spherically symmetric. The viscous dissipation rate is negligible and the pressure is uniform in space (low Mach number), but varies in time. The gas mixture is thermally perfect. Binary diffusion coefficients for each pair of species are taken to be equal, thermal mass diffusion is neglected. Mass diffusion and heat conduction are governed by Fick's and Fourier's law, respectively. The diffusion coefficient varies with temperature and pressure. The Lewis number is unity. Radiative heat transfer is neglected. The combustion chemistry is described by means of a single-step, irreversible reaction of the volatiles with the oxygen, and Arrhenius-type kinetics with non-unity exponents for fuel and oxygen concentrations apply. The equations also include coupling terms accounting for mass, momentum and energy exchanges between the gas phase and particle phase.

In the simplified treatment of the 'particle phase', a coal particle was represented by a sphere containing ash, fixed carbon and volatiles in specified initial fractions. The particle was considered to remain spherical and conserve its volume. The temperature was considered uniform in the particle, including its surface. The transport processes in the gas film next to the particle were assumed to be quasi-steady, and the thermophysical properties of the air/fuel vapour mixture were assumed uniform and evaluated at a conveniently averaged value of the temperature in the gas film. The fuel vapour production rate was assumed to depend on the particle temperature and global composition only. During the particle heat-up, the volatiles were assumed to be released according to a simple one-step Arrhenius pyrolysis reaction. Due to the highly transient character of the particle history in this kind of phenomena, surface oxidation reactions were not considered. This eliminated the need for considering the mass transfer processes in the film. All of the volatiles released by the particle are immediately available in the gas phase. The model accounts for the effects of the convective transport caused by the gas-particle relative motion by means of correction factors to the spherically symmetric stagnant film situation.

Ignition was induced by introducing a heat source of a given intensity in the energy equation for a limited time. The model then predicted the particle heat-up, devolatilization and ignition of the volatiles, and the subsequent flame propagation through the spherical volume.

Figure 4.24 shows an example of computations for laminar explosions of coal dusts of various diameters in air, in a spherical vessel of 0.10 m diameter.

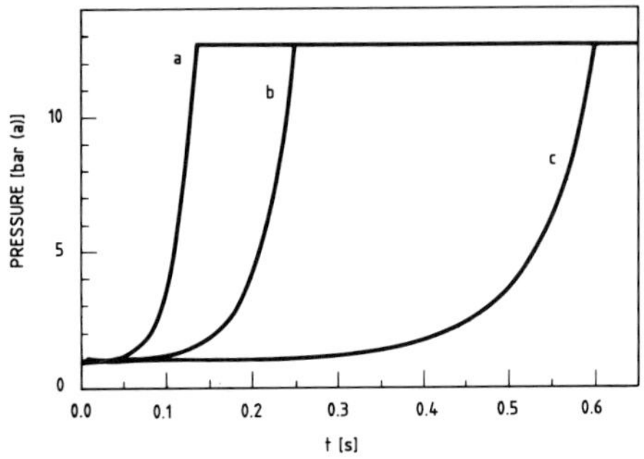

Figure 4.24 *Computed pressure-versus-time patterns for spherical explosions of coal dust of various particle diameters in air in a vessel of 0.10 m diameter: a) 30 μm; b) 50 μm; c) 100 μm (From Continello, 1988)*

The predicted final pressure of about 12.5 bar absolute is close to the maximum theoretical adiabatic pressure. This is much higher than maximum pressures found in experiments. The reasons are that the model neither accounts for heat losses, nor for endothermic dissociation in the burnt mixture.

Continello (1988a) expressed some important view points concerning the use of computer models for simulating dust explosions. A space resolution of the order of a few μm is necessary for a detailed description of particle-scale phenomena. On the other hand, the typical thickness of a real dust flame is of the order of 10 mm or more, whereas the physical dimensions of process units in which dust explosions take place is of the order of 1–10 m. This means that the ratio of the various length scales involved covers up to seven orders of magnitude. Therefore detailed comprehensive modelling considering all of the relevant mechanisms across all seven orders of magnitude is not really feasible even by means of extensive numerical computing. Besides, such a model would require information about a number of microscopic characteristics of the dust particles and of their interaction with heat and gas flows, which can only be acquired by complex, extensive experimentation. Furthermore, as discussed in Chapter 3, the mechanics of generation of dust clouds is very complex, and very small particles of the order of 1 μm diameter may not become dispersed into individual primary particles, but appear as considerably larger agglomerates constituting the effective particles in the dust cloud.

The optimal simulation model should include the minimum level of detail necessary to reproduce the significant features of the explosion development with sufficient accuracy. The specific interpretation of this statement may vary with the objective of the simulation. From an industrial safety point of view, the upper range of the length scale is most important, whereas for studies of the combustion process as such, for example for predicting chemical conversion, the smaller scales may be of greater interest.

No matter what the objective is, it is beyond doubt that computer simulation is the future tool for predicting dust explosion development in industrial practice. However, it is then necessary to include some other important factors in addition to those considered by Continello (1988), in particular turbulence and aspects of entrainment and dispersion of dust particles as discussed in Chapter 3 (see also Section 4.4.8).

4.2.6 MINIMUM AND MAXIMUM EXPLOSIBLE DUST CONCENTRATIONS

4.2.6.1 The problem

The existence of well-defined minimum and maximum explosible concentrations of fuel in air is well established for various gases and vapours. At the outset it would be reasonable to expect that such limits of explosible fuel concentrations also exist for combustible dusts. However, as shown by Makris *et al.* (1988), who considered the minimum explosible concentration, there is substantial disagreement between experimental data for a given dust. For example, reported values for maize starch in air range from 8 g/m^3 to 400 g/m^3. The disagreement arises from considerable differences in apparatus and interpretation of data. Because of the extremely energetic pyrotechnical ignitor used, it is not surprising that the exceptionally low value of 8 g/m^3 was determined by Siwek (1977) using a 20 litre spherical bomb. On the other hand, it canot be excluded that there were some real differences between the dusts used. Although the primary grains of maize starch have a fairly uniform size of 10–15 μm diameter, commercial maize starch qualities often contain considerable fractions of stable agglomerates that will behave as large single particles, as shown by Eckhoff and Mathisen (1977/1978). Furthermore, the moisture contents of the maize starches investigated were often not reported and may have varied.

One basic problem in all experimental determination of explosibility limits is the definition of an 'explosion'. It has been customary to relate this definition to either direct observation of a self-sustained flame through the dust cloud, at constant pressure, or to the increase in pressure that results if flame propagation occurs at constant volume in a closed vessel. If the dust concentrations are in the middle of the explosible range, the observation of explosion is simple, irrespective of the criterion chosen. Both extensive flame propagation and extensive pressure build-up will result. The problems arise when the dust concentration is approaching the lower or upper explosibility limits, and flame propagation and pressure rise become marginal. Because of the inherent inhomogeneity of real dust clouds and the corresponding comparatively poor reproducibility of repeated apparently identical experiments, it is necessary to choose some arbitrary criterion of a minimal explosion, either in terms of a minimal extent of flame propagation, or a minimal magnitude of pressure rise. Unfortunately there seems to be no really basic scientific criterion that defines the 'right' choice.

In their analysis, Makris *et al.* (1988) concluded that any meaningful criterion of a minimum explosible dust concentration must be related to a distinct flame propagation in the dust/air mixture at constant pressure. They claimed that it is not possible to decide whether or not such flame propagation occurs in constant volume experiments, and they therefore did not consider that results from closed bombs had any fundamental significance. This argument may not be fully justified, but it is necessary to account for the fact that in any closed-bomb experiment the unburnt mixture will start to become compressed right from the onset of flame propagation.

4.2.6.2
Experimental determination of minimum explosible dust concentration

Selle and Zehr (1957) described a closed-bomb method that utilized a flame propagation criterion of explosion. A spherical glass bomb of volume 1.4 litres was used, in which a given quantity of dust, placed in a small hemispherical cup, was dispersed into a cloud by means of a blast of compressed air and exposed to an ignition source at the sphere centre. The concentration of dust was gradually lowered in a series of consecutive experiments until flame did no longer propagate throughout the entire volume of the bomb. This means that Selle and Zehr had chosen the requirement of a fully developed flame within the bomb as their criterion of explosion. The size of the flame was recorded on a photographic film, and this facilitated an objective decision of whether the flame had actually filled the entire volume of the bomb. Nevertheless, the explosion criterion itself was the result of a subjective choice.

Selle and Zehr observed that flames that occupied only part of the bomb volume, were not necessarily located in the vicinity of the ignition source. Due to inhomogeneities in the dust concentration throughout the volume of the explosion bomb, flame propagation could be restricted to local, almost detached ‘pockets’ in the dust cloud.

This kind of non-homogeneous structure is an inherent feature of real dust flames in general. This clearly complicates the interpretation of marginal flame propagation in small-scale apparatus in terms of minimum explosible concentration in large, industrial-scale systems. Therefore, experiments have also been conducted in fairly large industrial-scale equipment. The work of Palmer and Tonkin (1971) is a good example. Their apparatus is illustrated in Figure 4.25.

The dust was introduced at the top of the tube by a screw feeder and dropped into a vibrating 20 cm diameter and 15 cm high dispersing cylinder hanging immediately underneath the screw exit. After having passed the perforated bottom of the cylinder, the dispersed dust settled freely under gravity through the entire length of the tube until finally being collected in a bin at the bottom end. Dust concentration and flame propagation could not be measured in the same test, but had to be determined in separate tests at nominally identical dust cloud generation conditions, i.e. rotating speed of the feeding screw conveyor and vibration mode of the dust disperser. The dust concentration was measured gravimetrically. A manually operated sliding tray was inserted into the tube like a gate valve about 3.5 m from the tube top. By simultaneously closing the tube at the top by a conventional sliding gate valve, the volume of dust cloud between the top valve and the tray was trapped. By dividing the amount of dust finally settled out on the tray by the volume 0.182 m^3 between the tray and the top valve, the average dust concentration in this section of the tube was obtained.

Immediately before performing an explosion test the dust feed was stopped and the bottom end of the tube closed by a gate valve located just below the ignition zone. The ignition source was a propane flame generated by injecting a small pocket of a propane/air

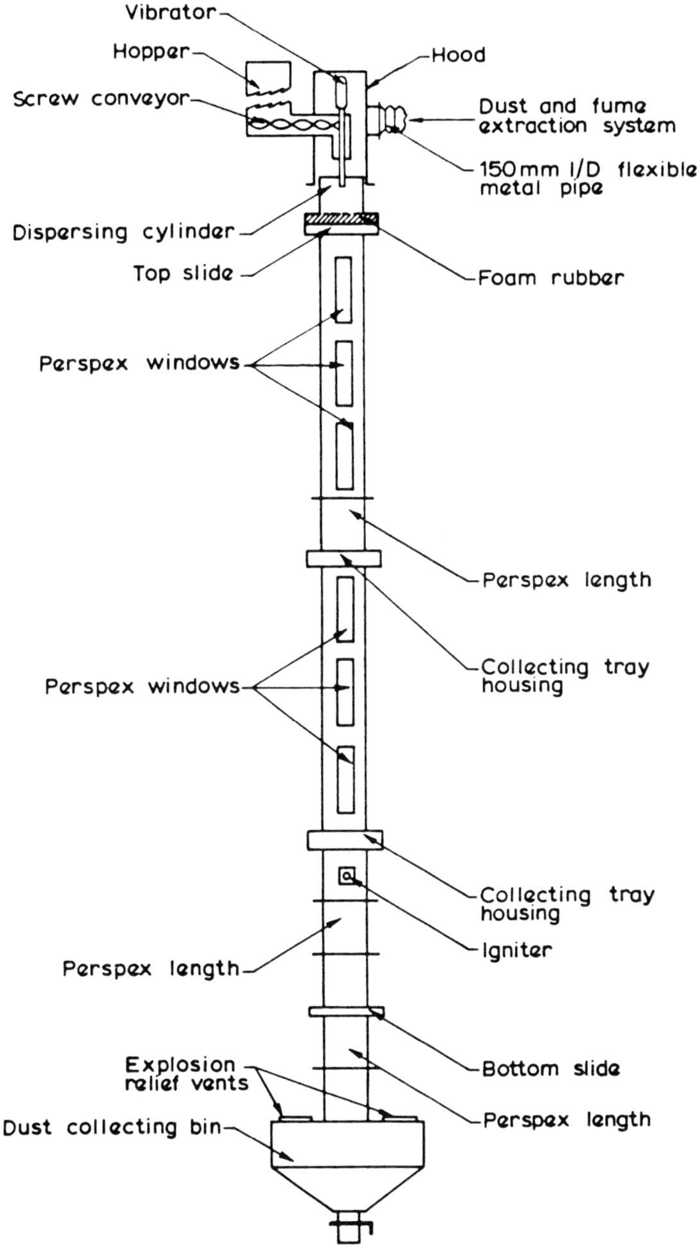

Figure 4.25 *Vertical large-scale explosion tube facility for flame propagation studies. Internal tube diameter 0.25 m. Total tube length 5.2 m. Ignition point 1.5 m above bottom (From Palmer and Tonkin, 1971)*

mixture into the bottom region of the explosion tube and igniting by means of an electric spark located at the tube axis.

This facility was made available to Eckhoff and Fuhre (1975) for determination of the minimum explosible concentration of a wheat grain dust of 10% moisture content, taken from a dust extraction filter in a grain silo plant. Due to poor flow properties of the dust, a constant rotation speed of the dust feeding screw did not always result in a constant dust feed. For this reason several dust concentration measurements had to be performed during a test series at a given screw rotation speed, and some scatter had to be accepted. Only flame propagation lengths of more than about 0.5 m upwards in the tube were considered significant. Propagation lengths of about 0.5–1.0 m were classified as 'marginal' propagation.

The results of eight test series are summarized in Figure 4.26. Each series, run at a given set of nominal dust cloud generation conditions, comprised from three to six consecutive experiments for measurement of dust concentration or flame propagation. Figure 4.26 gives the actual average dust concentration values determined in individual experiments in each series and the corresponding flame propagation results.

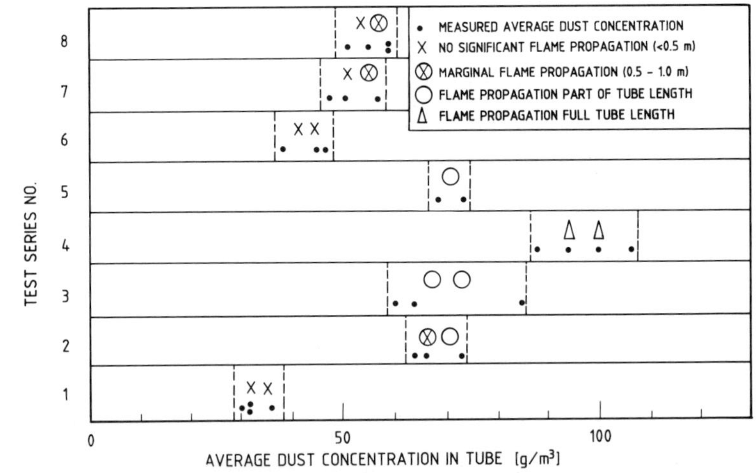

Figure 4.26 *Results from determination of the minimum explosible concentration in air of a wheat grain dust from a grain silo plant, containing 10% moisture, using the vertical large-scale dust explosion tube of diameter 0.25 m and height 5.2 m developed by Palmer and Tonkin (1971) (From Eckhoff and Fuhre, 1975)*

For dust concentrations below 50 g/m^3 no significant flame propagation was observed, whereas marginal propagation was observed in the range 50–60 g/m^3. From 60–80 g/m^3 flame propagated over part of the tube length, whereas full tube length propagation required dust concentration of at least 90–100 g/m^3.

This gradual increase of the extent of flame propagation with dust concentration over a considerable range was also observed by Palmer and Tonkin (1971) and was typical for this facility. This illustrates that realistic dust clouds are never perfectly homogeneous and that a sharp minimum explosible concentration value is therefore non-existent. However, some numerical value may be required in practice and in the present case, a conservative figure would be 50 g/m^3.

The absence of a sharp minimum explosible concentration seems to be common also for experiments in smaller scale. Therefore, the specification of a given value of the minimum explosible concentration for a given dust inevitably implies the use of some arbitrary criterion of explosion, as a finite minimum pressure rise at constant volume or a minimum finite minimum extent of flame propagation at constant pressure. A transition range representing a factor of two of average dust concentrations, from the first sign of self sustained flame, to extensive flame propagation is probably typical of many experiments.

Another aspect that needs consideration is the influence of the settling of particles due to gravity on the minimum explosible dust concentration. Burgoyne (1963), discussing the minimum explosible concentration of clouds of liquid droplets, distinguished between the 'static' and the 'kinetic' minimum explosible concentrations C_s and C_k. If the drops are sufficiently large for their gravitational sedimentation valocities v_t to be significant, and S_u is the upwards burning velocity in the drop cloud, then C_s and C_k differs according to:

$$C_k = \left(\frac{S_u + v_t}{S_u}\right) C_s \tag{4.65}$$

This equation should also be applicable to solid particles that volatilize/pyrolyze in the pre-heating zone of the flame front, i.e. organic materials and coals.

Figure 3.10 in Chapter 3 shows that for a density of 1 g/cm^3, a particle diameter of 10 μm gives v_t = 0.004 m/s, which means that for a limit value of S_u of about 0.1 m/s, C_k and C_s differs by only 4%. However, for particle diameters of 50 and 100 μm, v_t = 0.09 and 0.3 m/s, which for S_u = 0.1 m/s gives C_k = 1.9 C_s and 4.0 C_s respectively. This indicates that due to gravitational settling, flame propagation through clouds in air of volatilizing/pyrolyzing particles of the order of 50–100 μm diameter can take place at considerably lower 'static' concentrations C_s than for particles of negligible v_t. Burgoyne converted independent experimental data for C_s and C_k for mists and sprays of organic liquids, to the corresponding C_k and C_s values, using Equation (4.65) and a limit value of S_u of 0.46 m/s for negligible v_t estimated by assuming that S_u and C_k were the same for upwards and downwards flame propagation. The results, shown in Figure 4.27, indicate that Equation (4.65) is in accordance with reality.

As pointed out by Burgoyne (1963), Equation (4.65) also applies to downwards flame propagation, but then v_t, being numerically the same as for upwards propagation, becomes negative. As a consequence, C_s for downwards propagation becomes larger than C_k, and

$$C_{s,\ \text{upwards}} < C_k < C_{s,\ \text{downwards}} \tag{4.66}$$

Hartmann and Nagy (1944) introduced an arbitrary pressure criterion when determining the minimum explosible dust concentration using the 1.2 litre Hartmann tube. The top of the tube was closed by a paper membrane of bursting strength about 0.2 bar(g). The smallest quantity of dispersed dust that generated at least this pressure rise, divided by the volume of the tube, was taken as the minimum explosible dust concentration.

The continued use of this criterion in the extensive later investigations by the US Bureau of Mines, was confirmed by Dorsett *et al.* (1960). However, Cashdollar and Hertzberg (1985) reconsidered the original USBM method and suggested their new 20 litre closed explosion vessel test as an alternative. The explosion criterion chosen was

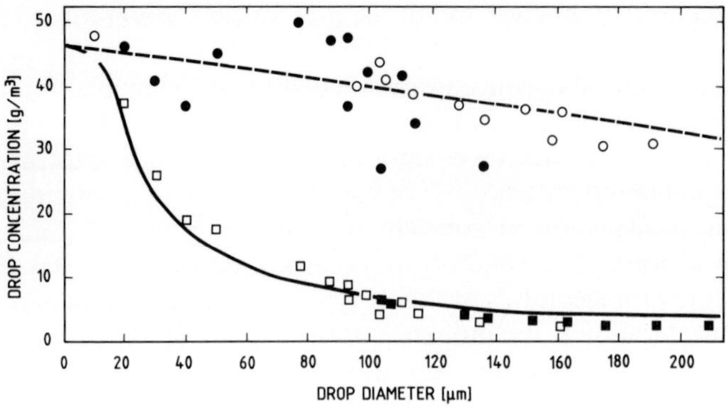

Figure 4.27 *Correlation of 'static' and 'kinetic' concentrations at the lower limit of flammability of tetralin/air suspensions with varying drop diameter*
□ = C_s *experimental* ■ = C_s *calculated*
○ = C_k *experimental* ● = C_k *calculated*
(From Burgoyne, 1963)

$P_m/P_0 \geqslant 2$, where P_m is the maximum absolute explosion pressure in the experiment and P_0 the initial pressure, both corrected for the pressure rise due to the 2.5–5 KJ chemical ignitors used. By adopting this method, Cashdollar *et al.* (1989) identified the minimum explosible concentration of Pittsburgh coal to about 90 g/m^3, in contrast to the earlier value of 135 g/m^3 found in an 8 litre bomb and reported by Hertzberg *et al.* (1979). Cashdollar *et al.* (1988) correlated minimum explosible dust concentrations of coal dusts measured in the USBM 20 litre bomb with values from large-scale mine experiments, and found good agreement.

Hertzberg *et al.* (1987) postulated that flames in low-concentration clouds of organic dusts and coal dusts of small particle sizes are essentially premixed gas flames. This is because the burning velocity close to the minimum explosible concentration is so low that each particle becomes completely devolatilized and the volatiles mixed with air in the pre-heating zone of the flame front before combustion gets under way.

Following this line of thought, Cashdollar *et al.* (1989) determined the minimum explosible concentrations for various coals and mixtures of graphite and polyethylene dust as a function of the content of volatiles. Figure 4.28 shows the resulting correlation. It is worth noting that the value of 33 g/m^3 for polyethylene, which devolatilizes completely, is close to the minimum explosible concentration of methane in air.

This is further in good agreement with the results of Eckhoff and Pedersen (1988) for polyester and epoxy dusts, using a method issued by Nordtest (1989). Their results are given in Figure 4.29.

The straight line through the square points is approximately horizontal, indicating that the minimum explosible concentration of combustible material is in fact almost constant and independent of pigment content. The chemical composition of the combustible substance does not seem to influence its minimum explosible concentration. It is interesting to observe the close agreement between this value of 31–35 g/m^3 and the value 33 g/m^3 found for polyethylene by Cashdollar *et al.* (1989). It is also of interest to compare the value of 31–35 g/m^3 with published explosibility limits for gaseous hydrocarbon-in-air

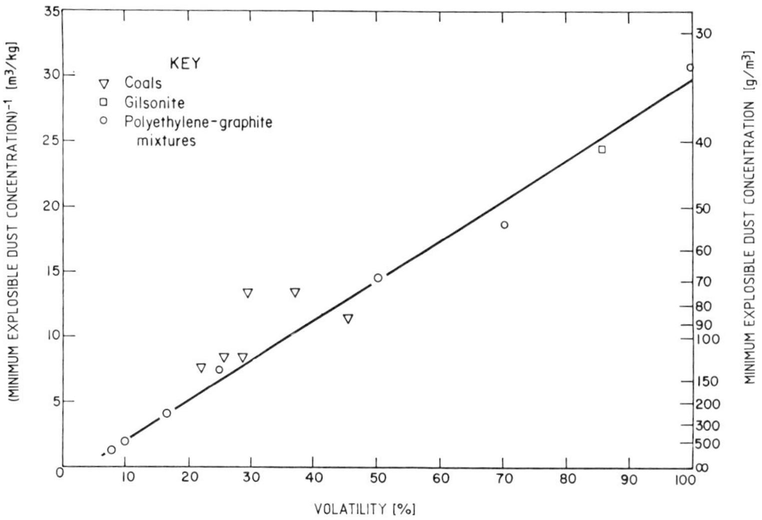

Figure 4.28 *Inverse and minimum explosible dust concentration versus content of volatiles for various dusts (From Cashdollar, Hertzberg and Zlochower, 1988)*

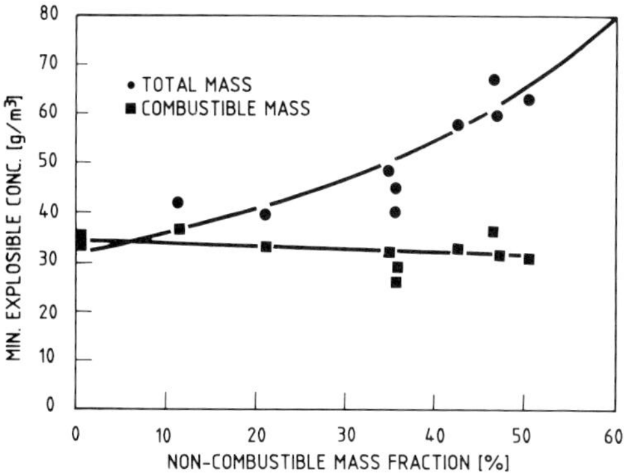

Figure 4.29 *Minimum explosible dust concentration versus non-combustible mass fraction for polyester and epoxy resins (From Eckhoff and Pedersen, 1988)*

mixtures. For methane-in-air and propane-in-air, the limits are approximately 5.0 vol% and 2.0 vol%, respectively. Converted to mass concentrations, these equal 33 g/m³ and 36 g/m³ respectively (at 25°C), which is close to the measured minimum explosible concentrations of combustible material for the polyester/epoxy powders. This supports the view that the flame propagation through the dust clouds at the limiting concentration is similar to that through a premixed gas, i.e. flame propagation takes place in the combustible gas evolved from the particles in the pre-heating zone just ahead of the flame. In accordance with this model, p being the mass fraction of non-combustibles in percent, the minimum explosible dust concentration (MEC) of this category of dusts is:

$$MEC = 32 \text{ g/m}^3 \, (100/(100 - p)) \tag{4.67}$$

This relationship gives the curved line in Figure 4.29, and it is seen that the agreement with the experimental points is reasonable. Approximate estimates of *MEC* for various contents of non-combustible material can be obtained by this relationship. However, undue extrapolation beyond the experimental points will give physically meaningless results because $MEC \rightarrow \infty$ when $p \rightarrow 100$.

Buksowicz and Wolanski (1983) studied flame propagation near the minimum explosible dust concentration, in a 5.5 litre vertical cylinder of 150 mm diameter. By choosing a minimum relative pressure rise P_m/P_0 of 1.5 as their explosion criterion, they obtained minimum explosible dust concentrations in close agreement with those based on self sustained flame propagation in a long tube of 100 mm diameter. Buksowicz and Wolanski demonstrated by direct photography that near the minimum explosible dust concentration the dust flame is fragmented into detached zones of burning particle clusters. They also emphasized the need for using a sufficiently energetic ignition source, when studying propagation of lean limit flames.

Schläpfer (1951) measured minimum explosible concentrations of various dusts in air in a laboratory-scale vertical tube of diameter 30 mm and with a vertical distance of 0.6 m from the ignition source to the open top end of the tube. Dust suspensions of known concentration were conveyed upwards in the tube at a laminar velocity of 0.6 m/s. Propagation of dust flame at least 3/4 of the distance of 0.6 m from the ignition source to the tube top was used as the explosion criterion. The ignition source was an electrically ignited 0.2 mm thick and 7 mm long aluminium wire. The results in Table 4.8 were obtained.

Table 4.8 Minimum explosible dust concentrations measured at laminar flow conditions in a 30 mm diameter vertical tube (From Schläpfer, 1951)

Dust type	Min. explosible concentration $[g/m^3]$
Aluminium flakes, mean flake thickness 0.5 μm	90
Lignin, 100% finer than 120 μm	48
Phenol resin, brown, 100% finer than 120 μm*	45
Phenol resin, grey, 100% finer than 120 μm*	36

*The chemical difference between the two resins was not given, but is probably primarily due to different colouring additives.

Hertzberg *et al.* (1987) found that the minimum explosible dust concentration for polymethylmethacrylate was about 80 g/m^3 and independent of the particle diameter up to 100 μm. For Pittsburgh seam bituminous coal the value of about 90 g/m^3 was found to apply from 2 μm to 60 μm particle diameter. However, when the particle diameter increased towards 200 μm, a substantial increase in the minimum explosible concentrations was found for both dusts. This influence of particle size agrees with the earlier results of Ishihama (1961) for various particle size fractions of coals of volatile matter contents in the range 46–49%. This worker also found that the minimum explosible dust concentration decreased with decreasing mean particle diameter down to about 60 μm. For a finer fraction, of mean size 29 μm, the minimum explosible concentration was only slightly lower than for the 60 μm fraction.

However, the actual minimum explosible concentration values found by Ishihama were only half of those found by Hertzberg *et al.* (1987). This can in part be explained by the higher content of volatiles in the coals used by Ishihama, but the major factor is probably the different experimental methods and explosion criteria used.

Minimum explosible dust concentrations were determined in a comparative test series amongst four different laboratories in different countries. Three different methods were used, namely the 20 litre sphere method developed by Siwek (1977) (see also ASTM (1988)), the 1 m^3 method specified by the International Standardization Organization (1985), and the Nordtest Fire 011 (1989) method. The results are shown in Table 4.9.

Table 4.9 Minimum explosible dust concentration [g/m^3] determined by three different methods in four different laboratories (From Eckhoff, 1988)

Test method / Dust	20-litre sphere					1 m^3	Nordtest Fire 011
	Lab. 1	Lab. 2	Lab. 3	Lab. 4	Arithm. mean	Lab. 1	Lab 3.
Lycopodium	30 (0.5)	20 (0.2)	15 (0.1)	40 (0.5)	26	30 (0.4)	35 ± 6*
Maize starch (11-12% moisture)	60 (0.5)	50 (0.3)	90 (0.2)	90 (0.2)	73	80 (0.3)	130 ± 14* [76 ± 4]
Light protecting agent 'Tinuvin'	30 (1.3)	10 (0.2)	20 ((0.2)	20 (0.1)	20	40 (0.3)	27 ± 3*
Spanish coal (36% volatiles)	40 (0.6)	30 (0.4)	25 (0.3)	30 (0.1)	31	70 (0.2)	98 ± 20* [73 ± 5]
Zinc	600 (0.9)	600 (0.3)	400 (0.3)	400 (0.1)	500	650 (0.8)	565 ± 65*

*Standard deviation.

Figures in brackets ( ) in Table 4.9 indicate measured maximum explosion pressure in bar(g) minus that due to the ignitor (assumed to be 1.3 bar(g) for the 20 litre sphere).

The 20 litre Siwek test has been based on a rather weak and vaguely defined pressure rise criterion. Besides, the very strong 10 KJ pyrotechnical ignitor may cause combustion of dust even if the dust concentration is below that required for self-sustained flame propagation at constant pressure. Therefore it is not unexpected that the 20 litre minimum explosible dust concentrations were generally lower than for the two other methods. The work of Continillo *et al.* (1986) with the 20 litre Siwek sphere indicates significant pressure rise for coal dust/air clouds even at dust concentrations as low as 50 g/m^3, in accordance with the low minimum explosible concentration for coal dust for this apparatus in Table 4.9. Furthermore, the real, local dust concentration in the region of the ignition source was not known. The problem of generating non-homogeneous distributions of dust concentration in small-scale experiments has been emphasized by Eggleston and Pryor (1967).

The 1 m^3 method also involves a 10 KJ ignition source and a pressure rise criterion, but because of the large size of the vessel, the net influence of the 10 KJ ignitor on the pressure rise is small. However, the distribution of dust concentration is not known.

The Nordtest Fire 011 is essentially a constant pressure method, because the top of the 15 litre vessel is covered only by a weak paper diaphragm. The explosion criterion is independent, upwards flame propagation through the experimental dust cloud to an extent that the flame, as observed visually, is clearly detached from the ignition source. A special feature of this method is that the actual local dust concentration in the region of the ignition source is measured directly gravimetrically. Most of the Nordtest data in Table 4.9 are based on an earlier, quite restrictive criterion of explosion, requiring fairly extensive flame propagation. More recent data based on the present criterion of any flame propagation that is clearly detached from the ignition source, are given in brackets [ ].

Lovachev (1976) discussed some unrealistically low values for the minimum explosible concentration of some dusts published in USSR, and he emphasized the necessity of observing self-sustained flame propagation through the dust cloud, beyond the influence of the ignition source (see also Section 7.13).

4.2.6.3 Experimental determination of maximum explosible dust concentration

The results of Palmer and Tonkin (1971) from the large-scale apparatus shown in Figure 4.25 give an indication of the maximum explosible concentration of a coal dust containing 36.4% volatiles on a dry, ash-free basis. Extrapolation of their data for mixtures of coal and sodium chloride to zero content of the latter, indicates a value of 2000–3000 g/m^3. This is of the same order as the value indicated by extrapolating the data from stabilized burner experiments with a similar coal dust (Pittsburgh) in air, presented by Smoot *et al.* (1977). These workers measured laminar burning velocities of more than 0.15 m/s even at 1800 g/m^3.

Slezak *et al.* (1986), using a tumbling horizontal explosion cylinder of 0.3 m diameter and 4.5 m length, estimated the maximum explosible concentration of Pittsburgh coal dust in air to be about 1500 g/m^3.

However, Cashdollar *et al.* (1988), using their closed 20 litre explosion vessel, were not able to detect any maximum explosive dust concentration for Pittsburgh coal up to 4000 g/m^3. They refer to other laboratory and large-scale experiments that confirm this result.

On the other hand, Ishihama *et al.* (1982) were able to determine maximum explosible concentrations of different non-cohesive coal dust fractions using a rotating drum apparatus in which the dust cloud was generated continually by being lifted up along the drum wall and subsequently falling freely under gravity. For the particle size fraction 35–50 μm the maximum explosible concentration in air was 2700 g/m^3 for a 45% volatiles coal, 2200 g/m^3 for 33% volatiles and 1400 g/m^3 for 22% volatiles. The maximum explosible concentration decreased with increasing particle size, and for the 45% volatiles coal it was 2400 g/m^3 for 50–75 μm and 1800 g/m^3 for 100–150 μm.

Ishihama *et al.* also investigated potato starch of mean particle size 50 μm and found a very high maximum explosible concentration of about 8000 g/m^3. It seems probable that the cohesive potato starch, as opposed to the free flowing coal dust fractions, only dispersed partly in the rotating drum apparatus, yielding a lower real concentration of dispersed dust than the nominal value.

Other data of maximum explosible dust concentration from direct experimental determination than these rather scattered and partly contradictory results, have not been traced. It is therefore of interest to consider more indirect determinations by Zehr (1959). He made the first order assumption that the conditions for flame propagation in a dust/air mixture only depends on the mass ratio dust/air, and is independent of air pressure and mean distance between particles. He then constructed the cylindrical combustion bomb illustrated in Figure 4.30 for determining the maximum explosible concentration of dusts.

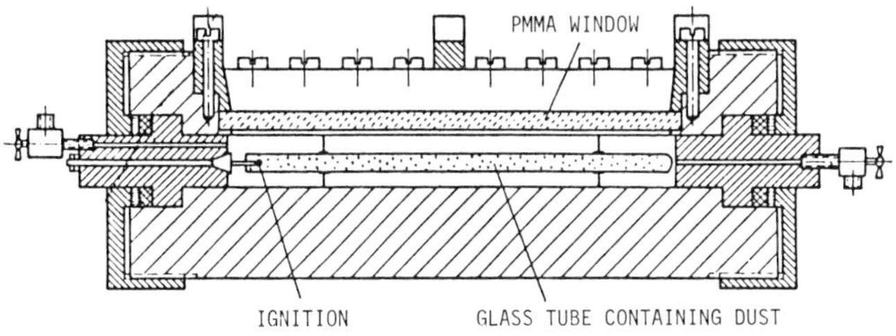

Figure 4.30 *Apparatus for indirect experimental determination of the maximum explosible concentration of dusts in air (From Zehr, 1959)*

The central 25 cm long glass tube of about 1 cm^2 cross section and the one end closed, is first filled completely with the dust to be tested, loosely packed. The glass tube is then inserted into the combustion bomb and the air pressure raised to the desired level. Because the bulk densities of loosely packed organic dusts are typically of the order of 500 kg/m^3 and maximum explosible concentrations of the order of 1 kg/m^3, air pressures up to the order of 500 bar were required for obtaining the same dust/air mass ratio in Zehr's combustion tube as in a dust cloud at the maximum explosible concentration at atmospheric pressure. At these high pressures the equation of state has to be corrected for non-ideal gas behaviour. Zehr (1959) gives a detailed description of the computational procedure used for conversion of the actual high pressure conditions to atmospheric pressure conditions. After having achieved the desired initial conditions, the dust was ignited at the open end of the glass tube and it was observed, through a narrow uncovered slit of the 'Perspex' window, whether combustion propagated along the tube towards its closed end. Some of Zehr's results are summarized in Table 4.10.

The experiments with kieselguhr mixed into the combustible dust were performed when the propagation of the combustion in the combustible dust only could not be clearly identified. However, although the kieselguhr facilitated distinction between propagation and no propagation, the maximum explosible dust concentrations estimated from the experiments with kieselguhr were much lower than they would be expected to be in the combustible dust alone, as illustrated by the data for polyvinyl alcohol in Table 4.10.

There might be potential for improving Zehr's method by increasing the glass tube diameter and using thermocouples at various locations in the tube for detecting propagation of combustion, rather than relying on visual observation. However, due to the very high temperatures to be expected, the method may not be suitable for metal dusts such as silicon, aluminium and magnesium.

Table 4.10 Maximum explosible concentrations of dusts in air determined by an indirect experimental method of Zehr (1959)

Dust	Maximum explosible dust concentration [g/m³]
Coal, < 10 μm particle size, high content of volatiles	> 800*
Wheat and rice flour	>1000*
Peat, dried	2600
Cotton	1800-2400
Cork	1400-1500
Cork	2000-2500**
Polyurethane	1000
Polyvinylalcohol	2000-2300
Polyvinylalcohol	> 750- 900*
Sulphur	> 900*

* Value obtained with 50% kieselguhr mixed into dust.
** Standard glass tube replaced by one of two times larger diameter.

4.2.6.5
Theories of minimum and maximum explosible dust concentrations

The first attempt to predict the minimum and maximum explosible concentrations for dust clouds theoretically was probably made by Jaeckel (1924), who considered the one dimensional heat transfer from a plane flame front to the adjacent unburnt layer of dust cloud.

The minimum explosible concentration will then be, according to Jaeckel, the minimum amount of dust, per unit volume of dust cloud, which by complete combustion liberates enough energy to heat the next unit volume of dust cloud to the ignition temperature. This means that the assumption of the existence of such a temperature is as basic in Jaeckel's theory, as it is in the classical flame propagation theory of Mallard/le Chatelier (1883).

According to Jaeckel, the maximum explosible concentration arises from the fact that the air contains a limited amount of oxygen, which is totally consumed by the complete combustion of a given amount of dust, the stoichiometric concentration C_s. A further increase in the dust concentration thus merely has the effect that more energy is required for heating the next volume to the ignition temperature, since the excess dust acts only as a coolant or heat sink.

Jaeckel (1924) formulated the condition for self-sustained flame propagation through the dust cloud of concentration $C < C_s$ at constant volume as:

$$C \times Q \geq L + (T_i - T_0)(C \times c_d + d_g \times c_v) \qquad (4.68)$$

where

c_v is the specific heat at constant volume of the gas
d_g is the density of the gas
Q is the heat of combustion of the dust
c_d is the specific heat of the dust particles
T_0 is the initial temperature of the dust cloud

T_i is the ignition temperature of the dust cloud
L is the heat losses, by radiation and conduction

By equating the two sides and rearranging, one obtains the expression for the minimum explosible concentration C_l:

$$C_l = \frac{L}{(Q - c_d(T_i - T_0))} + \frac{(T_i - T_0)d_g \times c_v}{(Q - c_d(T_i - T_0))} \tag{4.69}$$

For dust concentrations above the stoichiometric concentration the heat production is constant and equal to $Q \times C_s$, whereas the heat consumption increases with the dust concentration. In this case the condition for self-sustained flame propagation will be:

$$C_s \times Q \geqslant L + (T_i - T_0)(C \times c_d + d_g \times c_v) \tag{4.70}$$

By rearranging, Jaeckel's theoretical upper explosible limit this becomes equal to:

$$C_u = \frac{C_s Q - L}{(T_i - T_0)c_d} - \frac{d_g \times c_v}{c_d} \tag{4.71}$$

Jaeckel considered a constant volume explosion. In a typical real case, a dust explosion is probably neither a pure constant pressure nor a pure constant volume process, since pressure will gradually build up in the unburnt cloud, although the flame may not be fully confined in volume.

As can be seen from Equations (4.69) and (4.71), a substitution of c_v by c_p increases C_l and decreases C_u. The loss L is difficult to estimate, and Jaeckel suggested, as a first approximation, that the loss factor L be neglected. If this is done, and c_v is replaced by c_p, Equations (4.69) and (4.71) can be written:

$$C_l = \frac{(T_i - T_0)d_g \times c_p}{Q - c_d(T_i - T_0)} \tag{4.72}$$

$$C_u = \frac{C_s Q}{(T_i - T_0)c_d} - \frac{d_g \times c_p}{c_d} \tag{4.73}$$

If the left-hand sides of the Equations (4.68) and (4.70), representing the heat production, are denoted H_p, it is seen that for $0 < C < C_s$, H_p is a linear function of C, and for $C > C_s$ it is constant and independent of dust concentration.

If the ignition temperature is considered to be independent of dust concentration and the loss L is neglected, and the right-hand sides of the equations (4.68) and (4.70) representing the heat consumption, are denoted H_c, H_c becomes a linear function of the dust concentration. According to Jaeckel's simple model, the condition of self-sustained flame propagation is:

$$H_p \geqslant H_c \tag{4.74}$$

Zehr (1957) suggested that Jaeckel's theory be modified by replacing the assumption of an ignition temperature of finite value by the assumption that dust flames of concentrations near the minimum explosible limit have a temperature of 1000 K above ambient. Zehr further assumed that the combustion is adiabatic and runs completely to products of the highest degree of oxidation, and that the dust particles are so small that the dust cloud can be treated as a premixed gas. The resulting equations for the minimum explosible concentration in air are:

$$C_l = \frac{1000\ M}{107\ m + 2.966[Q_m - \Sigma\Delta I]}\ [\text{g/m}^3] \qquad (4.75)$$

for constant pressure, and

$$C_l = \frac{1000\ M}{107\ m + 4.024[Q_m - \Sigma\Delta U]}\ [\text{g/m}^3] \qquad (4.76)$$

for constant volume. Here M is the mole weight of the dust and m the number of moles of O_2 required for complete oxidation of 1 mole of dust. Q_m is the molar heat of combustion of the dust, $\Sigma\Delta I$ the enthalpy increase of the combustion products and $\Sigma\Delta U$ the energy increase of the combustion products.

Schönewald (1971) derived a simplified empirical version of Equation (4.75) that also applies to dusts containing a mass fraction $(l - \alpha)$ of inert substance, α being the mass fraction of combustible dust:

$$C_l^* = \frac{C_l/\alpha}{1 - 2.966(1 - \alpha)c_p C_l/\alpha} \qquad (4.77)$$

where the minimum explosible dust concentration without inert dust is $C_l = -1.032 + 1.207 \times 10^6/Q_0$, Q_0 being the heat of combustion per unit mass (in J/g), as determined in a bomb calorimeter. As can be seen from Freytag (1965), Equations (4.75) and (4.76) were used in F. R. Germany for estimating minimum explosible dust concentrations, but in more recent years this method has been replaced by experimental determination.

Table 4.11 gives examples of minimum explosible dust concentrations calculated from Equations (4.75) and (4.76), as well as some experimental results for comparison. The calculated and experimental results for the organic dusts polyethylene, phenol resin and starch are in good agreement. This would be expected from the assumptions made in Zehr's theory. However, the result for graphite clearly demonstrates that Zehr's assumption of complete combustion of any fuel as long as oxygen is available, is inadequate for other types of fuel. The results for bituminous coal and the metals also reflect this deficiency.

Buksovicz and Wolanski (1983) postulated that at the minimum explosible concentration, flames of organic dusts have the same temperature as lower limit flames of premixed hydrocarbon gas/air. They then proposed the following simple semi-empirical correlation between the heat of combustion (calorific value) Q [KJ/kg] of the dust, and the minimum explosive concentration C_l [g/m^3] in air at normal pressure and temperature:

$$C_l = 1.55 \times 10^7 \times Q^{-1.21} \qquad (4.78)$$

The assumptions implied confine the applicability of this equation to the same dusts to which Zehr's Equations (4.75) and (4.76) apply. For starch, Equation (4.78) gives $C_l = 114$ g/m^3, which is somewhat higher than the value of 70 g/m^3 found experimentally by Proust and Veyssiere (1988), but close to that calculated by Zehr for constant pressure. For polyethylene, Equation (4.78) gives 36 g/m^3, in close agreement with both experiments and Zehr's calculations.

Table 4.11 Minimum explosible dust concentrations calculated by the theory of Zehr (1957). Most data published by Freytag (1965). Comparison with experimental data

Dust type	Calculated minimum explosible dust concentrations [g/m³]		Experimental minimum explosible dust concentration [g/m³]
	Constant volume	Constant pressure	
Aluminium	37	50	90, constant pressure [Schläpfer (1951)]
Graphite	36	45	Flame propagation in graphite/ air at normal conditions not observed
Magnesium	44	59	
Sulphur	120	160	
Zinc	212	284	500-600, constant pressure, constant vol. [Eckhoff (1988)]
Zirconium	92	123	
Polyethylene	26	35	33, constant volume [Cashdollar, Hertzberg and Zlochower (1988)]
Polypropylene	25	35	
Polyvinyl alcohol	42	55	
Polyvinyl chloride	63	86	
Phenol resin	36	49	36-45, constant pressure [Schläpfer (1951)]
Maize starch	90	120	70, constant pressure [Proust and Veyssiere (1988)]
Dextrin	71	99	
Cork	44	59	50, constant pressure [Essenhigh and Woodhead (1958)]
Lignite	49	68	
Bituminous coal	35	48	70-130, constant volume [Cashdollar, Hertzberg and Zlochower (1988)]

Lunn (1988) also investigated this group of materials and obtained further support for the hypothesis that the minimum explosible concentration of organic dusts that burn more or less completely in the propagating flame, is primarily a function of the heat of combustion of the dust.

Shevchuk *et al.* (1979), being primarily concerned with metal dusts, advocated the view that a discrete approach, considering the behaviour and interaction of individual particles, is necessary for producing an adequate theory for the minimum explosible dust concentration. They analysed the distribution of a heat wave in a dilute suspension of monosized solid fuel particles in a gas, assuming no relative movement between particles and gas, no radiative heat transfer, and that the rate of heat production q_p during combustion of a single particle of mass m_p was constant during the entire burning lifetime t_b of the particle, and equal to $q_p = Qm_p/t_p$, where Q is the heat of combustion of the particle material. The resulting equation for the minimum explosible dust concentration, assuming that the average flame temperature equals the ignition temperature T_i of the dust cloud as determined in a heated-wall furnace, was:

$$C_l = (T_i - T_0)c_g\rho_g/(FQ - c_d(T_i - T_0)) \qquad (4.79)$$

Here T_0 is the ambient temperature, c_g and c_d the heat capacities of gas and dust material, ρ_g the gas density and F a special particle distribution factor resulting from this particular analysis, and which causes Equation (4.79) to differ from Jaeckel's Equation (4.72). Using T_i data from Jacobson *et al.* (1964), Shevchuk *et al.* compared Equations (4.72) and (4.79) as shown in Table 4.12.

Table 4.12 Minimum explosible concentrations of metal powders in air (From Shevchuk *et al.*, 1979)

Powder type	TI [K]	Eqn. (4.72) [g/m^3]	Eqn. (4.79) [g/m^3]
Aluminium	920	25	51
Magnesium	890	29	62
Titanium	600	21	44
Iron	590	52	107
Manganese	730	62	129

Reliable experimental data for metal dusts are scarce. However, Schläpfer (1951) found a value of 90 g/m^3 for fine aluminium flakes, which indicates that both equations underestimate the minimum explosible concentration considerably, Equation (4.72) by a factor of nearly four and (4.79) by a factor of nearly two. One main reason for this is probably the use of the ignition temperature T_i as a key parameter.

Mitsui and Tanaka (1973) derived a theory for the minimum explosible concentration using the same basic discrete microscopic approach as adopted later by Nomura and Tanaka (1978) for modelling laminar flame propagation in dust clouds, and discussed in Section 4.2.4.4. Working with spherical flame propagation, they defined the minimum explosible dust concentration in terms of the time needed from the moment of ignition of one particle shell to the moment when the air surrounding the particles in the next shell has been heated to the ignition temperature of the particles. If this time exceeds the total burning time of a particle, the next shell will never reach the ignition temperature. Because this heat transfer time increases with the mean interparticle distance, it increases with decreasing dust concentration. By using some empirical constants, the theory reproduced the trend of experimental data for the increase of the minimum explosible dust concentration of some synthetic organic materials with mean particle size in the coarse size range from 100–500 μm particle diameter.

Nomura, Torimoto and Tanaka (1984) used a similar discrete theoretical approach for predicting the maximum explosible dust concentration. They defined this upper limit as the dust concentration that just consumed all available oxygen during combustion, assuming that a finite limited quantity of oxygen, much less than required for complete combustion, was allocated for partial combustion of each particle. Assuming that oxygen diffusion was the rate controlling factor, they calculated the total burning time of a particle in terms of the time taken for all the oxygen allocated to the particle to diffuse to the particle surface. In order for the flame to be transmitted to the next particle shell, the particle burning time has to exceed the heat transfer time for heating the gas surrounding the next particle shell to the ignition temperature. Equating these two times defines the maximum explosible dust concentration. Two calculated values were given, namely 1400 g/m^3 for terephthalic acid of 40 μm particle diameter and 4300 g/m^3 for aluminium of 30 μm particle diameter. The ignition temperatures for the two particle types were taken as 950 K and 1000 K respectively.

Bradley *et al.* (1989) proposed a chemical kinetic theoretical model for propagation of flames of fine coal dust near the minimum explosible dust concentration. It was assumed that the combustion occurred in premixed volatiles (essentially methane) and oxidizing gas, the char particles being essentially chemically passive. The predicted minimum explosive concentrations were in good agreement with experimental values (about 100 g/m^3 for 40% volatile coal, and 500 g/m^3 for 10–15% volatiles)

4.3 NON-LAMINAR DUST FLAME PROPAGATION PHENOMENA IN VERTICAL DUCTS

This section will treat some transitional phenomena that are observed under conditions where laminar flames could perhaps be expected. This does not include fully turbulent combustion, which will be discussed in Section 4.4.

Buksowicz *et al.* (1982) and Klemens and Wolanski (1986) describe experiments with a lignite dust of 52% volatiles, 6% ash and < 75 μm particle size, in a 1.2 m long vertical duct of rectangular cross section of width 88 mm and depth 35 mm. The duct was closed at the top and open at the bottom. Dust was fed at the top by a calibrated vibratory feeder yielding the desired dust concentration. The ignition source (an electric spark of a few J energy or a gas burner flame) was located near the open bottom end. Flame propagation and flame structure were recorded through a pair of opposite 80 mm × 80 mm glass windows. Diagnostic methods included Mach-Zehnder interferometry, high-speed framing photography, and high-frequency response electrical resistance thermometry. Figure 4.31 shows a compensation photograph of a lignite dust/air flame propagating upwards in the rectangular duct. The heterogeneous structure of the flame, which is typical for dust flames in general, is a striking feature. This is reflected by the marked temperature fluctuations recorded at fixed points in the flame during this kind of experiments, as shown in Figure 4.32.

The amplitudes of the temperature oscillation with time are substantial, up to 1000 K. The very low temperature of almost ambient level at about 1.1 s in Figure 4.32b shows that at this location and moment there was probably a pocket of cool air or very dilute, non-combustible dust cloud. Klemens and Wolanski (1986) were mainly concerned with quite low dust concentrations. From quantitative analysis of their data they concluded that the thickness of the flame front was 11–12 mm, whereas the total flame thickness could reach 0.5 m due to the long burning time (and high settling velocities) of the larger particles and particle agglomerates. The flame velocities relative to unburnt mixture of 0.5–0.6 m/s were generally about twice the velocity for lean methane/air mixtures in the

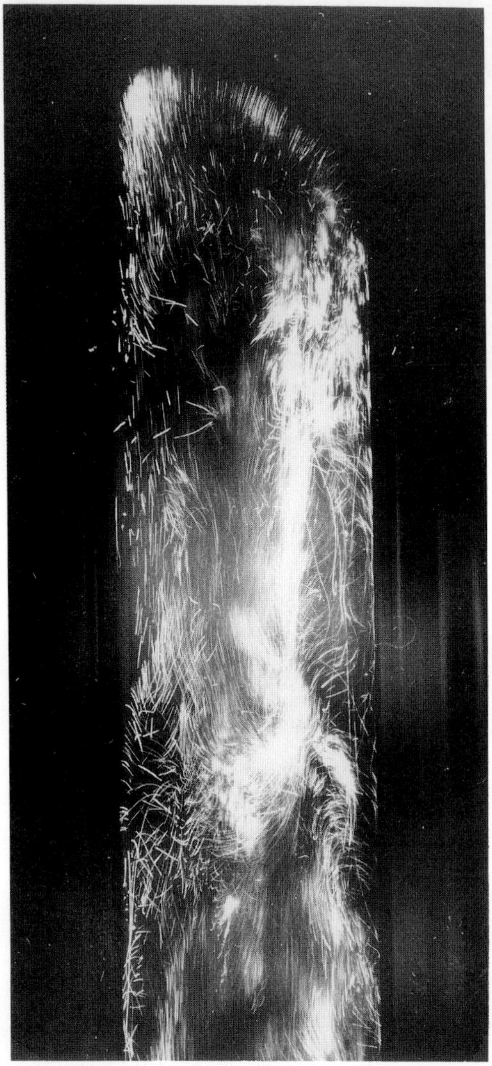

Figure 4.31 *Compensation photograph of a 80 g/m³ lignite dust/air flame in a vertical rectangular duct of width 88 mm (From Buksowicz, Klemens and Wolanski, 1982)*

same apparatus. This was attributed to the larger flame front area for the dust/air mixture, and to the intensification of the heat and mass exchange processes in the dust/air flame. Even for Reynolds' numbers of less than 2000 (calculated as proposed by Zeldovich *et al.* (1980)) eddies, generated by the non-uniform spatial heat generation rate caused by the non-uniform dust cloud, could be observed in the flame front.

Gmurczyk and Klemens (1988) conducted an experimental and theoretical study of the influence of the non-uniformity of the particle size distribution on the aerodynamics of the combustion of clouds of coal dust in air. It was suggested that the non-homogeneous particle size, amplified by imperfect dust dispersion, produces a non-homogeneous heat release process, and leads to the formation of vortices.

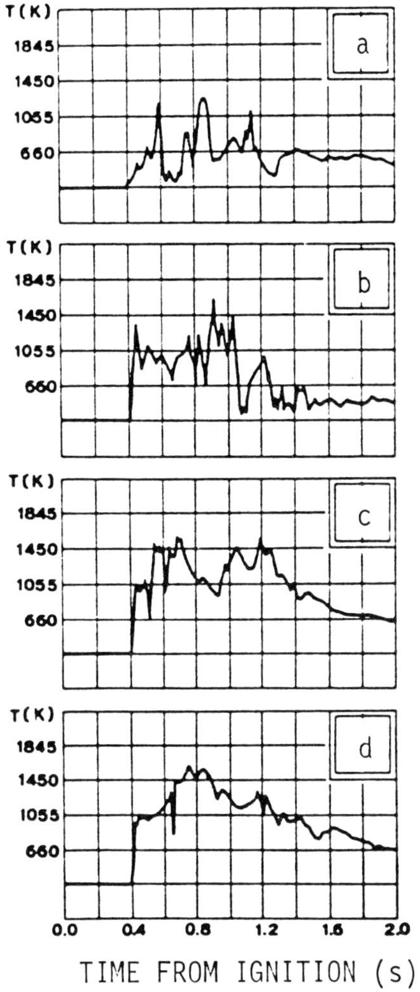

Figure 4.32 *Temperature variation with time at four fixed locations in a 103 g/m³ lignite/air dust flame propagating in a vertical duct of 88 mm × 35 mm rectangular cross section. Temperature probe locations: a = 2 mm from duct wall; b = 6 mm from duct wall; c = 26 mm from duct wall; d = 44 mm from duct wall (= duct centre) (From Klemens and Wolanski, 1986)*

Deng Xufan *et al.* (1987) and Kong Dehong (1986) studied upwards flame propagation in airborne clouds of Ca-Si dust and coal dust, in a vertical cylindrical tube of i.d. 150 mm and length 2 m. The tube was open at the bottom end and closed at the top. The Ca-Si dust contained 58% Si, 28% Ca, and 14% Fe, Al, C etc. and had a mean particle diameter of about 10 μm. The Chinese coal dust from Funsun contained 39% volatiles and 14% ash and had a median particle diameter by mass of 13 μm. The dust clouds were generated by vibrating a 300 μm aperture sieve, mounted at the top of the combustion tube and charged with the required amount of dust, in such a way that a stationary falling dust cloud of constant concentration existed in the tube for the required period of time. The dust concentration was measured by trapping a given volume of the dust cloud in the tube between two parallel horizontal plates that were inserted simultaneously, and weighing the trapped dust. Ignition was accomplished by means of a glowing resistance wire coil at the tube bottom, after 10–20 s of vibration of the sieve. Upwards flame velocities and flame thicknesses were determined by means of two photodetectors positioned along the tube. For the Ca-Si dust, flame velocities were in the range 1.3–1.8 m/s, and the total thickness of the luminous flame extended over almost the total 2 m length of the tube. The net thickness of the reaction zone was not determined. Figure 4.33 shows a photograph of a Ca-Si dust flame propagating upwards in the 150 mm diameter vertical tube.

Figure 4.34 gives the average upwards flame velocities in clouds of various concentrations of the Chinese coal dust in air.

On average these flame velocities for coal/air are about half those found for the Ca-Si under similar conditions. The data in Figure 4.34 indicate a maximum flame velocity at about 500 g/m^3. If conversion of these flame velocities to burning velocities is made by

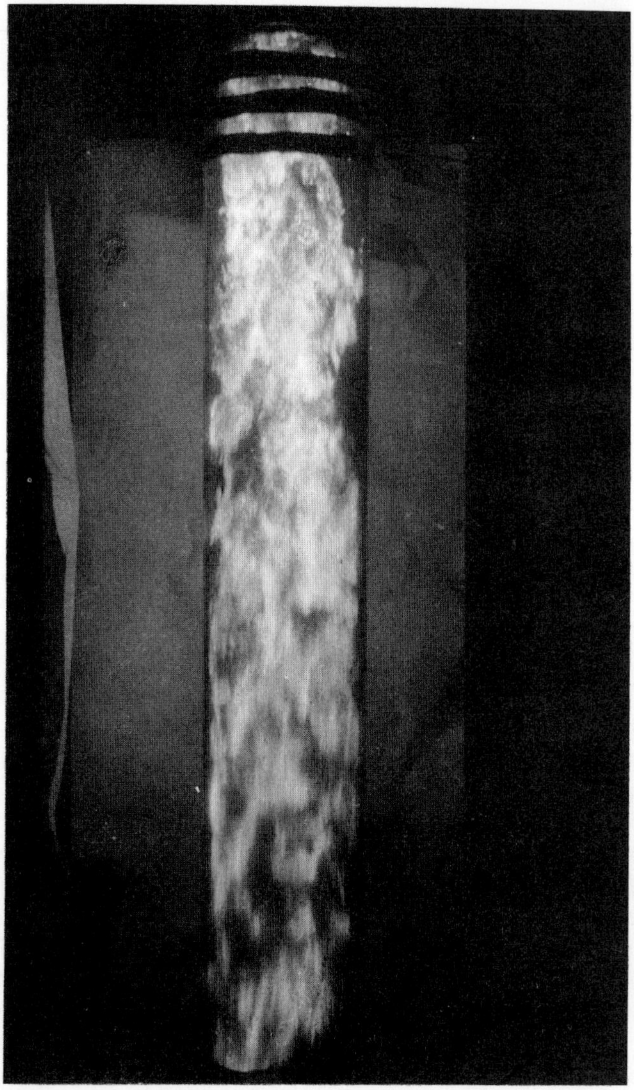

Figure 4.33 *Photograph of upwards flame propagation in a Ca-Si dust cloud in the 150 mm i.d. vertical combustion tube (From Deng Xufan* et al., *1987)*

assuming some smooth convex flame front shape, the resulting estimates are considerably higher than the expected laminar values. This agrees with the conclusion of Klemens and Wolanski (1986) that this kind of dust flames in vertical tubes will easily become non-laminar due to non-homogeneous dust distribution over the tube volume.

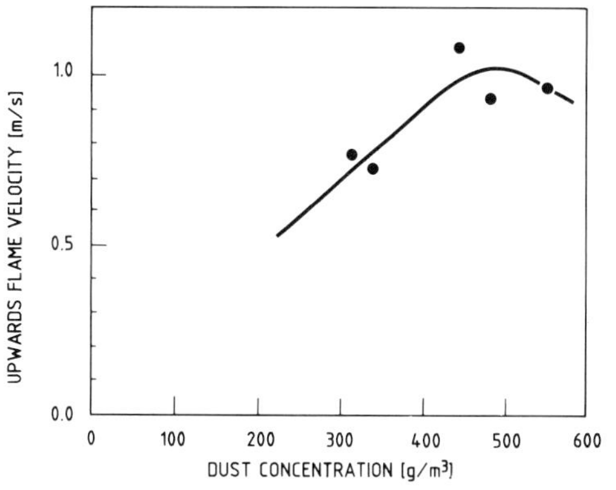

Figure 4.34 *Upwards flame velocity versus concentration of dry coal dust in air in vertical tube of i.d. 150 mm, open at bottom and closed at top. Coal dust from Funsun in P. R. China, 39% volatiles and 14% ash. Median particle diameter by mass 13 μm, and particle density 2.0–2.5 g/cm^3 (Data from Kong Dehong, 1986)*

In the initial phase of the experiments of Proust and Veyssiere (1988) in the vertical tube of 0.2 m × 0.2 m square cross section, non-laminar cellular flames as shown in Figure 4.35 were observed. In these experiments the height of the explosion tube was limited to 2 m. Over the propagation distance explored, the mean flame front velocity was about 0.5 m/s, as for the proper laminar flame, but careful analysis revealed a pulsating flame movement of about 60 Hz. A corresponding 60 Hz pressure oscillation, equal to the fundamental standing wave frequency for the one-end-open 2 m long duct, was also recorded inside the tube. Further, a characteristic sound could be heard during the propagation of the cellular flames. Proust and Veyssiere, referring to Markstein's (1964) discussion of cellular gas flames, suggested that the observed cellular flame structure is closely linked with the 60 Hz acoustic oscillation. However, there seems to be no straightforward relationship between the cell size and the frequency of oscillation.

It is of interest to relate Proust and Veyssiere's discussion of the role of acoustic waves to the maize starch explosion experiments of Eckhoff *et al.* (1987) in a 22 m long vertical cylindrical steel silo of diameter 3.7 m, vented at the top. Figure 4.36 shows a set of pressure-versus-time traces resulting from igniting the starch/air cloud in the silo at 13.5 m above the silo bottom, i.e. somewhat higher up than half-way.

This kind of exaggerated oscillatory pressure development occurred only when the ignition point was in this region. The characteristic frequency of 4–7 Hz agrees with the theoretical first harmonic standing wave frequency in a 22 m long one-end-open pipe (22 m = $\frac{1}{4}$ wave length). The increase of frequency with time reflects the increase of the average gas temperature as combustion proceeds. It is interesting to note that the peak amplitude occurs at about 2 s after ignition. The pulsating flow probably gradually distorts the flame front and increases the combustion rate. The oscillatory nature of this type of explosion could be clearly seen on video recordings. 'Packets' of flames were ejected at a frequency matching exactly that of the pressure trace. Similar oscillations were also generated in experiments in the 236 m^3 silo when the vent was moved from the silo roof to the cylindrical silo wall, just below the roof (Eckhoff *et al.*, 1988).

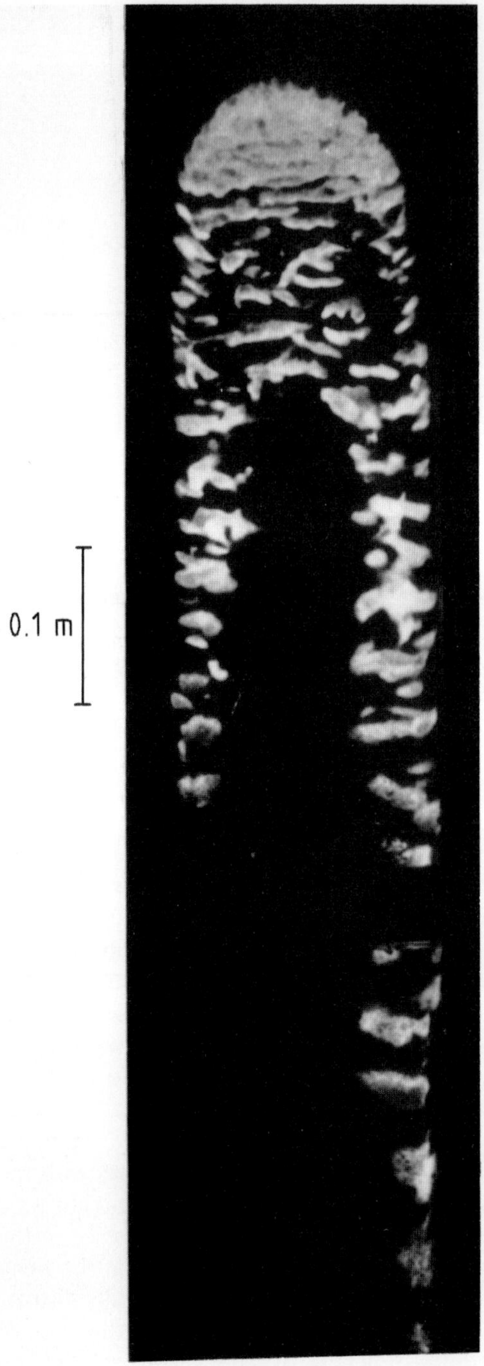

Figure 4.35 *Photograph of a typical cellular flame in 150 g/m³ maize starch in air, at 1.5 m above the ignition point. Upwards propagating flame in a vertical duct of 0.2 m × 0.2 m cross section (From Proust and Veyssiere, 1988)*

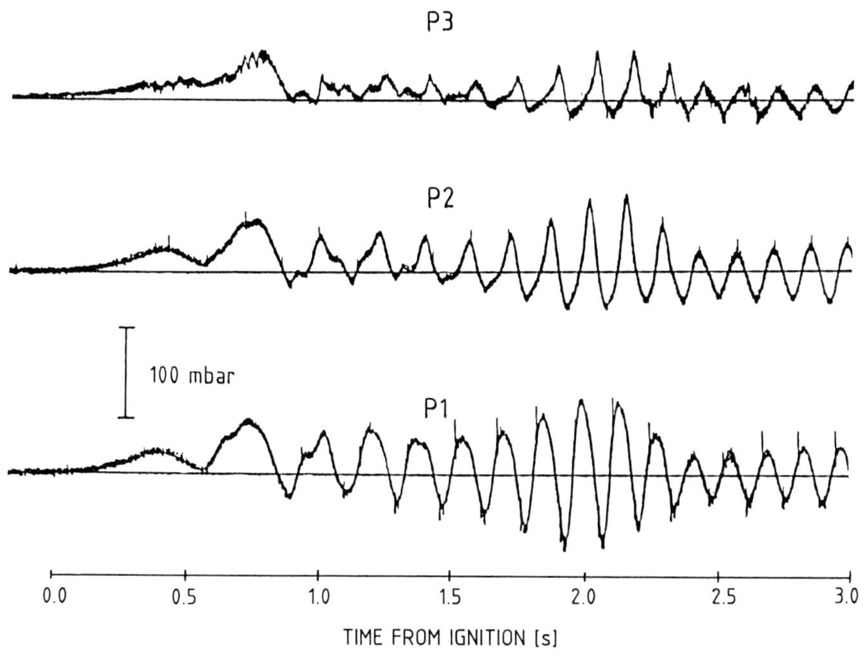

Figure 4.36 *Maize starch/air explosion in a vertical cylindrical silo of height 22 m and diameter 3.7 m and with an open 5.7 m^2 vent in the roof. Oscillatory pressure development resulting from ignition in upper half of silo (13.5 m above bottom).Oscillations persisted for about 5 s. Dust concentration 400–600 g/m^3. P_1, P_2 and P_3 were located at 3, 9 and 19.5 m above silo bottom respectively (From Eckhoff* et al., *1987)*

Artingstall and Corlett (1965) analysed the interaction between a flame propagating outwards in a one-end-open duct, and reflected shock waves, making the simplifying assumptions that:

- The initial shock wave and the flame are immediately formed when the ignition takes place and immediately have constant velocities.
- The burning velocity, i.e. the speed of flame relative to the unburnt reactants, is constant.
- Friction can be neglected.
- The effect of having to disperse the dust can be neglected.

They realized that the three first assumptions are not in accordance with realities in long ducts, where extensive flame acceleration is observed, but they indicated that their theoretical analysis can be extended to accelerating flames by using numerical computer models. It is nevertheless interesting to note that the simplified calculations predict the kind of oscillation shown in Figure 4.36. The calculations in fact showed that before the flame reached the open end, the air velocity at the open end could become negative, i.e. the air would flow inwards. Further reflections would cause the flow to reverse again. Artingstall and Corlett suggested that this theoretical result could help to explain the pulsating flow observed in some actual dust explosions in experimental coal mine galleries.

It is of interest to mention in this context that Samsonov (1984) studied the development of a propagating gas flame in an impulsive acceleration field generated by a free falling explosion chamber being suddenly stopped by a rubber shock absorber. He observed flame folding phenomena typical of those resulting from Taylor instabilities. These phenomena were also similar to those resulting from passage of a weak shock wave through a flame.

Essenhigh and Woodhead (1958) used an apparatus similar to that used by Schläpfer (1951), but of a large scale, for investigating flame propagation in clouds of cork dust in air in a one-end-open vertical duct. The duct was 5 m long and of diameter either 760 or 510 mm. They studied both upwards and downwards propagating flames, and ignition at the closed as well as the open end. With ignition at the open end and upwards flame propagation, constant flame velocities of 0.4–1.0 m/s were measured. For upwards propagation and the top end open, the maximum flame speeds were about 20 m/s. Some of this difference was due to the expansion ratio burnt/unburnt, but some was also attributed to increased burning rate.

Photographs of the flames were similar to Figures 4.31 and 4.33. Total flame thicknesses were in the range 0.2–1.2 m. The minimum explosible concentration of cork dust in air was found to be 50 ± 10 g/m^3 independent of median particle size by mass in the range 150–250 μm.

Phenomena of the kind discussed in the present section are important for the explanation of moderate deviations from ideal laminar conditions. However, the substantial deviations giving rise to the very violent explosions that can occur in industry and coal mines, are due to another mechanism, namely combustion enhancement due to flow-generated turbulence.

4.4 TURBULENT FLAME PROPAGATION

4.4.1 TURBULENCE AND TURBULENCE MODELS

Before discussing combustion of turbulent dust clouds, it is appropriate to include a few introductory paragraphs to briefly define and explain the concept of turbulence. A classical source of information is the analysis by Hinze (1975). His basic theoretical definition of turbulent fluid flow is 'an irregular condition of flow in which the various quantities show a random variation with time and space coordinates, so that statistically distinct average values can be discerned'. Turbulence can be generated by friction forces at fixed walls (flow through conduits, flow past bodies) or by the flow of layers of fluids with different velocities past or over one another. There is a distinct difference between the kinds of turbulence generated in the two ways. Therefore it is convenient to classify turbulence generated and continuously affected by fixed walls as 'wall turbulence' and turbulence in the absence of walls as 'free turbulence'.

In the case of real viscous fluids, viscosity effects will result in the kinetic energy of flow being converted into heat. If there is no continual external source of energy for maintaining the turbulent motion, the motion will decay. Other effects of viscosity are to make the turbulence more homogeneous and to make it less dependent on direction. The turbulence is called isotropic if its statistical features have no preference for any direction, so that perfect disorder exists. In this case, which is seldom encountered in practice, no average shear stress can occur and, consequently, no gradient of the mean velocity. The mean velocity, if any, is constant throughout the field.

In all other cases, where the mean velocity shows a gradient, the turbulence will be non-isotropic (or anisotropic). Since this gradient in mean velocity is associated with the occurrence of an average shear stress, the expression 'shear-flow turbulence' is often used to designate this class of flow. Most real turbulent flows, such as wall turbulence and anisotropic free turbulence fall into this class.

If one compares different turbulent flows, each having its distinct 'pattern', one may observe differences, for instance, in the size of the 'patterns'. Therefore, in order to describe a turbulent motion quantitatively, it is necessary to introduce the concept of scale of turbulence. There is a certain scale in time and a certain scale in space. The magnitude of these scales will be determined by the geometry of the environment in which the flow occurs and the flow velocities. For example, for turbulent flow in a pipe one may expect a time scale of the order of the ratio between pipe diameter and average flow velocity, i.e. the average time required for a flow to move a length of one pipe diameter, and a space scale of the order of magnitude of the diameter of the pipe.

However, it is insufficient to characterize a turbulent motion by its scales alone, because neither the scales nor the average velocity tell anything about the violence of the motion. The motion violence is related to the fluctuation of the momentary velocity, not to its average value. If the momentary velocity is:

$$V = \overline{V} + v \tag{4.80}$$

where $\overline{V}$ is the average velocity and v the momentary deviation, $\overline{v}$ is zero per definition. However, $\overline{v^2}$ will be positive and it is customary to define the violence of the turbulent motion, often called the intensity of the turbulence by

$$v' = (\overline{v^2})^{1/2} \tag{4.81}$$

The relative turbulence intensity is then defined by the ratio $v'/\overline{V}$.

As discussed by Beer, Chomiak and Smoot (1984) in the context of pulverized coal combustion, it is customary to distinguish between three main domains of turbulence, namely large-scale, intermediate-scale and small-scale. The large-scale turbulence is closely linked to the geometry of the structure in which the flow exists. Large-scale turbulence is characterized by strong coherence and high degree of organization of the turbulence structures, reflecting the geometry of the structure. For plane flow the coherent large-scale structures are essentially two-dimensional vortices with their axes parallel with the boundary walls. For flow in axi-symmetric systems, concentric large-scale vortex rings are formed. The theoretical description of the three-dimensional large-scale vortex structures encountered in practice presents a real challenge. Also experimental investigation of such structures is very difficult. According to Beer, Chomiak and Smoot, the lack of research in this area is the most serious obstacle to further advances in turbulent combustion theory.

On all scale levels turbulence has to be considered as a collection of long-lasting vortex structures, tangled and folded in the fluid. This picture is quite different from the idealized hypothetical stochastic fluctuation model of isotropic turbulence. Beer, Chomiak and

Smoot argue against the common idea that the small-scale structures are randomly distributed 'little whirls'. According to these authors it is known that the fine-scale structures of high Reynolds number turbulence become less and less space filling as the scale size decreases and the Reynolds number increases.

According to Hinze (1975) Kolmogoroff postulated that if the Reynolds number is infinitely large, the energy spectrum of the small-scale turbulence is independent of the viscosity, and only dependent on the rate of dissipation of kinetic energy into heat, per mass unit of fluid, ϵ. For this range Kolmogoroff arrived at his well-known energy spectrum law for high Reynolds numbers:

$$E(\alpha,t) = A \times \epsilon^{2/3} \times \alpha^{-5/3} \qquad (4.82)$$

$E(\alpha,t)$ is called the 'three-dimensional energy spectrum function of turbulence'. α is the wave number $2\pi n/\overline{V}$, where n is the frequency of the turbulent fluctuation of the velocity, and $\overline{V}$ is the mean global flow velocity. A is a constant, and ϵ is the rate of dissipation of turbulent kinetic energy into heat per unit mass of fluid.

Figure 4.37 illustrates the entire three-dimensional energy spectrum of turbulence, from the largest, primary eddies via those containing most of the kinetic energy, to the low-energy range of very high wavenumbers (or very high frequencies). Figure 4.37 includes the Kolmogoroff law for the universal equilibrium range.

In the range of low Reynolds numbers other theoretical descriptions than Kolmogoroff's law are required. In principle the kinetic energy of turbulence is identical to the integral of the energy spectrum curve $E(\alpha, t)$ in Figure 4.37 over all wave numbers.

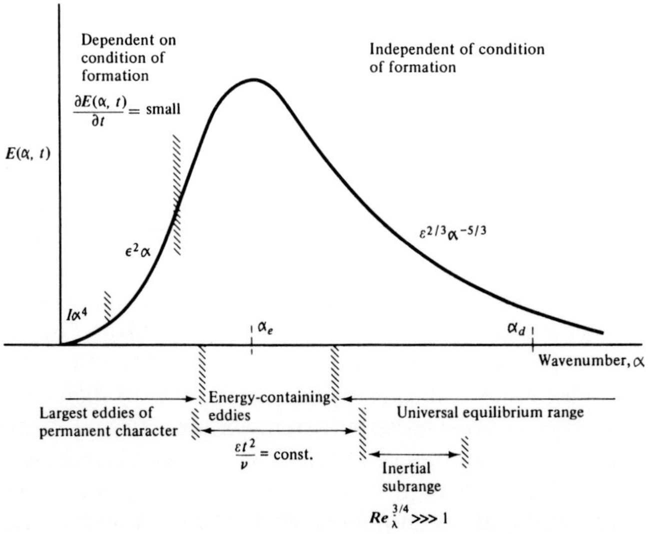

4.37 *Illustration of the three-dimensional energy spectrum* E(α, t) *in the various wave number ranges.* I *is Loitsianskii's integral,* $\in$ *is eddy viscosity,* ϵ *is dissipation of turbulent energy into heat per unit time and mass, and* υ *is kinematic viscosity.* Re_λ *is defined as* $v'\lambda_g/\upsilon$, *where* v' *is the turbulence intensity as defined by Equation (4.81), and* λ_g *is the lateral spatial dissipation scale of turbulence (Taylor micro-scale) (From Hinze, 1975)*

A formally exact equation for ϵ may be derived from the Navier-Stokes equations. However, the unknown statistical turbulence correlations must be approximated by known or calculable quantities. Comprehensive calculation requires extensive computational capacity, and it is not yet a realistic approach for solving practical problems. Therefore simpler and more approximate approaches are needed. One widely used approximate theory, assuming isotropic turbulence, is the $k - \epsilon$ model by Jones and Launder (1972, 1973), where k is the kinetic energy of turbulence, and ϵ the rate of dissipation of the kinetic energy of turbulence into heat. The $k - \epsilon$ model contains Equation (4.82) as an implicit assumption. The approximate equations for k and ϵ proposed by Jones and Launder were:

$$\rho u \frac{\partial k}{\partial x} + \rho v \frac{\partial k}{\partial y} = \frac{\partial}{\partial y}\left[\left(\mu + \frac{\mu_T}{\sigma_k}\right)\frac{\partial k}{\partial y}\right] + \mu_T \left(\frac{\partial u}{\partial y}\right)^2 - \rho\epsilon - 2\mu\left(\frac{\partial k^{1/2}}{\partial y}\right)^2$$

$$\rho u \frac{\partial \epsilon}{\partial x} + \rho v \frac{\partial \epsilon}{\partial y} = \frac{\partial}{\partial y}\left[\left(\mu + \frac{\mu_T}{\sigma_\epsilon}\right)\frac{\partial \epsilon}{\partial y}\right] + c_1 \frac{\epsilon}{k} \mu_T \left(\frac{\partial u}{\partial y}\right)^2 - \frac{c_2 \rho \epsilon^2}{k} + 2\frac{\mu \mu_T}{\rho}\left(\frac{\partial^2 u}{\partial y^2}\right)^2 \quad (4.83)$$

Here ρ is the fluid density, and u and v the mean fluid velocities in streamwise and cross-stream directions respectively. μ is molecular viscosity and μ_T turbulent viscosity. σ_k and σ_ϵ are turbulent Prandtl numbers for k and ϵ respectively and c_1 and c_2 are empirical constants or functions of Reynolds number. Both equations are based on the assumption that the diffusional transport rate is proportional to the product of the turbulent viscosity and the gradients of the diffusing quantity. Jones and Launder (1973) emphasized that the last terms of the two equations were included on an empirical basis to bring theoretical predictions in reasonable accordance with experiments in the range of lower Reynolds numbers, where Equation (4.82) is not valid. They foresaw future replacements of these terms by better approximations. The $k - \epsilon$ model has been used for simulating turbulent combustion of gases and turbulent gas explosions. More recently, as will be discussed in Section 4.4.8, it has also been adopted for simulating turbulent dust explosions.

Whilst the $k - \epsilon$ theory has gained wide popularity, it should be pointed out that it is only one of several theoretical approaches. Launder and Spalding (1972) gave a classical review of mathematical modelling of turbulence, including stress transport models, which is still relevant.

When the structure of turbulent dust clouds is to be described, further problems have to be addressed. Some of these have been discussed in Chapter 3. Beer, Chomiak and Smoot (1984) pointed out that there are two aspects of the turbulence/particle interaction problem. The first is the influence of turbulence on the particles, the second the influence of particles on the turbulence. With regard to the influence of turbulence on the particles in a burning dust cloud, two effects are important, namely mechanical interactions associated with particle diffusion, deposition, coagulation and acceleration, and convective interactions associated with heat and mass transfer between gas and particles, which influence the particle combustion rate. Beer, Chomiak and Smoot (1984) discussed available theory for the various regimes of Reynolds number (see Chapter 3) for the particle motion in the fluid. They emphasized that turbulence is a rotational phenomenon, and therefore the motion of the particles will also include a rotational component. Consequently one can define a relaxation time for the particle rotation τ_{pr} as well as one for the translatory particle motion, τ_p. Both relaxation times are proportional to the square of the particle diameter and hence decrease markedly as the particles get smaller.

When $\tau_p \gg \tau_L$, where τ_L is the characteristic Lagrangian time of the turbulent motion, the particle is not convected by the turbulent fluctuations and its motion is fully determined by the mean flow. However, when $\tau_p \ll \tau_L$, the particle adjusts to the instantaneous gas velocity. If the particle follows the turbulent fluctuations, its turbulent diffusivity is equal to the gas diffusivity. If the particle does not follow the turbulence, its diffusivity is practically equal to zero. An interesting but most complicated case occurs when the characteristic relaxation times and turbulence times are of the same order. In this case, the particle only partially follows the fluid and its motion depends partially on Lagrangian interaction with the fluid and partially on Eulerian interaction over the distance which it travels outside the originally surrounding fluid.

The effects of particles on the turbulence structure are complex. The simplest effect is the introduction of additional viscous-like dissipation of turbulent energy caused by the slip between the two phases. This effect is substantial in the range of explosible dust concentrations. Even small changes in dissipation can have a strong influence on the turbulence level. This is because turbulence energy is the result of competition between two large and almost equal sources of production and dissipation.

Beer, Chomiak and Smoot (1984) state that the change in turbulence intensity and structure caused by the increased dissipation will affect the mean flow parameters and in turn the turbulence production terms, so that the outcome of the chain of changes is difficult to predict even when the most advanced techniques are used. The difficulties are enhanced by a lack of reliable experimental data. For example, some experiments demonstrate dramatic effects of even minute admixtures of particles on turbulent jet behaviour. Others demonstrate smaller effects even for high dust concentrations. (See Section 3.8 in Chapter 3.)

4.4.2
TURBULENT DUST FLAMES. AN INTRODUCTORY OVERVIEW

The literature on turbulent dust flames and explosions is substantial. This is because it has long been realized that turbulence plays a primary role in deciding the rate with which a given dust cloud will burn, and because this role is not easy to evaluate either experimentally or theoretically. There are close similarities with turbulent combustion of premixed gases, as shown by Bradley *et al.* (1988), although the two-phase nature of dust clouds adds to the complexity of the problem. Hayes *et al.* (1983) mentioned two predominant groups of theories of turbulent burning of a premixed fluid system of a fuel and an oxidizer:

1. The laminar flame continues to be the basic element of flame propagation. The essential role of turbulence is to increase the area of the flame surface that burns simultaneously.
2. Turbulence alters the nature of the basic element of flame propagation by increasing rates of heat and mass transport down to the scale of the 'elementary flame front', which is no longer identical with the laminar flame.

In their comprehensive survey Andrews, Bradley and Lwakamba (1975) emphasized the importance of the turbulent Reynolds number $R_\lambda = v'\lambda/v$ for the turbulent flame propagation, where v' is the turbulence intensity defined by equation (4.81), λ the Taylor microscale and ν the kinematic viscosity. They suggested that for $R_\lambda > 100$, a wrinkled laminar flame structure is unlikely and that turbulent flame propagation is then associated with small dissipative eddies. A supplementary formulation is that laminar flamelets can only exist in a turbulent flow if the laminar flame thickness is smaller than the Kolmogoroff microscale of the turbulence. Bray (1980) gave a comprehensive discussion of the two physical conceptions and pointed out that the Kolmogoroff micro-scales and laminar flame thicknesses are difficult to resolve experimentally in a turbulent flame. Because of the experimental difficulties, the real nature of the fine structure of premixed flames in intense turbulence is still unknown.

Abdel-Gayed *et al.* (1989) proposed a modified Borghi diagram for classifying various combustion regimes in turbulent premixed flames, using the original Borghi parameters L/δ_l and u'/u_l as abscissa and ordinate. Here L is the integral length scale, δ_l the thickness of the laminar flame, u' the rms turbulent velocity and u_l the laminar burning velocity. The diagram identifies regimes of flame propagation and quenching, and the corresponding values of the Karlovitz stretch factor, the turbulent Reynolds number, and the ratio of turbulent to laminar burning velocity.

Spalding (1982) discussed an overall model that contains elements of both of the physical conceptions 1 and 2 of a turbulent flame defined above. An illustration is given in Figure 4.38. Eddies of hot, burnt fluid and cold unburnt fluid interact with the consequences that both fluids become mutually entrained.

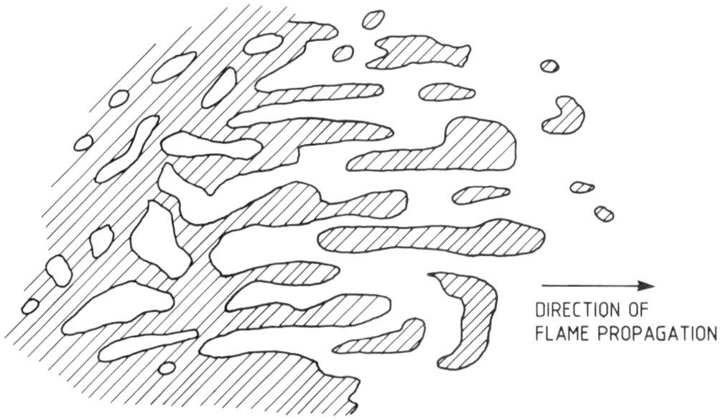

Figure 4.38 *Postulated micro-structure of burning turbulent fluid. Shaded areas represent burnt fluid, unshaded unburnt (From Spalding, 1982)*

Entrainment of burnt fluid into unburnt and vice versa is the rate controlling factor as long as the chemistry is fast enough to consume the hot reactants as they appear. In other words: The instantaneous combustion rate per unit volume of mixture of burnt and unburnt increases with the total instantaneous interface area between burnt and unburnt per unit volume of the mixture. Spalding introduced the length l as a characteristic mean dimension of the entrained 'particles' of either burnt or unburnt fluid, and l^{-1} as a measure of the corresponding specific interface surface area. He then assumed a differential equation of the form:

$$\frac{d(l^{-1})}{dt} = M + B + A \tag{4.84}$$

where M represents the influence of mechanical processes such as stretching, breakage, impact and coalescence. B represents the influence of the burning, whereas A represents influences of other processes such as wrinkling, smoothing and simple interdiffusion. Spalding indicated tentative equations for M, B and A, but emphasized that the identification of expressions and associated constants that correspond to physical reality over wide ranges, 'is a task for the future'.

It is nevertheless clear that the strong enhancing effect of turbulence on the combustion rate of dust clouds and premixed gases, is primarily due to the increase of the specific interface area between burnt and unburnt fluid by turbulence, induced by mutual entrainment of the two phases. The circumstances under which the interface itself is a laminar flame or some thinner, elementary flame front, remains to be clarified.

When discussing the specific influence of turbulence on particle combustion mechanisms, Beer, Chomiak and Smoot (1984) distinguished between micro-scale effects and macro-scale effects. On the micro-scale, turbulence directly affects the heat and mass transfer and therefore the particle combustion rate. They discussed the detailed implications of this for coal particle combustion, assuming that CO is the only primary product of heterogeneous coal oxidation. On the macro-scale there is a competition between the devolatilization process and turbulent mixing. Concerning modelling of turbulent combustion of dust clouds, these authors stressed that three-dimensional microscopic models are too detailed to allow computer simulation without use of excessive computer capacity and computing time. They therefore suggested alternative methods based on theories like the $k - \epsilon$ model, adopting the Lagrangian Escimo approach proposed by Spalding and co-workers (Ma *et al.* 1983), or alternative methods developed for accounting for the primary coherent large-scale turbulence structures (Ghoniem *et al.*, 1981).

Lee (1987) suggested that the length scale that characterizes the reaction zone of a turbulent dust flame is at least an order of magnitude greater than that of a premixed gas flame. For this reason dust flame propagation should preferably be studied in large-scale apparatus. It should be emphasized, however, that from a practical standpoint, large or full scale is not an unambiguous term. For example, a dust extraction duct of diameter 150 mm is full industrial scale, and at the same time of the scale of laboratory equipment. On the other hand, the important features of an explosion in a large grain silo cell of diameter 9 m and height 70 m are unlikely to be reproduced in a laboratory silo model of 150 mm diameter.

It should be mentioned here that Abdel-Gayed *et al.* (1987) identified generally applicable correlations in terms of dimensionless groups, enabling prediction of acceleration of flames in turbulent premixed gases. A similar approach might in some cases offer a means of scaling even of dust explosions. The role of radiative heat transfer in dust flames then needs to be discussed, as done by Lee (1987). His conclusion was that conductive and convective heat transfer are probably more important than radiative transfer. This may be valid for coal and organic dusts, but probably not for metal dusts like silicon and aluminium.

Amyotte *et al.* (1989) reviewed more than a hundred publications on various effects of turbulence on ignition and propagation of dust explosions. They considered the influence of both initial and explosion induced turbulence on flame propagation in both vented and fully confined explosions. They suggested two possible approaches towards an improved understanding. First, concurrent investigations of dust and gas explosions, and secondly direct measurement of turbulent scales and intensities in real experiments as well as in industrial plants.

4.4.3
EXPERIMENTAL STUDIES OF TURBULENT DUST FLAMES IN CLOSED VESSELS

4.4.3.1
Common features of experiments

The majority of the published experimental studies of turbulent dust explosions in closed vessels have been conducted in apparatus of the type illustrated in Figure 4.39.

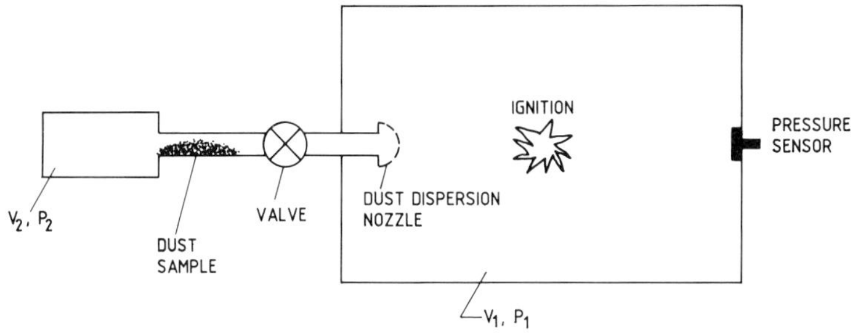

Figure 4.39 *Illustration of the type of apparatus commonly used in closed-vessel turbulent dust explosion experiments*

The closed explosion vessel of volume V_1 and initial pressure P_1 is equipped with a dust dispersion system, a pressure sensor and an ignition source. In most equipment the dust dispersion system consists of a compressed-air reservoir of volume $V_2 \ll V_1$, at an initial pressure $P_2 \gg P_1$. In some apparatuses the dust is initially placed on the high-pressure side of the dispersion air valve, as indicated in Figure 4.39, whereas in other apparatus it is placed downstream of the valve. Normally, the mass of dispersion air is not negligible compared with the initial mass of air in the main vessel. This causes a significant rise of the pressure in the main vessel once the dispersion air has been discharged into the main vessel. In some investigations this is compensated for by partial evacuation of the main vessel prior to dispersion so that the final pressure after dispersion completion, just prior to ignition, is atmospheric. This is important if absolute data are required, because the maximum explosion pressure for a given dust at a given concentration is approximately proportional to the initial absolute air pressure. Both the absolute sizes of V_1 and V_2 and the ratio between them vary substantially from apparatus to apparatus. The smallest V_1 used are of the order of 1 litre, whereas the largest that has been traced is 250 m^3. The design of the dust dispersion system varies considerably from apparatus to apparatus. A number of different nozzle types have been developed with the aim to break up agglomerates and ensure homogeneous distribution of the dust in the main vessel. The ignition source has also been a factor of considerable variation. In some of the earlier investigations, continuous sources like electric arcs or trains of electric sparks, and glowing resistance wire coils were used, but more recently it has become common to use short-duration sources initiated at a given time interval after opening of the dust dispersion valve. These sources vary from electric sparks, via exploding wires to various forms of electrically triggered chemical ignitors.

An important inherent feature of all apparatus of the type illustrated in Figure 4.39 is that the dispersion of the dust inevitably induces turbulence in the main vessel. The level of turbulence will be at maximum during the main phase of dust dispersion. After the flow of dispersion air into the main vessel has terminated, the turbulence decays at a rate that decreases with increasing V_1. (Compare time scales of Figures 4.41 and 4.42.)

In view of this it is clear that both the strength of the dispersion air blast and the delay between opening of the dust dispersion value and ignition have a strong influence on the state of turbulence in the dust cloud at the moment of ignition, and consequently also on the violence of the explosion. The situation is illustrated in Figure 4.40.

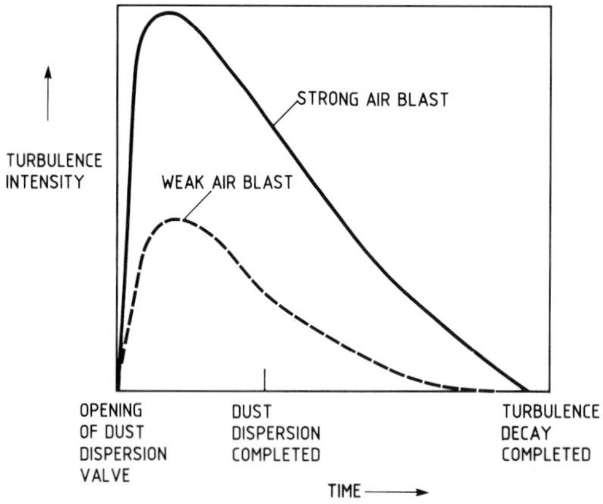

Figure 4.40 *Illustration of generation and decay of turbulence during and after dispersion of dust in an apparatus of the type illustrated in Figure 4.39. Note: A common way of quantitifying turbulence intensity is the rms turbulent velocity*

4.4.3.2
Experimental investigations

The data from Eckhoff (1977) given in Figure 4.41 illustate the influence of the ignition delay on the explosion development in a cloud of lycopodium in air in a 1.2 litre Hartmann bomb. As can be seen there is little difference between the maximum explosion pressure obtained with a delay of 40 ms and of 200 ms, whereas the maximum rate of pressure rise is drastically reduced, from 430 bar/s to 50 bar/s, i.e. by a factor of almost ten. There is little doubt that this is due to the reduced initial turbulence in the dust cloud at the large ignition delays. With increasing ignition delay beyond 200 ms, the maximum explosion pressure is also reduced as the dust starts to settle out of suspension before the ignition source is activated.

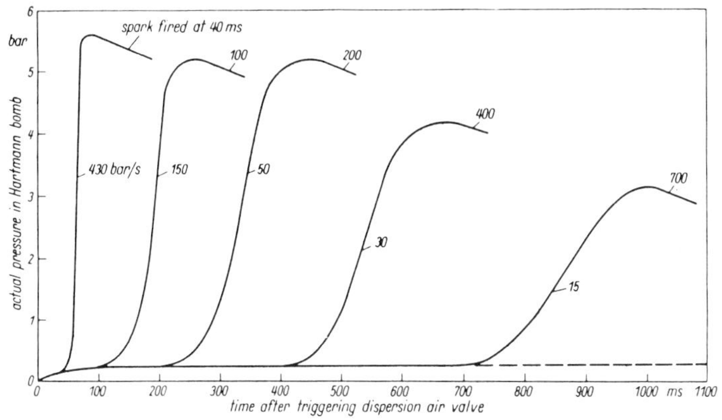

Figure 4.41 *Influence of ignition delay on development of lycopodium/air explosion in a 1.2 litre Hartmann bomb. Ignition source 4 J electric spark of discharge time 2–3 ms. Dust concentration 420 g/m^3. Initial pressure in 60 cm^3 dispersion air reservoir 8 bar(g) (From Eckhoff, 1977)*

As would be expected, the same kind of influence of ignition delay as shown in Figure 4.41 is in fact found in all experiments of the type illustrated in Figure 4.39. One of the first researchers to observe this effect was Bartknecht (1971). Some of his results for a 1 m^3 explosion vessel are given in Figure 4.42. As the ignition delay is increased from the lowest value of about 0.3 s to about 1 s, there is marked decrease of $(dP/dt)_{max}$, whereas P_{max} is comparatively independent of the ignition delay for both dusts. If the ignition delay is increased further, however, there is a marked decrease even in P_{max} for the coal. The 1 m^3 apparatus used by Bartknecht in 1971 is in fact the prototype of the standard test apparatus specified by the International Organization for Standardization (1985).

In this standard an ignition delay of 0.6 is prescribed. As Figure 4.42 shows, this is not the worst case, because a significantly higher level of initial turbulence and resulting rates of pressure rise exist at shorter ignition delays, down to 0.3 s. The delay of 0.6 s was chosen as a standard because at approximately this moment the dust dispersion was completed, i.e. pressure equilibrium between V_1 and V_2 in Figure 4.39 was established. In view of this there is no logical argument for claiming that an ignition delay of 0.6 s corresponds to 'worst case'. One can easily envisage situations in industry where dust injection into the explosion space is continued after ignition.

As shown by Eckhoff (1976), the data from experiments of Nagy *et al.* (1971) in closed-bombs of various volumes confirm the arbitrary nature of $(dP/dt)_{max}$ values from closed-bomb tests. This was re-emphasized by Moore (1979), who conducted further comparative tests in vessels of different volumes and shapes.

Dahn (1991) studied the influence of the speed of a stirring propeller on the rate of pressure rise, or the derived burning velocity, during lycopodium/air explosions in a 20 litre closed vessel. The purpose of the propeller was to induce turbulence in addition to

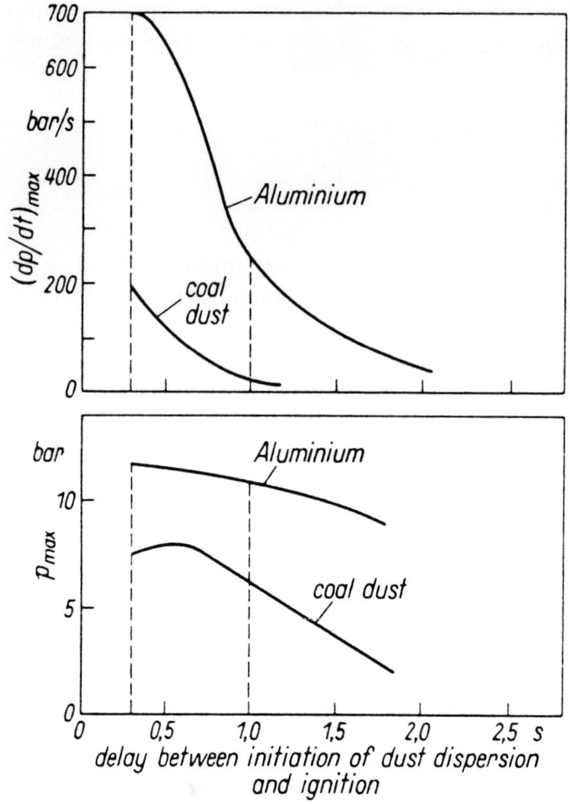

Figure 4.42 *Results from explosions of aluminium/air and coal dust/air in a closed 1 m³ vessel. Ignition source: chemical ignitor at vessel centre (Data from Bartknecht, 1971)*

that generated by the dust dispersion air blast. $(dP/dt)_{max}$ typically increased by a factor of 2–2.5 when the propeller speed increased from zero to 10 000 rpm.

The implication of the effects illustrated by Figures 4.40–4.42 for predicting explosion violence in practical situations in industry was neglected for some time. The strong influence of turbulence on the rate of combustion of a dust cloud is also indeed of significance in practical explosion situations in industry (see Chapter 6).

In the past sufficient attention was not always paid to the influence of the ignition delay on the violence of experimental closed-bomb dust explosions. Often continuous ignition sources, like flowing resistance wire coils, were used, as opposed to short-duration sources being active only for a comparatively short interval of time, allowing control of the moment of ignition. Some consequences of using a continuous ignition source were investigated by Eckhoff and Mathisen (1977/78). They disclosed that a correlation between $(dP/dt)_{max}$ and dust moisture content found by Eckhoff (1976) on the basis of Hartmann bomb tests, using a glowing resistance wire coil ignition source, was misleading. The reason is that a dust of a higher moisture content ignites with a longer delay than a comparatively dry dust. This is because the ignitability of a moist dust is lower than for a dried dust. Therefore ignition of the moist dust with a continuous source is not possible until the turbulence has decayed to a sufficiently low level, below the critical level for ignition of the dried dust. In other words: As the moisture content in the dust increases, the ignition delay also increases. Therefore the strong influence of moisture content on $(dP/dt)_{max}$ found earlier, was in fact a combined effect of increasing dust moisture and decreasing turbulence.

Eckhoff (1987) has discussed a number of the closed-bomb test apparatuses used for characterizing the explosion violence of dust clouds in terms of the maximum rate of pressure rise. It is clear that the $(dP/dt)_{max}$ from such tests are bound to be arbitrary as long as the test result is not associated with a defined state of initial turbulence of the dust cloud. In view of this the direct measurements of the rms (root mean square) turbulence as a function of time after opening the dispersion air valve in a Hartmann bomb, by Amyotte and Pegg (1989), and their comparison of the data with the data from Hartmann bomb explosion experiments by themselves and Eckhoff (1977), are of considerable interest. The results of Amyotte and Pegg's Laser-Doppler velocimeter measurements, obtained without dust in the dispersion system, are shown in Figure 4.43. It is seen that a decay by a factor of almost ten of the turbulence intensity occurs within the same time frame of about 40 to 200 ms as a corresponding decay of $(dP/dt)_{max}$ in Eckhoff's (1977) experiments (Figure 4.41). It is also seen that the turbulence intensity increases systematically with the initial pressure in the dispersing air reservoir, i.e. increasing strength of the air blast, in accordance with the general picture indicated in Figure 4.40.

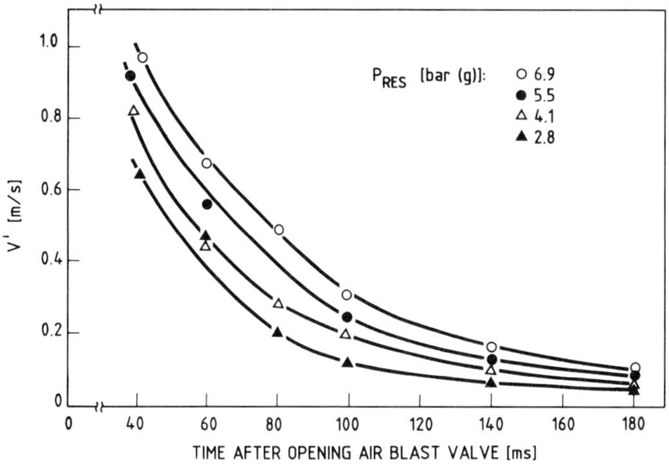

Figure 4.43 *Variation of rms turbulence velocities within 5 ms 'windows' in a Hartmann bomb with time after opening of air blast valve, and with initial pressure in dispersion reservoir. Air only, no dust (From Amyotte and Pegg, 1989)*

Kauffman *et al.* (1984) studied the development of turbulent dust explosions in the 0.95 m^3 spherical explosion bomb illustrated schematically in Figure 4.44. The bomb is equipped with six inlet ports and eight exhaust ports, both sets being manifolded and arranged symmetrically around the bomb shell. Dust and air feed rates were set to give the desired dust concentration and turbulence level. The turbulence level generated by a given air flow was measured by means of a hot-wire anemometer. The turbulence intensity v', assuming isotropic turbulence, was determined from the rms (root mean square) and mean velocities extracted from the hot-wire signal in the absence of dust. As pointed out by Semenov (1965), a hot-wire probe senses all velocities as positive, and therefore a positive mean velocity will be recorded even if the true mean velocity is zero. In agreement with the suggestion by Semenov, Kauffman *et al.* therefore assumed that $v' = (1/2)^{1/2} \times [(\text{rms velocity})^2 + (\text{mean velocity})^2]^{1/2}$. This essentially is a secondary rms of two different mean velocities, namely the primary rms and the arithmetic mean of the hot wire signal.

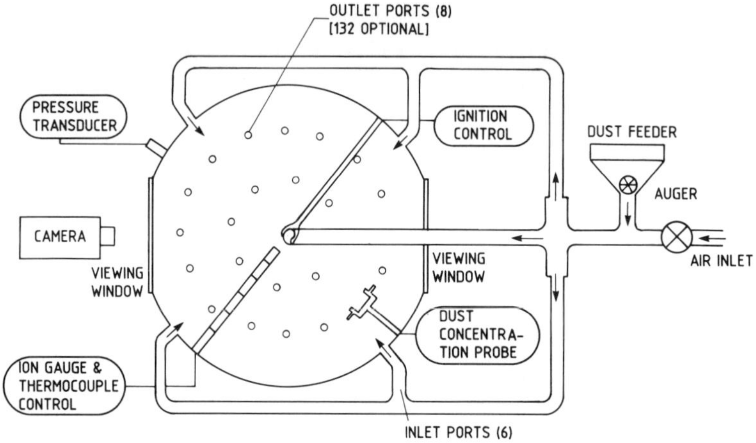

Figure 4.44 *0.95 m³ spherical closed bomb for studying combustion of turbulent dust clouds (From Kauffman* et al.*, 1984)*

Kauffman *et al.* were aware of the complicating influence of dust particles on the turbulence structure of the air, but they were not able to account for this. It was found that the turbulence intensity, in the absence of dust, was reasonably uniform throughout the 1 m^3 vessel volume.

When a steady-state dust suspension of known concentration had been generated in the 0.95 m^3 sphere, all inlet and exhaust openings were closed simultaneously and the dust cloud ignited at the centre. The rise of explosion pressure with time was recorded and $(dP/dt)_{max}$ and P_{max} determined. Figures 4.45 and 4.46 show a set of results for maize starch.

The marked increase of $(dP/dt)_{max}$ with turbulence intensity v' in Figure 4.45 was expected and in agreement with the trend in Figures 4.41–4.43. However, as shown in Figure 4.46, v' also had a distinct influence on P_{max}. At the first glance this conflicts with the findings of Eckhoff (1977) and Amyotte and Pegg (1989) in the 1.2 litre Hartmann bomb, where there was little influence of the ignition delay on P_{max} up to 200 ms delay. However, Eckhoff (1976) discussed the effect of initial dispersion air pressure on the development of explosion pressure in the Hartmann bomb. He found a comparatively steep rise of both P_{max} and $(dP/dt)_{max}$ with increasing dispersion pressure, and suggested that this was probably due to a combined effect of improved dust dispersion and increased

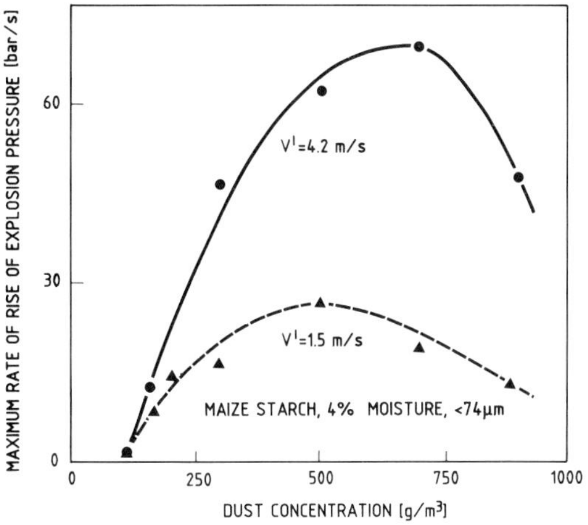

Figure 4.45 *Effect of turbulence on maximum rate of rise of explosion pressure in a 0.95 m³ spherical closed bomb (From Kauffman* et al.*, 1984)*

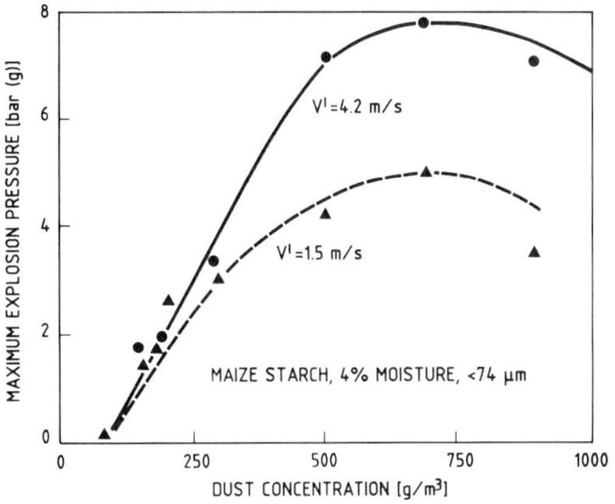

Figure 4.46 *Effect of turbulence on maximum explosion pressure in a 0.95 m³ spherical closed bomb (From Kauffman* et al.*, 1984)*

initial turbulence. A similar distinct influence on P_{max} of the intensity of the air blast used for dispersing the dust was also found by Amyotte and Pegg (1989). This could be interpreted in terms of improved degree of dust dispersion or deagglomeration, rather than degree of turbulence, being responsible for more effective combustion and thus higher P_{max}. Therefore, the primary effect on P_{max} of increasing v' in Kauffman *et al.*'s (1984) experiments could be improved degree of dust dispersion.

The rms turbulence intensities in Amyotte and Pegg's (1989) investigation were determined by means of a Laser-Doppler velocimeter, whereas Kauffman *et al.* (1984) used a hot-wire anemometer. Therefore the two sets of v' values may not be directly comparable. Amyotte and Pegg's values were generally lower than those of Kauffman *et al.*

Tezok *et al.* (1985) extended the work of Kauffman *et al.* (1984) to measurement of turbulent burning velocities in the 0.95 m^3 spherical explosion bomb. Radial turbulent burning velocities of 0.45–1.0 m/s were measured for mixed grain dust/air and 0.70–3.3 m/s for maize starch/air in the range of turbulence intensities of 1.5–4.2 m/s and dust concentrations between 50 and 1300 g/m^3. The ratio of turbulent to laminar burning velocity was found to correlate well with the ratio of the rms turbulence velocity to laminar burning velocity as well as with the Reynolds number. Some data from experiments with < 74 μm maize starch of 4% moisture content are shown in Figure 4.47. The laminar burning velocities S_L were the same as those derived by Kauffman *et al.* (1984) by extrapolating measured burning velocities in the 0.95 m^3 bomb to zero turbulence intensity. The S_L value of 0.7 m/s for 700 g/m^3 is, however, considerably higher than the highest value of 0.27 m/s arrived at for maize starch/air at constant pressure by Proust and Veyssiere (1988).

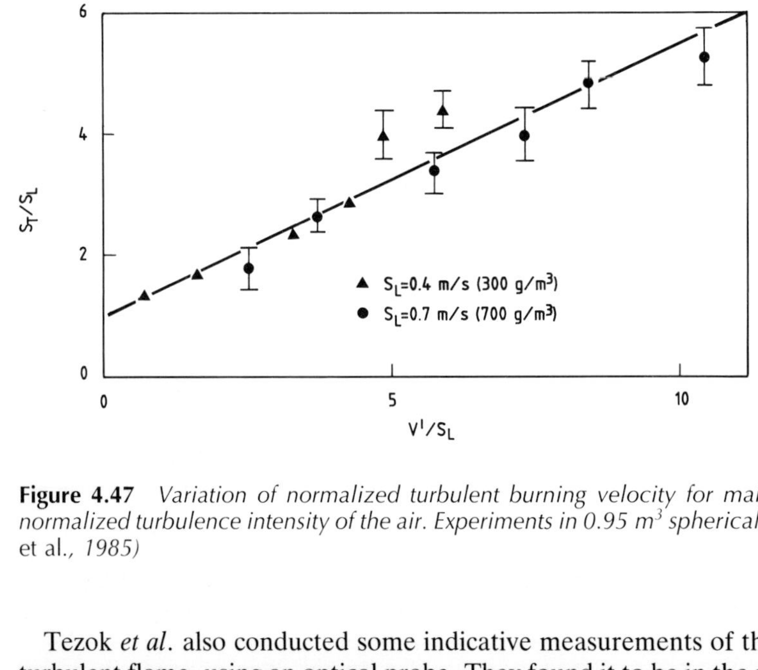

Figure 4.47 *Variation of normalized turbulent burning velocity for maize starch/air clouds, with normalized turbulence intensity of the air. Experiments in 0.95 m^3 spherical closed bomb (From Tezok* et al., *1985)*

Tezok *et al.* also conducted some indicative measurements of the total thickness of the turbulent flame, using an optical probe. They found it to be in the range of 0.15 to 0.70 m, and increasing with increasing turbulence intensity and dust concentration. This would mean that the total flame thickness was of the same order as the dimensions of the experimental vessel.

It should be mentioned that Lee *et al.* (1987) studied some further aspects of the influence of turbulence on $(dP/dt)_{max}$ and P_{max} in closed-bomb dust explosions.

In an investigation following up the work of Tezok *et al.* (1985), Tai *et al.* (1988) used laser Doppler anemometry for studying turbulent dust explosions in the 0.95 m^3 explosion vessel. It was found that the dust had little effect on the turbulence intensity, as compared to that in pure gas under the same conditions of turbulence generation. Turbulent burning velocities were determined for a range of dusts at turbulence intensities up to 3.3 m/s. Laminar burning velocities were estimated by extrapolating to zero turbulence intensity. The effect of turbulence and dust concentration on flame thickness was also studied.

Bradley *et al.* (1988) measured turbulent burning velocities in clouds of well-dispersed maize starch in air, in a fan-stirred 22 litre explosion bomb. Turbulence was varied by varying the fan speed. Isotropic turbulence in the central measurement region of the bomb was created by using four fans. Turbulent velocities and integral length scales corresponding to different conditions of stirring were measured in stirred air, in the absence of dust, by laser Doppler velocimetry. It was found that the correlation of the ratio of turbulent to laminar burning velocities with the ratio of effective rms turbulent velocity to laminar burning velocity and the Karlovitz flame stretch factor was similar to that obtained in stirred premixed gas explosions (methane/air).

Further comparative investigations of turbulent dust and gas explosions are discussed in section 4.4.5.

4.4.3.3
K_{St} and the 'cube-root-law'

The K_{St} concept was introduced by Bartknecht (1971, 1978). He claimed (1978) that the so-called 'cube-root-law':

$$(dP/dt)_{max} \times V^{1/3} = \text{constant} \equiv K_{St} \qquad (4.84)$$

had been confirmed in experiments with numerous dusts in vessel volumes from 0.04 m^3 and upwards. The K_{St} value [bar m/s], being numerically identical with the $(dP/dt)_{max}$ [bar/s] in the 1 m^3 standard ISO test (International Standardization Organization (1985)), was denoted 'a specific dust constant', which has led to some confusion. From what has been said in Sections 4.2.5.1, 4.4.3.1 and 4.4.3.2, the 'cube-root-law' is only valid in geometrically similar vessels, if the flame thickness is negligible compared to the vessel radius, and if the burning velocity as a function of pressure and temperature is identical in all volumes. Furthermore, the flame surface must be geometricallay similar (for example spherical). In view of the relationships in Figures 4.40 to 4.43, it is clear that K_{St} is bound to be an arbitrary measure of dust explosion violence, because the state of turbulence to which it refers, is arbitrary. As pointed out by Eckhoff (1984/85), this fact has sometimes been neglected when discussing K_{St} in relation to industrial practice, and may therefore need to be brought into focus again. Table 4.13 shows an arbitrary selection of K_{St} values for maize starch dust clouds in air, determined in various apparatus. The values range from 5–10 bar m/s to over 200 bar m/s, corresponding to a factor of more than 20. Some of the discrepancies can probably be attributed to differences in moisture content and effective particle size of the starch, and to different data interpretation (peak or mean values). However, differences in the turbulence of the dust clouds probably play the main role.

When using K_{St} values for sizing of vent areas and other purposes according to various codes, it is absolutely essential to use only data obtained from the standard test method specified for determining K_{St}. Normally this will be the method of the International Standardization Organization (1985), or a smaller-scale method that has been calibrated against the ISO-method. In addition it is necessary to appreciate the relative and arbitrary nature even of these K_{St} values (see Chapter 7).

It should be mentioned that Bradley *et al.* (1988) were able to express K_{St} in terms of a 'mass burning rate' and the initial and final pressure. The K_{St} concept was then defined by Equation (4.84).

Table 4.13 K_{St} values measured for clouds of maize starch dust in air in different closed vessels and arranged according to vessel volume. $K_{St} = (dP/dt)_{max}\ V^{1/3}$ (Extended and modified version of table from Yi Kang Pu, 1988)

Investigator	$(dP/dt)_{max}$ [bar/s]	Volume V of apparatus [m^3]	K_{St} [bar·m/s]
Bartknecht (1978)	680	0.0012	73
Nagy and Verakis (1983)	612	0.0012	66
Eckhoff et al. (1987)*	220	0.0012	23
Nagy and Verakis (1983)	413	0.009	86
Aldis et al. (1983)	320	0.020	87
Eckhoff et al. (1987)*	365	0.020	100
Yi Kang Pu (1988)	10-20	0.026	3-6
Yi Kang Pu (1988)	60-80	0.026	20-25
Nagy and Verakis (1983)	272	0.028	83
Bond et al. (1986)	50	0.33	34
Kauffman et al. (1984)	72	0.95	71
Kauffman et al. (1984)	20	0.95	20
Nagy and Verakis (1983)	136	3.12	200
Nagy and Verakis (1983)	110	6.7	209
Nagy and Verakis (1983)	55	13.4	131

*Arithmetic mean values, 11% moisture in starch.

4.4.4 TURBULENT FLAME PROPAGATION IN PARTLY OR FULLY UNCONFINED GEOMETRIES

The important work of Tamanini (1989), and Tamanini and Chaffee (1989) will be discussed more extensively in Chapter 6 on Venting of Dust explosions. In the present context it should only be briefly mentioned that explosion experiments were conducted in a 64 m^3 vented vessel at a series of different, known turbulence intensities at the moment of ignition. The turbulence intensities were measured by means of a bi-directional impact probe. For a given dust, dust concentration and vent characteristics, the maximum pressure in the vented explosion increased systematically with increasing initial turbulence intensity in the experimental range 2–12 m/s.

Hayes *et al.* (1983) investigated the influence of the speed of four shrouded axial fans mounted above the channel floor, on the dust flame speed in a horizontal channel of 1.5 m length and 0.15 × 0.15 m square cross section, open at both ends. A cloud of dried wheat flour of mean particle size 100 μm was produced in the channel and ignited by a propane/air flame while the fans were running. Some results are shown in Figure 4.48.

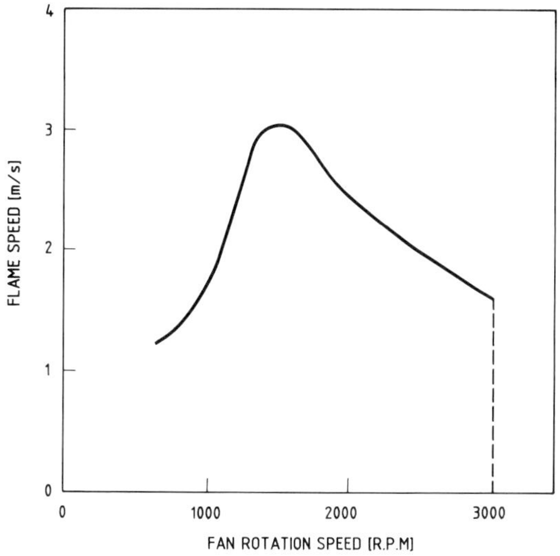

Figure 4.48 *Variation of dust flame speed in a horizontal channel with open ends, with rotational speed of four fans located in the channel, and 300 g/m*^{3} *of dried wheat flour in air (From Hayes* et al., *1983)*

It was anticipated that the flame speed would increase markedly with fan speed, and this was also observed up to a fan speed of about 1500 rpm. However, as the fan speed was increased further, the flame speed exhibited a marked decrease, to about 3000 rpm beyond which ignition of the dust cloud by the propane flame was no longer possible. Referring to the work by Chomiak and Jarosinski (1982) on quenching of turbulent gas flames by turbulence, Hayes *et al.* (1983) attributed the fall-off of flame speed in the region 1500 rpm to 3000 rpm to quenching by excessive turbulence. Turbulent flame quenching occurs when the induction time for onset of combustion exceeds the characteristic lifetime of the turbulence eddies, so that an eddy composed of hot combustion products and unburnt fluid dissipates before the unburnt gas has become ignited. Hayes *et al.* did not discuss whether dust could have been separated out at high fan speeds in regions of non-random circulation flow in the channel (cyclone effect). It was confirmed, by means of hot-wire anemometry, that the degree of turbulence was proportional to the fan speed. For this reason Hayes *et al.* used a fan Reynolds number as a relative measure of the degree of turbulence in the experimental channel.

Klemens *et al.* (1988) investigated the influence of turbulence on wood and coal dust/air flame propagation in the laboratory scale flow-loop shown in Figure 4.49.

The flow was first streamlined by being passed through a battery of stator blades upstream of the measurement section. Turbulence was then induced in the first part of the measurement section by a number of cylindrical rods or rods of V-profiles, mounted with their axes perpendicular to the main flow direction. The electric spark ignition source was located immediately downstream of the turbulizing zone, and turbulent flame propagation was observed in the remaining part of the measurement section. Experiments were conducted with two types of brown coal, a maize dust, and a wood dust, all dusts being finer than 75 μm particle size. Figure 4.50 shows the average turbulent burning velocity for maize dust/air in the loop as a function of the average normalized turbulence intensity.

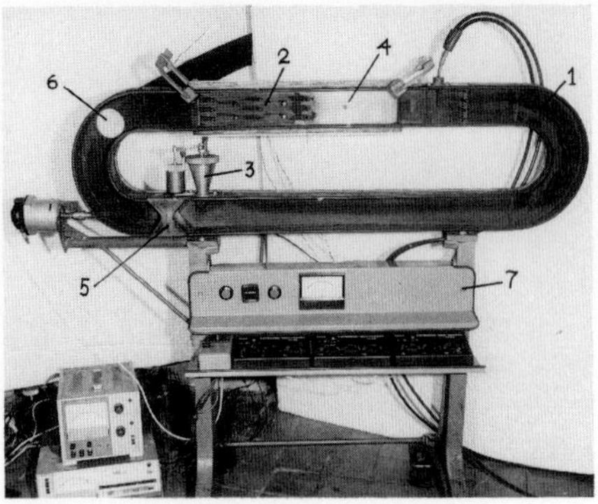

Figure 4.49 *Laboratory-scale flow loop for studying influence of turbulence on the propagation of dust/air flames:*
1 – Flow channel of cross section 80 mm × 35 mm
2 – Measurement section of 0.50 m length
3 – Dust feeder
4 – Ignition spark electrodes
5 – Fan
6 – Bursting membrane
7 – Automatic control system
(From Klemens et al.*, 1988)*

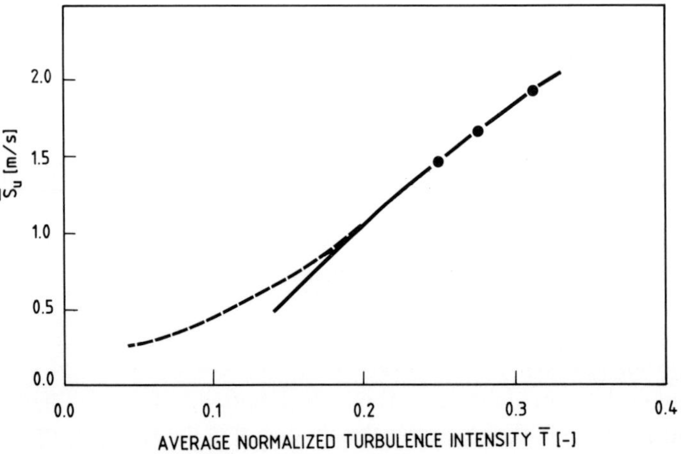

Figure 4.50 *Average turbulent burning velocity $\overline{S}_u$ in a cloud of maize dust in air as a function of the average normalized turbulence intensity $\overline{T}$, both quantities being averaged over the 80 mm height of the channel cross section. T = $1/V\,(V_x^2 + V_y^2 + V_z^2)^{1/2}$, where V is the overall flow velocity at a given location in the channel cross section, and V_x, V_y and V_z are the turbulence velocities in the three main directions at the same location (From Klemens* et al.*, 1988)*

Klemens *et al.* (1988) observed that their turbulent maize dust flame had the same characteristic non-homogeneous structure as observed by Proust and Veyssiere (1988) for turbulent maize starch/air flames in a vertical duct.

Shevchuk *et al.* (1986) studied flame propagation in unconfined clouds of aluminium dust in air at various levels of pre-ignition turbulence. The clouds were generated from a set of four dust dispersers driven by a short blast of compressed air. Each disperser was charged with 1 kg–10 kg of dust. After completion of dust dispersion, the dust cloud was ignited after a desired delay. The highest level of pre-ignition turbulence existed immediately after completion of the dispersion. As the ignition delay was increased, the turbulence decayed, and after a sufficiently long delay, the dust cloud was essentially quiescent. Figure 4.51 gives some results.

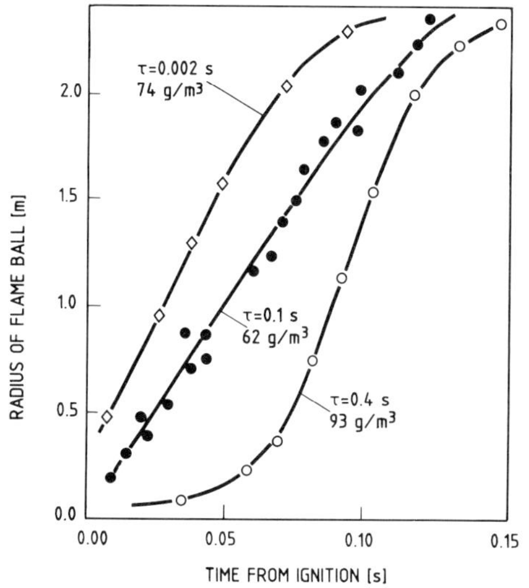

Figure 4.51 *Radius of flame ball as a function of time from central ignition of unconfined cloud of 10 μm diameter aluminium flakes in air. τ is the delay between completion of dust dispersion and effective ignition of the dust cloud. Dust concentrations are nominal averages (quantity of dust dispersed divided by visually estimated volume of dust cloud at moment of ignition) (From Shevchuk* et al.*, 1986)*

The data points for $\tau = 0.1$ s and 62 g/m^3 are from three different but nominally identical experiments. Figure 4.51 shows that the initial radial flame speed decreased systematically with increasing ignition delay, or decreasing initial turbulence, from about 30 m/s at $\tau = 0.002$ s via 20 m/s at $\tau = 0.1$ s to about 1 m/s at $\tau = 0.4$ s. The ignition delay of 0.4 s was probably sufficiently long to render the dust cloud practically laminar at the moment of ignition. However, after about 0.05 s the flame was no longer laminar, and accelerated rapidly to about 40 m/s over the very short period 0.05 to 0.07 s. Shevchuk *et al.* suggested that this 'switch' from laminar to turbulent conditions is triggered by flame instabilities due to non-homogeneous dust concentration, which is inevitable in a real dust cloud. They defined a special Reynolds number for establishing a criterion for the laminar-to-turbulent transition:

$$Re^* = \begin{matrix}\text{(Radius of flame ball}\\ \text{at transition point)}\end{matrix} \times \begin{matrix}\text{(Flame speed}\\ \text{at transition point)}\end{matrix} \Big/ \begin{matrix}\text{(Kinematic}\\ \text{viscosity of air)}\end{matrix}$$

and found that the transition generally occurred at Re^* in the range 10^4 to 10^5.

4.4.5 SYSTEMATIC COMPARATIVE STUDIES OF TURBULENT GAS AND DUST EXPLOSIONS

The dramatic influence of turbulence on gas explosions has been studied extensively. The investigations by Moen *et al.* (1982) and Eckhoff *et al.* (1984) are examples of fairly large scale experiments with obstacle- and jet-induced turbulence. It has been suggested, for example by Nagy and Verakis (1983), that there may be similarities between the influence of turbulence on gas and dust explosions. One of the first systematic comparative studies of turbulence influence on dust and gas explosions was conducted by Bond *et al.* (1986). They concluded that the relative burning rate variations caused by turbulence were equal in a 300 g/m^3 maize starch-in-air cloud and in premixed 7.5 vol% methane-in-air. However, they also emphasized the need for further work.

Yi Kang Pu (1988) and Yi Kang Pu *et al.* (1988) made further comparison of turbulent flame propagation in premixed methane-in-air and in clouds of maize starch in air, in identical geometries and at identical initial turbulence intensities. The experiments under turbulent conditions were conducted in closed vertical cylindrical vessels of 190 mm diameter and length either 0.91 m or 1.86 m. All experiments were conducted with initial turbulence generated by the blast of air used for dispersing the dust. The influence of ignition delay on the flame propagation and pressure development was studied. In the case of the gas experiments, the initial turbulence was generated by a blast of compressed methane/air, from the same reservoir as used for the compressed air for dust dispersion in the dust cloud experiments. In some experiments a battery of concentric ring obstacles were mounted in the tube for studying the influence of the additional turbulence generated by the expansion-induced flow of the unburnt gas or dust cloud past the obstacles.

A comparable set of Yi Kang Pu's results are shown in Figures 4.52 (gas) and 4.53 (dust). On average the combustion of the gas is twice as fast as that in the dust cloud. The laminar burning velocity of 550 g/m^3 maize starch in air, as determined by Proust and Veyssiere (1988), is about 0.20 m/s. Extrapolation of Zabetakis' (1965) data for methane in air to 5.5 vol% methane gives lower values, in the range of 0.15 m/s or less. It is therefore clear that the higher average turbulent flame speeds found by Yi Kang Pu for the 5.5 vol% methane-in-air cannot be attributed to a higher laminar burning velocity.

As the methane/air flame approached the end of the tube, the average flame speed $\overline{w}$ had reached the same value of 60–70 m/s irrespective of the ignition delay (initial turbulence), which means that the obstacle-induced turbulence played the main role in the latter part of the combustion. In the dust cloud, however, the high final flame speed of about 70 m/s is only reached in the case of high initial turbulence. The role of possible dust concentration inhomogeneities causing this discrepancy is not clear.

The maximum explosion pressures were in the range 4–5 bar(g) for the gas and somewhat higher, 5–7 bar(g) for the dust.

Yi Kang Pu's work indicates that there may not exist a simple one-to-one relationship between the response to flow-induced turbulence of gas and dust flames. There is little doubt that more research is needed in this area.

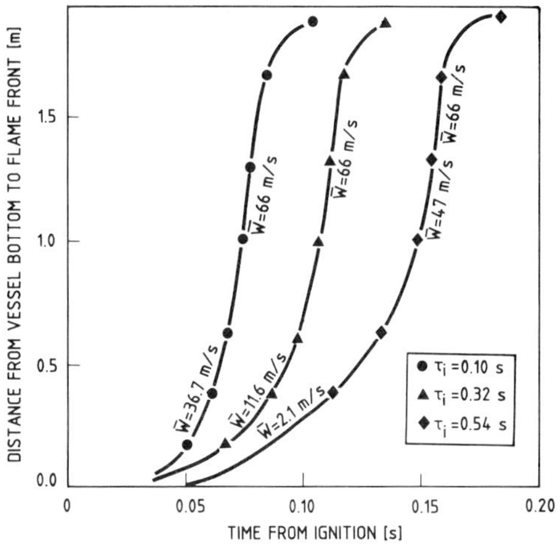

Figure 4.52 *Pressure rise and flame front location during combustion of 5.5 vol% methane/air in a 1.86 m long closed vertical tube of diameter 190 mm, as a function of time, under the influence of obstacle-induced turbulence. Three different ignition delay times τ_i. Ignition at tube bottom (From Yi Kang Pu, 1988)*

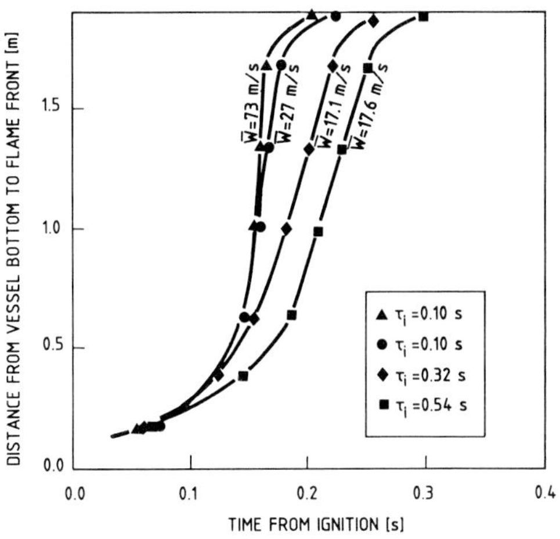

Figure 4.53 *Pressure rise and flame front location during combustion of a cloud of 550 g/m³ maize starch in air in a 1.86 m long closed tube of diameter 190 mm, as a function of time, under the influence of obstacle-induced turbulence. Three different ignition delay times, τ_i. Ignition at tube bottom (From Yi Kang Pu, 1988)*

4.4.6 MAXIMUM EXPERIMENTAL SAFE GAP (MESG) FOR DUST CLOUDS

The maximum experimental safe gap (MESG) can be defined as the largest width of a slot that will just prevent transmission of a flame in a gas or dust cloud inside an enclosure to a similar gas or dust cloud on the outside. This definition is somewhat vague and raises several questions. It neither defines the length of the slot, nor the explosion pressure inside or the volume of the enclosure. Therefore, MESG is not a fixed constant for a given explosible cloud, but depends on the actual circumstances. However, MESG is of importance in practice, and therefore needs to be assessed. In general it is smaller than the laminar quenching distance. This is because of the forced turbulent flow of the hot combustion products through the slot due to the pressure build-up inside the primary enclosure. Therefore the conditions of flame transmission are in the turbulent regime and should be discussed in the context of turbulent flame propagation.

Jarosinski *et al.* (1987), as part of their work to determine laminar quenching distances of dust clouds, also measured MESG under certain experimental conditions. The experiments were performed in a vertical tube of diameter 0.19 m and length 1.8 m, with a battery of parallel quenching plates of 75 mm length half-way up in the tube. Laminar quenching distances were determined at constant pressure, with ignition at the open bottom end of the tube and the top of the tube closed. MESG were determined with bottom ignition but both tube ends closed. This means that unburnt dust cloud was forced through the parallel plate battery as soon as significant expansion of the combustion products in the lower ignition end of the tube had started. Turbulence would then be generated in the flow between the parallel plates by wall friction and transmitted to the unburnt cloud immediately downstream of the plates. When the upwards propagating flame reached the plate battery, hot combustion products would be transmitted through the slots between the parallel plates, and re-ignition may or may not occur downstream of the plates. Under those circumstances MESG for 600 g/m^3 maize starch in air was found to be 1.5–2.2 mm, depending on the location of the primary ignition source. The lowest values were obtained with ignition at the tube bottom, the highest values with ignition just below the parallel plate battery. These values of MESG are not universal for 600 g/m^3 maize starch in air, but relate to the actual experimental conditions.

Schuber (1988, 1989) investigated the influence of various parameters on MESG. The apparatus is shown in Figures 4.54 and 4.55. Explosible dust clouds of desired concentrations were generated simultaneously in both vessels from compressed dust reservoirs and the cloud in the primary vessel subsequently ignited. It was then observed whether the cloud in the secondary vessel was ignited by the flame jet being transmitted through the annular gap in the wall of the primary vessel. Examples of flame jets of maize starch/air that are or are not capable of igniting the secondary cloud, are shown in Figure 4.56(a) and (b) respectively.

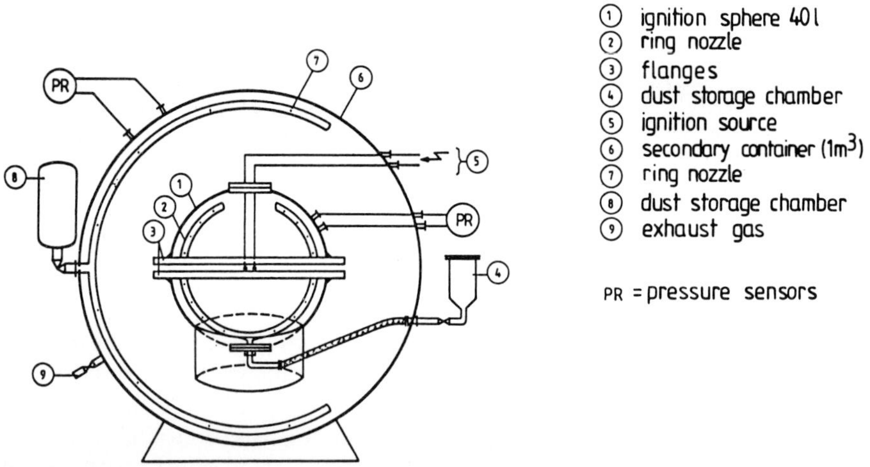

Figure 4.54 *Schematic illustration of primary 40 litre explosion sphere inside the secondary 1 m^3 vessel for determination of MESG of dust clouds. The annular gap for possible flame transmission is between the two flanges (3) (From Schuber, 1989)*

The dusts used in Schuber's investigation are listed in Table 4.14 together with their ignitability and explosibility properties. Table 4.14 does not contain metal dusts such as aluminium and silicon, and Schuber emphasized that his results are limited to organic dusts and coals.

In general Schuber found that MESG decreased with decreasing initial turbulence in the dust clouds. This is in harmony with the decrease of the minimum electric spark energy for ignition of both gases and dust clouds with decreasing turbulence. In order to ensure conservative results, Schuber's experiments to establish correlations between MESG and

Figure 4.55 *Photograph of actual assembly of primary 40 litre and secondary 1 m^3 vessels (From Schuber, 1989)*

Figure 4.56 *Visible flame jets of maize starch/air transmitting from the primary to the secondary dust cloud. (a) Flame jet too weak for igniting secondary cloud; (b) Flame jet will ignite secondary cloud (From Schuber, 1989)*

dust properties were conducted with comparatively low initial turbulence in the dust clouds. Schuber correlated his experimental MESG values with the product of minimum electric spark ignition energy and the dimensionless minimum ignition temperature

Table 4.14 Ignitability and explosibility properties of dusts used for determining MESG for dust clouds in air (From Schuber, 1988/1989)
M = Median particle diameter by mass.
P_{max} = Maximum explosion pressure according to International Standardization Organization (1985).
K_{St} = Normalized maximum rate of pressure rise according to International Standardization Organization (1985).
MIE = Minimum net electric spark energy for ignition of dust cloud with 1 mH inductance in capacitive discharge circuit.
TI = Minimum ignition temperature of dust cloud determined in BAM-furnace (see Chapter 7).

Dust type	M [mm]	P_{max} [bar(g)]	K_{St} [bar·m/s]	MIE [mJ]	TI [°C]	$MIE \cdot \frac{TI + 273}{273}$ [mJ]
Wettable sulphur*	50	4.3	86	< 1	260	< 1
Lycopodium	30	7.6	179	2	370	4.7
Benzanthrone	27	7.2	175	2	580	6.2
Light stabilizer	< 10	8.0	270	5	410	12.5
Polyethylene	123	5.5	55	10	410	25
Maize starch	< 10	7.9	186	20	400	49
Pea flour	54	7.4	95	100	410	250
Coal dust I	20	7.2	141	100	500	280
Coal dust II	18	7.8	130	2000	540	5960
Saar coal	54	6.4	84	5000	500	14200

*Wettable sulphur = 80% sulphur + 20% lignium sulphate.

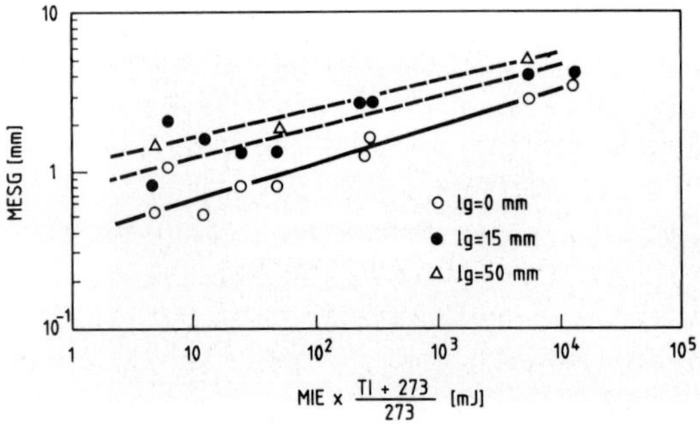

Figure 4.57 *Correlations between MESG and ignition sensitivity of dust clouds for various gap lengths (From Schuber, 1989)*

(column furthest to the right in Table 4.14) and the result is shown in Figure 4.57, where lg is the length of the gap (width of the flanges (3) in Figure 4.54).

There is an increase of MESG with increasing gap length from 0 to 50 mm by a factor of two to three. For a constant gap length there is a fair correlation between MESG and the ignition sensitivity parameter used. A closer examination of this parameter reveals that $(TI + 273)/273$ is in the range two to three for most of the dusts in Table 4.14, which means that the double-logarithmic correlation in Figure 4.57 is essentially between MESG and MIE. Schuber found that MESG approached a constant value as lg approached 50 mm. This value was considerably smaller than the laminar quenching distance. For example, Schuber's value for maize starch/air at lg = 50 mm was 1.8 mm, whereas the laminar quenching distance found independently by Jarosinski *et al.* (1987) and Proust and Veyssiere (1989) was 6–7 mm. Schuber's asymptotic value of 1.8 mm agrees well with the MESG of 1.5–2.2 mm found for maize starch/air by Jarosinski *et al.* (1987).

An important general conclusion from Schuber's work (1988) is that for a fairly long gap length of 25 mm, MESG for both gases, vapours and organic and sulphur dusts in air can be correlated with MIE $(TI + 273)/273$ in one single empirical equation:

$$\text{MESG [mm]} = (MIE\ [mJ] \times (TI + 273)/273)^{0.157} \qquad (4.85)$$

Equation (4.85) could in principle be refined by incorporating the gap length as a further parameter. For short gap lengths of a few mm this would give a reduction of MESG as compared to values from Equation (4.85) by a factor of two to three or more.

Schuber regarded the transmission of the flame through the slot as being primarily a process of ignition of the dust cloud downstream of the slot by the turbulent jet of hot combustion products being expelled from the slot, rather than flame propagation through the slot. He attributed the strong correlation between MESG and ignition sensitivity to this. On the other hand, it is well known that a strong correlation exists between laminar quenching distances and minimum ignition energies for gases.

The original motivation for Schuber's work was the uncertainty related to the ability of rotary locks to prevent transmission of dust explosions. He investigated results from experiments in the apparatus shown in Figure 4.58, where a rotary lock was mounted between two vessels in which dust clouds could be generated simultaneously.

The dust cloud on the one side was then ignited and it was observed if transmission of flame occurred to the extent that the dust cloud on the other side was also ignited. Figure 4.59 shows the essential parameters of the rotary lock.

On the basis of numerous experiments, Schuber (1989) proposed the nomograph in Figure 4.60 as a basis for predicting maximum permissible gaps W between the edges of the rotary lock blades and the housing. The radial gap W_r is defined in Figure 4.59. W_a is the axial gap. The gap that needs consideration depends somewhat on the details of the rotor construction. N_V is the number of consecutive rotor blades that form consecutive gaps. For example, on the right-hand side of the rotor, as viewed in Figure 4.59, $N_V = 3$.

Schuber emphasized that the nomograph does not apply to metal dusts, and that it is assumed that the rotor blades do not deform during the explosion. Figure 4.60 illustrates the use of the nomograph for maize starch/air for $N_V = 2$ and the two gap lengths 3 mm and 10 mm. The maximum permissible clearances are 0.4 mm and 1.1 mm respectively. For $N_V = 1$ the corresponding values would be about 0.1 mm and 0.25 mm, i.e. considerably smaller than the values 0.9 mm and 1.1 mm given for MESG for maize starch/air at zero and 15 mm gap lengths in Figure 4.57. This discrepancy could be due to integration of a safety margin in the nomograph. On the other hand, one would expect that much larger primary explosion volumes than 40 litres would be able to push larger quantities of burnt dust cloud through the slot and therefore create more favourable conditions for ignition of the dust cloud downstream of the slot.

Figure 4.58 *Arrangement for investigating the ability of rotary locks to prevent transmission of dust explosions (From Schuber, 1988)*

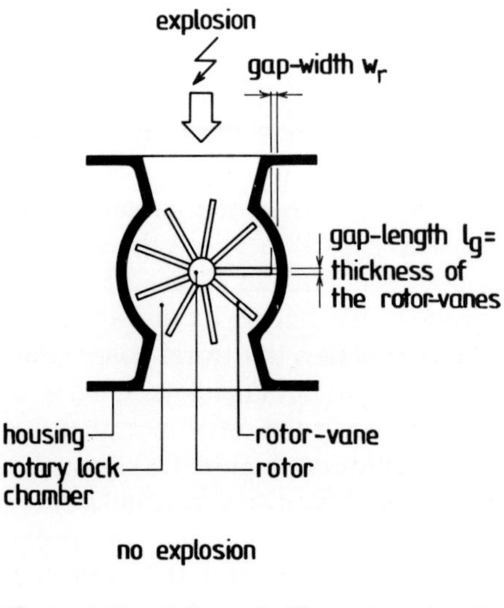

Figure 4.59 *Schematic illustration showing gap width and gap length related to explosion transmission through rotary locks (From Schuber, 1989)*

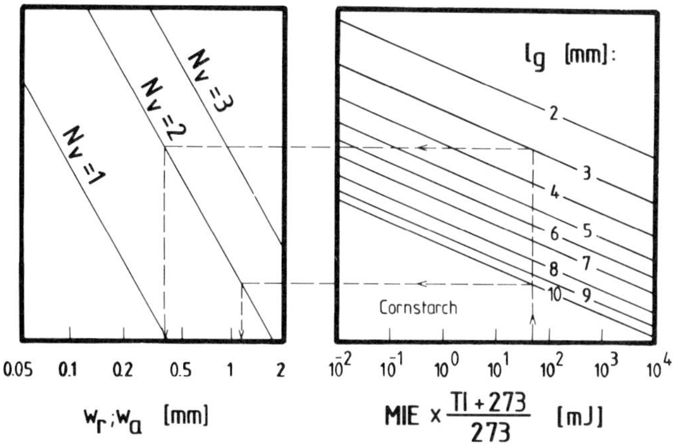

Figure 4.60 *Nomograph for estimating maximum permissible clearance W between rotor blades and housing for prevention of transmission of dust explosions through rotary locks (From Schuber, 1989)*

4.4.7 ACCELERATION OF TURBULENT DUST EXPLOSIONS IN ENCLOSURES OF LARGE L/D (DUCTS, PIPES, GALLERIES ETC.)

Coal mines essentially consist of long galleries of large length-to-diameter ratio (L/D). Since the onset of systematic research on the propagation of coal dust explosions in mines, large-scale experimental galleries have been a main tool of investigation. According to Cybulski (1975), Hall's experiments in coal mines in UK about 1890 was probably the first of this kind. Some years later, Taffanel (1907) published the results of his pioneering large-scale gallery experiments in France. These experiments were conducted as a consequence of the disastrous coal dust explosion in the Courriers mine in 1906, where 1099 miners lost their lives. Similar work was subsequently initiated in Poland, Russia, Germany and USA.

Greenwald and Wheeler (1925) used a horizontal explosion tube of internal diameter 2.3 m and length 230 m, i.e. $L/D = 100$ in their experiments. One end was normally closed, the other fully open. A pulverized nut coal, ground to 85% by mass < 74 μm particle size, and containing 33% volatiles, was used. The ignition source was 800 g of black powder igniting a primary cloud of 10 kg of coal dust. The main quantity of coal dust was spread along the gallery floor from the point of ignition at 61 m to the full opening of the gallery at 230 m (see Figure 4.61). The quantity of dust spread on the gallery floor was about 1500 g per m of gallery length, corresponding to a nominal dust concentration in a fully dispersed state of 360 g/m^3. No dust was spread out in the 61 m long section between the normally closed upstream end of the gallery and the ignition point.

The main purpose of Greenwald and Wheeler's experiments was to investigate the influences of the location and size of vents on the development of dust explosions in the gallery. As Figure 4.61 shows, flame speeds of up to 800 m/s were generated with the upstream end of the gallery fully closed. Whether the plateau of constant flame speed at 800 m/s beyond 165 m indicates detonation, is unclear. Lindstedt and Michels (1989) observed violent, constant-velocity deflagrations supported by wall-friction induced turbulence for alkanes in air. Similar steady combustion phenomena may also exist for dust explosions in long tubes and ducts. The flame speed will then be somewhat lower than for a proper detonation. (Detonation of dust clouds is discussed in Section 4.5.)

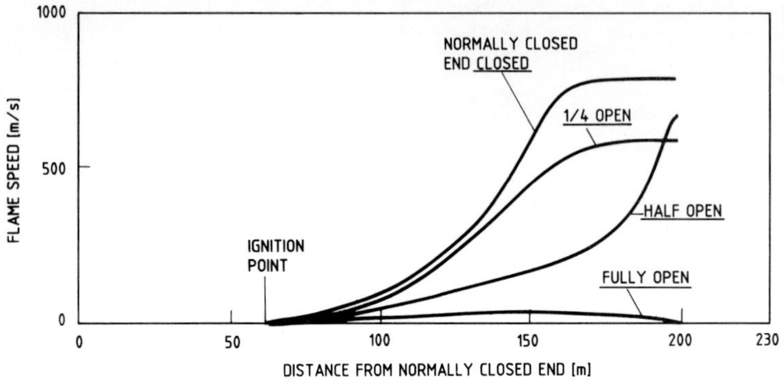

Figure 4.61 *Acceleration of coal dust explosions in a horizontal gallery of internal diameter 2.3 m and length 230 m. Effect of venting at upstream, normally closed end (From Greenwald and Wheeler, 1925)*

Figure 4.61 demonstrates that venting at the upstream, normally closed, end reduced the acceleration of the explosion appreciably. With a fully open upstream end, comparatively weak explosions of maximum flame speeds around 50 m/s resulted. In this case Greenwald and Wheeler made some interesting observations. The flame motion was markedly vibratory, and the column of dust and air preceding the flame was expelled from the gallery exit in puffs instead of in a continuous stream. The flame itself could be seen to issue from one of the openings two or three times, with a slight in-rush of air occurring between each flame appearance. This finding is in agreement with Chapman and Wheeler's (1926) observations of vibratory premixed gas flames in a laboratory tube open at both ends. They found that the 'periodicity of the vibrations was that of the fundamental tone of the tube'. As already discussed, and illustrated in Figure 4.36, Eckhoff *et al.* (1987) observed the same phenomenon during dust explosions in a large vertical silo of diameter 3.7 m and height 22 m, and vented at the top, provided the ignition point was in the upper part of the silo. Greenwald and Wheeler (1925) also measured explosion pressures at various locations in the large gallery. The maximum values recorded by the low-frequency-response manometers available at that time were 5.0 bar(g), 4.8 bar(g), 3.3 bar(g) and 0.14 bar(g) for the normally fully closed end fully closed, 1/4 open, half-open and fully open respectively. Pressure recordings further upstream were lower than this and decreased systematically with increasing distance to the down-stream exit.

Fischer (1957) reported results from coal dust explosion experiments in a 260 m long experimental coal mine gallery of equivalent-circle cross sectional diameter of 3.2 m, i.e. L/D of about 80. The main purpose of these experiments was to investigate whether deposits of stone dust on shelves in the upper part of the gallery cross section would prevent the propagation of coal dust explosions in the gallery. However, it appeared that under certain circumstances this stone dust had little effect, and flame acceleration phenomena of the same violent type as found by Greenwald and Wheeler (1925) were observed, as shown in Figure 4.62.

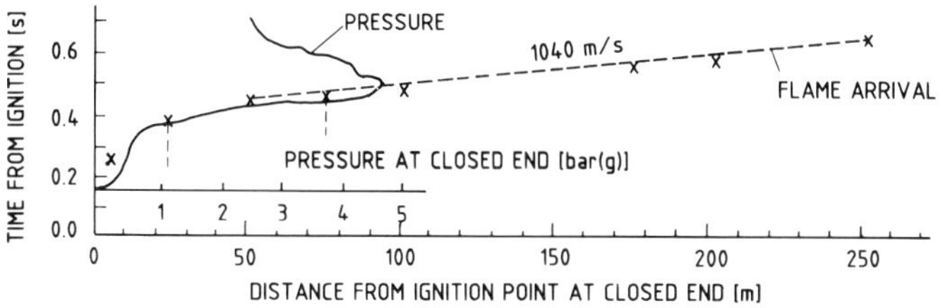

Figure 4.62 *Time of arrival of bituminous coal dust/air flame as a function of distance from ignition point at closed end of gallery of length 260 m and diameter 3.2 m. Pressure at closed end as a function of time. Nominal average dust concentration 500 g/m^3 (From Fischer, 1957)*

The coal dust explosion was initiated by an explosion of 40 m^3 methane/air at the upstream closed end of the gallery. The gas was ignited by black powder probably ensuring violent combustion of the gas. The blast from the gas explosion in turn swept up the coal dust layer of 4 kg per m length of gallery on the floor and initiated the self sustained dust explosion down the entire length of the gallery. The most striking feature of Figure 4.62 is the very constant flame speed of 1040 m/s measured from about 50 m from the closed end right to the open tube end 200 m further down. Fischer associated this with 'some kind of detonation' (see Section 4.5). The pressure versus time was recorded only at the upstream closed end of the gallery, because the explosion was so violent that all the measurement stations further down the gallery were destroyed. As can be seen, the peak pressure at the closed end was about 5 bar(g). It would be anticipated that the pressures further down the gallery were considerably higher.

Jost and Wagner (in Freytag (1965)) have illustrated the various characteristic phenomena occurring during acceleration of premixed gas flames in long one-end-open tubes. There are good reasons for assuming that their overall picture, as reproduced in Figure 4.63, also applies to dust clouds. The only major difference is that a dust cloud needs to be generated by raising dust deposits into suspension. This means that stage one and possibly also stage two in Figure 4.63, the ignition and laminar propagation of the initial flame, may not be relevant for dust flames. As already discussed, Greenwald and Wheeler (1925) used black powder for stirring up and igniting the primary dust cloud, whereas Fischer (1957) used a turbulent gas flame. However, once the primary dust explosion gets under way, the blast wave generated by it will entrain dust further downstream as already discussed. Therefore all stages of Figure 4.63, from stage three and downwards, will apply even to dust clouds. The essential reason for the flame acceleration is turbulence generated in the unburnt cloud ahead of the flame due to wall friction when the cloud is pushed towards the open tube end by the expansion of the part of the cloud that has burnt. When the flame front reaches the turbulent unburnt cloud, the combustion rate increases. This in turn increases the expansion rate of the combustion products and therefore also the

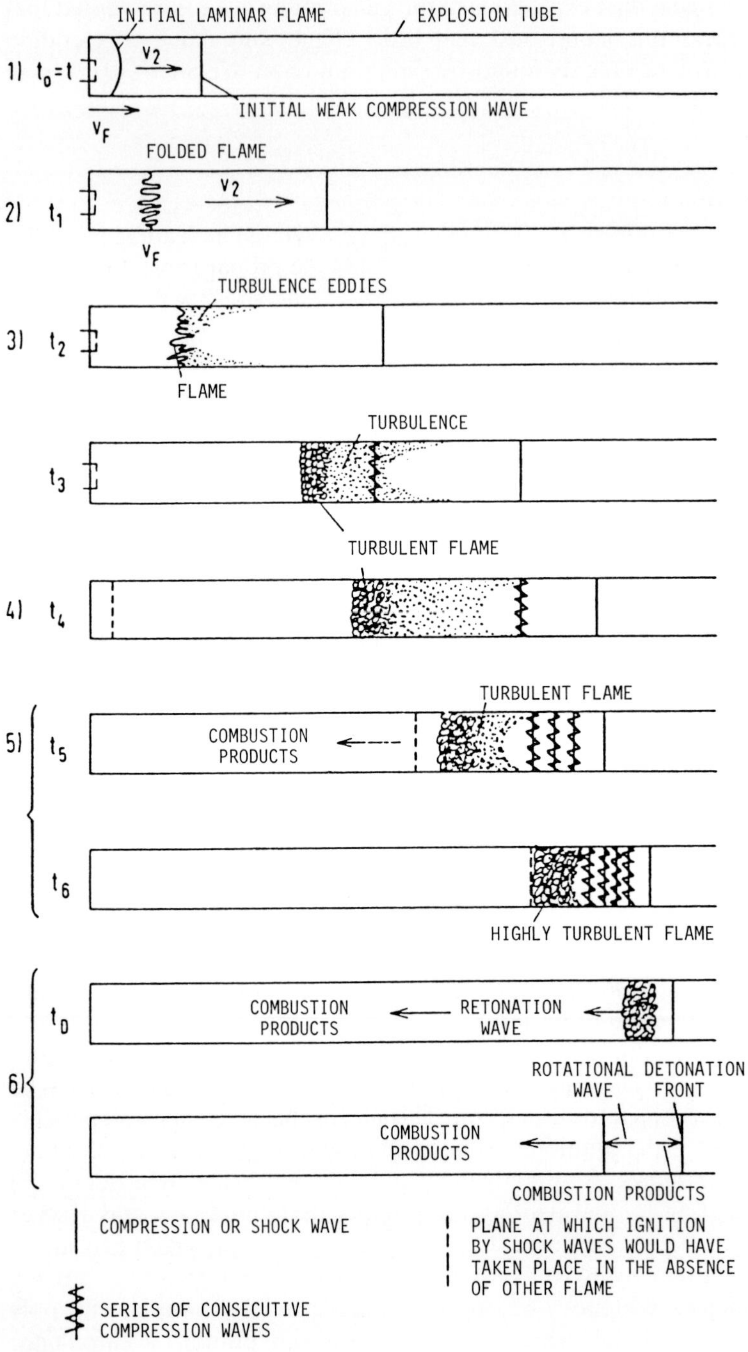

Figure 4.63 *Characteristic phenomena during acceleration of gas or dust flames in one-end-open long tubes, from laminar combustion via turbulent combustion to detonation. V_F is the flame speed; V_2 is the velocity of the unburnt gas or dust cloud ahead of the flame (From Jost and Wagner, in Freytag, 1965)*

flow rate of the unburnt cloud ahead. The result is an even higher turbulence level and further increase of the combustion rate. During all these stages, compression waves will be emitted and propagate towards the open tube end. Because of heating of the cloud ahead of the flame due to adiabatic compression, each wave will propagate at slightly higher velocity than the previous one. Ultimately, therefore, they will all catch up with the initial wave and form a strong leading shock front. The turbulent flame front will also, due to the positive feedback mechanism of combustion rate → flow rate → turbulence → enhanced combustion rate, eventually catch up with the leading shock wave. If the leading shock is sufficiently strong, a switch can occur in the mechanism of flame propagation. Instead of heat being transferred by turbulent diffusion behind the leading shock wave, the dust cloud may become ignited in the highly compressed state inside the leading shock. If the induction time of ignition is sufficiently short, the chemical reaction zone and the propagating shock wave then become closely coupled and propagate through the cloud at constant velocity. This is detonation. (see Section 4.5.) However, as already mentioned, flame propagation at a constant high speed will not necessarily have to be a classical detonation, but can also be a high-speed turbulent deflagration supported by wall friction induced turbulence.

Figure 4.63 shows a tube with a comparatively smooth internal wall. However, if the wall roughness is increased, the positive feedback loop of combustion acceleration becomes more effective, and acceleration up to detonation occurs over a shorter distance. Gas explosion experiments have been conducted in tubes in which the 'wall friction' was increased systematically by inserting in the tube a number of equally spaced narrow concentric rings in contact with the wall. Such experiments were in fact carried out by Chapman and Wheeler (1926) in a small laboratory-scale tube of diameter 50 mm and length 2.4 m, and open at both ends. For methane/air, flame speeds of up to 420 m/s were measured as opposed to 1.2 m/s without the rings. Chapman and Wheeler were fully aware of the essential role played by flow-generated turbulence. Similar dramatic effects of such equally spaced rings were found by Moen *et al.* (1982) for methane/air explosions in a one-end-open large-scale tube of 2.5 m diameter and 10 m length.

These investigations are of considerable interest in relation to dust explosions in coal mines, where the supporting structures of the mine galleries would seem to have the same type of turbulence increasing effect as the concentric rings in tubes (Fischer, 1957). In the process industry, the legs of bucket elevators are in fact long ducts with repeated obstacles.

Rae (1971) analysed coal dust explosion experiments in various large-scale tubes and galleries of lengths in the range 100–400 m, conducted in the time period 1911 to 1971. He pointed out that the initiating explosion causes events analogous to those observed in shock tubes. The initial thin turbulent dust flame entrains deposited dust and develops into the more extensive main explosion, which may in turn lead either to detonation-like phenomena including strong shock waves, or to oscillating flames, depending on various circumstances.

Bartknecht (1971) used an external dust dispersion system by which he avoided the use of a primary explosion for initiating dust entrainment and flame propagation. He generated a dust cloud of the most explosible concentration along the whole tube length by simultaneously injecting dust from a number of equally spaced external pressurized reservoirs. (This is essentially the same dust dispersion method as specified in the 1 m^3 test approved by the International Standardization Organization (1985).) The dust cloud was ignited by a strong chemical ignitor, or a pocket of exploding methane/air as soon as it had been generated. On the one hand, Bartknecht's experiments were clean and well-defined. On the other hand, they differed from conditions often met in mines and other industry, where the dust is initially deposited as layers that are dispersed by the air blast preceding the flame as the explosion propagates. There may be situations, however, where Bartknecht's dispersion method corresponds to reality, for example in pneumatic transport of explosible dust concentrations.

Figure 4.64 gives some of Bartknecht's results from experiments in 0.40 m diameter horizontal one-end open pipes of various lengths. As can be seen, there is close correlation between the K_{St} value, as determined in agreement with the recommendation by the International Standardization Organization (1985) and the violence of the explosions in the tubes.

The aluminium dust was comparatively coarse, having a median particle diameter on a mass basis of 30 μm, with 10% > 100 μm and 10% < 20 μm. Nevertheless, a maximum flame speed of 2200 m/s and maximum explosion pressure of 25 bar(g) was measured in a 20 m long pipe, with enlarged diameter in the ignition zone for increasing the initial 'push' and establishing a high level of turbulence and burning rate at an early stage. The explosion pressures were measured by piezoelectric sensors and were those acting normal to the tube wall, i.e. normal to the direction of propagation. There are reasons to believe

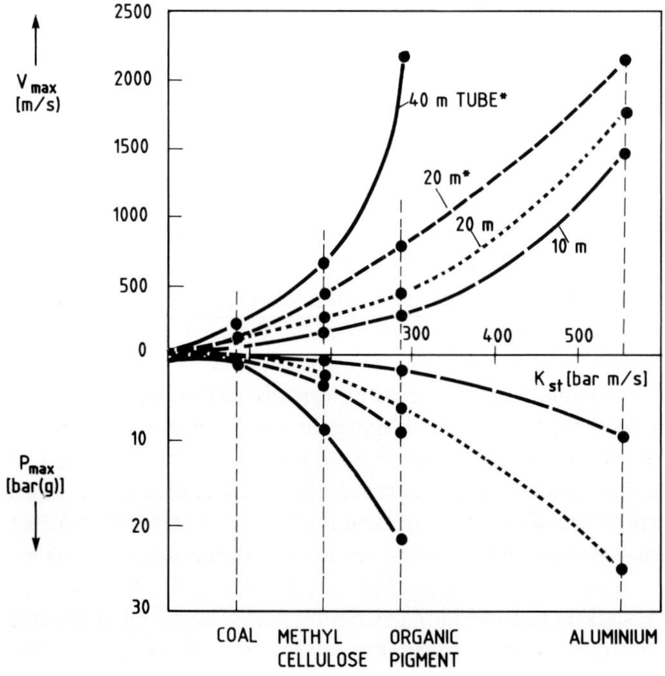

Figure 4.64 *Dust explosions in 0.40 m diameter horizontal one-end-open pipes of various lengths. Maximum flame speeds and maximum explosion pressures as functions of tube length and K_{St} value of dusts. *denotes enlarged pipe diameter in the ignition zone at the closed end (From Bartknecht, 1971)*

that the 2200 m/s phenomenon observed was in fact a proper detonation. (See Section 4.5.)

The coal dust only produced maximum flame speeds of up to 250 m/s and maximum explosion pressures of the order of 1 bar(g). The median particle size was 22 μm, with 10% > 60 μm and 10% < 5 μm (extrapolation of data). The volatile content was not specified.

Bartknecht attributed the comparatively slow coal dust explosions to the relatively small tube diameter of 0.4 m. He also conducted coal dust explosion experiments in a much larger one-end-open tube, of diameter 2.5 m and length 130 m, and with ignition at the closed end by a pocket of methane/ air. With 250 g/m^3 dust, maximum flame speeds of up to 500 m/s were measured. With 500 g/m^3, the maximum flame speeds were 700 m/s or more.

Bartknecht further conducted experiments where the dust was spread as a layer along the tube floor in a quantity corresponding to 250 g/m^3 if dispersed homogeneously over the whole tube cross section. When using the same ignition source (turbulent methane/air explosion at closed tube end) as with the pre-dispersed clouds that he normally worked with, he found lower flame speeds and explosion pressures than with pre-dispersed clouds. However, Figures 4.61 and 4.62 show that the layer-spreading technique can indeed give very high flame speeds if only the initiating blast is sufficiently violent. This illustrates that choosing conditions of experimentation that correspond to the actual industrial hazard is an important aspect of applied dust explosion research.

Pineau and Ronchail (1982) and Pineau (1987) described experimental research on the propagation of wheat and wood dust explosions in ducts of diameters from 25 mm to 700 mm. They pointed out the fact that in any industrial installation where dust extraction or pneumatic transport of powdered material is used, there will be a number of ducts connected either to blowers, to fans or to pumps. In addition, the arrangements may include cyclones, bag filters, hoppers and bins, and other process equipment, some of which may be inter-connected by pressure balance ducts. It is therefore essential, in the case of explosible powders and ducts, to understand the mechanisms by which dust explosions may propagate in dusts. In addition to straight ducts, ducts containing bends also need to be considered, because such bends are frequent in the process industry.

In one series of experiments reported by Pineau and Ronchail (1982), straight tubes of diameters from 250 mm to 700 mm and lengths from 12 m to 42 m were used. The tubes were either closed at both ends, closed at one end and fully or partly open at the other, or they were fully or partly open at both ends. In some experiments the ignition point was at a closed tube end, in others near an open end. In one experiment it was midway down the tube. The dust was initially distributed as a layer along the tube floor, the quantity of dust per unit length of duct corresponding to the desired nominal dust concentration. Ignition was sufficiently powerful to start dust entrainment and flame propagation through the dust cloud, but subsequent propagation was dependent on whether a sufficiently strong flow field was generated ahead of the flame for entraining the dust further downstream. This was, as expected, dependent on the extent to which the tube ends were closed or open and on the location of the ignition point. Some examples are given in Figure 4.65.

The results for the 700 mm diameter tube show that the maximum explosion pressures were low and nearly the same, i.e. 1 bar(g) for the one-end-open tube, and 1.5 bar(g) for the fully closed one. In the case of the closed tube, the low pressure means that the flame speeds and associated gas velocities were too low to cause entrainment and dispersion of

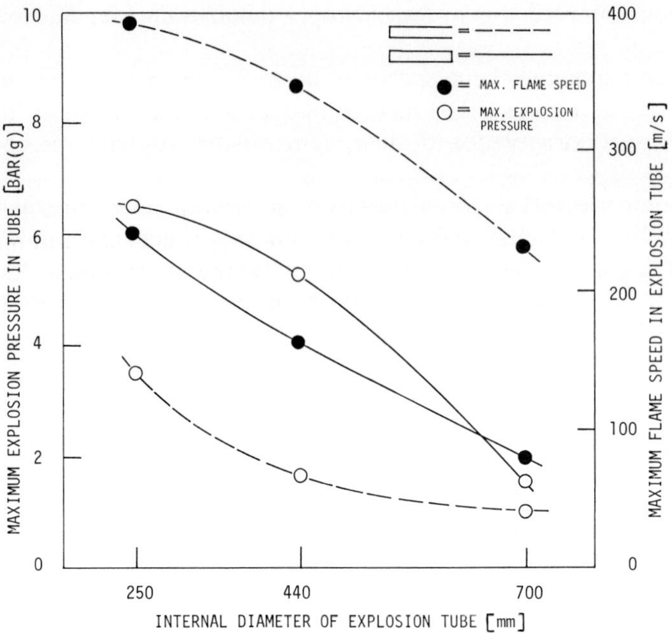

Figure 4.65 *Maximum explosion pressures and maximum flame speeds during wheat flour/air explosions in one-end-open or fully closed tubes of lengths 42 m and three different diameters. Nominal average dust concentration 470 g/m^3. Ignition at closed tube end (Replot of data from Pineau and Ronchail, 1982)*

the majority of the dust. In the case of the one-end-open tube, the speed of the blast ahead of the flame was much higher due to the venting at the open end. This in turn entrained the still unburnt part of the dust in the tube and gave rise to a sufficiently high combustion rate in the resulting dust cloud to generate 1 bar(g) pressure in spite of generous venting. As the tube diameter decreased, the maximum flame speed increased, even in the closed tube, and more dust was entrained and burnt, and a higher maximum explosion pressure resulted.

Pineau and Ronchail (1982) also conducted a number of experiments in tubes of smaller diameters in the range 25–100 mm connected to a vessel in which the explosion was initiated. The tube lengths varied between 10 m and 40 m and the volume of the vessel was either 1 m^3 or 0.1 m^3. The influence of a 90° bend in the duct was also investigated. Furthermore the effect of venting, either of the vessel or at the bend was studied. Numerous results were produced for various configurations and locations of the ignition point. Generally the trends found can be understood on a qualitative basis in terms of increasing turbulence, dust entrainment, combustion rate and venting, with increasing flow rate in the system. However, the complex pattern of results re-emphasizes the need for a unified theoretical dust explosion model suitable for computer simulation of the course of explosions in complex, integrated systems for which specific experimental data do not exist.

Pineau and Ronchail (1982) found that powders having $K_{St} > 200$ bar m/s (International Standardization Organization (1985)) can generate detonations in tubes of diameters 25–100 mm and up to 40 m length. Such detonations are associated with maximum pressures of more than 20 bar(g) and flame speeds of about 2000 m/s. This, for example, occurred with wood dust in a 25 m long tube of 100 mm diameter, connected to a 1 m^3 vessel in which the explosion was initiated. The inclusion of a 90° bend 6 m from the vessel, i.e. 19 m from the open tube exit, reduced the explosion violence somewhat, but still detonation resulted in one experiment in a series of eight.

Radandt (1989) emphasises that in industrial practice, as in dust extraction and pneumatic conveying systems, the initial dust clouds in ducts or tubes are not stagnant, but flow at a considerable velocity, typically in the range 15–25 m/s. He therefore conducted a comprehensive series of dust explosion experiments with a maize starch of K_{St} = 220 bar m/s being conveyed at various concentrations and velocities, using the experimental loop illustrated in Figure 4.66.

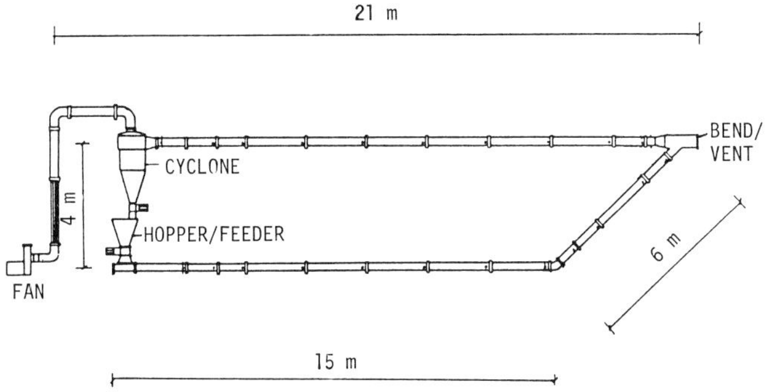

Figure 4.66 *Loop for dust explosion experiments, consisting of a dust feeder, 35 m of 200 mm diameter tube, a cyclone recycling the dust to the feeder, and a suction fan for establishing the air flow (From Radandt, 1989)*

The air was sucked into the system through the fully open tube end at the dust feeding point by the underpressure generated by the suction fan to the very left in Figure 4.66. A vent arrangement is indicated at the sharp 45° bend. Experiments were conducted both with a vent at this point, and with just a closed, smooth bend. The dust concentration was varied in the range 100–450 g/m^3 and the mean air velocity in the tube prior to ignition in the range 15–25 m/s. The ignition point was also varied from immediately downstream of the dust feeder to a number of other locations along the tube. A number of pressure and flame detectors were located at various strategic points. In most of the experiments the vent at the sharp bend reduced the maximum explosion pressures in the tube as compared with pressures generated with a smooth, closed bend. However, if the dust cloud was ignited near the dust feeding point, both the maximum pressure and the flame speed were higher with venting than without. This can be explained in terms of the higher flow velocity in the tube, due to the explosion, with a vent than without. Following ignition close to the dust feeder, the vent will open when the flame has propagated only part of the distance to the vent. The result is a sudden increase of the flow rate of the unburnt cloud ahead of the flame, and a corresponding increase of the turbulence in this cloud. Consequently, when the flame reaches these turbulent regions, the combustion rate will increase markedly. Under such circumstances the flow out of the vent can easily become choked, and very high explosion pressures can result. The combustion rate will also increase because the pressure of the unburnt cloud ahead of the flame increases. Radandt's investigation produced much valuable empirical data, which, however, again re-emphasizes the need for a unified computer-based model that accounts for the coupling between gas dynamics and turbulent combustion in complex systems.

Tamanini (1983) investigated the propagation of dust explosions in a large-scale gallery illustrated in Figure 4.67. A central objective was to determine the minimum quantity of dust, spread as a layer on the gallery floor, per unit gallery length, that was able to propagate a dust explosion sweeping along the gallery. A second objective was to investigate whether venting of a primary explosion in a confined space could prevent the development of secondary explosions in adjacent areas by reducing the expansion velocities and hence the dust entrainment potential of the primary explosions in those areas. The experiments showed that a dust flame propagated down the gallery even if the mass of the dust layer, per unit length of gallery, was considerably smaller than that corresponding to the minimum explosible concentration if dispersed uniformly over the whole gallery cross section. This is because the dust was dispersed only in the lower part of

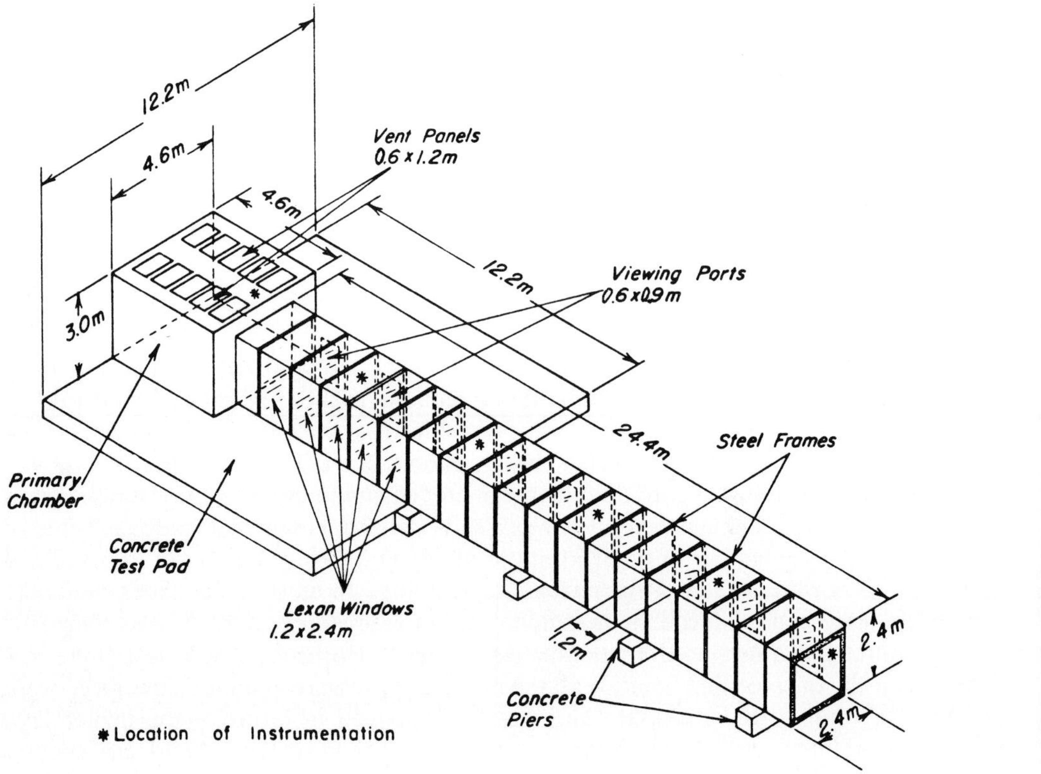

Figure 4.67 *Large-scale gallery used for investigating entrainment of dust layers and propagation of secondary explosions in a gallery due to a primary explosion in an adjacent chamber (From Tamanini, 1983)*

the gallery volume and therefore gave real dust concentrations that were higher than the nominal values. In accordance with this it was observed that the dust flame thickness was in fact considerably smaller than the height of the gallery. Such secondary dust flames were found to sweep along the gallery floor all the way down to the exit even if the dust layer on the floor was only 0.25 mm thick, representing a nominal dust concentration referred to the entire gallery volume, of only 77 g/m^3 of maize starch, i.e. at the limit for upwards laminar flame propagation.

Typical explosion pressures in the gallery were 0.2–0.4 bar(g) if the gallery was unvented and 0.07 bar(g) with vents close to the primary explosion chamber.

The fact that the dust entrained from the floor was distributed only in the lower part of the gallery may throw light on the results from Fischer's (1957) experiments, where stone dust barriers in the upper half of the gallery cross-section under certain conditions proved entirely ineffective in damping the propagation of the coal dust explosion. Fischer suggested that the primary turbulent torus sweeping down the gallery entrained the coal dust in the lower part of the gallery cross section and the stone dust in the upper part, with little mixing of the two.

Experiments of the type conducted by Tamanini (1983) and also by the other workers who used a primary explosion to initiate dust entrainment and the main explosion, depend very much on the nature of the primary explosion. Therefore, few generally valid quantitative conclusions can be drawn from such experiments until the various processes have been theoretically coupled.

Kauffman *et al.* (1984) studied the propagation of dust explosions in a horizontal tube of length 36.6 m and internal diameter 0.30 m, i.e. L/D = 122. One main objective of the experiment was similar to the one of Tamanini (1983), i.e. to identify the minimum quantity of dust deposited as a layer on the internal tube wall that was able to propagate a dust explosion sweeping down the tube. The exhaust end of the tube terminated with a 90° bend of 2 m radius leading into a 2.5 m long tube with a number of vents in the wall, but with the far downstream end closed. The ignition source, located at the far upstream end of the main tube, consisted of a 2.4 m long 50 mm diameter tube filled with stoichiometric hydrogen/oxygen. In the first 3.6 m of the main tube a dust layer was placed in a V-channel running inside the tube parallel with the tube axis. This dust could be dispersed into a primary cloud by air blasts from a series of nozzles at the bottom of the V-channel. In the remaining 33 m of the main tube the dust layer was resting directly on the tube wall, either as strips of widths 12.5 mm or 90 mm along the tube bottom, or as a thin layer around the whole tube wall. The explosions were initiated by first dispersing the dust in the V-channel, and then igniting the hydrogen/oxygen mixture, which would then in turn ignite the dispersed dust. The blast from this violent primary explosion would then sweep down the main tube and entrain and disperse the dust from the layer on the tube wall, as in the experiments of Greenwald and Wheeler (1925), Fischer (1957), Pineau (1987), Tamanini (1983), and in the other investigations discussed by Rae (1971).

Kauffman *et al.* found that in general, for a given mass of dust layer per unit length of tube, a uniform layer around the entire tube wall gave the most violent explosions. The dusts tested included a maize starch, a mixed natural organic dust, a wheat grain dust and an oil shale dust. Various ranges of particle sizes and moisture contents were investigated.

The strength of the primary explosion was varied by varying the initial pressure of the hydrogen/oxygen mixture and the initial quantity of dust dispersed from the V-channel. It was generally found that for a given mass of dust per unit length of the main tube, the maximum pressure, temperature and flame speed of the secondary explosion increased with the strength of the primary explosion. Figure 4.68 shows how the nominal minimum and maximum explosible concentrations (mass of dust layer/m^3 tube) varied with the strength of the primary explosion in terms of its maximum overpressure. Assuming a bulk density of the dust layer of 0.5 g/cm^3, a nominal concentration of 1000 g/m^3 corresponds to layer thicknesses of 0.15 mm if all the tube wall is covered, and 1.6 mm and 11 mm for 90 mm and 12.5 mm layer widths respectively.

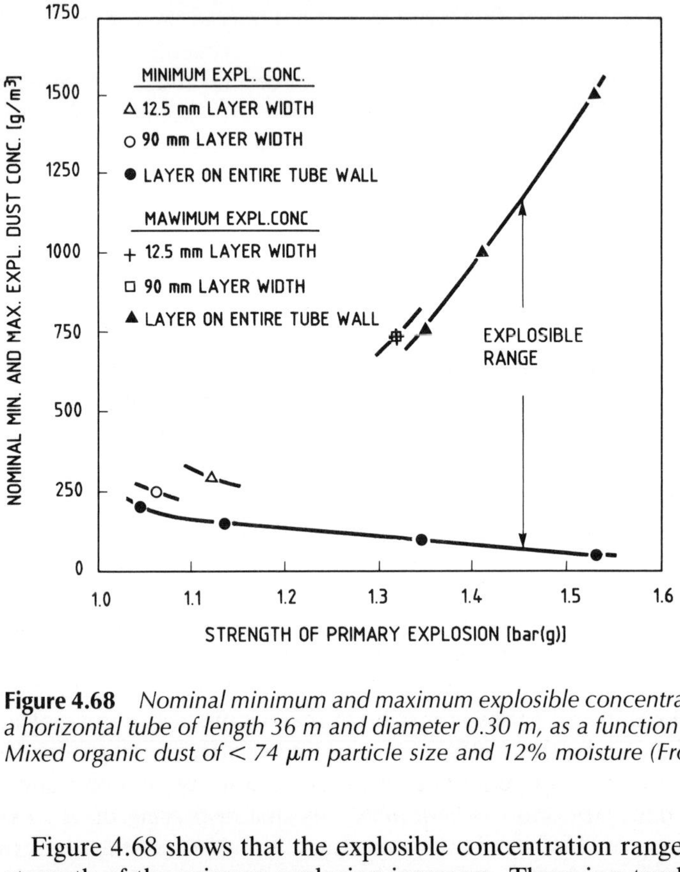

Figure 4.68 *Nominal minimum and maximum explosible concentrations for secondary explosions in a horizontal tube of length 36 m and diameter 0.30 m, as a function of strength of primary explosion. Mixed organic dust of < 74 μm particle size and 12% moisture (From Kauffman* et al., *1984)*

Figure 4.68 shows that the explosible concentration range expands at both ends as the strength of the primary explosion increases. There is a tendency of the range to become shifted towards higher nominal concentrations as the dust layer becomes concentrated in a narrow strip. The minimum explosible concentration of 50 g/m^3 at an explosion strength of 1.53 bar(g) is close to the value measured by Eckhoff and Fuhre (1975) for wheat grain dust of 12% moisture, in an entirely different large-scale apparatus.

In a subsequent study using the same 36 m tube facility as used by Kauffman *et al.* (1984), Srinath *et al.* (1985) determined the maximum overpressure and maximum flame speed for a dried mixed natural organic dust, and found 5.4 bar(g) and 607 m/s respectively. A numerical code developed by Chi and Perlee (1974) for premixed gas explosions was used for solving the one-dimensional compressible flow equations for a flame propagating through a tube of the same dimension as used experimentally. The code did not, however, close the loop connecting flow, turbulence and combustion rate, and an empirically based relationship between the turbulent burning velocity and flame propagation distance derived from the actual dust explosion experiments had to be employed. Under these circumstances the code predicted a maximum explosion pressure of 5.89 bar(g) and a maximum flame speed of 607 m/s, in good agreement with experiments.

However, as will be discussed in the next section, improved, more comprehensive numerical code concepts for dust explosion simulation are being developed.

4.4.8
THEORIES OF FLAME PROPAGATION IN TURBULENT DUST CLOUDS COMPUTER MODELS

4.4.8.1
Background

The discussion of flame propagation mechanisms in the previous sections, in particular turbulent propagation where turbulence is generated *in situ* by flow produced by the explosion itself, has demonstrated the vast complexity of the turbulent flame propagation process. Simple experimental correlations are not sufficient for predicting explosion development in complex practical situations. A theory is needed that can unify all these experiments. In its comprehensive form the theory should include the mechanisms of dust entrainment and dispersion (Chapter 3) as an integrated element in the complex feedback interaction between combustion of dust cloud, expansion of combustion products, gas flow ahead of flame, turbulence in gas flow ahead of flame, intensified entrainment and dispersion of dust ahead of flame, – and back to intensified combustion. Increasing availability of computational power has facilitated considerable progress over the last 10 to 15 years. It seems reasonable to expect that comprehensive computer codes for predicting dust explosion propagation even in complex industrial geometries will become available in a not too distant future. Encouraging progress has been made in the prediction of gas explosion propagation in congested geometries as for example in modules on offshore oil and gas production platforms, or compact onshore refineries and petrochemical plants. The pioneering work by Hjertager (1982, 1984, 1986), which makes use of the two-equation $k - \epsilon$ model of turbulence by Jones and Launder (1972, 1973) and the combustion model of Magnussen and Hjertager (1976), should be mentioned specifically. More recently Cant and Bray (1989) developed a theoretical model of turbulent combustion of premixed gases in closed bombs, which may also be a useful starting point for dust cloud explosion simulation.

However, in the case of dust clouds, the two-phase nature of the problem adds considerably to the complexity. The previous sections of this chapter give some elements of the present experimental and theoretical knowledge of the complex physics and chemistry involved, and which must be accounted for in a comprehensive dust explosion model.

In his model of gas explosions, Hjertager (1982, 1984, 1986) used the induction time for ignition, as determined in shock tube ignition experiments, as a global measure of the chemical kinetics for the combustion reaction. As reviewed by Eckhoff (1987) similar experiments have been conducted with dust clouds, and induction times for various types of dust are available. A more recent example is the induction time determination for aluminium and magnesium dust clouds in oxygen of 0.1 to 1.0 bar(abs.) initial pressure, by Boiko *et al.* (1989).

4.4.8.2
Simplified model by Ogle *et al.* for aluminium dust/air

Ogle *et al.* (1988) developed their model for numerical simulation of turbulent spherical aluminium/air explosions in a closed bomb, assuming spatially uniform pressure at any instant. Due to lack of computational power, Ogle *et al.* were unable to use the $k - \epsilon$ model or an equivalent model for describing the turbulence. Instead they adopted the empirical Abdel-Gayed eddy diffusivity correlation for confined turbulent combustion of premixed gases for obtaining first order approximate values of the turbulent diffusivities of heat and mass:

$$e/\upsilon = 11\ Re_\lambda^{0.56} \tag{4.86}$$

Here e is the eddy diffusivity and υ the kinematic viscosity of the gas. Re_λ is the turbulent Reynolds number defined as $Re_\lambda = v'\lambda/\upsilon$, where v' is the turbulence intensity, or characteristic fluctuating velocity component and λ is the characteristic micro scale of the turbulence. When comparing theoretical predictions with dust explosion experiments in a spherical bomb of diameter 0.34 m, Ogle *et al.* fixed the turbulence intensity at 0.1 m/s and the large-scale eddy size at 0.1 m in all the computations.

The model was formulated for aluminium dust/air explosions, and corresponding experiments were conducted in the 0.34 m diameter spherical bomb with a range of aluminium powders of different particle sizes and shapes. In the model the influence of particle size and dust concentration was accounted for by assuming that the rate of oxidation of the aluminium particles in the cloud was proportional to the surface area of the particles per unit volume of dust cloud. On the assumption that Al(liquid) + AlO $\rightarrow$ Al_2O is the rate controlling reaction in the combustion of aluminium, Ogle *et al.* reformulated the expression for the combustion rate in terms of consumption of molecular species, adopting the standard Arrhenius form of the reaction rate coefficient.

4.4.8.3
Model by Kjäldman for peat/air

Kjäldman (1987) used the $k - \epsilon$ turbulence model for homogeneous gas flow in his finite volume simulation of dust explosions in closed and vented vessels. Referring to explosion experiments in the 20 litre Siwek-sphere (Siwek 1977) and to the turbulence measurements of Kauffman *et al.* (1984) in a 0.95 m^3 spherical vessel, the values of k and ϵ at the moment of ignition in the 20 litre sphere were taken as $k = 10^{-2}(m/s)^2$ and $\epsilon = 2(m/s)^2/s$ respectively, corresponding to a turbulent time scale $k/\epsilon = 5$ ms. The particles were treated as a second hypothetical continuous phase interacting with the gas phase and having the microscopic properties of monosized peat particles of diameter d. Comparatively simple sub-models of particle drying, pyrolysis/devolatilization, gas combustion and char combustion were incorporated. The two continuous phases were assumed to interact by transport of material from the particle phase to the gas phase and transport of heat in either directions depending on whether the gas or the char was burning. The rate of the chemical gas phase reaction was assumed to be controlled by turbulent diffusion, i.e. by ϵ/k. The fuel consumption under these circumstances was calculated using the expression proposed by Magnussen and Hjertager (1976). Kjäldman estimated the role of thermal radiation to be small for the actual type of particles and used a simplified treatment to account for this effect.

Table 4.15 gives a set of corresponding experimental and computed data for peat dust explosions in a 20 litre explosion vessel extracted from Kjäldman's report. The experiments were conducted separately by Weckman (1986).

Table 4.15 Comparison of experimental and computed pressure development during peat dust/air explosions in a closed 20 litre spherical vessel (From Kjäldman, 1987)

Particle diam. [μm]	Dust concentration [g/m³]	Moisture content [weight per cent]	Time from ignition to pressure peak [ms]		P_{max} [bar(g)]		$(dP/dt)_{max}$ [bar/s]	
			Exp.	Comp.	Exp.	Comp.	Exp.	Comp.
38	500	14	46	21	8.4	8.1	513	670
54	500	0	35	18	8.4	8.1	610	700
72	500	34	60	35	7.2	9.0	248	370
96	500	0	47	22	7.8	8.6	413	570
100	1000	13	59	34	7.8	7.2	350	280
165	500	0	45	29	7.7	9.1	395	390

The data in Table 4.15 show good correlation between experimental and computed $(dP/dt)_{max}$ values for four or five of the six powders. An exception is the 100 μm powder, for which the computed value is comparatively low.

This may in part be due to problems with dispersing all the dust in this experiment (20 litre Siwek sphere), which means that the real dust concentration was probably lower than the nominal one of 1000 g/m^3. The maximum pressures, both experimental and computed, are all within 7–9 bar(g), but the correlation between experiments and computations within this narrow pressure range is rather poor. On the other hand, the correlation between experimental and computed times from ignition to pressure peak is good, although there is a systematic deviation by a factor of about two. It should be emphasized that the experiments were conducted with peat dusts of comparatively wide particle size distributions, whereas the computations were for monosized dusts of particle diameter equal to the mass average particle diameter of the real dust.

Kjäldman's contribution constitutes a further valuable step towards development of comprehensive computer models for simulation of dust explosions. The employment of the $k - \epsilon$ model of turbulence represents a significant step forward, but in future it may be necessary to replace even the $k - \epsilon$ model with better approximations in the range of low Reynolds numbers, where it is known that $k - \epsilon$ model may fall short in reproducing reality. Furthermore, the assumption of isotropic turbulence, which is inherent in the $k - \epsilon$ model, may not be acceptable for real dust clouds. Dust explosions in industry are often comparatively slow, in particular in the initial stages, and the turbulence levels correspondingly low.

It should be mentioned that Kjäldman also used his model for some introductory computations of the pressure development in vented explosions, but experimental data were not available for comparison. Kjäldman (1989) also extended the application of his numerical model, in a slightly modified form, to simulating pulverized peat dust combustion in a 5 MW furnace for heat production. Good agreement between the experimental and the computed furnace temperature distributions was obtained.

4.4.8.4
The Clark/Smoot model for accelerating coal dust flames

Explosions in one-end-open ducts with ignition at the closed end, as illustrated in Figure 4.63, constitutes a case where the positive feedback from combustion via expansion, flow and turbulence and back to combustion, is particularly strong. At the same time this case is of primary practical significance in mine gallery explosions. This was the motivation for the development of a numerical model of accelerating coal dust flames in long ducts undertaken by Clark and Smoot (1985).

Like Ogle *et al.* (1988), Clark and Smoot adopted a comparatively simple sub-model for the coupling between flow, turbulence and combustion rate. They used an empirical correlation of the ratio between turbulent and laminar burning velocity, and the turbulent Reynolds number, based on the gas explosion data of Andrews *et al.* (1975), and data from Richmond and Liebman's (1975) and Richmond *et al.*'s (1978) large-scale gallery coal dust explosion experiments. The correlation was:

$$S_t/S_l = C\, Re_\lambda \tag{4.87}$$

where S_t and S_l are the turbulent and laminar burning velocities, C an empirical constant, and Re_λ the turbulent Reynolds number defined as for Equation (4.86). With the eddy viscosity μ_e equal to $v'l$, where v' is the turbulence intensity and l the macro-scale of the turbulence, and introducing further correlations and assumptions, Clark and Smoot expressed Re_λ as:

$$Re_\lambda = 7(\mu_e/\upsilon)^{1/2} \tag{4.88}$$

and

$$Re_\lambda^2 = 24.3[(1 + 0.096\, f(Re)^2) - 1]\Big/\left[1 + \frac{\rho_d}{\rho_g}\right]^{1/2} \tag{4.89}$$

Here υ is the kinematic viscosity, f the Fanning friction factor of the gallery wall, Re the Reynolds number for the overall flow in the gallery, and ρ_d and ρ_g the dust cloud density and gas-phase density respectively.

Clark and Smoot's work confirmed that the increasing level of turbulence in the accelerating fluid is a major driving force behind the flame acceleration in coal dust flames in coal mine galleries.

4.5
DETONATIONS IN DUST CLOUDS IN AIR

4.5.1
QUALITATIVE DESCRIPTION OF DETONATION

Detonation is a singular, extreme mode of propagation of a flame through a premixed gas or a dust cloud. The transfer of heat from the burning to the unburnt cloud by molecular or turbulent diffusion, which is characteristic of the deflagration mode of explosion propagation discussed so far, is replaced by direct ignition by extreme compression of unburnt cloud in a shock wave being driven through the cloud at supersonic speed by the explosion itself. As will be mentioned in Section 4.5.3.2, the detailed mechanism of ignition and combustion inside the shocked detonation front is still a subject of research.

The necessary condition for self-sustained detonation propagation is that the shock wave is sufficiently strong for the volume inside it to become ignited and react chemically before the shock wave has travelled a significant distance away. In this way the shock wave and the chemical reaction zone remain closely coupled, and the shock wave speed and strength is maintained. Typical maximum detonation velocities in premixed hydrocarbon gas/air and dust/air mixtures at normal pressure and temperature and optimum fuel concentrations are in the range 1500–2000 m/s. This is of the order of five times the velocity of sound in the unburnt and uncompressed premixed gas/air or dust cloud in air, and hence the unburnt mixture obtains no gas dynamic signal from the approaching detonation front until being caught by the front itself. Therefore, reducing the maximum explosion pressure of a detonation by venting, is impossible.

It follows from what has been said above, that a detonation in a premixed gas or a dust cloud can only be initiated by a sufficiently strong shock wave. This can either be supplied by an explosive charge or similar external means of generating the initial shock, or by gradual build-up of a strong shock by turbulent acceleration of the explosion itself as illustrated in Figure 4.63. (Deflagration-to-Detonation-Transition, DDT.)

Wolanski (1987) gave a comprehensive review of experimental evidence and theory of dust cloud detonations up to that time.

4.5.2
EXPERIMENTAL EVIDENCE OF DETONATIONS IN DUST CLOUDS IN AIR

4.5.2.1
Experiments in ducts and large-scale galleries

Figure 4.63 illustrates how detonation may develop in ducts of large L/D via enhanced combustion due to flow-generated turbulence. As already mentioned in Section 4.4.7, both Greenwald and Wheeler (1925) and Fischer (1957) reported that coal dust flames in one-end-open large-scale galleries, with ignition near the closed end, accelerated up to a point whereafter a high, constant flame speed was maintained during the remaining length of the gallery. In the case of Greenwald and Wheeler this steady flame speed was about 800 m/s, whereas Fischer reported 1040 m/s as a maximum value in his experiments. These velocities are lower than the Chapman/Jouquet detonation velocities (see Section 4.5.3) that would be expected for coal dust in air. Therefore, Greenwald/Wheeler and Fischer may have observed the kind of constant high-velocity turbulent deflagrations described by Lindstedt and Michels (1989). However, such high-turbulence deflagrations can be nearly as violent as proper detonations. One indication of this is that in Fischer's experiments the pressure measurement stations in the region of the gallery of the constant, high flame speeds were destroyed by the explosion.

Similar evidence of steady high-speed turbulent deflagrations of dust clouds in large-scale galleries was found by Cybulski (1952), Bartknecht (1971) and Rae (1971).

However, both Pineau and Ronchail (1982) and Bartknecht (1971) found clear evidence of proper dust detonations in ducts of smaller diameters. In these cases steady flame speeds of the order of 2000 m/s and high peak pressures of the order of 20 bar(g) were measured, as mentioned in Section 4.4.7 and illustrated in Figure 4.64.

On this background the contribution by Kauffman *et al.* (1982, 1984a) is important. They demonstrated that a steady detonation wave could propagate in clouds of oats and wheat grain dust in air, in a vertical laboratory scale duct of square cross section 6.35 cm by 6.35 cm and length 6 m. The dust was charged into the tube at the top at a mass rate giving the desired dust concentration during gravity settling down the tube. The main dust explosion was initiated by a local hydrogen/oxygen explosion at the bottom tube end.

Using a laser Schlieren technique, it was observed that the shock front was followed closely by an induction zone, which was in turn followed by a reaction zone, as would be expected in a proper detonation wave. The leading shock caused intense dispersion of the particle agglomerates into an optically dense cloud of primary particles within a few mm behind the shock front, where the particles ignited and burnt. After combustion the mixture was again optically transparent. The combustion process was nearly completed 0.5 m behind the shock front, corresponding to a time interval of about 0.3 ms. At an oats dust concentration of 250–270 g/m^3, slightly lower than the stoichiometric one of 300 g/m^3, the measured detonation wave velocity was 1540 m/s, which is somewhat lower than the theoretical Chapman-Jouguet velocity at stoichiometric concentration of 1800 m/s. It would be expected, however, that inevitable energy losses in a dust detonation would cause the real detonation velocity to be lower than the ideal C-J velocity. The highest measured peak pressure was about 24 bar, quite close to the theoretical C-J pressure at stoichiometric concentration, 22.4 bar.

Kauffman *et al.* (1984a) also investigated the upper and lower dust concentration limit for detonations of oats dust in air in their laboratory-scale vertical tube. They found that detonations could only be initiated, even with very vigorous ignition sources, within the narrow concentration range of approximately 200–450 g/m^3.

Further important evidence demonstrating detonations in dust clouds in air has been provided by Gardner *et al.* (1986). The dusts used were coals and included a fine British coal fraction of 87% by mass < 71 μm particle size, containing 33.5% volatiles and 3.5% moisture, and an equally fine US sub-bituminous coal fraction of 41.3% volatiles and 17.3% moisture. Coarser particle size fractions of the two coals were also tested.

The experimental arrangement consisted of a 20 m^3 ignition chamber connected to a 42 m long straight test duct of diameter 0.6 m, which was essentially open at the downstream end. Air was blown through the system at a rate giving 20–30 m/s in the duct, and dust was fed into the air stream just upstream of the 20 m^3 chamber to give the desired dust concentration, ranging from 30 g/m^3 to 850 g/m^3, in the explosion chamber and the 42 m long duct. The dust cloud was ignited in the 20 m^3 chamber by a flame jet or a chemical ignitor. The main results are summarized in Figure 4.69.

Figure 4.69 also gives the theoretical relationships obtained by Artingstall (1961) by solving the concentration equations for a steady state coal dust/air deflagration. The experimental relationship found by Bartknecht (1971, 1978) is also included. Gardner *et al.*'s results are in good agreement with Artingstall's deflagration theory, whereas, on average, the Chapman-Jouguet detonation pressure calculated by Artingstall is significantly lower than Gardner *et al.*'s experimental pressures at the calculated C-J velocity of about 2350 m/s (see Section 4.5.3). The extreme experimental peak pressure value of 81 bar(g) is remarkable. However, Gardner *et al.* refer to Bull's argument that at the onset of detonation there is always a regime in which the combustion wave is overdriven before settling down to the C-J conditions. During this transient period, the detonation pressure can exceed the C-J value considerably.

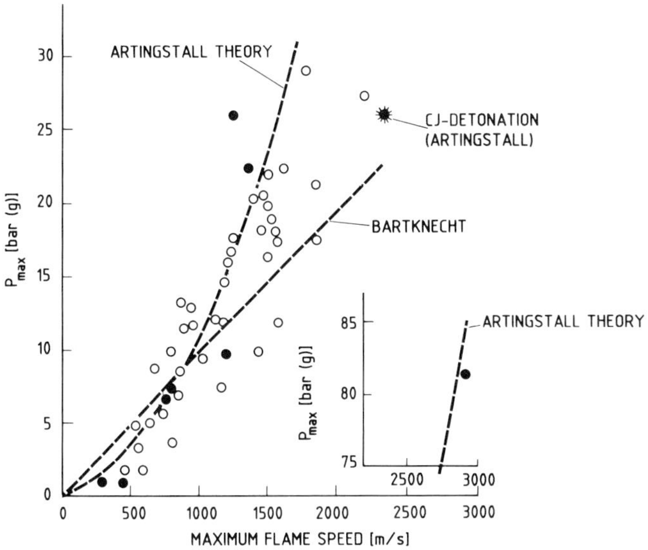

Figure 4.69 *Maximum explosion pressure versus maximum flame speed during coal dust/air explosions in a 42 m long duct of 0.6 m diameter. Particle size: 87% (mass) < 71 μm.* • *US sub-bituminous coal;* ○ *UK coal (From Gardner* et al.*, 1986)*

Gardner *et al.*'s contribution supports the view that proper detonations can also occur in dust clouds, and that such detonations can be brought about by *in situ* transition of fast deflagrations to detonations via turbulent flame acceleration (DDT), as in premixed gases.

4.5.2.2
Unconfined dust cloud detonations

As discussed by Lee (1987), Borisov *et al.* (1983) came to the conclusion that unconfined dust clouds may be considered as practically impossible to detonate. This was because the ignition delay times of clouds in air of wheat flour and similar materials are at least one order of magnitude greater than for methane/air, which requires at least 22 kg of high explosives to detonate in the unconfined state. By assuming that the minimum detonation charge for dust clouds is proportional to the cube of the induction time, as for premixed gases, one finds that a minimum of 20 tons of high explosive would be required for direct initiation of detonation in an unconfined cloud of wheat flour in air. However, the induction time for dust clouds decreases with decreasing particle size, or increasing specific surface area, and therefore unconfined detonations in clouds of very fine dusts become less unlikely than in wheat flour/air.

The only direct experimental observation of a self-sustained detonation wave in an unconfined dust cloud in air that has been traced, was made by Tulis and Selman (1984) and Tulis (1984). They used aluminium dusts of various finenesses, and found that detonation could only be initiated with a very fine aluminium flake powder of specific surface area 3–4 m^2/g, corresponding to spherical particles of diameter smaller than 1 μm. In the first phase of this work, Tulis and Selman (1984) worked with an unconfined dust cloud of approximately cylindrical shape, 6 m in diameter and 1 m in height, containing 4.5 kg of the fine aluminium powder, corresponding to an average nominal dust concentration of 160 g/m^3. The centrally located initiator charge was 2.3 kg of high explosive. Although indications of self-sustained detonations were demonstrated in these experiments, the size of the dust cloud was too small for elimination of the influence of the initiation charge on the detonation wave propagation. Therefore, as reported by Tulis (1984) a further experiment was conducted, using three simultaneously generated aluminium flake clouds of the same size and concentration as described above, forming one large elongated flat unconfined cloud of length 10 m.

When this cloud was initiated at one end with a 2.3 kg high explosive charge, stable, self-sustained detonation was achieved. The average velocity of the detonation wave was 1750 m/s, and the peak pressures in excess of 28 bar. The corresponding calculated C-J values were 1850 m/s and 26 bar. This close agreement between experiment and theory seems to confirm that the phenomenon observed was in fact a proper, unconfined dust cloud detonation.

4.5.3
THEORIES OF DETONATION

4.5.3.1
The Chapman-Jouguet theory

As pointed out by Lee (1987), there is no a priori reason for assuming that the classical Chapman-Jouguet theory for detonation, which has been successfully applied to premixed gases and solid and liquid explosives, does not apply even to explosible dust clouds. This theory was developed independently by Chapman (1899) and Jouguet (1905, 1906), and predicts detonation front velocities, temperatures, pressures and concentrations of reaction products. Davis (1987) discussed the slightly different approaches taken by Chapman and Jouguet. Chapman simply postulated that a detonation front is a shock wave precipitating in its wake chemical reactions that supply the energy required for maintenance of the steady propagation of the shock wave through the explosible gas. In that case the theory of shock wave propagation through a gas could be used to describe detonation by replacing the unreacted gas behind the shock by the products of the combustion reaction and adding the heat of reaction. The resulting theory predicted a specific minimum velocity for self sustained detonation for any given explosible mixture, which Chapman found to be in excellent agreement with the velocities measured in the gas explosion experiments conducted by Dixon (1893). Chapman therefore simply postulated that the minimum velocity predicted by his theory was the detonation velocity of the system considered.

Jouguet (1905, 1906) had been working along similar lines, but his slightly different approach revealed the important additional conclusion that the detonation wave velocity equals the velocity of sound in the hot, compressed reaction products immediately behind the shock front. The Chapman-Jouguet theory is only concerned with the simple system consisting of a homogeneous unreacted gas at a set of initial conditions, and the corresponding shocked reaction products, separated by an infinitely thin, plane of discontinuity. The theory results from the three equations for conservation of mass, momentum and energy across the discontinuity, and the equation of state, as shown by e.g. Jost and Wagner (in Freytag (1965)), Glassman (1977) and Nettleton (1987). Nettleton refers to computer codes that can be used for calculating C-J parameters for various explosible gas mixtures.

As pointed out by Kuchta (1985) the detonation peak pressure for gaseous mixtures is approximately twice the maximum pressure for adiabatic constant-volume combustion of the same mixture (absolute pressures). Kuchta also gave the following equation, which relates the C-J detonation front pressure ratio to the detonation front velocity V_D:

$$\frac{P_2}{P_1} = \frac{1 + \gamma_1 (V_D/C_1)^2}{1 + \gamma_2} \tag{4.90}$$

Here P_2 is the detonation front pressure, P_1 the pressure in the unreacted gas ahead of the detonation front, γ_1 and γ_2 the specific heat ratios of the unreacted gas mixture and the reaction products, and C_1 the sound velocity in the unreacted gas mixture.

As long as the reaction zone is very thin, as it will be for many explosible gas/air mixtures if the composition does not deviate too much from the stoichiometric one, the predicted C-J parameters agree with experiments within a few per cent. However, when the composition approaches the limits of detonation where the thickness of the reaction zone becomes significant, this is no longer the case. The Chapman-Jouguet theory is only concerned with the initial and final states of the gas, and not with the route from the one state to the other. Nettleton (1987) pointed out that close to the limits of ability to detonate the predicted C-J detonation velocities are significantly higher, by 20% or more, than those actually measured. The discrepancies between predicted and measured pressures and densities of the flow just behind the shock front are also correspondingly large in such mixtures. Therefore more refined theories were required.

4.5.3.2
The ZND-theory

The first significant steps towards explaining the details of how chemical reactions are initiated by shock compression and how the resulting energy is transferred to the flow of reaction products were taken independently by Zeldovich (1940), von Neumann (1942) and Döring (1943). As pointed out by Nettleton (1987), the resulting original ZND model of gaseous detonations assumed that the leading shock wave generated a flow of the density and temperature required to initiate exothermic chemical reactions not far behind the shock (1–10 mm).

However, Lee (1987) pointed out that the one-dimensional ZND structure of the detonation front in homogeneous gaseous or liquid explosives has been found to be unstable theoretically, and that the ZND structure has in fact never been observed experimentally in self-sustained gaseous detonations, which rather have a cellular structure. Lee proposed that intense turbulence generated in the shear layers at the cell boundaries causes rapid mixing of unburnt mixture and combustion products and therefore plays a main role in causing ignition just behind the leading shock.

The need for accounting for the role of turbulence in detonation wave propagation was also emphasized by Davis (1987). By doing this, it may also be possible to describe DDT (deflagration-to-detonation-transition) within a unified theory for turbulent flame propagation.

4.5.3.3
Dust clouds

Wolanski *et al.* (1984) were concerned with the detailed structure of dust cloud detonation waves and developed a first order model for the reaction zone, accounting for both two-phase flow effects and wall losses. The flow in the reacton zone was assumed to be one-dimensional and steady, the dust particles were assumed to be spherical and of the same temperature as the surrounding gas. A simplified dust combustion rate model was adopted, assuming heterogeneous reaction. After tuning of the constants of the model against experimental pressure profile data from detonation of wheat dust/air, the reaction zone profiles of particle and gas velocities, temperature, and density could be computed. The calculated detonation velocities were in good agreement with experimental values.

Kulikovskii (1987) discussed the existence of convergent cylindrical and spherical Chapman-Jouguet waves in dust clouds. The theoretical analysis revealed that the ratio between two dimensionless parameters determines the influence of the particles on the detonation wave structure. The first parameter is the ratio of solid particle volume to total dust cloud volume, the second is the product of the mean curvature of the cylindrical or spherical wave and the characteristic particle dimension. If the first parameter is much smaller than the second, the particles have negligible influence. If, however, the first parameter is of the same order or greater than the second, the particles begin to play an important role by significantly altering the flow behind the C-J wave, and the range of its existence.

In another theoretical investigation, Ischii (1983) analysed the influence of the size distribution of the dust particle on the flow structure behind a shock front passing through a dust cloud. A main conclusion was that the particle size distribution is important, and that the assumption of monosized particles, which is often adopted in theoretical work, can lead to poor predictions if the size distribution is in reality comparatively wide.

Lee (1987) anticipated some operational problems in applying the C-J theory to dust clouds because of difficulties with defining the relevant final states after compression and chemical reaction. The assumption of complete chemical equilibrium may differ significantly from actual detonation wave characteristics.

Wolanski (1988) also emphasized the complexity of the dust detonation wave, using coal dust as an example. The measured ignition delays are of the order of ten times those of premixed hydrocarbons. This indicates that release of volatiles from the particles is the rate controlling factor. Volatiles will mix with the oxidizer gas and ignite as soon as they have been released. One cannot consider the induction period as consisting of two distinctly separable consecutive steps, namely devolatilization and subsequent combustion of volatiles. It would be expected that a similar argument applies to dusts of natural and synthetic organic materials (see also Wolanski, 1987).

Fan and Sichel (1988) developed a comprehensive model of the structure of dust cloud detonations, comprising both the induction and the reaction zone, without separating the two. The oxidation of the particles was treated as a heterogeneous surface reaction. Conductive heat transfer within the particles, convective heat transfer between the particles and the gas and reaction heat release within the particles were included in the model. However, because of lack of kinetic data, some constants in the model had to be estimated by fitting theoretical predictions to experimental data. Transverse cellular structure was not accounted for by the model. According to Fan and Sichel the existence of such structure in dust cloud detonations remains to be demonstrated.

Fan Zhang (1989) investigated detonation propagation in maize starch/oxygen clouds in a horizontal tube of 140 mm internal diameter and 17.4 m length. The stoichiometric concentration of maize starch in oxygen at 1 bar(abs) is 1110 g/m^3. For an initial pressure of 1 bar(abs) stable detonation was observed over the dust concentration range from 300 to 9000 g/m^3. The highest stable detonation velocity of 1988 m/s occurred at 2000 g/m^3, and the highest detonation peak pressure of 66.9 bar(abs.) at 3000 g/m^3. The corresponding values at 300 g/m^3 were 1766 m/s and 35.8 bar(abs.), and at 9000 g/m^3 1795 m/s and 43.4 bar(abs.). Fan Zhang concluded, however, that the observed stable detonations could not generally be regarded as classical C-J detonations. This is because of the comparatively long total reaction time, which makes the detonation propagation dependent on the apparatus.

REFERENCES

Abdel-Gayed, R. G., Bradley, D., and Lawes, M. (1987) Turbulent Burning Velocities: A general Correlation in Terms of Straining Rates. *Proc. Roy. Soc.* Lond. A 414, pp. 389–413

Abdel-Gayed, R. G., Bradley, D., and Lung, F. K.-K. (1989) Combustion Regimes and the Straining of Turbulent Premixed Flames. *Combustion and Flame* **76** pp. 213–218

Aldis, D. F., Lee, R. S., and Lai, F. S. (1983) Pressure Development in Explosions of Cornstarch, Wheat Flour and Lycopodium Dust Clouds in a 20 Litre Spherical Chamber. *Pacific Regional Meeting of Fine Particle Society*, (August), Honolulu

Alekseev, A. G., and Sudakova, I. V. (1983) Flame Propagation Rate in Air Suspensions of Metal Powders. *Fizika Goreniya i Vzryva* **19** No. 5 pp. 34–36 (Translated by Plenum Publishing Corporation, 1984)

Amyotte, P. R., and Pegg, M. J. (1989) Lycopodium Dust Explosions in a Hartmann Bomb: Effects of Turbulence. *J. Loss Prev. Process Ind.* **2** pp. 87–94

Amyotte, P. R., Chippett, S., and Pegg, M. J. (1989) Effects of Turbulence on Dust Explosions. *Prog. Energy Combust. Sci.* pp.293–310

Andrews, G. E., and Bradley, D. (1972) The Burning Velocity of Methane/Air Mixtures. *Combustion and Flame* **19** pp. 275–288

Andrews, G. E., Bradley, D., and Lwakamba, S. B. (1975) Turbulence and Turbulent Flame Propagation, A Critical Appraisal. *Combustion and Flame* **24** pp. 285–304

Artingstall, G. (1961) On the Relation Between Flame and Blast in Coal-Dust Explosions. Research Report No. 204, Safety in Mines Research Establishment, Washington

ASTM (1988) *Annual Book of ASTM Standards*. Section 14 General Methods and Instrumentation, ASTM, Philadelphia pp. 692–698

Ballal, D. R. (1983) Flame Propagation through Dust Clouds of Carbon, Coal, Aluminium and Magnesium in an Environment of Zero Gravity. *Proc. R. Soc.* London A385 pp. 21–51

Bartknecht, W. (1971) Brenngas- und Staubexplosionen. Forschungsbericht F45. Bundesinstitut für Arbeitsschutz, Koblenz, F. R. Germany

Bartknecht, W. (1978) *Explosionen, Ablauf und Schutzmassnahmen,* Springer-Verlag

Beer, J. M., Chomiak, J., and Smoot, L. D. (1984) Fluid Dynamics of Coal Combustion: A Review. *Prog. Energy Combust. Sci.* **10** pp. 177–208

BIA/BVS/IES (1987) *Brenn- und Explosionskenngrössen von Stäuben.* Erich Schmidt Verlag GmbH, Bielefeld

Boiko, V. M., Lotov, V. V., and Papyrin, A. N. (1989) Ignition of Gas Suspensions of Metallic Powders in Reflected Shock Waves. *Combustion, Explosion and Shock Waves*, **25** pp. 193–199

Bond, J. F., Knystautas, R., and Lee, J. H. S. (1986) Influence of Turbulence on Dust and Gas Explosions in Closed Vessels. *AIAA Progress in Astronautics and Aeronautics* **106** pp. 571–584

Borisov, A. A., Gelfand, B. E., Timofeev, E. I., *et al.* (1984) Ignition of Dust Suspensions behind Shock Waves. *Proc. of 9th ICDERS*, Poitier, France (1983), published in *Progress in Astronautics and Aeronautics* **94** pp.

Bradley, D., and Mitcheson, A. (1976) Mathematical Solutions for Explosions in Spherical Vessels. *Combustion and Flame* **26** pp. 202–217

Bradley, D., Habik, S. El-Din, and Swithenbank, J. R. (1986) Laminar Burning Velocities of CH_4-Air-Graphite Mixtures and Coal Dusts. *Proc. of 21st Symposium (Internat.) on Combustion.* The Combustion Institute pp. 249–256

Bradley, D., Chen, Z., and Swithenbank, J. R. (1988) Burning Rates in Turbulent Fine Dust/Air Explosions. *Proc. of 22nd Symp. (Int.) on Combustion*, The Combustion Institute, Pittsburgh USA, pp. 1767–1775

Bradley, D., Dixon-Lewis, G., and Habik, S. El-Din, (1989) Lean Flammability Limits and Laminar Burning Velocities of CH_4-Air-Graphite Mixtures and Fine Coal Dusts. *Combustion and Flame* **77** pp. 41–50

Bray, K. N. C. (1980) Turbulent Flows with Premixed Reactants, Chapter in *Turbulent Reaction Flows* by P. A. Libby and F. A. Williams. Springer-Verlag, pp. 115–183

Bryant, J. J. (1971) The Combustion of Premixed Laminar Graphite Dust Flames at Atmospheric Pressure. *Combustion Science and Technology* **2** pp. 389–399

Buksowicz, W., Klemens, R., and Wolanski, P. (1982) An Investigation of the Structure of Dust-Air-Flames. *Proc. of 2nd Int. Specialist Meeting of the Combustion Institute of Oxidation*, Budapest, Hungary

Buksowicz, W., and Wolanski, P. (1983) Flame Propagation in Dust/Air Mixtures at Minimum Explosive Concentration. *Progr. Astronautics and Aeronautics* **87** pp. 414–425

Burgoyne, J. H. (1963) The Flammability of Mists and Sprays. *Proc. of 2nd Symp. Chem. Process Hazards*. (B1) Inst. Chem. Engrs., UK

Cant, R. S., and Bray, K. N. C. (1989) A Theoretical Model of Premixed Turbulent Combustion in Closed Vessels. *Combustion and Flame* **76** pp. 243–263

Cashdollar, K. L., and Hertzberg, M. (1983) Infrared Temperatures of Coal Dust Explosions. *Combustion and Flame* **51** pp. 23–35

Cashdollar, K. L., Sapko, M. J., Weiss, E. S., *et al.* (1988) Laboratory and Mine Dust Explosion Research at the Bureau of Mines, Special Technical Publication No. 958 *Industrial Dust Explosions*, ASTM, Phildadelphia

Cashdollar, K. L., Hertzberg, M., and Zlochower, I. A. (1989) Effect of Volatility on Dust Flammability Limits for Coals, Gilsonite and Polyethylene, *22nd Internat. Symp. on Combustion* The Combustion Institute

Cassel, H. M., Das Gupta, A. K., and Guruswamy, S. (1949) Factors Affecting Flame Propagation through Dust Clouds. *Proc. of 3rd Symp. Combustion, Flame, Explosion Phenomena*. Williams & Wilkins Co.

Cassel, H. M., and Liebman, I. (1959) The Cooperative Mechanism in the Ignition of Dispersions. *Combustion and Flame* **3** p. 467

Cassel, H. M., and Liebman, I. (1963) Combustion of Magnesium Particles II – Ignition Temperatures and Thermal Conductivities of Ambient Atmospheres. *Combustion and Flame* **7** pp. 79–81

Cassel, H. M. (1964) Some Fundamental Aspects of Dust Flames. Rep. Inv. 6551, US Bureau of Mines, Washington

Chamberlain, C. T., and Gray, W. A. (1967) Combustion of Coal in Oxygen. *Nature* **216** p. 1245

Chapman, D. L. (1899) On the Rate of Explosion in Gases. *Philosophical Magazine*, Series 5, **47** pp. 90–104

Chapman, W. R., and Wheeler, R. V. (1986) The Propagation of Flame in Mixtures of Methane and Air. Part IV. The Effect of Restrictions in the Path of the Flame. *Chemical Society London Journal* **37** pp. 2139–2147

Chi, D. N., and Perlee, H. E. (1974) Mathematical Study of a Propagating Flame and its Induced Aerodynamics in a Coal Mine Passageway, Rep. Inv. 7908, US Bureau of Mines, Washington

Chomiak, J. and Jarosinski, J. (1982) Flame Quenching by Turbulence. *Combustion and Flame* **48** pp. 241–249

Clark, D. P., and Smoot, L. D. (1985) Model of Accelerating Coal Dust Flames. *Combustion and Flame* **62** pp. 255–269

Continillo, G., Crescitelli, S., Fumo, E., *et al.* (1986) Misure di explodibilita di carbone in autoclave sferica. *La Rivista dei Combustibili* **40** pp. 77–83

Continillo, G. (1988) Numerical Study of Coal Dust Explosions in Spherical Vessels. *Proc. of 195 ACS National Meeting and 3rd Chemical Congress of North America* (June) **33** No. 2 pp. 188–196

Continillo, G. (1988a) A Two-Zone Model and a Distributed Parameter Model of Dust Explosions in Closed Vessels. *Proc. of 3rd Internat. Colloquium on Dust Explosions*, Szczyrk, Poland

Cybulski, W. (1952) Explosibility of Coal Dust of Very High Fineness. Paper No. 36. *Proc. of 7th Internat. Conf. of Directors of Safety in Mines Research*, Dortmund

Cybulski, W. (1975) *Coal Dust Explosions and their Suppression* (English translation from Polish) pubished by the Foreign Scientific Publications Department of the National Center for Scientific, Technical and Economic Information, Warsaw, Poland

Dahn, C. J. (1991) Effects of Turbulence on Dust Explosion Rate of Pressure Rise. *Archivum Combustionis* Warsaw, Poland

Davis, W. C. (1987) The Detonation of Explosives. *Scientific American* pp. 98–105

Deng Xufan, Kong Dehong, Yen Ginzen, *et al.* (1987) Ignitability and Explosibility of Ca-Si Dust Clouds and Some of their Predictions. *Archivum Combustionis* **7** pp. 19–31

Dixon, H. B. (1893) The Rates of Explosions of Gases. *Philos. Trans. Roy. Soc.* (London) Ser. A 184, 97

Dixon-Lewis, G., and Islam, S. M. (1982) Flame Modelling and Burning Velocity Measurement, *Proc. of 19th Symposium (Internat.) on Combustion*. The Combustion Institute pp. 283–291

Dorsett, H. G., Jacobson, M., Nagy, J., *et al.* (1960) Laboratory Equipment and Test Procedures for Evaluating Explosibility of Dusts. Rep. Inv. 5624, US Bureau of Mines, Washington

Döring, W. (1943) Ueber den Detonationsvorgang in Gasen. *Ann. Physik* **43** p. 421

Eckhoff, R. K., and Fuhre, K. (1975) Investigations Related to the Explosibility of Clouds of Agricultural Dusts in Air. (Part 3), Report No. 72001, (May) Chr. Michelsen Institute, Bergen, Norway

Eckhoff, R. K. (1976) *A Study of Selected Problems Related to the Assessment of Ignitability and Explosibility of Dust Clouds*, Beretninger XXXVIII, 2. John Grieg Chr. Michelsen Institute, Bergen, Norway

Eckhoff, R. K. (1977) The Use of the Hartmann Bomb for Determining K_{St} Values of Explosible Dust Clouds. *Staub-Reinhalt. Luft* **37** pp. 110–112

Eckhoff, R. K., and Mathiesen, K. P. (1977/1978) A Critical Examination of the Effect of Dust Moisture on the Rate of Pressure Rise in Hartmann Bomb Tests. *Fire Research* **1** pp. 273–280

Eckhoff, R. K., Fuhre, K., Guirao, C. M., *et al.* (1984) Venting of Turbulent Gas Explosions in a 50 m^3 Chamber. *Fire Safety Journal* **7** pp. 191–197

Eckhoff, R. K. (1984/1985) Use of $(dP/dt)_{max}$ from Closed-Bomb Tests for Predicting Violence of Accidental Dust Explosions in Industrial Plants. *Fire Safety Journal* **8** pp. 159–168

Eckhoff, R. K. (1987) Measurement of Explosion Violence of Dust Clouds, *Proc. Internat. Symp. Explosion Hazard Classification of Vapours, Gases and Dusts*. Publication NMAB-447, National Academy Press, Washington DC. (Also: Report No. 863350–2 (1986) Michelsen Institute, Bergen, Norway

Eckhoff, R. K., Fuhre, K., and Pedersen, G. H. (1987) Dust Explosion Experiments in a Vented 236 m^3 Silo Cell. *Journal of Occupational Accidents* **9** pp. 161–175

Eckhoff, R. K. (1988) Determination of the Minimum Explosible Dust Concentration by Laboratory-Scale Tests. Report No. 88/02102–1 Chr. Michelsen Institute, Bergen, Norway

Eckhoff, R. K., and Pedersen, G. H. (1988) Ignitability and Explosibility of Polyester/Epoxy Resins for Electrostatic Powder Coating. *Journal of Hazardous Materials* **19** pp. 1–16

Eckhoff, R. K., Alfert, F., Fuhre, K., *et al.* (1988) Maize Starch Explosions in a 236 m^3 Experimental Silo with Vents in the Silo Wall. *Journal of Loss Prevention in the Process Industries* **1** pp. 16–24

Eggleston, L. A., and Pryor, A. J. (1967) The Limits of Dust Explosibility. *Fire Technology* **3** pp. 77–89

Ellis, de C. O. C. (1928) Flame Movement in Gaseous Explosive Mixtures. *Fuel* **7** pp. 5–12

Ellis, de. C. O. C., and Wheeler, R. V. (1928) Explosions in Closed Vessels. The Correlation of Pressure Development with Flame Movement. *Fuel* **7** pp. 169–178

Elsner, Th., Köneke, D., and Weinspach, P.-M. (1988) Thermal Radiation of Gas/Solid Mixtures. *Chem. Eng. Technol.* **11** pp. 237–243

Ermakov, V. A., Razdobreev, A. A., Shorik, A. I., *et al.* (1982) Temperature of Aluminium Particles at the Time of Ignition and Combustion. *Fizika Goveniya i Vzryva* **18** No. 2, pp. 141–143 (Translated by Plenum Publishing Corporation)

Essenhigh, R. H., and Woodhead, D. W. (1958) Speed of Flame in Slowly Moving Clouds of Cork Dust. *Combustion and Flame* **2** pp. 365–382

Essenhigh, R. H. (1961) Combustion Phenomena in Coal Dusts. Part I. *Colliery Engineering* pp. 534–539

Essenhigh, R. H., Froberg, R., and Howard, J. B. (1965) Combustion Behaviour of Small Particles. *Industrial and Engineering Chemistry* **57** pp. 32–43

Essenhigh, R. H., Misra, M. K., and Shaw, D. W. (1989) Ignition of Coal Particles: A Review. *Combustion and Flame* **77** pp. 3–30

Fan Bao-Chun, and Sichel, M. (1988) A Comprehensive Model for the Structure of Dust Detonations. *Proc. 22nd Symp. (Int) on Combustion*, pp. 1741–1750. The Combustion Institute, Pittsburgh, USA

Fan Zhang (1989) Phänomene von Wellen in Medien. Part II: Stabile Detonationen in einer Zweiphasenströmung aus reaktiven Teilchen und Gas. *Doctorate Thesis*, (Dec.), Technical University of Aachen, Germany

Faraday, M., and Lyell, Chas. (1845) Report on the Explosion at the Haswell Collieries, and on the Means of Preventing Similar Accidents. *Philosophical Magazine* **26** pp. 16–35

Fernandez-Pello, A. C. (1982) An Analysis of the Forced Convective Burning of a Combustible Particle. *Combustion Science and Technology* **28** pp. 305–313

Field, M. A. (1969) Rate of Combustion of Size-Graded Fractions of Char from low-rank Coal between 1200 and 2000 K. *Combustion and Flame* **13** p. 237

Fischer, K. (1957) Ablauf von Kohlenstaubexplosionen. *Zeitschrift für Elektrochemie* **61** pp. 685–692

Florko, A. V., Zolotko, N. V., Kaminskaya, N. V., *et al.* (1982) Spectral Investigation of the Combustion of Magnesium Particles. *Fizika Goveniya i Vzryva* **18** pp. 17–22 (Translated by Plenum Publishing Corporation)

Florko, A. V., Kozitskii, S. V., Pisarenko, A. N., *et al.* (1985) Study of Combustion of Single Magnesium Particles at Low Pressure. *Fizika Goveniya i Vzryva* **22** pp. 35–40(Translated by Plenum Publishing Corporation, 1986)

Freytag, H. H. (1965) *Handbuch der Raumexplosionen,* Verlag Chemie GmbH, Weinheim/ Bergstrasse

Friedman, R., and Macek, A. (1962) Ignition and Combustion of Aluminium Particles in Hot Ambient Gases. *Combustion and Flame* **6** pp. 9–19

Friedman, R., and Macek, A. (1963) Combustion Studies of Single Aluminium Particles. *Proc. of 9th Symp. (Internat.) on Combustion.* Academic Press, pp. 703–712

Fristrom, R. M., Avery, W. H., Prescott, R., *et al.* (1954) Flame Zone Studies by the Particle Track Technique. *J. Chem. Phys.* **22** pp. 106–109

Froelich, D., Corbel, S., Prado, G., *et al.* (1987) Experimental Study and Modelling of the Combustion of a Coal Particle. *International Chemical Engineering* **27** pp. 66–69

Frolov, Yu.V. Pokhil, P. F., and Logachev, V. S. (1972) Ignition and Combustion of Powdered Aluminium in High-Temperature Gaseous Media and in a Composition of Heterogeneous Condensed Systems. *Combustion, Explosion and Shock Waves* **8** pp. 213–236

Gardiner, D. P., Caird, S. G., and Bardon, M. F. (1988) An Apparatus for Studying Deflagration through Electrostatic Suspensions of Atomized Aluminium in Air. *Proc. 13th Int. Pyrotech. Sem.* pp. 311–326

Gardner, B. R., Winter, R. J., and Moore, M. J. (1986) Explosion Development and Deflagration-to-Detonation Transition in Coal Dust/Air Suspensions. *Proc. of 21st Symp. (Internat.) on Combustion.* The Combustion Institute, pp. 335–343

Ghoniem, A. F., Chorin, A. J., and Oppenheim, A. K. (1981) Numerical Modelling of Turbulent Combustion in Premixed Gases. *Proc. of 18th Symp. (Internat.) on Combustion.* The Combustion Institute, pp. 1375–1383

Ghosh, B., Basu, D., and Roy, N. K. (1957) Studies of Pulverized Coal Flames. *Proc. of 6th Symposium (Internat.) on Combustion.* The Combustion Institute, Reinhold, New York, pp. 595–602

Gieras, M., Klemens, R., and Wojcicki, S. (1985) Ignition and Combustion of Coal Particles at Zero Gravity. *Acta Astronautica* **12** pp. 573–579

Gieras, M., Klemens, R., Wolanski, P., *et al.* (1986) Experimental and Theoretical Investigation into the Ignition and Combustion Processes of Single Coal Particles under Zero and Normal Gravity Conditions. *Proc. of 21st Symp. (Internat.) on Combustion.* The Combustion Institute, pp. 315–323

Glassman, I. (1977) Combustion. Academic Press

Gmurczyk, G., and Klemens, R. (1988) Effect of Non-Homogeneous Coal Particle Distribution on Combustion Aerodynamics. *Progress in Astronautics and Aeronautics* **113** pp. 102–111

Gomez, C. O., and Vastola, F. J. (1985) Ignition and Combustion of Single Coal and Char Particles. A Quantitative Differential Approach. *Fuel* **64** pp. 559–563

Goral, P., Klemens, R., and Wolanski, P. (1988) Mechanism of Gas Flame Acceleration in the

Presence of Neutral Particles. *Progress in Astronautics and Aeronautics* **113** pp. 325–335
Graaf, J. G. A. de (1965) Ueber den Mechanismus der Verbrennung von festen Kohlenstoff. *Brennstoff-Wärme-Kraft* **17** pp. 227–231
Greenberg, J. B., and Goldman, Y. (1989) Volatilization and Burning of Pulverized Coal, with Radiation Heat Transfer Effects, in a Counter Flow Combustor. *Combust. Sci. Tech.* **64** pp. 1–17
Greenwald, H. P., and Wheeler, R. V. (1925) *Coal Dust Explosions. The Effect of Release of Pressure on their Development.* Safety in Mines Research Board. Paper No. 14. His Majesty's Stationery Office, London
Grigorev, A.I., and Grigoreva, I. D. (1976) Ignition of Metal Particles. *Fizika Goreniya i Vzryva* **12** pp. 208–211 (Translated by Plenum Publishing Corporation)
Guenoche, H. (1964) Flame Propagation in Tubes and Closed Vessels, Chapter in *Non Steady Flame Propagation* by G. H. Markstein, Pergamon Press, Oxford
Hartmann, I., and Nagy, J. (1944) Inflammability and Explosibility of Powders Used in the Plastics Industry. Rep. Inv. 3751, US Bureau of Mines, Washington
Hayes, T., Napier, D. H., and Roopchand, D. R. (1983) Effect of Turbulence on Flame Propagation in Dust Clouds. *Proc. of Spring Technical Meeting*. The Combustion Institute, Canadian Section
Held, E. F. M. (1961) The Reaction between a Surface of Solid Carbon and Oxygen. *Chem. Eng. Sci.* **14** pp. 300–313
Helwig, N. (1965) Untersuchungen über den Einfluss der Korngroesse auf den Ablauf von Kohlenstaubexplosionen. D.82 (Diss. T. H. Aachen). Mitteilungen der Westfählischen Berggewerkschaftskasse. Heft 24
Hertzberg, M., Cashdollar, K. L., and Opferman, J. J. (1979) The Flammability of Coal Dust/Air Mixtures. Lean Limits, Flame Temperatures, Ignition Energies, and Particle Size Effects. Rep. Inv. 8360, US Bureau of Mines, Washington
Hertzberg, M. (1982) The Theory of Flammability Limits, Radiative Losses and Selective Diffusional Demixing. Rep. Inv. 8607, US Bureau of Mines, Washington
Hertzberg, M., Cashdollar, K. L., Daniel, L. NG., *et al.* (1982) Domains of Flammability and Thermal Ignitability for Pulverized Coals and Other Dusts: Particle Size Dependences and Microscopic Analyses. *Proc. of 19th Symp. (Internat.) on Combustion*. The Combustion Institute, pp. 1169–1180
Hertzberg, M., Zlochower, I. A., and Cashdollar, K. L. (1986) Volatility Model for Coal Dust Flame Propagation and Extinguishment. *Proc. of 21st Symp. (Internat.) on Combustion*. The Combustion Institute, pp. 325–333
Hertzberg, M., Zlochower, I. A., Conti, R. S., *et al.* (1987) Thermokinetic Transport Control and Structural Microscopic Realities in Coal and Polymer Pyrolysis and Devolatilization. Their Dominant Role in Dust Explosions. Prepr. Pap. *Am. Chem. Soc. Div. Fuel Chem.* **32**(3) pp. 24–41
Hinze, J. O. (1975) *Turbulence,* (2nd Edition), McGraw-Hill
Hjertager, B. H. (1982) Simulation of Transient Compressible Turbulent Reactive Flows. *Combustion Sci. Technol.* **27** pp. 159–170
Hjertager, B. H. (1984) Influence of Turbulence on Gas Explosions. *J. Hazard. Materials* **9** pp. 315–346
Hjertager, B. H. (1986) Three-Dimensional Modelling of Flow, Heat Transfer and Combustion, Chapter 41 of *Handbook of Heat and Mass Transfer*. Gulf Publishing Company, Houston pp. 1303–1350
Horton, M. D., Goodson, F. P., and Smoot, L. D. (1977) Characteristics of Flat, Laminar Coal Dust Flames. *Combustion and Flame* **28** pp. 187–195
Howard, J. B., and Essenhigh, R. H. (1966) Combustion Mechanism in Pulverized Coal Flames. *Combustion and Flame* **10** pp. 92–93
Howard, J. B., and Essenhigh, R. H. (1967) Pyrolysis of Coal Particles in Pulverized Fuel Flames:

Industrial and Engineering Chemistry. *Process Design and Developm.* **6** pp. 74–84
International Standardization Organization (1985) Explosion Protection Systems. Part I: Determination of Explosion Indices of Combustible Dusts in Air. ISO 6184/1, ISO, Geneva
Ishihama, W. (1961) Studies on the Critical Explosion Density of Coal Dust Clouds. *Proc. of 11th Internat. Conf. of Directors of Safety in Mines Research*, (Oct.) Warsaw
Ishihama, W., Enomoto, H., and Sekimoto, Y. (1982) Upper Explosion Limits of Coal Dust/Methane/Air Mixtures. *Journal of the Association of the Japanese Mining Industry* (in Japanese) pp. 13–17
Ishii, R. (1983) Shock Waves in Gas-Particle Mixtures. *Faculty of Engineering Memoirs* **45** Kyoto University pp. 1–16.
Jacobson, M., Cooper, A. R., and Nagy, J. (1964) Explosibility of Metal Powders. Rep. Inv. 6516, US Bureau of Mines, Washington
Jaeckel, G. (1924) Die Staubexplosionen. *Zeitschrift für technische Physik* pp. 67–78
Jarosinski, J. (1984) The Thickness of Laminar Flames. *Combustion and Flame* **56** pp. 337–342
Jarosinski, J., Lee, J. H. S., Knystautas, R., *et al.* (1987) Quenching Distance of Self-Propagating Dust/Air Flames. *Archivum Combustionis* **7** pp. 267–278
Johnson, G. R., Murdoch, P., and Williams, A. (1988) A Study of the Mechanism of the Rapid Pyrolysis of Single Particles of Coal. *Fuel* **67** pp. 834–842
Jones, W. P., and Launder, B. E. (1972) The Prediction of Laminarization with a Two-Equation Model of Turbulence. *Int. J. Heat Mass Transfer* **15** pp. 301–314
Jones, W. P., and Launder, B. E. (1972) The Calculation of Low-Reynolds-Number Phenomena with a Two-Equation Model of Turbulence. *Int. J. Heat Mass Transfer* **16** pp. 1119–1130
Jouguet, M. (1905) Sur la propagation des reactions chimiques dans les gaz. Chapitre I et II. *Journal de Mathematiques pures et appliquées* **1** Series 61, pp. 347–425
Jouguet, M. (1906) Sur la propagation des reactions chimiques dans les gaz. Chapitre III. *Journal de Mathematiques pures et appliquées* **2** Series 61, pp. 5–86
Kaesche-Krischer, B., and Zehr, J. (1958) Untersuchungen an Staub/Luft-Flammen. *Zeitschrift für Physikalische Chemie Neue Folge* **14** 5/6
Kaesche-Krischer, B. (1959) Untersuchungen an vorgemischten, laminar Staub/Luft-Flammen. Staub **19** pp. 200–203
Kauffman, C. W., Ural, E., Nichols, J. A. *et al.* (1982) Detonation Waves in Confined Dust Clouds. *Proc. of 3rd Internat. School of Explosibility of Dusts*, (5–7 Nov.), Turawa, Poland
Kauffman, C. W., Srinath, S. R., Tezok, F. I., *et al.* (1984) Turbulent and Accelerating Dust Flames. *Proc. of 20th Symp. (Internat.) Combustion.* The Combustion Institute pp. 1701–1708
Kauffman, C. W., Wolanski, P., Arisoy, A., *et al.* (1984a) Dust, Hybrid and Dusty Detonations. *Progress in Astronautics and Aeronautics* **94** pp. 221–240
Kawakami, T., Okajima, S., and Tinuma, K. (1988) Measurement of slow burning velocity by zero-gravity method. *Proc. 22nd Symp. (Int.) on Combustion*, The Comb. Inst. pp. 1609–1613
Khaikin, B. I., Bloshenko, V. N., and Merzhanov, A. G. (1970) *Fizika Goreniya i Vzryva* **5** No. 4
Kjäldman, L. (1987) Numerical Simulation of Peat Dust Explosions. Research Report No. 469, Technical Research Centre of Finland, Espoo
Kjäldman, L. (1989) Modelling of peat dust combustion. *Proc. of 3rd Internat. PHOENICS User Conference.* (Aug./Sept.) Dubrovnik
Klemens, R., and Wolanski, P. (1986) Flame Structure in Dust/Air and Hybrid/Air Mixtures near Lean Flammability Limits. *Progr. Astronautics and Aeronautics* **105** pp. 169–183
Klemens, R., Kotelecki, M., Malanovski, P., *et al.* (1988) An Investigation of the Mechanism of Turbulent Dust Combustion. Private communication to Eckhoff
Kong Dehong (1986) Study of Flame Propagation in a Laminar Dust Cloud. M. Eng. Thesis. Dept. of Metallurgical and Physical Chemistry, Northeast University of Technology, Shenyang, P. R. China
Krazinski, J. L., Backius, R. O., and Krier, H. (1977) Modelling Coal-Dust/Air Flames with

Radiative Tansport. *Proc. Spring Meeting Central States Section.* (March) The Combustion Institute, Cleveland, Ohio
Krazinski, J. L., Backius, R. O., and Krier, H. (1978) A Model for Flame Propagation in Low Volatile Coal-Dust/Air Mixtures. *J. Heat Transfer* **100** pp. 105–111
Kuchta, J. M. (1985) *Investigation of Fire and Explosion Accidents in the Chemical, Mining and Fuel-Related Industries – A Manual.* Bulletin 680, US Bureau of Mines, Washington
Kulikovskii, V. A. (1987) Existence of convergent Chapman-Jouguet Detonation Waves in Dust-Laden Gas. *Fizika Goreniya i Vzryva* **23** pp. 35–41 (Translated by Plenum Publishing Corporation)
Launder, B. E., and Spalding, D. B. (1972) *Mathematical Models of Turbulence.* Academic Press
Lee, J. H. S. (1987) Dust Explosions: An Overview. *Proc. Internat. Symp. Shock Tubes and Waves*, Aachen, F. R. Germany, 16: pp. 21–38
Lee, J. H. S., Yi Kang Pu, and Knystautas, R. (1987) Influence of Turbulence in Closed Volume Explosion of Dust/Air Mixtures. *Archivum Combustionis* **7** pp. 279–297
Lee, J. H. S. (1988) Dust Explosion Parameters, their Measurement and Use. *VDI-Berichte* **701** pp. 113–122
Leuschke, G. (1965) Beitrage zur Erforschung des Mechanismus der Flammenausbreitung in Staubwolken. *Staub* **25** pp. 180–186
Levendis, Y. A., Flagan, R. C., and Gavals, G. R. (1989) Oxidation Kinetics of Monodisperse Spherical Carbonaceous Particles of Variable Properties. *Combustion and Flame* **76** pp. 221–241
Lewis, B., and von Elbe, G. (1961) *Combustion, Flames and Explosion of Gases.* 2nd Ed., Academic Press
Liebman, I., Corry, J., and Perlee, H. E. (1972) Ignition and Incendivity of Laser Irradiated Single Micron-Size Magnesium Particles. *Combustion Science and Technology* **5** pp. 21–30
Lindstedt, R. P., and Michels, H. J. (1989) Deflagration to Detonation Transitions and Strong Deflagrations in Alkane and Alkane/Air Mixtures. *Combustion and Flame* **76** pp. 169–181
Ma, A. S. C., Spalding, D. B., and Sun, L. T. (1982) Application of 'Escimo' to Turbulent Hydrogen/Air Diffusion flame. *Proc. of 19th Symp. (Internat.) on Combustion.* The Combustion Institute pp. 393–402
Magnussen, B. F., and Hjertager, B. H. (1976) On Mathematical Modelling of Turbulent Combustion with Special Emphasis on Soot Formation and Combustion. *Proc. of 16th Symp. (Internat.) on Combustion.* The Combustion Institute, Pittsburgh pp. 719–729
Mallard, E., and le Chatelier, H. L. (1883) Recherches Experimentales et Theoretiques sur la Combustion des Mélanges Gazeux Explosifs. *Annales des Mines* **4** p. 379
Malte, P. C., and Dorri, B. (1981) The Behaviour of Fuel Particles in Wood-Waste Furnaces. *Proc. of Spring Meeting, Western States Section*, (April), Combustion Institute, Washington State University
Mason, W. E., and Wilson, M. J. G. (1967) Laminar Flames of Lycopodium Dust in Air. *Combustion and Flame* **11** pp. 195–200
Matalon, M. (1982) The Steady Burning of a Solid Particle. *SIAM J. Appl. Math.* **42** pp. 787–803
Mitsui, R., and Tanaka, T. (1973) Simple Models of Dust Explosion. Predicting Ignition Temperature and Minimum Explosive Limit in Terms of Particle Size. *Ind. Eng. Chem. Process Des. Develop.* **12** pp. 384–389
Moen, I., Lee, J. H. S., and Hjertager, B. H. (1982) Pressure Development due to Turbulent Flame Propagation in Large-Scale Methane/Air Explosions. *Combustion and Flame* **47** pp. 31–52
Moore, P. E. (1979) Characterization of Dust Explosibility: Comparative Study of Test Methods. *Chemistry and Industry* **7** p. 430
Nagy, J., Seiler, E. C., Conn, J. W., *et al.* (1971) Explosion Development in Closed Vessels. Rep. Inv. No. 7507, US Bureau of Mines, Washington
Nagy, J., Conn, J. W., and Verakis, H. C. (1969) Explosion Development in a Spherical Vessel. Rep. Inv. 7279, US Bureau of Mines, US Dept. Interior, Washington

Nagy, J., and Verakis, H. C. (1983) *Development and Control of Dust Explosions.* Marcel Dekker, Inc.
Nelson, L. S., and Richardson, N. L. (1964) The Use of Flash Heating to Study the Combustion of Liquid Metal Droplets. *The Journal of Physical Chemistry* **68** No. 5 pp. 1269–1270
Nelson, L. S. (1965) Combustion of Zirconium Droplets Ignited by Flash Heating. *Pyrodynamics* **3** pp. 121–134
Nettleton, M. A. (1987) *Gaseous Detonations: Their Nature, Effects and Control,* Chapman and Hall, London
Neumann, J. von (1942) Progress Report on the Theory of Detonation Waves. Report No. 549, OSRD
Nomura, S.-I., and Tanaka, T. (1978) Theoretical Discussion of the Flame Propagation Velocity of a Dust Explosion. The Case of Uniform Dispersion of Monosized Particles. Heat Transfer. *Japanese Research* **7** pp. 79–86
Nomura, S.-I., and Tanaka, T. (1980) Prediction of Maximum Rate of Pressure Rise due to Dust Explosion in Closed Spherical and Non-Spherical Vessels. *Ind. Eng. Chem. Process Des. Dev.* **19** pp. 451–459
Nomura, S.-I., Torimoto, M., and Tanaka, T. (1984) Theoretical Upper Limit of Dust Explosions in Relation to Oxygen Concentration. *Ind. Eng. Chem. Process Des. Dev.* **23** pp. 420–423
Nordtest (1989) Dust Clouds: Minimum Explosible Dust Concentration. NT Fire 011, Nordtest, Helsinki
Nusselt, W. (1924) Der Verbrennungsvorgang in der Kohlenstaubfeuerung. *Zeitschrift Ver. Deutscher Ingenieure* **68** pp. 124–128
Ogle, R. A., Beddow, J. K., and Vetter, A. F. (1983) Numerical Modelling of Dust Explosions: The Influence of Particle Shape on Explosion Intensity. Powder and Bulk Solids Handling and Processing, Technol. Progr., Internat. Powder Science Institute
Ogle, R. A., Beddow, J. K., Vetter, A. F. (1984) A Thermal Theory of Laminar Premixed Dust Flame Propagation. *Combustion and Flame* **58** pp. 77–79
Ogle, R. A., Beddow, J. K., Chen, L. D. (1988) An Investigation of Aluminium Dust Explosions. *Combustion Science and Technology* **61** pp. 75–99
Palmer, K. N., and Tonkin, P. S. (1971) Coal Dust Explosions in a Large-Scale Vertical Tube Apparatus. *Combustion and Flame* **17** pp. 159–170
Pineau, J. P., and Ronchail, G. (1982) Propagation of Dust Explosions in Ducts. *Proc. of International Symposium: The Control and Prevention of Dust Explosions*, (November) Organized by Oyes/IBC, Basle
Pineau, J. P. (1987) Dust Explosions in Pipes, Ducts and Galleries. A State-of-the-Art Report with Criteria for Industrial Design. *Proceedings of Shenyang International Symposium on Dust Explosions*, Sept. 14–16, NEUT, Shenyang, P. R. China
Prentice, J. L. (1970) Combustion of Pulse-Heated Single Particles of Aluminium and Beryllium *Combustion Science and Technology* **1** pp. 385–398
Proust, C., and Veyssiere, B. (1988) Fundamental Properties of Flames Propagating in Starch Dust/Air Mixtures. *Combustion Science and Technology* **62** pp. 149–172
Radandt, S. (1989) Explosionsabläufe in Abhängigkeit von Betriebsparametern. *VDI-Berichte* **701,** Volume 2. VDI-Verlag, Düsseldorf pp. 801–817
Rae, D. (1971) Coal Dust Explosions in Large Tubes. *Proc. of 8th International Shock Tube Symposium*, (July), London
Razdobreev, A. A., Skorik, A. I., and Frolov, Yu.V. (1976) Ignition and Combustion Mechanism of Aluminium Particles. *Fizika Goreniya i Vzryva* **12** No. 2 pp. 203–208 (Translated by Plenum Publishing Corporation)
Richmond, J. K., and Liebman, I. (1975) A Physical Description of Coal Mine Explosions. *Proc. of 15th Symp. (Internat.) on Combustion*. The Combustion Institute, Pittsburgh, USA pp. 115–126
Richmond, J. K., Liebman, I., Bruszak, A. E., *et al.* (1978) A Physical Description of Coal Mine

Explosions. Part II. *Proc. of 17th Symp. (Internat.) on Combustion*. The Combustion Institute, Pittsburgh, USA pp. 1257–1268

Samsonov, V. P. (1984) Flame Propagation in an Impulsive Acceleration Field. *Fizika Goreniya i Vzryva* **20** No. 6 pp. 58–61 (Translated by Plenum Publishing Corporation)

Schläpfer, P. (1951) Ueber Staubflammen und Staubexplosionen. *Schweiz. Verein von Gas- und Wasserfachmännern Monatsbulletin* No. 3, **31** pp. 69–82

Scholl, E. W. (1981) Brenn- und Explosionsverhalten von Kohlenstaub. *Zement-Kalk-Gips* No. **5** pp. 227–233

Schuber, G. (1988) Zünddurchschlagverhalten von Staub-/Luft-Gemischen und Hybriden-Gemischen. Publication Series: *Humanisierung des Arbeitslebens*, **Vol. 72** VDI-Verlag, Düsseldorf

Schuber, G. (1989) Ignition Breakthrough Behaviour of Dust/Air and Hybrid Mixtures through Narrow Gaps. *Proc. of 6th Internat. Symp. Loss Prev. Safety Prom. Proc. Ind.*, Oslo

Schönewald, I. (1971) Vereinfachte Methode zur Berechnung der unteren Zündgrenze von Staub/Luft-Gemischen. *Staub-Reinhalt. Luft* **31** pp. 376–378

Selle, H., and Zehr, J. (1957) Experimentaluntersuchungen von Staubverbrännungsvorgängen und ihre Betrachtung von reaktionsdynamischen Standpunkt. *VDI-Berichte* **19** pp. 73–87

Semenov, E. S. (1965) Measurement of Turbulence Characteristics in a Closed Volume with Artificial Turbulence. *Combustion, Explosion and Shock Waves* **1** No. 2 pp. 57–62

Semenov, N. N. (1951) Tech. Memo. No. 1282, NACA

Shevchuk, V. G., Kondrat'ev, E. N., Zolotko, A. N., *et al.* (1979) Effect of the Structure of a Gas Suspension on the Process of Flame Propagation. *Fizika Goreniya i Vzryva* **15** No. 6 pp. 41–45 (Translated by Plenum Publishing Corporation)

Shevchuk, V. G., Bezrodnykh, A. K., Kondrat'ev, E. N., *et al.* (1986) Combustion of Airborne Aluminium Particles in Free Space. *Fizika Goreniya i Vzryva* **22** No. 5 pp. 40–43 (Translated by Plenum Publishing Corporation)

Siwek, R. (1977) 20 Liter Laborapparatur für die Bestimmung der Explosionskennzahlen brennbarer Stäube. MSc. Thesis, Winterthur Engineering College, Wintherthur

Slezak, S. E., Buckius, R. O., and Krier, H. (1986) Evidence of the Rich Flammability Limit for Pulverized Pittsburgh Seam Coal/Air Mixtures. *Combustion and Flame* **63** pp. 209–215

Smith, I. W. (1971) Kinetics of Combustion of Size-Graded Pulverized Fuels in the Temperature Range 1200–2270 K. *Combustion and Flame* **17** pp. 303–4

Smoot, L. D., and Horton, M. D. (1977) Propagation of Laminar Coal-Air Flames. *Progr. Energy Combust. Sci.* **3** pp. 235–258

Smoot, L. D., Horton, M. D., and Williams, G. A. (1977) Propagation of Laminar Pulverized Coal-Air Flames. *Proc. of 16th Symp. (Internat.) on Combustion*. The Combustion Institute, pp. 375–387

Spalding, D. B. (1957) Predicting the Laminar Flame Speed in Gases with Temperature-explicit Reaction Rates. *Combustion and Flame* **1** pp. 287–295

Spalding, D. B., Stephenson, P. L., and Taylor, R. G. (1971) A Calculation Procedure for the Prediction of Laminar Flame Speeds. *Combustion and Flame* **17** p. 55

Spalding, D. B. (1982) Representations of Combustion in Computer Models of Spark Ignition. Report CFD/82/18, Computational Fluid Dynamic Unit, Imperial College of Science and Technology, London

Specht, E., and Jeschar, R. (1987) Ermittlung der geschwindigkeitsbestimmenden Mechanismen bei der Verbrennung von dichten Kohleteilchen. *VDI-Berichte* **645** pp. 45–56

Srinath, R. S., Kauffman, C. W., Nicholls, J. A., *et al.* Flame Propagation due to Layered Combustible Dusts. *Proc. of 10th International Colloquium on Dynamics of Explosions and Reactive Systems*, (August), Berkeley, USA

Taffanel, M. J. (1907) Premiers Essais sur l'Inflammabilité des Poussieres, Rapport publique par la Comité Central des Houillères de France, Aout

Tai, C. S., Kauffman, C. W., Sichel, M., *et al.* (1988) Turbulent Dust Combustion in a Jet-Stirred

Reactor. *Progress in Astronautics and Aeronautics*, **113** pp. 62–86
Tamanini, F. (1983) Dust Explosion Propagation in Simulated Grain Conveyor Galleries, Technical Report FMRC J.I. OFIR2.RK, (July), Prepared for National Grain and Feed Association, Washington DC
Tamanini, F. (1989) Turbulence Effects on Dust Explosion Venting. *Proc. of AIChF Loss Prevention Symposium*, (April 2–6), Session 8, Plant Layout, Houston
Tamanini, F., and Chaffee, J. L. (1989) Large-Scale Vented Dust Explosions – Effect of Turbulence on Explosion Severity. Technical Report FMRC J.I. OQ2E2.RK, (April), Factory Mutual Research
Tanford, C., and Pease, R. N. (1947) Theory of Burning Velocity. II. The Square Root Law for Burning Velocity. *J. Chemical Physics* **15** pp. 861–865
Tulis, A. J., and Selman, J. R. (1984) Unconfined Aluminium Particle Two-Phase Detonation in Air. *Progress in Astronautics and Aeronautics* **94** pp. 277–292
Tulis, A. J. (1984) Initiation and Propagation of Detonation in Unconfined Clouds of Aluminium Powder in Air. *Proc. of 9th Int. Semin. Pyrotechnics*
Ubhayakar, S. K., and Williams, F. A. (1976) Burning and Extinction of a Laser-Ignited Carbon Particle in Quiescent Mixtures of Oxygen and Nitrogen. *Journ. Electrochem. Society* **123** pp. 747–756
Vareide, D., and Sönju, O. K. (1987) Theoretical Predictions of Char Burn-Off. Report No. STF15 A87044 SINTEF, Trondheim, Norway
Wagner, R., Schulte, A., Mühlen, H.-J., *et al.* (1987) Laboratoriumsuntersuchungen zum Zünden und Abbrandgeschwindigkeit bei der Verbrennung einzelner Kohlekörner. *VDI-Berichte* **645** pp. 33–43
Weber, R. O. (1989) Thermal Theory for Determining the Burning Velocity of a Laminar Flame, Using the Inflection Point in the Temperature Profile. *Combust. Sci. and Tech.* **64** pp. 135–139
Weckman, H. (1986) Safe Production and Use of Domestic Fuels. Part 4. Fire and Explosion Properties of Peat. Research Report No. 448. Technical Research Centre of Finland, Espoo
Wolanski, P. (1977) Numerical Analysis of the Coal Dust/Air Mixtures Combustion. *Archivum Termodynamiki i Spalania* **8** pp. 451–458
Wolanski, P., Lee, D., Sichel, M., *et al.* (1984) The Structure of Dust Detonations. *Progress in Astronautics and Aeronautics* **94** pp. 242–263
Wolanski, P. (1987) Detonation in Dust Mixtures, *Proc. Shenyang Internat. Symp. Dust Expl.* NEUT, Shenyang, P. R. China, pp. 568–598
Wolanski, P. (1988) Oral Statement made at *3rd Internat. Coll. on Dust Explosions*, (Oct. 24–28) Szczyrk, Poland
Yi Kang Pu: (1988) Fundamental Characteristics of Laminar and Turbulent Flames in Cornstarch Dust/Air Mixtures. (January), Ph.D. Thesis, Dept. Mech. Eng., McGill University
Yi Kang Pu, Mazurkiewicz, J., Jarosinski, J. *et al.* (1988) Comparative Study of the Influence of Obstacles on the Propagation of Dust and Gas Flames. *Proc. 22nd Symp. (Int.) on Combustion* The Combustion Institute pp. 1789–1797 Pittsburgh, USA
Zabetakis, M. G. (1965) *Flammability Characteristics of Combustible Gases and Vapors.* Bulletin 627, US Bureau of Mines, Washington
Zehr, J. (1957) Anleitung zu den Berechnungen über die Zündgrenzwerte und die maximalen Explosionsdrücke. *VDI-Berichte* **19** pp. 63–68
Zehr, J. (1959) Die Experimentelle Bestimmung der oberen Zündgrenze von Staub/Luft-Gemischen als Beitrag zur Beurteilung der Staubexplosionsgefahren. *Staub* **19** pp. 204–214
Zeldovich, Ya.B. (1940) On the Theory of the Propagation of Detonation in Gaseous Systems. *J. Exp. Theor. Phys. USSR* **10** p. 524. (Translation: NACA Tech. Memo No. 1261, (1950) pp. 1–50)
Zeldovich, Ya.B., Istratov, A. G., Kidin, N. I., *et al.* (1980) Flame Propagation in Tubes: Hydrodynamics and Stability. *Combustion Science and Technology* **24** pp. 1–13

Chapter 5

Ignition of dust clouds and dust deposits: further consideration of some selected aspects

5.1 WHAT IS IGNITION?

The word 'ignition' is only meaningful when applied to substances that are able to propagate a self-sustained combustion or exothermal decomposition wave. Ignition may then be defined as the process by which such propagation is initiated.

Ignition occurs when the heat generation rate in some volume of the substance exceeds the rate of heat dissipation from the volume and continues to do so as the temperature rises further. Eventually a temperature is reached at which diffusion of reactants controls the rate of heat generation, and a characteristic stable state of combustion or decomposition is established.

The characteristic dimension of the volume within which ignition/no ignition is decided, is of the order of the thickness of the front of a self-sustained flame though the mixture. This is because self-sustained flame propagation can be regarded as a continuing ignition wave exposing progressively new parts of the cloud to conditions where the heat generation rate exceeds the rate of heat dissipation. A similar line of thought applies to propagation of smouldering fires in powder deposits and layers, as discussed in Section 5.2.2.4.

In the ignition process the concepts of stability and instability play a key role. Thorne (1985) gave an instructive simplified outline of some basic features of the instability theory of ignition, which will be rendered in the following. In most situations diffusion, molecular as well as convective, plays a decisive role in the ignition process. Systems that can ignite, may be characterized by a dimensionless number D_a, the Damköhler number, which is the ratio of the rate of heat production within the system due to exothermic chemical reactions, to the rate of heat loss from the system by conduction, convection and radiation. Often D_a is expressed as the ratio of two characteristic time constants, one for the heat loss and one for the heat generation:

$$D_a = \tau_L/\tau_G \quad (5.1)$$

The influence of temperature on the rate of chemical reactions is normally described by the exponential Arrhenius law:

$$k = f\exp(-E/RT) \quad (5.2)$$

where k is the rate constant, f the pre-exponential factor or frequency factor, E the activation energy, R the gas constant and T the absolute temperature.

In general the chemical rate of a combustion reaction may be written:

$$R_C = kC_f^p\ C_{OR}^q \tag{5.3}$$

where $p + q = m$ is the order of the reaction, and C_f and C_{OR} the concentration of fuel and oxygen in the reaction zone. In the case where the fuel is non-depleting and $q = 1$, one gets:

$$R_C = kC_{OR} \tag{5.4}$$

The rate of diffusion of oxygen from the surroundings into the reaction zone is:

$$R_D = D(C_{OS} - C_{OR}) \tag{5.5}$$

where D is the thermal diffusion rate constant and C_{OS} the oxygen concentration in the surroundings.

As the temperature in the reaction zone increases, the thermal reaction rate increases according to Equations (5.2) and (5.4), and a point is reached where the rate is controlled by the diffusional supply of oxygen to the reaction zone. Then $R_C = R_D$ and the right-hand sides of Equations (5.4) and (5.5) are equal.

$$kC_{OR} = D(C_{OS} - C_{OR}) = C_{OS} \times \beta \tag{5.6}$$

where

$$\beta = kD/(k + D) \tag{5.7}$$

is called the Frank-Kamenetskii's overall rate constant, and k is as defined in Equation (5.2). By introducing the heat of reaction Q, the rate of heat generation can, according to Equation (5.6), be expressed as:

$$R_G = Q \times C_{OS} \times \beta \tag{5.8}$$

By inserting Equation (5.2) into (5.7) and substituting for β in (5.8), one gets:

$$R_G = \frac{QC_{OS}Df\exp\left(-E/RT\right)}{D + f\exp\left(-E/RT\right)} \tag{5.9}$$

The general expression for the heat loss from the system considered is:

$$R_L = U(T - T_0)^n,\ n \geqslant 1 \tag{5.10}$$

where U and n are characteristic constants for the system, T the temperature in the reaction zone and T_0 the ambient temperature.

Figure 5.1 illustrates the stability/instability conditions in a system that behaves according to Equations (5.9) and (5.10). Figure 5.1 reveals three intersections between the S-shaped R_G curve and the heat loss curve R_L. In the figure, R_L is a straight line, corresponding to $n = 1$, which applies to heat loss by conduction only. For convection, n is 5/4 and for radiation 4. The upper (3) and lower (1) intersections are stable, whereas the intermediate one (2) is unstable. A perturbation in T at this point either leads to cooling to the lower intersection (1), or to a temperature rise to the upper intersection (3). If the heat loss decreases due to changes of the constants in Equation (5.10), the heat loss curve R_L shifts to the right, and the intersection points (1) and (2) approach each other and finally merge at the critical point of tangency (4). At the same time intersection point (3), which determines the stable state of combustion, moves to higher temperatures.

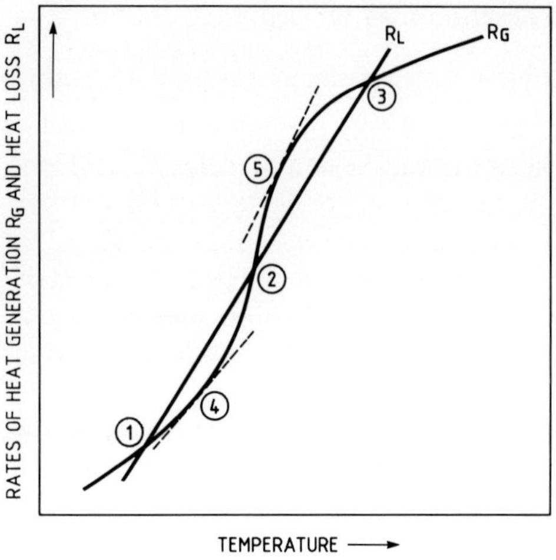

Figure 5.1 *Heat generation and heat loss as functions of temperature in the reaction zone. Explanation of the various features (1)–(5) are given in the text (From Thorne, 1985)*

If U in Equation (5.10) increases, another critical point of tangency (5) is reached. If U increases further, ignition becomes impossible.

If the temperature rise ΔT of the system described by Figure 5.1 is plotted as a function of the Damköhler number as defined in Equation (5.1), a stability/instability diagram as illustrated in Figure 5.2 is obtained. The intersection and tangency points (1) to (5) in Figure 5.1 are indicated.

The lower branch in Figure 5.2 is stable and corresponds to a slow, non-flaming reaction. The upper branch is also stable and corresponds to steady propagation of the combustion or decomposition wave. The intermediate branch is unstable. The system temperature can be raised from ambient temperature without significant increase in the reaction rate until the ignition point (2) has been passed. Then the system jumps to the

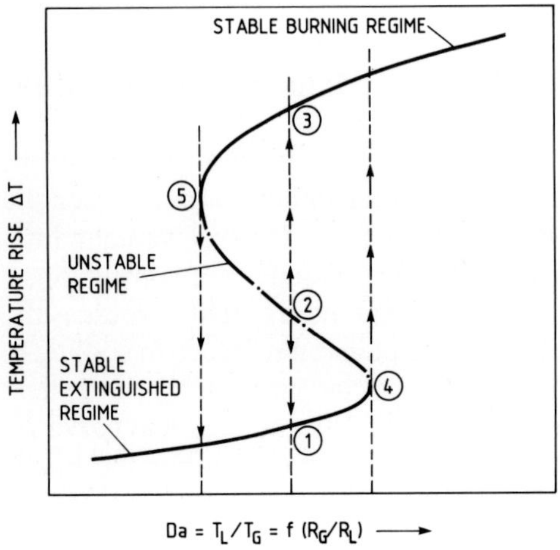

Figure 5.2 *Stability/instability diagram for a combustible system. The features of the points (1)–(5) are explained in the text (From Thorne, 1985)*

upper stable flame propagation branch. Upon cooling, i.e. increasing U or n or both in Equation (5.10), the rate of reaction is reduced. However, the reaction continues right down to (5) in Figure 5.2, from which the system temperature drops to a stable condition in the extinguished regime.

The scheme illustrated in Figures 5.1 and 5.2 is quite general and applicable to different types of systems. More extensive treatments of general ignition/combustion-stability theory were given for example by Gray and Lee (1967), Gray and Sherrington (1977) and Bowes (1981). The classical basis for this type of analysis was established by Semenov (1959) and Frank-Kamenetzkii (1969). The book by Bowes (1984) provides a unique, comprehensive overview of the field of self-heating and ignition, not least in solid materials including dust layers and heaps.

Although the basic considerations implied in Figures 5.1 and 5.2 to some extent provide a satisfactory general definition of ignition, the precise theoretical definition has remained a topic of scientific discussion. One example is the dialogue between Lermant and Yip (1984, 1986) and Essenhigh (1986).

5.2 SELF-HEATING AND SELF-IGNITION IN POWDER DEPOSITS

5.2.1 OVERVIEWS

Bowes (1984) gave the state of the art of experimental evidence and theory up to the beginning of the 1980s. Considerable information was available, and theory for predicting self-heating properties of powders and dusts under various conditions of storage had been developed.

There were nevertheless some gaps in the quantitative knowledge, one of which is biological heating. In vegetable and animal materials such as feed stuffs and natural fibres, self-heating may be initiated by biological activity, in particular if the volume of material is large, its moisture content high and the period of storage long. However, because the micro organisms responsible for the biological activity cannot survive at temperatures above about 75°C, biological heating terminates at this temperature level. Further heating to ignition, therefore, must be due to non-biological exothermic oxidation, for which theory exists. It is possible, however, that the long-term biological activity in a real industrial situation may generate chemically different starting conditions for further self-heating than the conditions established in laboratory test samples heated artificially to 75°C by supply of heat from the outside. Further research seems required in this area.

Starting with the extensive account by Bowes (1984), Beever (1988) highlighted the theoretical developments that she considered most useful for assessing the self-heating and ignition hazards in industrial situations. In spite of many simplifying assumptions, the models available appeared to agree well with experimental evidence. However, extrapolating over orders of magnitude, from laboratory scale data to industrial scale, was not recommended. Biological activity was not involved in the self-heating processes considered.

5.2.2
SOME EXPERIMENTAL INVESTIGATIONS

5.2.2.1
Isoperibolic experiments

In the isoperibolic configuration, the outside of the dust deposits is kept at a constant temperature while the temperature development at one or more points inside the deposit is monitored. The dust sample may either be mechanically sealed from the surroundings, or air may be allowed to penetrate it, driven by the buoyancy of heated gases inside the dust sample or by external over-pressure or suction.

Leuschke (1980, 1981) conducted extensive experimental studies of the critical parameters for ignition of deposits of various combustible dusts under isoperibolic conditions, with natural air draught through the sample, driven by buoyancy. Figure 5.3 shows a plot of the minimum ambient air temperature for self-ignition of deposits of cork dust samples of various shapes and sizes as a function of the volume-to-surface ratio of the sample.

This correlation can be interpreted in terms of the critical Frank-Kamenetzkii parameter for self-ignition (Equation (5.11) below), which was discussed extensively by Bowes (1984). Note that the abscissa scale in Figure 5.3 is linear with the logarithm of the

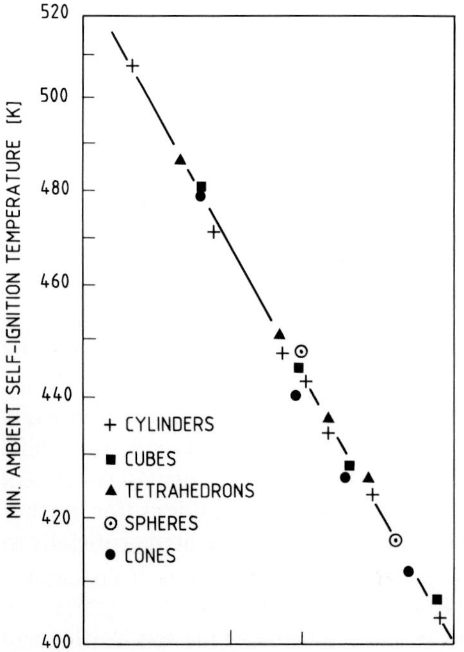

Figure 5.3 *Minimum ambient air temperature for self-ignition of cork dust deposits of various sizes and shapes as a function of the volume/surface area ratio (From Leuschke, 1981)*

volume-to-surface area, whereas the ordinate axis is linear with the reciprocal of the temperature [K].

Some further experimental results produced by Leuschke (1980, 1981) are mentioned in Section 5.2.3.2.

Hensel (1987), continuing the line of research initiated by Leuschke, investigated the influence of the particle size of coal on the minimum self-ignition temperature. Some of his results are given in Figure 5.4.

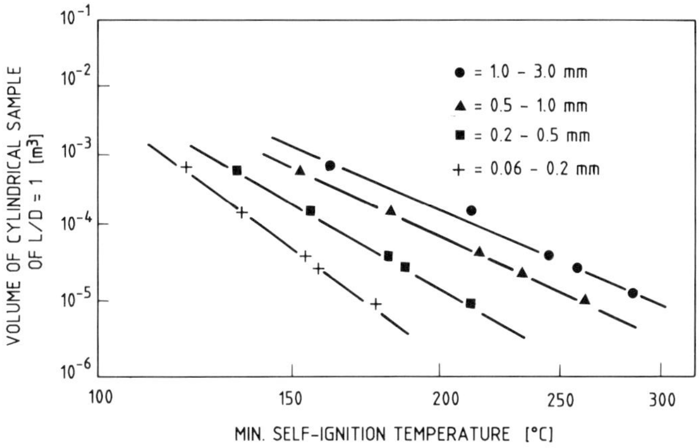

Figure 5.4 *Influence of particle size of coal of 28 wt% volatiles and 6.4 wt% ash on the minimum self-ignition temperature in a heated chamber for various sample volumes (From Hensel, 1987)*

The abscissa axis is linear with the reciprocal of the absolute temperature, which means that $1/T_{min} = A \times \log(\text{sample volume}) + B$, where A and B are constants depending on the particle size. As shown by Hensel (1987), these data also gave linear Arrhenius plots, from which apparent activation energies could be extracted, using the Frank-Kamenetzkii parameter:

$$\delta = Er^2Q\rho f\exp(-E/RT_a)/RT_a^2\,\lambda \tag{5.11}$$

as the theoretical basis. Here E is the activation energy, R the universal gas constant, f the pre-exponential factor, r the characteristic linear dimension of the dust sample, T_a the ambient temperature (temperature of the air surrounding the dust sample in the furnace), Q heat of reaction per unit mass, ρ bulk density of the dust sample, and λ the thermal conductivity of the dust sample.

In a further contribution Hensel (1989) confirmed that data of the type shown in Figure 5.3, for various sample shapes, could in fact be correlated with a good fit using the Frank-Kamenetzkii parameter (Equation (5.11)). The linear dimension r was defined as the shortest distance from the centre of the powder sample to its surface.

Heinrich (1981), being primarily concerned with self-ignition in coal dust deposits, produced a nomograph from which the minimum ambient air temperature for self-ignition in the deposit could be derived from measured values for the same dust and bulk density at two different known volume-to-surface ratios. Although attractive from the practical point of view, however, extrapolating laboratory-scale data to large industrial scale may, as pointed out by Beever (1988), yield misleading results.

Guthke and Löffler (1989) nevertheless proposed that reliable prediction of induction times to ignition in large scale can be obtained from activation energies derived from laboratory-scale self-heating experiments under adiabatic conditions.

5.2.2.2 Dust deposit on hot surface at constant temperature

Miron and Lazzara (1988) determined minimum ignition temperatures for dust layers on a hot surface, for several dust types, using the method recommended by the International Electrotechnical Commission, and described in Chapter 7. The materials tested included dusts of coal and three oil shales, lycopodium spores, maize starch, grain dust and brass powder. For a few of the dusts the effects of particle size and layer thickness on the minimum ignition temperatures were examined. The minimum hot-surface ignition temperatures of 12.7 mm thick layers of these dusts, except grain dust and maize starch, ranged from 160°C for brass to 190°C for oil shale. Flaming combustion was observed only with the brass powder. The minimum ignition temperatures decreased with thicker layers and with smaller particle sizes. Some difficulties were encountered with the maize starch and grain dusts. During heating, the starch charred and expanded, whereas the grain dust swelled and distorted. The test was found acceptable for the purpose of determining the minimum layer ignition temperature of a variety of dusts.

Tyler and Henderson (1987) conducted a laboratory-scale study in which the minimum hot-plate temperatures for inducing self-ignition in 5–40 mm thick layers of sodium dithionite/inert mixtures were determined. The kinetic parameters for the various mixing ratios were determined independently using differential scanning calorimetry (DSC) in both scanning and isothermal modes, and by isothermal decomposition tests. This allowed measured minimum hot-plate temperatures for ignition to be compared with corresponding values calculated from theory, using a modified version of the Tyler/Jones computer simulation code. The code did not require any approximation of the temperature dependence, and reactant consumption was accounted for assuming first order kinetics.

Tyler and Henderson found that the minimum hot-plate temperatures for ignition were significantly affected by the air flow conditions at the upper boundary, as predicted by theory. This must be allowed for when interpreting or extrapolating experimental data. It was further found that the simple model of Thomas and Bowes can be used to interpret experimental results even when appreciable reactant consumption occurs.

Henderson and Tyler (1988) observed that for certain types of dust different experimental routes for the determination of the minimum ignition temperature of a dust layer can lead to widely differing experimental values. For sodium dithionite, experiments starting at a high temperature and working down led to an apparent minimum ignition temperature of nearly 400°C compared to a value of about 190°C when experiments started at a low temperature, working up. The cause of this behaviour was the two stage decomposition of sodium dithionite, and the problems with preparing the dust layer on the hot-plate fast enough for the first stage temperature rise to be observable at high plate temperatures in the range 350–400°C. Similar behaviour may be expected from some other materials.

5.2.2.3
Constant heat flux ignition source in dust deposit

As pointed out by Beever (1984) situations may arise in industry where hot surfaces on which dust accumulates should be described as constant heat flux surfaces rather than as surfaces at constant temperature. Beever mentioned casings of electric motors, high-power electric cables and light bulbs that have become buried in powder or dust, as examples. Practical situations where the temperature of the hot surface is not influenced by the thermal insulation properties of dust accumulations may, in fact, be comparatively rare.

In her constant heat flux ignition experiments, Beever (1984) used samples of wood flour contained in a cylindrical stainless steel wire mesh basket of 0.8 m length and 0.1 m diameter. The ignition source was an electrically heated metal wire coinciding with the axis of the basket. In order to generate different ratios of the radius of the central cylindrical hot surface and the thickness of the cylindrical dust-sample, the heating wire was enveloped by ceramic tubes of different diameters. Some essential properties of the wood flour are given in Table 5.1.

Table 5.1 Properties of wood flour used in self-ignition experiments reported by Beever (1984).

Bulk density, ρ	220 ± 10 kg/m^3
E/R	$1.275 \cdot 10^4$ K
Thermal conductivity, λ	0.346 kJ/hmK
$\frac{\rho \cdot Q \cdot f}{\lambda} \cdot \frac{E}{R}$	$7.678 \cdot 10^{20}$ K/m^2

Here E is the activation energy of the exothermic chemical reaction, R the gas constant, Q the heat of reaction, and f the pre-exponential frequency factor.

Figure 5.5 shows some of Beever's experimental results for a hollow cylindrical wood flour deposit surrounding a cylindrical hot-surface ignition source. A curve predicted from an approximate theory is also shown. The agreement of the theoretical predictions, using a step function approximation, with the experimental results is reasonable, except when the radius of the hot-surface is very small in relation to the thickness of the dust layer.

5.2.2.4
Ignition of dust layers by a small electrically heated wire coil source: propagation of smouldering combustion in dust layers

Leisch, Kauffman and Sichel (1984) studied ignition and smouldering combustion propagation of dust layers in a wind tunnel where the top surface of the dust layer could be subjected to a controlled air flow.

The ignition source was a coil of 0.33 mm diameter platinum wire on a ceramic support. A constant power P was dissipated in the coil for a given period of time Δt, the dissipated energy being $P\Delta t$. Both P and Δt were varied systematically and the minimum dissipated energy for ignition was determined as a function of dissipated power per unit area of the

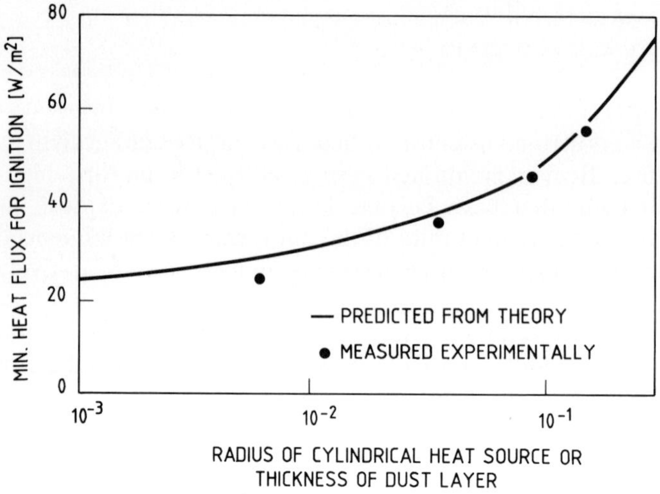

Figure 5.5 *Minimum heat flux for ignition of a centrally heated infinitely long cylindrical wood flour deposit (From Beever, 1984)*

dust envelope in contact with the ignition source. Some results are shown in Figure 5.6. The points in Figure 5.6 are experimental results, whereas the dotted curve is the expected trend in the low power end. The vertical dashed line indicates the value of power/area at which the rate of energy input is equal to the rate of heat loss from the layer. The experimental data in Figure 5.6 indicate that for the higher values of power/area, more energy was needed to ignite the dust layer than in the lower range. According to Leisch, Kauffman and Sichel, this is because at the higher values of power/area, the combustion rate was oxygen diffusion limited and therefore much of the heat transferred to the layer was lost by dissipation into the surroundings. At very low values of power/area, represented by the expected dotted curve, much of the energy furnished to the layer was conducted away before the reaction rate had increased significantly.

Leisch, Kauffman and Sichel (1984) also studied the propagation of smouldering combustion in layers of wood and grain dust. The studies revealed that the smouldering combustion wave had a definite structure, and could be divided into four distinct regions. The initial part of the wave was characterized by discoloration of the unburnt material due to pyrolysis. Pyrolysis occurred when the temperature of the unburnt material reached a minimum value characteristic of that particular material. The pyrolysis products were gaseous volatiles and solid char. The volatiles escaped to the surroundings while the char remained in the layer, forming the second region of the combustion wave, the combustion zone. Oxygen from the atmosphere diffused into this zone, oxidizing the hot char, thereby releasing heat. In the case of forced air flow over the dust layer surface, the combustion zone could contain a visibly glowing subregion. The products of the combustion reaction were CO, CO_2, H_2O vapour, and solid ash. If the combustion was incomplete, some unburnt char would also remain. The ash and any unburnt char would then form the third region of the combustion wave. The final region of the combustion wave was termed the 'cavity'. Only gases (air plus combustion products) were present in this region. However,

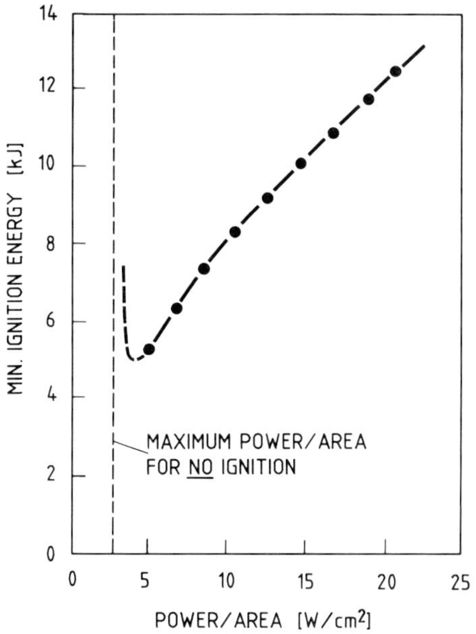

Figure 5.6 *Influence of dissipated power in a hot platinum wire coil, embedded in a layer of grain dust, per unit area of the dust in contact with the coil, on the minimum dissipated energy required for initiating smouldering combustion in the dust layer. Thickness of dust layer 102 mm. Ignition source located 12.7 mm below dust surface. No forced air flow past the dust surface (From Leisch, Kauffman and Sichel, 1984)*

it was shown to constitute an important part of the wave structure in the presence of forced air flow.

Some results from the experiments by Leisch, Kauffman and Sichel are given in Table 5.2 together with values predicted by using a numerical model developed by the same authors.

Table 5.2 Comparison of results from numerical modelling of smouldering combustion in wood dust (pine) layers, with results from experiments (From Leisch, Kauffman and Sichel, 1984)

Property	Experiment	Model
Combustion wave velocity [mm/s]	0.011 to 0.032	0.029
Char temperature [K]	500	485
Maximum temperature in combustion zone [K]	865	770
Reaction zone thickness [mm]	17	13

The data in Table 5.2 refer to experiments with no forced air flow past the surface of the dust layer. With an air flow of 4 m/s the combustion wave velocity was in the range 0.02 to 0.07 mm/s, i.e. about a factor of two higher than without forced air flow. For grain dust layers the combustion wave velocity was 0.0035 to 0.008 mm/s without forced air flow and two to two and one half times higher for 4 m/s air flow. These values are lower than those for wood dust by a factor of three or four.

5.2.2.5
Heat conductivity of dust/powder deposits

As Figure 5.1 illustrates, the rate of heat loss plays an important role as to whether self-heating will result in self-ignition. The heat conductivity of the powder deposit is a central parameter in the heat loss process. It is of interest, therefore, to consider this property more closely. Table 5.3 gives some thermal data for dust/powders published by Selle (1957).

Table 5.3 Specific heats and heat conductivities of some combustible materials in solid and powdered form (Data from Selle, 1957)

Substance	Density [g/cm³]		Powder porosity [vol.%]	Specific heat [J/gK]	Heat conductivity [kJ/mhK]	
	Solid	Powder			Solid	Powder
Air	0.0012	-	0	1.0	0.088	-
Water	1.0	-	0	4.2	2.0	-
Aluminium	2.7	0.31	88.5	0.88	730	0.230
Sulphur	2.1	0.67	67.5	0.75	0.96	0.042
Sugar	1.6	0.65	59	1.25	2.20	0.063
Wood	0.55*	0.15	90	2.5	*** 0.50 to 1.25	0.059
Cork	-	0.074	95	2.5	-	0.033
Brown Coal	1.16**	0.39	74	1.05	0.61	0.067

* 63 vol.% pores in solid
** 33 vol.% pores in solid
*** Depending on orientation of fibres

The heat conductivities in Table 5.3 for the powders, except for aluminium, are very low, and in fact lower than for air. Selle did not describe the method of measurement and further analysis of his data is not possible.

However, in recent years, John and Hensel (1989) developed a hot wire cell allowing more accurate measurement of the heat conductivity of powder and dust deposits. The cell was a vertical cylinder of diameter about 50 mm and height about 200 mm. The heat source was a straight electrically heated resistance wire coinciding with the cell axis, and generating a constant power. The temperature was measured as a function of time at a point in the powder midway between the hot wire and the cell wall. John and Hensel used the Fourier-type equation:

$$\lambda = \frac{q}{4\pi} \times \frac{ln(t_2/t_1)}{T_2 - T_1} \tag{5.12}$$

for calculating the heat conductivity of the powder from two measured temperatures T_1 and T_2 at times t_1 and t_2. Here λ is the heat conductivity and q the heat generated by the hot wire per unit time and wire length. This is a valid approach as long as the two measured temperatures are within a range where the temperature is a linear function of the logarithm of time. A set of data from measurements with this cell are given in Table 5.4.

Table 5.4 Heat conductivities of deposits of some combustible powders and dusts determined from measurements in a hot wire cell, using Equation (5.12) (From John and Hensel, 1989)

Dust type	q [kJ/mh]	t_2 [h]	t_1 [h]	T_2 [°C]	T_1 [°C]	λ [kJ/mhK]
Bituminous coal	59	0.67	0.28	44.9	33.3	0.35
Cork dust	62	0.67	0.25	53.6	40.7	0.38
Wheat flour	53	0.50	0.22	35.3	27.0	0.43
Lycopodium	56	0.57	0.28	35.4	28.8	0.47
Methyl cellulose	54	0.67	0.25	37.4	30.5	0.61
Iron powder	58	0.67	0.25	29.8	23.9	0.77

Faveri *et al.* (1989) presented a theory for the heat conduction in coal piles, using the following expression for the heat conductivity λ in a powder, developed for porous oxides by Ford and Ford (1984):

$$\lambda = \lambda_s(1 - (1 - a\lambda_g/\lambda_s)\epsilon)/(1 + (a - 1)\epsilon) \tag{5.13}$$

where

$$a = \frac{3\lambda_s}{2\lambda_s + \lambda_g}$$

and λ_s and λ_g are the heat conductivity for the solid and gas respectively and ϵ is the porosity of the powder deposit (see Chapter 3). As long as $\lambda_s \gg \lambda_g$, Equation (5.13) reduces to:

$$\lambda = \lambda_s(1 - \epsilon)/(1 + \epsilon/2) \tag{5.14}$$

If this equation is applied to Selle's data in Table 5.3 for powdered sugar, the heat conductivity becomes 0.70 kJ/mhK, and for aluminium and sulphur 58 and 0.23 kJ/mhK respectively. All these values are considerably higher than those given by Selle. For cork dust of porosity 0.95, assuming a value of 2.2 kJ/mhK for λ_s (same as for sugar), Equation (5.14) yields the value 0.074 kJ/mhK, which is lower than for air and therefore must be wrong. The reason is that the simplified Equation (5.14) yields $\lambda = 0$ for $\epsilon = 1$, whereas according to physical reality $\lambda = \lambda_g$. This requirement is satisfied by the more comprehensive Equation (5.13), which, when applied to the cork data, yields a value of 0.16 kJ/mhK. This differs only by a factor of two from the experimental value reported for cork dust by John and Hensel (Table 5.4). If John and Hensel worked with a significantly lower porosity than 0.95, this could explain the difference.

Liang and Tanaka (1987a) used the following formula to account for the influence of temperature on the heat conductivity of cork dust:

$$\lambda = 6.45 \times 10^{-4}\, T + 0.1589 \text{ [kJ/mhK]} \tag{5.15}$$

For T = 300 K, this gives λ = 0.35 kJ/mhK, which is close to the experimental value in Table 5.4. For T = 500 K, Equation (5.15) gives λ = 0.48 kJ/mhK.

Duncan *et al.* (1988) reviewed various theories for the heat conductivity of beds of spherical particles, and compared predicted values with their own experimental results for 2.38 mm diameter spheres. They found that none of the theories tested were fully adequate. In particular, the experiments revealed that gas conduction in the pores between the particles had a significant and predictable effect on the bed conductivity. For a loosely packed bed of aluminium spheres the experimental heat conductivity was 20 and 9 kJ/mhK in nitrogen at atmospheric pressure, and in vacuum respectively. For aluminium and a porosity ϵ of 0.35, Equation (5.14) yields a bed conductivity of about 400 kJ/mhK, which exceeds the experimental values substantially.

Duncan *et al.* found that the heat conductivity of beds of aluminium spheres in nitrogen increased by a factor of 1.5–2.0 when the bed was exposed to a compacting pressure of about 1 MPa. This effect, which was practically absent in beds of spheres of non-ductile materials, is probably due to enlargement of the contact areas between the particles in the bed by plastic deformation.

It seems that a generally applicable theory for reliable estimation of heat conductivities of powder deposits does not exist. Therefore one must rely on experimental determination, e.g. by the method developed by John and Hensel (1989).

5.2.3 FURTHER THEORETICAL WORK

5.2.3.1 The Biot number

The dimensionless Biot number is an important parameter in the theory of self-heating and self-ignition of dust deposits. It is defined as

$$Bi = hr/\lambda \tag{5.16}$$

where h is the heat transfer coefficient at the boundary between the dust deposit and its environment, r is half the thickness, or the radius of the dust deposit, and λ its thermal conductivity. The Biot number expresses the ease with which heat flows through the interface between the powder deposit and its surroundings, in relation to the ease with which heat is conducted through the powder. A Biot number of zero means that the heat conductivity in the powder is infinite and the temperature distribution uniform at any time. $Bi = \infty$ implies that the resistance to heat flow across the boundary is negligible compared to the conductive resistance within the powder.

As pointed out by Bowes (1981) and Hensel (1989), the classical work of Semenov (1935) rests on the assumption that $Bi = 0$, whereas Frank-Kamenetzkii assumed $Bi = \infty$. Thomas (1958) derived steady-state solutions of the basic partial differential heat balance equation for finite plane slabs, cylinders and spheres from which the Frank-Kamenetzkii parameter (Equation (5.11)) could be calculated for Biot numbers $0 < Bi < \infty$.

Liang and Tanaka (1987) found that the fairly complex approximate relationships between the critical condition for ignition and the Biot number originally proposed by Thomas, could be replaced by much simpler formulae based on the Frank-Kamenetzkii approximate steady-state theory. Improved accuracy was obtained by adjusting the formulae to closer agreement with the more exact general numerical solutions for non-steady state.

5.2.3.2 Further theoretical analysis of self-ignition processes: computer simulation models

Liang and Tanaka (1987a, 1988) used the experimental results of Leuschke (1980, 1981) from ignition of cylindrical cork dust samples under *isoperibolic conditions* as a reference for comprehensive computer simulation of the self-heating process in such a system. They assumed that heat did not flow in the axial direction, only radially, and arrived at the following partial differential equation for the heat balance, considering heat generation by zero-order chemical reaction and heat dissipation by radial conduction:

$$\rho C \frac{\partial T}{\partial \theta} = \frac{\lambda}{r} \frac{\partial}{\partial r}\left(r \frac{\partial T}{\partial r}\right) + Qfe^{-E/RT} \qquad (5.17)$$

where

r = radial distance in cylindrical coordinates [m]
ρ = density of the sample [kg m^{-3}]
C = specific heat of the sample [J kg^{-1} K^{-1}]
θ = storage time [h]
λ = thermal conductivity of the sample [J m^{-1} h^{-1} K^{-1}]
Q = heat of reaction [J kg^{-1}]
f = frequency factor of chemical reaction rate [kg m^{-3} h^{-1}]
E = activation energy [J mol^{-1}]
R = universal gas constant [J mol^{-1} K^{-1}]
T = temperature [K]

In order to compare predictions by Equation (5.17) with the data from Leuschke (1980, 1981) for cork dust, the appropriate boundary conditions had to be specified, including a combined heat transfer coefficient of heat dissipation by natural convection and radiation from the cylindrical wall of the cork dust sample. Temperature profiles of cylindrical cork dust samples at any time could then be calculated at various ambient temperatures by solving Equation (5.17) using the finite element method. The predicted radial temperature distributions at any time, the minimum self-ignition temperature, as well as the induction time to ignition, for various sample sizes, agreed well with the experimental data reported by Leuschke (1981), except at extremely high ambient temperatures.

Figure 5.7 gives a set of predicted temperature profiles for cork dust samples of 0.16 m diameter, at three different ambient air temperatures. The predictions were in good agreement with the corresponding experimental data reported by Leuschke (1980, 1981).

At very low ambient air temperatures, close to the minimum for ignition (about 412 K for the 0.16 m diameter sample), ignition starts at the sample axis, whereas at high temperatures it starts at the periphery. This is also in complete agreement with the experimental findings of Leuschke (1980).

Figure 5.8 shows the minimum self-ignition temperature as a function of sample volume for cylindrical cork dust samples, as determined experimentally by Leuschke (1981) and by computer simulation by Liang and Tanaka (1987a, 1988).

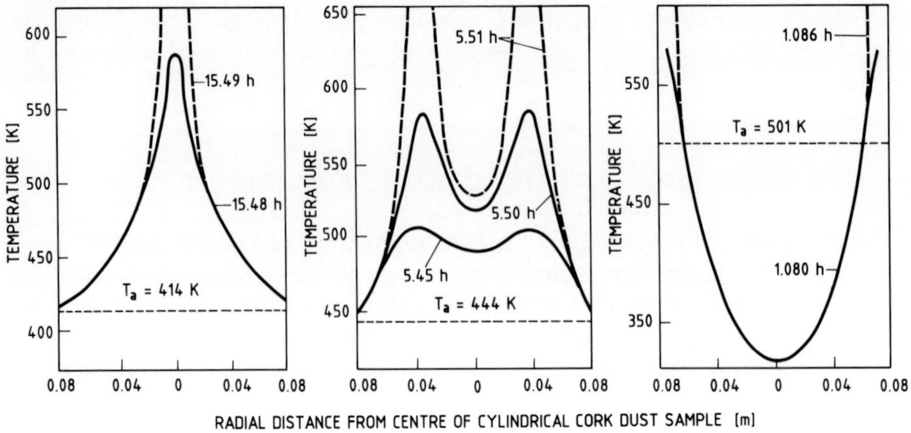

Figure 5.7 *Temperature distributions in a cylindrical cork dust sample of diameter 0.16 m just before ignition (solid lines) and just after (dotted lines), for three different ambient air temperatures* T_a. *Theoretical predictions by Liang and Tanaka (1987a)*

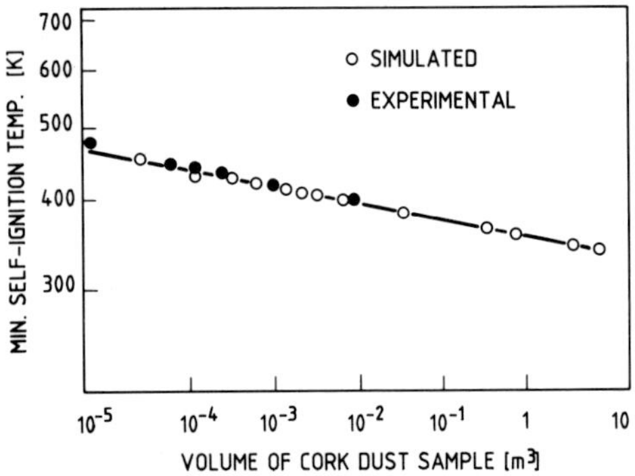

Figure 5.8 *Dependence of minimum self-ignition temperature for cylindrical cork dust samples on sample volume. Experimental data from Leuschke (1981) and computer simulation results from Liang and Tanaka (1987a, 1988)*

Figure 5.9 shows the increase of the induction time to ignition, i.e. the time from introducing the dust sample into air of temperature T_a to ignition of the sample, with increasing sample volume and decreasing T_a.

Leuschke (1981) did not provide data for cork dust corresponding to the simulation results in Figure 5.9. However, he gave a set of experimental data for another natural organic dust, which exhibit trends that are very close to those of the results in Figure 5.9.

The induction time to ignition is an important parameter from the point of view of industrial safety, because it specifies a time frame within which precautions may be taken to prevent self-ignition. This in particular applies to large volumes at comparatively low ambient temperatures, for which the induction times may be very long.

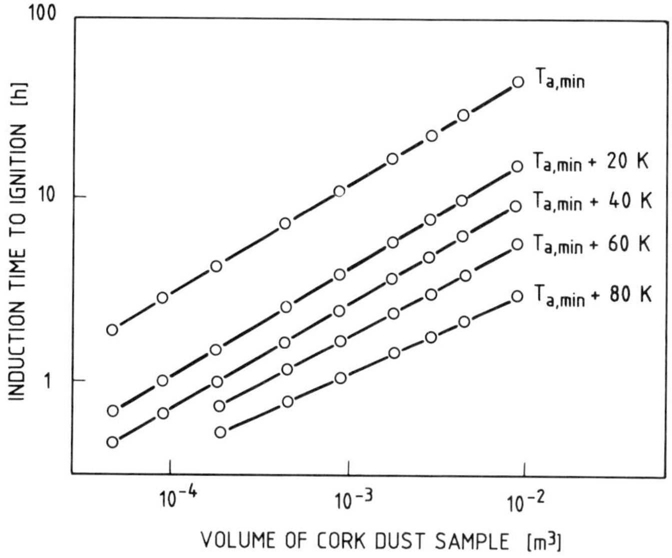

Figure 5.9 *Influence of dust sample volume and ambient air temperature on the induction time to self-ignition of cylindrical deposits of cork dust.* $T_{a,min}$ *is the minimum ambient air temperature for self-ignition. Computer simulation results (From Liang and Tanaka, (1987a)*

The finite element computer simulation approach offers a possibility for analysing self-ignition hazards in a wide range of other geometrical configurations than cylinders. Dik (1987) proposed the use of the thermal impedance method for numerical prediction of critical conditions for self-ignition for various boundary conditions.

Adomeit and Henriksen (1988) developed a computer model addressing the same problem as the model used by Tyler and Henderson (1987), i.e. simulation of self-ignition in dust layers on hot surfaces. It was assumed that the combustion was mainly controlled by homogeneous gas phase reactions, following an initial step of pyrolysis of the solid fuel. The system described by the model is composed of three zones as illustrated schematically in Figure 5.10.

The model implied the following overall picture of the various steps in the ignition process:

1. Formation of a thin gas layer close to the hot surface due to initial pyrolysis of the dust. Reduction of temperature of dust closest to the hot surface due to thermal insulation by the gas.
2. At a given minimum gas layer thickness a homogeneous gas phase reaction starts in a rich premixed zone close to the hot surface.
3. Formation of a second diffusion flame zone between the burning premixed zone and the hot surface, receiving fuel via further pyrolysis caused by the rich primary burning zone.
4. Extinction of diffusion flame due to lack of oxidizer. Drop in pyrolysis rate due to cooling by extinguishing gas.
5. Stabilization of premixed flame close to dust/gas interface.

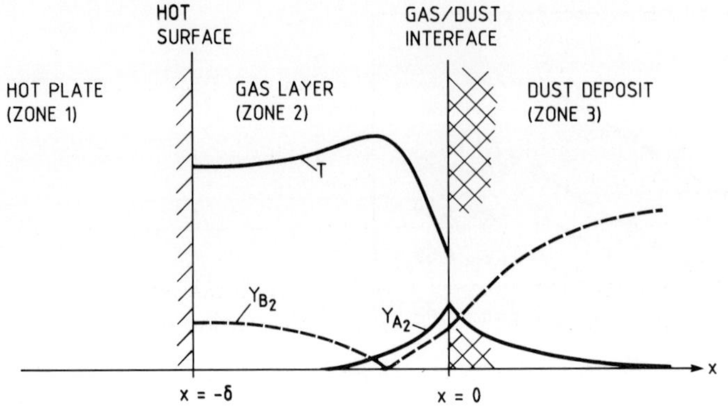

Figure 5.10 *Schematical illustration of system described by computer simulation model for self-ignition of dust layers on hot surfaces.* Y_{A2} *and* Y_{B2} *are the mass fractions of fuel and oxidizer in the gas phase,* T *the gas temperature,* δ *the thickness of the gas layer and* x *the distance from the dust/gas interface (From Adomeit and Henriksen, 1988)*

This model seems to address the case of comparatively high hot-surface temperatures and thin dust layers. Self-ignition in comparatively thick dust layers resting on hot surfaces of quite low temperatures often occurs inside the layer rather than at the hot surface.

Beever (1984) applied the classical self-ignition theory to a dust deposit exposed to a hot surface at constant heat flux boundary conditions. She adopted the step-function approximation devised by Zaturska (1978) and found good agreement between values of the critical Frank-Kamenetzkii parameter for ignition calculated by the approximate theory and values obtained analytically by Bowes, for self-heating in a plane dust slab with constant heat flux on one face. As shown in Section 5.2.2.3, Beever also found good agreement between the predicted minimum heat flux for ignition and experimental results for cylindrical dust deposits heated by an internal concentric cylindrical constant flux heat source.

Leisch, Kauffman and Sichel (1984) were primarily interested in the propagation of a one-dimensional smouldering combustion wave in a dust layer. They obtained a numerical solution of the conservation equations for this process in good agreement with experimental results. (See Section 5.2.2.4). The theoretical model also gave temperature and density profiles within the combustion wave similar to those observed experimentally.

5.2.4 APPLICATIONS TO DIFFERENT POWDER/DUST TYPES: A BRIEF LITERATURE SURVEY

5.2.4.1 Coal dust

Elder *et al.* (1945) studied the relative self-heating tendencies of 46 different coal samples of particle sizes finer than 6 mm, using an adiabatic calorimeter and a rate-of-oxygen-consumption meter. It was found that:

- The self-heating tendency increased with decreasing coal rank.
- The self-heating tendency increased with storage temperature.
- The self-heating tendency decreased with increasing pre-oxidation of the coal prior to the test.
- The rate of heat generation due to oxidation was proportional to the vol.% oxygen in the air in contact with the coal, raised to the power of 2/3.
- The rate of heat generation due to oxidation was proportional to the cube root of the specific surface area of the coal.
- Increasing the ash content in the coal, decreased the self-heating tendency.
- An appreciable moisture content in the coal decreased the self-heating tendency.

Guney and Hodges (1969) reviewed the various experimental methods used up to that time for determining the relative self-heating tendencies of coals. They concluded that only isothermal and adiabatic methods would give consistent results. Shea and Hsu (1972) used an adiabatic method for studying self-heating of various dried coals and petroleum cokes at 70°C in atmospheres of oxygen or nitrogen saturated with water vapour, or in dry oxygen. In a completely dry system there was no appreciable self-heating, even in pure oxygen. The absorption of water from humid atmospheres by dry carbonaceous materials was the major origin of the primary temperature rise from 70 to 90°C.

Chamberlain and Hall (1973) discussed the various chemical and physical properties of coals that influence their oxidizability. Continuous measurement of gases produced during the oxidation process showed that carbon monoxide gives the earliest indication of spontaneous heating.

Heinrich (1981) provided a nomograph from which minimum ambient air temperatures for self-ignition in coal dust deposits may be determined from laboratory-scale measurements of the minimum self-ignition temperatures for two powder samples of different volume to surface ratios. (See also Section 5.2.2.1.)

Heemskerk (1984), using both isothermal and adiabatic test methods, investigated the relationship between the rate of self-heating in coal piles and the oxygen content in the atmosphere in the range 0–20 vol.% oxygen. A systematic decrease of the self-heating rate with decreasing oxygen content was found. Addition of sulphuric acid and iron salts to coal piles stimulated self-heating. Smith *et al.* (1988) investigated the effectiveness of ten different additives, applied as solutions in water, to inhibit self-heating in deposits of a coal of high self-ignition potential, using an adiabatic heating oven. Sodium nitrate, sodium chloride and calcium carbonate were found to be the most effective inhibitors, whereas sodium formate and sodium phosphate stimulated the self-heating process.

Enemoto *et al.* (1987) studied the process leading to a fire in a new bag house installed with a cyclone separator in a pneumatic transport system for pulverized coal. By using classical Frank-Kamenetzkii type theory and appropriate values for the thermal conductivity of the very fine coal dust (2.3 μm) and for the kinetic parameters, it was confirmed that the fire was most probably caused by self-ignition in a dust deposit in the bag house.

Bigg and Street (1988) developed a mathematical computer model for simulation of spontaneous ignition and combustion of a bed of activated carbon granules through which heated air was passed. The model simulated the temporal development of temperature and gas species concentration. The model was validated against the experimental data of Hardman *et al.* (1983) and good agreement was found.

Brooks *et al.* (1988) formulated a mathematical model for evaluating the risk of spontaneous combustion in coal stock piles, using a personal computer. The model predicts expected trends with change in various parameters, but comprehensive validation against experiments was not reported.

Tognotti *et al.* (1988) studied self-ignition in beds of coal particles experimentally, using various cylindrical-shaped beds of diameters 17–160 mm and heights 10–80 mm. Theoretical thermal ignition models were used for interpreting and extrapolating the data from the small-scale experiments. Results from additional isothermal experiments were compared with the small-scale ignition tests. The boundary conditions (Biot number) played an important part in deciding whether ignition would occur.

Takahashi *et al.* (1989) simulated the temperature rise with time in a coal deposit due to spontaneous oxidation, using a numerical computer model. The maximum temperature occurred at the centre of the bed when the oxygen concentration inside the bed was not reduced due to the oxidation reaction, whereas it occurred near the bed surface when the oxygen concentration in the bed decreased due to the consumption. The rate of temperature rise was significantly affected by the activation energy and frequency factor of the coal oxidation. Measurement of the moisture absorbed on the oxidized coal samples showed that the loss in mass due to oxidation increased markedly at temperatures above 120°C. By assuming that the limiting temperature for significant self-heating in coal storage is 120°C, a maximum permissible size of stored coal deposit to prevent self-ignition was estimated for various types of coal.

Hensel (1988) was concerned with a similar problem, namely predicting maximum permissible storage periods for large coal piles. He extrapolated empirical laboratory-scale correlations between the volume/surface area ratio of the dust deposit and the induction time to ignition. An induction time of 10 years was predicted for some 20-year-old, large coal piles in Berlin, in which self-ignition had been observed repeatedly over the last years. By extrapolating the laboratory-scale data, Hensel also confirmed that the size of the actual coal piles was larger than the maximum permissible size for preventing self-ignition at average ambient air temperatures in the Berlin region.

5.2.4.2 Natural organic materials

Raemy and Löliger (1982) used a heat flow calorimeter for studying the thermal behaviour of cereals above 20°C. When the samples were heated in sealed measuring cells, strong exothermic reactions were observed at about 170°C. These reactions were attributed mainly to carbonization of the carbohydrates in the cereals. Raemy and Lambelet (1982) conducted a similar heat flow calorimetric study of self-heating in coffee and chicory above 20°C.

In a study of the thermal behaviour of milk powders, Raemy, Hurrell and Löliger (1983) used both heat flow calorimetry and differential thermal analysis. They found that four main types of reactions are involved in the thermal degradation of milk powders. In order of increasing temperature they are crystallization of amorphous lactose, Maillard reactions, fat oxidation and lactose decomposition.

Self-ignition properties of fish meals were studied by Alfert and co-workers at CMI, Bergen, Norway, by storing the samples, supported by metal gauze baskets, in air at constant temperatures in the range 100–250°C. Some results were reported by Höstmark (1989). For 1- and 2-litre samples the minimum ambient air temperatures for self-ignition were 140 and 130°C respectively. The corresponding induction times to ignition were 5–6 and 8 hours. At ambient air temperatures exceeding 200–240°C, the dust samples ignited close to the surface after induction times of the order of 2 hours. (See trend in Figure 5.7.)

5.2.4.3
Corrosion of direct-reduced iron

Birks and Alabi (1986, 1987, 1988) were concerned with the special problem of self-ignition in piles of direct-reduced iron when exposed to water. The problem arose because direct-reduced iron is stored and transported in charges of considerable size, and it had been observed that the bulk material has a tendency to oxidize to an extent leading to self-ignition. Birks and Alabi investigated the various chemical reactions operating when direct-reduced iron reacts with water and the oxygen in the air.

5.2.4.4
Self-ignition in dust deposits in bag filters in steel works

This problem was studied by Marchand (1988). Two specific cases were discussed to illustrate how hot-spots and smouldering combustion can develop in fabric filter plants in steel works. The cause of accumulation of deposits of very fine dust fractions in the clean section of some filters, and the various possibilities of ignition were analysed. The dusts in question contained a large fraction of combustible material, including carbon, various organic compounds and metallic iron. The typical ignition sources were burning metal droplets expelled from the molten metal and conveyed to the filter.

5.3
IGNITION OF DUST CLOUDS BY ELECTRIC SPARK DISCHARGES BETWEEN TWO METAL ELECTRODES

5.3.1
HISTORICAL BACKGROUND

Holtzwart and von Meyer (1891) were probably the first scientists to demonstrate that dust clouds could be ignited by electric sparks. They studied the explosibility of brown coal dusts in a small glass explosion vessel of 50 m^3 capacity, fitted with a pair of platinum electrodes, between which an inductive spark could pass.

A few years later Stockmeier (1899), who investigated various factors affecting the rate of oxidation of aluminium powder, was able to demonstrate that aluminium dust, shaken up in a glass bottle, ignited in the presence of an electric spark.

Since the publication of these pioneering papers, numerous contributions to the published literature on the spark ignition of dust clouds have been produced. Indeed they have confirmed that ignition of dust clouds by electric discharges is a real possibility and the cause of many severe dust explosions during the years, in mines as well as in industrial plants.

5.3.2 THE OHMIC RESISTANCE OF A SPARK CHANNEL BETWEEN TWO METAL ELECTRODES

Ohm's law can be applied to a spark channel just as well as to any other current-carrying conductor. However, the resistance per unit length of channel is not a constant, but depends on the extent to which the gas in the gap between the electrodes is ionized. This in turn depends on the energy dissipation in the gap per unit time, which determines the temperature in the ionized zone. If equilibrium has been established, one would, for a given gas at a given temperature and pressure, expect a consistent relationship between the gap resistance per unit length and the current flowing through the gap. This has in fact been investigated for air at atmospheric pressure and normal temperature by several workers, as summarized in Figure 5.11.

If it is assumed that the spark resistance for a given current is proportional to the spark gap length, the data from Rose (1959) for a 1.1 mm gap length should be shifted upwards

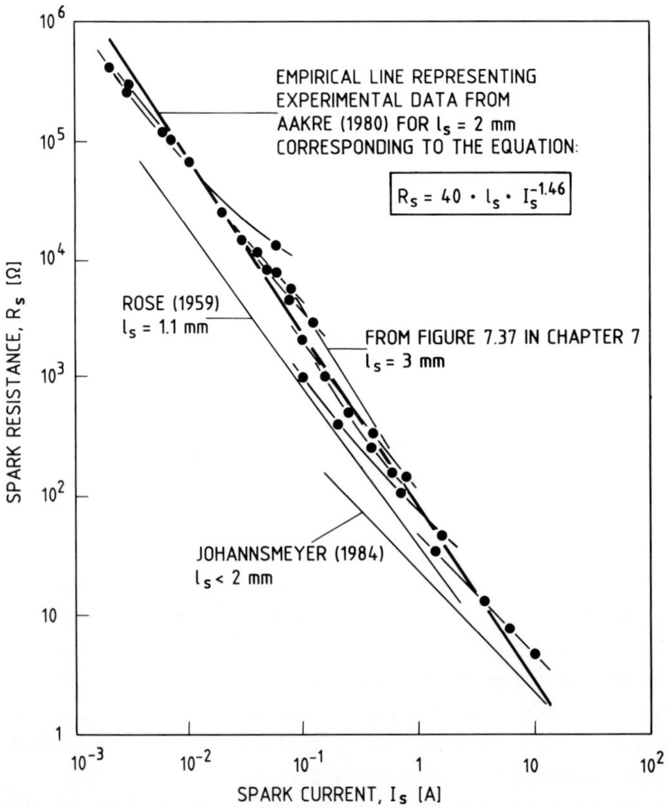

Figure 5.11 *Spark gap resistance R_s as a function of spark current I_s for capacitive spark discharges across a 2 mm spark gap in air at normal pressure and temperature. (Data from Aakre, 1980. Comparison with data for shorter gaps from Rose, 1959 and Johannsmeyer, 1984 and for a 3 mm gap from Figure 7.37 in Chapter 7)*

by a factor of 1.8, and the data from Figure 7.37 in Chapter 7 downwards by a factor of 1.5. The gap length for Johannsmeyer's (1984) data is not known, but it is shorter than 2 mm. It therefore seems as all the data tend to group reasonably well round the data from Aakre (1980), if adjusted to a gap length of 2 mm.

The empirical correlation of all the data in Figure 5.11 yields:

$$R_s = 40 \times l_s \times I_s^{-1.46} \tag{5.18}$$

Normally the ohmic energy dissipation E_s in the spark gap, often called the 'net spark energy', is determined experimentally by simultaneous measurement of the spark current I_s and the spark gap voltage V_s as functions of time during the discharge, and subsequent calculation using the equation:

$$E_s = \int_0^{t_{max}} I_s \times V_s dt \tag{5.19}$$

However, assuming that Ohm's law is valid at any time:

$$V_s = R_s \times I_s \tag{5.20}$$

and substitution of (5.18) into (5.20) yields:

$$V_s = 40 \times l_s \times I_s^{-0.46} \tag{5.21}$$

which, when substituted into (5.19) gives:

$$E_s = 40 \times l_s \int_0^{t_{max}} I_s^{0.54}\, dt \tag{5.22}$$

This equation offers a possibility for determining the ohmic energy dissipation in the spark gap, i.e. the net spark energy, by measuring the spark current $I_s(t)$ only. Figure 5.12 shows a correlation of net spark energies determined from Equations (5.19) and (5.22) using the experimental data from Aakre (1980). As can be seen, the agreement is within a factor of two for $E > 0.1$ mJ. It remains to be seen whether Equation (5.18) is a reasonable approximation even outside the range covered by the data in Figure 5.11.

Equations (5.18) and (5.20) can also be used to express E_s as an integral of V_s instead of I_s.

5.3.3 INFLUENCE OF SPARK DISCHARGE DURATION ON THE MINIMUM ELECTRIC SPARK IGNITION ENERGY FOR DUST CLOUDS

5.3.3.1 Displacement of dust particles by blast wave from spark discharge

The strong influence of the spark discharge duration on the minimum spark energy for ignition of dust clouds was probably first discovered by Boyle and Llewellyn (1950). They were able to demonstrate that the minimum capacitor energy 1/2 CV^2, C being the capacitance and V the initial capacitor voltage, capable of igniting clouds of various powders in air, decreased quite considerably when a series resistance was introduced in

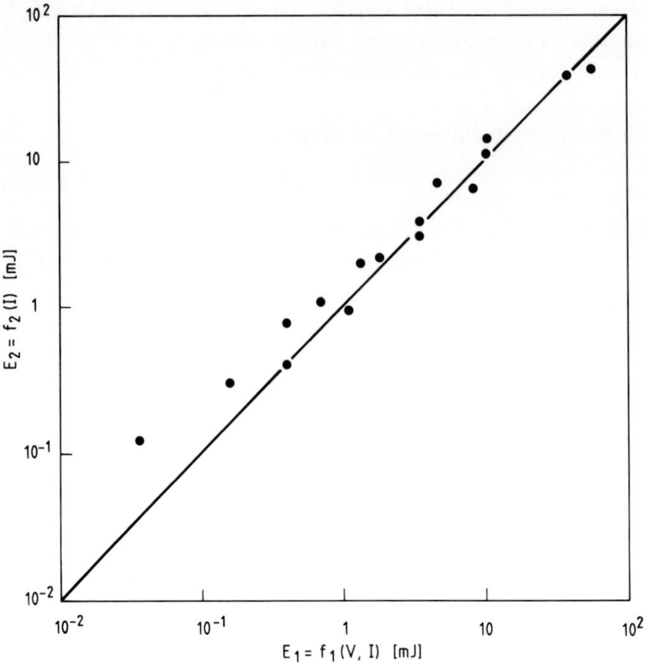

Figure 5.12 *Correlation between spark energies E_1 determined from simultaneous measurements of spark gap voltage V_s and spark current I_s as functions of time, and energies E_2 determined from the spark current measurements only, using the empirical correlation of spark current and spark resistance per unit length of spark gap in Figure 5.11*

the discharge circuit. Some results obtained by Boyle and Llewellyn are shown in Figure 5.13.

As can be seen, the minimum 1/2 CV^2 for ignition decreased by a factor of about ten both for granular aluminium and magnesium, when a series resistance of 10^4 to 10^5 Ω was added to the discharge circuit. Similar trends were also found by these workers for dust clouds of ferromanganese, zinc, silicon, and sulphur.

Boyle and Llewellyn expressed their results in terms of stored capacitor energy 1/2 CV^2. However, a large series resistance in the spark discharge circuit will, during discharge, absorb a significant fraction of the capacitor energy, so that the energy delivered in the spark gap will be considerably lower than 1/2 CV^2. This fraction has been determined experimentally by various workers, as shown by Eckhoff (1975). From independent investigations it can be concluded that with the capacitances used by Boyle and Llewellyn and using a series resistance in the range of 10^4 to 10^7 Ω, the net spark energies were only of the order of 5 to 10% of the stored capacitor energy 1/2 CV^2.

This, in turn, means that in the experiments of Boyle and Llewellyn, an inclusion of a series resistance of 10^4 to 10^5 Ω in the discharge circuit, reduced the minimum net spark energy for ignition to only 1%, or perhaps even less, of the energy required without additional series resistance.

In a later investigation, Line *et al.* (1959) ignited steady-state wall-free and wall-confined 25 mm and 50 mm diameter columns of lycopodium spores in air by electric sparks. Some results for 25 mm columns are shown in Figure 5.14.

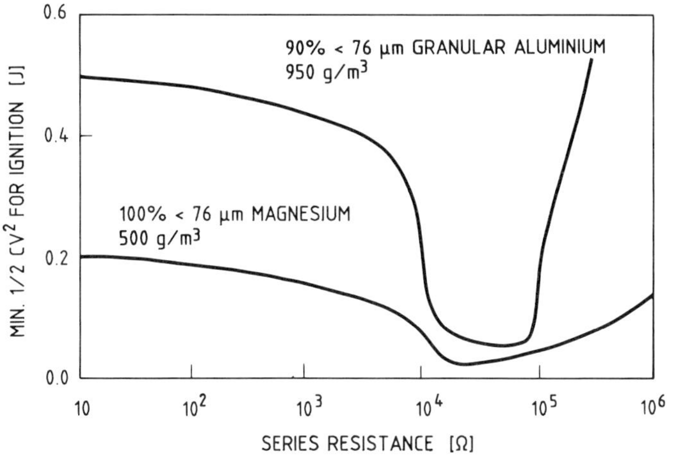

Figure 5.13 *Results from ignition of dust clouds by capacitive sparks, using an additional series resistance in the discharge circuit (Data from Boyle and Llewellyn, 1950)*

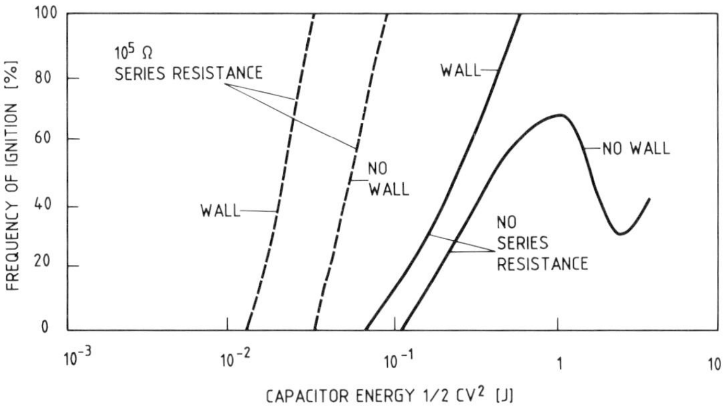

Figure 5.14 *Frequency of ignition of a 25 mm diameter stationary column of 80 g/m³ of lycopodium in air, as a function of capacitor energy. Effect of wall confinement and additional series resistance in the discharge circuit (Data from Line* et al., 1959)

As can be seen both for wall free and confined dust columns, the stored capacitor energies required for producing a given probability of ignition, decreased roughly by a factor of ten if a series resistance of 10^5 Ω was included in the discharge circuit. Both the order of magnitude of the decrease, as well as the order of magnitude of the series resistance giving this maximum decrease, agree with the corresponding figures found by Boyle and Llewellyn for other powders.

Line *et al.* attributed the dramatic influence of spark discharge time to decreasing disturbance of the dust cloud by the blast wave from the spark discharge as the discharge time increased and the spark energies decreased. In the case of high stored capacitor energies and short discharge times, using high-speed filming, they were able to observe the formation of a dust-free zone round the spark before ignition got under way.

Smielkow and Rutkowski (1971) conducted an independent study of the influence of spark discharge duration on the minimum ignition energy of dust clouds. As in the work of Line *et al.* (1959), the spark discharge duration was increased by either adding a very large series inductance (0.1–1.0 H) or a large series resistance (0.45–0.90 MΩ). Reductions in the minimum ignition energies (1/2 CV^2) of the order of a factor of ten was observed, as by Line *et al.*

Eckhoff (1970) conducted further studies of the ignition of clouds of lycopodium spores in air by capacitor sparks of comparatively high energies and short discharge times. Some results are given in Figure 5.15.

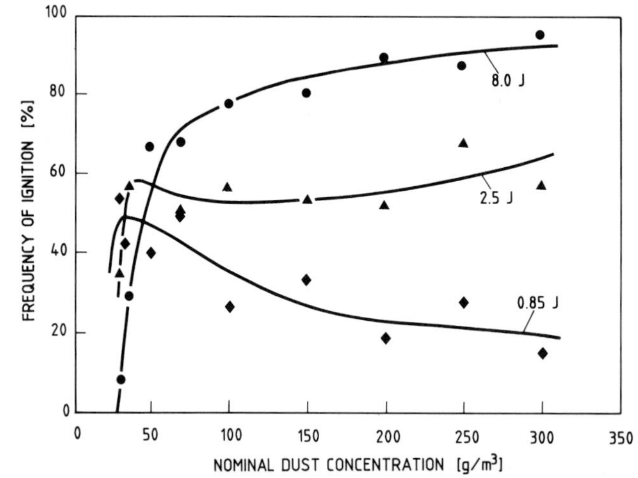

Figure 5.15 *Probability of electric spark ignition of clouds of lycopodium in air as a function of dust concentration, for three different spark energies. Spark discharge duration 5–10 μs. External circuit resistance 0.01 Ω. Circuit inductance 2 μH. Spark gap length 2.7 mm (From Eckhoff, 1970)*

The results in Figure 5.15 are in accordance with those for no wall and no series resistance in Figure 5.14. Even with spark energies of nearly 10 J the frequency of ignition is lower than 100%. The most probable reason for this is that there is a 'knife edge' competition between the heat transfer from the spark to the surrounding dust, which promotes ignition, and the mechanical separation of the dust from the hot spark kernel by the blast wave, which counteracts ignition. The results in Figure 5.14 even show a drop in the frequency of ignition as the spark energy increases from 1 J to 3 J. Eckhoff and Enstad (1976) demonstrated that the blast wave from capacitive discharges of durations of the order of 1 μs and energies of 100–200 mJ, could push a 4 × 5 mm paper piece pendulum, supported by thin threads, an appreciable distance from the spark. The results are given in Table 5.5.

Table 5.5 Displacement distances of a 4 × 5 mm paper piece pendulum due to blast waves from capacitive spark discharges. Initial distance between paper piece and spark gap 1 mm (From Eckhoff and Enstad, 1976)

Spark energy [mJ]	Length of spark gap [mm]	Displacement distance [mm]	
		'short' spark	'long' spark
10	0.1	< 0.5	0
25	0.2	2.5	0
100	1.0	12	< 0.5
300	2.0	35	1

Table 5.5 clearly demonstrates that as the spark energy increased beyond 100 mJ, the displacement of the paper piece by 'short' sparks was appreciable. On the other hand, as the spark energy decreased below 10 mJ, the displacement was practically negligible even for the 'short' sparks, which means that the minimum ignition energy may not necessarily increase with decreasing discharge duration in the range of low spark energies below 10 mJ. This was confirmed by the results of Parker to be discussed below. However, first the theoretical analysis by Enstad (1981) of the interference of the blast from a 'short' spark discharge with the surrounding dust particles, will be outlined. Enstad made the following assumptions:

- The spark discharge time is very short, i.e. less than 0.1 μs for a 1 J spark and less than 0.01 μs for a 1 mJ spark. This means that the spark discharge is completed before any significant expansion of the hot gas has taken place.
- The maximum temperature, i.e. the temperature immediately after completion of the very short heating period and prior to the onset of the subsequent expansion of the hot gas, is estimated at 60 000 K, based on the peak temperature of 50 000 K in a 2 μs, 1 J spark found experimentally by Krauss and Krempl (1963).
- The initial spark is spherical, and the rapid expansion of the hot gas sphere to ambient pressure, following the discharge, is adiabatic, and a rectangular radial temperature distribution in the hot gas is maintained throughout this process. The equation of state for ideal gases, and the expressions $C_v = 5/2\ R$ and $C_p/C_v = 1.5$ apply.
- After completion of the rapid expansion, the hot gas is cooled to ambient temperature by heat conduction into the surrounding gas. This process, involving diffusion of both heat and mass, is described by the equation:

$$\frac{\partial u}{\partial \theta} = u^3 \left\{ \frac{\partial^2 u}{\partial x^2} + \frac{2}{x} \times \frac{\partial u}{\partial x} \right\} \tag{5.23}$$

where u is a dimensionless function of the spark temperature, x a dimensionless expression of the distance from the spark centre, and θ a dimensionless expression of the time.

- The upward movement of the hot gas due to buoyancy is neglected.
- The radial distribution of gas pressure is assumed rectangular throughout the supersonic expansion of the hot gas to ambient pressure.
- The particles are first accelerated by the extremely rapid passage of the shock front through the particle, and by the rapid outward flow of expanding gas following the shock front. At a certain point the particle velocity, because of the inertia, will overtake the gas velocity, and from this stage on the particle velocity will gradually decrease.
- Depending on the Reynolds number, *Re*, either the laminar drag:

$$K_l = \frac{24}{Re} \times \frac{\rho}{2} V^2 \times A_p \tag{5.24}$$

or the turbulent drag:

$$K_t = \frac{\rho}{2} V^2 \times A_p \tag{5.25}$$

acts on the particles during the acceleration as well as during the subsequent retardation process.

The theoretical treatment by Enstad confirmed that a dust free zone, separating the dust cloud from the hot gas core, may in fact be established. As an example, the theoretical results for a 'short' 1.5 J spark discharge in a cloud of lycopodium in air are summarized in Figure 5.16. The distance of a dust particle from the spark centre is given as a function of the time after spark discharge, and the initial position of the particle. Beyond a given instant, depending on the initial particle position, the particle to spark centre distance decreases with time. This is because beyond this point the settling velocity of the particles in quiescent air ($\approx$ 2 cm/s for lycopodium) will dominate, and the particles above the spark will approach the hot gas core.

The 1000 K and 700 K radii of the hot gas sphere as functions of time are also given in Figure 5.16. The minimum ignition temperature of lycopodium clouds in air at atmospheric pressure, as determined in the standard Godbert-Greenwald furnace, is about 700 K. From Figure 5.16 it thus follows that a dust free, cool zone, separating the dust cloud from the incendive part of the hot gas core, is gradually formed from 100 μs after the spark discharge and onwards, making ignition impossible. Figure 5.16 indicates that from less than 1 μs to about 100 μs after the spark discharge, particles with initial positions 2 to 5 mm from the spark centre will be trapped in the spark. However, this is unlikely to cause ignition, because the induction period for 'long' spark ignition of lycopodium clouds in air, as shown by high speed photography by Line *et al.* (1959), is of the order of 1 ms.

It is of interest to note that the radius of the dust free zone at 2 ms after spark discharge, as predicted by Figure 5.16, is in close agreement with the experimental value of about 10 mm found by Line *et al.* (1959) for the same spark energy, type of dust and instant after spark discharge.

The Schlieren flash photograph of a rising hot spark kernel in Figure 5.17 may suggest that Enstad's assumption that the buoyancy of the spark kernel can be neglected, may not be entirely valid.

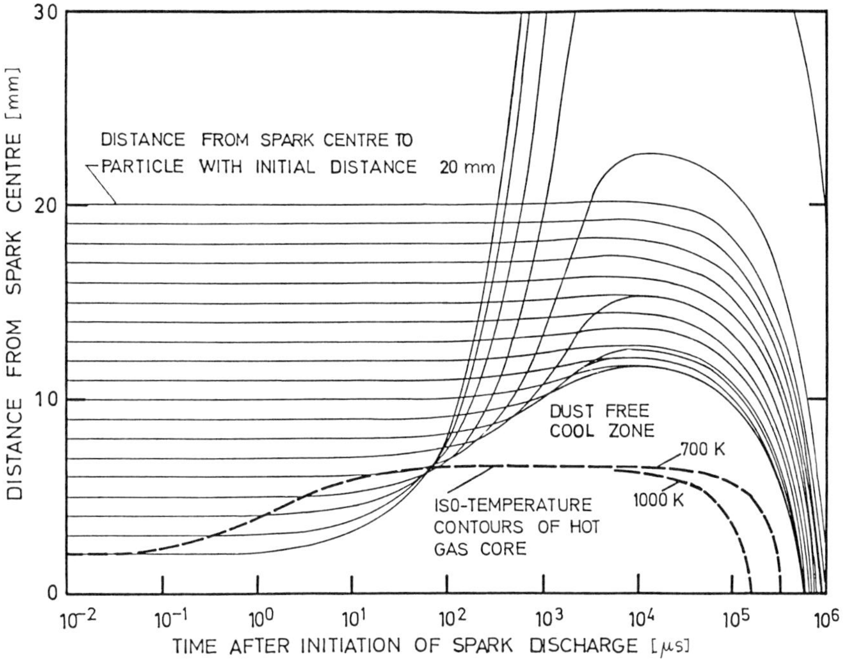

Figure 5.16 *Summary of theoretical prediction of positions of dust particles and radius of the hot gas kernel following a 'short' discharge of a 1.5 J electric spark in a cloud of lycopodium in air (From Enstad, 1981)*

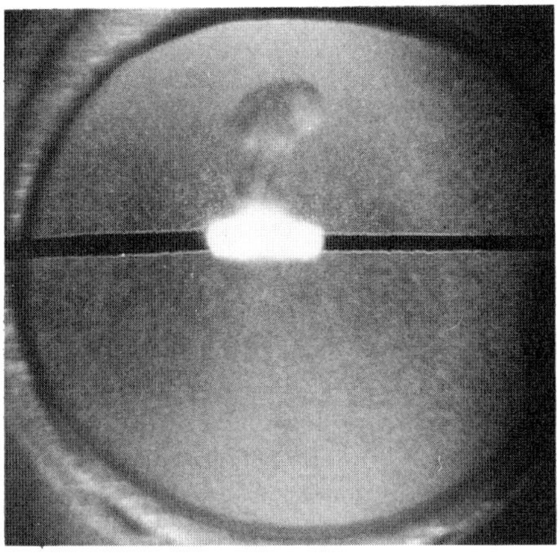

Figure 5.17 *Open-shutter Schlieren flash photo of rising hot spark kernel following an electric spark discharge in a cloud of lycopodium in air. Spark energy 4.3 mJ. Spark discharge duration 28 μs. Delay from initiation of spark discharge to flash 500 μs. Spark gap width 4 mm. Electrode diameter 0.5 mm. The luminous spark image is due to self-radiation during the discharge 472–500 μs before the Schlieren flash. The discharge did not ignite the dust cloud, but some individually burning dust particles are visible just above the luminous spark channel (Courtesy of S. J. Parker)*

5.3.3.2
Optimum spark discharge duration for ignition

A specific study of this aspect was performed by Matsuda and Naito (1983). For lycopodium and < 105 μm cork dust in air, they found the lowest minimum ignition energies for spark durations in the range 0.1–1.0 ms.

The current in an overdamped R-C-L series discharge circuit, after the initial rapid rise to its maximum value, is given by the equation:

$$I = \frac{V_0}{R} \exp(-t/RC) \tag{5.26}$$

where V_0 is the initial capacitor voltage, R the total circuit resistance, C the capacitance, and t the time. By defining the discharge duration as the time required for the current to decrease to one per cent of the initial value at $t = 0$, Equation (5.26) yields:

$$t = 4.6 \times R \times C \tag{5.27}$$

The values of R and C that gave the most incendiary sparks in the investigation by Boyle and Llewellyn (1950) and Lines *et al.* (1959), indicate that the lowest minimum ignition energies were found for discharge durations in the rate 0.1–1.0 ms. Furthermore, the optimum duration appeared to decrease with decreasing minimum net spark ignition energy.

The influence of discharge duration on the minimum electric spark ignition energy of dust clouds was studied systematically by Parker (1985). He used a method of electric spark generation by which the energy and duration of the uni-directional spark discharges could be varied independently in a controlled manner. Parker investigated four different dusts in air, and the results are summarized in Figure 5.18.

For two of the dusts (lycopodium and PAN) there seemed to be a fairly distinct region of optimal discharge durations. For shorter durations, the minimum ignition energy increased markedly. For the two other dusts, however, this effect was absent. As indicated in Figure 5.18, an optimum discharge duration line may be drawn through the results for the four powders. For comparison the spark duration/spark energy characteristic of the CMI discharge circuit (see Chapter 7) has also been included in Figure 5.18. This refers to an R-C-L circuit of inductance L $\geqslant$ 1 mH, for which the discharge will normally be a damped oscillation. The discharge time may then be defined as the time needed for the exponential damping factor to decrease to one per cent of the initial maximum value. The discharge duration then equals:

$$t = 9.2 \times L/R \tag{5.28}$$

which corresponds to Equation (5. 27) for the overdamped case.

As Figure 5.18 shows (Parker, 1985), there is a limit to the combination of spark discharge duration on spark energy which can be realized in practice. This is because a stable arc phase cannot exist unless the degree of ionization of the gas, which is determined by the spark current, exceeds a certain minimum level.

In Chapter 7 the concept of electric spark ignition sensitivity profile is discussed in connection with a standard test for ignition of dust layers by electric sparks (Figure 7.33). In fact, Parker's results for the four dusts in cloud form, as presented in Figure 5.18, are electric spark ignition sensitive profiles. The influence of the spark discharge duration on

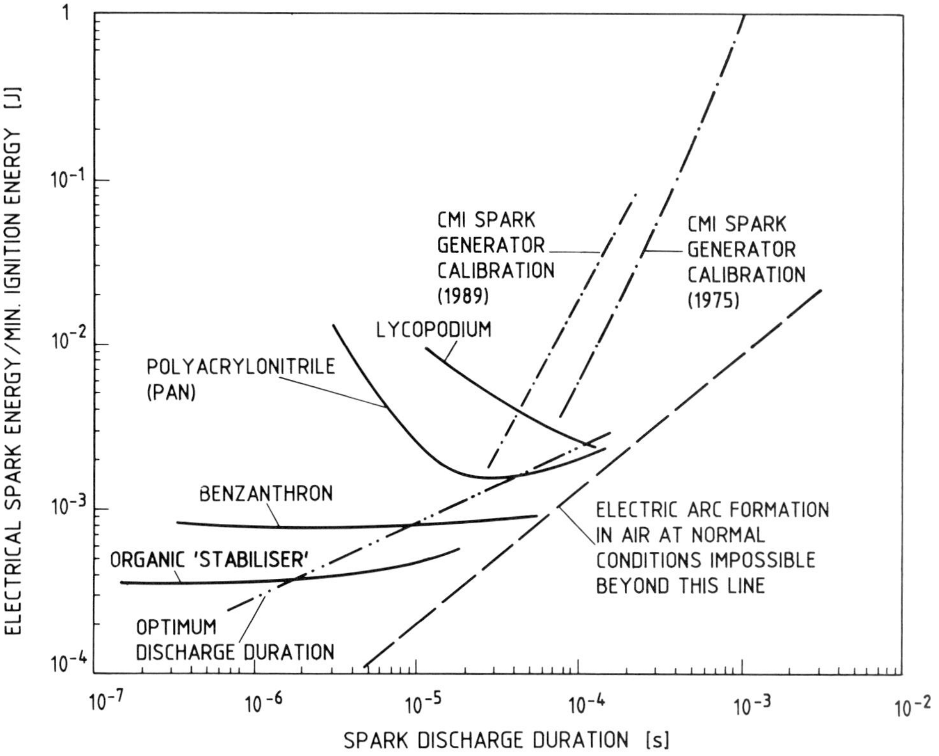

Figure 5.18 *Minimum electric spark ignition energies approx. 5% probability of ignition for four powders as functions of spark discharge duration, as determined by Parker (1985). Electric energy of sparks from CMI-spark generator, as a function of spark discharge-duration (From Eckhoff, 1975)*

the minimum ignition energy is important for adequate use of test data in practice. For example, very low minimum ignition energies determined by the standard discharge circuits of $L \geqslant 1$ mH discussed in Chapter 7 may not be relevant for assessing the electrostatic spark ignition hazard in industrial plant. This is because high inductance values are unlikely to occur in accidental electrostatic discharge circuits in industry.

As discussed in Section 1.1.4.6 in Chapter 1, there are several kinds of electrostatic discharges in air that do not occur across two well-defined, sharp electrodes and therefore do not have such a well-defined shape as the discharge in Figure 5.17. In such cases, which will not be discussed any further in the present context, one could expand the concept of an ignition sensitivity profile to that of an ignition sensitivity surface for a given dust cloud, by adding a spark geometry dimension, as illustrated in Figure 5.19. The definition of an appropriate geometry parameter would, however, require careful consideration.

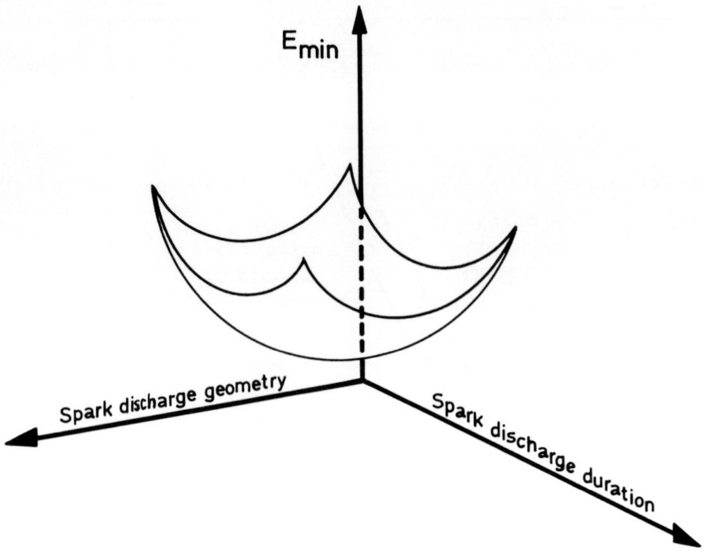

Figure 5.19 *Schematic illustration of the possible concept of electric spark ignition sensitivity surface of an explosible dust cloud*

5.3.4
INFLUENCE OF SOME FURTHER PARAMETERS ON THE MINIMUM IGNITION ENERGY OF DUST CLOUDS

5.3.4.1
Movement/turbulence of dust cloud

The marked increase of the minimum ignition energy for premixed gases with the turbulence intensity of the gas mixture has been demonstrated by various workers, including Ballal and Lefebvre (1977), and Bradley and Lung (1987). One would expect a similar influence of the turbulence intensity of dust clouds on their minimum ignition energies, as indicated by Figure 1.40 in Chapter 1. Figure 5.20 gives some supplementary data by Smielkow and Rutkowski (1971).

These workers dispersed a given quantity of dust from a small cup into the spark gap region by means of an air jet of known velocity. The minimum ignition energies of three dusts, using a 0.1–1.0 H inductance in the capacitive spark discharge circuit, were measured as functions of the estimated velocity of the dispersed dust cloud through the spark gap region. As can be seen from Figure 5.20, a systematic increase of the energy required for ignition, with the dust/air velocity, was found.

5.3.4.2
Spark gap length

This effect was studied by Ballal (1979), using quasi-laminar dust clouds of various materials. A set of results are given in Figure 5.21, which indicate a systematic increase of

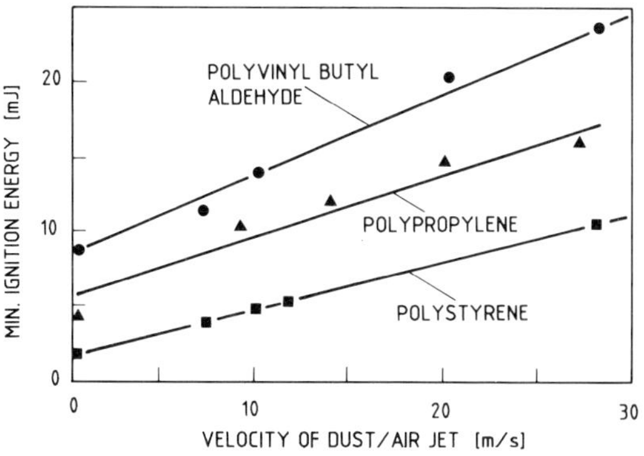

Figure 5.20 *Influence of velocity of dust cloud through spark gap region on the minimum electric spark ignition energy for three plastic dusts of particle size < 75 μm (From Smielkow and Rutkowski, 1971)*

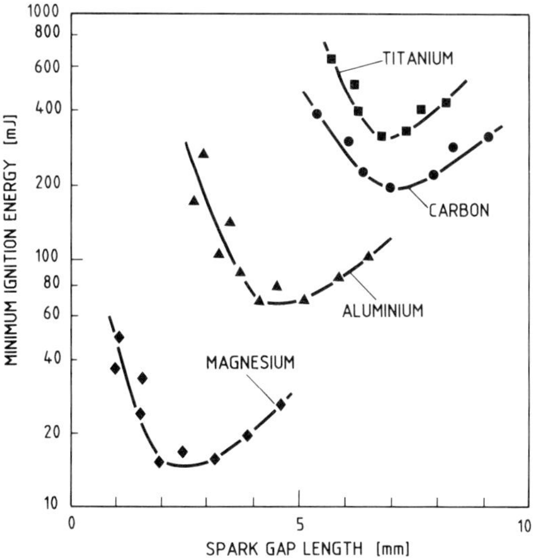

Figure 5.21 *Influence of electrode gap length on the minimum ignition energy of clouds of four metal dusts in air at atmospheric pressure. Dust concentration corresponding to equivalence ratio 0.65. Surface/volume mean particle size 40 μm. (From Ballal, 1979)*

the optimum spark gap length for ignition with increasing minimum ignition energy at the optimum gap length. This is consistent with the general picture for premixed gases, for which a close correlation between quenching distance and minimum ignition energy has been established.

Norberg *et al.* (1988) found that the optimum spark gap length for igniting clouds in air of various easily ignitable powders was in the range 6 to 8 mm. The capacitive sparks were of the short-duration type (low series inductance and resistance). The minimum ignition energies were in the range 1–6 mJ.

5.3.5
THEORIES OF ELECTRIC SPARK IGNITION OF DUST CLOUDS

Smielkow and Rutkowski (1971) derived a semi-empirical equation for the minimum electric spark ignition energy of dust clouds. Their experiments disclosed the following empirical relationship:

$$E_{min} = AS_f^{-3.56} \tag{5.29}$$

where E_{min} is the minimum ignition energy [mJ] and S_f the spatial laminar flame front speed [cm/s) of the dust cloud in question, and A is a constant.

The semi-empirical equation was obtained by inserting a Mallard/le Chatelier type expression for S_f (see Section 4.2.1 in Chapter 4) into Equation (5.29).

In their theoretical analysis, Kalkert and Schecker (1979) used the basic equation in the Jost-theory for ignition of premixed gases

$$\lambda \left(\frac{\partial^2 T}{\partial r^2} + \frac{2 \partial T}{r \partial r} \right) = \rho c \frac{\partial T}{\partial t} \tag{5.30}$$

as the point of departure. Here λ is the heat conductivity of the gas, T the temperature, r the radius, ρ the gas density, c the specific heat of the gas at constant pressure, and t the time.

By making several simplifying assumptions, they were able to derive the following equation for E_{min}

$$E_{min} = (4\pi\kappa)^{3/2} \rho c \left[\frac{ln2}{12} \frac{\rho_s c_s}{\lambda} \right]^{3/2} T_f \, d^3 \tag{5.31}$$

where $\kappa = \lambda/(\rho \times c)$ is the 'temperature conductivity' of air, d the diameter of the dust particles (monosized), ρ_s and c_s the density and specific heat of the particle material, and T_f the flame temperature (defined as 1300 K).

One central feature of Equation (5.31) is that $E_{min} \sim d^3$. Figure 1.30 in Chapter 1 shows the close agreement between predictions by Equation (5.31) and experimental values for polyethylene dust. (Note: E_{min} and MIE are interchangeable notations for the minimum electric spark ignition energy.)

Klemens and Wojcicki (1981) were specifically interested in modelling the electric spark ignition of coal dust clouds in air. They were able to validate their model predictions against unique experimental evidence of the development of the spark kernel and subsequent establishment of self-sustained flame propagation through the dust cloud away from the spark. An example is shown in Figure 5.22.

The overall physical picture of the ignition process on which the model of Klemens and Wojcicki was based, is as follows: During and following the spark discharge, the dust particles and the air in the vicinity of the spark kernel are heated. As a consequence, volatiles are evolved from the particles, mix with air and the mixture ignites. As the temperature increases further, the oxidation of the solid particle phase (coke) begins.

The temperature in the spark kernel and its close vicinity decreases with time due to heat drain. However, if ignition occurs, a flame front appears at the same time, at a certain distance from the spark axis, and starts to propagate outwards at the laminar flame speed of the coal dust/air cloud in question.

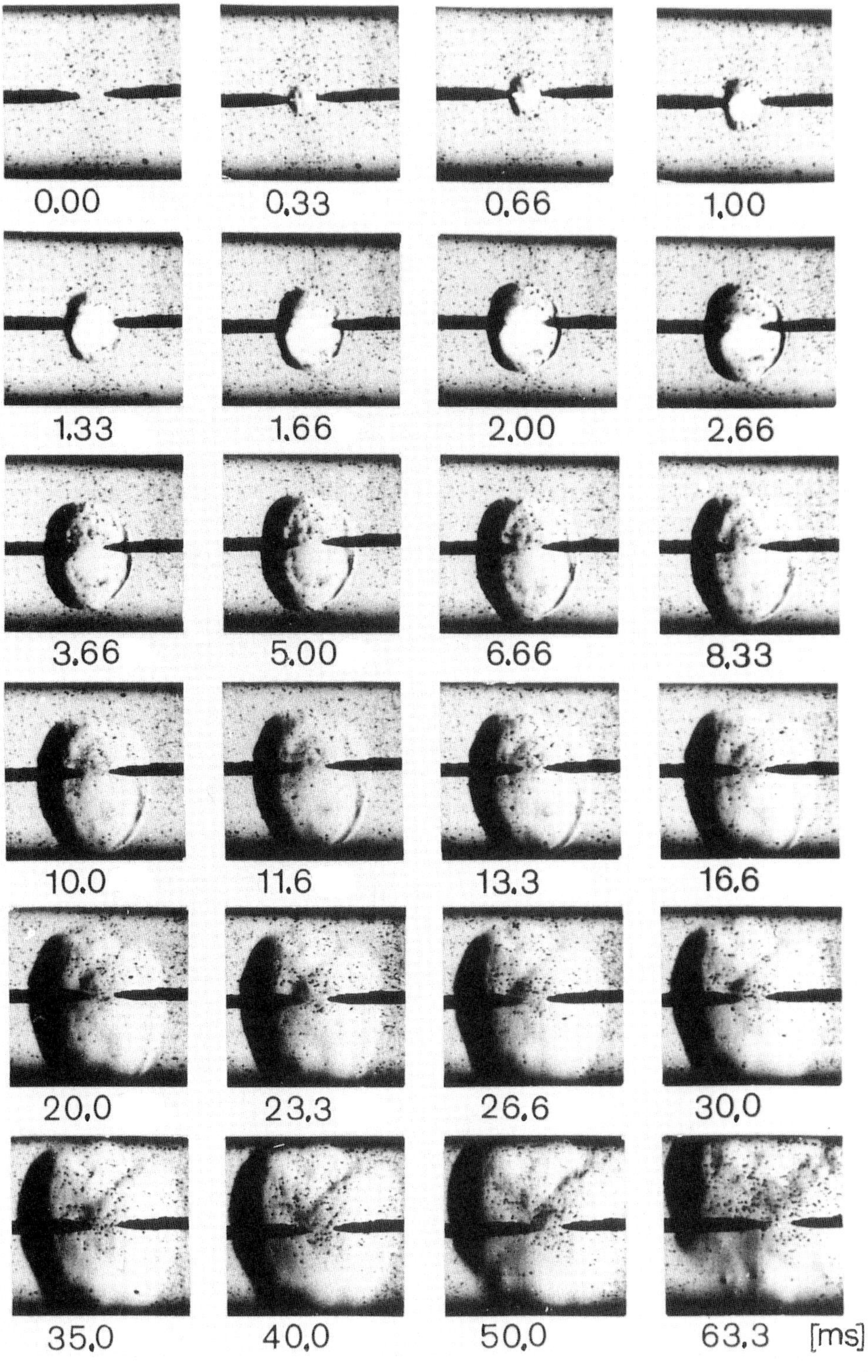

Figure 5.22 *Electric spark ignition of a cloud of lignite dust in air. Dust concentration 106 g/m*^{3}. *Spark energy 3.0 J. Spark discharge duration 0.10 ms (From Klemens and Wojcicki, 1981)*

The rate of energy delivery to the spark channel during spark discharge was taken into account in the mathematical model. Typically the duration of a 50 mJ spark would be about 0.10 ms. It was assumed that the energy density along the radius of the spark channel was linear at any instant. Both plane, cylindrical and spherical models were formulated.

Numerical simulations, using the model, were carried out, employing the establishment of a flame front propagating at a defined speed, as the criterion of ignition. In other words, whenever the spark energy exceeds the minimum ignition energy, the region over which the temperature rises is not limited to the spark region, but spreads into the mixture at the speed corresponding to the fundamental burning velocity of the dust cloud.

Gubin and Dik (1986) developed a mathematical model assuming that the oxidation occurred as a heterogeneous reaction between oxygen from the gas phase and the particle surface. They further assumed that the spark discharge initially generated a certain quantity of heat located within a narrow channel in the spark gap. The heat drain from the channel to the surroundings was assumed to occur essentially by conduction, radiation and covection being neglected. The basic heat balance equation was of the same form as that derived by previous workers. As in the case of other workers, the ignition criterion was the establishment of self-sustained laminar flame propagation in the dust cloud. It would appear that Gubin and Dik may not have been aware of the other investigations mentioned above.

5.4 IGNITION OF DUST CLOUDS BY HEAT FROM MECHANICAL RUBBING, GRINDING OR IMPACT BETWEEN SOLID BODIES

5.4.1 BACKGROUND

Whether or not metal sparks or hot-spots from accidental impacts, rubbing operations etc. between solid bodies can initiate dust explosions, has remained a controversial issue for a long time. Many attempts have been made at resolving the puzzle by analysing past accidents with the objective to identify the ignition sources. A summary with reference to the grain, feed and flour industry is given in Table 5.6.

Table 5.6 Percentage of dust explosions in the grain, feed and food industry assumed to be initiated by 'friction sparks' or unknown sources (From Pedersen and Eckhoff, 1987)

Investigation No.	Period	Number of explosions	% ignited by 'friction sparks'	% unknown	% 'friction sparks' + unknown
1	1860-1973	535	20	46	66
1	1949-1973	128	17	27	44
2	1941-1945	91	54	18	72
3	1958-1975	137	9	62	71
4	1965-1980	83	28	5	30-35

As can be seen, 'friction sparks' are claimed to play a significant part. If it is further taken into account that it is often tacitly implied that a substantial part of the 'unknowns' may have been initiated by some untraceable sources such as metal sparks and electrostatic discharges, the 'friction spark' becomes the most suspect of all the potential ignition sources.

The situations in which metal sparks and hot-spots can be generated in an industrial process plant fall into two main categories. The first is grinding and cutting operations, by which continuous, dense showers of sparks are produced and comparatively large hot-spots may be generated. The second is single accidental impacts.

5.4.2 SPARKS/HOT-SPOTS FROM RUBBING, GRINDING AND MULTIPLE IMPACTS

The ability of metal sparks or hot-spots from grinding operations to ignite dust clouds, has been demonstrated by several researchers. The experiments by Leuschke and Zehr (1962) are probably the first ones in which dust clouds were ignited by grinding wheel metal sparks. However, no clouds of organic dusts ignited. Zuzuki *et al.* (1965) ignited different coal dusts using both sparks and hot-spots from a piece of steel in contact with a grinding wheel rotating at 23–47 m/s peripheral velocity. Allen and Calcote (1981) conducted similar experiments in which metal sparks were generated by pressing a steel rod against a rotating grinding wheel. By retarding and focusing the spark stream, it was possible to ignite clouds of natural organic dusts such as maize starch and wheat grain dusts.

Kachan *et al.* (1976) studied the ignition of clouds of coal dust by metal sparks or hot spots generated by the cutters of a coal cutting machine, when cutting pyrite and sandstone at a speed of 1.5–2.0 m/s. The coal dust contained 24% volatiles or more, and 85% was finer than 75 μm. The dust concentration was 100 g/m^3. In the case of pyrite containing more than 35% sulphur, and with a load per cutter of 1–3 kW, the probability of ignition was practically 100%. However, the coal dust cloud did not ignite until after 15–80 s of continuous cutting with sparking, depending on the load. This long delay suggests that the ignition source was not the spark shower, but a hot-spot generated either at the cutter tip or on the pyrite surface just behind the cutter.

Ritter (1984, 1984a) and Müller (1989) conducted extensive studies of ignition of dust clouds by sparks/hot surfaces generated by scratching, grinding and multiple-impact processes. They used the concept of equivalent electric spark energy for characterizing the ignition potential of the various scratching/grinding/impact sources studied. This was done by first determining the lowest concentration of a given dust in air at which an essentially quiescent dust cloud could be ignited by the heat source investigated. The minimum electric spark ignition energy at this particular dust concentration was then determined and taken as the equivalent electric spark energy of that particular heat source.

Ritter and Müller found linear correlations between the minimum ignition temperature of the dust cloud determined by the BAM furnace (Chapter 7), and the logarithm of the equivalent minimum electric spark ignition energy, for the various ignition sources investigated, as illustrated in Figure 5.23.

The example indicated by dotted lines says that a dust cloud of minimum ignition temperature 615°C cannot be ignited by flint sparks from grinding or scratching unless its

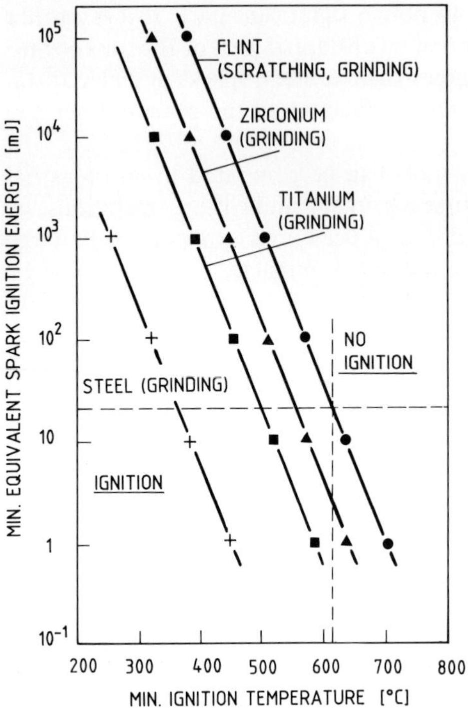

Figure 5.23 *Correlation between minimum ignition temperature of dust clouds (BAM furnace) and minimum equivalent electric spark ignition energy for various scratching/grinding ignition sources (From Müller, 1989)*

minimum ignition energy is lower than about 20 mJ. Similar correlations were found for sparks/hot-spots from multiple impacts.

Unfortunately the type of relationships illustrated in Figure 5.23 are not generally applicable because grinding/scratching/impact processes in practice may differ from the specific ones used in the experiments of Ritter and Müller. However, their approach is an interesting attempt at resolving a very complex matter.

Dahn and Reyes (1987), using a 20 litre explosion vessel, studied ignition of transient dust clouds by grinding sparks generated by forcing a metal rod against a rotating grinding wheel located within the vessel.

One striking feature is the negative result obtained with the two aluminium rods. This is in accordance with the discussion in Section 1.1.4.5 in Chapter 1.

5.4.3 SPARKS, HOT-SPOTS AND FLASHES FROM SINGLE ACCIDENTAL IMPACTS

Pedersen and Eckhoff (1987) studied the ignition of clouds of maize starch and grain dust in air by sparks, hot-spots and thermite flashes from single accidental impacts, using the

Table 5.7 Ignition of dust clouds by metal sparks/hot spots generated by forcing a metal rod against a rotating grinding wheel. Rod diameter 6.3 mm. Contact force 13.2 N. Estimated contact time between rod and wheel before ignition < 1 s (From Dahn and Reyes, 1987)

Dust cloud properties			Minimum peripheral grinding wheel speed for ignition using various metal rod materials [m/s]			
Dust tested	Particle size [μm]	Dust concentration [g/m3]	1018 Mild Steel	316 Stainless steel	3003-H14 Aluminium	6061-T6 Aluminium
M6 (propellant)	< 75	250	12.4	13.5	NI*	NI
M30AI (propellant)	< 75	410	11.2	13.5	NI	NI
M31AI (propellant)	< 150	320	10.4	14.0	NI	NI
CBI (igniter)	< 150	410	8.8	14.0	NI	NI
Black powder	< 75	250	11.6	14.0	NI	NI
Pittsburgh coal dust	< 75	350	10.0	NI	NI	NI
Maize starch	< 75	350	8.0	14.0	NI	NI

*NI = No ignition up to a peripheral grinding wheel speed of 20 m/s

apparatus described in Section 7.12.2 in Chapter 7 and illustrated in Figures 7.40 and 7.41. They investigated impacts of net energies up to 20 J and tangential velocities of approach from 10 m/s to 25 m/s. Table 7.2 in Chapter 7 gives some results from ignition with titanium impacting against rusty steel (thermite flash ignition). A positive correlation between the frequency of ignition and the minimum electric spark ignition energy is indicated.

Single impacts with steel as spark-producing material generated a very low number of sparks as compared to the number produced by titanium under the same impact conditions. The temperatures of individual steel sparks, however, could reach the same level as those of titanium sparks (~2500°C).

Impacts of standard quality aluminium against rusty steel did not generate any sparks or any other luminous reaction at all, only a thin smear of aluminium on top of the rust (see Section 1.1.4.5 in Chapter 1). Impacts with hard aluminium-containing alloys were not investigated.

In most cases, ignition by titanium sparks (e.g. from titanium against concrete) was observed very close to the point of impact. However, occasionally ignition was also observed 10–30 cm downstream of the impact point. Ignition by one single metal spark was never observed. A fairly dense cluster of sparks seemed to be necessary for igniting the clouds of maize starch.

Any moving object in the dust cloud will reduce the ignition sensitivity of the cloud in the vicinity of the object by inducing turbulence. The experiments showed that at a given net impact energy, the ignition frequency dropped when the impact velocity increased. Thus, at a given net impact energy, objects generating low turbulence represent a greater ignition hazard than objects generating high turbulence. This was illustrated by an experiment in which the impacting object on the spring-loaded arm (Figure 7.40) was withdrawn slightly, allowing it, once the arm was released, to pass just above the anvil without touching it. Instead an electric spark was discharged at the point where the impact would normally have occurred, and the frequency of ignition measured as a function of electric spark energy for various tangential arm-tip velocities. The results are shown in Figure 5.24.

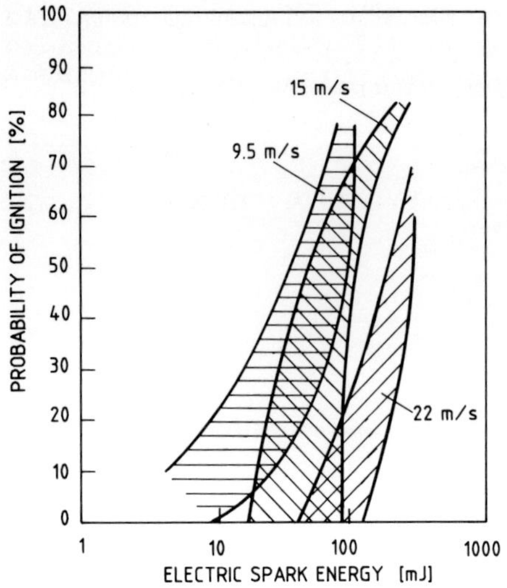

Figure 5.24 *Effect of tangential 'impact' velocity on ignition sensitivity of clouds of lycopodium in air. Delay between 'impact' and electric spark discharge 0.3–10 ms. Envelopes embrace the experimental points (From Pedersen and Eckhoff, 1987)*

The microscopic nature of the anvil surface is decisive for the spark formation process. For example, impacts against worn concrete surfaces exposing naked stone and gravel faces produced considerably more sparks than impacts against a fresh concrete surface covered with cement.

The overall practical conclusion of the investigation by Pedersen and Eckhoff (1987) is that up to net impact energies of 20 J, tangential accidental single impacts between various types of steel, and between steel and rusty steel or concrete, are unable to ignite clouds of grain and feed dust, or flour, even if the dusts are dry. Impacts of standard quality aluminium against rusty steel will not even generate any visible sparks. In the case of titanium, the sparks produced can initiate explosions in clouds of dried maize starch, but not in clouds of starch containing 10% moisture or more, not even in the case of thermite flashes. However, for net impact energies $\gg$ 20 J the situation may be different.

5.5 IGNITION OF DUST CLOUDS BY HOT SURFACES

5.5.1 EXPERIMENTAL STUDY OF THE INFLUENCE OF SIZE OF HOT SURFACE

The decrease of the minimum ignition temperatures of explosible gas mixtures with increasing hot surface size has been known for a long time. One classic investigation of this subject is that of Silver (1937). A similar dependence of the minimum ignition temperature on the area of the hot surface would be expected for dust clouds. Figure 5.25 gives some experimental data from Pinkwasser (1989) confirming this expectation.

The three smallest surfaces were 10 mm long pieces of heated wire of thickness 0.7, 1.2 and 6.0 mm respectively, bent as a U. The largest surface of 1000 mm^2 was obtained by coiling 50 mm of the 6.0 mm diameter wire. Figure 5.25 also gives the BAM furnace minimum ignition temperatures of the three dusts, assuming a hot surface area of about 2000 mm^2.

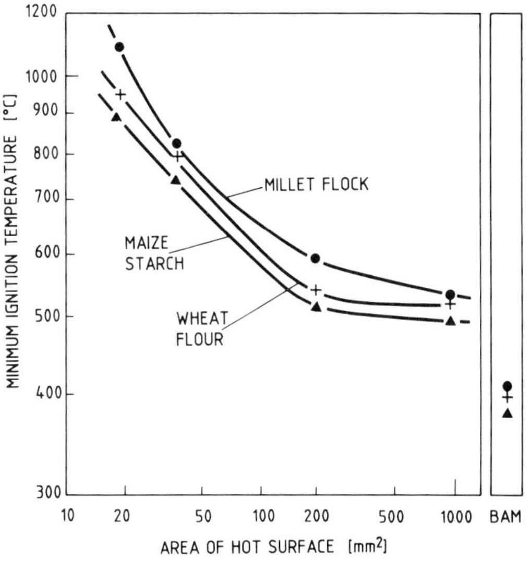

Figure 5.25 *Influence of area of hot surface on the minimum ignition temperature of clouds in air of natural organic dusts. Comparison with results from BAM furnace tests (hot surface area approx. 2000 mm^2) (From Pinkwasser, 1989)*

5.5.2
THEORIES FOR PREDICTING THE MINIMUM IGNITION TEMPERATURES OF DUST CLOUDS

In their theoretical investigation, Mitsui and Tanaka (1973) focused on the influence of particle size on the minimum ignition temperature. They considered a spherical dust cloud inside which heat was generated by combustion, and from which heat was lost due to convection and radiation. They then assumed a combustion rate with Arrhenius type exponential temperature dependence, and being proportional to the total particle surface area in the spherical dust cloud. The critical ignition condition was specified as the rate of heat generation due to chemical reaction being equal to the rate of heat loss. The resulting equation seemed to predict an influence of particle size in good agreement with experimental results when using a tuning constant depending on the dust chemistry.

A similar study, focusing particularly on the geometry of the Godbert-Greenwald furnace (see Chapter 7), was undertaken by Takigawa and Yoshizaki (1982). They investigated natural organic dusts and found a reasonably good agreement between measured and numerically predicted dependence of minimum ignition temperature on particle size. The numerical model calculations further revealed that the residence time of the dust particles in the furnace largely affects the ignition process. It was concluded that the steady state solution of the minimum ignition temperature is not applicable to the ignition process in the Godbert-Greenwald furnace apparatus. Comparison of model predictions with experimental data from other workers confirmed the validity of the predicted effect of the residence time of the dust particles in the furnace on the minimum ignition temperature.

Nomura and Callott (1986) modified the Cassel/Liebman theory to make it account for the influence of the residence time of the dust particles in the hot furnace. The theory suggested that it is possible for the ignition temperature of mono-sized coal particles of about 50 μm diameter to be minimal even for a limited residence time.

The theory was extended to dust clouds with a distribution of particle sizes. It was shown that there exists a range of size distributions for which the possibility of ignition is at a maximum. The calculated results were presented as Rosin-Rammler charts indicating the size distributions that are most sensitive to ignition.

Higuera, Linan and Trevino (1989) analysed the heterogeneous ignition of a cloud of spherical monodisperse coal particles injected instantaneously in the space between two parallel isothermal walls. They focused on the range of large gas/particles thermal capacity ratios, for which the temperature difference between the particles and the gas is important. Radiative heat transfer was accounted for by using the Eddington differential approximation, and heat conduction between the particles and the gas was also included in the model. Heat release was assumed to occur at the surface of the particles through the heterogeneous reaction $C(s) + 1/2\ O_2 \rightarrow CO$, obeying an Arrhenius law with large activation energy. Critical conditions for ignition were determined on the basis of a quasi-steady treatment. The effects of the ratio of gas temperature to wall temperature, the conduction/radiation parameter, and the size of the reacting dust cloud relative to the optical length was explained.

Tyler (1987) was concerned with the problem of scaling ignition temperatures of dust clouds from laboratory test apparatus to industrial scale. In particular he focused on the Godbert-Greenwald furnace (see Chapter 7). As pointed out by Tyler, there does not seem to be any single physical/chemical pattern for ignition of a dust cloud. In the case of substances as sulphur and polyethylene, the minimum ignition temperatures are high enough to allow complete evaporation or pyrolysis to form gaseous fuels. At the other extreme there are metals of minimum ignition temperatures at which neither the metal nor its oxide will vaporize fully. In the first case the exothermic oxidation process almost certainly takes place in the gas phase, whereas in the second it occurs at the surface of or within the particle. (See also Section 4.1 in Chapter 4.) However, these differences may not necessarily be important in the establishment of the unstable state of ignition which precedes a fully developed flame.

Tyler developed a Semenov-type mathematical model of the ignition of a dust cloud in a heated environment (furnace). However, validation of the model was difficult. No reliable activation energies were found in the literature which could be definitely attributed to the heat release reaction that occurs at the ignition temperature, and Tyler pointed out that the activation energy could be quite different from that associated with a fully fledged flame; indeed the dominant mechanism could well be different in the two situations.

Nevertheless useful parametric studies could be performed. For example, the model predicted comparatively large changes of the minimum ignition temperature with furnace diameter. The Godbert-Greenwald furnace has a diameter of 37 mm. For a furnace diameter of 300 mm and a dust with a Godbert-Greenwald value of 1000 K the model predicted a minimum ignition temperature at least 150°C lower than the Godbert-Greenwald value.

However, few experimental data were traced for the influence of increased furnace diameter on the minimum ignition temperature except when comparing data from the new US Bureau of Mines furnace (see Chapter 7) and the Godbert-Greenwald furnace. The ratio of the two furnace diameters is 2.7, and therefore significant differences in the minimum ignition temperatures from the two apparatuses would be expected. However, the picture offered by existing data was inconclusive. For some dusts the experimental Godbert-Greenwald value was even lower than that from the new furnace.

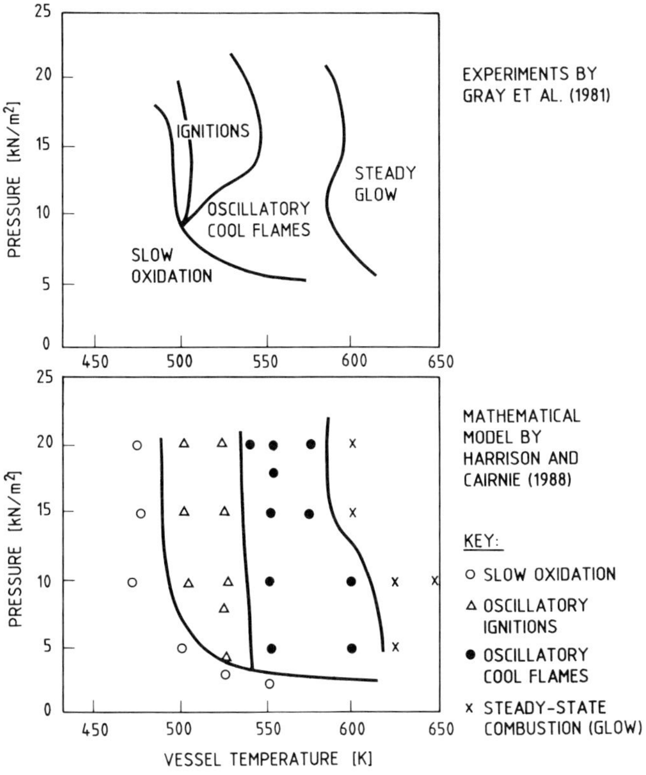

Figure 5.26 *Comparison of experimentally observed and numerically predicted ignition diagrams for acetaldehyde/air in a continuously stirred 0.5 litre glass bulb (From Harrison and Cairnie, 1988)*

Tyler concluded that there was no theoretical model by which data from Godbert-Greenwald furnace tests could be transformed to minimum ignition temperatures in more complex practical situations in industry. He suggested that stirred reactor ignition experiments, as performed successfully for combustible gas mixtures, could provide a more fundamental understanding of dust cloud ignition processes. Such experiments may yield appropriate activation energies for the ignition processes, which may be used to scale minimum ignition temperatures more reliably. The study of ignition of acetaldehyde/air mixtures, by Harrison and Cairnie (1988) and Harrison *et al.* (1988), taking this approach, is an excellent example of its potential. Figure 5.26 shows a comparison of experimentally determined and theoretically predicted ignition diagrams for the acetaldehyde/air system.

REFERENCES

Aakre, M. (1980) Maaling av laagaste gnistenergi for tenning av pulverlag, Report No. 790705–1, (April), Chr. Michelsen Institute, Bergen, Norway

Adomeit, P., and Henriksen, K. (1988) Zeitlicher Verlauf der Entflammung eines brennbaren Schüttguts an einer heissen Oberfläche. *Staub-Reinhalt. Luft* **48** pp. 371–377

Allen, J., and Calcote, H. F. (1981) Grain Dust Ignition by Friction Sparks. Report SMS-81-049. National Grain and Feed Association, Washington DC

Ballal, D. R., and Lefebvre, A. H. (1977) Ignition and Flame Quenching in Flowing Gaseous Mixtures. *Proc. Roy. Soc.* A **357** London, pp. 163–181

Ballal, D. R. (1979) Ignition and Flame Quenching of Quiescent Dust Clouds of Solid Fuels. *Proc. Roy. Soc.*, Series A, London (July)

Beever, P. (1984) Self-Heating of Powders in Theory and Practice. *VDI-Berichte* **494,** VDI-Verlag GmbH, Düsseldorf, pp. 151–155

Beever, P. F. (1988) Self-Heating and Spontaneous Combustion. Chapter 1–22 in *SFPE Handbook of Fire Protection Engineering*, National Fire Protection Association, USA

Bigg, S., and Street, P. J. (1989) Predicting Spontaneous Ignition and Combustion in Fixed Beds of Activated Carbon. *Combust. Sci. and Tech.*,**65** pp. 245–262

Birks, N., and Alabi, A. G. F. (1986) Reactions Leading to the Autoignition of Direct Reduced Iron. *Proc. 5th Int. Iron & Steel Congress*, Washington DC, April pp. 83–91

Birks, N., and Alabi, A. G. F. (1987) Mechanisms in Corrosion-Induced Autoignition of Direct Reduced Iron. *Proc. of 70th Steelmaking Conf.*, (March/April), Pittsburgh PA, USA

Birks, N., and Alabi, A. G. F. (1988) The Role of Corrosion in the Autoignition of Direct Reduced Iron. *Proc. W. O. Philbrook Memorial Symp.* pp. 61–65

Bowes, P. C. (1981) A General Approach to the Prediction and Control of Potential Runaway Reaction. *Proc. of I. Chem. E. Symposium Series* No. 68, pp. 1/A:1–1/A:35. The Institution of Chemical Engineers, UK

Bowes, P. C. (1984) *Self-Heating: Evaluating and Controlling the Hazards,* Elsevier Publ. Co., Amsterdam

Boyle, A. R., and Llewellyn, F. J. (1950) The Electrostatic Ignitability of Dust Clouds and Powders. *J. Soc. Chem. Ind. Trans.*, **69** pp. 173–181

Bradley, D., Lung, F. K.-K. (1987) Spark Ignition and the Early Stages of Turbulent Flame Propagation. *Combustion and Flame*, **69** pp. 71–93

Brooks, K., Svanas, N., and Glasser, D. (1988) Evaluating the Risk of Spontaneous Combustion in Coal Stockpiles. *Fuel*, **67** pp. 651–656

Chamberlain, E. A. C., and Hall, D. A. (1973) The Liability of Coals to Spontaneous Combustion. *Colliery Guardian* (February) pp. 65–72

Dahn, C. J., and Reyes, B. N. (1987) Determination of Metal Sparking Characteristics and the Effects on Explosive Dust Clouds. *Industrial Dust Explosions*, (eds K. L. Cashdollar and M. Hertzberg, Special Technical Publication 958, ASTM, Philadelphia, USA pp. 324–332

Dik, I. G. (1987) Solution by the Thermal Impedance Method to the Problem of Critical Conditions

for Self-Ignition. In *Combustion, Explosion and Shock Waves*, Plenum Publishing Corporation, pp. 495–501
Duncan, A. B., Peterson, G. P., Fletcher, L. S. (1988) Effective Thermal Conductivity within Packed Beds of Spherical Particles. *ASME HTD* **104** pp. 77–88
Eckhoff, R. K. (1970) The Energy Required for the Initiation of Explosions in Dust Clouds by Electric Sparks. M.Phil. Thesis, University of London
Eckhoff, R. K. (1975) Towards Absolute Minimum Ignition Energies for Dust Clouds? *Combustion and Flame*, **24** pp. 53–64
Eckhoff, R. K., and Enstad, G. G. (1976) Why are 'Long' Electric Sparks More Effective Dust Explosion Initiators than 'Short' Ones? *Combustion and Flame*, **27** pp. 129–131
Elder, J. L., Schmidt, L. D., Steiner, W. A., *et al.* (1945) Relative Spontaneous Heating Tendencies of Coals. US Dept. Interior Technical Paper 681, Washington
Enemoto, H., Nifuku, M., Deguchi, M., *et al.* (1987) On the Ignition Source of Fire in a Bag Separator in the Pneumatic Transport System for Pulverized Coal. *Safety Engineering*, (Japanese) **26**, No. 3 pp. 153–160
Enstad, G. G. (1981) Effect of Shock Wave Emitted from Electric Spark Discharges on the Energy Required for Spark Ignition of Dust Clouds, Report No. 813101-2, (February), Chr. Michelsen Institute, Bergen, Norway
Essenhigh, R. H. (1986) Comment on the Definition of Ignition Used by Lermant and Yip. *Combustion and Flame* **63**, pp. 303–304
Faveri, D. M. de, Zonato, C., Vidili, A., *et al.* (1989) Theoretical and Experimental Study on the Propagation of Heat Inside Deposits of Coal. *J. Hazardous Materials*, **21** pp. 135–142
Ford, L. F., and Ford, J. D. (1984) Thermal Diffusivity of Some Porous Oxides. *Can. J. Chem. Eng.*, **62** pp. 125–134
Frank-Kamenetzkii, D. A., (1969) Diffusion and Heat Transfer in Chemical Kinetics, 2nd Edition. Translated by J. P. Appleton, Plenum Press, New York/London
Gray, P., and Lee, P. R. (1967) Thermal Explosion Theory. In *Oxidation and Combustion Reviews*, **2** pp. 1–183, (ed by C. F. H. Tipper), Elsevier Publ. Comp.
Gray, P., and Sherrington, M. E. (1977) Self-Heating, Chemical Kinetics and Spontaneous Unstable Systems. Specialist Periodical Reports. *Gas Kinetics and Energy Transfer* **2** Chapter 8, The Chem. Soc., London
Gray, P., Griffiths, J. E., Hasko, S. M., *et al.* (1981) *Proc. Roy. Soc.* (London), A 374:313
Gubin, E. I., and Dik, I. G. (1986) Ignition of a Dust Cloud by a Spark. *Combustion, Explosion and Shock Waves*, **22** 2, pp. 135–141
Guney, M., and Hodges, D. J. (1969) Adiabatic Studies of the Spontaneous Heating of Coal. *Colliery Guardian*, **217** pp. 105–109
Hardman, J. S., Lawn, C. J., and Street, P. J. (1983) Further Studies of the Spontaneous Ignition Behaviour of Activated Carbon. *Fuel*, **62** p. 632
Harrison, A. J., and Cairnie, L. R. (1988) The Development and Experimental Validation of a Mathematical Model for Predicting Hot-Surface Autoignition Hazards Using Complex Chemistry. *Combustion and Flame*, **71** pp. 1–21
Harrison, A. J., Furzeland, R. M., Summers, R., *et al.* (1988) An Experimental and Theoretical Study of Autoignition on a Horizontal Hot Pipe. *Combustion and Flame*, **72** pp. 119–129
Heemskerk, A. H. (1984) Self-Heating of Sub-Bituminous Coal. Report PML 1984–C35, Prins Maurits Laboratorium, TNO, Rijswijk, NL
Henderson, D. K., and Tyler, B. J. (1988) Dual Ignition Temperatures for Dust Layers. *J. Hazardous Materials*, **19** pp. 155–159
Heinrich, H.-J. (1981) Grundlagen für die Einstufung von Kohlenstäuben in die Gefahrklasse 4.2 der Beförderungsvorschriften. *Amts- und Mitteilungsblatt der BAM* **11** No. 4, pp. 326–330
Hensel, W. (1987) The Dependence of Self-Ignition Temperatures of Dusts upon Grain Size. *Archivum Combustionis*, **7** No. 1–2, pp. 45–57

Hensel, W. (1988) Schwelbrände in Steinkohlen-Bevorratungslagern des Senators für Wirtschaft und Arbeit von Berlin. *Amts- und Mitteilungsblatt der BAM* **18** pp. 377–384

Hensel, W. (1989) Entzündung abgelagerter Stäube. *VDI-Berichte* **701,** VDI-Verlag GmbH, Düsseldorf, pp. 143–166

Higuera, F. J., Linan, A., and Trevino, C. (1989) Heterogeneous Ignition of Coal Dust Clouds. *Combustion and Flame*, **75** pp. 325–342

Holtzwart, R., and Meyer, E. von (1891) One the Causes of Explosions in Browncoal Briquette Works. *Dinglers Journal* **280** pp. 185–190 and 237–240

Höstmark, Ö (1989) Selvantenning og stöveksplosjonsfaren for fiskemel. Meldinger fra SSF No. 2, (October), pp. 22–26, Sildolje- og Silemelindustriens Forskningsinstitutt, Bergen, Norway

Johannsmeyer, U. (1984) Zündung explosionsfähiger Gemische durch Kurzzeitige Schliessfunken in kapazitiven Stromkreisen für die Zündschutzart Eigensicherheit. Dr.-Ing. Thesis, Technical University, Carolo-Wilhelmina, Braunschweig

John, W., and Hensel, W. (1989) Messung der Wärmeleitfähigkeit abgelagerter Stäube mit einem Heizdraht-Messgerät. *Staub-Reinhalt. Luft* **49** pp. 333–335

Kachan, V. N., Kocherga, N. G., Petrukhin, P. M., *et al.* (1976) Ignition of Coal Dust by Frictional Sparks. Translated from *Fizika Goreniya i Vzryva* **12** pp. 302–304 Plenum Publishing Corporation, New York

Kalkert, N., and Schecker, H.-G. (1979) Theoretische Überlegungen zum Einfluss der Teilchengrösse auf die Mindestzündenergie von Stäuben. *Chem.-Ing.-Tech.* **51** pp. 1248–1249

Klemens, R., and Wojcicki, S. (1981) Model of Ignition of Dust/Air Mixtures by a Low-Energy Electric Spark. Copy of unpublished manuscript given to R. K. Eckhoff by R. Klemens, Warsaw Technical University, Poland

Krauss, L., and Krempl, H. (1963) Über das Temperaturabklingen in Funken 50 000 bis 5000 Grad. *Zeitschrift für angewandte Physik* **16** pp. 243–247

Leisch, S. O., Kauffman, C. W., and Sichel, M. (1984) Smouldering Combustion in Horizontal Dust Layers. *Proceedings of 20th Symp. (Internat.) on Combustion*, The Combustion Institute pp. 1601–1610

Lermant, J.-C., and Yip, S. (1984) Generalized Semenov Model for Thermal Ignition ... *Combustion and Flame*, **57** pp. 41–54

Lermant, J.-C., and Yip, S. (1986) Response to R. H. Essenhigh, 'A Comment on the Definition of Ignition Used by Lermant and Yip'. *Combustion and Flame*, **63** p. 305

Leuschke, G., and Zehr, J. (1962) Zündung von Staublagerungen und Staub/Luftgemischen durch mechanisch erzeugte Funken. *Arbeitsschutz*, **6** p. 146

Leuschke, G. (1980) Induction Times of Dust Deposits Stored at and above Self-Ignition Temperatures. *Proc. of 3rd Int. Symp. Loss Prev. Safety Prom. Proc. Ind.*, Basle **Vol. 2** pp. 8/647–8/656

Leuschke, G. (1981) Experimental Investigations on Self-Ignition of Dust Deposits in Hot Environments. *Proc. of I. Chem. E. Symp.*, Series No. 68, Institution of Chemical Engineers, UK

Liang, H., and Tanaka, T. (1987) Effect of Biot Number on the Critical Condition for Spontaneous Ignition. *Kagaku Kogaku Ronbunshu* **13** pp. 847–849

Liang, H., and Tanaka, T. (1987a) The Spontaneous Ignition of Dust Deposits. Ignition Temperature and Induction Time. Kona. *Powder Science and Technology in Japan*, No. 5 pp. 25–32

Liang, H., and Tanaka, T. (1988) Simulation of Spontaneous Heating for the Evaluation of the Ignition Temperature and Induction Time of a Combustible Dust. *International Chemical Engineering* **28** pp. 652–660

Line, L. E., Rhodes, H. A., and Gilmer, T. E. (1959) The Spark Ignition of Dust Clouds. *Journal of Physical Chemistry* **63** pp. 290–294

Marchand, D. (1988) Entstehung von Glimmnestern bzw. Schwelbränden und deren Auswirkungen auf die Betriebssicherheit von Tuchfiltern. *Staub-Reinhalt. Luft* **48** pp. 369–370

Matsuda, T., and Naito, M. (1983) Effects of Spark Discharge Duration on Ignition Energy for

Dust/Air Suspensions. In *Particulate Systems, Technol & Fundam.*, (ed. J. K. Beddow.) Hemisphere Publ. Corp./McGraw Hill Int. Book Co., pp. 189–198

Miron, Y., and Lazzara, C. P. (1988) Hot-Surface Ignition Temperatures of Dust Layers. *Fire and Materials* **12** pp. 115–126

Mitsui, R., and Tanaka, T. (1973) Simple Models of Dust Explosion. Predicting Ignition Temperature and Minimum Explosive Limit in Terms of Particle Size. *Ind. Eng. Chem., Proc. Des. & Devel.* **12** p. 384–389

Müller, R. (1989) Zündfähigkeit von mechanisch erzeugten Funken und heissen Oberflächen in Staub-/Luftgemischen. *VDI-Berichte* **701,** VDI-Verlag GmbH, Düsseldorf pp. 421–466

Nomura, S., and Callcott, T. G. (1986) Calculation of the Ignition Sensitivity of Dust Clouds of Varying Size Distributions. *Powder Technology* **45** pp. 145–154

Norberg, A., Xu, D., and Zhang, D. (1988) Powder Ignition Energy Measured Utilizing a New Fluidized Bed Ignition Chamber. Report ISSN 0349–8352. (September), Institute for High Voltage Research, Uppsala, Sweden

Parker, S. J. (1985) Electric Spark Ignition of Gases and Dusts. Ph.D. Thesis, (August) City University, London. Dust part also published as Report No. 853351–4, (December) Chr. Michelsen Institute, Bergen, Norway

Pedersen, G. H., and Eckhoff, R. K. (1987) Initiation of Grain Dust Explosions by Heat Generated During Simple Impact Between Solid Bodies. *Fire Safety Journal* **12** pp. 153–164

Pinkwasser, Th. (1989) Influence of Size of Hot Surface on the Minimum Ignition Temperature of Dust Clouds. Private Communication to R. K. Eckhoff from Th. Pinkwasser, Bühler, Switzerland

Raemy, A., and Lambelet, P. (1982) A Calorimetric Study of Self-Heating in Coffee and Chicory. *J. Food Technology* **17** pp. 451–460

Raemy, A., and Löliger, J. (1982) Thermal Behaviour of Cereals Studied by a Heat Flow Calorimeter. *Cereal Chemistry* **59** pp. 189–191

Raemy, A., Hurrell, R. F., and Löliger, J. (1983) Thermal Behaviour of Milk Powders Studied by Differential Thermal Analysis and Heat Flow Calorimetry. *Thermochimica Acta* **65** pp. 81–92

Ritter, K. (1984) Die Zündwirksamkeit mechanisch erzeugter Funken gegenüber Gas/Luft- und Staub/Luft-Gemischen. Doctorate Thesis, University of Karlsruhe (TH), F. R. Germany

Ritter, K. (1984a) Mechanisch erzeugte Funken als Zündquellen. *VDI-Berichte* **494,** VDI-Verlag GmbH, Düsseldorf, pp. 129–144

Rose, H. E. (1959) Über die Initialzündung von Explosionen durch elektrische Funken. *Staub* **19** pp. 215–220

Selle, H. (1957) Die chemischen und physikalischen Grundlagen der Verbrennungsvorgänge von Stäuben. *VDI-Berichte* **19** pp. 25–36

Semenov, N. N. (1935) *Chemical Kinetics and Chain Reactions,* The Clarendon Press, Oxford

Semenov, N. N. (1959) *Some Problems of Chemical Kinetics and Reactivity,* Vol. **2** Translated by J. E. S. Bradley, Pergamon Press, London/New York

Shea, F. L., and Hsu, H. L. (1972) Self-Heating of Carbonaceous Materials. *Ind. End. Chem. Prod. Res. Develop.* **11** pp. 184–187

Silver, R. S. (1937) The Ignition of Gaseous Mixtures by Hot Particles. *Phil. Mag. Series 7.* **23** No. 156 Suppl. (April), pp. 633–657

Smielkow, G. I., and Rutkowski, J. D. (1971) Badania Zjawiska Zaplonu Mieszanin Pylowo-powietrznych wywolanego wyladonaniami iskrowymi. *Chemia Stosowana* **XV** 3 pp. 283–292

Smith, A. C., Miron, Y., and Lazzara, C. P. (1988) Inhibition of Spontaneous Combustion of Coal. Rep. Inv. 9196, US Bureau of Mines, Washington

Stockmeier, H. (1899) Über die Ursachen der Explosion bei der Bereitung des Aluminiumbronze-pulvers. *Zeitschr. f. Untersuchung d. Nahr.- und Genussmittel. Freie Verein Bayer. Vertr. angew. Chemie* **2** pp. 49–61

Takahashi, H., Obata, E., Takeuchi, T., *et al.* (1989) Simulation of a Maximum Storage Amount of Coal for Preventing Spontaneous Combustion and Degradation in Quality. *Kona* No. **7** pp. 89–96

Takigawa, T., and Yoshizaki, S. (1982) Research of Several Problems Related to the Minimum Ignition Temperature of Agricultural Dusts. *J. Soc. Powder Technol. Japan* **19** No. 10, pp. 582–591

Thomas, P. H. (1958) On the Thermal Conduction Equation for Self-Heating Materials with Surface Cooling. *Trans. Faraday Soc.* **54** pp. 60–65

Thorne, P. F. (1985) The Physics of Fire Extinguishment. *Phys. Technol.* **16** pp. 263–268

Tognotti, L., Petarca, L., and Zanelli, S. (1988) Spontaneous Combustion in Beds of Coal Particles. *Proceedings 21st Symp. (Internat.) on Combustion.* The Combustion Institute, pp. 201–210

Tyler, B. J. (1987) Scaling the Ignition Temperatures of Dust Clouds. Report No. F3/2/347, (June) Fire Research Station, UK.

Tyler, B. J., and Henderson, D. K. (1987) Spontaneous Ignitions in Dust Layers: Comparison of Experimental and Calculated Values. *I. Chem. E. Symp.* Series No. 102, Institute of Chemical Engineers, UK, pp. 45–59

Zaturska, M. B. (1978) An Elementary Theory for the Prediction of Critical Conditions for Thermal Explosion. *Combustion and Flame* **32** pp. 277–284

Zuzuki, T., Takaoka, S., and Fujii, S. (1965) The Ignition of Coal Dust by Rubbing, Frictional Heat and Sparks. *Proc. Restricted Int. Conf. Directors Safety Min. Res.*, (July), Safety in Mines Res. Establishment, UK

CHAPTER 6

Sizing of dust explosion vents in the process industries: further consideration of some important aspects

6.1 SOME VENT SIZING METHODS USED IN EUROPE AND USA

6.1.1 VENT RATIO METHOD

This and other methods were reviewed by Schofield (1984), Lunn (1984) and Field (1984). The vent ratio method requires that $P_{red} \leq 0.14$ bar(g) and that the opening pressure and inertia of the vent cover are small. The vent ratio is defined as vent area per unit of enclosure volume. Originally a fixed vent ratio, determined by the maximum rate of pressure rise in the 1.2 litre Hartmann bomb (see Chapter 7), was specified for a specific dust. However, as the enclosure volume gets larger, the required vent area increases more than if geometrical similarity is used for scaling, and unreasonably large vents result. For example, with a vent ratio of 1 m^2/6 m^3, a 6 m^3 spherical vessel would need only 6% of the sphere surface for venting, whereas a huge sphere of 24 000 m^3 volume would need the entire surface. Because of this the vent ratio method has been modified by reducing the required ratio as the enclosure volume increases. It has not been possible to trace the experimental basis for the vent ratio method. The method has, however, been widely used for example in UK.

6.1.2 'NOMOGRAPH' METHOD

This method was originally designed by Verein deutscher Ingenieure (1979), but later it was also adopted for USA by National Fire Protection Association (1988).

The nomograph method rests partly on the results of extensive large-scale experimentation by Donat (1971) and Bartknecht (1978) and partly on theoretical analysis by Heinrich (1974). A self-contained system for vent sizing has been developed. It consists of the 1 m^3 ISO standard test (see Chapter 7) for determination of a maximum rate of pressure rise used as the characteristic of the explosion violence of a given dust (K_{St}-value), and a series of nomographs from which vent areas can be estimated, using K_{St}, the enclosure volume, the maximum explosion pressure P_{red}, and the static opening pressure of the vent cover

P_{stat} as parameters. The relevance of the 'nomograph' method is limited by the nature of the large-scale experiments on which it rests. These experiments were conducted with dust clouds generated by blowing the dust into the experimental enclosures from pressurized reservoirs through narrow nozzles ensuring uniform, well-dispersed, and highly turbulent dust clouds. Consequently the burning rate for a given dust (see Chapter 4) was very high, in fact too high to be representative of the dust clouds in most industrial situations. In spite of this, the nomograph method has been widely used. However, when Verein deutscher Ingenieure (1991) issue a draft revision of their venting guide line, they may allow for considerable reduction in the vent area requirements for some industrial situations.

Lunn *et al.* (1989) extended the VDI 3673 nomographs from 1979 to K_{St}-values as low as 10 bar m/s, and P_{red} values down to 0.05 bar(g). However, the experiments on which this extension was based were of the same kind as those forming the basis of the 1979 VDI 3673 nomograph, i.e. of very high turbulence and degree of dust dispersion. Therefore the extended nomographs are subject to the same basic limitations as the original VDI nomographs. This is illustrated in Figure 6.1, in which maximum pressure/vent area correlations predicted by the extended nomographs are shown together with experimental data from the investigation by Brown and Hanson (1933). It is felt that the experiments of Brown and Hanson were considerably closer to the reality most often encountered in industry than those on which the nomograph method was based. This exemplifies the fact that much excellent experimental work that was performed in the past on various aspects

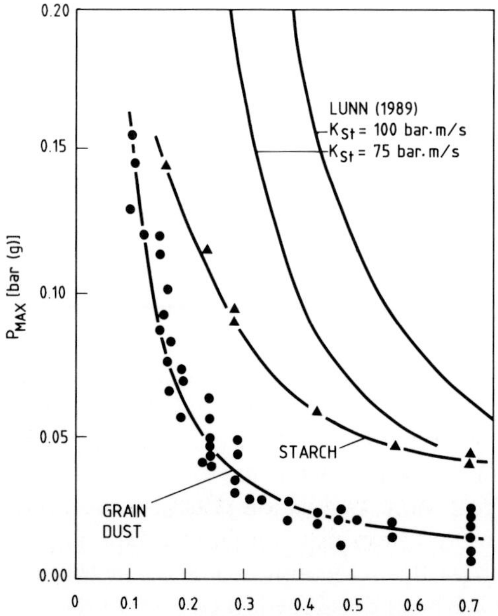

Figure 6.1 *Experimental correlations between vent area and maximum explosion pressure for grain dust and starch explosions in a 2.8 m³ cubical vessel. Vents covered with hinged metal doors or glass panes in pivoted frames (From Brown and Hanson, 1933). Comparison with correlations suggested by Lunn, 1989)*

of the dust explosion problem often seems to be overlooked by some more recent investigators. The paper of Brown and Hanson indeed deserves to be read carefully even today.

The dusts used by Brown and Hanson were dried to less than 8% moisture, and it seems reasonable to assume that their K_{St} values would not have been lower than the values 75 and 100 bar m/s used for comparison with the Lunn-predictions. If the starch was from maize, a K_{St} of 100 bar m/s is a low estimate.

Comparisons between 'nomograph'-method predictions and data from more recent realistic experiments are given in Section 6.2.

6.1.3 THE SWEDISH METHOD

In this method, issued by Danielson (1981), the explosion violence classification of the dust is by $(dP/dt)_{max}$ from standard 1.2 litre Hartmann bomb tests (see Chapter 7), as in the case of the vent ratio method. Dusts are classified in three groups:

Group 1: $(dP/dt)_{max} \leqslant 300$ bar/s
Group 2: 300 bar/s $< (dP/dt)_{max} <$ 600 bar/s
Group 3: $(dP/dt)_{max} \geqslant 600$ bar/s

The required vent area per unit enclosure volume is specified for each of the three groups, but the value decreases with increasing enclosure volume. For Group 1 dusts and hinged vent panels of maximum mass 20 kg/m^2, the tabular relationship between vent area A[m^2], volume V[m^3] and P_{red} [bar(g)] can be approximated by the equation:

$$A = \frac{0.019}{P_{red}^{0.5}} \times V^{0.635} \qquad (6.1)$$

For Group 2 dusts and using hinged vent covers of maximum mass 12 kg/m^2, the Swedish method corresponds approximately to the equation:

$$A = \frac{0.044}{P_{red}^{0.5}} \times V^{0.685} \qquad (6.2)$$

For enclosures of $L/D > 3$, the Swedish method requires that the enclosure be divided in the minimum number of fictitious subvolumes needed for all of these to have $L/D \leqslant 3$. The required vent area for the actual enclosure is then the sum of the areas calculated for all the fictitious subvolumes.

The official Swedish guide line for dust explosion prevention and mitigation is currently being revised (in 1991), and changes are expected, also in the venting section.

6.1.4 THE NORWEGIAN METHOD (MODIFIED DONAT METHOD)

The method most often used in Norway and described by Eckhoff (1988) is a slightly modified version of the method of Donat (1971), based on Hartmann bomb assessment of

$(dP/dt)_{max}$. The modification consists in the use of continuous graphs, obtained by interpolation and extrapolation of Donat's tabulated data. For elongated enclosures of length-to-equivalent-diameter ratios exceeding 4, the enclosure should be divided in the number of fictitious subvolumes required for L/D of each subvolume to be $\leqslant 4$. The vent area for each subvolume is assessed individually and the sum of all the areas is taken as the total area required for venting the enclosure.

The official Norwegian regulations for prevention and mitigation of industrial dust explosions are to be revised in the near future. This will also apply to the venting section.

6.1.5 THE RADANDT SCALING LAW FOR VENTED SILO EXPLOSIONS

Bartknecht (1987) indicated that Equation (6.3), derived by Radandt:

$$P_{red} = (b \times V^c)/(A - a \times V^c) \qquad (6.3)$$

could be used for scaling vent areas for silos. In this equation, which was also presented by Radandt (1989), A [m^2] is the vent area, P_{red} [bar(g)] the maximum explosion pressure in the vented silo, V [m^3] the silo volume, and a, b and c empirical constants depending on the K_{St} value of the dust. P_{stat} is assumed to equal 0.1 bar(g), and P_{red} must not exceed 2 bar(g). For K_{St} = 200 bar m/s, i.e. the upper limit of the dust explosion class St 1, the constants are $a = 0.011$, $b = 0.069$ and $c = 0.776$, based on results from experiments in a 20 m^3 silo with direct injection of dust from a conventional pneumatic transport line. It is not clear, however, how the volume scaling constant was obtained.

Eckhoff (1991) investigated Equation (6.3) by comparing data from silo explosions of twice the linear scale used by Radandt, with Radandt's data. As shown by Eckhoff (1987), the violence of explosions in vented large-scale silos of $L/D = 6$ are strongly dependent on the location of the ignition point. For this and other reasons it appears that Radandt's Equation (6.3) may not be entirely satisfactory as a general scaling law for silo vent areas.

6.1.6 OTHER VENT SIZING METHODS

Some further methods that have been suggested for sizing of dust explosion vents are also discussed by Schofield (1984) and Lunn (1984). They include the K-factor method investigated by Gibson and Harris (1976), the Schwab and Othmer Nomograph, the equivalence coefficient method by Maisey (1965), and the method by Rust (1979), the latter three being based on $(dP/dt)_{max}$ from the 1.2 litre Hartmann bomb.

6.2
COMPARISON OF DATA FROM RECENT REALISTIC FULL-SCALE VENTED DUST EXPLOSION EXPERIMENTS, WITH PREDICTIONS BY VARIOUS VENT SIZING METHODS

6.2.1
EXPERIMENTS IN LARGE SILOS OF L/D ≤ 4

The experiments by Matusek and Stroch (1980) in a 500 m^3 silo of $L/D = 3$ should be mentioned. Unfortunately, however, explosibility data for the dusts used were not provided, and the results are therefore difficult to analyse in a general context.

A series of experiments in a 500 m^3 silo of $L/D = 4$ was reported by Eckhoff and Fuhre (1984). A cross section of the silo is shown in Figure 6.2.

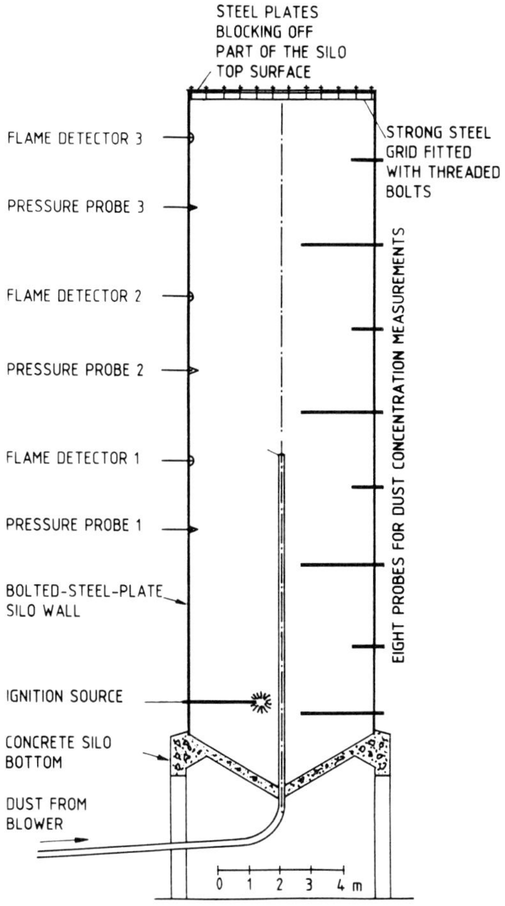

Figure 6.2 *Cross section of 500 m³ silo used in vented dust explosion experiments in Norway (From Eckhoff and Fuhre, 1984)*

The silo was one of a complex of several partly condemned bolted steel plate silos that were made available for dust explosion experiments. Pneumatic injection of dust through a 200 mm Ø pipeline was used for generating explosible dust clouds in the silo. With wheat grain dust, about 300 kg of dust was blown into the silo per experiment. With maize starch the quantity was somewhat less, about 200 kg. A larger quantity was required for the wheat grain dust because dispersion was not complete due to fibrous particles, and lumps settled to the silo bottom before all the dust had been injected. The ignition point was close to the silo bottom. In all experiments but one, dust injection was terminated a few seconds before ignition, allowing the dust cloud to become comparatively quiescent. The experimental results for wheat grain dust as well as maize starch explosions are shown in Figure 6.3 together with predictions based on some of the vent sizing methods discussed or mentioned in Section 6.1.

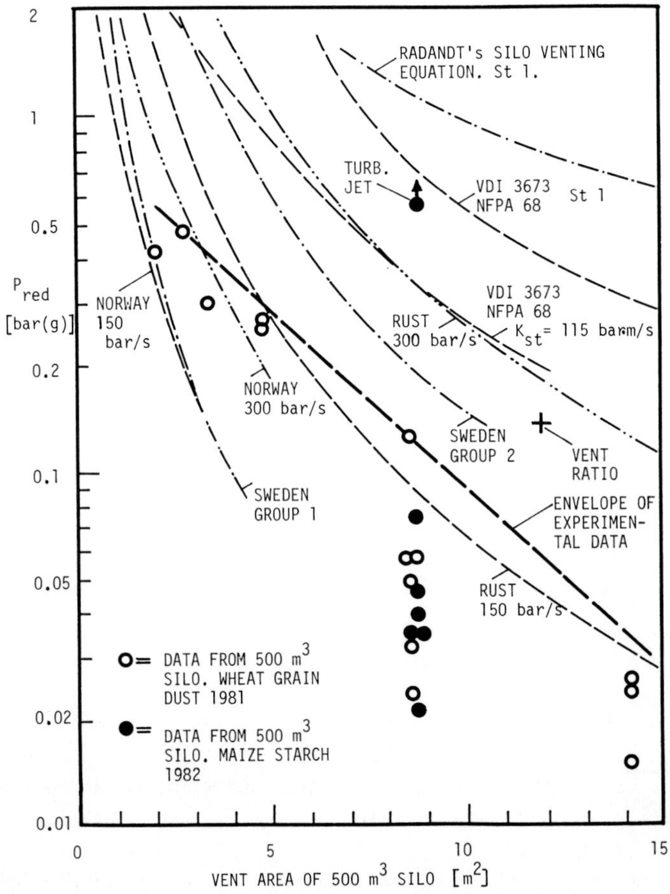

Figure 6.3 *Results from vented maize starch and wheat grain dust explosions in 500 m^3 silo cells in Norway. Comparison with predicted P_{red}/vent area correlations by various vent sizing methods in current use. P_{red} means the maximum pressure in the vented enclosure during the explosion (From Eckhoff and Fuhre, 1984, with adjustments by Eckhoff, 1990)*

The vent cover used in the experiments was a sheet of plastic with a low static opening pressure of the order of 0.01–0.02 bar(g). The final, exceptional maize starch explosion named 'turbulent jet' was so violent that the silo wall (nailed steel plates) ruptured, at about 0.6 bar(g), as shown in Figure 6.4. As seen in Figure 6.5, this produced an additional vent of about 50 m^2, which prevented the pressure from rising even further.

Figure 6.4 *High-turbulence maize starch explosion in 500 m^3 bolted steel plate silo at Vaksdal, Norway, in April 1982 (Photographer: A. M. Fosse, Vaksdal). For a much clearer picture see colour plate 5*

Figure 6.5 *Damaged 500 m^3 silo after explosion shown in Figure 6.4.*

In this exceptional experiment 300 kg of starch had been charged to the dust feeding system, and dust was still being injected at full rate, when the ignition source was activated. Such a configuration, with dust being blown upwards in the silo from a point about half-way up on its axis, is probably not a realistic situation in an industrial silo. However, the dramatic result demonstrates that even the conservative VDI 3673 from 1979 may, in certain cases of particularly high turbulence levels, specify too small vent openings.

All the other experimental results in Figure 6.3 show that in the case of a large enclosure of 500 m^3 and L/D = 4, VDI 3673 from 1979 and NFPA 68 from 1988 oversize the vent by at least a factor of 2–3 if the K_{St} value of 115 bar m/s for the maize starch used in the experiments is applied, and at least by a factor of 5 if the common St 1 nomograph is used.

The Rust method agrees well with the wheat grain dust experiments (150 bar/s in the Hartmann bomb), but oversizes the vents for the maize starch (300 bar/s) by a factor of 3. The 'vent ratio' method, which simply requires half the silo top as vent for keeping P_{red} equal to maximum 0.14 bar(g), oversizes the vent in this particular case by a factor of 2:3. Radandt's equation gives exceptionally large vents, but the small length to diameter ratio of four of the 500 m^3 silo may be too small to make Radandt's constants for 'silos' applicable.

Both the Swedish and the Norwegian vent sizing methods are comparatively liberal, suggesting P_{red} values that are in fact lower than the experimental maximum values. An argument in defence of such a liberal approach is that vent sizing involves risk-analytical considerations, rather than being a fully deterministic problem. (See Section 6.6.) In the 500 m^3 silo explosion experiments it was attempted to create a worst-case situation in terms of both dust concentration and location of the ignition point.

6.2.2 EXPERIMENTS IN SLENDER SILOS OF L/D $\approx$ 6

Sizing of vents for large, slender silos of the kind frequently used in the grain, feedstuffs and food industries, remains a controversial subject. However, during the last decade, valuable information has been gained through large-scale experiments. Two experimental programmes were run in parallel, namely one in Norway, reported by Eckhoff, Fuhre and Pedersen (1987) and Eckhoff *et al.* (1988), and one in Switzerland/F. R. Germany, reported by Radandt (1985, 1989), and Bartknecht (1987). Both silos had *L/D* close to 6, but the volume of the Norwegian silo was 236 m^3, whereas that of the Swiss/German one was only 20 m^3.

A cross section of the Norwegian silo is shown in Figure 6.6, and a picture of the experimental site in Figure 6.7.

The vent was located in the silo roof, and the silo was instrumented with dust concentration probes of the type shown in Figure 1.76 in Chapter 1, and pressure sensors, and the ignition point could be shifted over the entire length, from top to bottom. Dust was injected pneumatically either from the bottom or the top as indicated in Figure 6.6. Air of the desired flow rate was supplied by a Roots blower, and dust was fed into the air flow at the desired mass flow rate. Before ignition the air flow was stopped and the dust cloud allowed to calm down for a few seconds. Figure 6.8 shows venting of a maize starch explosion in the 236 m^3 silo.

In the experiments by Eckhoff *et al.* (1988) the vent opening was located in the silo wall near the top, as opposed to in the silo roof. This, in fact, reduced the maximum explosion pressures somewhat, but otherwise the results were similar to those from roof venting.

Radandt apparently assumed that effects observed in the comparatively small 20 m^3 silo are also representative of larger silos. However, results from the 236 m^3 silo experiments suggest that this is not a valid assumption in general. A more detailed discussion of this problem is given by Eckhoff (1991).

One set of results from the two silo sizes for which the vent areas were fairly close to geometric similarity, is given in Figure 6.9. The hatched envelope of all experimental data for the 236 m^3 silo shows a dramatic decrease of the maximum explosion pressure when

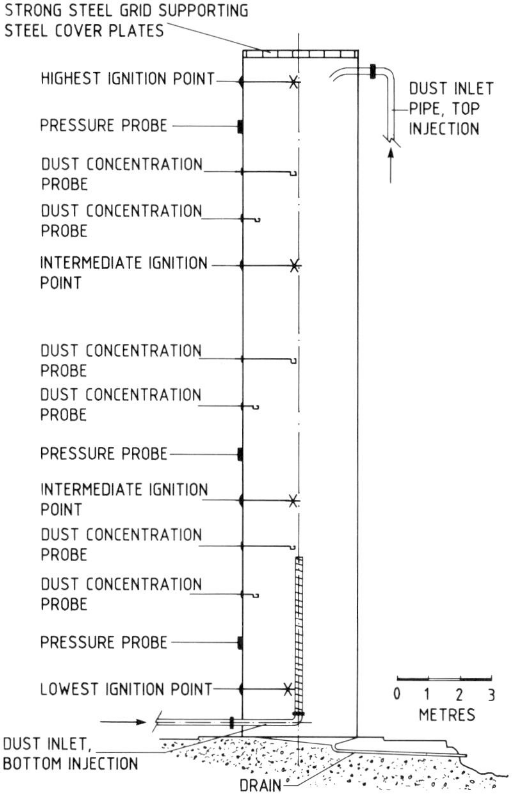

Figure 6.6 *236 m³ silo in Norway for dust explosion experiments. L/D = 6 (From Eckhoff, Fuhre and Pedersen, 1987)*

the ignition point was moved upwards along the silo axis, whereas the hatched strip enveloping the data for the 20 m^3 silo shows very little of this effect. The discrepancy is most probably due to the different ways in which the dust clouds were generated in the silos and to the difference between the silo heights (10 m and 22 m), as discussed by Eckhoff (1991).

In the case of the 236 m^3 silo, the dust cloud was essentially quiescent at the moment of ignition. There is little doubt that the marked increase of the maximum explosion pressure as the ignition point was shifted downwards in the silo, was caused by flame acceleration due to expansion-induced flow. This is in complete accordance with what has been found in the past in numerous dust and gas explosion experiments in one-end-open tubes, ducts and galleries.

In case of the 20 m^3 silo, the strong, turbulent dust jet, extending from the top of the silo and several metres downwards, was maintained during the ignition and explosion process. This most probably caused very rapid propagation of any flame initiated at the top of the silo, to the central parts. This would explain why top ignition and central ignition gave almost the same explosion pressures. In the case of bottom ignition, the initial flame propagation was probably comparatively slow. But as soon as the flame front reached the turbulent dust zone in the central parts of the silo, a much more rapid flame propagation pattern, similar to that generated by central ignition, probably developed.

Figure 6.7 *Experimental site outside Bergen, Norway, with 236 m^3 steel silo, dust injection system and instrumentation cabins. Enclosed winding staircase along the silo wall to the left. For a much clearer picture see colour plate 6*

In Figure 6.10, predictions by Radandt's Equation (6. 3) are compared with the results of the experiments in the 236 m^3 silo.

The dotted curves are obtained by means of Equation (6.3) for 20 m^3 and 236 m^3 volumes, using the constants given by Radandt for dust class St 1. The upper limit of St 1 is K_{St} = 200 bar m/s, and it was assumed that the dotted curves apply exactly to this K_{St} value. In the absence of further information, it was then assumed that the ratios between the vent areas for K_{St} = 200 bar m/s and 100 bar m/s for the 20 m^3 silo and a given maximum pressure also apply to the 236 m^3 silo (geometrical similarity). This makes it possible to estimate theoretical Radandt-predictions for K_{St} = 100 bar m/s even for 236 m^3, by shifting the Radandt curve for 200 bar m/s and 236 m^3 to the left by Δ (log A).

The actual experimental maximum explosion pressures found in the 236 m^3 silo of $L/D = 6$ are partly considerably higher, partly considerably lower than the estimated Radandt-value for K_{St} = 100 bar m/s. On the other hand, even the highest experimental pressure of 1.2 bar(g) is significantly lower than the Radandt value of 1.75 bar(g) for K_{St} = 200 bar m/s (St 1 dusts).

It must be concluded that so far the relevance of, and experimental and theoretical basis for, the simple scaling law suggested by Radandt have not been fully substantiated.

All the dust clouds in the large-scale silo experiments reported by Eckhoff and co-workers were generated by pneumatic pipeline injection, in accordance with typical industrial practice. After a series of experiments using the VDI-method for dust cloud generation, as described by Radandt (1983), Bartknecht and Radandt decided to adopt pneumatic pipeline injection even in their 20 m^3 silo experiments, as discussed by

Figure 6.8 *Vented maize starch explosion in 236 m^3 steel silo in Norway. For a much clearer picture see colour plate 7*

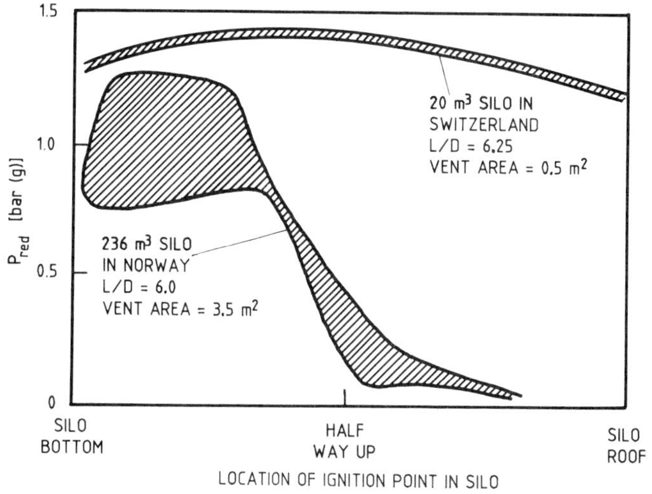

Figure 6.9 *Influence of location of the ignition point in the silo on maximum vented explosion pressure. Comparison of trends in a 20 m^3 and a 236 m^3 silo (From Eckhoff, 1990)*

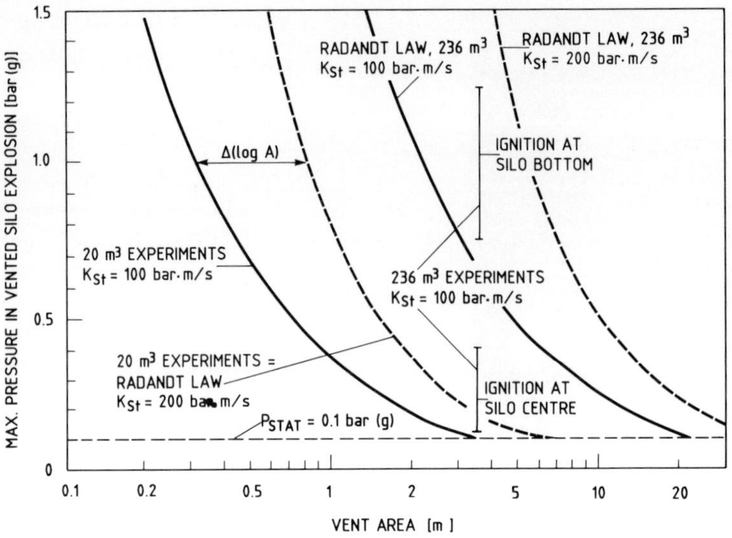

Figure 6.10 *Comparison of Radandt's scaling law for silo vent sizing and experiments in a 236 m^3 vented silo in Norway (From Eckhoff, 1990)*

Bartknecht (1988). This was a significant decision, reflecting the appreciation of the need for conducting experiments in accordance with reality in industry. In fact, Bartknecht and Radandt took a further, most relevant step, by adding experiments in which the dust was not injected directly into the silo, but via a cyclone at the silo top. In this way the dust cloud in the silo, generated by discharge of dust from the cyclone bottom via a rotary lock, was neither well dispersed, nor very turbulent. The published results from using the latter method which are traced, are limited to one data point for maize starch shown in Figure 6.11. For a vent area of 1.3 m^2, the maximum explosion pressure generated by dropping the dust from a cyclone at the silo top via a rotary lock, was only 0.2 bar(g), whereas direct pneumatic injection gave about 0.5 bar(g), and the traditional, artificial VDI-method (discharge of dust from pressurized bottles) about 0.75 bar(g).

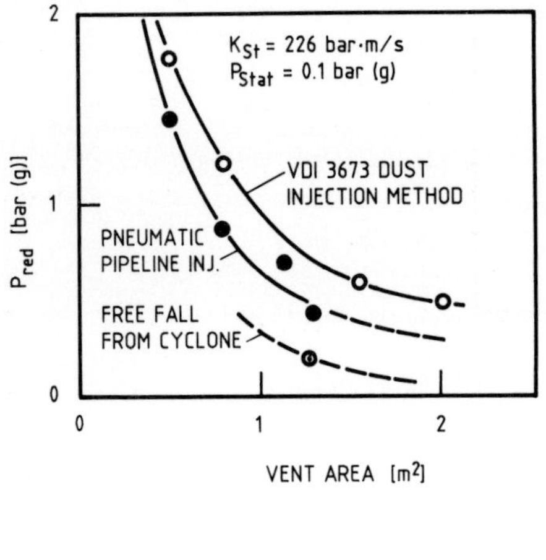

Figure 6.11 *Results from vented maize starch explosions in a 20 m^3 silo, demonstrating the marked influence of the mode of dust cloud generation on the maximum pressure P_{red} in the silo during the vented explosion (Data from Bartknecht, 1987, 1988)*

Figure 6.11 illustrates the need both for applying a differentiated approach to sizing of dust explosion vents and for further full-scale experiments under realistic industrial conditions.

The pneumatic injection experiments in the 20 m^3 silo gave substantial scatter in the experimental maximum explosion pressures, as discussed by Bartknecht (1988). The P_{red} data in Figure 6.11 are the highest values obtained for each vent area, which means that the curve for direct pneumatic injection is quite conservative. This kind of scatter, which is also apparent in Figure 6.3, emphasizes the relevance of applying risk-analytical considerations in vent sizing. (See Section 6.6)

6.2.3
PNEUMATIC PIPELINE INJECTION EXPERIMENTS IN VESSELS OF SMALL L/D

It was noted with considerable interest when the German/Swiss 'school' of dust explosion venting research started to use industrial pneumatic pipeline transport systems for generating dust clouds even in vessels of small *L*/*D*. Siwek (1989, 1989a) discussed a series of explosion experiments in which the dust clouds were generated in this way in vented enclosures of 10, 25 and 250 m^3 respectively. Some results for maize starch are reproduced in Figure 6. 12 together with the VDI 3673 recommendations from 1979. Experimental results for two quite high dust cloud injection velocities are given. For all three enclosure volumes the dust entered the vessel through a 90 mm ∅ nozzle at the vessel top, which generated a strong dust jet vertically downwards into the vessel. For all three volumes the predicted VDI 3673 areas are substantially larger than those found experimentally for any given P_{red} (note that the vent area scale in Figure 6.12 is logarithmic). The discrepancy increases systematically with increasing enclosure volume. For example for P_{red} = 0.5 bar(g), the VDI 3673 vent area for the 10 m^3 vessel is 3.7 times the experimental value even for the highest injection velocity of 78 m/s. For the 25 m^3 vessel the corresponding factor is 5.4 times, and for the 250 m^3 vessel as large as 8.3 times. The reason for this trend could be the following: The relative influence of the dust jet with respect to inducing turbulence in the vessel and thereby increasing the combustion rate, must necessarily decrease with vessel volume. In the experiments forming the basis of VDI 3673 from 1979, the systems for dispersing the dust into the vessels were scaled up with the vessel size until the desired, very high turbulence level had been reached even in large volumes. Such experimental conditions are extremely conservative and must lead to grossly oversized vents for large empty volumes of moderate *L*/*D*.

The systematic trend in Figures 6.12 of 78 m/s yielding more violent explosions than 39 m/s is probably due to increase of both the degree of turbulence and the degree of dust dispersion (de-agglomeration) with the velocity of the dust cloud jet.

It is important to note that the experimental data in Figure 6.12 are not generally valid for pneumatic injection of the maize starch used. Upwards injection from the bottom, or sideways injection might give different results. Furthermore, the written account indicates that Siwek (1989) did not investigate lower maize starch concentrations in the feeding pipe than 2–3 kg/m^3. In the closed 1 m^3 standard ISO vessel (see Chapters 4 and 7), this particular starch gave the most violent explosions for 0.6–0.7 kg/m^3. It could well be, therefore, that more violent explosions would have resulted in Siwek's experiments, if the dust concentration in the feeding pipe had been in this range.

Siwek (1989) also used wheat flour (K_{St} = 75 bar m/s) and 'Technocel' (K_{St} = 170 bar m/s) in his experiments. He found a systematic increase of P_{red} with K_{St} for the three dusts, for the test conditions investigated.

It is expected that evidence of the type illustrated in Figure 6.12 will be included in the new VDI 3673 guideline (draft probably 1991).

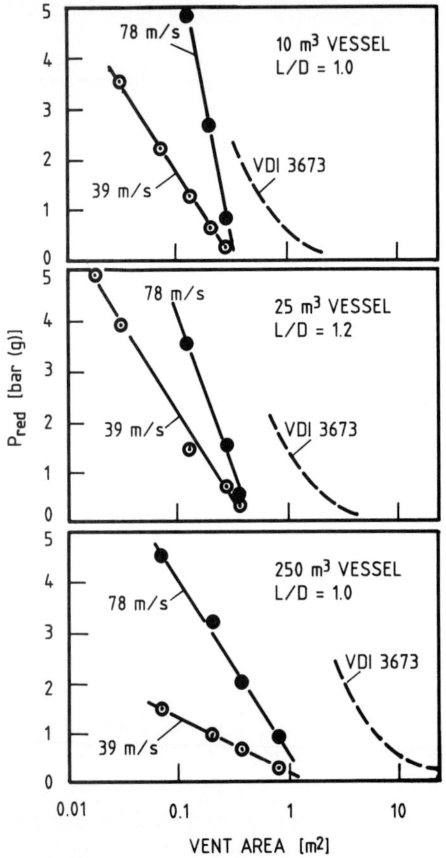

Figure 6.12 *Results from vented maize starch explosion experiments (low-moisture starch of K_{St} = 226 bar m/s) in vessels of various volumes. Pneumatic injection of dust into vessels through a 90 mm ∅ nozzle located at the vessel apex, pointing vertically downwards. Dust jet velocities 78 and 39 m/s. Dust concentration in jet 12.5 kg/m³, P_{stat} = 0.1 bar(g). Comparison with VDI 3673 (1979 edition) predictions (Experimental data from Siwek, 1989)*

6.2.4 EXPERIMENTS IN JAPAN ON VENTING OF DUST EXPLOSIONS IN A 0.32 m³ CYCLONE

Vented dust explosion experiments in a 1.2 m³ cyclone under realistic industrial conditions, were reported by Tonkin and Berlemont (1972) and Palmer (1973). As shown by Eckhoff (1986), the vent area requirements in these realistic experiments were considerably smaller, by a factor of five, than those prescribed by VDI 3673 (1979).

Similar cyclone explosion experiments were conducted in Japan more recently by Hayashi and Matsuda (1988). Their apparatus is illustrated in Figure 6.13.

The volume of the cyclone vessel was 0.32 m^3, its total height 1.8 m and the diameter of the upper cylindrical part 0.6 m. Dust clouds were blown into the cyclone through a 150 mm diameter duct. The desired dust concentration was acquired by independent control of the air flow through the duct (suction fan at downstream end of system), and the dust feeding rate into the air flow. The dust trapped in the cyclone dropped into a 0.15 m^3 dust collecting chamber bolted to the bottom outlet. The exhaust duct of 0.032 m^2 cross section and 3 m length ended in a 0.73 m^3 cubical quenching box fitted with two vents of 0.3 m^2 and 0.1 m^2 respectively. The venting of the cyclone itself was through the 0.032 m^2

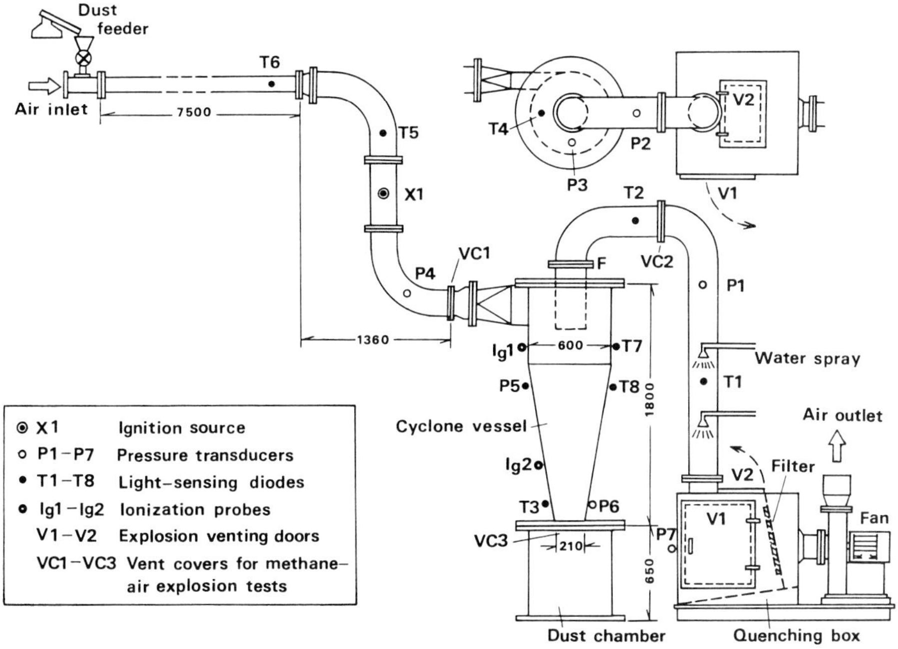

Figure 6.13 *Experimental cyclone plant for studying dust explosion development under realistic industrial conditions (From Hayashi and Matsuda, 1988)*

exhaust duct and the almost 10 m long 0.008 m^2 dust feeding duct. During explosion experiments two water spraying nozzles for flame quenching were in operation in the exhaust duct in order to protect the fan just outside the quenching box. The ignition source was a 5 kJ chemical igniter located in the dust feeding duct about 2 m upstream of the cyclone. Two different polymer dusts were used in the experiments, namely an ABS resin dust of median particle size 180 μm, and an ethylene-vinyl acetate copolymer dust (EVA) of median particle size 40 μm.

In addition to the realistic 'dynamic' explosion experiments, Hayashi and Matsuda (1988) conducted a series of experiments with the same two dusts, using an artificial 'static' dust cloud generation method, very similar to that used in the experiments being the basis of the VDI 3673 (1979 edition). As illustrated in Figure 6.14, the dust feeding duct was then blocked at the entrance to the cyclone, which reduced the effective vent area slightly, to 0.032 m^2.

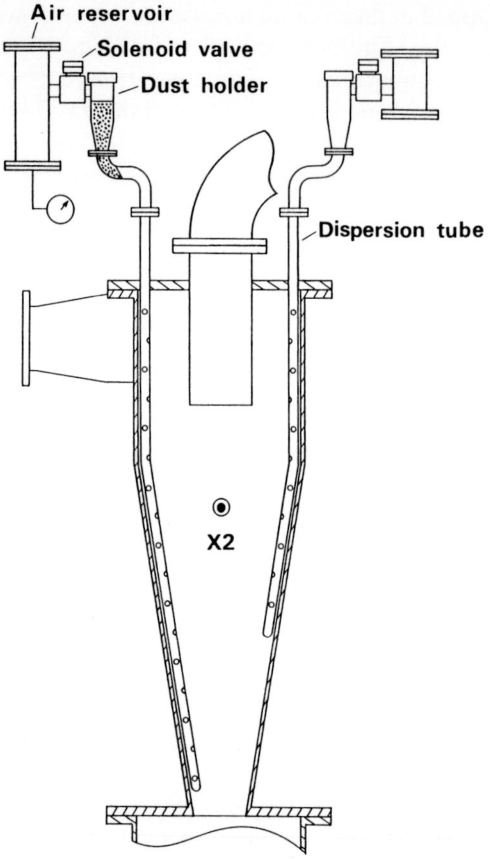

Figure 6.14 *0.32 m^3 cyclone modified for generation of dust clouds by high-pressure injection through perforated dust dispersion tubes (From Hayashi and Matsuda, 1988)*

A system of two pressurized dust reservoirs and perforated tube dispersion nozzles were employed for generating the dust clouds. The 5 kJ ignition source was located inside the cyclone, half way up on the axis (indicated by X2). The ignition source was activated about 100 ms after onset of dust dispersion.

Envelopes embracing the results of both series of experiments are given in Figure 6.15. As can be seen, the artificial 'static' method of dust dispersion gave considerably higher maximum explosion pressures in the cyclone, than the realistic 'dynamic' method. This is in accordance with the results of the earlier realistic cyclone experiments of Tonkin and Berlemont (1972). It is of interest to compare the 'static' results in Figure 6.15 with predictions by VDI 3673 (1979 edition). A slight extrapolation of the nomographs to 0.32 m^2 vent area, assuming St 1 dusts, gives an expected maximum overpressure of about 2.5 bar(g), which is of the same order as the highest pressures of 1.5 bar(g) measured for

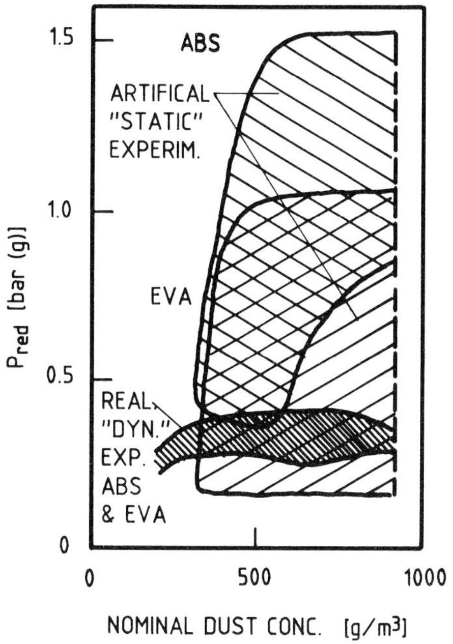

Figure 6.15 *Results from vented dust explosions in a 0.32 m³ cyclone using two different polymer dusts and two different methods of dust cloud generation. 0.03–0.04 m² open vents with ducts. Data from Hayashi and Matsuda (1988) (From Eckhoff, 1990)*

the artificial 'static' dust dispersion method, and much higher than the pressures measured in the realistic experiments.

The NFPA 68 (1988 edition) includes an alternative nomograph which covers all St 1 dusts that do not yield higher P_{max} in standard closed bomb tests than 9 bar(g). This nomograph gives much lower P_{red} values than the standard nomograph, in particular for small volumes. In the case of the 0.32 m^3 cyclone with a 0.032 m^2 vent, the alternative nomograph gives P_{red} equal to 0.50 bar(g), which in fact is close to the realistic experimental values. This alternative nomograph originates from Bartknecht (1987), and represents a considerable liberalization, by a factor of two or so, of the vent area requirements for most St 1 and St 2 dusts. However, the scientific and technical basis for this liberalization does not seem to have been fully disclosed in the open literature.

6.2.5 REALISTIC EXPERIMENTS IN BAG FILTERS

6.2.5.1 Vented explosions in a 6.7 m^3 industrial bag filter unit in UK

Lunn and Cairns (1985) reported on a series of dust explosion experiments in a 6.7 m^3 industrial bag filter unit. The experiments were conducted during normal operation of the filter, which was of the pulsed-air, self-cleaning type. Four different dusts were used, and their K_{St} values were determined according to ISO (1985) (see Chapter 7). The ignition source was located in the hopper below the filter bag section. In the experiments of main interest here, the vent was in the roof of the filter housing. Hence, in order to get to the vent, the flame had to propagate all the way up from the hopper and through the congested filter bag section. The results from the experiments are summarized in Figure 6.16, together with the corresponding VDI 3673 (1979 edition) predictions.

Figure 6.16 first shows that the P_{red} in the actual filter explosions were mostly considerably lower than the corresponding VDI 3673 predictions and close to the theoretical minimum value 0.1 bar(g) at which the vent cover ruptured. Secondly, there is no sensible correlation between the VDI 3673 ranking of expected pressures according to the K_{St} values, and the ranking actually found.

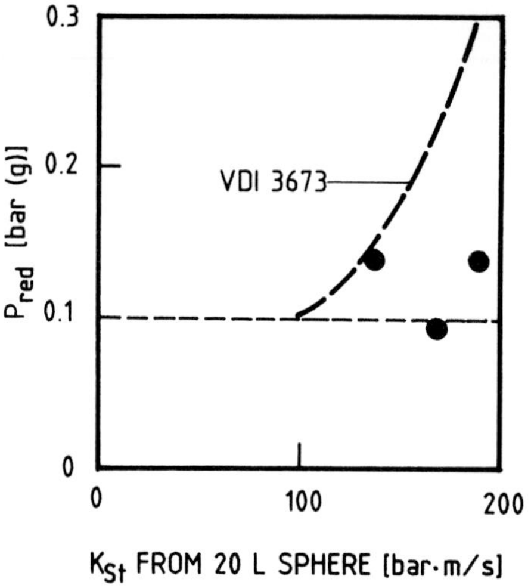

Figure 6.16 *Maximum explosion pressures P_{red} measured in dust explosions in an industrial 6.7 m^3 bag filter unit in normal operation. P_{stat} = 0.1 bar(g). Data from Lunn and Cairns (1985). Comparison with VDI 3673 (1979 edition) (From Eckhoff, 1990)*

Lunn and Cairns (1985) also reported on a series of dust explosion experiments in a generously vented 8.6 m^3 empty horizontal cylindrical vessel of $L/D \approx 6$. The same dusts were used as in the filter experiments, but the dust clouds were generated 'artificially' by injection from pressurized reservoirs as in the standard VDI 3673 method. In spite of the similarity between the dust dispersion method used and the VDI 3673 dispersion method, there was no correlation between P_{red} and K_{St}.

6.2.5.2
Vented explosions in a 5.8 m^3 bag filter in Norway

These experiments were reported in detail by Eckhoff, Alfert and Fuhre (1989). A perspective drawing of the experimental filter is shown in Figure 6.17 and a photograph of a vented maize starch explosion in the filter in Figure 6.18.

Dust explosions were initiated in the filter during normal operation. A practical worst-case situation was realized by blowing dust suspensions of the most explosible concentration into the filter at 35 m/s and igniting the cloud in the filter during injection. Four dusts were used, namely, maize starch and peat dust, both having K_{St} = 115 bar m/s, and polypropylene and silicon dusts, both having K_{St} = 125 bar m/s. Considerable effort was made to identify worst-case conditions of dust concentration, and ignition-timing. At these conditions, experimental correlations of vent area and P_{red} were determined for each dust.

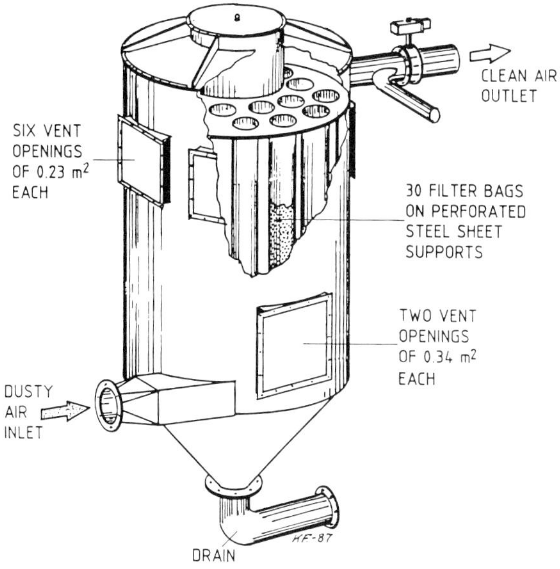

Figure 6.17 *5.8 m³ experimental bag filter in Norway (From Eckhoff, Alfert and Fuhre, 1989)*

Figure 6.18 *Maize starch explosion in 5.8 m³ experimental bag filter unit in Norway. Vent area 0.16 m². Static opening pressure of vent cover 0.10 bar(g). Maximum explosion pressure 0.15 bar(g). For a much clearer picture see colour plate 8*

As shown in Figure 6.19, the peat dust gave significantly lower explosion pressures than those predicted by VDI 3673 (1979), even if the predictions were based on the volume of the dusty filter section (3.8 m^3) only.

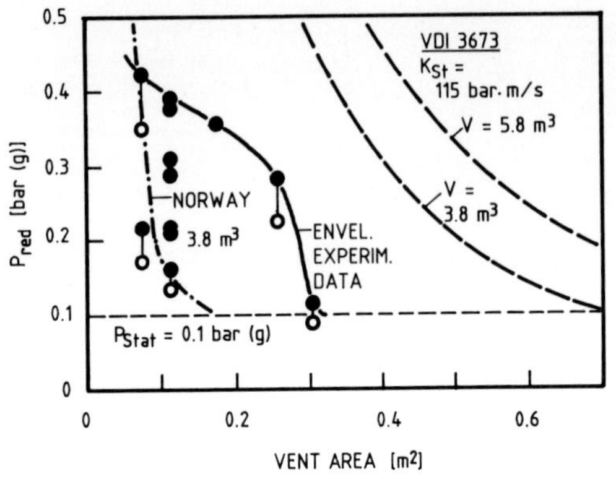

Figure 6.19 *Results from vented peat dust explosions in a 5.8 m^3 filter at P_{stat} = 0.1 bar(g). Comparison with VDI 3673 (1979 edition) and vent sizing method used in Norway (Eckhoff (1988)). Injected dust concentration 600 g/m^3. ● = dusty section of filter, ○ = clean section of filter (From Eckhoff, 1990)*

Figure 6.20 summarizes the results for all the four dusts. As can be seen, the explosion pressures measured were generally considerably lower than those predicted by VDI 3673 (1979 edition) for all the four dusts as long as the ignition source was a nitrocellulose flame. However, the singular result obtained for silicon dust ignited by a silicon dust flame emphasizes the different nature of initiation and propagation of metal dust flames, as compared with flames of organic dusts. (See discussion by Eckhoff, Alfert and Fuhre (1989), and Chapter 4.)

As illustrated by Figure 6.19, P_{red} scattered considerably, even when the nominal experimental conditions were identical. This again illustrates the risk-analytical aspect of the vent sizing problem (see Section 6.6). Figure 6.19 suggests that VDI 3673 is quite conservative, whereas the method used in Norway is quite liberal, in agreement with the picture in Figure 6.3.

In Figures 6.20 and 6.21 the 5.8 m^3 filter results for all four dusts are plotted as functions of K_{St} from 1 m^3 ISO standard tests, and $(dP/dt)_{max}$ from Hartmann bomb tests. (See Chapter 7.)

Predictions by various vent sizing methods have also been included for comparison. The data in Figure 6.20 show poor correlation between the maximum explosion pressures measured in the filter at a given vent area, and the maximum rates of pressure rise determined in standard laboratory tests. Although the K_{St} values of the four dusts were very similar, ranging from 115 to 125 bar m/s, the P_{red} (nitrocellulose flame ignition) for the four dusts varied by a factor of two to three.

In the case of the Hartmann bomb Figure 6.21 indicates a weak positive correlation between P_{red} and $(dP/dt)_{max}$ for nitrocellulose ignition, but it is by no means convincing. Figure 6.21 also gives the corresponding correlations predicted by three different vent sizing methods based on Hartmann bomb tests. Both the Swedish and the Norwegian methods are quite liberal. The Rust method oversizes the vents for the organic dusts excessively for $(dP/dt)_{max}$ > 150 bar/s. There is, however, fair agreement with the data for silicon dust ignited by a silicon dust flame.

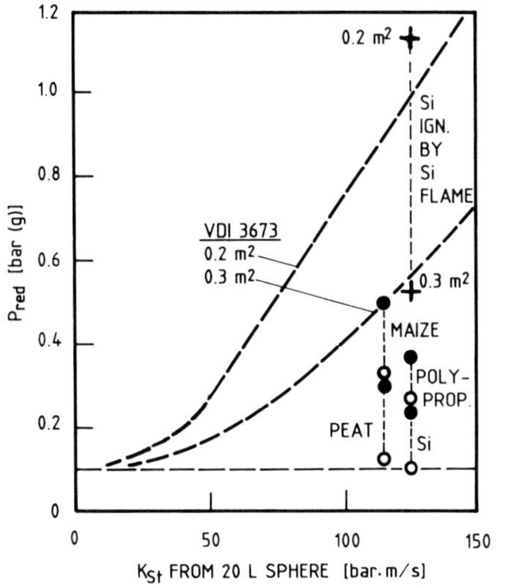

Figure 6.20 *Maximum explosion pressures for four dusts in a vented 5.8 m³ filter at two vent areas, as functions of K_{St} determined by the 20 litre Siwek sphere.*

● = 0.2 m² vent area ⎫ nitrocellulose flame ignition
○ = 0.3 m² vent area ⎭

+ = silicon dust flame ignition of silicon dust
P_{stat} = 0.1 bar(g)

Comparison with VDI (1979 edition) predictions for 3.8 m³ volume (dusty section of filter) (From Eckhoff, 1990)

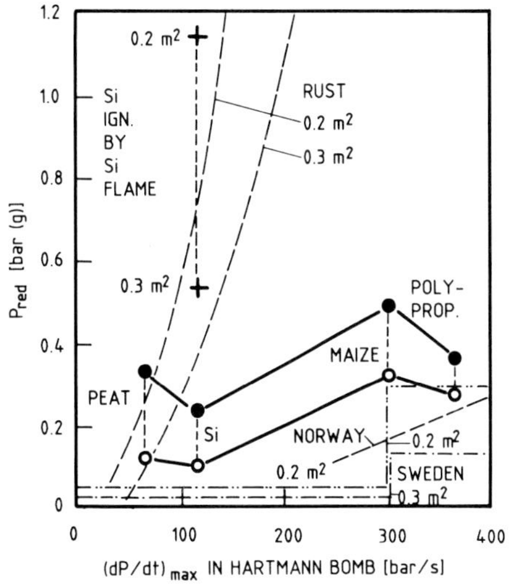

Figure 6.21 *Maximum explosion pressures for four different dusts in a vented 5.8 m³ filter at two vent areas, as functions of $(dP/dt)_{max}$ determined by the Hartmann bomb.*

● = 0.2 m² vent area ⎫ nitrocellulose flame ignition
○ = 0.3 m² vent area ⎭

+ = silicon dust flame ignition of silicon dust
P_{stat} = 0.1 bar(g)

Comparison with maximum explosion pressures predicted for 3.8 m³ volume (dusty section of filter) by three different methods (From Eckhoff, 1990)

The use of closed-bomb tests for predicting the violence of accidental dust explosions in industrial plants was discussed by Eckhoff (1984/85). (See also Chapter 7.)

6.2.6 OTHER LARGE-SCALE EXPERIMENTS RELEVANT TO INDUSTRIAL PRACTICE

Some quite early work that is still of considerable interest and practical value deserves attention. The pioneering work of Greenwald and Wheeler (1925) on venting of coal dust explosions in long galleries is discussed in Section 4.4.7 in Chapter 4.

A set of results from the comprehensive investigation by Brown and Hanson (1933) on venting of dust explosions in volumes typical of the process industry were reproduced in Figure 6.1. The paper by Brown and Hanson describes a number of interesting observations and considerations including the effect of the location and distribution of the vents and the influence of the size and type of ignition source.

Brown (1951) studied the venting of dust explosions in a 1.2 m diameter, 17 m long horizontal tube with and without internal obstructions. The tube was either closed at one end and vented at the other, or vents were provided at both ends. In some experiments an additional vent was also provided in the tube wall midway between the two ends. The location of the ignition point was varied.

Brown and Wilde (1955) extended the work of Brown (1951) by investigating the performance of a special hinged vent cover design on the explosion pressure development in a 0.76 m diameter, 15 m long tube with one or more vents at the tube ends and/or in the tube wall.

Pineau, Giltaire and Dangreaux (1974, 1976), using geometrically similar vented vessels of L/D about 3.5 and volumes 1, 10 and 100 m^3, investigated the validity of the vent area scaling law $A_2 = A_1 (V_2/V_1)^{2/3}$. They concluded that this law, which implies geometrical similarity even of vent areas, was not fully supported by the experiments. However, as long as the dust clouds were generated in similar ways in all three vessel sizes, and the ignition points were at the vessel centres, the experiments were in agreement with the law $A_2 = A_1 (V_2/V_1)^{0.52}$.

Pineau, Giltaire and Dangreaux (1978) presented a series of experimentally based correlations for various dusts between vent area and vessel volume for open and covered vents, with and without vent ducts. Both bursting membranes and spring-loaded and hinged vent covers were used in the experiments.

Zeeuwen and van Laar (1985) and van Wingerden and Pasman (1988) studied the influence of the initial size of the exploding dust cloud in a given vented enclosure, on the maximum pressure developed during the vented explosion.

The investigation showed that the pressure rise caused by the explosion of a dust cloud filling only part of a vented enclosure is higher than would perhaps be intuitively expected. Even if the dust cloud is considerably smaller than the enclosure volume, it is usually necessary to size the vent as if the entire volume of the enclosure were filled with explosible cloud.

Gerhold and Hattwig (1989) studied the pressure development during dust explosions in a vented steel silo of rectangular cross section. The length-to-equivalent-diameter ratio could be varied between two and six. The explosion pressure and flame front propagation histories were measured using a measurement system similar to that illustrated in Figure 6.6. The influence of the key parameters of industrial pneumatic dust injection systems on the explosion development was investigated, in particular injection pipe diameter, air flow and dust-to-air ratio. The general conclusion was that the maximum pressures generated with realistic pneumatic injection were substantially lower than those predicted by the VDI 3673 (1979 edition) guideline.

6.3 VENT SIZING PROCEDURES FOR THE PRESENT AND NEAR FUTURE

6.3.1 BASIC APPROACH AND LIMITATIONS

As shown in Section 6.2, realistic vented dust explosion experiments, mostly conducted during the 1980s, have demonstrated that none of the vent sizing codes in use up to 1990 are fully adequate. It is proposed, therefore, that for the present and near future, sizing of dust explosion vents be primarily based on the total evidence from realistic experiments that is available at any time.

The following suggestions presuppose that the initial pressure in the enclosure to be vented is atmospheric. Furthermore, the vent covers must open completely within times comparable to the opening times of standard calibrated rupture diaphragms. In the case of heavier, and reversible, vent covers such as hinged doors with counterweights, or spring-loaded covers, additional considerations are required. The same applies to the use of vent ducts and the new, promising vent closure concept that relieves the pressure, but retains the dust and flame, thus rendering vent ducts superfluous. (See Section 1.4.6 in Chapter 1.)

6.3.2 LARGE EMPTY ENCLOSURES OF L/D < 4

As shown in Figure 6.3, a large empty enclosure of volume 500 m^3 and $L/D = 4$, in the absence of excessive dust cloud turbulence, requires considerably smaller vents than those specified by VDI 3673 (1979 edition) or NFPA 68 (1988 edition). This also applies to the more liberal St 1 nomograph for constant-volume pressures $P_{max} < 9$ bar(g), proposed by Bartknecht (1987). (Not included in Figure 6.3.) As shown in Figure 6.12, even more dramatic reductions in vent area requirements were found in a 250 m^3 spherical vessel. In this case the vent area actually needed was only one-eighth of that specified by VDI 3673 (1979 edition).

When sizing vents for large enclosures of $L/D \leqslant 4$, the exact vent area reduction factor as compared with VDI 3673 (1979 edition), has to be decided in each case, but it should certainly not be greater than 0.5. In some cases it may be as small as 0.2 to 0.1. The new edition of VDI 3673 (draft probably 1991) is likely to take this into account.

6.3.3 LARGE, SLENDER ENCLOSURES (SILOS) OF L/D > 4

The only investigation of vented dust explosions in vertical silos of $L/D > 4$ and volumes > 100 m^3 that has been traced, is that described in Section 6.2.2. The strong influence of the location of the ignition source on the explosion violence, as illustrated in

Figure 6.9, is a major problem. It is necessary, in each specific case, to analyse carefully what kind of ignition sources are likely to occur, and at what locations within the silo volume ignition has a significant probability (Eckhoff (1987)). For example, if the explosion in the silo cell can be assumed to be a secondary event, initiated by an explosion elsewhere in the plant, ignition will probably occur in the upper part of the silo by flame transmission through dust extraction ducts or other openings near the silo top. In this case a vent of moderate size will serve the purpose even if L/D of the silo is large. However, the analysis might reveal that ignition in the lower part of the silo is also probable, for example because the dust has a great tendency to burn or smoulder. In this case even the entire silo roof may in some situations be insufficient for venting, and more sophisticated measures may have to be taken in order to control possible dust explosions in the silo.

6.3.4 SMALLER, SLENDER ENCLOSURES OF L/D > 4

The data of Bartknecht (1988) and Radandt (1985, 1989) from experiments in the 20 m^3 silo constitute one useful reference point. Further data for a 8.7 m^3 vessel of $L/D = 6$ is found in the paper by Lunn and Cairns (1985). However, it is necessary to pay adequate attention to the way in which the dust clouds are generated in the various experiments and select experimental conditions that are as close as possible to the conditions prevailing in the actual industrial enclosure (see Figure 6.11). Depending on the way in which the dust cloud is generated in practice, vent area reduction factors, with reference to VDI 3673 (1979), may vary between 1.0 and 0.1.

6.3.5 INTERMEDIATE (10–25 m^3) ENCLOSURES OF SMALL L/D

The experimental basis is that of the VDI 3673 guideline (1979 edition) with highly homogeneous, well-dispersed and turbulent dust clouds, and the more recent results for much less homogeneous and less well-dispersed clouds (Figure 6.12). The vent area requirements identified by these two sets of experiments differ by a factor of up to 5. Adequate vent sizing therefore requires that the conditions of turbulence, dust dispersion and level and homogeneity of dust concentration for the actual enclosure be evaluated in each specific case.

6.3.6 CYCLONES

Two realistic investigations have been traced (Tonkin and Berlemont (1972) and Hayashi and Matsuda (1988)), and both suggest a significant vent area reduction in relation to VDI 3673 (1979 edition). The early investigation by Tonkin and Berlemont using a cyclone of 1.2 m^3, indicates an area reduction factor of 0.2. The more recent investigation by Hayashi and Matsuda, using a smaller cyclone of 0.32 m^3, indicates a factor of about 0.5.

Hence, for organic St 1 dusts ($K_{St} \leq 200$ bar m/s) there seems to be room for vent area reductions with reference to the VDI 3673 (1979 edition), by factors in the range 0.5–0.2. However, for metal dusts such as silicon, although there is no direct evidence from cyclone explosions with such dusts, the VDI 3673 (1979 edition) requirements should probably be followed as in the case of filters (see Section 6.3.7).

6.3.7 BAG FILTERS

The experimental basis is the evidence in Figures 6.16 and 6.19 to 6.21, produced by Lunn and Cairns (1985) and Eckhoff, Alfert and Fuhre (1989). If ignition inside the filter itself is the most probable scenario (no strong flame jet entering the filter nor any significant pressure piling prior to ignition), the vent area requirements of VDI 3673 (1979 edition) for St 1 dusts can be reduced by at least a factor of 0.5. If the dust concentration in the feeding duct to the filter is lower than the minimum explosive concentration, the vent area may be reduced even more.

However, in the case of some metal dusts such as silicon, primary ignition in the filter itself may be less probable and ignition will be accomplished by a flame jet entering the filter from elsewhere. In this case it is recommended that the vent area requirements of VDI 3673 (1979 edition) be followed.

6.3.8 MILLS

The level of turbulence and degree of dust dispersion in mills vary with the type of mill. The most severe states of turbulence and dust dispersion probably occur in air jet mills. The experimental technique for dust cloud generation used in the experiments on which VDI 3673 (1979 edition) is based, is likely to generate dust clouds similar to those in an air jet mill. For this reason it seems reasonable that VDI 3673 (1979 edition) be used without modifications for sizing vents for this type of mills. In the case of mills generating dust clouds that are less turbulent and less well dispersed, it should be possible to ease the vent area requirements, depending on the actual circumstances.

6.3.9 ELONGATED ENCLOSURES OF VERY LARGE L/D

This enclosure group includes galleries in large buildings, pneumatic transport pipes, dust extraction ducts, bucket elevators, etc. In such enclosures severe flame acceleration can take place because of the turbulence produced by expansion-generated flow in the dust cloud ahead of the flame. In extreme cases, transition to detonation can occur. (See Chapter 4.) The generally accepted main principles for venting of such systems should be followed. Either the enclosure must be made sufficiently strong to be able to sustain even a detonation, and furnished with vents at one or both ends, or a sufficient number of vents have to be installed along the length of the enclosure to prevent severe flame acceleration. Chapter 8 of National Fire Protection Association (1988) provides useful more detailed advice. Further evidence of how dust explosions propagate in long ducts under realistic process conditions was presented by Radandt (1989a), as discussed in Chapter 4.

6.3.10 SCALING OF VENT AREAS TO OTHER ENCLOSURE VOLUMES AND SHAPES, AND TO OTHER P_{red} AND DUSTS

The number of reported realistic vented dust explosion experiments is still limited. It may therefore be difficult to find an experiment described in the literature that corresponds sufficiently closely to the case wanted. A procedure for scaling is therefore needed. National Fire Protection Association (1988) suggests the following simple equation intended for scaling of vent areas for weak structures of $P_{red} \leq 0.1$ bar(g):

$$A = \frac{C \times A_s}{P_{red}^{0.5}} \tag{6.4}$$

Here A is the vent area, A_s is the internal surface area of the enclosure and P_{red} is the maximum pressure (gauge) in the vented explosion. C is an empirical constant expressing the explosion violence, based on experimental evidence. By using the internal surface area as the scaling parameter for the enclosure 'size', the enclosure shape is accounted for such that an elongated enclosure of a given volume gets a larger vent than a sphere of the same volume.

Equation (6.4) was originally intended for the low-pressure regime only, but its form presents no such limitations. Therefore, this equation may be adopted even for $P_{red} > 0.1$ bar(g) and used for first approximation scaling of vent areas from any specific realistic experiment, to other enclosure sizes and shapes, other P_{red} and other dusts. At the outset the constant C should be derived from the result of the closest realistic experiment, from which data are available. Subsequent adjustment of C should be based on additional evidence/indications concerning influence of dust type, turbulence, etc.

Most often this approach will imply extrapolation of experimental results, which is always associated with uncertainty. Therefore the efforts to conduct further realistic experiments should be continued.

6.3.11 CONCLUDING REMARK

Over the last decade our understanding of the dust explosion venting process has increased considerably. Unfortunately, however, this has not provided us with a simple, coherent picture. On the contrary, new experimental evidence gradually forces us to accept that dust explosion venting is a very complex process. What may happen with a given dust under one set of practical circumstances may be far apart from what will happen in others. Therefore the general plant engineer may no longer be able to apply some simple rule of thumb and design a vent in five minutes. This may look like a step backwards, but in reality it is how things have developed in most fields of engineering and technology. Increasing insight and knowledge has revealed that apparently simple matters were in fact complex, and needed the attention of somebody who could make them their specialities and from whom others could get advice and assistance.

On the other hand, some qualitative rules of thumb may be indicated on a general basis. One example is Figure 6.22, which shows how, for a given type of dust, the violence of the dust explosion, or the burning rate of the dust cloud, depends on the geometry of the enclosure in which the dust cloud burns. Turbulence and dust dispersion induced by flow is a key mechanism for increasing the dust cloud burning rate.

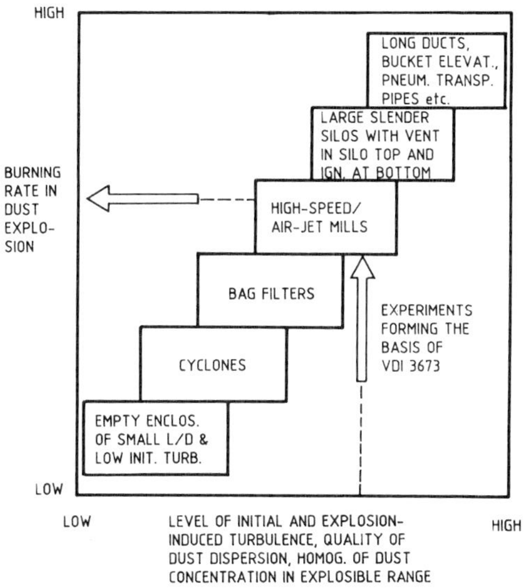

Figure 6.22 *Qualitative illustration of correlation between degree of dust dispersion, level of dust cloud turbulence and presence of homogeneous explosible dust concentration for a given dust in various industrial situations, and the burning rate of the dust cloud*

6.4 INFLUENCE OF ACTUAL TURBULENCE INTENSITY OF THE BURNING DUST CLOUD ON THE MAXIMUM PRESSURE IN A VENTED DUST EXPLOSION

This problem was studied specifically by Tamanini (1989) who conducted vented dust explosion experiments in a 64 m^3 rectangular enclosure of base 4.6 m $\times$ 4.6 m and height 3.0 m. The vent was a 5.6 m^2 square opening in one of the four 14 m^2 walls of the enclosure. Details of the experiments were given by Tamanini and Chaffee (1989).

The dust injection system essentially was of the same type as illustrated in Figure 4.39 and discussed in Section 4.4.3.1 in Chapter 4. It consisted of four pressurized-air containers, each of 0.33 m^3 capacity and 8.3 bar(g) initial pressure and being connected to four perforated dust dispersion nozzles. Two nozzle sets, i.e. eight nozzles, were mounted on each of two opposite walls inside the chamber. The dust was placed in four canisters, one for each of the pressurized air containers, located in the lines between the pressurized containers and the dispersion nozzles. On activation of high-speed valves, the pressurized air was released from the containers, entrained the dust and dispersed it into a cloud in the 64 m^3 chamber via the 16 nozzles. The high-speed valves were closed again when the pressure in the pressurized containers had dropped to a preset value of 1.4 bar(g).

As illustrated in Figures 4.40, 4.41 and 4.42 in Chapter 4, this type of experiment generates transient dust clouds characterized by a comparatively high turbulence intensity during the early stages of dust dispersion, and subsequent marked fall-off of the turbulence intensity with increasing time from the start of the dispersion. This means that the turbulence level of such a dust cloud at the moment of ignition can be controlled by controlling the delay between start of dust dispersion and activation of the ignition source.

Tamanini (1989) and Tamanini and Chaffee (1989) used this effect to study the influence of the turbulence intensity at the moment of ignition on the maximum pressure generated by explosion of a given dust at a given concentration in their 64 m^3 vented chamber. The actual turbulence intensity in the large-scale dust cloud at any given time was measured by a bi-directional fast-response gas velocity probe, in terms of the RMS (root-mean-square) of the instantaneous velocity.

However, Tamanini and Chaffee (1989) also found that during the dispersion air injection into the 64 m^3 chamber, a strong mean flow accompanied the turbulent fluctuations, at least in certain regions of the chamber. Furthermore, despite the injection of the air charge through a large number of distributed points, the flow field in the chamber was highly non-uniform, with the non-uniformity continuing during the decay part of the transient turbulence when the discharge of the air containers was complete. However, it was pointed out that the observed deviation of the flow field from uniformity is probably representative of the situation in actual process equipment, and complicates the application of flame velocity data obtained in homogeneous turbulence, to practical situations in industry. It also complicates the correlation of turbulence data with overall flame propagation characteristics.

In order to characterize the turbulence intensity in the 64 m^3 enclosure for a given small time interval by a single figure, the RMS-values found for that time interval at a large number of probe locations were averaged.

Figure 6.23 gives a set of data showing a clear correlation between the maximum pressure in the vented explosion and the average RMS of the instantaneous fluctuating turbulence velocity as measured by the pressure probes.

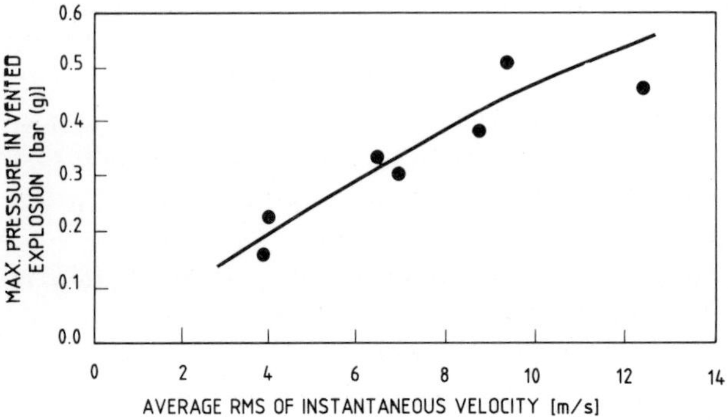

Figure 6.23 *Influence of turbulence intensity of burning dust cloud on maximum pressure in vented maize starch explosion in 64 m^3 rectangular chamber. Starch concentration 250 g/m^3 . Vent size 5.6 m^2. Ignition source 5 J chemical igniter at the chamber centre (From Tamanini, 1989)*

The contribution of Tamanini and co-workers is particularly valuable because it suggests that a quantitative link between systematic venting experiments, in which the turbulence is quantified, and the real industrial explosion hazard may be obtained via measurement of characteristic turbulence levels in dust clouds in industrial process equipment.

Tamanini and Chaffee (1989) encountered problems when trying to correlate maximum rates of pressure rise from 20 litre sphere tests with the maximum pressures in large-scale vented explosions. This is in agreement with the findings illustrated in Figures 6.20 and 6.21.

6.5 THEORIES OF DUST EXPLOSION VENTING

6.5.1 INTRODUCTORY OUTLINE

As described in Section 1.4.6.1 in Chapter 1, the maximum explosion pressure in a vented explosion, P_{red}, is the result of two competing processes:

- Burning of the dust cloud, which develops heat and increases the pressure.
- Flow of unburnt, burning and burnt dust cloud through the vent, which relieves the pressure.

In most cases the two processes are coupled via expansion-induced flow of the dust cloud ahead of the flame, which increases the turbulence of the unburnt dust cloud and hence its burning rate. In a comprehensive theory of dust explosion venting it will be necessary to include a mathematical description of this complex coupling. As discussed in Chapter 4, this has to some extent been possible in advanced modelling of gas explosions in complex geometries, where the turbulence is generated by flow past comparatively large geometrical obstacles. It is to be expected that the current rapid progress in gas and dust explosion modelling will soon result in comprehensive theories and computer simulation codes for conventional venting configurations in the process industry.

However, in the meantime several less comprehensive, more approximate theories are in use, in which it is assumed that the burning of the dust cloud and the flow out of the vent can be regarded as independent processes. In all the theories traced, it is assumed that the burning rate of the dust cloud in the vented enclosure can in some way or other be derived from the burning rate of the same dust in a standard closed-bomb test. The theories vary somewhat in the way in which this derivation is performed, but in general none of the existing venting theories seem to handle the complex burning rate problem satisfactorily. As Table 4.13 in Chapter 4 shows, K_{St} values from dust explosions with the same dust in closed bombs of various volumes and design can vary substantially, depending on dust concentration, degree of dust dispersion and dust cloud turbulence.

When using a given K_{St} value, or a maximum-rate-of-pressure-rise value, as input to the various existing theories, the relevance of the laboratory test conditions yielding the value, in relation to the dust cloud state in the actual industrial situation to be simulated, must be evaluated.

The second part of the venting theories, describing the flow out of the vent, is generally based on the classical, well-established theory for flow of gases through orifices.

A third common feature of existing theories is the use of the fact that at the maximum explosion pressure, P_{red}, in the vented enclosure, the first derivative of pressure versus time is zero. This means that the rate of expansion of the dust cloud inside the enclosure due to the combustion equals the rate of flow through the vent. An alternative formulation is that the incremental pressure rise due to combustion equals the incremental pressure drop due to venting.

In the general gas dynamics theory for venting of pressure vessels, one must distinguish between the two cases sub-sonic and sonic flow. If the ratio of internal to external pressure exceeds a certain critical value, the flow is governed by the upstream conditions only, whereas at lower pressure ratios the pressure drop across the orifice plays a main role. For a vent of small diameter compared with the vessel size (e.g. as in Figure 6.18), and neglecting friction losses, the critical pressure ratio equals

$$\frac{P_c}{P_0} = \left(\frac{\gamma + 1}{2}\right)^{\gamma/(\gamma-1)} \tag{6.5}$$

where γ is the ratio of the specific heat of the gas at constant pressure and volume. For air and most combustion gases generated in dust explosions in air this value is about 1.8–1.9, which corresponds to a pressure inside the vessel of 0.8–0.9 bar(g) at normal atmospheric ambient pressure. For most conventional process equipment the maximum permissible explosion pressure in the vented vessel will be lower than 0.8–0.9 bar(g), and in such cases the flow out of the vent is sub-sonic. However, in the case of quite strong process units, such as certain types of mills, the pressure ratio P_c/P_0 during the first part of the venting process may exceed the critical value, and the sonic flow theory will apply.

In the following sections only venting theories that were developed specifically for dust explosions are included. However, as long as the dust cloud is regarded as a combustible continuum, there is little difference between the theoretical treatment of a dust and a gas explosion, apart from the dust dispersion and initial turbulence problem. Therefore reference should be made at this point to some central publications on gas explosion venting, including Yao (1974), Anthony (1977/78), Bradley and Mitcheson (1978, 1978a), McCann, Thomas and Edwards (1985), Epstein, Swift and Fauske (1986) and Swift and Epstein (1987).

6.5.2 THEORY BY MAISEY

An early attempt to develop a partial theory of dust explosion venting was made by Maisey (1965). As a starting point he used a simple theory for laminar gas explosion development in a closed spherical vessel, with ignition at the centre. The radial laminar flame front speed was, as a first approximation, assumed to be a constant for a given fuel. For dusts it was estimated from Hartmann bomb test data (see Chapter 7). A central assumption was that the maximum pressure in a closed-bomb test is proportional to the laminar radial flame speed. However, Maisey fully appreciated the fact that in the Hartmann bomb test, as in any closed-bomb dust explosion test, the dust cloud is turbulent, and that turbulence increases the flame speed. He suggested that Hartmann bomb test data be converted to equivalent turbulent flame speeds, corresponding to the turbulence level in the test. However, because this turbulence level is probably higher than in dust clouds in most industrial plant, Maisey recommended a reduction of this equivalent Hartmann bomb flame speed, according to the actual industrial situation.

The second main part of the venting problem, the flow of gas and dust out of the vent opening, was not treated theoretically by Maisey, who instead used various experimental results to derive semi-empirical correlations between maximum vented explosion pressure and vent area for various enclosure volumes and closed-bomb flame speeds.

6.5.3 THEORY BY HEINRICH AND KOWALL

Heinrich and Kowall (1971), following the philosophy outlined in 6.5.1, and considering sub-sonic flow, arrived at the following expression for the pressure equilibrium at the maximum pressure P_{red}:

$$\left(\frac{dP_{ex}}{dt}\right)_{P_{red}\,V} = \alpha\,\frac{A}{V}\left(\frac{2RT}{M}\,P_{red}(P_{red} - P_0)\right)^{1/2} \tag{6.6}$$

where the left-hand side expresses the rate of rise of explosion pressure in the enclosure at the pressure P_{red}, had the vent been closed for an infinitely small interval of time, and

- A is the vent area [m^2]
- V is the volume of the vented enclosure [m^3]
- R is the universal gas constant = 8.31 J/(K mol)
- T is the temperature [K]
- M is the average molecular weight of the gas to be vented [kg]
- P_{red} is the maximum explosion pressure in vented enclosure [bar(abs)]
- P_0 is the ambient (normally atmospheric) pressure [bar(abs)]
- α is the vent coefficient [-], equal to 0.8 for sharp-edged vents

By rearranging Equation (6.6), the vent area A can be expressed as a function of the other parameters, including the hypothetical rate of pressure rise at P_{red}, had the vent been closed.

Heinrich and Kowall discussed the problems in quantifying the latter key parameter for dust explosions. They correlated results from actual dust explosion venting experiments, using vessel volumes up to 5 m^3, with maximum rate of pressure rise values from the 1.2 litre Hartmann bomb (see Chapter 7).

It was then assumed that the 'cube root law' (see Section 4.4.3.3 in Chapter 4) could be applied:

$$\left(\frac{dP_{ex}}{dt}\right)_{P_{red}\,V_{encl}} = \left(\frac{dP_{ex}}{dt}\right)_{P_{red}\,V_{Hartm}} \times \left(\frac{V_{Hartm}}{V_{encl}}\right)^{1/3} \tag{6.7}$$

It was concluded that the Hartmann bomb data could be correlated with the large-scale data via Equations (6.6) and (6.7) using correction factors in the range 0.5–1.0. However, Heinrich and Kowall encouraged the development of a new closed-bomb test method that would yield maximum rates of pressure rise closer to industrial reality.

In a subsequent investigation, Heinrich and Kowall (1972) discussed the influence on P_{red} of replacing the point ignition source normally used in the large-scale experiments, by a turbulent flame jet. Whereas flame-jet ignition caused a considerable increase of $(dP/dt)_{max}$ in closed vessel experiments, the increase of P_{red} in vented experiments was comparatively small. As discussed in Section 1.4.4..1 in Chapter 1, and illustrated in Figure 1.78, this conclusion can by no means be extended to flame jet ignition in general. In some cases, e.g. with strong jets from long ducts, appreciably higher P_{red} values than with point source ignition must be expected.

In his further studies, Heinrich (1974) incorporated experimental data from other workers and proposed a set of nomographs for calculating vent areas, using maximum rates of pressure rise from the 1 m^3 closed Bartknecht-vessel (subsequently made an ISO-standard) for identifying the combustion rate. The underlying assumption was a positive, monotonic correlation between $(dP_{ex}/dt)_{P_{red}}$ in the vented explosion and $(dP_{ex}/dt)_{max}$ in the closed bomb, which was indicated by some experimental data.

Heinrich's nomographs formed an essential part of the basis of the VDI 3673 (from 1979) and NFPA 68 (from 1988).

Heinrich (1980) subsequently gave a useful analysis of the theory of the flow of a compressed gas from a container into the surrounding atmosphere after a sudden provision of a vent opening. Both the adiabatic and the isothermal cases were considered. The gas dynamic analysis was also extended to two and three vessels coupled by ducting. Good agreement with experiments was demonstrated.

Lunn *et al.* (1988) and Lunn (1989) applied the Heinrich-Kowall theory for extending the Nomograph method for vent sizing to the region of low maximum explosion pressures.

6.5.4
THEORY BY RUST

Rust (1979) based his theory on considerations very similar to those of Heinrich and Kowall, using maximum rates of pressure rise from closed-bomb tests for assessing an average burning velocity in the vented explosion via the cube root law. The weakest point in Rust's theory, as in all theories of this category, is the assessment of the burning velocity of the dust cloud.

6.5.5
THEORY BY NOMURA AND TANAKA

The process studied theoretically by Nomura and Tanaka (1980), being identical with that considered by Yao (1974) for gases, is illustrated in Figure 6.24. They envisaged a boundary surface x – x that was sufficiently close to the vent for essentially all the gas in the vessel being to the left of the surface, and sufficiently apart from the vent for the gas velocity through the surface to be negligible. They then formulated a macroscopic energy balance equation for the flow system describing the venting process, assuming that all the pressure and heat energy was located to the left of the x – x line in Figure 6.24, and all the kinetic energy to the right.

Although the approach taken by Nomura and Tanaka is somewhat different from those of Heinrich and Kowall, and Rust, the basic features are similar and in accordance with what has been said in Section 6.5.1. It may appear as if Nomura and Tanaka were not aware of the fact that Heinrich and Kowall (1971) used Equation (6.7) for estimating the rate of pressure rise in the vented enclosure from standard closed-bomb test data.

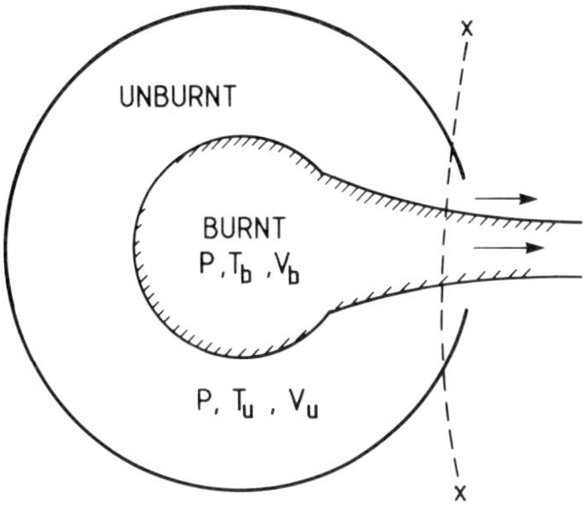

Figure 6.24 *Conceptual model of explosion venting (From Nomura and Tanaka, 1980)*

Nomura and Tanaka correlated their theoretical predictions with experimental data from various workers and found that the calculated vent areas were about three times the experimental ones. Their analysis confirmed that $A/V^{2/3}$ = constant seems to be a sensible scaling law for enclosures of length-to-diameter not much larger than unity.

6.5.6 THEORETICAL ANALYSIS BY NAGY AND VERAKIS

Nagy and Verakis (1983) first gave a comprehensive analysis of the physical process of venting of a vessel containing compressed air, applying classical gas dynamics theory, as also done by Heinrich (1980). Both the sonic and subsonic regimes were explored. They then formulated the theory of the thermodynamics of the combustion process, and finally discussed the combustion rate in more qualitative terms. The combustion part of the theory was of the same nature as that of closed vessel explosions reviewed in Section 4.2.5.1 in Chapter 4.

Nagy and Verakis first developed a one-dimensional theory for unrestricted sub-sonic venting of a dust explosion in a long cylinder with the vent at one end. Three cases were considered, namely ignition at the closed cylinder end, at the vent and at the centre. Turbulence generation due to flow of unburnt cloud towards the vent was not considered. The one-dimensional theory was then extended to the spherical configuration illustrated in Figure 6.24. The corresponding theory for sonic venting was also formulated.

The treatment by Nagy and Verakis provides a basis for formulating various equations connecting maximum pressure and vent area, assuming that $dP/dt = 0$ at the maximum pressure, using vessel shape, ignition point and flow regime as parameters.

However, Nagy and Verakis were not able to formulate a comprehensive burning rate theory. They applied the simplified 2–zone model of combustion, assuming a very thin flame and a burning velocity $S_u\alpha$, where S_u is the laminar burning velocity and $\alpha > 1$ a turbulence enhancement factor. The product $S_u\alpha$ was estimated from closed-bomb experiments with the dust of interest.

Nagy and Verakis also extended their theory to the case where the bursting pressure of the vent cover is significantly higher than the ambient pressure. Theoretical predictions were compared with experimental data from dust explosions in a 1.8 m^3 vented vessel.

6.5.7 THEORY BY GRUBER *ET AL.*

In their study, Gruber *et al.* (1987) applied the same basic gas dynamics considerations as previous workers to analyse the flow through the vent. The influence of the turbulence on the combustion rate was accounted for by multiplying the laminar burning velocity with a turbulence factor, as done by Nagy and Verakis (1983). Gruber *et al.* included a useful discussion of the nature and magnitude of the turbulence factor, by referring to more recent work by several workers. In particular, attempts at correlating empirical turbulence factors with the Reynolds number of the flow of the burning cloud were evaluated.

6.5.8 THEORY BY SWIFT

Swift (1988) proposed a venting equation implying that the maximum pressure in the vented vessel is proportional to the square of the burning velocity of the dust cloud. A turbulence factor, obtained from correlation with experimental data, was incorporated in the burning velocity, as in the case of Nagy and Verakis.

6.5.9 THEORY BY URAL

The special feature of this theory compared with those outlined above, is the assumption that the pressure rise in the unvented explosion can be described by the simple function shown in Figure 6.25.

This implies that the maximum rate of pressure rise in the unvented explosion equals:

$$\left(\frac{dP}{dt}\right)_{max} = \frac{\pi}{2t_{max}}(P_{max} - P_0) \tag{6.8}$$

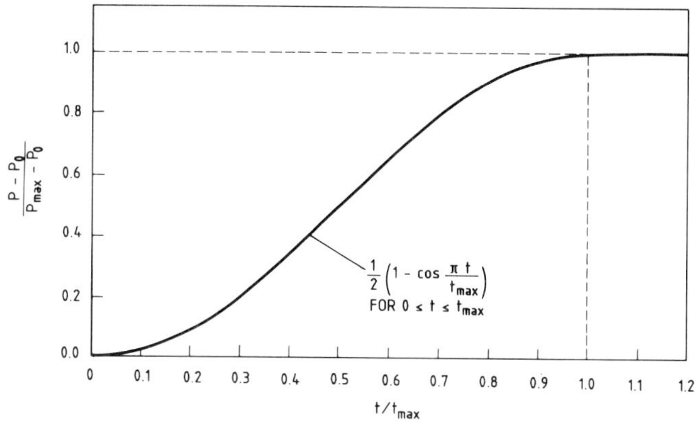

Figure 6.25 *Mathematical approximation for the shape of the pressure rise curve for the unvented explosion used in the venting theory by Ural (1989)*

where P_{max} and P_0 are the maximum and initial pressures and t_{max} is the time from ignition to when the maximum pressure has been reached. The explosion rate is then essentially characterized by the single parameter t_{max}. By means of the generalized form of Equation (6.7), experimental values of $(dP/dt)_{max}$ from closed-bomb tests may be converted to $(dP/dt)_{max}$ for the actual enclosure, without venting, and then to the corresponding t_{max} using Equation (6.8), which may be used in the venting theory for predicting maximum vented explosion pressures, P_{red}. It is then assumed that the rate of heat release in the vented explosion versus time is the same as in the unvented explosion.

As for the other theories discussed, a central requirement for obtaining reasonable predictions is that the state of the dust cloud in the closed-bomb test used for predicting the explosion violence corresponds to the state of the dust cloud in the vented explosion of concern.

6.5.10 CONCLUDING REMARK

In all the theories outlined above, the modelling of the burning rate of the dust cloud is incomplete. The situation may be improved by making use of systematic correlations of burning rates and initial dust cloud turbulence intensities determined experimentally in controlled explosion experiments, and measurements of typical turbulence intensities in various industrial plants. The studies of Tamanini and co-workers, discussed in Section 6.4, constitute a valuable step in this direction. The approach for the future is probably further development of the type of more comprehensive theories discussed in Section 4.4.8 in Chapter 4.

6.6 PROBABILISTIC NATURE OF THE PRACTICAL VENT SIZING PROBLEM

6.6.1 BASIC PHILOSOPHY

This aspect of the venting problem was treated by Eckhoff (1986). Section 1.5.1 in Chapter 1 gives a general overview of the probabilistic element in designing for dust explosion prevention and mitigation.

Consider a specific process unit being part of a specific industrial plant in which one or more specific combustible materials are produced and/or handled in powdered or granular form. The process unit can be a mill, a fluidized bed, a bucket elevator, a cyclone, a storage silo or any other enclosure in which explosible dust clouds may occur.

Assume that the plant can be operated for one million years from now, with no systematic changes in technology, operating and maintenance procedures, knowledge and attitudes of personnel, or in any other factor that might influence the distribution of ways in which dust clouds are generated and ignited. One can then envisage that a certain finite number of explosion incidents will occur during the one-million-year period. Some of these will only be weak 'puffs', whereas others will be more severe. Some may be quite violent. Because it is assumed that 'status-quo' conditions are re-established after each incident, the incidents will be distributed at random along the time axis from now on and a million years ahead.

The enclosure considered is equipped with a vent opening. The expected maximum pressure P_{max} generated in vented explosions in the enclosure, will by and large decrease with increasing vent size. This is illustrated in Figure 6.26. If the vent area is unnecessarily large, as A_1, the distribution of expected explosion pressures will be well below the maximum permissible pressure P_{red}. On the other hand, if the vent is very small, as A_3, a considerable fraction of all explosions will generate pressures exceeding the maximum permissible one. (Note that the A_2 and A_3 cases in Figure 6.26 illustrate the pressures that would have been generated had the enclosure been sufficiently strong to withstand even $P_{max} > P_{red}$.)

In the case of A_2, the vent size is capable of keeping a clear majority of all explosion pressures below P_{red}. If the fraction of the explosions that generates $P_{max} > P_{red}$ represents a reasonable risk, A_2 will constitute an adequate vent size for the case in question. However, the decision as to whether the fraction of expected destructive explosions is acceptable, depends on several considerations. The first is the expected total number of incidents of ignition of a dust cloud in the enclosure in the one-million-year period. This number is strongly influenced both by the standard obtained with respect to elimination of potential ignition sources and the standard of housekeeping. If these standards are comparatively low, the overall chance of cloud ignitions will be comparatively high. Consequently, it will be necessary to require that the fraction of all expected explosions that will not be taken care of by a vent, be comparatively small to ensure that the expected number of destructive explosions is kept at an acceptable level. On the other hand, if the probability of dust cloud ignition is low, one can rely on a smaller vent than if the standard of housekeeping and the efforts to eliminate ignition sources are inadequate. This is illustrated in Figure 6.27.

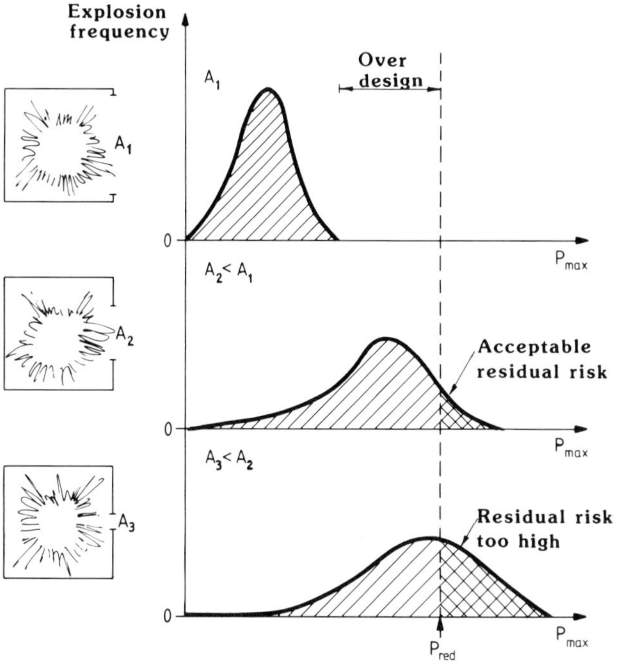

Figure 6.26 *Distributions of maximum explosion pressures generated in a given process unit, fitted with vents of different sizes, by the same one-million-year population of explosions. The unit of explosion frequency is number of explosions per million years per unit of pressure. The areas under the frequency curves then give the total number of explosions in one million years and are thus the same for the three cases*

Risk is often defined as the product of the expected number of a specific type of undesired event in a given reference period, and the consequence per event. When specifying the maximum acceptable number N of destructive explosions in the one-million-year period, i.e. the maximum acceptable number of explosions of $P_{max} > P_{red}$, it is therefore necessary to take into account the expected consequences of the destructive explosions. This comprises both possible threats to human life and health and possible damage to property.

In principle, the standard of explosion prevention can be so high that the total number of expected explosions in the one-million-year period is of the same order as the acceptable number of destructive explosions. In such cases it is questionable whether installing a vent would be advisable at all.

Figure 6.26 illustrates the 'random' variation of the expected combustion rate for a specific process unit in a specific plant handling a specific dust. However, if the dust chemistry or the particle size distribution is significantly changed, the distributions of P_{max} will also change. For example, if the particle size is increased and a systematic reduction of combustion rate results, all three distributions in Figure 6.26 will be shifted towards lower P_{max} values. The small vent area A_3 may then turn out to be sufficient. Alternatively, the average running conditions of the process could be altered in such a way that a significant systematic change in the dust cloud turbulence or concentration within the process unit

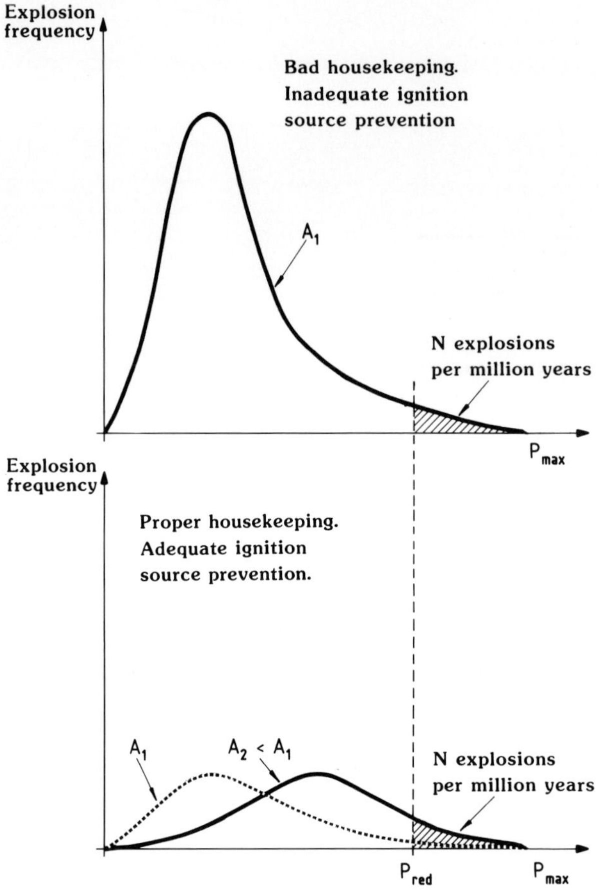

Figure 6.27 *Illustration of the reduction of necessary vent area resulting from reduction of the overall probability of dust cloud ignitions. N is the maximum acceptable number of destructive explosions per one million years*

would result. This would also cause the distributions in Figure 6.26 to change, rendering the original vent size either too small or unnecessarily large.

A general illustration of the consequence of any significant systematic change of this kind is given in Figure 6.28.

If the system is altered in such a way that the dust cloud combustion rates would generally be reduced (Modification I in Figure 6.28), the original vent size A would be unnecessarily large. On the other hand, if the alteration would generally lead to increased explosion violence (Modification II in Figure 6.28), the original vent area might turn out to be too small.

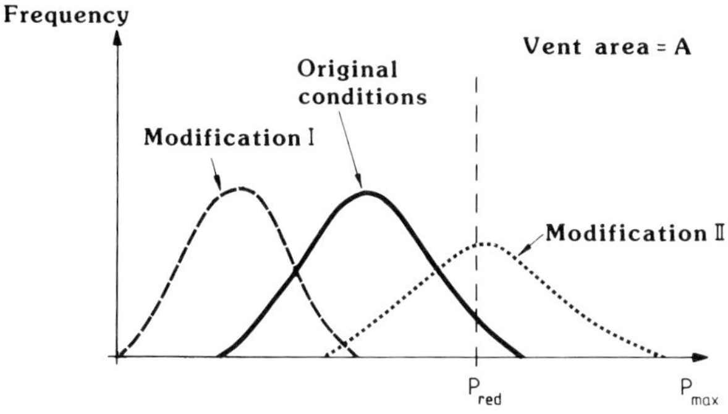

Figure 6.28 *Illustration of the influence of modifying dust properties or process design on the distribution of* P_{max}

6.6.2 THE 'WORST CREDIBLE EXPLOSION'

The discussion in Section 6.6.1 has exposed a central problem in prescribing an adequate vent size for a given purpose: Identification of the 'worst-case' explosion to be designed for. In some venting cases and guidelines, the choice of 'worst case' is rather conservative, both with respect to dust concentration, turbulence level and degree of dust dispersion. In defence of this approach, it has been argued that the venting code ensures safe venting under all circumstances encountered in practice. However, extreme conservatism may not be the optimal solution. Excessive overdesign of vents quite often imposes significant, unnecessary practical problems and costs both in finding a suitable vent location that does not conflict with other design criteria, and in designing excessive vent cover arrangements. Furthermore, providing a large vent opening may significantly reduce the strength of the process unit to be vented, necessitating complicating reinforcement for maintaining the original strength.

Conservative, rigid venting requirements may cause industry to conclude that venting is not applicable to their problem at all, and no vents are provided. This situation has been quite common in the case of large storage silos in the grain, feed and flour industry. The alternative venting philosophy outlined in Section 6.6.1 implies that even a modestly sized vent may add significantly to the safety standard of the plant by being capable of providing adequate relief for the majority of the expected explosions.

Results from realistic experiments of the kind discussed in Section 6.2, combined with proper knowledge about the actual industrial process and plant, constitute the existing basis for assessing the 'worst credible explosion'. In the future, systematic studies of different selected representative scenarios can probably be conducted by using comprehensive computer simulation models.

REFERENCES

Anthony, E. L. (1977/1978) The Use of Venting Formulae in the Design and Protection of Building and Industrial Plant from Damage by Gas or Vapour Explosions. *J. Hazardous Materials* **2** pp. 23–49

Bartknecht, W. (1978) *Explosionen,* Springer-Verlag

Bartknecht, W. (1987) *Staubexplosionen.* Springer-Verlag

Bartknecht, W. (1988) Massnahmen gegen gefährliche Auswirkungen von Staubexplosionen in Silos und Behälter: Explosionsdruckentlastung beim Befüllen durch Flugförderung. Abschlussbericht Project No. 01 HK 664 AO, Bundesministerium für Forschung und Technologie

Bradley, D., and Mitcheson, A. (1978) The Venting of Gaseous Explosions in Spherical Vessels. I – Theory. *Combustion and Flame* **32** pp. 221–236

Bradley, D., and Mitcheson, A. (1978a) The Venting of Gaseous Explosions in Spherical Vessels. II – Theory and Experiment. *Combustion and Flame* **32** pp. 237–255

Brown, H. R., and Hanson, R. L. (1933) Venting of Dust Explosions. *National Fire Protection Association Quarterly* **26** pp. 328–341

Brown, K. C. (1951) Dust Explosions in Factories: The Protection of Elevator Casings by Pressure Relief Vents, SMRE Res. Rep. No. 22, Safety in Mines Research Establishment, Sheffield, UK

Brown, K. C., and Wilde, D. G. (1955) Dust Explosions in Factories: The Protection of Plant by Hinged Explosion Doors, SMRE Res. Rep. No. 119, Safety in Mines Research Establishment, Sheffield, UK

Danielson, G. (1981) *Dammexplosioner.* Arbetarskyddsstyrelsens Författningssamling. AFS 1981:5. Liber Förlag, Stockholm

Donat, C. (1971) Auswahl und Bemessung von Druckentlastungseinrichtungen für Staubexplosionen. *Staub-Reinhalt. Luft* **31** pp. 154–160

Eckhoff, R. K., and Fuhre, K. (1984) Dust Explosion Experiments in a Vented 500 m^3 Silo Cell. *J. Occ. Acc.* **6** pp. 229–240

Eckhoff, R. K. (1984/1985) Use of $(dP/dt)_{max}$ from Closed-Bomb Tests for Predicting Violence of Accidental Dust Explosions in Industrial Plants. *Fire Safety J.* **8** pp. 159–168

Eckhoff, R. K. (1986) Sizing Dust Explosion Vents. The Need for a New Approach Based on Risk Assessment. *Bulk Solids Handling* **6** No. 5 (October)

Eckhoff, R. K. (1987) A Differentiated Approach to Sizing of Dust Explosion Vents: Influence of Ignition Source Location with Particular Reference to Large, Slender Silos. In *Industrial Dust Explosions*, *ASTM Special Techn. Publ.* 958, ed by K. L. Cashdollar and M. Hertzberg, ASTM, Philadelphia, USA, pp. 265–280

Eckhoff, R. K., Fuhre, K., and Pedersen, G. H. (1987) Dust Explosion Experiments in a Vented 236 m^3 Silo Cell. *J. Occ. Acc.* **9** pp. 161–175

Eckhoff, R. K. (1988) Beregning av trykkavlastningsarealer for støveksplosjoner (Sizing of Dust Explosion Vents). Report No. 88/02501–3, Chr. Michelsen Institute, Bergen, Norway

Eckhoff, R. K., Alfert, F., Fuhre, K., *et al.* (1988) Maize Starch Explosions in a 236 m^3 Experimental Silo with Vents in the Silo Wall. *J. Loss Prev. Process Ind.* **1** pp. 16–24

Eckhoff, R. K., Alfert, F., and Fuhre, K. (1989) Venting of Dust Explosions in a 5.8 m^3 Bag Filter under Realistic Conditions of Dust Cloud Generation. *VDI-Berichte* **701,** VDI-Verlag GmbH, Düsseldorf, pp. 695–722

Eckhoff, R. K. (1990) Sizing of Dust Explosion Vents in the Process Industries. Advances Made During the 1980s. *J. Loss. Prev. Process Ind.* **3** pp. 268–279

Eckhoff, R. K. (1991) Scaling of Vented Dust Explosions in Large Silos. (To be published) *Archivum Combustionis*, Warsaw

Epstein, M., Swift, I., and Fauske, H. K. (1986) Estimation of Peak Pressure for Sonic-Vented Hydrocarbon Explosions in Spherical Vessels. *Combustion and Flame* **66** pp. 1–8

Field, P. (1984) Dust Explosion Protection – A Comparative Study of Selected Methods for Sizing

Explosion Relief Vents. *J. Hazardous Materials* **8** pp. 223–238

Gerhold, E., and Hattwig, M. (1989) The Investigation of Dust Explosions in a Metal Silo. *Proceedings 6th Internat. Symp. Loss. Prev. Safety Prom. Process Ind.*, Norwegian Society of Chartered Engineers Vol. **IV** pp. 23–1 to 23–15

Gibson, N., and Harris, G. F. P. (1976) The Calculation of Dust Explosion Vents. *Chemical Engineering Progress* (November) pp. 62–67

Gruber, U., Puppich, P., Noll, E., *et al.* (1987) Zeitlicher Druckverlauf bei Explosionen als Grundlage zur Auslegung von Behältern und Apparaten. *Chem.-Ing.-Tech.* **59** pp. 917–926

Hayashi, T., and Matsuda, T. (1988) Dust Explosions and Their Venting in a Real Scale Cyclone. *3rd Int. Coll. On Dust Explosions*, Szczyrk, Poland, October 24–28, to be published in *Archivum Combustionis*

Heinrich, J.-J., and Kowall, R. (1971) Ergebnisse neuerer Untersuchungen zur Druckentlastung bei Staubexplosionen. *Staub-Reinhalt. Luft* **31** pp. 149–153

Heinrich, H.-J., and Kowall, R. (1972) Beitrag zur Kenntnis des Ablaufs druckentlasteter Staubexplosionen bei Zündung durch turbulente Flammen. *Staub-Reinhalt. Luft* **32** pp. 293–297

Heinrich, H.-J. (1974) Druckentlastung bei Staubexplosionen. *Arbeitsschutz* No. **11** pp. 314–318

Heinrich, H.-J. (1980) Beitrag zur Kenntnis des zeitlichen und örtlichen Druckverlaufs bei der plötzlichen Entlastung unter druckstehender Behälter und Behälterkombinationen. Forschungsbericht 75 (November), Bundesanstalt für Materialprüfung (BAM), Berlin

Lunn, G. A. (1984) *Venting Gas and Dust Explosions – A Review.* Inst. Chem. Engrs., UK

Lunn, G. A., and Cairns, F. (1985) The Venting of Dust Explosions in a Dust Collector. *Journal of Hazardous Materials* **12** pp. 87–107

Lunn, G. A., Brookes, D. E., and Nicol, A. (1988) Using the K_{St} Nomographs to Estimate the Venting Requirements in Weak Dust-Handling Equipment. *J. Loss Prev. Process Ind.* **1** pp. 123–133

Lunn, G. A. (1989) Methods for Sizing Dust Explosion Vent Areas: A Comparison when Reduced Explosion Pressures are Low. *J. Loss Prev. Process Ind.* **2** pp. 200–208

Maisey, H. R. (1965) Gaseous and Dust Explosion Venting. Part 1: *Chem. & Process Engng.*, (October), pp. 527–535 and p. 563

Maisey, H. R. (1965a) Gaseous and Dust Explosion Venting. Part 2: *Chem. & Process Engng.*, (December), pp. 662–672

Matusek, Z., and Stroch, V. (1980) Problematik der Staubexplosionen und Massnahmen gegen Explosionsgefahren in Grossraumbunker für Schüttgut. *Staub-Reinhalt. Luft* **40** pp. 503–510

McCann, D. P. J., Thomas, G. O., and Edwards, D. H. (1985) Gasdynamics of Vented Explosions. Part II: One-Dimensional Wave Interaction Model. *Combustion and Flame* **60** pp. 62–70

Nagy, J., and Verakis, H. C. (1983) *Development and Control of Dust Explosions.* Marcel Dekker, Inc.

National Fire Protection Association (1988) *Venting of Deflagrations*, NFPA 68, National Fire Protection Association, USA

Nomura, S.-I., and Tanaka, T. (1980) Theoretical Study of Relief Venting of Dust Explosions. *Journal of Chemical Engineering of Japan* **13** pp. 309–313

Palmer, K. N. (1973) *Dust Explosions and Fires.* Chapman and Hall, London

Pineau, J., Giltaire, M., and Dangreaux, J. (1974) Efficacité des évents d'explosion. Étude d'explosions de poussières en récipients de 1, 10 et 100 m^3. Note No. 881–74–74. Cahiers de Notes Documentaires, Paris

Pineau, J., Giltaire, M., Dangreaux, J. (1976) Efficacité des évents. Étude d'explosions de poussières en récipients de 1, 10 et 100 m^3: Influence de la nature de la poussière et de la présence d'une canalisation prolongeant l'event. Note No. 1005–83-76. Cahiers de Notes Documentaires, Paris

Pineau, J., Giltaire, M., and Dangreaux, J. (1978) Efficacité des évents dans le cas d'explosions de poussières Choix des surfaces d'event et de leurs dispositifs d'obturation. Note No. 1095-90-78.

Cahiers de Notes Documentaires, Paris

Radandt, S. (1983) Staubexplosionen in Silos. Untersuchungsergebnisse. Teil 2. *Symposium No.* 12, (November) Berufsgenossenschaft Nahrungsmittel und Gaststätten, Mannheim, F. R. Germany

Radandt, S. (1985) Staubexplosionen in Silos. Untersuchungsergebnisse. Teil 3. *Symposium No.* 14, (September) Berufsgenossenschaft Nahrungsmittel und Gaststätten, Mannheim, F. R. Germany

Radandt, S. (1989) Einfluss von Betriebsparametern auf Explosionsabläufe in Silozellen. *VDI-Berichte* **701,** VDI-Verlag GmbH, Düsseldorf, pp. 755–774

Radandt, S. (1989a) Explosionsabläufe in Rohrleitungen in Abhängigkeit von Betriebsparametern. VDI-Berichte 701, VDI-Verlag GmbH, Düsseldorf, pp. 801–818

Rust, E. A. (1979) Explosion Venting for Low-Pressure Equipment. *Chemical Engineering*, (November), pp. 102–110

Schofield, C. (1984) *Guide to Dust Explosion Prevention and Protection – Part 1: Venting,* Inst. Chem. Engrs., UK

Siwek, R. (1989) Druckentlastung von Staubexplosionen beim pneumatischen Befüllen von Behältern. *VDI-Berichte* **701,** VDI-Verlag GmbH, Düsseldorf, pp. 529–567

Siwek, R. (1989a) Dust Explosion Venting for Dusts Pneumatically Conveyed into Vessels. *Plant/Operations Progress* **8** pp. 129–140

Swift, I., and Epstein, M. (1987) Performance of Low-Pressure Explosion Vents. *Plant/Operations Progress* **6** pp. 98–105

Swift, I. (1988) Designing Explosion Vents Easily and Accurately. *Chemical Engineering*, (April), pp. 65–68

Tamanini, F. (1989) Turbulence Effects on Dust Explosion Venting. *AIChE Loss Prevention Symposium*, Paper 12a Session 8, (April), Houston

Tamanini, F., and Chaffee, J. L. (1989) Large-Scale Vented Dust Explosions – Effect of Turbulence on Explosion Severity. Technical Report FMRC J.I. OQ2E2.RK, (April) Factory Mutual Research

Tonkin, P. S., and Berlemont, F. J. (1972) Dust Explosions in a Large-Scale Cyclone Plant. Fire Research Note No. 942, (July) Fire Research Station, UK

Ural, E. A. (1989) Simplified Analytical Model of Vented Explosions. In Large-Scale Vented Dust Explosions – Effect of Turbulence on Explosion Severity, (Tamanini and Chaffee) Technical Report FMRC J.I. OQ2E2.RK, (April), FMRC

Verein deutscher Ingenieure (1979) Druckentlastung von Staubexplosionen. VDI-Richtlinie 3673. VDI-Verlag GmbH, Düsseldorf

Verein deutscher Ingenieure (1991) Druckentlastung von Staubexplosionen. VDI-Richtlinie 3673. (Draft of new version), VDI-Verlag GmbH, Düsseldorf

Wingerden, C. J. M. van, and Pasman, H. J. (1988) Explosion Venting of Partially Filled Enclosures. *Proc. of Conference on 'Flammable Dust Explosions'*, (November 2–4), St. Louis, Miss., USA

Yao, C. (1974) Explosion Venting of Low-Strength Equipment and Structures: Loss Prevention. CEP Technical Manual, Vol. 8, pp. 1–9, Am. Inst. Chem. Engineers, New York

Zeeuwen, J. P., and Laar, G. F. M. van (1985) Explosion Venting of Enclosures Partially Filled with Flammable Dust/Air Mixtures. *Proc. of Internat. Symp. Control of Risks in Handling and Storage of Granular Foods*. (April), APRIA, Paris

Chapter 7

Assessment of ignitability, explosibility and related properties of dusts by laboratory scale tests

7.1 HISTORICAL BACKGROUND

Since the beginning of this century considerable work has ben carried out in many countries on assessing the explosion hazard of various types of combustible dusts by laboratory testing. Palmer (1973) gave an informative account of the status in three or four different countries up to the beginning of the 1970s. The more recent summary of Field (1982) included some significant developments in the late 1970s and also work conducted in some further countries.

In the USA the US Bureau of Mines has, since its establishment in 1910, conducted studies of ignitability and explosibility of dusts. At the beginning, the investigations were mainly on coal dusts, but from 1936 the work was extended to all sorts of agricultural, industrial and other dusts (Jacobson *et al.* (1961), Jacobson *et al.* (1962), Jacobson *et al.* (1964), Nagy *et al.* (1965), Dorsett and Nagy (1968)). Equipment and procedures were developed to investigate the various ignitability and explosibility properties, as described by Dorsett *et al.* (1960). More recently new and more refined tests methods were developed by the US Bureau of Mines, as discussed by Hertzberg *et al.* (1979, 1985). Lee *et al.* (1982) proposed that some of the traditional US Bureau of Mines test methods be improved by including more refined diagnostic instrumentation. The Committee on Evaluation of Industrial Hazards (1979) suggested some additional methods for testing the ignitability and electrical resistivity of dust layers. Schwab (1968) focused on the central problem of interpreting the results of the laboratory-scale US Bureau of Mines tests in terms of the real industrial hazards and practical means of dust explosion prevention and mitigation.

In the UK systematic testing of dust ignitability and explosibility was undertaken by Wheeler at Safety in Mines Research Establishment (SMRE) from early in this century. However, in the 1960s, much of this work, except for coal dust explosion research and testing, was transferred to the Joint Fire Research Organization, now Fire Research Station, at Borehamwood. Raftery (1968) discussed the early work carried out by this organization on testing of dusts for ignitability and explosibility, and it appears that the experimental procedures and equipment were to a large extent similar to those of the US Bureau of Mines. More recently some of the test methods in the UK were modified or replaced by new ones, as discussed by Field (1983). Gibson (1972) described some further test methods used by the chemical industry in UK, whereas Burgoyne (1978) related the results of the various test methods to means of preventing and mitigating the industrial hazard.

In Germany, Selle (1957) gave an account of the quite extensive work on dust explosion testing that was carried out, in particular at the Bundesanstalt für Materialprüfung (BAM) in Berlin, in the first half of this century.

Leuschke (1966, 1967) gave updated comprehensive accounts of the various test methods used at the BAM, whereas Heinrich (1972) discussed some fundamental problems related to applying data from such methods in practical safety engineering. In a later paper, Leuschke (1979) discussed the problem of classifying the explosion hazard to be associated with a given dust on the basis of test data. Other more recent survey papers covering the scene of F. R. Germany include those of Ritter and Berthold (1979), Beck and Glienke (1985) and Hattwig (1987). In addition to BAM, BVS at Dortmund-Derne and the large chemical companies in F. R. Germany have conducted extensive research on development and assessment of test methods related to ignitability and explosibility of dusts. Verein deutscher Ingenieure (1988) summarized the status in F. R. Germany by the end of the 1980s.

An overview of comparatively early corresponding work conducted in the German Democratic Republic was given by Kohlschmidt (1972).

Zeeuwen (1982) and Zeeuwen and van Laar (1984) presented tests and methods of interpretation of test results used by TNO in the Netherlands. In Italy, work on test methods has been conducted by Stazione Sperimentale per i Combusibili (Milan) and in Spain by Laboratorio Oficial J. M. Madariaga (Madrid).

Poland has a long tradition in coal dust explosion research and testing. The work by Cybulski (1975) has gained international recognition. Much valuable work on initiation and propagation of dust explosions in industry has been conducted at The Technical University of Warsaw, and at other Polish universities.

Testing of dust ignitability and explosibility in France has mostly been carried out by CERCHAR near Paris. An account of the status on apparatuses and procedures by the end of the 1970s was given by Giltaire and Dangreaux (1978). It is interesting to note that a tensile strength test was used for assessing the cohesiveness of the powders/dusts (see Chapter 3).

In Switzerland, the extensive work by Ciba-Geigy AG has dominated the development of methods for testing the ignitability and explosibility properties of dusts during the last two or three decades. The pioneering contribution by Lütolf (1971) should be mentioned specifically. He described a complete system for testing ignitability and explosibility of dust clouds, as well as the flammability of dust layers. The system, which also incorporated some test methods developed by others than Ciba-Geigy AG, was designed to satisfy the requirement that all test results for a given powder or dust should be available within 24 hours from the sample having been received by the test laboratory. Lütolf's quick-tests still seem to be adequate for the purpose that they were intended to serve. Fairly recently comprehensive accounts of the test methods used by the Swiss process industries was given by Siwek and Pellmont (1986), Bartknecht (1987) and Siwek (1988).

Laboratory tests for dust ignitability and explosibility have been developed and investigated extensively by various organizations in USSR. The Research Institute of Material Science Problems in Kiev has played a key role in this respect. Nedin, Nejkov and Alekseev (1971) described some of the test methods in use at this institution by about 1970. Some supplementary information was provided by Eckhoff (1977). Much work has also been conducted by USSR Academy of Sciences in Moscow. Efimockin *et al.* (1984) produced an industrial standard for determination of the ignitability and explosibility parameters of dust clouds. Korolchenko and Baratov (1979) argued against the earlier practice in USSR, by which safety measures against dust explosions were specified on the basis of a measured value of the minimum explosible concentration only.

Significant work on testing of dust ignitability and explosibility has also been carried out at the University of Sydney in Australia, at the Indian Institute of Technology, Kharagpur, and the Central Building Research Institute, Roorkee, both in India, and at various universities in Japan.

Similar research and development is also being conducted at several universities in P. R. China, among which the Northeast University of Technology in Shenyang plays a central role.

In Scandinavia, Chr. Michelsen Institute in Bergen, Norway has been the central institution for ignitability and explosibility testing of dusts since about 1974. Eckhoff (1975) described the initial phase of the build-up of the laboratory, whereas Pedersen (1989) gave a recent summary of the test methods in use. During her stay at Chr. Michelsen Institute, Racke (1989) produced a summary of commercially available equipment for testing ignition sensitivity, thermal stability, and combustibility properties of reactive chemicals, including dusts. As part of a research programme on ignitability and explosibility of peat dust, the Central Research Laboratory of Finland established a laboratory comprising a limited range of test methods.

7.2 PHILOSOPHY OF TESTING IGNITABILITY AND EXPLOSIBILITY OF DUSTS: RELATIONSHIP BETWEEN TEST RESULT AND THE REAL INDUSTRIAL HAZARD

As discussed in Chapter 1, a dust explosion in industry may be initiated by a variety of different ignition sources, amongst which smouldering dust and powder, open flames, hot surfaces and electric sparks are perhaps the most important ones. The prevention of ignition may be accomplished by eliminating ignition sources, by inerting the dust cloud and in certain cases by maintaining the dust concentration below the lower explosible limit. Should an explosion nevertheless be initiated, damage may be prevented or limited by precautions such as the use of process units of small volumes separated by explosion chokes or fast-acting valves, by explosion suppression, by using explosion-proof equipment, by venting and by proper housekeeping.

The purpose of the various laboratory tests for ignitability and explosibility of dusts is to provide the quantitative data for the various hazards related to dust explosions and fires that are required for designing relevant safety precautions.

However, the relationship between the laboratory test conditions and real life in industry is not always straight-forward. The general situation is illustrated in Figure 7.

The test method produces a quantitative measure of some property of the dust, which is supposed to be related to the particular hazard in question. However, before statements can be made about the real hazard, the test result must be passed through an adequate theory of the industrial system and transformed to a useful statement of the behaviour of the system.

Figure 7.1 is a 'philosophical' model, which becomes useful only when the contents of the boxes are adequately specified. There are two extremes for the testing box to the left: The first is full-scale realistic testing in true copies of industrial plants, the other is measurements of basic behaviour of particles and molecules. In the first case there will be no need for the coupling theory, because what is measured in the left-hand box is per definition what happens in the box to the right. In the second case, however, a very detailed and comprehensive theory is required in order to transform the fundamental test data to real system performance.

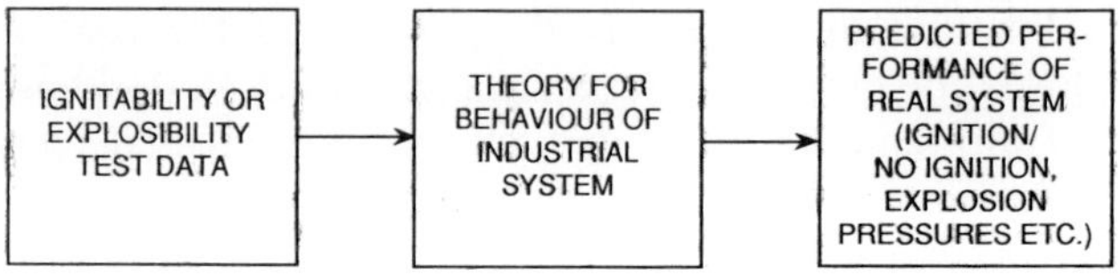

Figure 7.1 *Overall context of ignitability and explosibility testing.*

It could be argued that one should generally aim at testing on a fairly basic level and develop corresponding, complex theories. However, the rational approach seems to be to take a more balanced view. In order to make an optimal choice of the level of resolution, some questions need to be answered: How good are the available measurement techniques? How good are the theories? How much resolution is really needed for adequate design in practice?

Consider for example the ignition of dust clouds by electric sparks. In practice there are many kinds of sparks, as discussed in Section 1.1.4.6 and in Chapter 5. When electrically conducting wires are broken, break-flashes occur and the spark energy is determined by the self-induction of the system, and the current. In other situations the spark arises from capacitive discharge from non-earthed electrically conductive bodies. Further, there are brush discharges from non-conducting surfaces, corona discharges, propagating brush discharges, lightning discharges and discharges from powder heaps.

So, how should one assess the electric spark ignition hazard?

The actual measurements, symbolized by the left-hand box in Figure 7.1, can take many forms. For example, one could construct a full-scale copy of the industrial plant, introduce the powder or dust in a realistic way and see whether ignition results. However, as a general approach to hazard identification, this would not be very practical.

A more realistic approach would be to design a range of separate laboratory tests, one exposing the dust cloud to capacitive sparks from non-earthed electrical conductors, another to break-flashes, and further special tests to other kinds of electrostatic discharges. In addition one would have to visit the industrial plant and measure the relevant parameters such as capacities, voltages and inductivities, and estimate likely discharge energy levels from theory (intermediate box in Figure 7.1). By comparing these theoretical energies with the minimum ignition energies measured in the various test apparatus, one could determine whether the electric discharge ignition hazard in the plant would be significant.

A third, more fundamental approach would be to characterize the electric discharge ignition sensitivity of the dust in more basic terms, for example as a function of the distribution of spark energy in space and time, as discussed in Chapter 5. However, in this case the theory needed for relating the test result to real system behaviour would have to be considerably more detailed and complex, perhaps too complex to be manageable at present. Furthermore, the measurements would be very demanding in themselves.

Therefore, whenever a test method is designed in order to identify real, specific industrial hazards, one has to ask the basic strategic question: What is the most suitable level of resolution and generality of experiment and associated theory?

Figure 7.2 gives an introductory overview of the various test methods to be considered in the following sections.

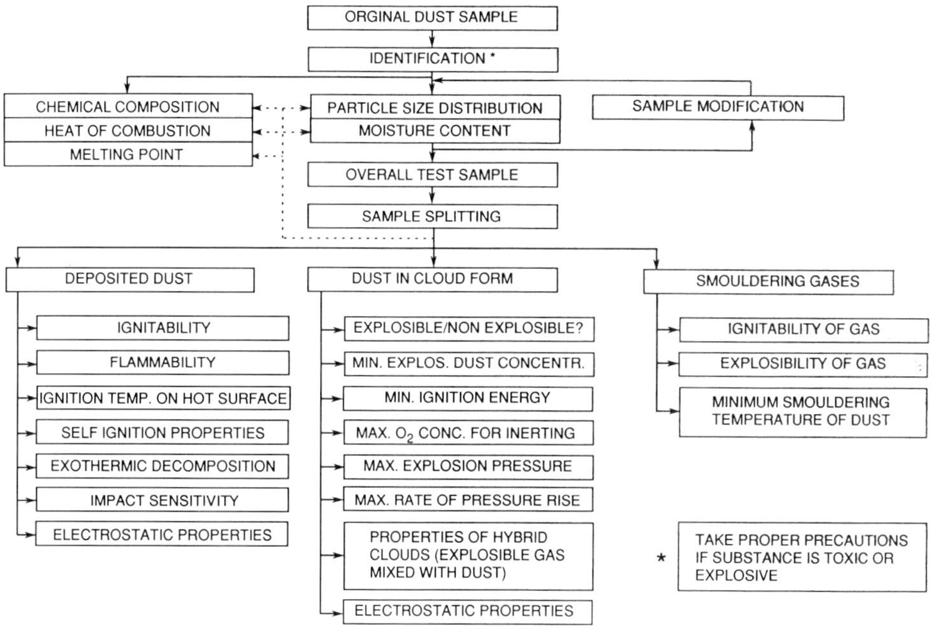

Figure 7.2 *Diagram of possible tests for assessing ignitability and explosibility properties of dusts (Slightly modified, translated version of original by Verein deutscher Ingenieure, 1988)*

7.3 SAMPLING OF DUSTS FOR TESTING

As part of a general philosophy of testing, a few words must also be said about the need for representative dust samples, Chemistry, including moisture content, and particle size and shape distributions, have a vital influence on both ignition sensitivity and explosibility. Therefore, if the dust sample tested is not representative of the dust or powder in the industrial process of concern, even the most perfect pair of test method and theory (Figure 7.1) will yield misleading assessments of the real industrial hazard.

It is useful to distinguish between two different levels of sampling. The first, and often most crucial, is the initial collection of dust in the industrial plant. Chemical composition, moisture content and particle size in samples taken from the main product stream, a dust filter, and a shelf or beam in the workroom, may vary substantially. It is important, therefore, that the initial sample be collected at a location and under operational conditions that ensures that it is really representative of the dust that creates the dust explosion hazard to be assessed.

Allen (1981) described methods for collecting dust/powder samples from bulk material in terms of:

- Sampling from a moving stream of powder.
- Sampling from a conveyor belt or chute.
- Sampling from a bucket conveyor.
- Bag sampling.
- Sampling spears.
- Sampling from wagons and containers.
- Sampling from heaps.

If the dust sample is to be collected from a gas stream, special precautions are required for ensuring representative sampling. As discussed by Allen, these relate both to the design of the sampling equipment and to the way in which the equipment is used. Whenever the sample is obtained by suction from a gas stream, it is important to ensure that the sampling conditions are iso-kinetic. Otherwise a sample of non-representative particle size distribution can result.

Once the main dust/powder sample has been collected, it remains to divide it into smaller sub-samples, down to the level required for a single dust ignitability or explosibility test. For some tests this can mean quantities as small as 0.1 g. If the initial main sample is a bag of 50 kg, in which significant segregation of particle sizes may exist, picking samples of 0.1 g directly from the bag may not ensure the required degree of representativeness of the samples. In order to solve the problem of producing representative sub-samples, various apparatuses and methods have been developed. Allen (1981) distinguished between:

- Scoop sampling.
- Coning and quartering.
- Table sampling.
- Chute splitting.
- The spinning riffler.
- Other devices.

In general, the spinning riffler has proven to yield the most homogeneous sub-samples, and this method is therefore recommended in most cases. Figure 7.3 shows a large spinning riffler used for splitting large bag-size samples of grain dust into a number of sub-samples. However, even a sub-sample of about 1 kg is very large compared with the very small quantities, down to 0.1 g, required for some ignitability or explosibility tests. Therefore, further sub-division may be necessary, and much smaller spinning rifflers than that shown in Figure 7.3 have been developed for this purpose.

Figure 7.3 *Large spinning riffler used at Chr. Michelsen Institute, Norway, for splitting large initial samples of grain dust into 26 sub-samples*

7.4 MEASUREMENT OF PHYSICAL CHARACTERISTICS OF DUSTS RELATED TO THEIR IGNITABILITY AND EXPLOSIBILITY

7.4.1 PARTICLE SIZE DISTRIBUTION AND SPECIFIC SURFACE AREA

Particle size analysis is a large field of research and development in itself, and the main purpose of the present section is to re-emphasize the major role played by particle size and shape and their distributions in deciding the ignitability and explosibility of a dust of a combustible material. (See Chapter 1.) The book by Allen (1981) is a main source of further information, both concerning the basic theory of particle size distributions, and the various experimental methods available. Allen grouped the various methods in the following main categories:

- Sieving (woven-wire and electroformed micro-mesh).
- Microscopy (light, TEM and SEM).
- Sedimentation in liquids (incremental and cumulative, gravitational and centrifugal).
- Electrical sensing zone.
- Light scattering.
- Permeametry and gas diffusion.
- Gas adsorption.

Some of the methods yield the full particle size distribution, others only a mean particle size or a specific surface area. Fast, computer-aided theoretical analysis of raw data can yield refined information. It is important to realize that the various groups of methods listed above detect different basic particle properties, and therefore the definitions of particle size, and hence also the size distributions derived for a given powder of non-spherical particles, differ for the various method groups. Nevertheless, as long as the particle shape is not extreme, such as long fibres or thin flakes, the discrepancies are normally moderate.

However, as discussed in detail in Chapter 3, and also in Section 1.3.3 in Chapter 1, powders/dusts of very small particles are difficult to disperse, particularly in a gas. Therefore, differing degrees of dispersion may give rise to considerable discrepancies in the apparent size distributions obtained for a given dust by various methods. For example, dry sieving of very fine and cohesive powders may leave significant residues of apparently coarse particles, which are in fact just agglomerates of very fine particles. Such agglomerates may be easily dispersed in liquids by using a suitable surfactant or ultrasonics, or both. Consequently, a method based on suspending and dispersing the powder in a liquid may yield a much finer size distribution than the dry-sieving method. The question is then which method gives the most realistic size distribution in relation to that of the dust clouds that may be generated in the industrial plant. (See Section 1.3.3 in Chapter 1 and Chapter 3.)

These circumstances should be kept in mind when selecting methods for particle size analysis in the context of assessment of ignitability and explosibility of dust clouds.

7.4.2 DISPERSIBILITY

The significance of this property of powders and dusts with respect to the ignitability and explosibility of clouds produced from them, has been discussed in Section 1.3.3 as well as in Chapter 3. However, neither the definition of dispersibility in practical terms, nor the development of adequate techniques for measuring this property are straight-forward tasks.

Eckhoff and Mathisen (1977/78), investigating the rate of pressure rise during dust explosions in the 1.2 litre Hartmann bomb, used the apparatus shown in Figure 7.4 for assessing the degree of dust dispersion generated by the standard dispersion system of the Hartmann bomb. Figure 7.5 shows the actual apparatus mounted on top of the Hartmann bomb dispersion cup, and dismantled. The main principle is that dispersed dust is collected on a double-stick tape mounted on a microscope slide fixed to the adjustable circular metal plate. The particles on the microscope slide are subsequently analysed with respect to size either by light microscope or SEM. The quantity of dust dispersed must not exceed the limit that still allows the individual particle units in the cloud, whether single primary particles or stable agglomerates, to appear as detached entities on the tape.

Figure 7.6 gives the number frequency distributions of maize starch agglomerates (see Figure 1.33 in Chapter 1) collected on the tape when dispersing 1.5 g of starch by the standard dispersion process of the Hartmann bomb. In this case there was no influence of moisture content, which means that the agglomerates were held together by other means than liquid bridges. The content of agglomerates was about 30% on a number basis and approximately 90% on the basis of mass, irrespective of moisture content.

Ural (1989) reviewed the literature on tests methods related to the dispersibility of dusts and powders. One of the classical methods was proposed by Professor Andreasen in Denmark in 1939. Two cubic centimetres of powder were poured through a narrow slit into a vertical tube of 2.5 m height and 45 mm diameter. The particles were separated to some extent as they fell through the air, and the percentage of the powder mass that had

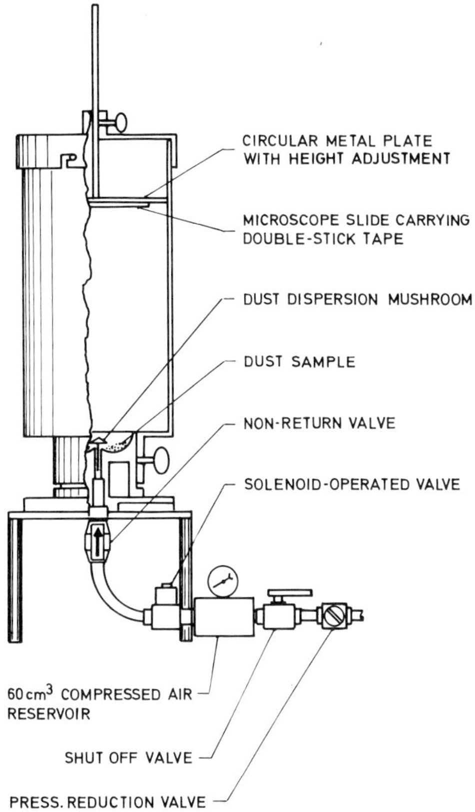

Figure 7.4 *Apparatus for assessing the degree of dispersion in dust clouds in the 1.2 litre Hartmann bomb (From Eckhoff and Mathisen, 1977/78)*

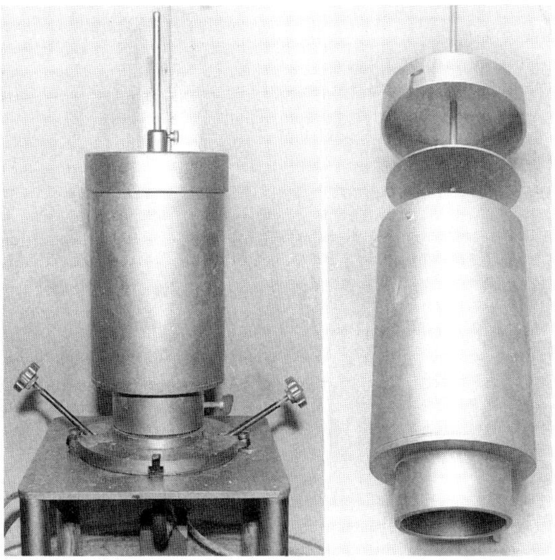

Figure 7.5 *Apparatus for assessing the degree of dispersion in dust clouds in the Hartmann bomb. Assembled on the dust dispersion cup of the Hartmann bomb (left), and dismantled (right)*

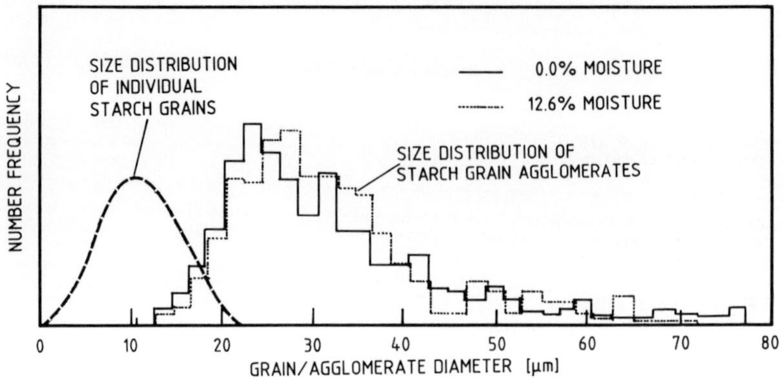

Figure 7.6 *Size distributions of maize starch grain agglomerates collected in the apparatus shown in Figures 7.4 and 7.5. Width and shape of number size distribution of individual starch grains indicated by dotted line (From Eckhoff and Mathisen, 1977/78)*

not settled to the bottom of the tube in six seconds was determined. Since the individual particles could not have reached the bottom in this time, Andreasen assumed that this figure represented the percentage of dispersed powder, which he called dispersibility. However, because some of the unsettled material could well be small agglomerates of tiny particles, this assumption may not have been entirely valid.

Another, semiquantitative test method was described by Carr (1965). The apparatus consisted of a vertical plastic tube of length 330 mm and internal diameter 100 mm, supported with its lower edge 100 mm above a 100 mm diameter watch glass. A 10 g sample of material was dropped 'en-masse' through the cylinder from a height of 600 mm above the watch glass. The material remaining on the watch glass was weighed, and the difference from the initial mass equalled the amount dispersed during the experiment.

Ural also quoted two ASTM (American Society for Testing and Materials) test methods related to dust dispersibility. One of these (Standard D547-41) is intended for determining an index of dustiness of coal and coke. The other (Standard D4331-84) serves the purpose of assessing the effectiveness of dedusting agents for powdered chemicals.

Ural (1989, 1989a) was specifically concerned with quantifying the ability of dust layers to become entrained by blasts from primary explosions and thus give rise to secondary dust explosions. His aim was to design experimental test methods that were simple and easy to perform, but should nevertheless measure fundamental quantities that could be used as input to mathematical models.

Two parameters were identified to play important roles in determining the dispersibility of powders affecting the severity of secondary explosions. The first was the settling velocity distribution of the dust and the second the entrainment threshold of a dust layer. Therefore two apparatus were built to classify powders according to these two properties.

The settling velocity apparatus is shown in Figure 7.7 and yields the settling velocity distribution of a powder sample dispersed by means of a reproducible and controllable aerodynamic disturbance.

A given quantity of dust is first placed inside the dust disperser located in the upper part of the vertical tube. Details of the disperser are shown in the expanded illustration in Figure 7.8.

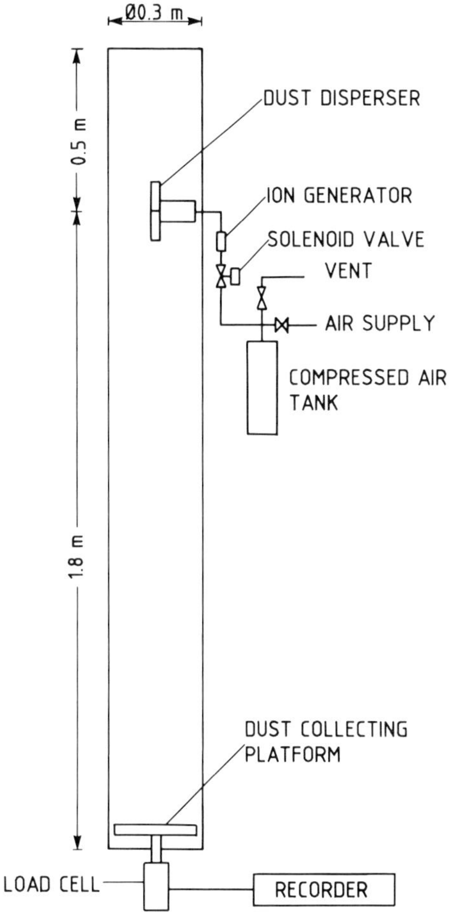

Figure 7.7 *A schematic view of the FMRC settling velocity apparatus (From Ural, 1989)*

The air pulse entrains the dust and forces the dust/air suspension through the narrow gap between the open ends of the two tubes constituting the main body of the disperser. The diffuser, consisting of a pair of flanges ensures that the dust cloud generated in the upper part of the settling velocity apparatus (Figure 7.7) does not have any significant vertical momentum. The purpose of the ion generator is to reduce the agglomeration due to electrostatic forces. The dust then settles under gravity and gradually accumulates on the bottom plate, which is supported by a load cell permitting continuous recording of the accumulated dust mass. Accumulated mass versus time is used as the primary quantification of the dispersibility of the powder. By varying the intensity of the air pulse and the width of the annular slot of the disperser, the dispersibility of a given powder can be determined as a function of basic dispersion parameters such as air velocities and viscous shear forces. The operating range of this apparatus in terms of particle size is from a few μm to 100 μm, i.e. in the range of primary interest in the context of dust explosions.

The second test method proposed by Ural (1989) was the FMRC lift-off apparatus for measuring the critical air velocity for lift-off of dust particles from a thin layer on a horizontal surface. A cross section of the basic apparatus is shown in Figure 7.9. The dust layer is spread evenly across a horizontal 380 mm ∅ circular plate. A second circular plate

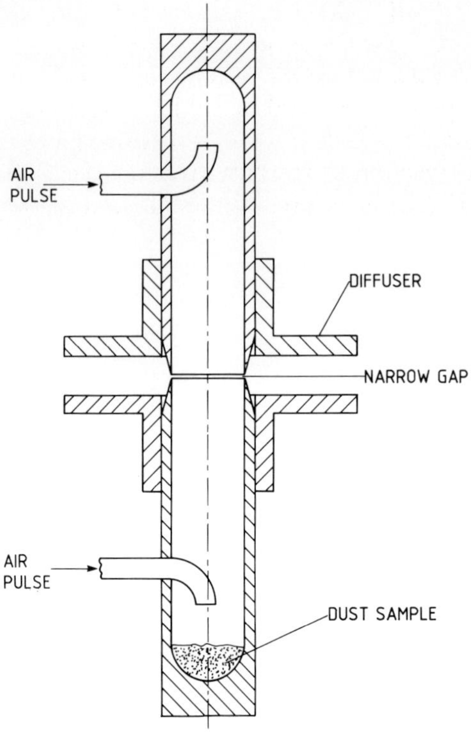

Figure 7.8 *Dust disperser for FMRC settling velocity apparatus (From Ural, 1989)*

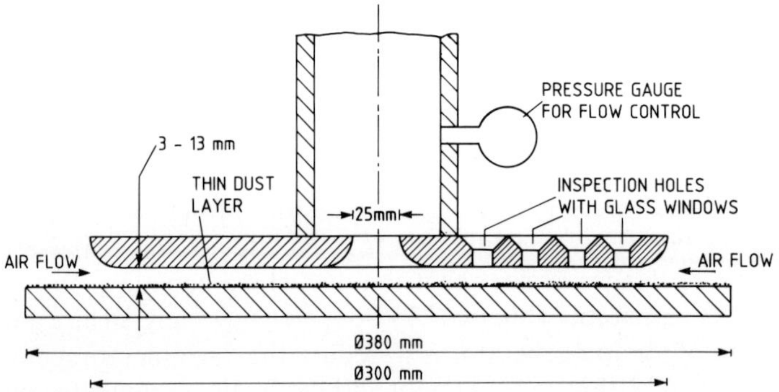

Figure 7.9 *The FMRC lift-off apparatus for assessing the dispersibility of thin dust layers (From Ural, 1989)*

of diameter 300 mm and a central hole of 25 mm Ø and rounded edges is mounted above and parallel with the dust layer plate, with a gap that can be varied from 3 mm to 13 mm. A given inwards air flow in the gap between the plates is established by creating an under-pressure in the vertical tube connected to the central hole in the upper plate. Because of diminishing flow cross section, the inwards radial flow velocity of the air increases with decreasing distance from the plate centre. When the air flow has been properly adjusted too the dust layer in question, the particles inside a circle of diameter smaller than 380 mm will be lifted off the plate by the air flow, whereas the particles outside this circle will remain in the dust layer. As long as this critical circle for lift-off can be reasonably well identified, a corresponding average critical air velocity for lift-off can also be identified. As would be expected from powder-mechanical considerations (see Chapter 3), the critical average linear air velocity for lift-off is not constant for a given cohesive dust, but depends on the degree of compaction of the dust layer. Figure 7.10 gives a set of results for maize starch layers generated by three different compaction methods.

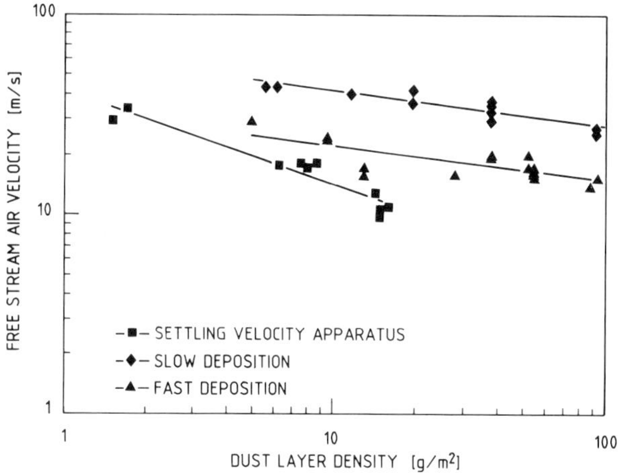

Figure 7.10 *Free stream air velocity required for 50% removal of maize starch as a function of dust layer thickness, for three different dust deposition methods (From Ural, 1989)*

The layer bulk density was highest with the fast deposition method and lowest when the layer was produced using the settling velocity apparatus illustrated in Figure 7.7. As Figure 7.10 shows, the average critical air velocity for lift-off decreased with decreasing bulk density and with increasing dust layer thickness.

Further development of the test procedures may make Ural's tests an attractive candidate for a standard method both for acquiring fundamental data and for relative ranking of dispersibility/dustability of dusts.

7.4.3
POWDER MECHANICAL PROPERTIES

The relationships between the mechanical bulk properties of a powder and the ease with which it can be dispersed into a dust cloud are discussed in detail in Chapter 3. Cohesion and tensile strength are two parameters often used for characterizing the cohesiveness of bulk powders. Test apparatuses in use are shown in Figures 3.5 and 3.6. However, both cohesion and tensile strength vary with the bulk density, or the degree of compaction, of the powder and therefore just a single figure may not be useful unless the method of preparing the powder sample is specified.

Other, relative powder mechanical test parameters that may be related to dispersibility include compressibility, angle of repose, angle of fall and angle of difference (Ural (1989)).

7.4.4 MOISTURE CONTENT

As discussed in Chapter 3 and illustrated by Figure 3.3, moisture in a powder can increase the powder cohesiveness considerably. Section 1.3.1 illustrates the strong influence of the dust moisture content on both ignitability and explosibility of the dust.

A range of measurement methods are available for determining the moisture content of dusts and powders. Cuckler (1987) gave a useful overview. The oldest and most common method of determining the moisture content consists of heating the sample to ensure complete drying. Moisture is calculated on the basis of weight difference between original and dried sample. The method is applicable to most powders and dusts and does not require unusual operator skill. Semiautomatic drying and weighing ovens are available. The moisture content is then indicated directly by the weighing scale built into the oven. Problems may arise if heating of the material also causes loss of volatile products other than water. It is essential that the material is completely dry at the time of final weighing, and that accurate weighings are made. The oven method is widely used and often serves as the primary standard for calibration of electrical and other indirect methods. Typical drying conditions for laboratory-scale samples are three to four hours at 105°C and subsequent cooling in a desiccator. Dried hygroscopic materials must be protected from humid atmosphere before weighing. Materials that pyrolyze at temperatures around 100°C may be dried at a lower temperature under vacuum. Materials that oxidize in air at normal drying temperatures may be dried in an inert atmosphere such as nitrogen.

Electrical conductivity methods are based on the relationship between dc resistance and moisture content for wood, textiles, paper, grain, and similar materials. Specific resistance plotted against moisture content results in an approximately straight line up to the moisture saturation point. Beyond the saturation point, conductivity methods are not reliable. This point varies from approximately 12% to 25% moisture, depending on the type of material.

Electrical capacitance methods are based on the difference in dielectric constant between dry and moist material. The dielectric constant of most vegetable organic materials is between two and five when dry, whereas water has a dielectric constant of 80. Therefore the addition of small amounts of moisture to these materials causes a considerable increase in the dielectric constant. The sample being measured forms part of a capacitance bridge circuit, which has radio-frequency power applied from an electronic oscillator. Electronic detectors measure bridge unbalance or frequency change, depending on the method employed.

Some instruments make use of the absorption of electromagnetic energy when passed through the material. Typical frequencies are below 10 MHz. The method gives the best results for materials composed of polar materials. The electromagnetic energy is passed through the polar material and the water molecules transform some of the energy into molecular motion.

Micro-wave absorption may also be used for measuring moisture content in powders. The principle of operation is based upon the fact that the water molecule greatly attenuates the transmitted signal with respect to other molecules in the material in the 'S' and 'X' band frequencies. In the case of the 'K' band microwave frequencies, the water molecule produces molecular resonance. There are no other molecules that respond to this particular resonant frequency, making this frequency most specific to moisture.

Some instruments for measuring moisture content are based on absorption of infrared radiation when such radiation is passed through the sample material. The water molecule becomes resonant at certain infrared frequencies and thus the amount of energy absorbed by the water absorption band is a measure of the moisture content.

More refined methods include the Karl Fischer technique and a special distillation technique.

When specifying the moisture content in terms of a percentage, it should be made clear whether this figure refers to the total mass, including the moisture, as 100%, or whether the 100% is the dry mass only.

7.4.5 ELECTRICAL RESISTIVITY

The significance of electrical resistivity of powders and dusts in the context of process safety is dual. First, the possibility of accumulating hazardous electrostatic charges and voltages in an industrial process increases with increasing electrical resistivity powder. Secondly, the chances that dusts which penetrate into electric and electronic equipment give rise to short circuits and equipment failure, increases as the dust resistivity decreases. From the point of view of the dust explosion hazard, both of these situations may lead to generation of ignition sources.

A test method for determination of the electrical resistivity of powders is being evaluated by the International Electrotechnical Commission (1988). The test cell, illustrated in Figure 7.11, consists of two metal electrodes resting on a non-conducting base plate (glass or PTFE).

The right-angled prismatic dust sample of length W and cross section H–L fills the gap between the electrodes. The actual dimensions are W = 10 cm, H = 1.4 cm and L = 1.25 cm. The recommended width of the electrodes is 3.3 cm and the thickness of the base plate 0.5–1.0 cm. Two glass bars of height 1.4 cm are placed across the ends of the electrodes to keep the dust sample in place. The dust to be tested shall be conditioned at a relative air humidity of 50 ± 5% and 20–25°C, and should normally pass a 71 μm test sieve. The moisture content of the dust, and any changes of it during the resistivity test must be reported.

During a test, the resistance R_0 of the empty test cell is first determined with the two glass bars in position across the electrode ends. Then a weighed amount of dust is poured into the cell and the excess dust scraped off and weighed, whereby the weight of the test sample and its bulk density are determined. The resistance R_s of the dust-filled cell is then

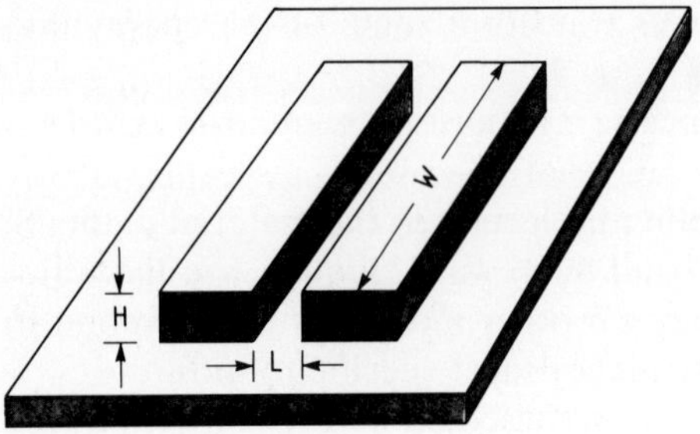

Figure 7.11 *Test cell for determination of the electrical resistivity of dust layers being evaluated by the International Electrotechnical Commission (1988)*

measured for a range of cell voltages from 110 to 2000 V. In general, $R_0 > 10\ R_s$, and the resistivity ρ of the dust is then approximately equal to:

$$\rho = R_s \times \frac{H \times W}{L} [\Omega \times \text{cm}] \qquad (7.1)$$

If $R_0 < 10\ R_s$,

$$\rho = R_s \times \frac{R_0}{R_0 - R_s} \times \frac{H \times W}{L} [\Omega \times \text{cm}] \qquad (7.2)$$

The resistivity normally varies somewhat with the applied voltage.

A dust is considered to be conductive, and thus capable of short-circuiting electrical and electronic equipment when admitted inside such equipment, if $\rho \leqslant 10^5\ \Omega$ cm ($10^3\ \Omega$ m).

7.5 CAN CLOUDS OF THE DUST GIVE EXPLOSIONS AT ALL? YES/NO SCREENING TESTS

Before embarking on more specific tests, it is sometimes considered desirable to perform an introductory test in order to determine whether the powder or dust in question can give a dust explosion at all. However, a very first screening should always be made by looking at the chemistry of the compound, which will tell whether or not it can produce significant quantities of heat by oxidation to stable products. If this is not the case, dust explosions can be excluded and testing is unnecessary.

In order to perform an introductory 'yes/no' test it is required to have a reasonably sized cloud of the dust, of concentration in the most explosive range (often 500–1000 g/m^3), and expose it to a sufficiently powerful ignition source. 'Sufficiently powerful' implies that whenever a dust cloud that is able to propagate a self-sustained flame is tested, ignition will take place.

The matter of how such a 'yes/no' test should preferably be designed is still being discussed. The tests used in various countries differ considerably, in particular with respect to the ignition source.

As described by Palmer (1973) and Field (1983), the ignition sources traditionally used in UK are electric sparks and glowing electrically heated metal wire coils. The decision as to whether the dust tested is explosible is based on visual observation of flame propagation following the dispersion of varying amounts of dust around a continuous train of electric sparks or a brightly glowing ignition coil in a vertical Perspex or glass tube of length 30 cm or 50 cm and diameter 6.4 cm. A dust is considered explosible if a dust flame becomes clearly detached from the ignition source. Normally the dust is first tested in the condition as received by the laboratory, apart from removal of particles larger than 1400 μm from the sample by sieving. However, if clouds of the dust 'as received' do not propagate a self-sustained flame, a dust sample is dried at 105°C in air for one hour and re-tested. If flame propagation still does not take place, the dry sample is sieved and individual size fractions tested, down to 25 μm. For some dusts, only the finest fractions, representing less than 1% of the bulk sample, are able to propagate dust flames. However, even in such cases, the dust will be regarded as explosible. This is because fine fractions may segregate out and become dispersed separately in an industrial situation.

In some countries in continental Europe, very powerful pyrotechnical igniters of energy about 10 kJ are used in 'yes/no' tests. Lee *et al.* (1983) discussed the production and performance of this kind of igniters. Closed 1 m^3 or 20 litre explosion bombs are often employed as test vessels (see Sections 7.15 and 7.16), and significant pressure rise is taken as an indication of explosion. However, some European countries have also adopted the modified Hartmann tube originally proposed by Lütolf (1971) for the 'yes/no' test. In this apparatus the ignition source is a comparatively weak electric spark. The argument put forward in defence of this approach is that none of the dusts that were classified as 'non-explosible' in the modified Hartmann tube test, ever caused any explosions in the chemical industry. Bartknecht (1978), on the other hand, warned against the use of the modified Hartmann tube test for 'yes/no' screening, unless the spark ignition source is replaced by a glowing resistance wire coil with a temperature of at least 1000°C.

In Norway, a pragmatic approach was taken, based on the experience that a welding torch flame seems to be among the strongest ignition sources encountered in industrial practice. The actual test apparatus is shown in Figure 7.12. A vertical tube of length 40 cm and diameter 14 cm, open at both ends, is fitted with a U-shaped dust dispersion tube and an acetylene/oxygen welding torch. A quantity of the powder is placed at the bottom of the dispersion tube, and a controlled blast from a compressed air reservoir disperses the dust into a cloud in the tube, which is then immediately exposed to the hot flame from the welding torch. The amount of powder and the dispersion air pressure are varied to produce optimal conditions for ignition. When a dust flame occurs, its maximum height, colour and apparent violence are assessed by the observer. Figure 7.13 shows a photograph of a welding-torch-flame ignition test.

The discussion of the 'yes/no' test problem is continuing, and a final, fully universal solution is not yet within sight. But, whenever the first screening is positive, i.e. the dust cloud catches fire, the screening test has fulfilled its objective, no matter which test it is. Then further, more specific testing may be required. The various methods used can be grouped in three main categories:

- Tests for ignition sensitivity of dust deposits and clouds.
- Tests for limiting conditions for flame or glow propagation in dust clouds and deposits.
- Tests for maximum rise and rate of rise of explosion pressure in dust clouds.

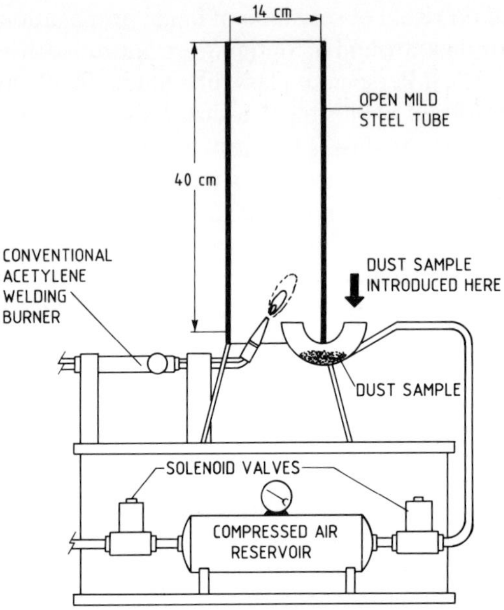

Figure 7.12 *Welding torch ignition test apparatus used in Norway (Chr. Michelsen Institute) for assessing whether or not a dust cloud is explosible*

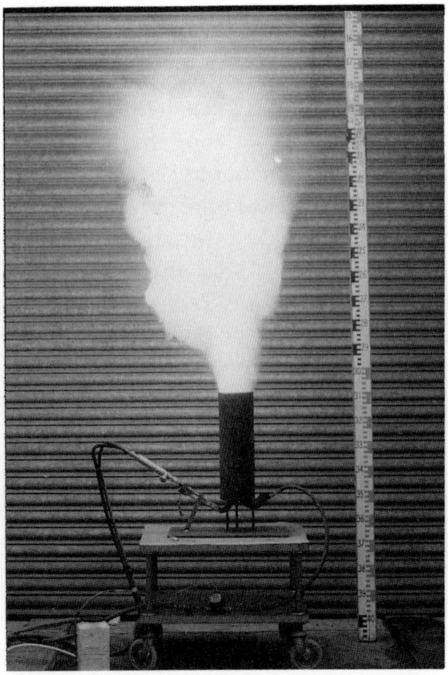

Figure 7.13 *Silicon dust explosion in the welding torch ignition test apparatus used in Norway. For a much clearer picture see colour plate 9*

Before discussing the various test methods, however, some special tests concerned with evolution of explosible gases from heated dusts should be mentioned.

7.6 CAN HAZARDOUS QUANTITIES OF EXPLOSIBLE GASES EVOLVE FROM THE DUST DURING HEATING?

7.6.1 THE INDUSTRIAL SITUATION

If the oxygen supply is limited, which will often be the case if smouldering takes place in large powder deposits in closed vessels, CO and other combustible gases can be produced, which can mix with the air and form an explosible gas cloud above the powder deposit. If the smouldering fire propagates to the surface of the powder deposit, the gas mixture can be ignited. (See Figure 1.9 in Chapter 1.) The primary gas explosion can then throw dust layers on walls, shelves, beams etc. into suspension, and give rise to considerably more extensive and severe secondary explosions, which can either be hybrid (mixture of explosible gas and dust) or pure dust explosions.

Some organic substances will decompose exothermally and release combustible gaseous products even in the absence of oxygen. This possibility represents a particular hazard.

7.6.2 LABORATORY TESTS

7.6.2.1 The BAM method

The apparatus developed by BAM (1974) and shown in Figure 7.14 illustrates the common main idea. A given quantity of dust is placed in a test tube, which is enclosed in a copper block that can be heated to any desired temperature up to 580°C. The test tube exit is connected to the bottom of a furnace with temperature control. When the upper furnace is kept at a temperature significantly higher than the minimum ignition temperature of the smouldering gases, it represents a severe ignition source. If the lower copper block furnace is then heated to a temperature where smouldering occurs, smouldering gases will leave the test tube and enter the upper furnace where they mix with the air in the furnace. If significant quantities of smouldering gases are evolved over some time, an explosible mixture with air will sooner or later occur in the upper furnace, and become ignited. If, on the other hand, the smouldering gas evolution is very small, explosible concentration may not be reached within a reasonable test period.

The BAM apparatus also permits quantitative determination of the minimum heating temperature for evolution of hazardous quantities of smouldering gases, as well as the minimum temperature in the upper furnace for ignition of explosible mixtures of such gases and air.

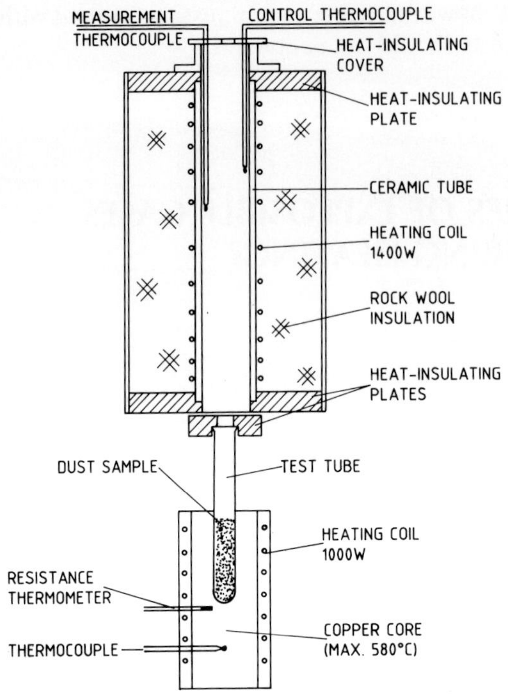

Figure 7.14 *Furnace for assessing the minimum temperature for evolution of combustible smouldering gases from dusts, and the minimum ignition temperature of mixtures of such gases and air (From BAM, 1974)*

7.6.2.2
The ASTM method

American Society for Testing and Materials (1989) prepared a test method for determination of temperature limits of flammability of chemicals in general, which also includes a procedure for testing powders and dusts. The sample of 50 cm^3 volume is placed at the bottom of a spherical 5 litre glass bottle which is kept at the desired elevated temperature by a flow of heated air sweeping through the casing in which the bottle is contained. A magnetic stirrer in the bottle ensures that the smouldering gas produced is mixed continuously with the main bulk of gas and air in the bottle. Hence, there is a reasonably homogeneous gas concentration throughout the 5 litres. The flammability of the gas/air mixture is tested by means of an electric spark discharged close to the centre of the bottle.

7.6.2.3
Lütolf's method

Figure 7.15 shows the quick-test method proposed by Lütolf (1978). Test tube (a) is for collecting the smouldering gases for other test purposes. Test tube (b) allows simple direct testing of whether ignitable quantities of smouldering gases are produced at the selected heating block temperature. Test tubes (c) and (d) allows detection of any exothermal decomposition of the dust/powder at the selected block temperature.

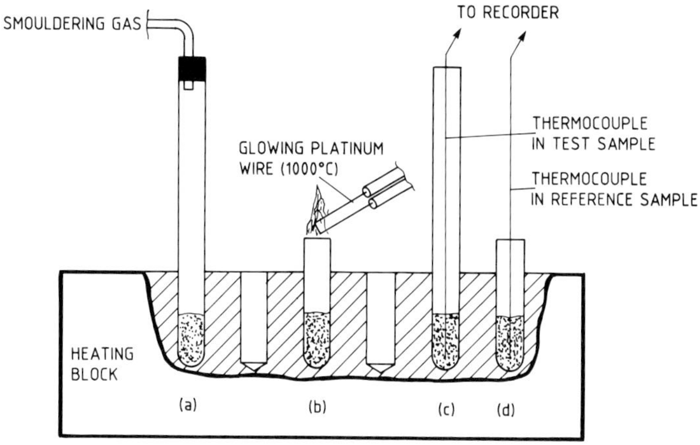

Figure 7.15 *Method for testing ignitability of smouldering gases from decomposition of dusts, and possible exothermal effects during decomposition (From Lütolf, 1978)*

7.7 IGNITION OF DUST DEPOSITS/LAYERS

7.7.1 THE INDUSTRIAL SITUATION

Smouldering combustion, or glow fires, in powder or dust deposits, can give rise to explosions in several ways. Smouldering lumps from the fire zone can be transported to areas containing explosible dust clouds and initiate dust explosions there. This may for example happen if the smouldering combustion takes place in a hopper that feeds powdered material to a larger storage silo via pneumatic transport. When powder is discharged from the hopper into the pneumatic line, for example through a rotary lock, the smouldering zone will eventually also reach the hopper outlet, and smouldering lumps will get into the pneumatic line and be transported to the larger silo. If the smouldering lumps are not quenched during transportation to the silo, and the silo contains an explosible dust cloud, the result can easily be a dust explosion.

Smouldering combustion can start as a slow, gentle process in the powder deposit at quite low temperatures, in some cases even at normal room temperature. Smouldering combustion can also be initiated by a hot object, which is either fully embedded in the dust deposit, or on which the deposit is lying. The hot object can be a piece of metal, for example a bolt or a nut that has loosened somewhere in the plant and been carried along with the process stream and heated up by repeated impacts against the internal walls of process equipment. Eventually it may come to rest embedded in a powder deposit in a silo, a bucket elevator boot or elsewhere in the plant. Alternatively, the hot object can be an overheated bearing or another larger hot object that has been covered with a layer of powder or dust.

Further details are given in Sections 1.1.4.2, 1.4.2.2, 1.4.2.4 and 1.4.2.5. See also Chapter 5.

7.7.2
LABORATORY TESTS

7.7.2.1
Semi-quantitative flammability test

The foundation of this method was laid by Lütolf (1971) and full descriptions were given by Siwek and Pellmont (1986) and Verein deutscher Ingenieure (1988). The apparatus and procedure are illustrated in Figure 7.16.

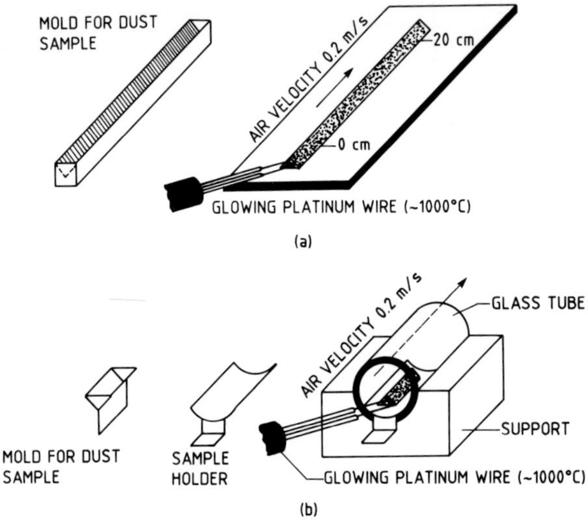

Figure 7.16 *Method for testing flame propagation ability of dust layers at ambient (a) and elevated (b) temperatures (From Siwek and Pellmont, 1986)*

For tests at ambient temperature, a ridge of the dust of triangular cross section is placed on a ceramic plate and one of the ends touched with a white-glowing platinum wire as shown in Figure 7.16 (a). For tests at elevated temperatures the sample holder shown in Figure 7.16 (b) is used and the sample placed in a glass tube heated to the desired temperature. The small air flow of about 0.2 m/s through the glass tube must be ensured.

Section A.1.2.9 in the Appendix describes the way in which results from tests with the apparatus shown in Figure 7.16 are classified, and Table A1 in the Appendix gives experimental results for a range of dusts.

7.7.2.2
Hot-plate test for minimum ignition temperature determination

The apparatus, which is shown in Figure 7.17, consists of a modified electric hot-plate, a temperature-control unit, three thermocouples and a 2-channel recorder.

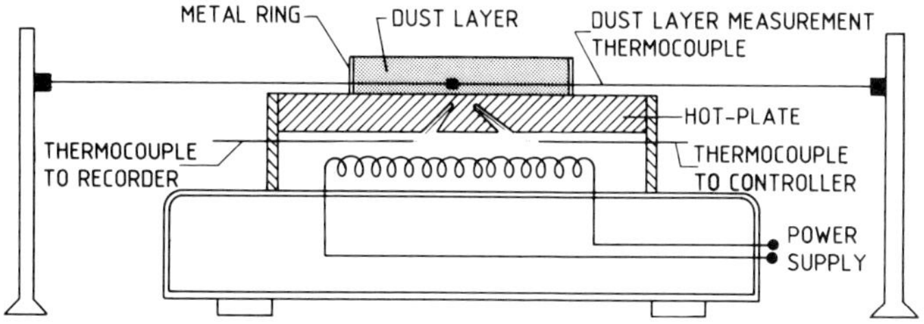

Figure 7.17 *Apparatus for determining the minimum temperature of a hot-plate that causes ignition of a dust layer on the hot-plate (Part of method being evaluated by the International Electrotechnical Commission, 1982)*

The hot-plate is kept at a given temperature, which is read by one of the thermocouples and displayed on one of the recorder channels.

On the surface of the plate is laid a metal ring, with a diameter of 100 mm and a height of either 5 mm or 15 mm. The powder sample to be tested is placed in the metal ring and carefully levelled off to the height of the ring. A thermocouple is placed in the sample through holes in the metal ring. The sample temperature is displayed on the second recorder channel. The third thermocouple is used for regulating the plate temperature.

The test procedure is specified in detail by the International Electrotechnical Commission (1982). Typical outcomes of a test are illustrated in Figure 7.18.

The temperature in the dust sample must exceed the hot-plate temperature by more than 20°C for the test to be recorded as ignition. Tests are conducted repeatedly until the minimum hot-plate temperature for ignition has been identified. This is defined as the lowest hot-plate temperature that gives ignition, rounded off to the nearest value in °C that is divisible by 10.

It is important to note that the minimum hot-plate igntion temperature decreases systematically with increasing dust layer thickness. If the values for two different layer

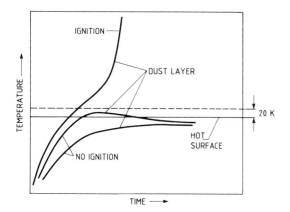

Figure 7.18 *Typical categories of results from measurements of temperature development in dust layers using the apparatus in Figure 7.17.*

thicknesses have been determined, simplified theory enables estimation of the values for other thicknesses, as shown by Bowes and Townshend (1962).

In the context of the possible IEC-test method, the similar German DIN test, using a dust layer thickness of 5 mm, should be mentioned. Data from this method are given in Table A1 in the Appendix.

7.7.2.3
Original US Bureau of Mines test

In the test originally used by US Bureau of Mines and described by Dorsett *et al.* (1960), a small basket of metal gauze is filled with the powder and placed in a furnace through which air of constant, known temperature is flowing at a slow, specified rate. The temperature within the powder bed is monitored continuously, and by increasing the air temperature in steps, a level is reached at which the temperature in the powder sample begins to rise above that of the surrounding air. This critical air temperature is taken as the minimum ignition temperature of the powder in question.

However, this temperature will not be a true powder constant but depend on the experimental conditions, in particular on the size of the powder sample tested, as shown in Section 7.7.2.5 below, and the air flow.

7.7.2.4
The Grewer-furnace test

Grewer (1971) developed a more refined version of the original US Bureau of Mines test. A cross section of the Grewer furnace is shown in Figure 7.19.

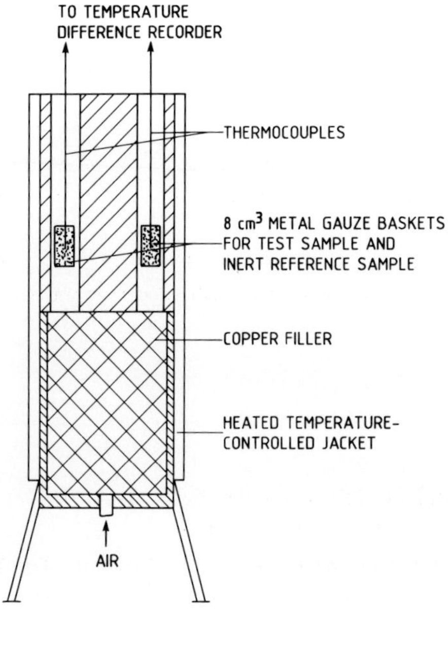

Figure 7.19 *Grewer-furnace for determination of minimum ignition temperatures of dust deposits (From Grewer, 1971)*

The furnace has six vertical cylindrical cavities, in which small metal gauze baskets are placed. One of the baskets contains an inert reference sample, the other five, test samples. The furnace can be programmed to give a specific rate of temperature rise, e.g. 1°C/min. The temperature at which a test sample temperature starts to rise faster than that of the inert reference sample, is taken as the minimum ignition temperature of that sample. Figure 7.20 shows an example of a set of results.

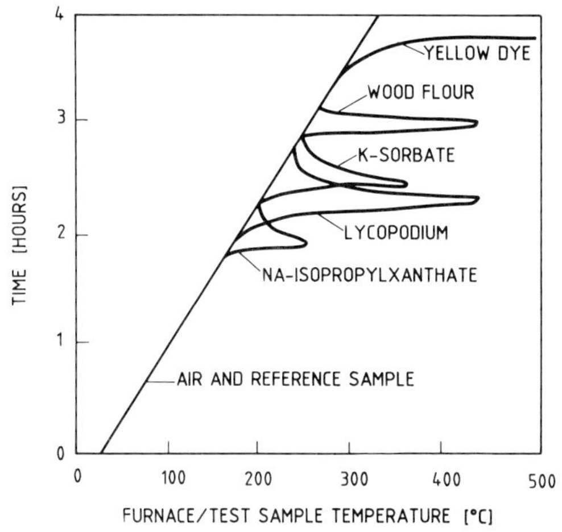

Figure 7.20 *Example of temperature recordings during a four-hour test of five different combustible powders in the Grewer furnace (From Grewer, 1971)*

7.7.2.5
Storage of bulk powder samples in heated atmosphere

The apparatus is shown in Figure 7.21. The sample is suspended in a metal gauze basket in a heating chamber through which a given flow of preheated air is circulated. The ambient air temperature and the temperature inside the powder sample are measured and the difference recorded. If the temperature in the powder sample rises beyond that of the air, but not higher than 400°C, the phenomenon is named self-heating. Temperature rises beyond 400°C are named self-ignition.

Figure 7.22 illustrates the typical linear relationship between the minimum ignition temperature and the sample size. Further details concerning self-heating and self-ignition processes are given in Chapter 5.

7.7.2.6
Other methods

Standard instruments for differential thermal analysis (DTA) have been used for fast screening of self-heating/self-ignition properties of dusts and powders.

Bowes (1984) gave an updated and comprehensive account of the state-of-the-art on theories and experiments on self-heating in powder deposits by 1983/1984. Since that date,

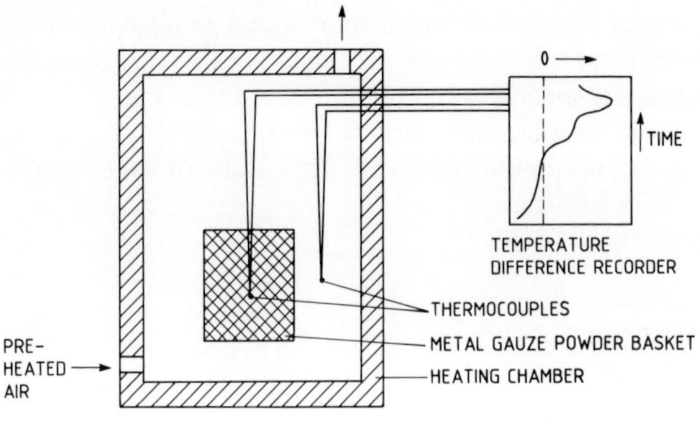

Figure 7.21 *Apparatus for storage of bulk powder samples in a heated atmosphere (From Verein deutscher Ingenieure, 1988)*

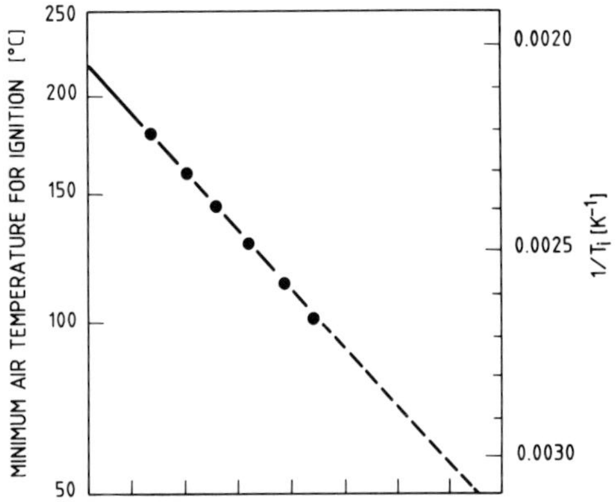

Figure 7.22 *Example of a typical series of results from determination of the minimum ambient air temperature* T_i *leading to ignition of different sizes of geometrically similar samples of a given powder (From Verein deutscher Ingenieure, 1988)*

further development has taken place, and new instruments for detailed studies of the rate of reaction as a function of temperature, under adiabatic conditions, are now commercially available. As described by Townsend and Tou (1980), such 'accelerating rate calorimeters' (ARC) are essentially highly computerized adiabatic calorimeters. During an ARC experiment, the sample is maintained in a near-to-perfect adiabatic condition, while time, temperature and pressure data are automatically collected and stored. The data can then be processed by computers. An ARC is illustrated in Figure 7.23.

In addition to ARC, differential scanning calorimetry (DSC) is also in use, as discussed by Snee (1987). DSC implies measurements of the rate at which heat must be transferred to or from the test sample in order to maintain it at the same temperature as an inert reference sample. The reference sample temperature is usually increased at a predetermined linear rate (constant temperature rise per unit time).

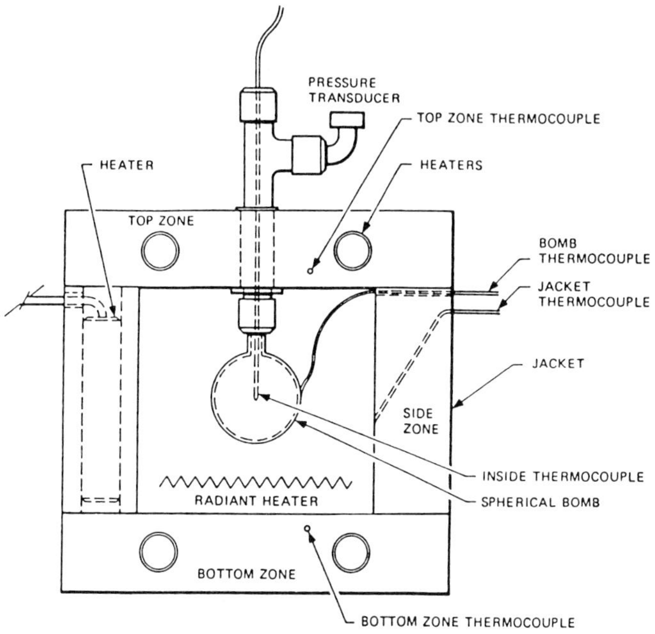

Figure 7.23 *Accelerating rate calorimeter for assessing exothermal behaviour of substances, including powders and dusts, under adiabatic conditions (From Townsend and Tou, 1980)*

Grewer *et al.* (1989) and Racke (1989) gave comprehensive reviews of instrumentation and procedures in use for assessing the exothermal behaviour of reactive chemicals including powders and dusts.

7.8 MINIMUM IGNITION TEMPERATURE OF DUST CLOUDS

7.8.1 THE INDUSTRIAL SITUATION

Hot surfaces capable of igniting dust clouds exist in a number of situations in industry, such as in furnaces and burners, and dryers of various kinds. In addition, hot surfaces can be generated accidentally by overheating of bearings and other mechanical parts.

If an explosible dust cloud is generated in some uncontrolled way in the proximity of a hot surface of temperature above the actual minimum ignition temperature, the result can be a dust explosion. It is important, therefore, to know the actual minimum ignition temperature and to take adequate precautions to ensure that temperatures of hot surfaces in areas where explosible dust clouds can occur, do not rise to this value.

However, the minimum ignition temperature is not a true constant for a given dust cloud, but depends on the geometry of the hot surface and the dynamic state of the cloud. (See Chapter 5.)

7.8.2
LABORATORY TESTS

7.8.2.1
Godbert-Greenwald furnace

In USA, as described by Dorsett *et al.* (1960), the ignition temperature of dust clouds in contact with a hot surface was traditionally determined in the Godbert-Greenwald furnace. In this apparatus the internal surface of a vertical cylindrical ceramic tube, open at the lower end, is kept at a known, constant temperature, and a sample of the powder is dispersed as a dust cloud into the tube from above by means of a blast of air. The automatically controlled temperature of the internal wall of the tube is changed in steps and the experiment repeated until the minimum temperature for ignition has been identified. In UK the same furnace has also been used for many years, as described by Raftery (1968) and Field (1983).

The International Electrotechnical Commission (IEC) investigated the performance of the Godbert-Greenwald furnace through several round-robin test series involving several central test laboratories in Europe and USA. The influences of a number of details of the apparatus itself and of the experimental procedure were studied and details of apparatus and procedure specified more closely. The resulting, improved Godbert-Greenwald furnace test was proposed as a standard for determining minimum ignition temperature of dust clouds. The essential details concerning both apparatus and procedure were given in a draft document from the International Electrotechnical Commission (1984). This includes details of the central ceramic tube, which is fitted with a special spiral groove for the heating element, and two holes for the two thermocouples. One of the holes is penetrating the wall, allowing the measuring thermocouple junction to be in direct contact with the internal wall of the ceramic tube. Specifications of the way of generating the air blast for dispersing the dust are also given. Figure 7.24 illustrates a version of the Godbert-Greenwald furnace that is in agreement with that being evaluated by the IEC on the essential points.

Figure 7.25 shows a photograph of a Godbert-Greenwald furnace test.

Griesche and Brandt (1976) used a Godbert-Greenwald furnace modified in such a way that dust clouds of known concentrations could be passed through the furnace at a desired constant velocity. They investigated the influence of the dust cloud velocity, or the mean residence time of the dust in the furnace, on the minimum ignition temperature. The results given in Figure 7.26 show that the minimum ignition temperature decreased quite significantly with increasing residence time. A conventional Godbert-Greenwald test on the same coal dust gave a minimum ignition temperature of 310°C. This is lower than all the data in Figure 7.26, but about 100°C higher than the very low value found for a residence time of > 1 s and 500 g/m^3 dust concentration. This evidence should be kept in mind when applying data from standard Godbert-Greenwald furnace tests in industrial practice.

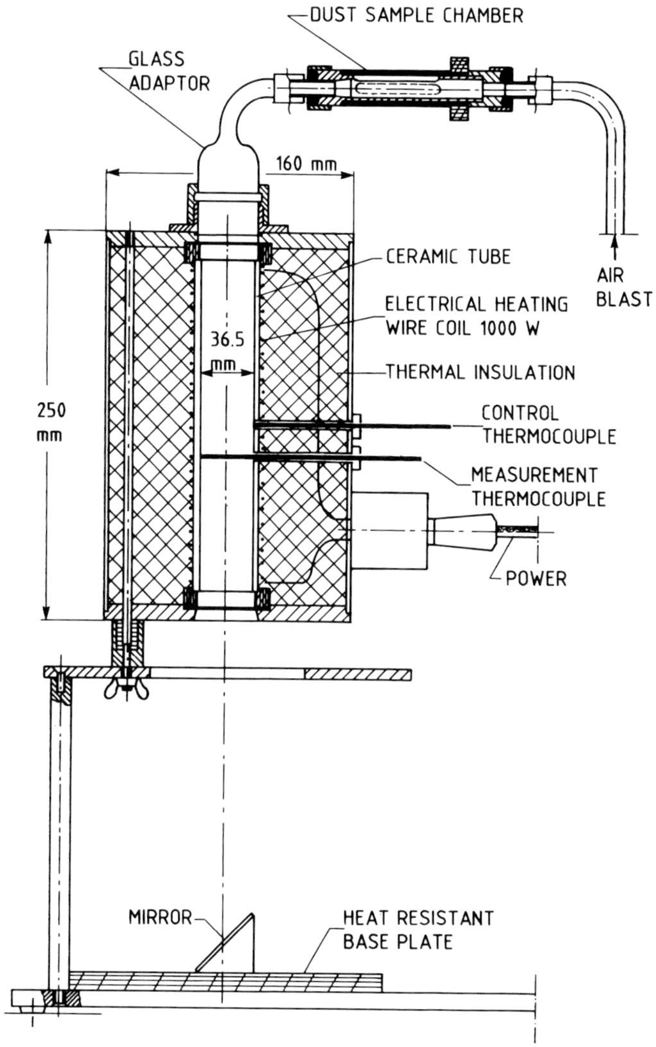

Figure 7.24 *Godbert-Greenwald furnace for determination of the minimum ignition temperature of dust clouds. Design being evaluated by International Electrotechnical Commission (1984)*

7.8.2.2
BAM furnace

In Germany an alternative furnace was developed by Bundesanstalt für Materialprüfung (BAM), as described by Leuschke (1966, 1966a). The furnace is illustrated in Figure 7.27. The experimental procedure is similar to that of the Godbert-Greenwald furnace, but the generation of the dust cloud is manual, by pressing a rubber bulb. The cloud is directed against a circular concave metal disc of about 20 cm^2 area and known temperature.

Figure 7.25 *Ignition of a dust cloud in the Godbert-Greenwald furnace. For a much clearer picture see colour plate 10*

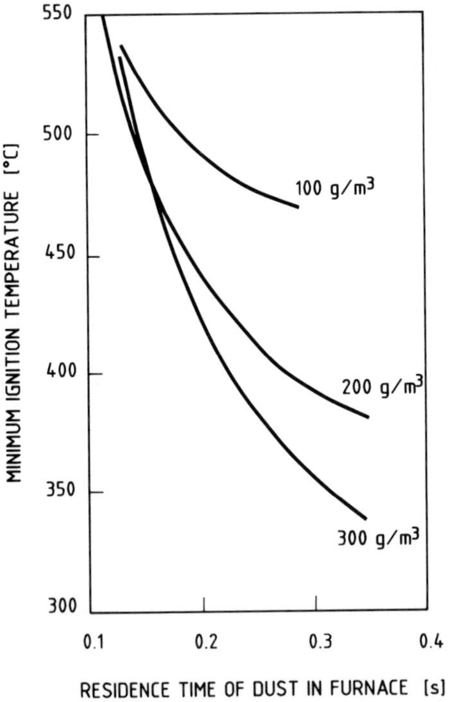

Figure 7.26 *Influence of the residence time of dust clouds in the Godbert-Greenwald furnace on the minimum ignition temperature. Three different concentrations of brown coal dust (From Griesche and Brandt, 1976).*

However, because of the horizontal geometry, the BAM furnace allows dusts that do not ignite directly in suspension, to settle on the hot internal bottom of the furnace. In this way smouldering gases can develop, which can subsequently ignite at a lower temperature than that required for direct ignition of the dust cloud. Ignition of smouldering gases normally occurs with a noticeable delay with respect to the dispersion of dust in the furnace. Because the BAM-furnace test method considers such delayed ignition of smouldering gases as equivalent to ignition of the dust cloud, the minimum ignition temperatures determined by this test method can be lower than those determined in the Godbert-Greenwald furnace for the same dusts. Figure 7.28 shows data from comparative tests of the same dust in the Godbert-Greenwald and the BAM furnace. In this case the difference is relatively small, about 20°C.

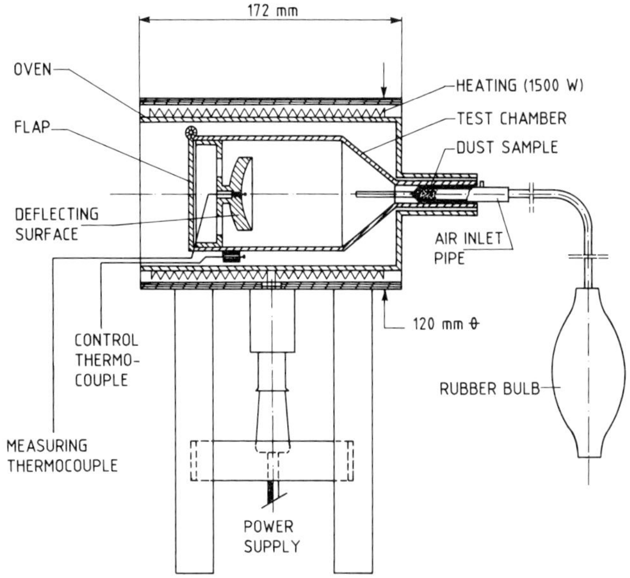

Figure 7.27 *BAM furnace for determination of the minimum ignition temperature of dust clouds (Courtesy of J. Lütolf, formerly Ciba-Geigy AG)*

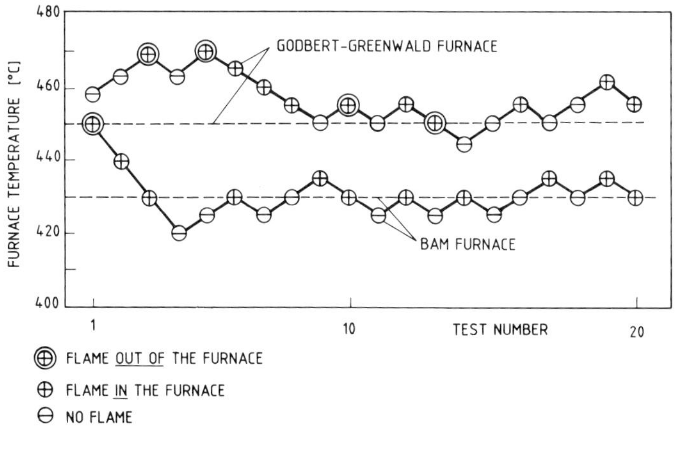

Figure 7.28 *Comparison of results from determination of the minimum ignition temperature of American lycopodium by two different furnaces. Dust sample dispersed in each test 1.6 cm*^{3} *(From Leuschke, 1975)*

7.8.2.3
New US Bureau of Mines furnace

This furnace, which was described in detail by Conti *et al.* (1983), is shown in Figure 7.29. The volume of the ceramic chamber is 1.2 litre. This apparatus was included as an equal possibility together with the Godbert-Greenwald and BAM furnaces in a draft by

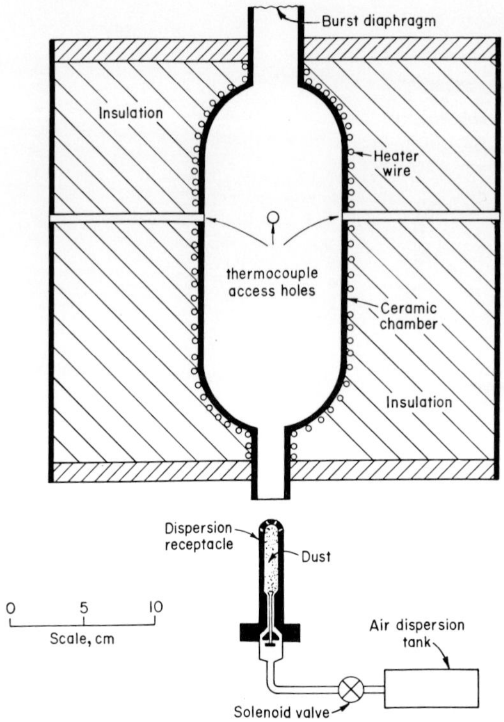

Figure 7.29 *New 1.2 litre US Bureau of Mines furnace for determination of the minimum ignition temperature of dust clouds (From Conti* et al., *1983)*

Rogerson (1989) for a possible new standard for measurement of the minimum ignition temperature of dust clouds. Conti *et al.* (1983) showed that for organic materials and coals the new furnace gave minimum ignition temperatures that were from 15°C to 90°C lower than those from the Godbert-Greenwald furnace. The agreement with the BAM furnace would be expected to be closer for some dusts.

7.8.2.4
Further comparison of test methods

Hensel (1984) compared the minimum ignition temperatures for a range of dusts, using four different apparatus, namely the original and a modified BAM furnace, and the original and a modified Godbert-Greenwald furnace. The BAM furnace was modified by replacing the rubber bulb for manual dust dispersion by an automatic system as used with the Godbert-Greenwald furnace. The Godbert-Greenwald furnace was modified by doubling its length, which increased the residence time of the dust in the furnace. The results are given in Table 7.1.

With the exception of powders 7, 8 and 16 in Table 7.1, the long version of the Godbert-Greenwald furnace gave the same values as, or lower values than the standard version. However, the differences were mostly moderate and not more than 50°C. The modification of the BAM furnace also led to a slight reduction of the minimum ignition temperatures.

Table 7.1 Comparison of minimum ignition temperatures of dust clouds in air determined in four different test furnaces (From Hensel, 1984)

Name of substance	Minimum ignition temperature of a dust cloud (°C)				Maximum temperature difference (K)
	G.G.-furnace		BAM-furnace		
	IEC	long	manual	autom.	
1 Sugar	420	420	360	340	80
2 Wheat flour	490	470	410	375	115
3 Dextran	400	380	370	340	60
4 Lycopodium	460	455	430	425	35
5 Tobacco additive	560	540	540	510	50
6 Painting powder (a)	570	570	530	490	80
7 Painting powder (b)	500	520	490	460	60
8 Painting powder (c)	470	480	460	450	30
9 Aluminium	560	515	510	510	50
10 Alloy (10% Zr)	430	380	380	370	60
11 Zircaloy fines	400	390	350	340	60
12 Brown coal	410	400	440	410	40
13 Pitch coal	590	560	560	560	30
14 Lignin	460	450	500	470	50
15 Pittsburgh coal	580	570	590	580	20
16 Plastic (a)	370	390	400	390	30
17 Plastic (b)	370	370	380	375	10
18 Plastic (c)	490	480	480	450	40
19 Plastic (d)	520	500	480	450	70
20 Plastic (e)	400	370	390	370	30
21 Chip board dust	510	480	450	450	60

7.9 MINIMUM ELECTRIC SPARK IGNITION ENERGY OF DUST LAYERS

7.9.1 THE INDUSTRIAL SITUATION

It is well known that explosible dust clouds can be ignited by electric sparks and arcs that occur in switches and motors, and in short-circuiting caused by damaged cables. In addition, some categories of electrostatic discharges may initiate dust explosions in industry, as discussed in Sections 1.1.4.6 and 1.4.2.7. Hazardous electrostatic discharges include capacitive sparks, propagating brush discharges and discharges from powder heaps.

The probability of a given dust layer or dust cloud becoming ignited by an electric spark not only depends on the spark energy, but indeed also on the distribution of this energy in time and space. This is discussed in greater detail in Chapter 5. In the case of dust layers the dependence on spark discharge duration is incorporated in the standard test procedure described in Section 7.9.2.2.

7.9.2
LABORATORY TESTS

7.9.2.1
Original US Bureau of Mines test

This method was described by Dorsett *et al.* (1960). The standard dust layer thickness tested was 1.6 mm. The layer was resting on a 25 mm diameter steel plate that also served as the negative electrode. The positive needle point electrode, connected to a capacitor bank charged to 400 V, was lowered by hand towards the surface of the dust layer until a spark discharge occurred. After an ignition occurred, the steel plate was cleaned, a new dust layer formed and the process repeated at progressively lower capacitance values until the lowest that gave at least one ignition in twenty trials was identified. The minimum ignition energy was defined as $1/2\ CV^2$, where C is the capacitance and V the charging potential of 400 volts.

7.9.2.2
'Nordtest Fire 016'

This method, described by Nordtest (1982), is primarily intended to be used for pyrotechnics and explosives in pulverized form. It may, however, also be applied to man-made and natural combustible materials, which, when distributed as a thin layer, resting on a flat metal surface, are able to propagate self-sustained combustion.

The test apparatus is illustrated in Figures 7.30 and 7.31.

The dust/powder is poured gently into the disc-shaped cavities formed by the slidable supporting plate/hole plate assembly and excess dust is removed by a scraper. Plane, circular dust/powder samples of thickness 2 mm and diameter 12 mm are thus obtained. The metal bottom of the cavities acts as one of the two electrodes forming the spark gap. A thin tungsten wire pointing downwards towards the dust/powder layer and with its tip just above the dust/powder surface, acts as the second electrode.

Electric sparks of the desired net energies and discharge times are passed through the sample, one at a time, and it is observed whether or not ignition occurs. An electric spark generator that permits independent variation of spark energy and spark duration is required. Figure 7.32 shows one type of generator used.

Twenty identical tests are carried out at each combination of net spark energy and discharge time, yielding a frequency of ignition in the range zero to 100%. After each spark discharge, the dust/powder sample is shifted horizontally to allow each spark to pass through fresh dust/powder not exposed to previous sparks. If ignition occurs, the sample tested is discarded and the test continued with a new sample.

The minimum electric spark ignition energy, defined as the net spark energy yielding an ignition frequency of 5%, is determined for various spark discharge durations Δt. The ultimate result of the test is an electric spark sensitivity profile $E_{min}(\Delta t)$, as illustrated in Figure 7.33.

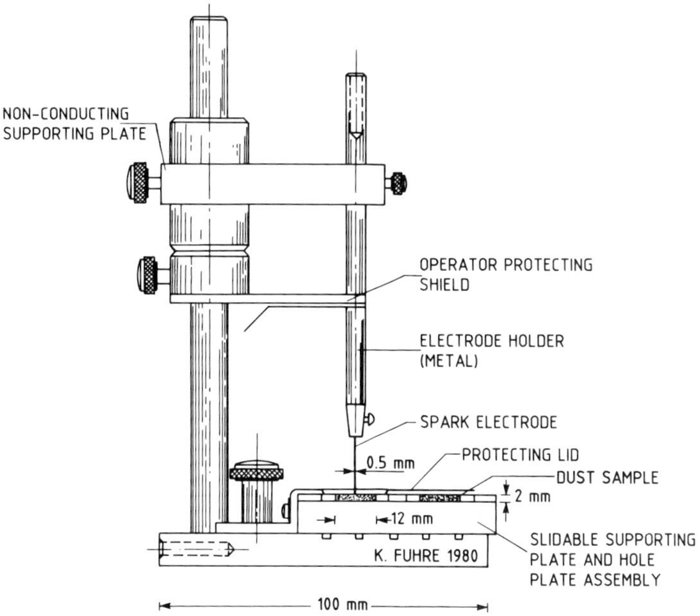

Figure 7.30 *Cross section of Nordtest apparatus for determination of electric spark ignition sensitivity of dust layers (From Nordtest, 1982)*

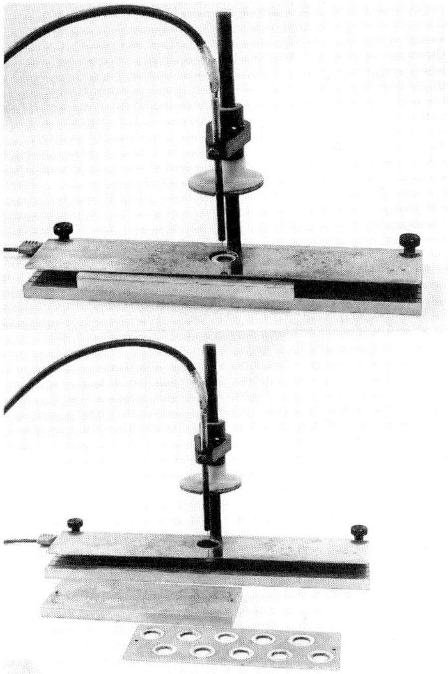

Figure 7.31 *Nordtest apparatus for determination of electric spark ignition sensitivity of dust layers. Top: Assembled as in Figure 7.30. Bottom: Slidable supporting plate and hole plate removed and shown separately*

Figure 7.32 *Electric spark generator used for determining electric spark ignition sensitivity profiles of dust layers*

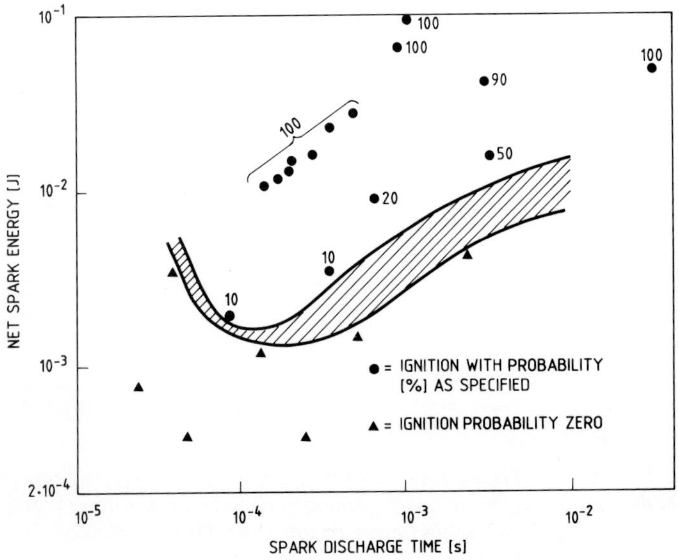

Figure 7.33 *Electric spark ignition sensitivity profile according to Nordtest (1982) for layers of a specific pyrotechnical product. Shaded area represents profile for ignition probabilities P [%] for which $0 < P < 10$*

7.10
MINIMUM ELECTRIC SPARK IGNITION ENERGY OF DUST CLOUDS

7.10.1
THE INDUSTRIAL SITUATION

Most of what has been said in Section 7.9.1 also applies to dust clouds. Whenever relating result from laboratory tests to practice, it is important to account for the influence of both the spatial and the temporal energy distribution in the discharge on the minimum spark energy for ignition. Relevant aspects are considered in Chapters 1 and 5.

7.10.2
LABORATORY TESTS

7.10.2.1
Original US Bureau of Mines method

The apparatus used by Dorsett *et al.* (1960) was essentially as illustrated in Figure 7.34. An appropriate quantity of dust was placed in the dispersion cup at the bottom of the 1.2 litre

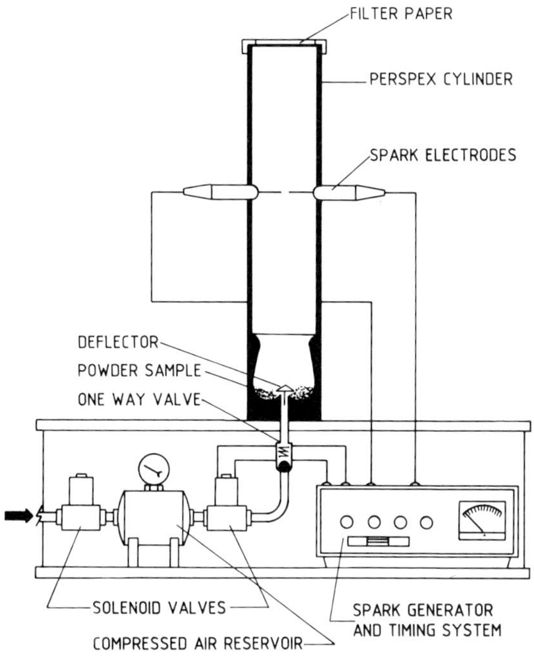

Figure 7.34 *Apparatus for determining the minimum ignition energy of dust clouds*

plastic cylinder and dispersed by a blast of air deflected by a conical 'hat' as indicated. A spark was discharged across the electrodes synchronously with the transient appearance of the dust cloud in the spark gap region. A photograph of a dust explosion in this type of apparatus is shown in Figure 7.35. However, the energies of the electric sparks used in the original US Bureau of Mines test were not satisfactorily well defined. This is due to the design of the spark discharge circuit, as shown in Figure 7.36 (a). The spark energy was generated by discharging the capacitor C at a DC voltage V through a step-up transformer. It was assumed that the spark energy equalled $1/2\ CV^2$, but some energy was inevitably lost in the transformer. A tentative correlation between the numerous $1/2\ CV^2$ values reported by US Bureau of Mines through the years, and the real electric spark energies is indicated in Figure A1 in the Appendix.

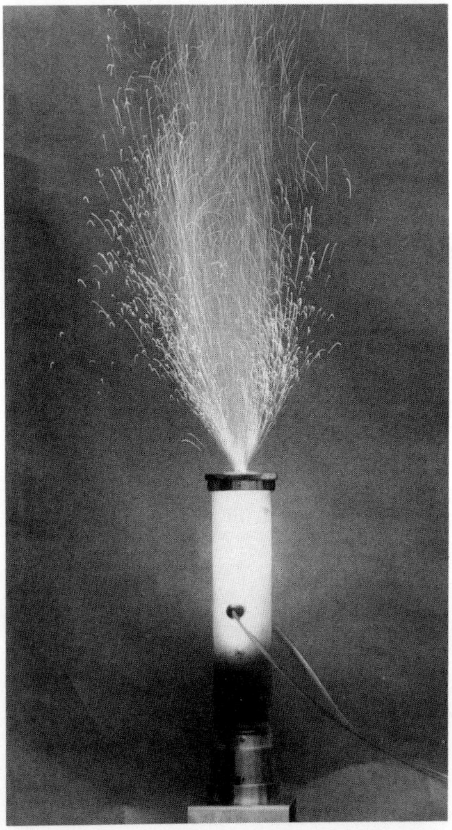

Figure 7.35 *Silicon dust explosion following electric spark ignition in an apparatus of the type illustrated in Figure 7.34*

7.10.2.2
Direct discharge of high-voltage capacitors

Direct discharge of capacitors at sufficiently high voltages for ensuring direct breakdown of the spark gap, as illustrated in Figure 7.36 (b), has also been used for test purposes. However, because of the high voltage, energy losses in the switch needed for synchroniz-

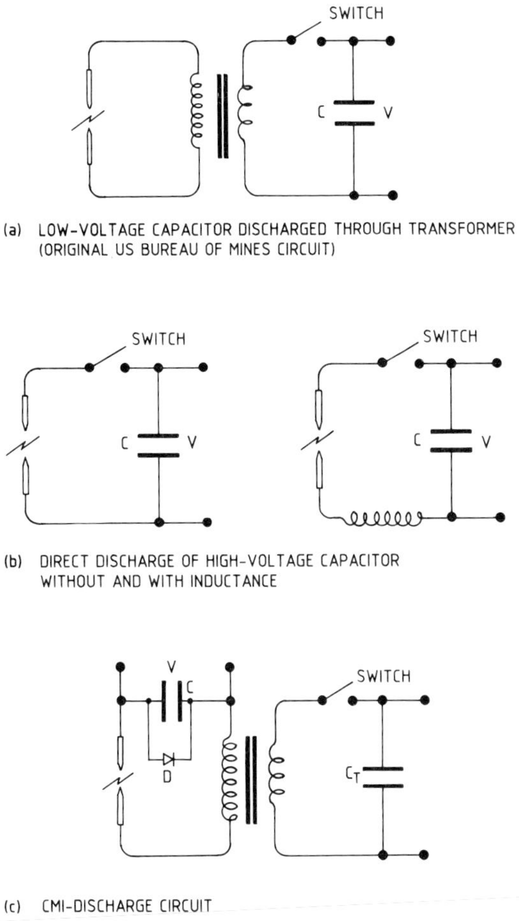

Figure 7.36 *Three different electric spark discharge circuits used for determining the minimum ignition energy of dust clouds*

ing the discharge with the dust cloud, may be appreciable. Sophisticated elements such as thyrathrons have been employed to solve this problem.

However, synchronization of spark and dust cloud can also be accomplished by incorporating a third, auxiliary spark electrode in the spark gap configuration. By discharging just a very small energy in the gap between one of the main electrodes and the auxiliary electrode, the main discharge is initiated. This method was used with success by Franke (1978).

Mechanical synchronization constitutes a further possibility. Prior to the experiment the capacitor is then charged to the high voltage required with the spark gap sufficiently long for breakdown to be impossible at that voltage. Pneumatically or spring-driven displacement of one of the spark electrodes towards a shorter spark gap, allowing spark-over, is synchronized with the occurrence of the transient dust cloud, for example via solenoids. Boyle and Llewellyn (1950) were probably amongst the first to use the electrode displacement method. Its drawback is that the actual spark gap distance at the moment of the discharge is not known.

One way of avoiding the synchronization problem is to work with a semi-stationary dust cloud and charge the high-voltage capacitor slowly until breakdown occurs naturally at the fixed spark gap distance chosen. Because of arbitrary variations, the actual voltage at breakdown will differ from trial to trial, and must be recorded for each experiment for obtaining the actual given spark energy $1/2\ CV^2$.

Figure 7.36 (b) illustrates two versions of the direct high-voltage discharge circuit, one without and one with a significant series inductivity, of the order of 1 mH. This difference can be significant with respect to the igniting power of sparks of similar energies. The induction coil makes the spark more effective as an ignition source by increasing the discharge duration of the spark. Such an induction coil is automatically integrated both in the original US Bureau of Mines circuit, and also in the CMI circuit, as shown in Figure 7.36 (a) and (c). (See Chapter 5 for further details concerning the influence of the spark discharge duration.)

If the test is to simulate a direct electrostatic discharge of an accidentally charged non-earthed electrically conducting object, the use of a discharge circuit with low inductance (left of Figure 7.36 (b)) seems most appropriate.

7.10.2.3
The CMI discharge circuit

The method for synchronization of dust cloud and spark discharge, which was developed by CMI (see Eckhoff (1975a)), is illustrated in Figure 7.36 (c). The method is similar to the 3-electrode technique in the sense that an auxiliary spark discharge is employed for breaking the spark gap down, but the use of a third electrode is avoided. The energy of the auxiliary spark is about 1–2 mJ. The CMI method requires that the spark energy be measured directly, in terms of the time integral of the electrical power dissipated in the spark gap. Figure 7.37 shows the traces of voltage and current for a spark of net electrical energy 13 mJ, produced by the CMI circuit. The spark discharge was completed after about 280 μs.

The general apparatus used by CMI was as otherwise shown in Figure 7.34, i.e. similar to that originally developed by US Bureau of Mines.

7.10.2.4
A new international standard method

As a part of its efforts to standardize safe design of electrical apparatus in explosible atmospheres, the International Electrotechnical Commission (1989) is considering a new test method for the minimum ignition energy of dust clouds. The draft is to a large extent based on work conducted by an international European working group and summarized by Berthold (1987).

The detailed design of the apparatus to be used in a possible IEC test method, in terms of explosion vessel, dust dispersion system, synchronization method, etc. was not specified, but some suitable apparatus were mentioned, including direct high-voltage discharge circuits as well as the CMI circuit. However, no matter which apparatus is chosen, the spark generating system must satisfy the following requirements:

- Inductance of discharge circuit $\geqslant$ 1 mH.
- Ohmic resistance of discharge circuit < 5 Ω.
- Electrode material: stainless steel, brass, copper or tungsten.
- Electrode diameter: 2.0 mm.
- Electrode gap: 6 mm.
- Capacitors: low-inductance type, resistant to surge currents.
- Capacitance of electrode arrangement: as low as possible.
- Insulation resistance between electrodes: sufficiently high to prevent significant leakage currents.

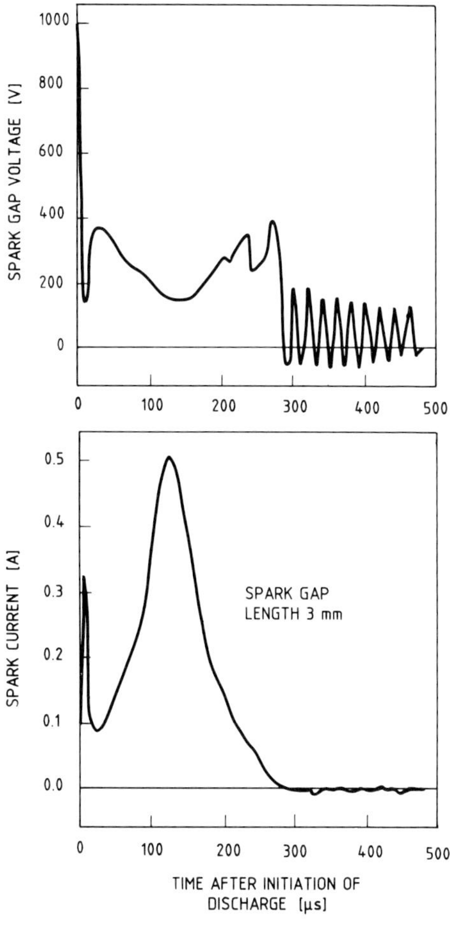

Figure 7.37 *Spark gap voltage and spark current versus time during discharge of a 13 mJ electric spark from the CMI spark generator. Spark discharge duration 280 μs. Energy of trigger spark (spike to the far left) is about 1–2 mJ*

It will be necessary to take account of the possible influences of dust concentration, dust cloud turbulence and degree of dust dispersion on the test result. Preliminary tests must be carried out to adjust the dust dispersion conditions and the ignition delay such that prescribed minimum ignition energies are actually measured for three specified reference dusts.

Starting with a value of spark energy that will reliably cause ignition of a given concentration of the dust being tested, the dust concentration being itself a variable, the test energy is successively halved until no ignition occurs in 10 successive tests. The minimum ignition energy is defined to lie between the highest energy at which ignition fails to occur in at least ten successive attempts to ignite the dust/air mixture, and the lowest energy at which ignition occurs within ten successive attempts.

7.11 SENSITIVITY OF DUST LAYERS TO MECHANICAL IMPACT AND FRICTION

7.11.1 THE INDUSTRIAL SITUATION

This hazard primarily applies to powders and dusts with explosive properties, i.e. which are able to react or decompose exothermally without oxygen supply from the air. Strong exothermal reactions may be initiated in layers of such materials if they are exposed to high mechanical stresses and fast heating by impact or rubbing, either accidentally or as part of an industrial process.

7.11.2 LABORATORY TESTS

7.11.2.1 Drop hammer tests

As summarized by Racke (1989), a number of impact/friction sensitivity test methods have been developed in several European countries, as well as in USA and Japan. The most common design concept for the impact test is the drop hammer, as illustrated in Figure 7.38.

Verein deutscher Ingenieure (1988) also mentioned the very similar test by Lütolf (1978) as a suitable standard method. In the Lütolf test the dust sample size is about 0.10 g and the theoretical maximum drop hammer impact energy 39 J (5 kg, 0.8 m). Up to ten trials are conducted and observations are made with respect to occurrence of explosion, flame, smoke or sparks. If all ten tests are negative, a new test series is conducted with the dust samples wrapped in thin aluminium foil (10 μm thickness), in case the aluminium should have a sensitizing effect on a possible exothermal reaction. If the tests with aluminium are positive, a new test series without aluminium is conducted.

The American Society for Testing and Materials (1988) adopted the US Bureau of Mines drop hammer method as their standard. Using a fixed drop hammer weight (2.0 or 3.0 kg), the drop height H_{50} giving 50% probability of a positive reaction, is determined. The lower H_{50}, the more sensitive the material is to impact ignition. In the test description

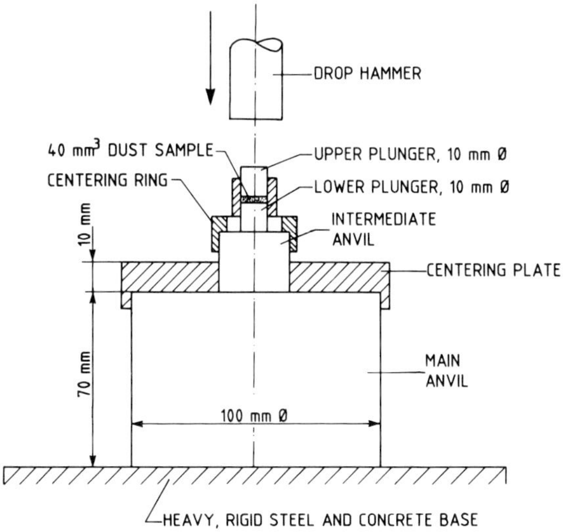

Figure 7.38 *Drop hammer test for dust layers by Koenen, Ide and Swart (1961). Drop hammer mass 5 kg and height of fall 1 m (From Verein deutscher Ingenieure, 1988)*

it is emphasized that the observation of the reaction of the sample is one of the difficult points in impact sensitivity testing. A positive test result is defined as an impact that produces one or more of the following phenomena: (a) audible reaction, (b) flame or visible light, (c) definite evidence of smoke (not to be confused with a dust cloud of dispersed sample), and (d) definite evidence of discolouration of the sample due to decomposition. The problem arises with reactions that yield no distinguishable audible response, no flame, and little sample consumption. The decision concerning reaction/no reaction in these cases must be based primarily upon the appearance of the sample after the test. The impact in most cases will compress the sample into a thin disc, portions of which may adhere to the striking tool surface, the anvil, or both. One should then inspect the tool and anvil surfaces and look for voids in the powder disc and discolouration due to decomposition in areas where voids occur. If there is discolouration from decomposition, the test trial is to be considered as positive. If there are small voids but no discolouration, the trial should be regarded as negative. In the case of doubt as to whether or not discolouration is present, the trial is to be regarded as negative. If the only evidence is a slight odour or a small amount of smoke, which may be a dust cloud from dispersed sample, the trial should also be considered negative.

7.11.2.2
Friction tests

As pointed out by Racke (1989), several different friction tests have been devised, including three described by Gibson and Harper (1981). One of these is illustrated in Figure 7.39.

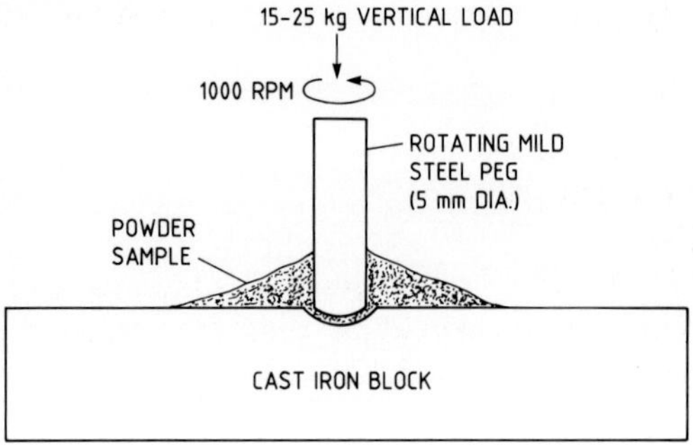

Figure 7.39 *Example of laboratory method for testing the sensitivity of powders to mechanical rubbing/friction (From Gibson and Harper, 1981)*

7.12 SENSITIVITY OF DUST CLOUDS TO IGNITION BY METAL SPARKS/HOT SPOTS OR THERMITE FLASHES FROM ACCIDENTAL MECHANICAL IMPACT

7.12.1 THE INDUSTRIAL SITUATION

Dense clouds of metal sparks, and also hot surfaces, are easily generated in grinding and cutting operations. Such operations are therefore generally to be considered as hot work, which should not be permitted in the presence of ignitable dusts or powders.

However, the evaluation of the ignition hazard to be associated with accidental impacts is less straight-forward. Such impacts can occur due to mis-alignment of moving parts in powder processing equipment, for example in grinders and bucket elevators. Or foreign bodies such as stones and tramp metal can get into the process line. Whether or not metal sparks/hot spots or thermite flashes from single accidental impacts between solid bodies, can in fact initiate dust explosions, has remained a controversial issue for a long time. It now seems that in the past 'friction sparks' have been claimed to be the ignition sources of dust explosions more often than one would consider as reasonable on the basis of more recent evidence. However, as long as necessary conditions for such impacts to be capable of initiating dust explosions have been unidentified, one has been forced to maintain the hypothesis that such sparks may be hazardous in general. This in turn has forced industry to take precautions that may have been superfluous, and caused fear that may have been unnecessary.

Generation of metal sparks/hot spots by accidental mechanical impacts is a complex process, involving a number of variables such as:

- Chemistry and structure of the material of the colliding bodies.
- Physical and chemical surface properties of the colliding bodies.
- Shapes of the colliding bodies.
- Relative velocity of the colliding bodies just before impact.
- Impact energy (kinetic energy transformed to heat in an impact).
- Single or repeated impacts?

Whether a given dust cloud will be ignited by a given impact not only depends on the specific dust properties, but also on:

- Dust concentration and dynamic state of the dust cloud.
- Composition, temperature and pressure of the gas phase.

In view of the great number of variables and the lack of an adequate theory, it is clear that the ignition experiments on the basis of which the practical hazard is to be assessed, should resemble the practical impact situation as closely as possible.

7.12.2 LABORATORY TESTS

No standardized test methods have so far been traced, but the ability of metal sparks/hot spots from grinding and cutting to ignite dust clouds has been demonstrated in laboratory tests by several researchers, including Leuschke and Zehr (1962), Zuzuki *et al.* (1965), Allen and Calcote (1981) and Ritter (1984). (See Chapter 5.)

Laboratory test methods for the incendivity of single accidental mechanical impacts seem to be less numerous. A test apparatus developed by Pedersen and Eckhoff (1987), is illustrated in Figure 7.40.

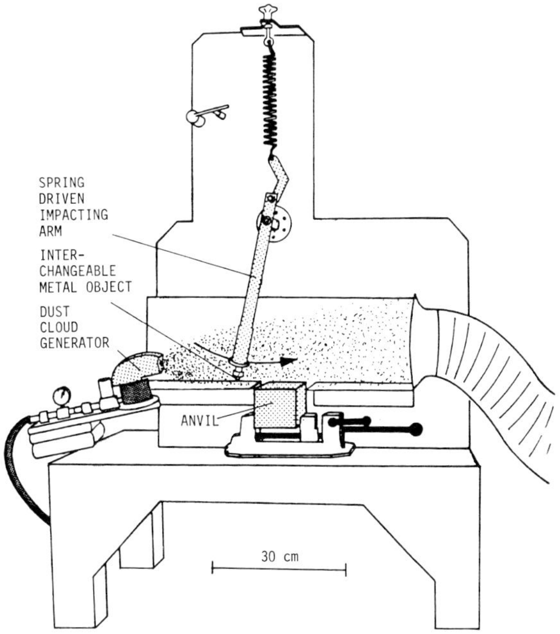

Figure 7.40 *Apparatus for determining the sensitivity of dust clouds to ignition by single accidental mechanical impacts (From Pedersen and Eckhoff, 1987)*

The basic principle of impact generation is that a spring-loaded rigid arm, which can swing around a fixed axis, and carries the test object at its tip, is released and hits a test anvil tangentially at a known velocity. Depending on the normal contact force during impact, the peripheral velocity of the tip of the arm will be more or less reduced. By knowing the mass distribution of the arm and the peripheral velocity of its tip just before and just after impact, the impact energy can be estimated in terms of loss of kinetic energy of the arm. The impact force is varied by varying the excess length of the arm compared with the distance from the arm axis to the anvil.

Figure 7.41 gives an expanded view of the test object holder at the arm tip. The dust cloud was generated by dispersing a given quantity of dust from a dispersion cup by a short blast of air. The dust concentration of the transient cloud near the point of impact, at the moment of impact, was measured by a calibrated light attenuation probe. (See Figure 1.76 in Chapter 1.)

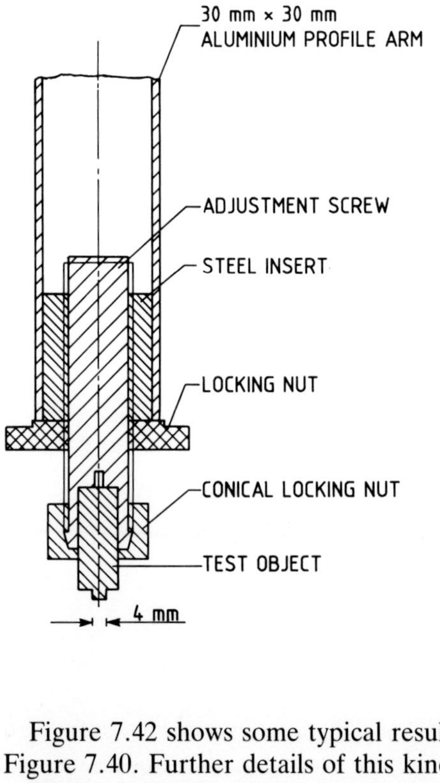

Figure 7.41 *Expanded view of test object holder of apparatus shown in Figure 7.40 (From Pedersen and Eckhoff, 1987)*

Figure 7.42 shows some typical results from experiments with the apparatus shown in Figure 7.40. Further details of this kind of experiments are discussed in Chapter 5.

Because of the lack of generally accepted test methods, it has been suggested that the sensitivity of a dust cloud to ignition by metal sparks/hot spots from accidental impacts may be correlated to the sensitivity of ignition by other sources, such as electric sparks. As discussed in Chapter 5, Ritter (1984) found a correlation involving both the minimum electric spark ignition energy and the minimum ignition temperature as determined by the BAM furnace. Table 7.2 indicates a correlation with the minimum electric spark ignition energy alone.

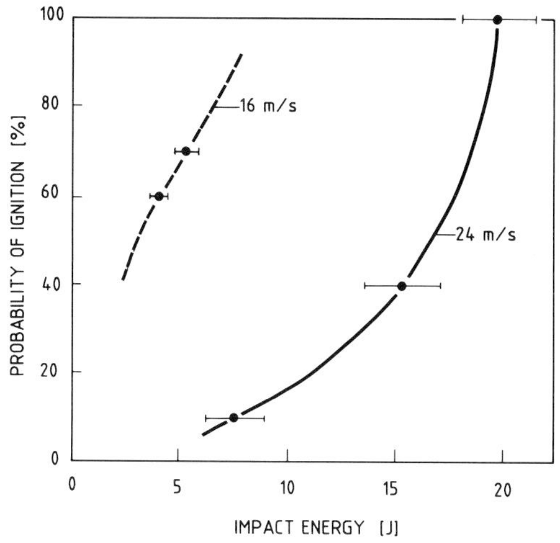

Figure 7.42 *Frequency of ignition of clouds of dried maize starch in air as a function of impact energy at 16 m/s and 24 m/s peripheral velocity of approach of the arm tip. Bars indicate ± 1 standard deviation. Impacts between titanium and rusty steel (thermite flashes) (From Pedersen and Eckhoff, 1987)*

Table 7.2 Results from single-impact ignition tests of dust clouds of different minimum electric spark ignition energies, using a 20 J thermite flash impact between titanium and rusty steel (From Pedersen and Eckhoff, (1987).

Dust	Minimum electric spark ign. energy [mJ]	Frequency of ignition in impact tests [%]
Maize starch, dried	< 4.5	100
Lycopodium	6	100
Barley protein	13	10
Barley starch	18-22	0
Maize starch, 10-11% moisture	27-36	0
Barley fibre	47-59	0

7.13 MINIMUM EXPLOSIBLE DUST CONCENTRATION

7.13.1 THE INDUSTRIAL SITUATION

For a given type of explosible dust, dispersed as a cloud in air, there is a reasonably well defined minimum quantity of dust per unit volume of air below which the dust cloud is not able to propagate a flame. (See Chapter 4 for full discussion.) In theory, therefore, one could eliminate the possibility of dust explosions by ensuring that the dust concentration does not exceed this minimum limit. In practice, however, most process equipment in plants where powders are manufactured and handled will always contain large quantities of powder, and hence this principle of preventing dust explosions is not practicable in general. There are, however, some types of process equipment to which the principle may be adapted in practice (see Section 1.4.3.2).

One example is dust extraction systems designed for the purpose of extracting a relatively small quantity of fine dust from a coarse main product, as in grain silo plants. In such cases the concentration of dust in the system can often be controlled to some extent by controlling the flow of air. It is then essential, however, that the air velocity is maintained sufficiently high to prevent dust from depositing on walls of ducting etc., since such deposits, if redispersed, may form clouds of explosible concentration.

Another type of equipment that can be protected by keeping the dust concentration sufficiently low, is systems for electrostatic powder painting. In such systems the concentration of particles in the air is relatively uniform and fairly easy to control. In fact, several countries have imposed specific maximum permissible average dust concentrations in the spraying booth, based on estimates of the minimum explosible dust concentration. (See Section 1.5.3.5.)

7.13.2 LABORATORY TESTS

Experimental determination of the minimum explosible dust concentration is discussed in detail in Section 4.2.6.2 in Chapter 4. This also includes comparisons between various test methods in use.

7.13.2.1 Tests developed in USA

In the standard test used in USA and UK for a number of years and described by Dorsett *et al.* (1960), a known quantity of the powder was dispersed as a cloud in a slim, vertical, cylindrical container of 1.2 litre volume and exposed to a continuous spark ignition source. Starting with very small powder quantities and repeating the test with steadily increasing amounts, a critical quantity was reached at which the dust cloud ignited. The critical mass of dust, divided by the volume of the test container, was taken as the minimum explosible dust concentration (MEC).

It was felt that the traditional test method was not fully satisfactory. On the one hand, the continuous ignition source was located in the lower part of the vertical, elongated explosion vessel, and this would allow the dust cloud, rising from the dispersion cup of the vessel bottom, to become ignited before having been fully dispersed throughout the entire vessel volume. Hence, the real concentration of dust in the cloud at the moment of ignition was likely to be higher than the nominal concentration estimated by dividing the mass of dust dispersed by the total vessel volume. This error would generally lead to underestimation of MEC. On the other hand, the traditional ignition source was a continuous train of relatively weak electric sparks that may not have been sufficiently energetic for igniting dust clouds of concentrations near the true limit for self-sustained flame propagation. This would generally yield overestimation of MEC. The effects of these two factors tend to cancel each other, and this may be the reason for the surprisingly good agreement that has in some cases been obtained between MEC values from the traditional small-scale lab test, and large-scale experiments. For example, Jacobson *et al.* (1961) found that various grain dusts and starches all gave MEC's of the order of 50 g/m^3 in the small lab-scale test, which compares favourably with the value 60 g/m^3 found for a typical wheat grain dust containing 10% moisture, in industrial-scale experiments by Eckhoff and Fuhre (1975).

However, such good agreement between the small-scale test and large-scale conditions would not be expected to be the general rule. For this reason, considerable efforts have been made in several countries during the last decade to develop an improved test for MEC.

In the USA, Hertzberg, Cashdollar and Opferman (1979) at the Bureau of Mines first developed an 8 litre explosion vessel in which transient dust clouds of quite homogeneous concentration distributions could be generated. One important conclusion from these studies was that determination of true MEC-values requires a strong ignition source. Therefore, Cashdollar and Hertzberg (1985) subsequently developed a 20 litre explosion vessel that would yield meaningful results even with quite strong ignition sources.

A cross section of the 20 litre vessel is shown in Figure 7.43. A photograph of the opened vessel, showing one of the light attenuation probes for measuring the dust concentration development in the transient dust cloud, is given in Figure 7.44.

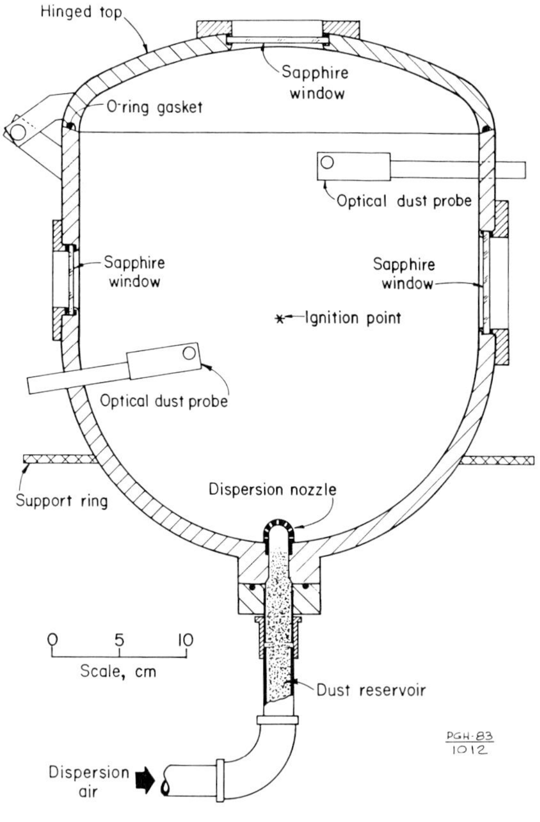

Figure 7.43 *Cross-section of US Bureau of Mines' 20 litre explosion vessel for determination of the minimum explosible concentration and other parameters of explosible dust clouds (From Cashdollar and Hertzberg, 1985)*

Figure 7.44 *Photograph of opened 20 litre US Bureau of Mines explosion vessel, showing one of the light attenuation probes for measuring dust concentration (Courtesy of K. L. Cashdollar, US Bureau of Mines, Pittsburgh, USA)*

Favourable agreement was obtained between minimum explosible concentrations found for coal dust in large-scale mine experiments and in the 20 litre vessel (Cashdollar *et al.*, 1987). The ignition source used in the 20 litre sphere was then a strong chemical igniter of calorific energy about 2500 J. The criterion of explosion was that the explosion pressure in the closed vessel should rise to at least twice the absolute initial pressure. For atmospheric initial pressure this means at least 1 bar(g). In addition the maximum rate of pressure rise should exceed 5 bar/s.

7.13.2.2
German/Swiss closed bombs

The 1 m^3 ISO vessel developed by Bartknecht and the 20 litre Siwek vessel are both discussed in Chapter 4, and further details are given in Sections 7.16 and 7.17. With the same ignition source and explosion criterion as used by Cashdollar and Hertzberg, the Siwek sphere should be able to yield comparable results. If, however, the 10 kJ igniter prescribed for the Siwek sphere for determining P_{max} and K_{St} values is used, too low minimum explosible concentration values would be expected for some dusts.

The 1 m^3 ISO vessel would be expected to yield the most reliable assessment of the minimum explosible concentration. Because of the large volume of the dust cloud, even a very strong ignition source of 10 kJ would not interfere with the main phase of dust cloud propagation. However, just because of its large size, the 1 m^3 test is not very suitable for routine testing, and smaller, laboratory-bench-scale methods are needed.

7.13.2.3
Nordtest Fire 011

The Nordtest (1989) method was designed specifically to meet the need of a reliable bench-scale test for the minimum explosible concentration of dust clouds. The apparatus consists of three main parts:

- A 15-litre explosion vessel with dust dispersion system.
- An ignition system.
- A dust concentration measurement system.

Figure 7.45 shows a maize starch explosion in the 15 litre Nordtest vessel.

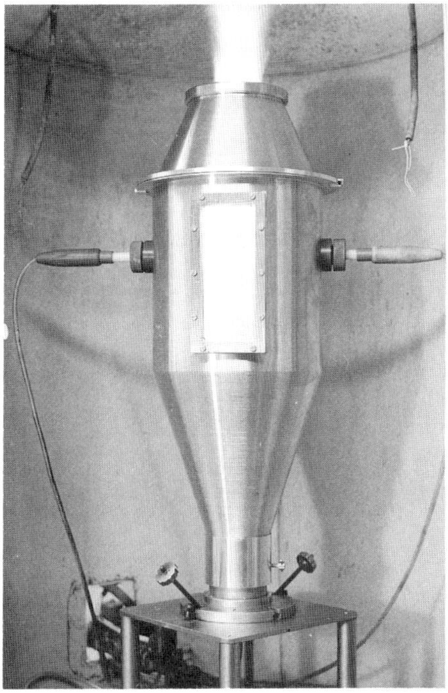

Figure 7.45 *Maize starch explosion in 15 litre Nordtest Fire 011 vessel. Ignition source: Strong electric arc between two thin metal electrodes*

The test procedure consists of two consecutive steps. First weighed quantities of the dust are dispersed into clouds in the 15 litre explosion vessel by means of a suitable, defined blast of air and exposed to an effective ignition source. The dispersion mushroom shown in Figure 7.46 is an essential part of the dust dispersion system.

The driving pressure and duration of the air blast are set to yield a reasonably homogeneous dust cloud in the vessel, as judged visually by the operator. Optimum dispersion conditions depend on particle size, shape, density and mass of dust to be dispersed. Immediately after completion of dispersion, the ignition source, positioned centrally within the cloud, is activated. By varying the dispersed mass of dust and

Figure 7.46 *Dispersion mushroom for Nordtest Fire 011 (right) compared with the IEC-version for the Hartmann bomb (left). Length of match approximately 50 mm*

conducting ten tests at each mass, the mass yielding a probability of explosion of 50% is estimated by interpolation.

The ignition source recommended for the test is a 200 W electric arc of 0.1 s duration. The arc is passed across a 3 mm spark gap between two 1.6 mm ∅ metal electrodes. The arc discharge is initiated by the closing action of the solenoid valve of the dust dispersion system. The ignition source must under no circumstances be less effective than this arc. However, in exceptional cases, the ignitability of the dust to be tested can be so low that a more effective ignition source may be required.

Explosion, i.e. a positive test result, is defined as independent flame propagation through the experimental dust cloud to the extent that the flame, as observed visually, is clearly detached from the ignition source.

In the second step of the test procedure the actual local dust concentration in the vicinity of the ignition source, at the same instant as the ignition source would be activated in the first step, is determined using the dust mass giving 50% of ignition, and exactly the same dust dispersion method as in the ignition tests. The arithmetic mean of five consecutive concentration measurements is taken as the minimum explosible dust concentration.

The version of the 15 litre vessel used in the second step is shown in Figure 7.47, and the basic principle of the traversing dust sampling cylinder is illustrated in Figure 7.48.

7.13.2.4
Possible international standard

The International Electrotechnical Commission (1990) is evaluating a test method based on the 20 litre Siwek (1988) sphere. Nordtest (1989) and the 1 m^3 vessel of the International Standardization Organization (1985) are specified as alternative methods.

The explosion criterion is that the maximum explosion pressure should be at least 1.5 bar(g). This includes the pressure of 1.1 ± 0.1 bar(g) generated by the powerful chemical 10 kJ igniter only, without dust. Tests are conducted with successively decreasing dispersed dust masses in steps of 0.2 g until a mass is reached at which the maximum pressure is lower than 1.5 bar(g) in three consecutive tests with the same dispersed dust mass. The minimum explosible concentration is then assumed to lie between the highest nominal concentration (dispersed mass divided by vessel volume) at which the maximum explosion pressure was less than 1.5 bar(g) in three successive tests, and the lowest nominal concentration at which the explosion pressure was 1.5 bar(g) or more in one of up to three successive tests.

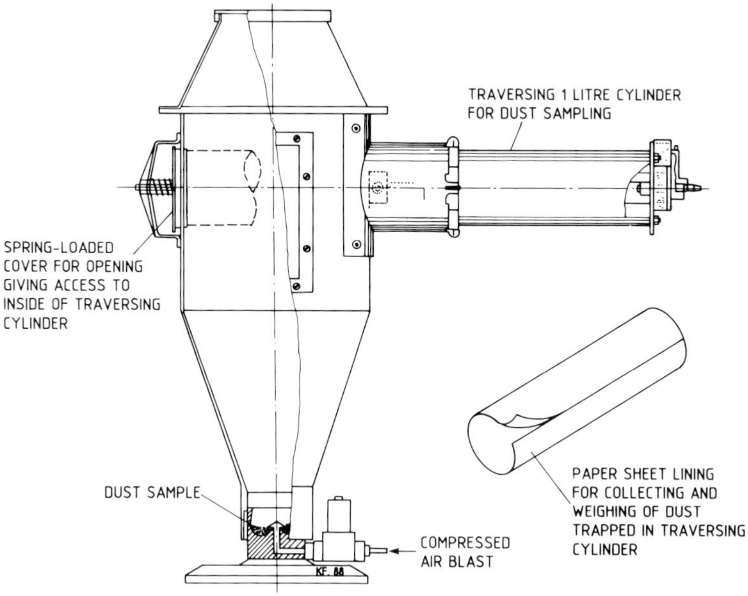

Figure 7.47 *A 15 litre Nordtest Fire 011 vessel equipped with traversing cylinder for measuring local dust concentration in the vicinity of the ignition source (From Nordtest, 1989).*

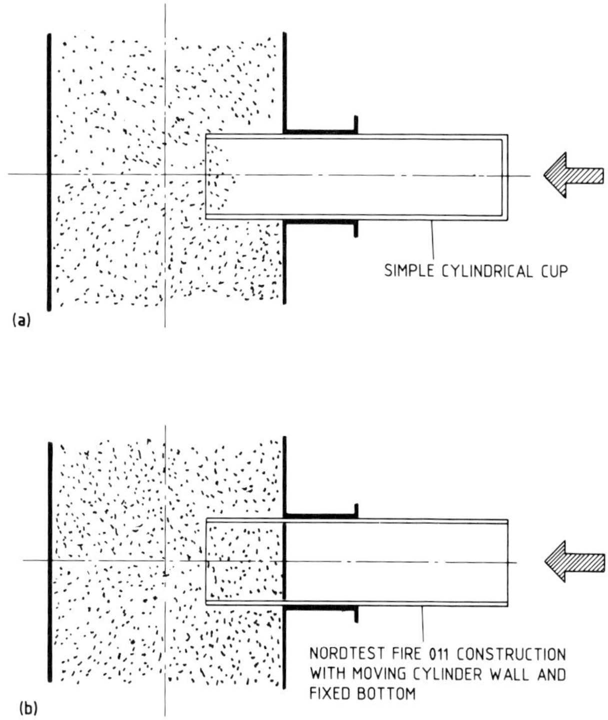

Figure 7.48 *Principle of Nordtest Fire 011 dust cloud sampling cylinder (b) compared with that of a simple cylindrical cup (a)*

As discussed in Section 4.26.2 in Chapter 4, there are indications of this test method yielding unexpectedly low minimum explosible dust concentrations for some dusts. This may be due to the use of the very energetic 10 kJ chemical ignition source that may support propagation of flames in dust clouds of lower concentrations than the true minimum explosible concentration.

This problem is avoided when using the ISO (1985) method, because the vessel of 1 m^3 volume is sufficiently large for ignition-source-independent flame propagation to be necessary for generation of significant explosion pressures.

Table 4.9 in Chapter 4 gives comparative data from tests with the three methods, using the same dusts.

7.14 MAXIMUM EXPLOSION PRESSURE AT CONSTANT VOLUME

7.14.1 THE INDUSTRIAL SITUATION

Most process equipment will not be sufficiently strong to withstand the typical pressures generated by unvented dust explosions. In principle, strengthening of the equipment can prevent it from bursting, but in general the structures required for achieving the sufficient strength will have to be so heavy that this approach is not generally recommendable, neither from the point of view of capital cost nor with respect to running and maintaining the plant. Exceptions are cylindrical dust extraction ducting, which can be made pressure resistant with reasonable wall thicknesses, and certain types of equipment which is heavy anyway, such as some mill types.

It nevertheless happens that the concept of fully pressure resistant process plant is adopted, e.g. when the powders are highly toxic and therefore in no circumstances can be admitted to outside the equipment. In such cases it is important to know the highest pressures to be expected, should a dust explosion occur within the equipment. As discussed in Section 3.3.8, the maximum explosion pressure (abs.) is generally proportional to the initial pressure (abs.), which must therefore be specified. In the case of a dust explosion in a fully confined, integrated system of various process items connected by comparatively narrow passages, pressure-piling may easily occur, as discussed in Section 1.4.4.1 in Chapter 1. This implies that a local explosion in one process unit may raise the pressure in the unburnt dust clouds elsewhere in the interconnected system. Should the flame then propagate into this pre-pressurized area, a considerably higher maximum pressure than if the initial pressure had been atmospheric, can result. Such pressure-piling, which may escalate in several stages, can give rise to local transient explosion pressures that are substantially higher than the adiabatic maximum explosion pressure at constant volume generated from normal atmospheric initial pressure. These possibilities must be considered carefully before adopting laboratory test data for the maximum explosion pressure, which are normally based on atmospheric initial pressure.

7.14.2
LABORATORY TESTS

7.14.2.1
Hartmann bomb

The Hartmann bomb, described by Dorsett *et al.* (1960), has been used throughout the world for assessing the maximum explosion pressure of dust clouds for nearly half a century. This apparatus, which is illustrated in Figures 7.49 and 7.50, basically consists of a closed vertical 1.2 litre stainless steel cylinder into which a known quantity of dust is dispersed as a cloud by a blast of air and exposed to an ignition source.

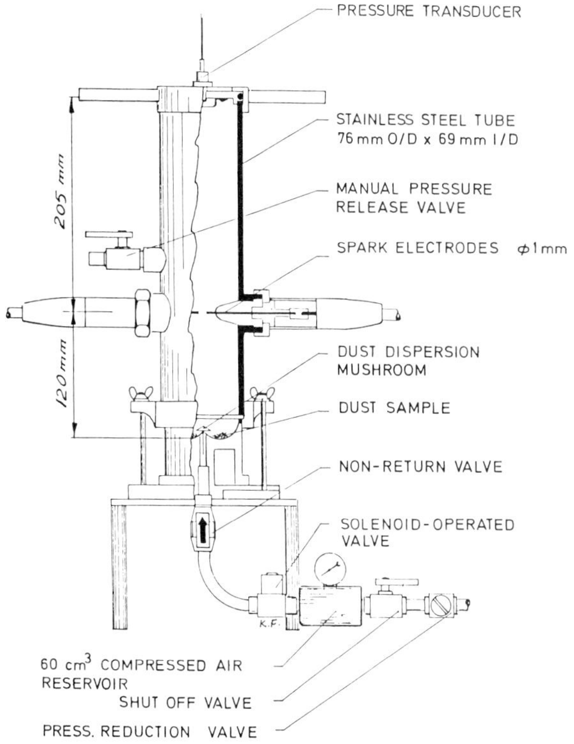

Figure 7.49 *A 1.2 litre Hartmann bomb for determination of pressure development in dust explosions at constant volume. Version developed during multinational cooperation and in all essentials adopted as a standard by the American Society for Testing and Materials (1988a)*

The dispersion mushroom design adopted in a multinational joint effort through the International Electrotechnical Commission, and shown in Figure 7.46 differs slightly from that included in the standard specified by the American Society of Testing and Materials (1988a).

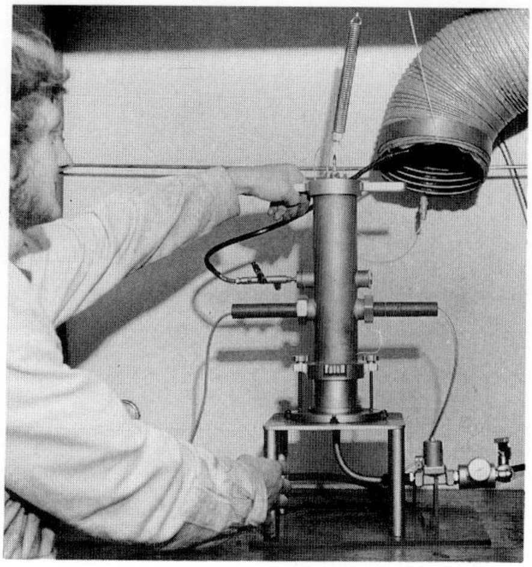

Figure 7.50 *Photograph of the version of the Hartmann bomb shown in Figure 7.49*

The ignition sources used include continuous trains of electric sparks, single synchronized sparks, synchronized chemical igniters and glowing resistance wire coils. Versions of the two latter are shown in Figure 7.51. For determination of maximum pressures the nature of the ignition source is not decisive because the maximum pressure is rather insensitive to the turbulence of the dust cloud at the moment of ignition. For the rate of pressure rise, however, turbulence is a key parameter and the moment of ignition must be exactly defined. (See Section 7.15.)

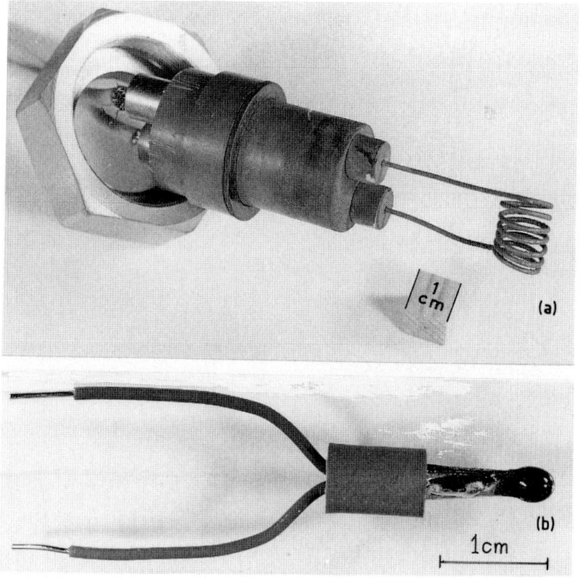

Figure 7.51 *Glowing resistance wire coil (continuous ignition source) (a) and chemical match head (instantaneous source) (b) used in Hartmann bomb tests (From Eckhoff, 1976)*

The development of explosion pressure as a function of time is recorded as illustrated in Figure 7.52 over a range of nominal dust concentrations (dispersed dust mass divided by bomb volume). Due to statistical scatter, several tests have to be determined at each nominal dust concentration. A typical set of results is shown in Figure 7.53. This figure also includes the maximum rate of pressure rise, i.e. the maximum value of the slope of the pressure-versus-time curve, which will be discussed separately in Section 7.15. In Norway it has been customary to take the highest 95% probability value as the result of the test. For the example in Figure 7.53 this means a maximum pressure of 6.8 bar(g).

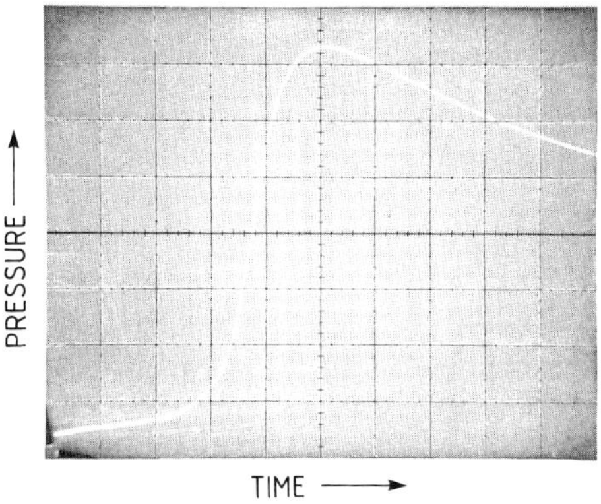

Figure 7.52 *Typical trace of pressure-versus-time during dust explosion in a closed vessel*

Because of the small volume of the Hartmann bomb and its elongated shape, the heat loss to the vessel wall during the explosion is significant. Therefore the maximum pressures measured are generally somewhat lower, typically by 25–30%, than those generated with the same dusts in larger vessels, such as the 1 m^3 ISO vessel and various 20 litre vessels. This is so in spite of the fact that the pressures measured in the Hartmann bomb are not corrected for the increased initial pressure due to the dust dispersion air.

The measurement of maximum constant-volume pressures generated by dust explosions in closed bombs is fairly straightforward. Apart from the wall cooling effects in small bombs, the results do not depend much on the details of the experiment as long as the dust cloud is reasonably well dispersed and the average nominal dust concentration is varied systematically to identify the worst case.

7.14.2.2
The 1 m^3 standard ISO vessel

Side and top views of this apparatus are illustrated schematically in Figure 7.54.

A container of approximately 5 litres capacity and capable of being pressurized with air to 20 bar is attached to the explosion chamber. The container is fitted with a 19 mm ∅ opening valve of 10 ms opening time. The container is connected to the explosion chamber via a 19 mm ∅ perforated semicircular spray pipe. The diameter of the holes in the pipe should be in the range 4–6 mm. The number of holes is chosen such that their total cross-sectional area is approximately 300 mm^2.

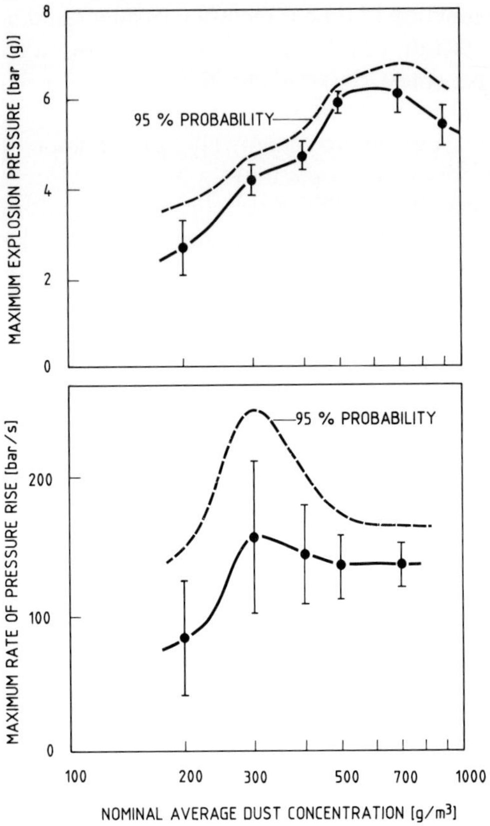

Figure 7.53 *Typical set of results from Hartmann bomb test of a given dust. Bars represent ± 1 st. dev. Dotted lines based on Gaussian distribution (mean + 1.65 st. dev.)*

The ignition source is a pyrotechnical igniter with a total energy of 10 kJ and arranged to fire after a fixed delay of 0.6 s after onset of dust injection. The mass of the pyrotechnical ignition source is 2.4 g, and it consists of 40% zirconium, 30% barium nitrate, and 30% barium peroxide. It is activated by an electric fuse head. The igniter is located at the geometric centre of the explosion chamber. Two pressure transducers, linked to a recorder, are fitted to measure the explosion chamber pressure development.

The way of determining the maximum explosion pressure is similar to that of the Hartmann bomb test, and Figures 7.52 and 7.53 also applies to the 1 m^3 test. However, due to the comparatively large size of the experiment, the amount of dust and the time required per experiment limit the number of tests that are normally performed.

Maximum explosion pressures measured with this apparatus would be expected to be relatively close to the theoretical maximum adiabatic pressures. Data for a range of dusts are given in Table A1 in the Appendix.

Figure 7.55 shows a 1 m^3 vessel that would most probably satisfy the ISO-standard requirement, if equipped with appropriate dust dispersion and ignition systems.

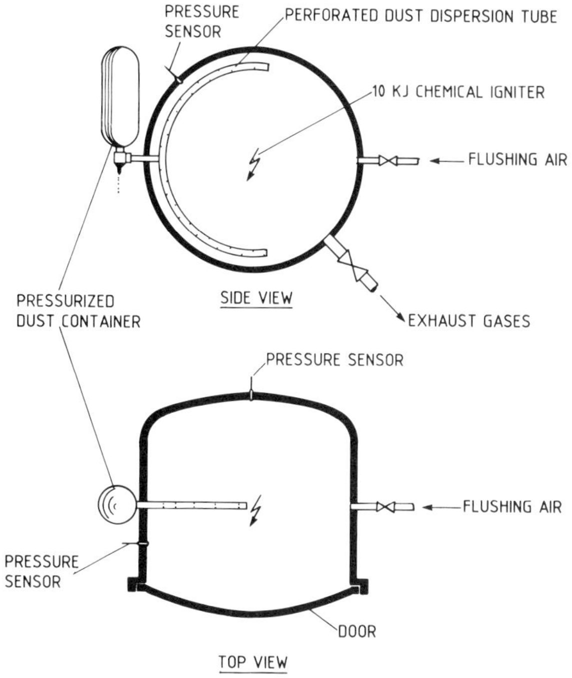

Figure 7.54 *1 m^3 closed vessel specified by the International Standardization Organization (1985) for determination of maximum explosion pressures and maximum rates of pressure rise of dust clouds in air (From Verein deutsche Ingenieure, 1988)*

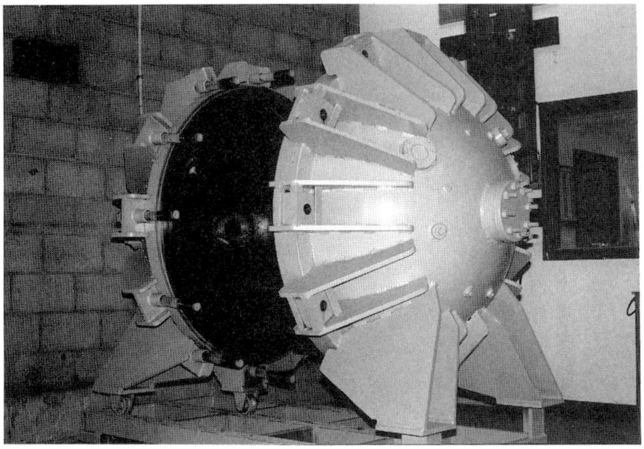

Figure 7.55 *1 m^3 spherical explosion vessel composed of two detachable hemispheres (Courtesy of Fike Corporation, USA)*

7.14.2.3
The Siwek 20 litre sphere

This vessel was developed by Siwek (1988) primarily with a view to obtain maximum explosion pressures and explosion rates in agreement with data from the 1 m^3 ISO vessel. The Siwek sphere is shown in Figure 7.56.

Figure 7.56 *A 20 litre Siwek sphere for determination of pressure development in dust explosions (Courtesy of R. Siwek, Ciba-Geigy AG, Switzerland)*

The sphere essentially is a small-scale version of the 1 m^3 ISO vessel. The original dust dispersion system was of the same type as that of the 1 m^3 ISO vessel, consisting of a pressurized dust reservoir, from which the dust was injected into the main vessel through a perforated tube, as illustrated in Figure 7.54. The experimental conditions required for obtaining agreement with the 1 m^3 ISO vessel were specified in a standard issued by the American Society for Testing and Materials (1988b). The ignition source has to be the same type of 10 kJ chemical igniter as used in the 1 m^3 ISO-test. The ignition delay is, however, shorter (60 ms) because of the smaller vessel size. For the determination of the rate of pressure rise (see Section 7.15) it is important to pay attention even to the design of the capsule containing the pyrotechnical mixture of the ignition source. Zhu *et al.* (1988) showed that igniters with metal capsules could give significantly different K_{St} values from those obtained for the same dusts with plastic capsules.

Under these circumstances, and testing dusts of small particle size, Siwek obtained quite good correlations between data from the 1 m^3 ISO vessel and his 20 litre sphere, as shown in Figure 7.57. (K_{St} is defined in Section 4.4.3.3 in Chapter 4.)

Experience in several laboratories disclosed, however, that many cohesive dusts, in particular of fibrous particles, can easily get packed and trapped inside the perforated dispersion tube of the original dust dispersion system, which is clearly unsatisfactory. This led to the development of an open nozzle system named a 'rebound' nozzle, shown in Figure 7.58, which has gradually replaced the original perforated ring. According to Siwek

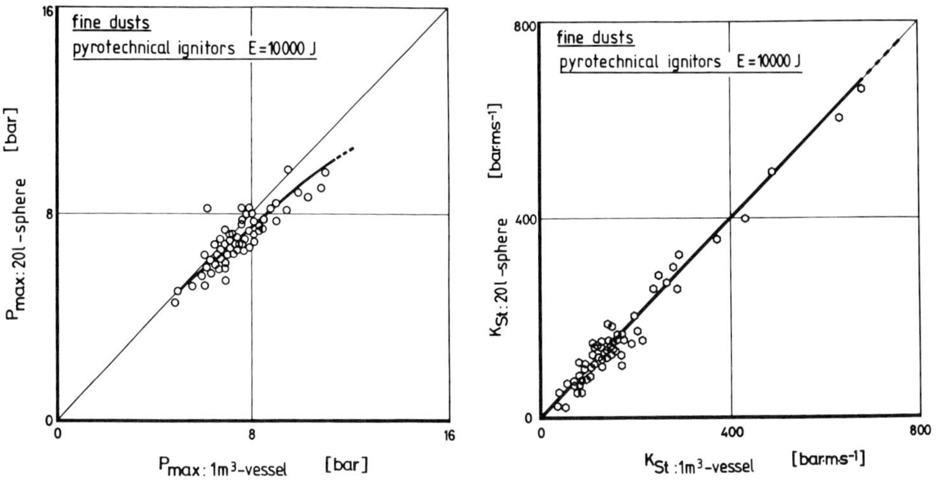

Figure 7.57 *Correlations of maximum explosion pressures and maximum rates of pressure rise from 1 m^3 ISO vessel and 20 litre Siwek sphere. Courtesy of R. Siwek, Ciba-Geigy AG, Switzerland)*

Figure 7.58 *New rebound nozzle for dispersing the dust in the 20 litre Siwek sphere (Courtesy of R. Siwek, Ciba-Geigy AG, Switzerland)*

(1988) the new nozzle produces both maximum pressures and K_{St} values in reasonable agreement with those generated by the original perforated-ring system.

7.14.2.4
Other 20 litre vessels

The US Bureau of Mines vessel described by Cashdollar and Hertzberg (1985) and shown in Figure 7.59, is a valid alternative to the Siwek vessel. An advantage, as demonstrated in Figure 7.44, is the large opening giving easy access to the inside of the vessel for cleaning, inspection, etc.

It would be expected that the US Bureau of Mines vessel would yield both maximum explosion pressures and rates of pressure rise in agreement with data from the Siwek sphere provided the dust dispersion and ignition conditions were the same in both vessels.

The 20 litre vessel system described by Burke (1988) was shown to be in accordance with the standard specified by American Society for Testing and Materials (1988b), for determination of both maximum explosion pressures and maximum rates of pressure rise.

Another complete 20 litre vessel test system is illustrated in Figure 7.60.

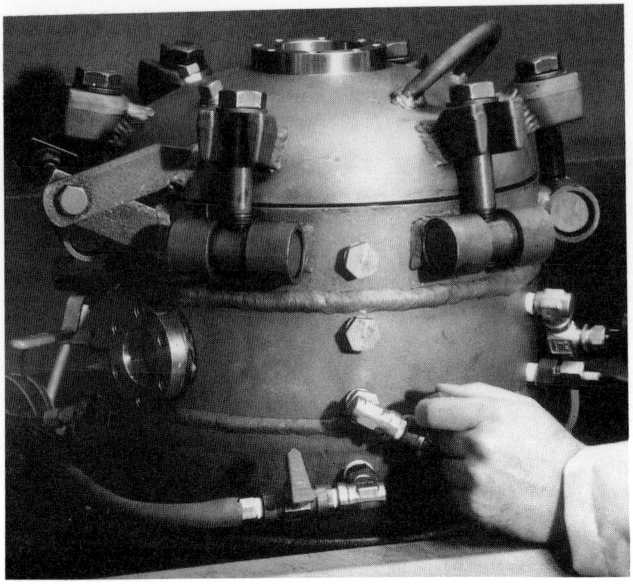

Figure 7.59 *Photo of the 20 litre US Bureau of Mines vessel with the lid on (Courtesy of C. L. Cashdollar, US Bureau of Mines, Pittsburgh, USA)*

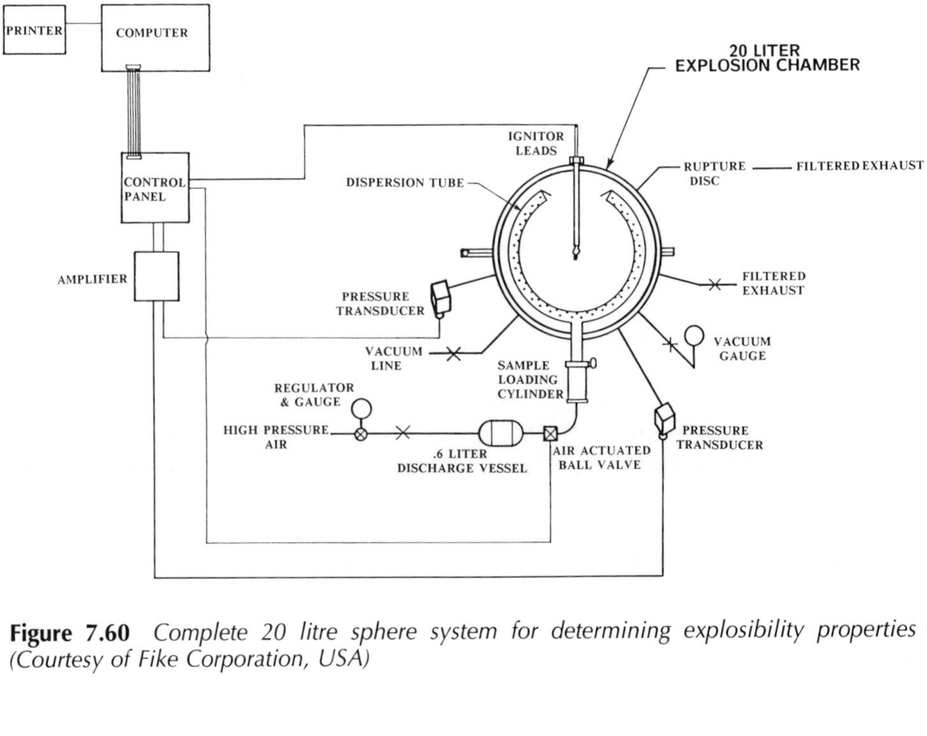

Figure 7.60 *Complete 20 litre sphere system for determining explosibility properties of dusts (Courtesy of Fike Corporation, USA)*

American Society for Testing and Materials (1988b) also indicate some further 20 litre test apparatuses and a 26 litre apparatus that are likely to satisfy the requirements of the standard.

7.15 MAXIMUM RATE OF RISE OF EXPLOSION PRESSURE AT CONSTANT VOLUME (EXPLOSION VIOLENCE)

7.15.1 THE INDUSTRIAL SITUATION

Industrial enclosures such as conventional process equipment, are normally far too weak to withstand the pressures exerted even by only partly developed, confined dust explosions. Consequently a primary objective of fighting an explosion after it has been initiated, is to prevent the build-up of destructive overpressures.

At least three techniques for preventing destructive overpressures are in current use in industry. The first and probably most widely used is venting. Another technique is automatic suppression. In case the explosion starts in an enclosure that is strong enough to withstand the explosion pressure, such as certain types of mills, isolation by high-speed valves to prevent the explosion from propagating to other, weaker enclosures constitutes a third means of protection.

Regardless of which protective technique is adopted, the violence of the dust explosion, i.e. the rate of heat generation inside the enclosure where the explosion is initiated, is a deciding factor as to whether a given protective system will perform adequately. In view of the fact that the combustion rate can vary substantially from dust cloud to dust cloud, it is important to base the design of industrial equipment on appropriate estimates of the explosion violence or combustion rate that will occur in practice.

7.15.2 LABORATORY TESTS

Maximum rates of pressure rise can be measured in all the closed vessels described in Section 7.14. Section 4.4.3 in Chapter 4 discusses the basic nature of such experiments, and it is shown that the maximum rate of pressure rise in closed bomb apparatuses of the type discussed in Section 7.14 is bound to be arbitrary. This also applies to K_{St} values of dusts (Section 4.4.3.3, equation (4.84)).

The method for determining K_{St} values of dusts specified by the International Standardization Organization (1985) is the same as for measurement of maximum explosion pressure and described in Section 7.14.2.2. Because the standard vessel has a volume of 1 m^3, the K_{St} values in bar m/s are numerically identical with the maximum rate of pressure rise in bar/s.

If smaller vessels, for example of 20 litres, are used for determining K_{St} values according to the ISO standard, the dust dispersion system, the ignition source strength and the ignition delay must be tuned in such a way that the products of the maximum rates of pressure rise measured and the cube roots of the vessel volumes equal the K_{St} values that would have been measured for the same dusts in the 1 m^3 ISO standard test. (See Equation (4.84) in Section 4.4.3.3 in Chapter 4.)

Through the years a considerable number of other non-standardized closed vessels have been used for assessing explosion violence. Thus Nagy *et al.* (1971) performed experiments in vessel volumes ranging from the 1.2 litre of the Hartmann bomb to 14 m^3. They normalized their results by multiplying all the measured maximum rates of pressure rise by the cube root of the vessel volume, the product being denoted *K*. With maize starch, all the three smallest vessels of volumes 1.2 litres, 8 litres and 28 litres gave close to identical *K*–values, whereas those of the three larger vessels of 3 m^3, 6.5 m^3 and 14 m^3 were all about twice as large. With coal dust, nearly identical K-values were obtained for the 8 litre and 28 litre vessels, and again these were about half values for the three larger vessels. However, in this case the value of the Hartmann bomb was only one third of that obtained in the 8 litre and 28 litre vessels. Hence, for some dusts the Hartmann bomb would yield *K*–values very similar to those generated in larger size vessels, whereas for other dusts the Hartmann bomb values were considerably smaller. A distinct, dust-independent increase of the *K*–value by a factor of two was observed when moving from the three laboratory-scale bombs to the closed vessels of industrial scale. This could be due to the use of a different type of dust cloud generation system in the large-scale experiments.

Moore (1979) performed a similar comparison of *K*–values obtained by testing the same dust in four different vessels. These were the Hartmann bomb, a 1.75 litre cylinder with L/D = 1, mounted on the standard dust dispersion unit of the Hartmann bomb, a 43 litre sphere, and the standard 1 m^3 ISO vessel. In general the Hartmann bomb gave the lowest *K*–values, but consistent correlation between values from the various vessels were difficult to establish. Moore interpreted the discrepancies in terms of different degrees of turbulence, different dust concentration distributions and different ignition source properties in the various tests.

Enright (1984) reported similar comparative experiments in three closed vessels of 1.2 litres, 8 litres and 20 litres respectively. The principle of the dust dispersion system was the same for all three vessels, namely an air blast from a dispersion mushroom impinging on a dust heap placed at the vessel bottom. However, both the gap between the dispersion mushroom and the vessel bottom, the volume of the dispersion air reservoir, and the ignition delay were increased somewhat arbitrarily with vessel volume. For all the three dusts tested, namely lycopodium, wheat starch and a '60 μm' aluminium powder, the lowest *K*–values were obtained with the 1.2 litre vessel and the highest with the 20 litre vessel.

This evidence re-emphasizes that even the K_{St} concept, as defined by ISO remains an arbitrary measure of the explosion violence. K_{St} is not a specific dust constant, but clearly also a function of the special test conditions in the ISO standard test.

On the other hand, the K_{St}, as defined by ISO, seems to provide a reasonable relative measure for ranking the explosion violence to be expected from various dusts in industrial dust explosions. However, the resolution must not be overrated. As shown in Chapter 6, four dusts of very similar K_{St} values in the narrow range 115–125 bar m/s gave maximum explosion pressures in a filter with a given vent, which varied by a factor of two to three.

It is important to keep in mind the various factors that influence the explosion rate of a dust cloud (see Chapter 4, Chapter 6, and Eckhoff (1987)) and to consider the extent to which they are the same in the standard test and the industrial situation of concern.

It is felt that other test methods for maximum rates of pressure rise, not complying with the ISO standard may also yield a reasonable relative ranking of dusts with respect to their explosion violence in practice. This includes the Hartmann bomb as standardized by the American Society for Testing and Materials (1988a).

7.15.3 FURTHER DEVELOPMENT OF ADEQUATE TEST METHODS FOR DUST EXPLOSION VIOLENCE ASSESSMENT

As already pointed out, the violence with which clouds of a given dust will explode in an industrial plant is not a specific dust property, but indeed also depends on the state of the dust cloud in the actual industrial situation. Test methods that would allow differentiation in test conditions could be designed by following at least three lines of approach (Eckhoff, 1987):

1. The first would be to retain one of the existing standard closed-bomb methods and add to it a differentiated procedure for interpreting results. For example, the measured $(dP/dt)_{max}$ or K_{St} value could be multiplied by one of a range of empirical correlation factors to match the particular industrial situation in question. This would allow existing nomograms for vent area assessment to be maintained. No matter which standard test method were to be chosen, it would be necessary to standardize extremely carefully both apparatus and experimental procedures. One could also make active use of the dependence of $(dP/dt)_{max}$ on dust concentration, which is in fact currently measured in the existing standard tests, as illustrated in Figure 7.53. In most cases it would seem justified to assume that worst-case-concentration throughout the cloud is rather unlikely.
2. A second possibility would be to retain one of the existing bombs, but change the experimental programme of the test. By including the ignition delay as a parameter, the reaction rate as a function of the relative turbulence level could be assessed experimentally. (See also Section 4.4.3 in Chapter 4.) This would correspond to varying the turbulence index T_u defined by the International Standardization Organization (1985). $(dP/dt)_{max}$ at various ignition delays would then represent the respective reaction rates corresponding to various situations in industry. One could then test at the turbulence level that would correspond to the actual industrial situation concerned. It could also be of interest to supplement the explosion test with a dust dispersibility test (see Section 7.4.2) to assess the degree of dust dispersion that would be expected from the dispersion process operating in the specific industrial situation of interest.
3. A third strategy would be to retain one of the existing bombs, but design a range of different 'plug-in' dust dispersion units to allow tests to be carried out with the unit producing the degree of dust dispersion and level of turbulence to be expected in practice. This would yield different correlations of $(dP/dt)_{max}$ versus ignition delay, depending on the intensity of the dispersion process. (See Figure 4.40 in Chapter 4.)

No matter which of these possibilities is pursued, it is necessary to conduct realistic full-scale dust explosion experiments to establish credible correlations between predicted dust cloud combustion rates, and those that will actually occur in the wide spectrum of situations in which dust clouds may burn in industry.

In the future the maximum pressure rise measurement is likely to be replaced by more basic parameters, such as the induction time of the dust cloud combustion reaction, which will be used as input to advanced computer simulation models for turbulent dust explosions (Eckhoff, 1987).

7.16 EFFICACY OF EXPLOSION SUPPRESSION SYSTEMS

Explosion suppression is discussed in Section 1.4.7 in Chapter 1. The International Standardization Organization (1985a) specified a test method for evaluating the effectiveness of explosion suppression systems against defined explosions in closed, or essentially closed vessels. The test does not cover explosions at elevated initial pressures. The method gives design criteria for apparatus for explosion suppression efficacy tests and criteria for defining the safe operating regime of an explosion suppression system.

The basic test apparatus is the 1 m^3 closed vessel described in Section 7.14.2.2 and shown in Figure 7.54, but other vessels may also be used provided that the volume is sufficiently large and the length-to-diameter ratio is less than two.

A complete test arrangement, with the suppression system to be tested mounted on the test vessel, is illustrated in Figure 7.61.

Figure 7.62 illustrates the type of pressure development that will be observed during a standard test.

Prior to initiation of the fully automated test, the 1 m^3 vessel is partially evacuated to compensate for the supply of air during the dust injection process, which constitutes the first step of the automatic test sequence. When all the explosible dust has been injected into the test vessel and atmospheric pressure has been restored, the explosible dust cloud is ignited after a pre-determined delay. The pressure detector of the suppression system under test has been pre-set at a given trigger level P_A, and when this explosion pressure is reached, suppressant injection starts. The efficacy of the suppression is reflected by the magnitude of the peak pressure P_{red}.

By varying the trigger level P_A of the pressure detector, and the K_{St} value of the dust, the efficacy of the specific suppression system under test can be assessed for a range of explosible cloud conditions.

The standard test method is unsuitable for predicting the performance of suppression systems if the industrial enclosure to be protected has one or more of the following features:

- Vessel aspect ratio greater than 2:1.
- Partially vented vessels.
- Container fitted with fixed or mobile apparatus which could impede the distribution of suppressant.
- Operating pressures and temperatures substantially higher or lower than normal atmospheric conditions.
- High levels of turbulence and/or dust/powder throughput.
- Vessel volumes substantially greater or smaller than those used in the efficacy test.

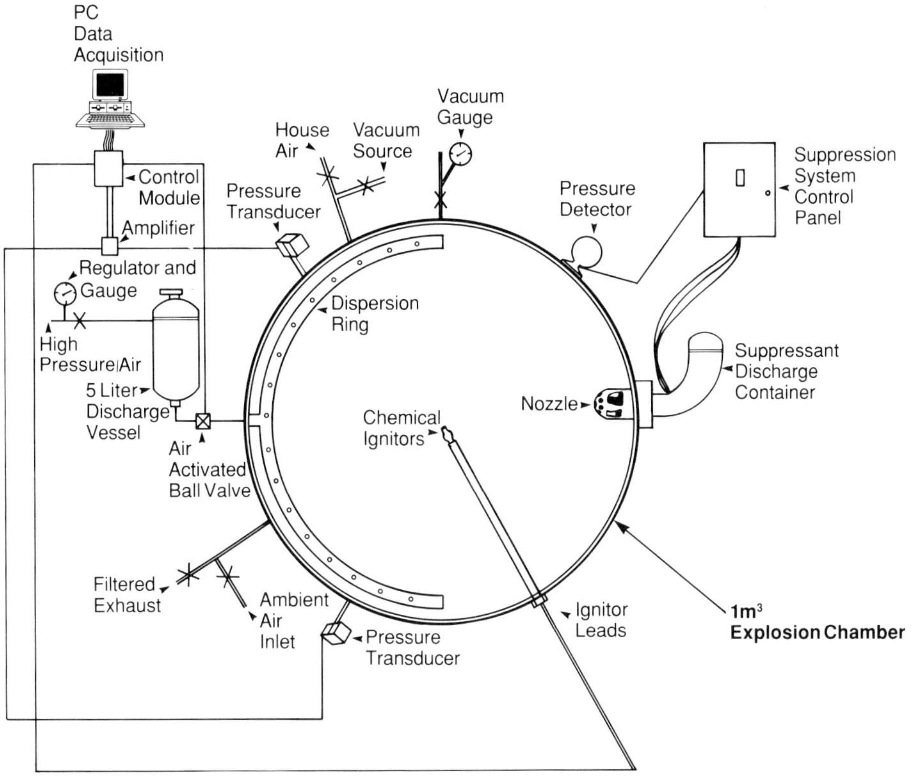

Figure 7.61 *Illustration of complete system for testing the efficacy of explosion suppression systems according to the International Standardization Organization (1985a) (Courtesy of Fike Corporation, USA)*

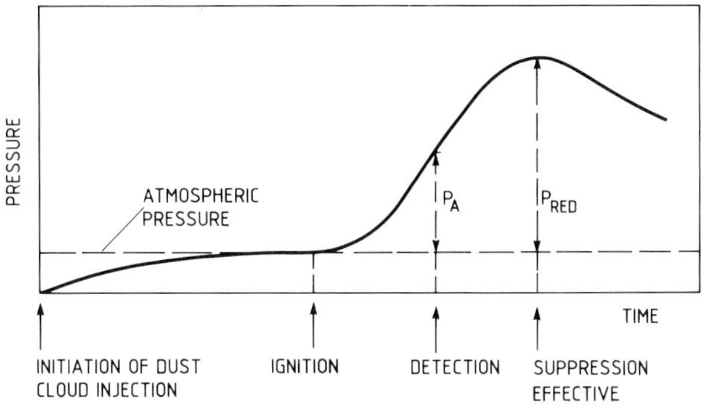

Figure 7.62 *Typical pressure-versus-time trace during a dust explosion suppression test in the standard 1 m^3 ISO apparatus*

Figure 7.63 illustrates a test arrangement relevant for testing suppression of larger volumes than 1 m^3. Design methods for systems for suppression of volumes of up to 250 m^3 is discussed in Section 1.4.7.

Figure 7.63 *A 10 m^3 test vessel for assessing the efficacy of suppression systems for larger volumes than 1 m^3 (Courtesy of Fike Corporation, USA)*

7.17 MAXIMUM EXPLOSION PRESSURE AND EXPLOSION VIOLENCE OF HYBRID MIXTURES OF DUST AND GAS IN AIR

The ignitability/explosibility of hybrid mixtures is discussed in Section 1.3.9 in Chapter 1. Such mixtures may be generated in industry in a number of ways, for example during drying of explosible dust containing organic solvents.

The International Standardization Organization (1985b) designed a test method for assessing the explosibility properties of explosible clouds other than dust/air and gas/air, based on the apparatus illustrated in Figure 7.64.

The method is primarily intended for hybrid mixtures of combustible dusts and gases in air, and mists of combustible liquids in air. However, it also seems to be suitable for investigating the explosibility of dusts in oxidizer gases of other oxygen contents than in air, as mentioned in Section 7. 19.

The procedure for testing hybrid gas/dust/air mixtures is as follows:

The gas/air mixture in the 1 m^3 chamber is prepared by the method of partial pressures or other suitable technique. It is important to ensure that the composition and homogeneity of the gas/air mixture is as required.

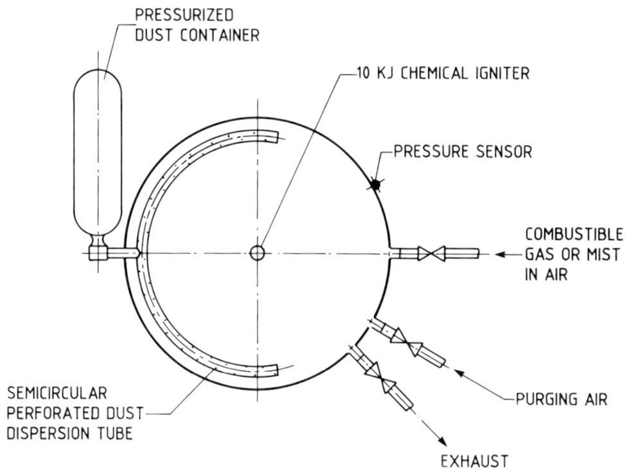

Figure 7.64 *A 1 m^3 closed vessel specified by the International Standardization Organization (1985b) for determining maximum explosion pressures and rates of pressure rise of other explosible clouds than dust/air and gas/air.*

Then the dust sample, of the mass required to obtain the appropriate cloud concentration, is placed in the 5 litre container which is subsequently pressurized with air to 20 bar. The pressure recorder is activated followed by activation of the dust sample container valve and the ignition source.

The flow of compressed dust/air suspension into the explosion chamber induces turbulence in the gas/air mixture. Therefore, choosing an appropriate ignition delay (turbulence level) is important. The influence of the compressed air from the dust reservoir on the final explosible gas concentration should be taken into account.

Tests are conducted for the range of total fuel concentrations and combustible gas/combustible dust ratios required.

7.18 TESTS OF DUST CLOUDS AT INITIAL PRESSURES AND TEMPERATURES OTHER THAN NORMAL ATMOSPHERIC CONDITIONS

Industrial processes are sometimes operated at higher initial pressures or temperatures, or both, than normal ambient conditions. In such cases, results from tests of ignitability and explosibility at normal ambient initial conditions may not be relevant. The general trends of the influences of initial pressure and temperature are outlined in Sections 1.3.7 and 1.3. 8 in Chapter 1.

Tests to elucidate specific problems are most conveniently conducted in closed bombs of the types described in Section 7.14, and fitted with adequate provisions for heating and pre-pressurization, and of sufficient strength. This applies both to ignition sensitivity tests and explosibility tests. The proportional increase of the maximum explosion pressure with initial pressure (Section 1.3.8) requires very strong bombs if the initial pressure is appreciable.

Bombs of the type in Figure 7.64 may be used if the gas phase differs from pure air.

7.19 INFLUENCE OF OXYGEN CONTENT IN OXIDIZING GAS ON THE IGNITABILITY AND EXPLOSIBILITY OF DUST CLOUDS

7.19.1 THE INDUSTRIAL SITUATION

Full and partial inerting is discussed in Sections 1.3.6 and 1.4.3 in Chapter 1.

The possibility of dust explosions in process equipment can in principle be effectively eliminated by substituting the air by a gas which makes flame propagation in the dust cloud impossible. Since the use of large quantities of inert gas in a plant can be expensive, it is important to limit the inert gas consumption to the extent possible. For most dusts it is not necessary to substitute the entire atmosphere in the actual area by e.g. nitrogen, carbon dioxide, or other inert gas to obtain inerting. Hence, it is essential to know the critical gas composition for inerting the dust in question. In some cases it may even be of interest to use smaller fractions of inert gas than required for completing inerting, because this will reduce both the ignition sensitivity of the dust cloud, and the maximum pressure and rate of pressure rise at constant volume.

7.19.2 LABORATORY TESTS

In USA, as described by Dorsett *et al.* (1960), two standard test methods were traditionally used. In both tests, the dust was dispersed in the appropriate gas mixture from above, into a fairly narrow vertical tube of i.d. 38 mm, and exposed to an ignition source. The apparatus is similar to the Godbert-Greenwald furnace described in Section 7.8. In the first test the ignition source was an electric spark, in the second test the hot tube wall. Usually the limiting gas compositions for flame propagation obtained for the same dust from the two tests, differed significantly, the hot surface test yielding lower critical permissible oxygen contents than the spark test.

Figure 7.65 shows a type of apparatus used by some laboratories for determining the maximum permissible oxygen content in the atmosphere for inerting of dust clouds.

An experimental procedure applicable to this apparatus is as follows: compressed air and inert gas are first mixed in the desired proportions in a mixing vessel by the partial pressure method. Once the powder to be tested has been placed in the dispersion cup, a quantity of 3 litres of the gas mixture is admitted gently into the explosion tube via the small reservoir and the thin flushing tube, with the filter paper in position at the top of the

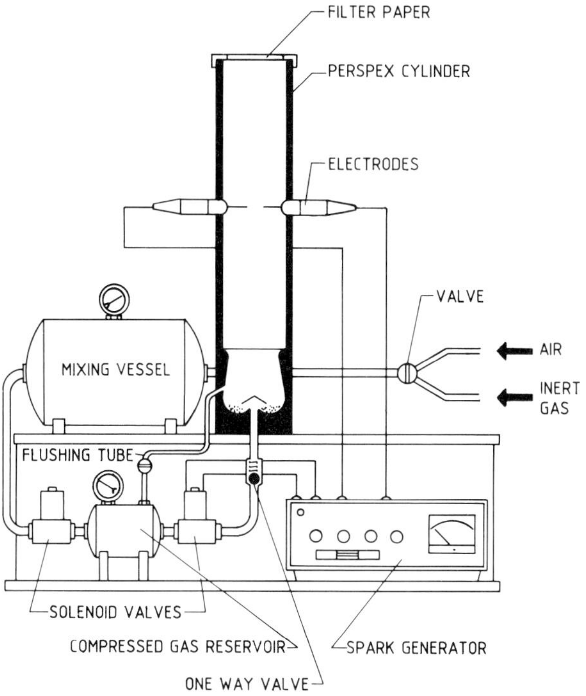

Figure 7.65 *Open 1.2 litre Hartmann tube apparatus for determining the influence of the oxygen content in the atmosphere on the ignitability of dust clouds*

Perspex cylinder. During this process the air that was originally in the Perspex cylinder leaks to the atmosphere. The small reservoir is now pressurized with the appropriate gas mixture to a pre-determined level, found in earlier trials to give the best dust dispersion conditions for ignition and flame propagation in air.

To initiate the test sequence, a push button on the electric spark generator opens the solenoid valve to disperse the powder. After a pre-set delay, a soft spark of approximately 3 J is discharged across the spark gap in the dust cloud in the tube. It is then observed whether ignition occurs. Ignition is defined as visual observation of a dust flame that is clearly detached from the spark. For each particular oxygen concentration 20 trials are carried out and the results plotted as a frequency-of-ignition versus oxygen-concentration graph. The maximum permissible oxygen content for inerting is then defined as lying between the lowest concentration at which at least one trial in 20 gave ignition, and the highest concentration at which no ignition occurred in 20 trials.

When applying the test result in industrial plant design, an appropriate safety margin must be incorporated. The method can be refined by actually measuring the oxygen concentration in the Perspex tube prior to each test. This may be necessary at low oxygen contents, of a few per cent and lower.

It is important that the ignition source is not the limiting factor for ignition. The situation in this respect is the same as for the minimum explosible dust concentration test. If the ignition source is too weak, apparently inert conditions will be found for oxygen concentrations that would in fact allow flame propagation once ignition had been accomplished by a sufficiently strong source. For some dusts a 3 J soft spark source may be too weak for identification of the true oxygen limit.

The apparatus in Figure 7.65 is also well suited for measurement of the minimum electric spark ignition energy as a function of oxygen concentration, which may be useful information for assessing the gain in safety obtained if partial inerting is used.

The apparatus in Figure 7.64 may be used for measuring maximum explosion pressure and rate of pressure rise as functions of the oxygen content in the atmosphere, which provides further information about the effect gained by partial inerting. The inlet for combustible gas/air would then instead be used for the mixture of inert gas and air.

7.20 INFLUENCE OF ADDING INERT DUST TO THE COMBUSTIBLE DUST, ON THE IGNITABILITY AND EXPLOSIBILITY OF DUST CLOUDS

Section 1.4.3.3 outlines the industrial situation. When required, all the test methods described in the present chapter can also be applied to mixtures of combustible and inert dust. One problem that could arise, however, is segregation of the two components during dust dispersion because of differences in particle properties (size, shape, density). If such segregation occurs, misleading results could arise because the ratio of inert to combustible dust in the region of the ignition source was either significantly above or below the assumed nominal average.

7.21 HAZARD CLASSIFICATION OF EXPLOSIBLE DUSTS

Attempts have been made in the past at classifying ignitability and explosibility of dusts by one or two dimensionless figures. Experience has shown that the usefulness of such indices is limited, because specific ignitability and explosibility properties are not necessarily correlated.

A draft for an alternative hazard classification system for combustible dusts is being evaluated by the International Electrotechnical Commission (1990a). The system is primarily concerned with the fire and explosion hazard if ignitable dusts are in contact with various kinds of electrical apparatus.

The system makes use of three symbols:

+ Indicates that there is a need for special attention in respect of the combustion property of the dust.

− Indicates that special attention is not required in the respect of the combustion property of the dust, or that the property cannot be measured by the test specified. (For example the minimum ignition temperature of a layer of the dust cannot be measured because it melts and runs off the hot surface of the test apparatus.)

? Indicates that information on the property is not available.

The + symbol is applied to the following four specific tests, using the sensitivity thresholds indicated:

- Electrical resistivity. $< 10^3\ \Omega$ m
- Minimum ignition temperature of dust cloud. $< 400°C$
- Minimum ignition temperature of dust layer of 5 mm thickness. $< 300°C$
- Minimum ignition energy of dust cloud. < 15 mJ

The symbol − is used for all data that equal or exceed these critical thresholds.

The thresholds chosen are bound to be arbitrary. On the other hand, the classification system allows a quick preliminary screening to identify dusts that are definitely hazardous. The system does not imply, however, that the hazard of dusts that appear with a negative sign for all four tests, can be neglected.

National Material Advisory Board (1988) discussed the specification of an explosion hazard classification system of dusts in relation to use of electrical equipment, with reference to the situation in USA. Two basic questions were considered essential:

- Is the dust combustible?
- Is the electrical resistivity of deposited dust higher or lower than $10^3\ \Omega$ m?

No final conclusions as to the details of a revised classification system were drawn, but some indications were given. The need for further research and international cooperation was emphasized.

REFERENCES

Allen, J., and Calcote, H. F. (1981) Grain Dust Ignition by Friction Sparks. SMS-81–049, National Grain and Feed Association, Washington DC

Allen, T. (1981) *Particle Size Measurement,* 3rd Edition. Chapman & Hall, London

American Society for Testing and Materials (1988) *ASTM Standard E 680–79 (Reapproved 1984) for Drop Weight Impact Sensitivity Testing of Solid-Phase Hazardous Materials.* Annual book of ASTM Standards, **Vol. 14.02,** pp.405–411, ASTM, Philadelphia, USA

American Society for Testing and Materials (1988a) ASTM Standard E 789–86 for Pressure and Rate of Pressure Rise for Dust Explosions in a 1.2 Litre Closed Cylindrical Vessel. *Annual book of ASTM Standards,* **Vol. 14.02,** pp. 488–506, ASTM, Philadelphia, USA

American Society for Testing and Materials (1988b) ASTM Standard E 1226–88 for Pressure and Rate of Pressure Rise for Combustible Dust. *Annual book of ASTM Standards,* **Vol. 14.02,** pp. 688–698 ASTM, Philadelphia, USA

American Society for Testing and Materials (1989) ASTM Standard E 1232 for Testing of Temperature Limit of Flammability of Chemicals. *Annual book of ASTM Standards,* **Vol. 14.02,** ASTM, Philadelphia, USA

BAM (1974) Apparatus for Determination of the Minimum Ignition Temperature of Smouldering Gases. Drawing produced by BAM, Berlin

Bartknecht, W. (1978) *Explosionen – Ablauf und Schutzmassnahmen.* Springer-Verlag

Bartknecht, W. (1987) *Staubexplosionen – Ablauf und Schutzmassnahmen.* Springer-Verlag

Beck, H., and Glienke, N. (1985) Verfahren zur Bestimmung brann- und explosionstechnischer Kenngrössen von Stäuben. *Staub-Reinhalt. Luft* **45** pp. 532–535

Berthold, W. (1987) *Bestimmung der Mindestzündenergie von Staub/Luft-Gemischen.* VDI-Fortschrittberichte. Reihe 3: Verfahrenstechnik Nr. 134. VDI-Verlag GmbH, Düsseldorf

Bowes, P. C., and Townshend, S. E. (1962) Ignition of Combustible Dusts on Hot Surfaces. *Brit. Journ. of Applied Physics* **13** p. 105

Bowes, P. C. (1984) *Self-Heating: Evaluating and Controlling the Hazards.* Elsevier Publ. Co., Amsterdam

Boyle, A. R., and Llewellyn, F. J. (1950) The Electrostatic Ignitability of Dusts and Powders. *Trans. Chem. Ind. (London)* **69** pp. 173–181

Burgoyne, J. H. (1978) The Testing and Assessment of Materials Liable to Dust Explosion or Fire. *Chemistry and Industry*, (4. February) pp. 81–87

Burke, R. (1988) Development of a 20 Litre Sphere Dust Explosion Test Facility. Technical Report FMRC J.I. OQ2E1. RK. (May) Factory Mutual Research, Norwood, Mass., USA

Carr, R. L. (1965) Evaluating Flow Properties of Solids. *Chemical Engineering,* (January) pp. 163–168

Cashdollar, K. L., and Hertzberg, M. (1985) 20 litre Explosibility Test Chamber for Dusts and Gases. *Rev. Sci. Instruments* **56** pp. 596–602

Cashdollar, K. L., Sapko, M. J., Weiss, E. S., *et al.* (1987) *Industrial Dust Explosions*, Special Techn. Publ. 958 Laboratory and Mine Dust Explosion Research at the Bureau of Mines., ASTM, Philadelphia, USA pp. 107–123

Committee on Evaluation of Industrial Hazards (1979) Test Equipment for Use in Determining Classifications of Combustible Dusts, Report NMAB 353–2, National Materials Advisory Board, Washington DC, USA

Conti, R. S., Cashdollar, K. L., Hertzberg, M., *et al.* (1983) Thermal and Electrical Ignitability of Dust Clouds. Rep. Inv. 8798. US Bureau of Mines, US Dept. Interior, Washington

Cuckler, L. E. (1987) Moisture Content Measurement. In *McGraw-Hill Encyclopedia of Science and Technology,* 6th Edition, **Vol. 11,** McGraw-Hill Book Co., pp. 303–306

Cybulski, W. (1975) *Coal Dust Explosions and their Suppression.* (English translation published by Foreign Scientific Publ.) Dept. of National Center for Scientific, Technical and Economical Information, Warsaw, Poland

Dorsett, H. G., Jacobson, M., Nagy, J., *et al.* (1960) Laboratory Equipment and Test Procedures for Evaluating Explosibility of Dusts. Rep. Inv. 5424, US Bureau of Mines

Dorsett, H. G., and Nagy, J. (1968) Dust Explosibility of Chemicals, Drugs, Dyes and Pesticides Rep. Inv. 7132, US Bureau of Mines, Washington

Eckhoff, R. K. (1975) Arbeider på feltet industrielle støveksplosjoner ved Chr. Michelsens Institutt 1970–1974. Et sammendrag. Report No. 72001–8/RKE, (August), Chr. Michelsen Institute, Bergen, Norway.

Eckhoff, R. K. (1975a) Towards Absolute Minimum Ignition Energies of Dust Clouds? *Combustion and Flame* **24** pp. 53–64.

Eckhoff, R. K., and Fuhre, K. (1975) Investigations Related to the Explosibility of Clouds of Agricultural Dusts in Air. Part 3. Report No. 72001, (May) Chr. Michelsen Institute, Bergen, Norway.

Eckhoff, R. K. (1976) *A Study of Selected Problems Related to the Assessment of Ignitability and Explosibility of Dust Clouds.* Chr. Michelsen Institute Beretninger XXXVIII-2. A/S John Grieg, Bergen, Norway.

Eckhoff, R. K. (1977) Some Notes Made During a Visit to the Research Institute of Material Science Problems, Kiev, USSR, (September 1977). Report No. 77002–2, (October), Chr. Michelsen Institute, Bergen, Norway

Eckhoff, R. K., and Mathisen, K. P. (1977/1978) A Critical Examination of the Effect of Dust Moisture on the Rate of Pressure Rise in Hartmann Bomb Tests. *Fire Research* **1** pp. 273–280.

Eckhoff, R. K. (1987) Measurement of Explosion Violence of Dust Clouds. In *Proceedings of Int. Symp. Explosion Hazard Classif. Vapors, Gases and Dusts.* Publ. NMAB-447, pp. 181–200. National Academy Press, Washington DC

Efimočkin, G. I., Kopickij, G. K., Fedotov, F. K., *et al.* (1984) Solid and Energetic Fuels: A Method for Determining Characteristics of Dust Explosibility. An Industrial Standard, OCT 34–70. Moscow

Enright, R. J. (1984) Experimental Evaluation of the 1.2, 8 and 20-litre Explosion Chambers. In Proceedings of First International Colloquium on Explosibility of Industrial Dusts, Baranow, Poland, November 8–10 p. 52

Field, P. (1983) Explosibility Assessment of Industrial Powders and Dusts. Department of the Environment, Building Research Establishment, Her Majesty's Stationery Office

Franke, H. (1978) Bestimmung der Mindestzündenergie von Kohlenstaub/Methan/Luft-Gemischen (Hybride Gemische). *VDI-Berichte* **304,** pp. 69–72. VDI-Verlag GmbH, Düsseldorf

Gibson, N. (1972) Characteristics of Powders Processed in the Chemical Industry. In *Proc. of Internat. ISSA Symp. Dust Expl. Risks in Mines and Industry,* Karlovy Vary, pp. 158–180

Gibson, N., and Harper, D. J. (1981) Friction and Localized Heat Initiation of Powders Capable of Fast Decomposition – An Exploratory Study. In Symp. Series No. 68, *Runaway Reactions, Unstable Products and Combustible Powders,* pp. 3/R:1–3/R:14. The Inst. Chem. Engrs., UK

Giltaire, M., and Dangréaux, J. (1978) Les poussieres explosibles. *Annales des Mines,* Jan./Febr. pp. 85–96

Grewer, T. (1971) Zur Selbstentzündung von abgelagertem Staub. *Staub-Reinhalt. Luft* **31** pp. 97–101

Grewer, T., Klusacek, H., Löffler, U., *et al.* (1989) Determination and Assessment of the Characteristic Values for the Evaluation of the Thermal Safety of Chemical Processes. *J. Loss Prev. Process Ind.* **2** pp. 215–223

Griesche, G., and Brandt, D. (1976) Einflussfaktoren auf die Zündtemperatur von Staub-Luft-Gemischen. *Die Technik* **31** pp. 504–507

Hattwig, M. (1987) Explosion Hazard Classification, Explosion Safety Parameters and Relevant Measurement Techniques Applied in the F. R. Germany. Part II. In *Proceedings of Int. Symp. Explosion Hazard Classif. Vapors, Gases and Dusts.* Publ. NMAB-447, pp. 159–178, National Academy Press, Washington DC

Heinrich, H. J. (1972) Methodische Probleme bei der Bestimmung sicherheits-technischer Kennzahlen brennbarer Stäube. In *Proc. of Internat. ISSA Symp. Dust Expl. Risks in Mines and Industry,* Karlovy Vary, pp. 51–83

Hensel, W. (1984) Methoden zur Bestimmung der Zündtemperatur von Staub/Luft-Gemischen an heissen Oberflächen. Eine vergleichende Untersuchung. BAM Jahresbericht pp. 86–88. BAM, Berlin

Hertzberg, M., Cashdollar, K. L. and Opferman, J. J. (1979) The Flammability of Coal Dust-Air Mixtures. Lean Limits, Flame Temperatures, Ignition Energies and Particle Size Effects. Rep. Inv. 8360. US Bureau of Mines, Washington

Hertzberg, M., Conti, R. S., and Cashdollar, K. L. (1985) Electrical Ignition Energies and Thermal Autoignition Temperatures for Evaluating Explosion Hazards of Dusts. Rep. Inv. 8988, US Bureau of Mines, Washington

International Electrotechnical Commission (1982) Electrical Apparatus for Use in the Presence of Ignitable Dust. Part 2: Test Methods. Sheet 2–1: Method for Determining the Minimum Ignition Temperature of Dust Layers on a Heated Surface at Constant Temperature. (Draft). Central Office of IEC, Geneva

International Electrotechnical Commission (1984) Electrical Apparatus for Use in the Presence of Ignitable Dust. Part 2: Test Methods. Sheet 2–2: Method for Determining the Minimum Ignition Temperature of Dust Clouds in a Furnace at a Constant Temperature. (Draft August). Central Office of IEC, Geneva

International Electrotechnical Commission (1988) Electrical Apparatus for Use in the Presence of Ignitable Dust. Part 2: Test Methods. Sheet 2–3: Method for Determining Electrical Resistivity of Dust in Layers. (Draft February), Central Office of IEC, Geneva

International Electrotechnical Commission (1989) Electrical Apparatus for Use in the Presence of Ignitable Dust. Part 2: Test Methods. Sheet 2–4: Method for Determining the Minimum Ignition Energy of Dust Clouds. (Draft), Central Office of IEC, Geneva.

International Electrotechnical Commission (1990) Electrical Apparatus for Use in the Presence of Ignitable Dust. Part 2: Test Methods. Sheet 2–5: Method for Determining the Minimum Explosible Concentration of Dust/Air Mixtures. (Draft), Central Office of IEC, Geneva

International Electrotechnical Commission (1990a) Electrical Apparatus for Use in the Presence of Ignitable Dust. Part 2: Test Methods. Sheet 2–6: Hazard Classification of Combustible Dusts. (Draft), Central Office of IEC, Geneva

International Standardization Organization (1985) Explosion Protection Systems. Part 1: Determination of Explosion Indices of Combustible Dusts in Air. International Standard ISO/DIS 6184/1, Geneva

International Standardization Organization (1985a) Explosion Protection Systems. Part 4: Determination of Efficacy of Explosion Suppression Systems. International Standard ISO/DIS 6184/4, Geneva

International Standardization Organization (1985b) Explosion Protection Systems. Part 3: Determination of Explosion Indices of Fuel/Air Mixtures other than Dust/Air and Gas/Air Mixtures. International Standard ISO/DIS 6184/3, Geneva

Jacobson, M., Nagy, J., Cooper, A. R., *et al.* (1961) Explosibility of Agricultural Dusts. Rep. Inv. 5753, US Bureau of Mines, Washington

Jacobson, M., Nagy, J., and Cooper, A. R. (1962) Explosibility of Dusts Used in the Plastic Industry. Rep. Inv. 5971, US Bureau of Mines, Washington

Jacobson, M., Cooper, A. R., and Nagy, J. (1964) Explosibility of Metal Powders. Rep. Inv. 6516, US Bureau of Mines, Washington

Koenen, H., Ide, K. H., and Swart, K.-H. (1961) Sicherheitstechnische Kenndaten explosionsfähiger Stoffe. *Explosivstoffe* **49** Nr. 1–3

Kohlschmidt, J. (1972) Bestimmung von Staubzündkennwerten im lagernden und schwebenden Zustand zur Einschätzung der Gefährlichkeits-eigenschaften. *Proc. of Internat. ISSA Symp. Dust Expl. Risks in Mines and Industry,* Karlovy Vary

Koril'chenko, A,Ya., and Baratov, A. N. (1980) Evaluating Fire and Explosion Danger of Industrial Dusts. (Translated by Plenum Publishing Corp.) *Fizika Goreniya i Vzryva* **15** (1979) pp. 146–147.

Lee, R. S., Aldis, D. F., Garrett, D.W., *et al.* (1982) Improved Diagnostics for Determination of Minimum Explosive Concentration, Ignition Energy and Ignition Temperature of Dusts. *Powder Technology* **31** pp. 51–62

Lee, R. S., Aldis, D., Lai, F. S., *et al.* (1983) A High-Energy Chemical Igniter for Dust Cloud Ignition. *Particulate Systems: Technology and Fundamentals.* (Ed. by J. K. Beddow), Hemisphere Publ., Washington DC pp. 267–280

Leuschke, G., and Zehr, J. (1962) Zündung von Staublagerungen und Staub/Luft-Gemischen durch mechanisch erzeugte Funken. *Arbeitsschutz* **6** p. 146

Leuschke, G. (1966) Über die Untersuchung brennbarer Stäube auf Brand- und Explosionsgefahren. *Staub-Reinhalt. Luft* **26** pp. 49–57

Leuschke, G. (1966a) Verfahren zur Ermittlung der Zündtemperaturen aufgwirbelter brennbarer Stäube. BAM Mitteilung, Materialprüfung, BAM, Berlin

Leuschke, G. (1967) Grundlagen und Auswirkungen von Staubbränden und Explosionen. Die Berufsgenossenschaft No. 10, (October) pp. 2–8

Leuschke, G. (1975) Private Communication to R. K. Eckhoff from G. Leuschke, BAM, Berlin

Leuschke, G. (1979) Über die Klassifizierung brennbarer Stäube. *Staub-Reinhalt. Luft* **39** pp. 326–332

Lütolf, J. (1971) Untersuchungen von Stäuben auf Brand- und Explosionsgefahr. *Staub-Reinhalt. Luft* **31** pp. 93–97

Lütolf, J. (1978) Kurzmethoden zur Prüfung brennbarer Stäube. *VDI-Berichte* **304,** pp. 39–46. VDI-Verlag GmbH, Düsseldorf

Moore, P. E. (1979) Characterization of Dust Explosibility: Comparative Study of Test Methods,

Chemistry and Industry, **7.** (July) pp. 430–434
Nagy, J., Dorsett, H. G., and Cooper, A. R. (1965) Explosibility of Carbonaceous Dusts. Rep. Inv. 6597, US Bureau of Mines, Washington
Nagy, J., Seiler, E. C., Conn, J. W., *et al.* (1971) Explosion Development in Closed Vessels. Rep. Inv. 7507, US Bureau of Mines, Washington
National Material Advisory Board (1988) *The Explosion Hazard Classification of Gases and Dusts Relative to Use of Electrical Equipment.* Publication NMAB-448, National Academy Press (USA)
Nedin, V. V., Nejkov, O. D., and Alekseev, A. G. (1971) Explosibility Characteristics of Dusts and Methods for their Determination. In *Prevention of Accidental Dust Explosions,* (Ed. by V. V. Nedin) Kiev
Nordtest (1982) Dust and Powder Layers: Electric Spark Sensitivity, Nordtest Method NT Fire 016 Nordtest, Espoo, Finland
Nordtest (1989) Dust Clouds: Minimum Explosible Dust Concentration, Nordtest Method NT Fire 011, Edition 2, Nordtest, Espoo, Finland
Palmer, K. N.(1973) *Dust Explosions and Fires.* Chapman and Hall, London
Pedersen, G. H., and Eckhoff, R. K. (1987) Initiation of Grain Dust Explosions by Heat Generated During Single Impact Between Solid Bodies. *Fire Safety Journal* **12** pp. 153–164
Pedersen, G. H. (1989) Dust Explosion Test Methods at the CMI Dust Explosion Laboratory. (April) Chr. Michelsen Institute, Bergen, Norway
Racke, D. (1989) Safety of Reactive Chemicals. Hazard Evaluation and Loss Prevention. *Diploma Thesis,* (November) University of Dortmund, F. R. Germany
Raftery, M. M. (1968) *Explosibility Tests for Industrial Dusts.* Fire Research Technical Paper No. 21, Her Majesty's Stationery Office, London
Ritter, K., and Berthold, W. (1979) Bedeutung sicherheitstechnischer Kenndaten für die Auswahl von Sicherheitsmassnahmen gegen Gas-, Staub- und Wärme-Explosionen. *Chem.-Ing. Tech.* **51** pp. 174–183
Ritter, K. (1984) Die Zündwirksamkeit mechanisch erzeugter Funken gegenüber Gas/Luft- und Staub/Luft-Gemischen. Dr.-Ing. Dissertation, Universität Fridericiana, Karlsruhe
Rogerson, J. (1989) Standard for Minimum Autoignition Temperature of Dust Clouds, (2nd Draft) Committee E-27, ASTM, Philadelphia, USA
Schwab, R. F. (1968) Interpretation of Dust Explosion Test Data. *AICh. E, Loss Prevention* **2** pp. 37–43
Selle, H. (1957) Die Grundzüge der Experimentalverfahren zur Beurteilung brennbarer Industriestäube. *VDI-Berichte* **19** pp. 37–48
Siwek, R., and Pellmont, G (1986) Safety Technical Indices: Methods of Determination and Factors Influencing Hazard Evaluation in Dust Handling Equipment. *Proc. of EUROMECH Colloquium 208*: Explosions in Industry, Göttingen, F. R. Germany, (April 14–16)
Siwek, R. (1988) Reliable Determination of Safety Characteristics in 20 Litre Apparatus. *Proc. of Conference on Flammable Dust Explosions,* (November 2–4) St. Louis, Miss., USA,
Snee, T. J. (1987) Incident Investigation and Hazard Evaluation Using Differential Scanning Calorimetry and Accelerating Rate Calorimetry. *J. Occupational Accidents* **8** pp. 261–271
Townsend, D. I., and Tou, J. C. (1980) Thermal Hazard Evaluation by an Accelerating Rate Calorimeter. *Thermochimica Acta* **37** pp. 1–30.
Ural, E. A. (1989) Dispersibility of Dusts Pertaining to their Explosion Hazard. Factory Mutual Research Report J.I. OQ2E3. RK, (April) Norwood, Mass., USA
Ural, E. A. (1989a) Experimental Measurement of the Aerodynamic Entrainability of Dust Deposits. In *Proc. of 12th Int. Coll. Dyn. Expl. React. Syst.* (July 24–28) Ann Arbor, Michigan, USA
Verein deutscher Ingenieure (1988) Untersuchungsmethoden zur Ermittlung von sicherheitstechnischen Kenngrössen von Stäuben. VDI Richtlinie 2263, (Draft) VDI-Verlag GmbH, Düsseldorf

Zeeuwen, J. P. (1982) Review of Current Research at TNO into Gas and Dust Explosions. In *Proceedings of Internat. Conf. on 'Fuel-Air-Explosions' in Montreal,* Canada (November 4–6) (1981) pp. 687–702. University of Waterloo Press, Canada

Zeeuwen, J. P., and Laar, G. F. M. von (1984) A Practical Look at Dust Explosion Characteristics. Explosion Protection in Practice. *Proceedings (Part 2) of 1st Internat. EuropEx Symp.* (16–19 April), EuropEx, Antwerp, Belgium

Zhu Hailin, Liu Xiangjun, and Li Hongquan (1988) Influence of Type of Chemical Igniter on Violence of Maize Starch and Aluminium Dust Explosions in a Closed 20 Litre Sphere. *Fire Safety Journal* **13** pp. 181–183

Zuzuki, T., Takaoka, S., and Fujii, S. (1965) The Ignition of Coal Dust by Rubbing, Frictional Heat and Sparks. *Proc. of Restricted International Conference of Directors of Safety in Mines Research*, (July) held at Safety in Mines Research Establishment, Sheffield, UK

APPENDIX

Ignitability and explosibility data for dusts from laboratory tests

A1. TABLES A1, A2 AND A3, AND COMMENTS, FROM BIA (1987)

A.1.1. LIMITATIONS TO THE APPLICABILITY OF THE DATA

A.1.1.1 Particle size and moisture content

The applicability of the data in Tables A1, A2 and A3 to other dusts of apparently identical materials is limited. In practice dusts of a given overall chemistry may differ widely in particle size, particle shape and sometimes also in particle surface reactivity. Furthermore, most ignitability and explosibility parameters are influenced by inherent features of the test method. Therefore, as a general rule, the tabulated data should only be used as indications, and not as the ultimate basis for design of actual safety measures in industry. On the other hand, data obtained using the same test method allows relative comparison of ignitability and explosibility of various dusts. It is always necessary, however, to account for any significant differences between the particle size distributions and particle shape of the actual dust of interest and those in Tables A1, A2 and A3.

For a given dust material, the maximum explosion pressure (P_{max}), and the maximum rate of pressure rise (K_{St}) increase systematically with decreasing particle size and moisture content. The minimum ignition energy (MIE) generally decreases with decreasing particle size and moisture content. Decreasing the moisture content and particle size can also give a decrease of both the minimum explosible dust concentration (C_{min}) and the minimum ignition temperature of a dust cloud (T_{min}). The dusts were tested 'as received', and general lack of information of the moisture content presents a further uncertainty concerning the specific applicability of the data. This in particular applies to the data for wood and cellulose, and food and feed stuffs. Such dusts often contain considerable fractions of moisture in the 'as received' state.

It is generally advisable to have the actual dust of interest tested in a professional laboratory.

A.1.1.2
Initial state and Composition of the Gas in which the Dust is Dispersed or Deposited

The data in Table A1 apply to

- Atmospheric pressure (from −0.2 to +0.2 bar(g))
- Oxygen content of air (from 18 to 22 Vol% 0_2)
- Normal ambient temperature (from 0 to 40°C)

In general P_{max}, and under certain conditions also $(dP/dt)_{max}$ or K_{St}, increase proportionally with the absolute initial pressure. Increased oxygen fraction in the atmosphere increases both the ignitability and the explosibility, whereas a lower oxygen content than in air reduces the hazard correspondingly. Increased initial temperature increases the ignition sensitivity (reduces MIE). Normally, data for conditions that deviate significantly from the standard test conditions, will have to be determined specifically in each particular case.

If the gas phase contains some combustible gas or vapour, even in concentrations considerably below the lower explosibility limit for the gas or vapour, hybrid effects can give rise to considerable increase of both ignition sensitivity and explosibility. In such cases, specific tests will definitely have to be conducted.

A.1.2.

COMMENTS TO THE VARIOUS ITEMS IN TABLE A1

A1.2.1
Selection and Identification of Dusts

The original table published in German by BIA (1987) contains nearly 1900 dusts. Therefore the selection of about 375 dusts in Table A1 constitutes about 20% of those in the original tables. When performing the selection, the samples of a given dust material that gave the most severe test data, were normally preferred. In addition, sequences for some given dust materials showing systematic effects of e.g. moisture content or particle size were included. Examples of this are data for peat and aluminium.

In the original German table the dusts are identified by a code number, which has been omitted in the present, condensed table. However, the sequence of the dusts in the condensed table is identical with that in the original table. If required, the dusts in the condensed table can be easily identified in the original German table by means of the particle size data and the ignitability and explosibility data.

A.1.2.2
Particle Size Distribution

Most of the dusts were tested as received. However, in some cases fractions passing a 63 μm sieve were tested.

A.1.2.3
Minimum Explosible Dust Concentration (C_{min})

Most of the tabulated data were determined in the standard closed 1 m^3 ISO vessel (1985) or in the closed 20 litre Siwek sphere. Experience has shown that the latter apparatus tends to give lower values than the 1 m^3 vessel, often by a factor of two. (Note: Another standard small-scale method approved by Nordtest (1989) seems to give data in somewhat closer agreement with those from the 1 m^3 ISO vessel.) The C_{min} values in brackets were determined in the modified 1.2 litre Hartmann apparatus in terms of the smallest dispersed dust quantity that gave flame propagation, divided by the vessel volume. These values are sometimes higher than the true C_{min}, because of the comparatively weak ignition source used.

A.1.2.4
Maximum Explosion Pressure (P_{max})

The maximum explosion pressures were obtained either in the standard 1 m^3 ISO vessel or in the 20 litre Siwek sphere. The data in brackets were obtained in the 20 litre sphere using a simplified test procedure due to limited amounts of dust for testing. The standard procedure requires at least three replicate tests at each dust concentration over a range of different concentrations.

A.1.2.5
Explosion Violence (K_{St}, St class)

K_{St} is defined as the maximum rate of pressure rise during a dust explosion in an equi-dimensional vessel, times the cube root of the vessel volume. K_{St} [bar m/s] is numerically equal to the maximum rate of pressure rise [bar/s] in the 1 m^3 standard ISO test (1985). The K_{St} data in the table were obtained either in the standard ISO test or in the 20 litre Siwek sphere, adopted by ASTM (1988), which has been calibrated to yield comparable K_{St} values.

The St class was determined using the modified Hartmann tube, with a hinged lid at the top. Brackets are used to indicate that this test method is not considered adequate in F. R. Germany for conclusive classification of St2 and St3 dusts. (St2 means that 200 bar m/s $\leqslant K_{St} <$ 300 bar m/s, and St3 that $K_{St} \geqslant$ 300 bar m/s.)

A.1.2.6
Minimum Ignition Temperature of Dust Clouds

These data were acquired using either the Godbert-Greenwald furnace or the BAM furnace. The data in brackets were obtained using a modified, elongated version of the Godbert-Greenwald furnace, yielding somewhat lower values than the version proposed as an IEC standard (International Electrotechnical Commission).

A.1.2.7
Minimum Ignition Energy (MIE)

In the original BIA (1987) publication, the MIE values appear in a separate table. However, because the dusts could be identified by their reference numbers, it was possible to incorporate the MIE values in Table A1. These values are determined using soft sparks (long discharge times) in agreement with the VDI method described by Berthold (1987). Down to net spark energies of about 1 mJ this method is in complete accordance with the CMI method described by Eckhoff (1976). The VDI and the CMI methods are the basis of the method for measuring MIE that is being evaluated by the IEC. The VDI and CMI methods differ from the earlier US Bureau of Mines method, in which an appreciable fraction of the $1/2\ CV^2$ quoted as MIE was lost in a transformer and never got to the spark. Therefore, the USBM MIE values are generally higher than those determined by the new method. A tentative correlation for transforming USBM data to equivalent VDI/CMI data is given in Figure A1.

A.1.2.8
Glow Temperature

These data were obtained with a 5 mm thick layer of dust resting on a hot-plate of known, controllable temperature (equivalent to proposed standard IEC method for determining the minimum ignition temperature of a dust layer on a hot surface).

A.1.2.9
Flammability

The dusts are classified according to their ability to propagate a combustion wave when deposited in a layer. Ignition is accomplished using either a gas flame or a glowing platinum wire at 1000°C. The test sample is a 2 cm wide and 4 cm long dust ridge resting on a ceramic plate. Ignition is performed at one end. The definitions are:

- Class 1: No self-sustained combustion.
- Class 2: Local combustion of short duration.
- Class 3: Local sustained combustion, but no propagation.
- Class 4: Propagating smouldering combustion.
- Class 5: Propagating open flame.
- Class 6: Explosive combustion.

The numbers in brackets refer to a modified test procedure according to which 20 weight% diatomaceous earth is mixed with the powder or dust to be tested. By this means some materials that would otherwise not propagate a flame because they melt, may show sustained flame propagation.

Table A1

COTTON / WOOD / PEAT	PARTICLE SIZE DISTRIBUTION								IGNITABILITY AND EXPLOSIBILITY OF DUST CLOUDS							DUST LAYERS	
									$1\ m^3$ or 20 l vessel			Mod. H.	G.G.	BAM	VDI	DIN	
	WEIGHT % < SIZE [µm]							MEDIAN	C_{min}	P_{max}	K_{St}	EXPLOS. < 63 µm	T_{min}		MIE	Glow temp.	Flamm. <250µm
DUST TYPE	500	250	125	71	63	32	20	µm	g/m³	bar(g)	bar•m/s	Class	°C	°C	mJ	°C	Class
Cotton			98	72		38	25	44	(100)	7.2	24	St. 1	560			350	3
Cellulose			92	71		20	3	51	60	9.3	66		500		250	380	5
Wood dust				90		47	7	33					500		100	320	
Wood dust	58		57	55		43	39	80					480		7	310	
Wood dust (chip board)				70		30		43	60	9.2	102	St. 2	490			320	3
Wood/cardboard/jute									30	5.8	26		610		245	360	5
Wood/card board/jute/resin									30	8.4	67		520		3	350	5
Lignin dust			96	85		66	57	18	15	8.7	208		470			> 450	5
Paper dust				91		83	73	< 10		5.7	18		580			360	
Paper tissue dust			75	58				54	30	8.6	52	(St. 2)	540			300	4
Paper (phenolresin treated)				100		90	25	23	30	9.8	190		490			310	
Peat (15% moisture)			84	58		26	3	58	60	10.9	157	St. 1	480			320	4
Peat (22% moisture)			82	65		40	15	46	125	8.4	69	St. 1	470			320	4
Peat (31% moisture)			87	76		43	20	38	125	8.1	64		500			320	
Peat (41% moisture)			88	76		40	18	39	No ignition				500			315	
Peat (from bottom of sieve)			78	48		22		74	125	8.3	51	St. 1	490			310	4
Peat (dust deposit)				66		33	11	49	60	9.5	144		(360)			295	
Paper pulp				93		76		29		9.8	168						

Table A1, continued

FOOD / FEED	PARTICLE SIZE DISTRIBUTION								IGNITABILITY AND EXPLOSIBILITY OF DUST CLOUDS							DUST LAYERS	
	WEIGHT % < SIZE [µm]							MEDIAN	1 m^3 or 20 l vessel			Mod. H.	G.G.	BAM	VDI	DIN	
									C_{min}	P_{max}	K_{St}	EXPLOS. < 63 µm	T_{min}		MIE	Glow temp.	Flamm. <250µm
DUST TYPE	500	250	125	71	63	32	20	µm	g/m^3	bar(g)	bar•m/s	Class	°C	°C	mJ	°C	Class
Gravy powder (21% starch)					100					5.1	12			500	>1000		
Citrus pellets					100				60	7.7	39			460	250		
Dextrose, ground			100		94	71		22				St. 2					2
Dextrose				38		5	4	80	60	4.3	18	St. 1	500			570	3
Fat/whey mixture	76		11	3				330		7.0	23		450		180	410	5
Fat powder (48% fat)		100	75		24	7		92	30	6.4	20	St. 2					2
Do.					100									430	> 100		
Fish meal	68		23		12			320	125	7.0	35		530				
Fructose (from filter)	99		39	17				150	60	9.0	102		430		< 1	melts	
Fructose	92		15					200	60	7.0	28		440		180	440q	
Fructose	81							400	125	6.4	27		530		>4000	melts	
Barley grain dust	79	51	25		8	3		240							100		4
Do.					100				125	7.7	83			400			
Oats grain dust	64		24		8			295	750	6.0	14	St. 1				350	3
Wheat grain dust				48		30		80	60	9.3	112	St. 1				290	3
Wheat grain dust	100	81	50		32	25		125				St. 2					3
Coffee (from filter)				100		99	89	< 10	60	9.0	90		470			> 450	
Coffee (refined)					100					6.8	11			460	> 500		4

Table A1, continued

FOOD FEED	PARTICLE SIZE DISTRIBUTION								IGNITABILITY AND EXPLOSIBILITY OF DUST CLOUDS							DUST LAYERS	
									1 m^3 or 20 l vessel			Mod. H.	G.G.	BAM	VDI	DIN	
	WEIGHT % < SIZE [µm]							MEDIAN	C_{min}	P_{max}	K_{St}	EXPLOS. < 63 µm	T_{min}		MIE	Glow temp.	Flamm. <250µm
DUST TYPE	500	250	125	71	63	32	20	µm	g/m^3	bar(g)	bar•m/s	Class	°C	°C	mJ	°C	Class
Cocoa bean shell dust					100				125	8.1	68				> 250		4
Cocoa/sugar mixture	53		20					500	125	7.4	43	St. 1	580			460	2
Potatoe granulate					100					6.4	21			440	> 250		3
Potatoe flour			86	53		26	17	65	125	9.1	69		480			> 450	
Lactose (from filter)				83		60	47	22	125	6.9	29		450		80	> 450	
Lactose (from cyclone)				97		70	41	23	60	7.7	81	St. 2	520			> 450	3
Maize seed waste (9% moisture)	98	67	40		23	16		165	30	8.7	117			440	> 10		
Milk powder			34	18				165	60	8.1	90		460		75	330	
Milk powder	98		15	8				235	60	8.2	75		450		80	320	
Milk powder (low fat, spray dried)	100	100	99		60	17		46	30	7.5	109				> 100		
Milk powder (full fat, spray dried)				30				88	60	8.6	83	St. 1	520			330	2
Whey fat emulgator	62		7	2				400		7.2	38		450		90	420	5
Olive pellets					100				125	10.4	74			470	>1000		
Rice flour					100				60	7.4	57			360	> 100		
Rye flour			94	76		58	15	29		8.9	79		490			> 450	
Soy bean flour				85		63	50	20	(200)	9.2	110	St. 1	620			280	2
Potatoe starch					100				30	7.8	43			420	>1000		
Potatoe starch				100		50	17	32		(9.4)	(89)		520		>3200	> 450	2

Table A1, continued

FOOD FEED	PARTICLE SIZE DISTRIBUTION								IGNITABILITY AND EXPLOSIBILITY OF DUST CLOUDS							DUST LAYERS	
									1 m^3 or 20 l vessel			Mod. H.	G.G.	BAM	VDI	DIN	
	WEIGHT % < SIZE [µm]							MEDIAN	C_{min}	P_{max}	K_{St}	EXPLOS. < 63 µm	T_{min}		MIE	Glow temp.	Flamm. <250µm
DUST TYPE	500	250	125	71	63	32	20	µm	g/m^3	bar(g)	bar•m/s	Class	°C	°C	mJ	°C	Class
Maize starch				99		98	94	< 10		10.2	128		520		300	> 450	2
Maize starch				94		81	60	16	60	9.7	158	St. 1	520			440	2
Rice starch (hydrolyzed)				29		15		120	60	9.3	190	(St. 2)	480			555	5
Rice starch				99		74	54	18					470		90	390	3
Rice starch				86		62	52	18		10.0	190	(St. 2)	530			420	3
Wheat starch						84	50	20	60	9.8	132	(St. 2)	500			535	3
Tobacco			81	64		29		49		4.8	12		470			280	
Tapioca pellets				61		42		44	125	9.0	53	St. 1	(450)			290	4
Tea (6% moisture)					100				30	8.1	68			510			≥3
Tea (black, from dust collector)			64	48		26	16	76	125	8.2	59	St. 1	510			300	4
Meat flour			69	52		31	21	62	60	8.5	106	St. 1	540			> 450	2
Wheat flour								50					500		540	> 450	
Wheat flour			97	60		32	25	57	60	8.3	87		430			> 450	
Do.					100									400	> 100		
Wheat flour 550				60		34	25	56	60	7.4	42		470		400	> 450	
Milk sugar				99		92	77	10	60	8.3	75		440		14	melts	5
Milk sugar				98		64	32	27	60	8.3	82	St. 1	490			460	2
Sugar (Icing-)				88		70	52	19					470			> 450	

Table A1, continued

COAL / COAL PRODUCTS	PARTICLE SIZE DISTRIBUTION								IGNITABILITY AND EXPLOSIBILITY OF DUST CLOUDS							DUST LAYERS	
									$1\ m^3$ or 20 l vessel			Mod. H.	G.G.	BAM	VDI	DIN	
	WEIGHT % < SIZE [μm]							MEDIAN	C_{min}	P_{max}	K_{St}	EXPLOS. < 63 μm	T_{min}		MIE	Glow temp.	Flamm. <250μm
DUST TYPE	500	250	125	71	63	32	20	μm	g/m^3	bar(g)	bar•m/s	Class	°C	°C	mJ	°C	Class
Activated carbon				99		80	55	18	60	8.8	44		790			> 450	
Activated carbon				88		64		22		No ignition			670			335	
Activated carbon (16% moisture)			84	65		38		46	125	8.4	67		(630)				
Brown coal			83	69		40	20	41		9.1	123		420		160	230	4
Brown coal (from electrostatic filter			75	60		27		55	60	9.0	143	St. 1	450			240	4
Brown coal (dust from grinding)			71	56		38	30	60		8.9	107		420		230	230	3
Brown coal/anthracite (80:20)				66		43	24	40	60	8.6	108		440		>4000	230	
Brown coal/anthracite (20:80)				91		85	80	< 10		0.4	1		590			280	
Brown coal coke	93		18	13				290	250	8.4	115	St. 1	560			> 450	3
Brown coal (graphitized)				82		55	35	28		No ignition			> 850			> 450	
Char coal				99		88	67	14	60	9.0	10	St. 1	520			320	4
Char coal				95		85	58	19	60	8.5	117		540			270	
Char coal	36							>500		No ignition			> 850			> 450	
Asphalt				83		54	32	29	15	8.4	117		550			melts	
Bituminous coal				97		93	85	< 10		9.0	55		590			270	
Bituminous coal (Petchora)			76	65		46	37	38	125	8.6	86		610			360	
Anthracite (dust from filter)				99		97	85	< 10		No ignition			> 850			360	
Bituminous coal (high volat.)							99	4	60	9.1	59		510			260	

Table A1, continued

OTHER NATURAL ORGANIC PRODUCTS	PARTICLE SIZE DISTRIBUTION								IGNITABILITY AND EXPLOSIBILITY OF DUST CLOUDS							DUST LAYERS	
									1 m^3 or 20 l vessel			Mod. H.	G.G.	BAM	VDI	DIN	
	WEIGHT % < SIZE [µm]							MEDI-AN	C_{min}	P_{max}	K_{St}	EXPLOS. < 63 µm	T_{min}		MIE	Glow temp.	Flamm. <250µm
DUST TYPE	500	250	125	71	63	32	20	µm	g/m^3	bar(g)	bar•m/s	Class	°C	°C	mJ	°C	Class
Cotton seed expellers	66		24	10				245	125	7.7	35	St. 1	(480)			350	3
Dextrin				57		26	5	55		8.8	109	St. 1	490			> 450	2
Wheat gluten (after mill)				78		28	13	48	30	8.7	105		540			melts	
Blood flour			93	61		27	5	57	60	9.4	85		610			> 450	1
Hops, malted	52		14	9				490		8.2	90		420			270	
Leather dust (from collector)									(100)			(St. 2)					5
Linen (containing oil)	63		21					300		6.0	17		(440)			230	
Lycopodium					100	91								410		280	
Oil shale dust				99		79	50	20	125	5.2	35		520			290	2
Oil shale dust				71		50	39	32		No ignition			610			> 450	
Grass dust	96		26					200	125	8.0	47		470			310	
Walnut shell powder									(100)			St. 1					4

Table A1, continued

PLASTICS RESINS RUBBER	PARTICLE SIZE DISTRIBUTION								IGNITABILITY AND EXPLOSIBILITY OF DUST CLOUDS							DUST LAYERS	
									1 m^3 or 20 l vessel			Mod. H.	G.G.	BAM	VDI	DIN	
	WEIGHT % < SIZE [µm]							MEDIAN	C_{min}	P_{max}	K_{St}	EXPLOS. < 63 µm	T_{min}		MIE	Glow temp.	Flamm. <250µm
DUST TYPE	500	250	125	71	63	32	20	µm	g/m^3	bar(g)	bar•m/s	Class	°C	°C	mJ	°C	Class
Acrylnitrile-Butadiene-Styrene-Co-polym.	79	37	24					200	60	9.2	147	(St. 2)	480			> 450	5
Epoxy resin (for powder coating)		100	82		58	28		55	(100)			(St. 2)					2
Cellulose-2, 5-Acetate				100		89	53	19	30	9.8	180		520			> 450	
Polyester resin with glass	92	91	89		80	72		14	(100)			(St. 2)					5
Rubber				93		45		34	(100)	7.4	106	(St. 2)					5
Rubber (dust from grinding)			78	43		12		80	30	8.5	138		500		13	230	5
Resin (from filter)				97		44		40	30	8.7	108		460			melts	
Epoxy resin (60% resin + 36% TiO_2)				99		67	43	23		7.8	155						
Epoxy resin				95		60	36	26	30	7.9	129	St. 1	510			melts	2
Epoxy resin with Al				90		46		34		8.9	208		570			melts	
Melamin resin				99		84	55	18	125	10.2	110	St. 1	840			> 450	2
Melamin resin				66		24	13	57	60	10.5	172	St. 1	470			> 450	2
Phenol resin				100		99	94	< 10	15	9.3	129	(St. 2)	610			> 450	2
Phenol Formaldehyde resin	100	98	81		50	30		60	(100)			St. 1					4
Polyamid resin				95		84	64	15	30	8.9	105		450			melts	
Polymethacrylate	56				100	33			15	8.0	199						(2)
Silicon resin	91		59	39		20	13	100	60	7.2	80		480			melts	
Caoutchouc			58	40		20		95	30	9.5	192		450			230	

Table A1, continued

PLASTICS RESINS RUBBER	PARTICLE SIZE DISTRIBUTION								IGNITABILITY AND EXPLOSIBILITY OF DUST CLOUDS							DUST LAYERS	
									1 m³ or 20 l vessel			Mod. H.	G.G.	BAM	VDI	DIN	
	WEIGHT % < SIZE [μm]							MEDIAN	C_{min}	P_{max}	K_{St}	EXPLOS. < 63 μm	T_{min}		MIE	Glow temp.	Flamm. <250μm
DUST TYPE	500	250	125	71	63	32	20	μm	g/m³	bar(g)	bar·m/s	Class	°C	°C	mJ	°C	Class
Synthetical caoutchouc			66	46		18	9	80	15	8.6	145	(St. 2)	450			240	5
Methylmethacrylate-Butadiene-Styrene			45	18				135	30	8.6	120		470		11	melts	5
Methylmethacrylate-Butadiene-Styrene			34	11				150	30	8.4	114		480		30	melts	5
Polyacrylamide (from filter)				100		95	81	10	250	5.9	12	St. 1	780			410	2
Polyacrylate (from filter)			100	63		11	1	62	125	6.9	38		460		>1800	420	5
Polyacrylnitrile (32% H_2O)			95	47		16		63	60	7.4	41						
Polyamide flock (3.3 dtex 0.5 mm)				100		25	3	37	30	9.8	93	St. 1	520			melts	2(3)
Polyester								< 10		10.1	194		570			melts	
Polyethylene			91	51		10		72		7.5	67		440			melts	
Polyethylene	82		8	2				280		6.2	20		470			melts	
Polyethylene (high-pressure)			98	93		65	10	26		8.7	104		490			> 450	
Polyethylene (low-pressure)						95	86	< 10	(30)	8.0	156	(St. 2)	420			melts	2(5)
Polyethylene (low-pressure)			36	10				150	125	7.4	54	St. 1	480			melts	3(5)
Polyethylene (low-pressure)	90		20	9				245	125	7.5	46	St. 1	460			melts	3(5)
Polymethacrylate (from filter)				90		70	48	21	30	9.4	269	(St. 2)	550			melts	5
Polymethacrylimide			45	15				105	30	9.6	125	(St. 2)	530			melts	5
Polypropylene				92		61	40	25	(30)	8.4	101	(St. 2)	410			melts	3(5)
Polypropylene	100		12					162	(200)	7.7	38	St. 1	440			melts	2(5)

Table A1, continued

PLASTICS / RESINS / RUBBER	PARTICLE SIZE DISTRIBUTION								IGNITABILITY AND EXPLOSIBILITY OF DUST CLOUDS							DUST LAYERS	
									1 m^3 or 20 l vessel			Mod. H.	G.G.	BAM	VDI	DIN	
	WEIGHT % < SIZE [µm]							MEDIAN	C_{min}	P_{max}	K_{St}	EXPLOS. < 63 µm	T_{min}		MIE	Glow temp.	Flamm. <250µm
DUST TYPE	500	250	125	71	63	32	20	µm	g/m³	bar(g)	bar•m/s	Class	°C	°C	mJ	°C	Class
Polystyrene (Copolymer)			32	11				155	30	8.4	110		450			melts	
Polystyrene (Hard-foam)	30		10	5				760		8.4	23						
Polyurethane					100	90		3	< 30	(7.8)	(156)						5
Polyvinylacetate (Copolymer)						83	50	20	60	8.7	86	St. 1	660			melts	2
Polyvinylalcohol				74		55	44	26	60	8.9	128	(St. 2)	460			melts	5
Polyvinylalcohol				57		29	9	56	60	8.3	83	St. 1	460			melts	5
Polyvinylchloride						100		< 10	30	8.4	168						
Polyvinylchloride			46	15				125	30	7.7	68		530			340	
Polyvinylchloride (Em., 97.5% PVC)				97		73	26	25	125	8.2	42		750		>2000	> 450	
Polyvinylchloride (Em., 97% PVC)				60		31	14	51	125	8.5	63		790		>2000	350	
Polyvinylchloride (Susp.)			66	23				105	125	7.7	45	St. 1	510			> 450	2
Polyvinylchloride (Susp.)			30					137	No ignition				> 800			> 450	
Urea-formaldehyde (mold.-form.)				99		91	75	13	60	10.2	136	St.1	700			390	2
Melamine-formaldehyde (mold.-form.)				93		86	70	14	60	10.2	189	St. 1	800			> 440	2
El.stat. coating powder (Epoxy)				100		70		29	30	8.9	100	(St. 2)	540			melts	2(3)
El.stat. coating powder (Polyurethane)				100		66	22	29	30	7.8	89	St. 1	490			melts	2(2)
Shellac					100	33			15	7.6	144	(St. 2)					
Wax (NN Ethylene distearamide)					100	95		10	15	8.7	269	(St. 2)					2(2)

Table A1, continued

PHARMACEUTICALS COSMETICS PESTICIDES	PARTICLE SIZE DISTRIBUTION								IGNITABILITY AND EXPLOSIBILITY OF DUST CLOUDS							DUST LAYERS	
									1 m^3 or 20 l vessel			Mod. H.	G.G.	BAM	VDI	DIN	
	WEIGHT % < SIZE [μm]							MEDIAN	C_{min}	P_{max}	K_{St}	EXPLOS. < 63 μm	T_{min}		MIE	Glow temp.	Flamm. <250μm
DUST TYPE	500	250	125	71	63	32	20	μm	g/m^3	bar(g)	bar•m/s	Class	°C	°C	mJ	°C	Class
Acetyl salicylic acid					100				15	7.9	217	(St. 2)		550			2(5)
Amino phenazone						100	98	< 10		10.3	238		330			> 450	
Ascorbic acid, L(+)-				93		75	61	14	60	6.6	48	(St. 2)	490			melts	2(2)
Ascorbic acid				92		38	15	39	60	9.0	111	(St. 2)	460			melts	2(2)
Coffein					100				30	8.2	165	(St. 2)		> 550		melts	2(5)
Cysteine hydrate				100		98	94	< 10	125	7.4	40		420		>2000	melts	
L-Cystin				100		95	69	15	60	8.5	142		400		40	melts	
Digitalis leaves				59		42		46	250	8.5	73						
Dimethylaminophenazone						100		< 10		10.0	337						
2-Ethoxybenzamide					100				15	8.6	214	(St. 2)		490		melts	2(5)
Fungicide (Captan)			100		99	93		5	(500)			St. 1					5
Fungicide (Org. zinc comp.)						99	96	< 10	60	9.0	154		480			300	
Fungicide (Maneb)				98		97	93	< 10					380		>2500	200	
Methionine				100		99	95	< 10	30	9.4	143		390		9	melts	5
Methionine				100		98	87	< 10	30	8.7	128		390		100	melts	5
Sodium - L(+) ascorbate				97		67	45	23	60	8.4	119	St. 1	380			380	2
Paracetamole					100				15	7.9	156	(St. 2)		> 550		melts	2(5)
Pesticide				99		98	95	< 10	60	8.6	151		410			320	

Table A1, continued

INTERMEDIATE PRODUCTS / AUXILIARY MATERIALS	PARTICLE SIZE DISTRIBUTION								IGNITABILITY AND EXPLOSIBILITY OF DUST CLOUDS								DUST LAYERS	
									1 m^3 or 20 l vessel			Mod. H.	G.G.	BAM	VDI	DIN	
	WEIGHT % < SIZE [µm]							MEDIAN	C_{min}	P_{max}	K_{St}	EXPLOS. < 63 µm	T_{min}		MIE	Glow temp.	Flamm. <250µm
DUST TYPE	500	250	125	71	63	32	20	µm	g/m^3	bar(g)	bar•m/s	Class	°C	°C	mJ	°C	Class
Adipinic acid				98		92	86	< 10	60	8.0	97	(St. 2)	580			melts	2(5)
Aging protective					100	67		< 32	15	8.2	256	(St. 2)					2(3)
Anthracene	89		20	7				235	15	8.7	231		600			> 450	
Anthrachinone						100		< 10		10.6	364						
Anthrachinone				100		90	75	12	30	9.1	91						
Azodicarbonamide						100		< 10		12.3	176						
Benzoic acid									(30)			(St. 2)					2(5)
Betaine hydrochloride				93		85	78	< 10	60	9.8	114	(St. 2)	400			> 450	3
Betaine monohydrate	34		4					710	60	8.2	63	St. 1	510			> 450	5
Diphenol ketylene				98		80	60	15		9.0	270						
Calcium acetate			74	41		25	17	92	500	5.2	9	St. 1	730			> 460	2
Casein				99		65	40	24	30	8.5	115		560			> 450	
Sodium caseinate (from filter)				100		99	77	17	60	8.8	117		560		740	> 450	
Carboxy methyl cellulose				97		89		< 15		9.2	184						
Carboxy methyl cellulose				50		20	12	71	125	8.9	127	St. 1	390			320	3
Methyl cellulose				96		87	30	22		10.0	157		400		12	380	
Methyl cellulose				100		69	10	29	60	10.0	152		400		105	> 450	5
Methyl cellulose				93		37	12	37	30	10.1	209		410		29	450	5

Table A1, continued

INTERMEDIATE PRODUCTS / AUXILIARY MATERIALS	PARTICLE SIZE DISTRIBUTION								IGNITABILITY AND EXPLOSIBILITY OF DUST CLOUDS								DUST LAYERS	
									1 m^3 or 20 l vessel			Mod. H.	G.G.	BAM	VDI	DIN	
	WEIGHT % < SIZE [µm]							MEDI-AN	C_{min}	P_{max}	K_{St}	EXPLOS. < 63 µm	T_{min}		MIE	Glow temp.	Flamm. <250µm
DUST TYPE	500	250	125	71	63	32	20	µm	g/m³	bar(g)	bar•m/s	Class	°C	°C	mJ	°C	Class
Ethyl cellulose				66		40		40		8.1	162		(330)			275	
Chloroacetamide	98	79	33		13	3		170	(200)			St. 1					2(2)
Cyanoacrylicacid methylester	69		20					260	30	10.1	269	(St. 2)	500			> 450	5
Dicyandiamide				99		98	97	< 10		3.7	9		> 850			> 450	
1.3-Diethyldiphenyl urea				98		93	83	< 10	15	8.8	163	(St. 2)	530			melts	2(5)
1.3-Diethyldiphenyl urea	8							1300	30	8.7	116	(St. 2)	600			melts	2(5)
Dimethyl terephtalate						60		27	30	9.7	247		460		2	> 450	
Diphenyl urethane	93		49	27				128	30	8.9	218	(St. 2)	660			melts	2
Diphenyl urethane	31							1100	30	7.6	51	(St. 2)	660			melts	2
Emulgator (50% CH, 30% fat)			89	50		11		71	30	9.6	167		430		17	390	
Ferrocene			71	33				95	15	8.3	267		500		5	> 450	5
Fumaric acid	100	75	24		15	11		215	(100)			(St. 2)					5
Epoxy resin hardener				97		85	60	17	60	10.0	64		> 850			melts	2
Urea	4	2	< 1					2900				St. 1					1(2)
Hexamethylene tetramine				100		69	42	27	30	10.5	286						
Hexamethylene tetramine	100		30	9				155		10.0	224		530			melts	
Cellulose ion exchange resin								< 10	60	10.0	91	(St. 2)	410			> 450	5
Cellulose ion exchange resin				27		9		112	30	9.4	112		(350)			> 465	

Table A1, continued

INTERMEDIATE PRODUCTS / AUXILIARY MATERIALS	PARTICLE SIZE DISTRIBUTION								IGNITABILITY AND EXPLOSIBILITY OF DUST CLOUDS							DUST LAYERS	
									1 m³ or 20 l vessel			Mod. H.	G.G.	BAM	VDI	DIN	
	WEIGHT % < SIZE [µm]							MEDIAN	C_{min}	P_{max}	K_{St}	EXPLOS. < 63 µm	T_{min}		MIE	Glow temp.	Flamm. <250µm
DUST TYPE	500	250	125	71	63	32	20	µm	g/m³	bar(g)	bar•m/s	Class	°C	°C	mJ	°C	Class
Condensation product (phenol)				92		74	50	20	15	8.2	171	(St. 2)	560			melts	2(5)
D(-)-Mannite				61		24	13	67	60	7.6	54	St. 1	460			melts	2
Melamine				98		95	88	< 10	1000	0.5	1	St. 1	> 850			> 450	2
Melamine peroxide				61		56	46	24	250	12.2	73	St. 1	> 850			380	2
Melamine phosphate					100	79		22				St.1					2
Melamine phtalate				99		89	65	16	125	8.1	52	St. 1	910			melts	2
Metal soap (Ba/Pb-stearate)					100	48			15	8.1	180	(St. 2)					2(2)
Metal soap (Zn-behenate)					100	80			15	8.1	119	(St. 2)					2(3)
Methacrylamide	42							580		8.5	113		530		180	> 450	
Naphtalene	89		66		35	12		95	15	8.5	178		660		< 1	> 450	(5)
Naphtalic acid anhydride						97	69	16	60	9.0	90		690		3	melts	
2-Naphtol				100		96	94	< 10		8.4	137		430		5	> 450	
Sodium amide									(200)			(St. 2)					2
Sodium cyclamate	97	52	13		5	2		260				St. 1					5
Sodium hydrogen cyanamide			95	90		28	8	40	125	7.0	47		460			melts	
Sodium ligno sulphonate			100		63	20		58	(200)			St. 1					2
Oil absorber (hydrophobic cellulose)			65	51		31	21	65	60	7.2	42		540			> 450	
Paraformaldehyde				89		65	41	23	60	9.9	178	(St. 2)	460			> 480	5

Table A1, continued

INTERMEDIATE PRODUCTS / AUXILIARY MATERIALS	PARTICLE SIZE DISTRIBUTION								IGNITABILITY AND EXPLOSIBILITY OF DUST CLOUDS							DUST LAYERS	
									$1\ m^3$ or 20 l vessel			Mod. H.	G.G.	BAM	VDI	DIN	
	WEIGHT % < SIZE [µm]							MEDIAN	C_{min}	P_{max}	K_{St}	EXPLOS. < 63 µm	T_{min}		MIE	Glow temp.	Flamm. <250µm
DUST TYPE	500	250	125	71	63	32	20	µm	g/m^3	bar(g)	bar•m/s	Class	°C	°C	mJ	°C	Class
Paraformaldehyde				86		58	37	27	60	10.7	222		460			> 450	
Pectin			86	61		21		59	60	9.5	162		460			300	
Pectinase				91		47	20	34	60	10.6	177		510		180	> 450	3
Pentaerythrite				100		98	86	< 10	30	9.6	120		470		< 1	melts	2(5)
Pentaerythrite (from filter)			90	33		6	3	85	30	9.1	188		490		6	melts	5
Pentaerythrite	86		47	36		20	12	135	30	9.0	158				27	melts	5
Phtalic acid anhydride									(100)			(St. 2)					5
Polyethylene oxide	99	83	53		29	14		115	(30)			(St. 2)					3(5)
Polysaccharide					100	78		23	(500)			St. 1					4
Propyleneglycol alginate			57	24				115	125	8.8	82		440			450	
Salicylic acid									(30)			(St. 2)					2(5)
Saponin				93		77	65	13		9.4	150	St. 1	440			> 450	3
Lead stearate			99	96		90	80	< 10					480		< 1	melts	
Lead stearate							90	12	30	9.2	152	(St. 2)	630			melts	5
Calcium stearate				99		92	84	< 10					520		9	melts	
Calcium stearate						92	80	< 10	30	9.2	99		580		16	> 450	
Calcium stearate	100		43	25				145	30	9.2	155		550		12	> 450	
Magnesium stearate									(100)			(St. 2)					2(2)

Table A1, continued

INTERMEDIATE PRODUCTS AUXILIARY MATERIALS	PARTICLE SIZE DISTRIBUTION								IGNITABILITY AND EXPLOSIBILITY OF DUST CLOUDS							DUST LAYERS	
									1 m³ or 20 l vessel			Mod. H.	G.G.	BAM	VDI	DIN	
	WEIGHT % < SIZE [μm]							MEDIAN	C_{min}	P_{max}	K_{St}	EXPLOS. < 63 μm	T_{min}		MIE	Glow temp.	Flamm. <250μm
DUST TYPE	500	250	125	71	63	32	20	μm	g/m³	bar(g)	bar•m/s	Class	°C	°C	mJ	°C	Class
Sodium stearate				92		67	45	22	30	8.8	123	St. 1	670			melts	2
Zinc stearate									(100)			(St. 2)					2(5)
Zinc stearate				95		86	72	13					520		5	melts	
Stearin/Lead				99		95	75	15	60	9.1	111		600		3	> 450	
Stearin/Calcium				100		89	64	16	30	9.3	133		620		25	> 450	
Stearic acid	12							1300	8	7.2	34	(St. 2)	500			melts	1(1)
Terephthalic acid dinitrile					100	78			< 30	8.8	260	(St. 2)					5
2.2-Thiodiacetic acid				48		27	18	75	30	6.5	72	St. 1	350			410	2
Thio urea	56		1					460	250	3.5	8	St. 1	440			melts	2(2)
Trimellitic anhydride	4							1250	30	6.8	33		740		>2500	melts	2(5)
Trisodium citrate	36	2	1					800				St. 1					2
Tyrosine (final product)	100		99			48		10				(St. 2)					5
Tyrosine (raw product)	99		96		91	74		15				(St. 2)					5
Viscose flock					100	94		13	(100)			St. 1					4
Tartaric acid	100	5	1					480				St. 1					2
Zinc cyanamide				99		96	94	< 10	No ignition				> 850			> 450	3
Zinc cyanamide	47	34			27	14		600		(4.8)	(53)	No. ign.					2
Zinc pyridine thione						100			(500)			St. 1					2(5)

Table A1, continued

OTHER TECHNICAL/ CHEMICAL PRODUCTS	PARTICLE SIZE DISTRIBUTION								IGNITABILITY AND EXPLOSIBILITY OF DUST CLOUDS							DUST LAYERS	
									1 m^3 or 20 l vessel			Mod. H.	G.G.	BAM	VDI	DIN	
	WEIGHT % < SIZE [μm]							MEDI-AN	C_{min}	P_{max}	K_{St}	EXPLOS. < 63 μm	T_{min}		MIE	Glow temp.	Flamm. <250μm
DUST TYPE	500	250	125	71	63	32	20	μm	g/m^3	bar(g)	bar•m/s	Class	°C	°C	mJ	°C	Class
Organic dyestuff (blue)				99		98	95	< 10		9.0	73		710			360	
Organic dyestuff (khaki)				86		29	11	44					690			450	
Organic dyestuff (red)								< 10	50	11.2	249		520			melts	
Organic dyestuff (red)				65		33	23	52	60	9.8	237	(St. 2)	470			> 450	5
Organic dyestuff (Azo,yellow)				100		98	95	< 10	60	11.0	288	(St. 2)	480			melts	2(5)
Organic dyestuff (Disp., brilliant pink)			91	73		25		46					610		>4000	450	
Organic dyestuff (brown)									(200)			St. 1					4
Organic dyestuff (Phthalocyanine)				96		86		< 10	(200)	8.8	73	St. 1	770			355	4
Fuchsin base				74		45	26	36		8.4	115		640			melts	
Bituminous hydrocarbon			23	11				260	30	7.6	63		500			melts	2
Light protection agent				97		92	83	< 10		8.9	214		530			> 450	
Light protection agent				100		93		< 15		10.0	310						
Soap								65	30	9.1	111		580			melts	
Surfacer (Epoxy based)					100	77		24	(200)			St. 1					2
Surfacer (Polyester based)					100	85		19	(500)			St. 1					2
Washing agent (Na-sulph.)	88		14					275	30	9.0	267	(St. 2)	330			melts	5
Wax raw material (Alkylaryl sulphonate)												St. 1					5
Wax raw material (Olefin sulphonate)			60	28				105	30	8.6	115		390			> 590	

Table A1, continued

METALS ALLOYS	PARTICLE SIZE DISTRIBUTION								IGNITABILITY AND EXPLOSIBILITY OF DUST CLOUDS							DUST LAYERS	
									1 m³ or 20 l vessel			Mod. H.	G.G.	BAM	VDI	DIN	
	WEIGHT % < SIZE [µm]							MEDIAN	C_{min}	P_{max}	K_{St}	EXPLOS. < 63 µm	T_{min}		MIE	Glow temp.	Flamm. <250µm
DUST TYPE	500	250	125	71	63	32	20	µm	g/m³	bar(g)	bar•m/s	Class	°C	°C	mJ	°C	Class
Aluminium powder				94		88	79	< 10	60	11.2	515		560			430	
Aluminium powder				98		70	45	22		12.5	400		650			270	
Aluminium powder				99		64	47	22	30	11.5	1100		500			> 450	
Aluminium powder				94		60	17	29	30	12.4	415	(St. 3)	710			> 450	4
Aluminium grit				100		96		23	30	11.0	320		850			> 450	
Aluminium grit				99		16	2	41	60	10.2	100		> 850			> 450	
Aluminium grit	92		26	6				170		No ignition			> 850			> 450	1
Aluminium shavings	80		35	20				190					620		>1800	> 450	
Aluminium shavings	79		29	17				240		No ignition			> 850			> 450	
Aluminium/Iron (50:50)				93		68	48	21	250	9.4	230		760			> 450	
Aluminium/Magnesium			47					130		10.4	52	St. 1	> 850			> 450	2
Aluminium/Nickel				95		86		< 10		11.4	300						
Aluminium/Nickel (50:50)				37		18		90		No ignition			> 850			> 450	
Bronze powder						97	60	18	750	4.1	31	St. 1	390			260	4
Calcium/Aluminium (30:70)						68	46	22		11.2	420		600			> 450	6
Calcium/Silicon (from cyclone)				94		75	48	21	60	9.8	200	(St. 2)	770			> 440	1
Calcium/Silicon				87		55		28					770		145	> 450	
Iron (from dry filter)				98		82	67	12	500	5.2	50		580			> 450	

Table A1, continued

METALS ALLOYS	PARTICLE SIZE DISTRIBUTION								IGNITABILITY AND EXPLOSIBILITY OF DUST CLOUDS							DUST LAYERS	
									1 m³ or 20 l vessel			Mod. H.	G.G.	BAM	VDI	DIN	
	WEIGHT % < SIZE [μm]							MEDIAN	C_{min}	P_{max}	K_{St}	EXPLOS. < 63 μm	T_{min}		MIE	Glow temp.	Flamm. <250μm
DUST TYPE	500	250	125	71	63	32	20	μm	g/m³	bar(g)	bar·m/s	Class	°C	°C	mJ	°C	Class
Iron carbonyl							96	< 10	125	6.1	111	(St. 2)	310			300	3
Ferrochromium				96		82	73	< 10	500	6.4	86		> 850			> 450	
Ferromanganese				99		97	90	< 10		6.8	84		730			> 450	
FeSiMg (22:45:26)				99		77	57	17		9.4	169		670		210	> 450	
Ferrosilicon (22:78)				97		70	47	21	125	9.2	87		> 850			> 450	
Hard metal (TiC, TiN, WC, VC, Mo)		100	95		68	40		43	(200)			St. 1					4
Co-Al-Ti (62:18:20)				92		61	41	25	500	7.4	134		730			> 450	
Magnesium				100		70		28	30	17.5	508						
Magnesium	99		1					240	500	7.0	12	(St. 2)	760			> 450	5
FeSiMg (24:47:17)				99		70	47	21		9.9	267		560		35	> 450	
Manganese (electrolyt.)				82		70	57	16		6.3	157		(330)			285	
Manganese (electrolyt.)				70		41		33		6.6	69						
Molybdenum				100		96	92	< 10	No ignition				> 850			390	
Niobium (6% Al)	87	44	24		9	3		250	(200)			St. 1					2
Silicon				99		98	97	< 10	125	10.2	126		> 850		54	> 450	3
Silicon (from filter)						100	99	< 10	60	9.5	116		> 850		250	> 450	1
Silicon (from dust extr.)				90		70	57	16	60	9.4	100		800			> 450	
Steel (100 Cr6) dust					100	74			(30)	(4.0)	(82)	(St. 2)					2

Table A1, continued

METALS ALLOYS	PARTICLE SIZE DISTRIBUTION								IGNITABILITY AND EXPLOSIBILITY OF DUST CLOUDS							DUST LAYERS	
									1 m³ or 20 l vessel			Mod. H.	G.G.	BAM	VDI	DIN	
	WEIGHT % < SIZE [μm]							MEDIAN	C_{min}	P_{max}	K_{St}	EXPLOS. < 63 μm	T_{min}		MIE	Glow temp.	Flamm. <250μm
DUST TYPE	500	250	125	71	63	32	20	μm	g/m³	bar(g)	bar•m/s	Class	°C	°C	mJ	°C	Class
Tantalum/Niobium				97		90	80	< 10		6.6	37		700			450	
Titanium				98		55	24	30					450			> 450	
Titanium (pre-oxidized)				77		46	26	35					380			400	
Ti/TiO_2 (dust deposit)	61	40	28		12	6		310	(100)			(St. 3)					5
Zinc (from zinc coating)				91		72	53	19		6.0	85	St. 1	800			> 450	2
Zinc (from zinc coating)				93		70		21	250	6.8	93		790			> 450	
Zinc (dust from collector)							99	< 10	250	6.7	125	(St. 2)	570			440	3
Zinc (dust from collector)				97		91	72	10	125	7.3	176	St. 1					2

Table A1, continued

OTHER INORGANIC PRODUCTS	PARTICLE SIZE DISTRIBUTION								IGNITABILITY AND EXPLOSIBILITY OF DUST CLOUDS							DUST LAYERS	
									1 m^3 or 20 l vessel			Mod. H.	G.G.	BAM	VDI	DIN	
	WEIGHT % < SIZE [μm]							MEDIAN	C_{min}	P_{max}	K_{St}	EXPLOS. < 63 μm	T_{min}		MIE	Glow temp.	Flamm. <250μm
DUST TYPE	500	250	125	71	63	32	20	μm	g/m^3	bar(g)	bar•m/s	Class	°C	°C	mJ	°C	Class
NH_4NO_3/Dicyanamide (66:34)				60		42	35	50	250	7.0	21		390			> 450	
Graphite (99.5% C)					100	97		7	< 30	5.9	71			> 600		680	1
Carbon fibres (99% C)									(100)			St. 1					2
Molybdenum disulphide				92		75	53	19	250	5.6	37	St. 1	520			320	4
Petroleum coke				93		75	59	15	125	7.6	47	St. 1	690			280	4
Petroleum coke			83	51		22	14	71	125	3.8	3		750			> 450	3
Petroleum coke (calcinated)			94	86		64	47	22	250	6.8	14		> 850			> 450	3
Phosphorus (red)				100		92	59	18		7.9	526		400			340	5
Soot							99	5	60	9.2	85		760			590	
Soot (from filter)								< 10	30	8.8	88		840			570	
Sulphur				97		85	71	12					240		< 1	250	
Sulphur				96		70	51	20	30	6.8	151	(St. 2)	280				5
Sulphur				86		23		40					330		3	270	
Sulphur			53			7		120					370		5	270	
Titanium carbide									(100)			(St. 2)					4
Titanium hydride									(200)			St. 1					2
Titanium monoxide									(200)			(St. 2)					4

Table A1, continued

OTHER MATERIALS	PARTICLE SIZE DISTRIBUTION								IGNITABILITY AND EXPLOSIBILITY OF DUST CLOUDS							DUST LAYERS	
									1 m³ or 20 l vessel			Mod. H.	G.G.	BAM	VDI	DIN	
	WEIGHT % < SIZE [μm]							MEDIAN	C_{min}	P_{max}	K_{St}	EXPLOS. < 63 μm	T_{min}		MIE	Glow temp.	Flamm. <250μm
DUST TYPE	500	250	125	71	63	32	20	μm	g/m³	bar(g)	bar•m/s	Class	°C	°C	mJ	°C	Class
Flyash (from electrofilter)			100		99	92		6	125	1.9	35	No.ign.					1
Ash concentrate				87		61	48	21	60	8.6	91		580			260	
Bentonite/Asphalt/Coal/ Org. (15:45:35:5)		90			55			54	(100)			St. 1					2
Bentonite/Coal (50:50)		98	86		69	41		42	(100)			St. 1					2
Bentonite der. + org. comp.				89		45	23	35	60	7.4	123		430			> 450	3
Pb and Ca stearate mixture		98			70			35	(100)			(St. 2)					2(2)
Break liner (grinding dust)				98		95	89	< 10	250	6.9	71		530			310	
Brush dust (Al-brushes)				99		74	30	25	30	11.4	360		590		< 1	450	
CaC/Diamide lime/Mg (72:18:10)		99		93		87	80	8	125	5.8	30						4
Mud from settling chamber			99	91		62	45	23	60	7.7	96		430			260	5
Dust from polishing (Al)			44	26				150		5.0	18		440			320	5
Dust from polishing (Zn)		60	35		15	2		190	(200)			St. 1		400		350	2
Dust from polishing (brass)									(100)			St. 1					4
Dust from grinding (Al)					100	85			(30)	(5.7)	(214)	(St. 2)					4
Dust from grinding (Zn)					100	67			(500)	(2.3)	(24)	St. 1		480			1
Dust from grinding (cardboard)	70	64	44		25	10		160	(100)			St. 1					5
Dust from grinding (polyester)				98		95	93	< 10	30	9.5	153		500			> 450	5
Dust from grinding (polyester)			97	84		60	41	25		9.4	237		550			> 450	5

Table A1, continued

OTHER MATERIALS	PARTICLE SIZE DISTRIBUTION								IGNITABILITY AND EXPLOSIBILITY OF DUST CLOUDS							DUST LAYERS	
									1 m^3 or 20 l vessel			Mod. H.	G.G.	BAM	VDI	DIN	
	WEIGHT % < SIZE [µm]							MEDIAN	C_{min}	P_{max}	K_{St}	EXPLOS. < 63 µm	T_{min}		MIE	Glow temp.	Flamm. <250µm
DUST TYPE	500	250	125	71	63	32	20	µm	g/m^3	bar(g)	bar•m/s	Class	°C	°C	mJ	°C	Class
Dust from grinding (Ti)	89	64	37		18	4		170	(100)			(St. 2)					2
Dust from grinding + polish. (polyester)				99		96	91	< 10					530		< 1	> 450	
Blasting dust (light metals)					100	82			15	7.6	242	(St. 2)		370		280	4
Immersion polishing agent	46							600	(30)	6.2	11	St. 1	580			340	2
Textile fibres (nat. + synth.)									(30)			St. 1					5
Toner							100	< 10	60	8.9	196			520	4	melts	
Toner							100	< 10	30	8.7	137			530	< 1	melts	5
Toner				100		96	48	21	60	8.8	134			530	< 1	melts	(3)
Toner				100		95	30	23	60	8.8	145			530	8	melts	(3)
Toner/iron powder				58		37		60	60	8.2	169			570		> 450	
Toner resin				98		78	55	18						580	< 1	> 450	(5)
Zinc stearate/Bentonite (90:10)									(100)			(St. 2)					3
Zinc stearate/Bentonite (20:80)												(St. 1)					2

Table A2 Maximum permissible O_2 concentration for inerting dust clouds in atmospheres of $O_2 + N_2$ (Data from BIA, 1987)

Dust type	Median particle diameter by mass [μm]	Maximum O_2 concentration for inerting by N_2 [Vol%]
Cellulosic materials		
Cellulose	22	9
Cellulose	51	11
Waste from wood cutting	130	14
Wood	27	10
Food and Feed		
Pea flour	25	15
Maize starch	17	9
Waste from malted barley	25	11
Rye flour 1150	29	13
Starch derivative	24	14
Wheat flour 550	60	11
Coals		
Brown coal	42	12
Brown coal	63	12
Brown coal	66	12
Brown coal briquette dust	51	15
Bituminous coal	17	14
Other materials		
Ground hops	500	17
Hops draff	490	18
Plastics, resins, rubber		
Resin	<63	10
Rubber powder	95	11
Polyacrylnitril	26	11
Polyacrylnitril	26	10
Polyethylene, h.p.	26	10
Pharmaceuticals, pesticides etc.		
Aminopheenazone	<10	9
Herbizide	10	12
Methionine	<10	12
Intermediate products, additives		
Barium stearate	<63	13
Benzoyl peroxide	59	10
Bisphenol A	34	9
Cadmium laurate	<63	14
Cadmium stearate	<63	12
Calcium stearate	<63	12
Methyl cellulose	29	15
Methyl cellulose	49	14
Methyl cellulose	70	10
Dimethyl terephtalate	27	9
Ferrocene	95	7
Bistrimethylcilyl-urea	65	9
Naphthalic acid anhydride	16	12
2-Naphthol	<10	9
2-Naphthol	<30	9
Sodium methallyl sulphonate	280	15

Table A2, continued

Dust type	Median particle diameter by mass [µm]	Maximum O_2 concentration for inerting by N_2 [Vol%]
Paraformaldehyde	23	6
Paraformaldehyde	27	7
Pentaerythrite	<10	11
Pentaerythrite	<10	11
<u>Other techn.chem. products</u>		
Blue dye	<10	13
Organic pigment	<10	12
<u>Metals, alloys</u>		
Aluminium	22	5
Aluminium	22	6
Calcium/Aluminium alloy	22	6
Ferrosilicon	17	7
Ferrosilicon	21	12
Magnesium alloy	21	3
<u>Other inorganic products</u>		
Soot	<10	12
Soot	<10	12
Soot	13	12
Soot	16	12
Soot desorbed from acetylene	86	16
Soot desorbed from acetylene	120	16
<u>Others</u>		
Bentonite derivative	43	12

Table A3 Inerting of dust clouds by mixing the combustible dust with inert dust. (1 m^3 standard ISO (1985) vessel. 10 KJ chemical igniter.) (From BIA, 1987)

Combustible dust		Inert dust		
Type of dust	Median particle size by mass [µm]	Type of dust	Median particle size by mass [µm]	Minimum mass% inert of total mass required for inerting
Methyl cellulose	70	$CaSO_4$	<15	70
Organic pigment	<10	$NH_4H_2PO_4$	29	65
Bituminous coal	20		14	65
Bituminous coal	20	$NaHCO_3$	35	65
Sugar	30	$NaHCO_3$	35	50

A.2
APPLICABILITY OF EARLIER USBM TEST DATA

A.2.1
BACKGROUND

US Bureau of Mines in Pittsburgh, PA, USA, developed a comprehensive set of laboratory test methods for characterizing ignitability and explosibility of dusts, and published a large number of test data, which have been widely used throughout the world. The test apparatuses and procedures were described by Dorsett *et al.* (1960). Test data for 220 agricultural dusts were published by Jacobson *et al.* (1961), for 314 dusts in the plastics industry by Jacobson *et al.* (1962), for 314 metal powders by Jacobson *et al.* (1964), for 241 carbonaceous dusts by Nagy *et al.* (1965), for 175 chemicals, drugs, dyes and pesticides by Dorsett and Nagy (1968), and for 181 miscellaneous dusts by Nagy *et al.* (1968), i.e. for 1445 dusts altogether.

In more recent years alternative test methods have been developed, and there is a need to indicate the extent to which the substantial amount of the earlier USBM data are compatible with more recent data, as for example those in Tables A1, A2 and A3.

A.2.2
MINIMUM IGNITION TEMPERATURE OF DUST CLOUD

The apparatus used was the original Godbert-Greenwald furnace, which is essentially the same apparatus as the Godbert-Greenwald furnace used for determining the data in Table A1. The earlier USBM data should therefore be compatible with those in Table A1.

A.2.3
MINIMUM IGNITION TEMPERATURE OF DUST LAYER

The earlier USBM method differs significantly from the hot-plate method used for producing the data in Table A1. 6 cm^3 of the dust was placed in a small stainless steel mesh basket that was kept suspended at the centre of the Godbert-Greenwald furnace, while a controlled, small flow of air was passed through the furnace. The temperature of the furnace was controlled and maintained at a predetermined value, and the temperature inside the dust sample was monitored by a thermocouple. Ignition was defined as a distinct increase of the dust temperature beyond that of the furnace within 5 minutes. The minimum ignition temperature was defined as the lowest furnace temperature at which ignition occurred.

As would be expected, the USBM layer ignition temperatures are generally significantly lower, by 100° or more, than the 'glow temperatures' of Table A1 for similar dusts.

A.2.4 MINIMUM IGNITION ENERGY OF DUST CLOUD (MIE)

Due to the design of the electric spark generator used earlier by USBM, part of the stored capacitor energy 1/2 CV^2 was lost in a high-voltage transformer, and therefore the net spark energy was smaller than the nominal 1/2 CV^2 quoted as the spark energy. However, when comparing MIE data for similar dusts, determined by the earlier USBM method and the more recent methods described by Eckhoff (1976) and Berthold (1987), an approximate empirical correlation is indicated, as shown in Figure A1. Note that the correlation cannot be extrapolated beyond the range in Figure A1.

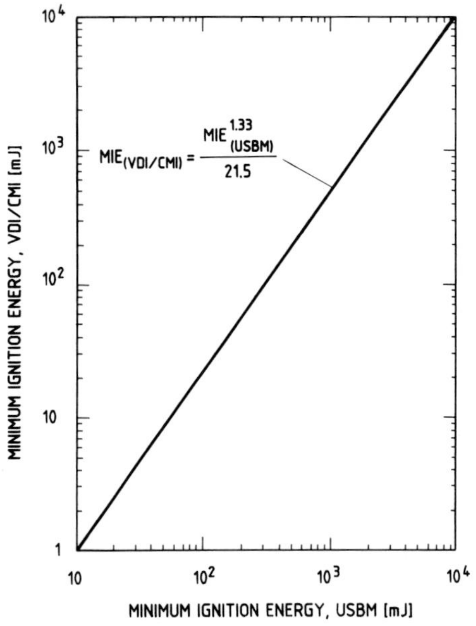

Figure A1 *Approximate empirical correlation between minimum ignition energies measured by the earlier USBM method described by Dorsett* et al. *(1960), and values generated by the more recent methods described by Eckhoff (1976) and Berthold (1987), and the method being evaluated by the IEC (see Chapter 7).*
Note: The correlation must be used as an indication only, and must not be extrapolated

A.2.5 MINIMUM EXPLOSIBLE DUST CONCENTRATION

The earlier USBM method was based on the 1.2 litre open Hartmann tube, with its top opening covered by a paper diaphragm. A comparatively weak continuous induction spark source was used for ignition. The dust concentration was defined as the quantity of dust dispersed, divided by the 1.2 litre volume of the tube. In spite of several probable sources of error, this method often yielded reasonable values as compared with more recent methods such as Nordtest (1989). This is probably because the effect of some of the sources of error partly cancel each other. However, data from the early USBM method must be regarded as indicative only.

A.2.6
MAXIMUM EXPLOSION PRESSURE

The early USBM data were determined in the original version of the closed 1.2 litre Hartmann bomb. Due to incomplete combustion and cooling by the walls, the maximum explosion overpressures in the Hartmann bomb are generally considerably lower, by up to 50%, typically 25–30% than those generated by the same dusts in larger vessels, such as the 1 m^3 standard ISO (1985) vessel and the 20 litre Siwek sphere. It does not seem advisable to indicate any general relationship between P_{max} from the early USBM tests and more recent data from larger vessels.

A.2.7
MAXIMUM RATE OF PRESSURE RISE

These data were determined in the same Hartmann bomb experiment as the maximum explosion pressures. However, there seems to be some justification for indicating the following tentative correlation between $(dP/dt)_{max}$ in the closed Hartmann bomb and the K_{St} from the 1 m^3 standard ISO (1985) method:

K_{St} [bar m/s] $= 0.35$ m $(dP/dt)_{max,\ Hartm.}$

Table A4 Examples of correlation of rates of pressure rise

(dP/dt)max,Hartm. [bar/s]	KSt [bar·m/s]
100	35
200	70
400	140
800	280
1600	560
3200	1120

Note: For quite coarse powders (non-homogeneous dust concentration distribution in Hartmann bomb) and for very fine, cohesive powders (poor dust dispersion in Hartmann bomb) this correlation can be substantially in error

A.2.8
MAXIMUM PERMISSIBLE O_2 CONCENTRATION FOR INERTING

USBM used two methods, namely an open glass tube with electric spark ignition, and the Godbert-Greenwald furnace at 850°C. As would be expected the latter method gave considerably lower limiting O_2 concentrations for inerting than the former.

Generally the values of Table A2 fall somewhere between the two USBM values for similar dusts. The arithmetic mean of the two USBM values would then perhaps be compatible with the data in Table A2.

REFERENCES

ASTM (1988) Standard Test Method for Pressure and Rate of Pressure Rise for Combustible Dusts. ASTM E 1226–88. *ASTM Annual Handbook* **Vol. 14.02** pp. 688–698

Berthold, W. (1987) Bestimmung der Mindestzündenergie von Staub/ Luft-Gemischen. Vortschritt-berichte Series 3, No. 134, VDI. VDI-Verlag, Düsseldorf

BIA (1987) Brenn- und Explosions-Kenngrössen von Staüben. Berufsgenossenschaftliches Institut für Arbeitssicherheit. Bergbau-Versuchsstrecke. Institut für Explosionsschutz und Sprengtechnik. Erich Schmidt Verlag GmbH & Co., Bielefeld

Dorsett, H. G., Jacobson, M., Nagy, J., *et al.* (1960) Laboratory Equipment and Test Procedures for Evaluating Explosibility of Dusts, Rep. Inv. 5624, US Bureau of Mines, Washington

Dorsett, H. G., and Nagy, J. (1968) Dust Explosibility of Chemicals, Drugs, Dyes and Pesticides. Rep. Inv. 7132, US Bureau of Mines, Washington

Eckhoff, R. K. (1975) Towards Absolute Minimum Ignition Energies for Dust Clouds? *Combustion and Flame* **24** pp. 53–64

International Standardization Organization (1985) *Explosion Protection Systems. Part 1: Determination of Explosion Indices of Combustible Dusts in Air.* ISO 6184/1 ISO, Geneva

Jacobson, M., Nagy, J., Cooper, A. R., *et al.* (1961) Explosibility of Agricultural Dusts. Rep. Inv. 5753, US Bureau of Mines, Washington

Jacobson, M., Nagy, J., and Cooper, A. R. (1962) Explosibility of Dusts Used in the Plastics Industry. Rep. Inv. 5971, US Bureau of Mines, Washington

Jacobson, M., Cooper, A. R., and Nagy, J. (1964) Explosibility of Metal Powders. Rep. Inv. 6516, US Bureau of Mines, Washington

Nagy, J., Dorsett, H. G., and Cooper, A. R. (1965) Explosibility of Carbonaceous Dusts. Rep. Inv. 6597, US Bureau of Mines, Washington

Nagy, J., Cooper, A. R., and Dorsett, H. G. (1968) Explosibility of Miscellaneous Dusts. Rep. Inv. 7208, US Bureau of Mines, Washington

Nordtest (1989) *Dust Clouds: Minimum Explosible Dust Concentration.* NT Fire 011, 2nd Edition, Nordtest, Helsinki

Index